Sources and Studies in the History of Mathematics and Physical Sciences

Sources and Studies in the History of Mathematics and Physical Sciences was inaugurated as two series in 1975 with the publication in Studies of Otto Neugebauer's seminal three-volume History of Ancient Mathematical Astronomy, which remains the central history of the subject. This publication was followed the next year in Sources by Gerald Toomer's transcription, translation (from the Arabic), and commentary of Diocles on Burning Mirrors. The two series were eventually amalgamated under a single editorial board led originally by Martin Klein (d. 2009) and Gerald Toomer, respectively two of the foremost historians of modern and ancient physical science. The goal of the joint series, as of its two predecessors, is to publish probing histories and thorough editions of technical developments in mathematics and physics, broadly construed. Its scope covers all relevant work from pre-classical antiquity through the last century, ranging from Babylonian mathematics to the scientific correspondence of H. A. Lorentz. Books in this series will interest scholars in the history of mathematics and physics, mathematicians, physicists, engineers, and anyone who seeks to understand the historical underpinnings of the modern physical sciences. All works are reviewed to meet the highest standards of scientific literature.

Titles from this series are indexed by Scopus.

Dale Martin Johnson

Theories of Dimension from Ancient to Modern Times

Springer

Dale Martin Johnson
Reston, VA, USA

ISSN 2196-8810 ISSN 2196-8829 (electronic)
Sources and Studies in the History of Mathematics and Physical Sciences
ISBN 978-3-032-02681-1 ISBN 978-3-032-02682-8 (eBook)
https://doi.org/10.1007/978-3-032-02682-8

Mathematics Subject Classification (2020): 01-02

This Springer imprint is published by the registered company Springer Nature Switzerland AG
The registered company address is: Gewerbestrasse 11, 6330 Cham, Switzerland

To the memories of my teachers and mentors:

Karl Popper

Imre Lakatos

Hans Freudenthal

Euclid alone has looked on Beauty bare.
Let all who prate of Beauty hold their peace,
And lay them prone upon the earth and cease
To ponder on themselves, the while they stare
At nothing, intricately drawn nowhere
In shapes of shifting lineage; let geese
Gabble and hiss, but heroes seek release
From dusty bondage into luminous air.
O blinding hour, O holy, terrible day,
When first the shaft into his vision shone
Of light anatomized! Euclid alone
Has looked on Beauty bare. Fortunate they
Who, though once only and then but far away,
Have heard her massive sandal set on stone.

— Edna St. Vincent Millay

Preface

This book is the result of several years of research that I carried out in the 1970s and the early 1980s. The result then was a series of three articles [Johnson, 1977] [Johnson, 1979] [Johnson, 1981] and a short partial summary [Johnson, 1987]. Given all the research that has gone on in the intervening years, I have considerably revised my earlier work and added a great deal of new material in writing this book.

The inspiration for this work comes from three sources, first, my time of study at the London School of Economics in the Department of Philosophy, Logic and Scientific Method with Professor Sir Karl Popper and Professor Imre Lakatos; second, my time working closely with Professor Dr. Hans Freudenthal at Utrecht on the topological work of L.E.J. Brouwer during a sabbatical year and after; and, third, my close association and correspondence with Professor Dr. Dirk van Dalen, Utrecht.

Acknowledgements

I would like to acknowledge the considerable help I received over many years from Hans Freudenthal, formerly professor at Universiteit Utrecht, now deceased. He was my mentor in all things Brouwer. He was a research assistant to Brouwer in the 1930s and knew Brouwer's topological work very well. He edited volume 2 of Brouwer's *Collected Works* that covered topology. As I worked closely with him he became a good friend.

I would like to acknowledge the help I received from Karl Menger, formerly professor at the Illinois Institute of Technology, Chicago, now deceased. As one of the founders of dimension theory he was able to give me many important insights into its history. We had many fruitful discussions together.

I must acknowledge the incredible work of Dirk van Dalen, Emeritus Professor at Universiteit Utrecht. He has done so much to assemble the Brouwer Archive and to publish the Brouwer correspondence. The Archive and published correspondence are invaluable sources of information concerning Brouwer. His biography of Brouwer in two editions is also an incredible source of detailed information about Brouwer. All of this material was of great value to me in my research. We have known one another for many years. He encouraged me to write this history. I am grateful for his help and encouragement.

I wish to express my thanks to Roger Cooke, Professor of Mathematics Emeritus, University of Vermont. His assistance with the ancient Greek texts referenced and translated was invaluable. Also, his assistance with the translation of the Russian text of Alexandroff related to Urysohn was invaluable. Moreover, he provided me with his careful review of chapters of the book, which I have found very useful. I value his friendship of long standing.

I wish to express my thanks to Brendon Lasell, Tutor at St. John's College, Annapolis, Maryland, for some clarifying discussions and interchanges on Greek mathematics and philosophy. His comments and suggestions were extremely useful to me for improving Chapter 2.

I wish to express my thanks to Jeremy Gray, Professor at the Open University in the UK. Jeremy has made many comments and corrections to help me improve my work. I value his friendship of long standing.

I wish to express my thanks to Donald Gillies, Emeritus Professor at University College London, who reviewed philosophical aspects of my work and offered valuable comments. We have known one another for many years.

I wish to express my thanks to Elías Fuentes Guillén of the Institute of Philosophy of the Czech Academy of Sciences, who reviewed my work and offered valuable comments.

I wish to express my thanks to Paul Rusnock, Professor of Philosophy at the University of Ottawa. He provided me with valuable comments and corrections for my analysis of the work on Bolzano on dimension and related philosophy.

I wish to express my thanks to David Eisenbud, Professor of Mathematics at the University of California, Berkeley. He provided me with important comments related to dimension in topology and in algebraic geometry and commutative algebra.

I wish to express my thanks to my Springer editor, Rémi Lodh. He has been a constant source of encouragement in my work on this book.

Finally, I wish to express my deep gratitude to my wife, Sidney. She has been my constant support in this work.

Contents

List of Figures

Chapter 1
Introduction to Theories of Dimension

> The great significance of specific problems for the progress of mathematics in general and the important role that they play in the work of individual researchers is undeniable. As long as a branch of science offers an abundance of problems, it remains alive; a lack of problems foreshadows extinction or the cessation of independent development. Just as every human undertaking strives towards goals, so mathematical research needs problems. The power of the researcher is tempered through the solution of problems; by them he discovers new methods and points of view and opens up a wider and more expansive horizon.[Ch1, 1]
>
> – Hilbert, 'Mathematische Probleme' [Hilbert, 1901], [Hilbert, 1902a][1]

1.1 Introduction to Problems of Dimension

Problems, above all, drive the progress of mathematics throughout its history. Mathematicians seek to solve problems—often very hard ones, whose complete or even partial solutions may take years, decades, or in some cases centuries to find. The results are mathematical theories or parts of larger theories or bodies of mathematics. David Hilbert, in opening his famous talk, 'Mathematische Probleme' ('Mathematical Problems') (see the quotation above) given at the International Congress of Mathematicians in 1900, emphasizes the importance of problems and problem solving to developing mathematics. In the talk he describes 23 critical and very difficult open problems, and he challenges mathematicians to seek to solve them.

In this work I am concerned with the problems, definitions of concepts, and resulting theories of dimension. I seek to analyze the conceptual and historical developments related to dimension. This problem domain has a long and fascinating history. Three problems have been fundamental to the growth of ideas of dimension

[1] References such as [Hilbert, 1901] or [Riemann, 1990, pages 573–590] with an author's name and date and possibly a page or page range or section denote items in the bibliography and page numbers or sections in these items. I have put the original language texts that are translated in the main body of the book in Appendix A of Original Language Texts of Quotations at the end of the book, and I have used raised endnote reference markings at end of the translated texts to refer to these texts.

D. M. Johnson, *Theories of Dimension from Ancient to Modern Times*, Sources and Studies in the History of Mathematics and Physical Sciences, https://doi.org/10.1007/978-3-032-02682-8_1

that eventually developed into a set of mature mathematical theories in the twentieth and twenty-first centuries: (1) the problem of defining the concept of dimension itself and some closely related concepts, such as the concepts of curve and surface; (2) the problem of explaining the number of dimensions of physical space (perhaps including time); and (3) the problem of proving the invariance of the dimension numbers of mathematical spaces under some restricted type of mapping, e.g., homeomorphism. The first and third problems, mathematical in nature, have had the most important direct influence on the growth of theories of dimension, mainly the topological theories of dimension. The second, a problem of physics or cosmology, has provided an indirect but significant motivation for the development of the theories from outside the domain of mathematics. In the next paragraphs I examine these three problems briefly as an introduction to the problem situation surrounding the history covered more fully in this work. From a longer-term historical perspective each of these problems encompasses a problem set, whose members, character, and treatment changed over the course of history.

The problem of defining dimension has ancient roots to a limited extent as suggested in brief passages in the writings of Greek philosophers and mathematicians. However, dimension theory is primarily a modern subject; its main historical roots start in the nineteenth century. At the very beginning of that century Bernard Bolzano examined several facets of the definition problem and proposed some insightful tentative solutions to it. After Bolzano a few nineteenth-century mathematicians took up the definition problem, but they did not resolve it satisfactorily. Interest in the problem heightened through the creation of the mathematical theory of higher-dimensional spaces and multi-dimensional geometries and the consequent methodological changes in the whole field of geometry. In the twentieth century Henri Poincaré was the first to construct a definition of dimension, one based on 'cuts' [Poincaré, 1903b] [Poincaré, 1912]. L.E.J. Brouwer soon gave a sympathetic reading and critique of it and proposed to replace it by an improved version [Brouwer, 1913b]. In the early 1920s Paul Urysohn (Pavel Uryson) and Karl Menger published independently of one another further definitions of dimension that turned out to be equivalent and that progressed a step beyond Brouwer's definition. These definitions of Brouwer, Urysohn, and Menger became the foundation for a widely accepted topological theory of dimension that began a rapid development in the 1920s and 1930s, yielding many significant consequences and theorems. In the early decades of the twentieth century mathematicians proposed other definitions of dimension with accompanying theories, not all of which were topological in character.

The problem of defining dimension and related concepts is surely the broadest of the three problems, having engendered the broadest solutions and theories. Beyond topology, there is a measure-theoretic concept of dimension, Hausdorff dimension. Felix Hausdorff laid down this notion and Abram Besicovitch, among others, developed it further, leading to fractal dimension and fractal geometry. There is also a need for a concept of dimension in algebraic geometry and related commutative algebra. The preferred modern concept in this case, Krull dimension, has both an algebraic aspect and a topological aspect. I shall cover some of this work in Chapter 16. Evidence of the breadth and richness of concepts of dimension is given in the nLab 'quasi-taxonomy' or list, `https://ncatlab.org/nlab/show/dimension`.

The cosmological problem of explaining the dimension number of physical space (or space with time), like the definition problem, arose in the speculations of the ancient Greeks, and over the centuries many thinkers have puzzled over it. In modern times Immanuel Kant considered it in his very first published work. For Poincaré this problem was the chief motive to propose his cut definition of dimension. There can be little doubt that this problem has had a significant role in the development of the mathematical theory of dimension.[2]

The origin of the invariance problem has a precise date, 1877, the year when Georg Cantor discovered—to his own astonishment—that geometrical figures of different dimension numbers can have their points arranged in one-one correspondence. Cantor's paradoxical result is a direct blow to our naive concept of dimension. When he and Richard Dedekind discussed it in an exchange of letters, they formulated the invariance problem for dimension, that is, the problem of showing that for certain continuous mappings dimension remains an invariant concept. Solution of the invariance problem quickly became regarded as the way to explain and resolve Cantor's strange result. As soon as Cantor published his paradoxical correspondence in 1878 a number of mathematicians tried to demonstrate the invariance, but the problem proved to be refractory. Giuseppe Peano dealt another blow to the naive concept of dimension in 1890 when he published the first example of a space-filling curve, a continuous mapping of a line segment onto an entire square. By the turn of the present century no one had found a general proof of dimensional invariance. With hindsight one can see why the problem was so difficult to solve. At the time there were virtually no adequate tools for the task. Topology was a very young subject. At last L.E.J. Brouwer achieved success by constructing a full and rigorous proof in 1910, which he published in 1911. Two years later he published another proof that he based on his definition of dimension. Undoubtedly the key to Brouwer's success was the set of new and fruitful ideas and tools that he introduced into topology.

The two problems of defining dimension and proving its invariance were the primary influences on the creation and growth of modern mathematical dimension theory. They are the foci of its history. However, the problem of explaining the dimension number of space was also at times a motive from outside mathematics for the development of the theory. Other problems as well had some influence on its development and, when the main topological theory came to maturity in the 1920s and 1930s, further problems became the center of attention. However, in the period up to about 1925 the twin problems of definition and invariance were the central motive forces with the cosmological problem as a pervading background force.

This book is about the history of concepts and theories of dimension related primarily to these problems. In the course of mathematical history the problems of interest change—the context changes, the formulations change, related background theories come into play and are modified and may become supporting frameworks for the mathematics of the problems. Also, the related problems may be solved, or

[2] I have excluded a treatment of dimension as found in string theory from the scope of this book.

new theories may be developed, and these may facilitate the solution of the main problems of interest. Essentially, the problem situation changes over time (to use a preferred phase of the philosopher Karl Popper). One should consider the problem situation in historical context with respect to the three problems of dimension to gain a deeper understanding of the problems and the various ways they may be solved.

When the term 'dimension theory' is used in this work in a modern context it refers primarily to the main topological theory for separable metric spaces and possibly more general topological spaces. In this statement the modern framework of topology is being used. Of course, before the twentieth century this framework was not available.

As noted, there are other topological dimension theories for other types of spaces, and there are theories that are not topological in character, such as the measure-theoretic theory. This work will present the history of most of these, at least to some extent. The aim of this book is to cover broadly the history of concepts and theories of dimension through the centuries.

1.2 Philosophical Perspectives

To gain some perspective on this history and to offer some interpretations of the mathematical developments, I shall initially consider some philosophical viewpoints concerning the nature of scientific, primarily mathematical, problems, concepts, definitions, and theories, and their development and growth. Then I shall offer some philosophical comments and insights as appropriate in the various chapters. Finally, I shall make some concluding comments in the last chapter. The philosophical commentary is not intended to be a substitute for the history or a rational reconstruction of it. To begin, I shall consider three philosophical viewpoints in brief, those of Rudolf Carnap (1891–1970), Karl Popper (1902–1994), and Imre Lakatos (1922–1974). I shall also comment on mathematical practices.

1.3 Rudolf Carnap: Words, Concepts, Explications

Philosophers since ancient times have analyzed in various ways the general nature of concepts and definitions as well as specific instances of concepts and their characterization and definition [Gupta & Mackereth, 2023]. In the twentieth century Rudolf Carnap put forward an interesting notion of *explication*, which has gained wide acceptance and which has often been used (in many cases without attribution to Carnap) [Carnap, 1956, pp. 7–8] [Carnap, 1962, Chapter 1] (cf. [Gupta & Mackereth, 2023, pp. 7–9], [Leitgeb & Carus, 2023, Supplement D. Methodology], [Murzi, 2024b], and [Cordes & Siegwart, 2024]. The last reference includes an extensive history and pre-history). Carnap states [Carnap, 1962, page 3]:

> By the procedure of *explication* we mean the transformation of an inexact, prescientific concept, the *explicandum*, into a new exact concept, the *explicatum*. Although the explicandum cannot be given in exact terms, it should be made as clear as possible by informal explanation and examples.
>
> The task of ***explication*** consists in transforming a given more or less inexact concept into an exact one or, rather, in replacing the first by the second. We call the given concept (or the term used for it) the ***explicandum***, and the exact concept proposed to take the place of the first (or the term proposed for it) the ***explicatum***. The explicandum may belong to everyday language or to a previous stage in the development of scientific language. The explicatum must be given by explicit rules for its use, for example, by a definition which incorporates it into a well-constructed system of scientific either logicomathematical or empirical concepts.

Carnap puts forward explication as a method to bring informal concepts into sharp focus as new exact concepts by providing them with exact formulations, the explicata (plural of explicatum)—in essence, formal definitions. Explication is a formalization process. It is part of his formal philosophy.

In the development of modern dimension theories mathematicians, such as Brouwer, Urysohn, and Menger, began by proposing formal definitions of the dimension concept. They took the informal concept and offered various replacement concepts in geometrical, set-theoretic, topological, or measure-theoretic frameworks, using the rigor that seemed to be appropriate or necessary to enable further mathematical development. Such exercises in defining formal concepts suggest something like the process of explication.

Carnap [Carnap, 1962, page 7] elaborates his idea in the following way:

> If a concept is given as an explicandum, the task consists in finding another concept as its explicatum which fulfills the following requirements to a sufficient degree.
>
> 1. The explicatum is to be *similar to the explicandum* in such a way that, in most cases in which the explicandum has so far been used, the explicatum can be used; however, close similarity is not required, and considerable differences are permitted.
> 2. The characterization of the explicatum, that is, the rules of its use (for instance, in the form of a definition), is to be given in an *exact* form, so as to introduce the explicatum into a well-connected system of scientific concepts.
> 3. The explicatum is to be a *fruitful* concept, that is, useful for the formulation of many universal statements (empirical laws in the case of a nonlogical concept, logical theorems in the case of a logical concept).
> 4. The explicatum should be as *simple* as possible; this means as simple as the more important requirements (1), (2), and (3) permit.

These requirements are essentially criteria for appraising good explicata resulting from the explication process. They are criteria for judging explications for their quality, 'goodness,' or value.

In this connection, Carnap mentions dimension theory and cites an earlier article by Karl Menger, a founder of modern dimension theory, 'What is dimension?' [Menger, 1943], which covers aspects of definition in general and implicitly explication (without using the term that Carnap later added to the philosophical vocabulary). As background he might also have mentioned a part of Menger's book *Dimensionstheorie* (*Dimension Theory*) [Menger, 1928b, II. Der Dimensionsbegriff, pp. 74–91] (English translation in part in [Menger, 1979a, pp. 207–210]). See Section 17.4 for more on the connection with Menger.

Carnap emphasizes *concepts that are formally developed* in his notion of explication. However, while dimension theories start with definitions, the real mathematical development normally occurs in the *theories* with results expressed as theorems supported by proofs. At first sight, it appears that in Carnap's treatment *concepts* are emphasized over *theories*. This emphasis may not be his intention. The larger aim of his book [Carnap, 1962] is to develop an alternative concept of probability, and with it derive results according to a specific interpretation of probability theory. However, Carnap is fundamentally formalistic in philosophy. He does not express a great interest in historical development.[3]

1.4 Karl Popper: Problems and Theories, Conjectures and Refutations, and the Growth of Knowledge

In contrast to Carnap, another philosopher, Karl Popper, in his writings strongly emphasizes theories over concepts (for example, see [Popper, 1962, passim] or later editions including [Popper, 2002, passim], especially [Popper, 2002, page 19, diagram]; cf. also [Thornton, 2023] and [Brendan, 2024]). Theories embody scientific explanations. One should view a well-developed theory as a logically ordered set of statements or propositions providing the heart of the meaning for a given scientific field. Statements can be true or false. In the case of mathematics the body of a given branch will start with definitions and axioms that are then developed into a set of interesting theorems. These deductive consequences are where the real meaning or value lies.

Popper also strongly emphasizes historical and methodological development. He argues that science develops through solving problems, and the solutions become tentative theories. In turn, these may be corrected through error elimination with further investigation and criticism. He claims that this form of development is how science progresses. Popper is a fallibilist—theories may in the course of time be refuted and then will need to be replaced with better ones. They may be revised and corrected. Hence, he thought in terms of 'conjectures and refutation'—theories

[3] Cf. also [Boniolo, 2003].

change over time as correcting criticism modifies theories, or transforms them into new theories. In his book *Objective Knowledge* [Popper, 1972, page 243] Popper uses a simple diagram to encapsulate his idea:

$$P_1 \rightarrow TS \rightarrow EE \rightarrow P_2,$$

where P_1, P_2 designate problems that change over time as new solutions are proposed and eliminated, TS designates tentative solution, and EE designates error elimination. He elaborates this diagram to cover multiple tentative solutions,

$$TS_1, TS_2, \ldots, TS_n.$$

This sequence of symbols designates a sequence of competing alternatives to solving the problem, leading to error eliminations and new resulting problems.

Popper held an objectivist view of the development of science. Philosophy of science was his primary interest, not so much philosophy of mathematics.

1.5 Imre Lakatos: Proofs and Refutations, and the Growth of Mathematical Knowledge

Imre Lakatos transformed many of Popper's ideas into the realm of the history and philosophy of mathematics in espousing his 'proofs and refutations' characterization of the growth of mathematics [Lakatos, 1964] [Lakatos, 1976] [Lakatos, 2015]. He enhanced his characterization with ideas from George Pólya on the role of heuristics and conjectural thinking in mathematics [Pólya, 2004] [Pólya, 1954] [Pólya, 1968]. In uniting these and other background elements Lakatos put forward an original and distinctive view of mathematical development, examining the interplay of concept formation and critical reformulation in close conjunction with proposed theorems and proofs, yielding theories, followed by critical refutations and reformulations. Examining these interactions in the longer term gives rise to a lively view of the development of mathematical theories—not as eternal truths, but as conjectural and potentially needing revision. 'Rigorous proofs' may not be as solid as they may seem at first sight. There can be significant revisions and rethinking as mathematicians grapple with problems, concepts, proofs, and theories during the course of a development of a domain.[4]

As I consider the theories of dimension and analyze them in historical context, I shall try to highlight philosophical indications of the perspectives of Carnap, Popper, and Lakatos, where possible. Although this work is intended primarily as a history, that history needs interpretations, and these philosophical views should help in formulating them. I shall return to these perspectives in the final Chapter 17.

[4] On Lakatos' work in history and philosophy of mathematics cf. [Musgrave & Pigden, 2023] [Gillies, 2023].

1.6 Recent Philosophical Views of Mathematical Practices

In the last few years in philosophy of mathematics there has been a new emphasis on mathematical practices—on how mathematicians create and develop their concepts and theories and how the results and theories interact with one another. Some philosophers of mathematics have paid less attention to foundational questions or to questions of the existence of types of mathematical objects (ontology) or to the way we might attain genuine mathematical knowledge (epistemology), although these deep questions still have a place for consideration. The more pragmatic view of mathematical practices accords very much with the way I have conceived this history of dimension theories. Among the writings belonging to this way of thinking about mathematical practices, I may mention: Poalo Mancosu's edited collection *The Philosophy of Mathematical Practice* [Mancosu, 2008], with ten contributors, Mancosu's earlier companion work *Philosophy of Mathematics & Mathematical Practice in the Seventeenth Century* [Mancosu, 1999], and José Ferreirós' recent book, *Mathematical Knowledge and the Interplay of Practices* [Ferreirós, 2015]. In this connection I also want to mention Barry Mazur's thought-provoking paper on 'Conjecture' [Mazur, 1997], giving an emphasis to the role of conjectures in mathematics.[5] I shall return to these philosophical perspectives in the final Chapter 17.

1.7 Mathematical Background

The reader of this book should have some basic background in point-set or general topology and a little background in algebraic topology. The following books provide ample background in topology, considerably more than is necessary:

- Witold Hurewicz and Henry Wallman, *Dimension Theory* [Hurewicz & Wallman, 1948]. This classic and very beautiful presentation of the theory also reviews some of the basics of topology.
- James R. Munkres, *Topology* [Munkres, 2000]. This is an excellent, well-written book introducing the main topics of topology, both general and algebraic.
- Stephen Willard, *General Topology* [Willard, 2004]. This a good reference in general topology and is available as a Dover book.
- Paul Sergeevich Alexandroff, *Einführung in die Mengenlehre und die Theorie der reellen Funktionen* (*Introduction to Set Theory and the Theory of Real Functions*) [Alexandroff, 1967]. This book is published in German translation and provides a well-motivated introduction to metric spaces and topology, and related topics in set theory and real function theory. Alexandroff is one of the main mathematicians figuring in this book.

[5] The Wikipedia article on conjecture contains some interesting references on mathematical conjectures of various sorts: `https://en.wikipedia.org/wiki/Conjecture`.

- Paul Sergeevich Alexandroff and Heinz Hopf, *Topologie I* (*Topology I*) [Alexandroff & Hopf, 1935] [Alexandroff & Hopf, 2013]. This book is a classic written in German by two important mathematicians, Alexandroff and Hopf, who figure prominently in the history of twentieth-century topology and dimension theory.

1.8 Summary of Later Chapters

The rest of the book is organized as follows. Chapter 2 deals with ancient Greek ideas touching on dimension. Chapter 3 takes the story of dimension from the middle ages to the end of the eighteenth century. Chapter 4 is devoted to the investigations of Bernard Bolzano into geometry and dimension. Chapter 5 looks at theories of higher-dimensional spaces in the nineteenth century. Chapter 6 examines the development of set theory with Georg Cantor and Richard Dedekind and the discovery of an important and surprising paradox of dimension. Chapter 7 treats early efforts to prove the invariance of dimension, 1878–1879. Chapter 8 surveys the early development of point-set theory and point-set topology, including Peano's space-filling curve. Chapter 9 covers the work of Henri Poincaré on dimension. Chapter 10 deals with other approaches to dimension that Frigyes Riesz, René-Louis Baire, and Maurice Fréchet took in the early years of the twentieth century. Chapter 11 covers L.E.J. Brouwer's approach to topology and his first proof of invariance of dimension. Chapter 12 deals with new ways to prove invariance of dimension and the dispute between L.E.J. Brouwer and Henri Lebesgue. Chapter 13 treats set-theoretic approaches to problems of curves and dimension in the early twentieth century. Chapter 14 covers Paul Urysohn's theory of curves and dimension. Chapter 15 covers Karl Menger's theory of curves and dimension. Chapter 16 concerns the further development of theories of dimension and curves in the twentieth century. Finally, Chapter 17 considers some historical, mathematical, and philosophical assessments and conclusions.

Chapter 2
Ideas Touching on Dimension in Ancient Greek Philosophy and Mathematics

2.1 Introduction

Several ancient Greek philosophers and mathematicians over many centuries considered and analyzed ideas akin to dimension. They grappled with basic concepts and implicit theories that touch on dimension. They put forward and examined fundamental ideas related to the dimension concept as embodied in the four basic geometrical figures of point, line, surface, and solid, their hierarchical ordering, and their mutual relations. From a modern point of view these figures are low-dimensional objects, i.e., objects of 0, 1, 2, and 3 dimensions, though the ancient Greek thinkers would not have identified them in this way. This chapter contains a summary of the ancient Greek ideas that touch on dimension. The main Greek ideas derive from Pythagoras (ca.570–ca.495 BCE) and the Pythagoreans who followed him, Aristotle (384–322 BCE), Euclid (ca.325–ca.265 BCE), and Claudius Ptolemy (ca.100–ca.170 CE). It is important to analyze these ideas related to dimension in their own historical context, including the related problems and problem situations. It is possible to view these early ideas to a very limited extent as precursors to the development of dimension theory, since the full range of broader Greek arithmetical and geometrical ideas have had a lasting effect on later mathematics.

The Greeks are rightly viewed as a principal ancient source of the science of mathematics, though by no means the only one.[1] Ancient Greek philosophers and mathematicians became expert in developing mathematical theories, both informal and more formal and axiomatic. Modern practitioners of the science have a continuing debt to these ancient philosophers and mathematicians for laying down methods of theory analysis and development that still can be valued.

Among the extant ancient Greek philosophical and mathematical texts there are distinct traces of thinking that relates to dimension. However, the Greeks did not formulate detailed theories of dimension; rather their 'dimension theories' were implicit. They did think and argue about conceptual problems related to dimension,

[1] The ancient Babylonians and Egyptians must be regarded as important to the early development of mathematics. The ancient Chinese also have a place in the earliest development of mathematics.

D. M. Johnson, *Theories of Dimension from Ancient to Modern Times*, Sources and Studies in the History of Mathematics and Physical Sciences, https://doi.org/10.1007/978-3-032-02682-8_2

and they did implicitly produce informal theories about dimension that have had a lasting influence on later generations of mathematicians. Aristotle, Euclid, and Ptolemy all thought about concepts touching on dimension. Even before these, the oldest sect of philosophers and mathematicians, the Pythagoreans, examined dimension-related problems in the context of their cosmogony, their theory of the world and its development related in numbers.

2.2 Pythagoras and the Pythagoreans

Pythagoras and his early followers demonstrated a strong interest in mathematics. This interest continued with later Pythagoreans over many centuries in ancient times. Everything could be explained by or 'reduced to' numbers, i.e., the positive integers (and ratios of integers—at least until the discovery of the problem of incommensurables concerning, e.g., the incommensurability of the side and diagonal of the square). The Pythagorean philosophy of nature and the world based in number was hugely influential and extended over centuries. Pythagorean ideas strongly influenced Plato (428/427 or 424/423–348/347 BCE). In later ancient times books and manuals on numbers and corresponding mathematics with a Pythagorean heritage continued to be written. The influence of Pythagoras and the Pythagoreans extended from the time of the philosopher over the following centuries to at least medieval and very early modern times.

Tracing and characterizing the writings of Pythagoras and the Pythagoreans have been a scholarly pursuit for some time. My exposition mainly relies on the books of Guthrie [Guthrie, 1962], Kirk and Raven [Kirk & Raven, 1957], [Kirk et al., 1983], Burkert [Burkert, 1972], and van der Waerden [Waerden, 1954]. In particular, Guthrie considers the difficulties of assessing the sources for information concerning Pythagoras and the Pythagoreans [Guthrie, 1962, pp. 148–172]. Indeed, Burkert devotes much of his book [Burkert, 1972] to dealing with that problem. B.L. van der Waerden [Waerden, 1954] offers an especially good analysis of the influence of Pythagoras and the Pythagoreans on the whole development of Greek mathematics.

There are many fragments concerning mathematics and philosophy ascribed to the Pythagoreans as well as many passages from other authors, such as Aristotle, dealing with them. Two ancient manuals on mathematics treat Pythagorean views on mathematics, even though they were written long after Pythagoras lived. The first is the *Theologumena Arithmeticae* or *The Theology of Arithmetic: On the Mystical, Mathematical and Cosmological Symbolism of the First Ten Numbers*, which is attributed to Iamblichus (ca.245–ca.325 CE), but which has passages attributed to earlier authors [Iamblichus, 1988]. The second is the *Introduction to Arithmetic* by Nicomachus of Gerasa (ca.60–ca.120 CE) [Nicomachus of Gerasa, 1926].

When trying to solve their fundamental cosmogonical problem of explaining the generation of physical things from numbers (i.e., the positive integers, and their ratios) in accordance with their distinctive mathematical philosophy, the early Pythagoreans described how the basic geometrical figures, the point, line, surface, and solid, could be produced from numbers as an intermediate stage in a genetic process. It was natural for the Pythagoreans to touch upon the dimension concept implicitly, inasmuch as their problem and resulting theory involved a discussion of things having different dimensions (to use the modern term) and the relations these things have to one another. According to modern scholars, it is necessary to distinguish two versions of the intermediate stage, an older one and a more recent one, which probably constituted an improvement on the first. Although the evidence for these pre-Socratic theories is fragmentary, the modern rational reconstructions of them has reasonable plausibility. Consider the primitive version of the intermediate stage. The most pertinent evidence comes in the following fragments. First there is an extensive passage on the Pythagorean ideas found in Iamblichus' *Theologumena Arithmeticae* (*The Theology of Arithmetic*) [Iamblichus, 1988], a passage that is ascribed to Speusippus[2] [Thomas, 1939, pp. 80–83]:

> For 1 is a point, 2 is a line, 3 is a triangle and 4 is a pyramid; all these are elements and principles of the figures like to them. In these numbers is seen the first of the progressions, that in which the terms exceed by an equal amount, and they have 10 for their sum. In surfaces and solids there are the elements—point, line, triangle, pyramid. . . . The same result is seen in their generation. For the first principle of magnitude is point, the second is line, the third is surface, the fourth is solid.[Ch2, 1]

Complementing this text are two passages referring to the Pythagoreans from Aristotle (*Metaphysics*, VII.2, 1028b16; XIV.3, 1090b5; quoted in [Guthrie, 1962, page 259]):

> Some think that the limits [boundaries] of bodies, such as surface and line and point or unit, are substances, rather than body and the solid.[Ch2, 2a]

and

> There are some who, because the point is the limit and end of a line, the line of a surface and the surface of a solid, hold it to be inescapable that such natures exist.[Ch2, 2b]

The intermediate stage of the theory may be described more fully as follows. There is a natural progression from point to line to surface to solid that is the pattern of generation. In particular, there are the special figures, the point, the line segment, the triangle, the tetrahedron, and these are associated with the numbers 1, 2, 3, 4, in arithmetic progression. In fact, the geometrical figures can be described by the minimum number of independent points required to contain them: one or the unit for the point itself, two for the line segment, three for the triangle, and four for the tetrahedron. According to the passages from Aristotle the key concept is limit (or extremity, boundary).

---

[2] Speusippus was the son of Plato's sister and his immediate successor as head of the Academy. He was strongly influenced by Pythagorean philosophy.

The neo-Pythagorean author Nicomachus of Gerasa provides an embroidered account of the old Pythagorean theory in his *Introduction to Arithmetic*. In this later account the dimension concept is brought to the fore with an informal philosophical gloss (or implicit analysis). Modern translations of D'Ooge and Guthrie translate the Greek word 'διάστημα' as both 'dimension' and 'interval;' in other contexts the same word may be translated (according to Liddell, Scott and Jones' *Greek-English Lexicon*) as 'difference' or 'extension.' [3] The text in translation is as follows [Nicomachus of Gerasa, 1926, Book II, chapter 6, pp. 237–238; Book II, chapter 7, page 239] [Guthrie, 1962, page 261]:

> Unity, then, occupying the place and character of a point, will be the beginning of intervals and of numbers, but not itself an interval or a number, just as the point is the beginning of a line, or an interval, but is not itself a line or an interval. Indeed, when a point is added to a point, it makes no increase, for when a non-dimensional thing is added to another non-dimensional thing, it will not thereby have dimension. . . .
>
> Unity, therefore, is non-dimensional [dimensionless, directionless, non-directional] and elementary, and dimension first is found and seen in 2, then in 3, then in 4, and in succession in the following numbers; for 'dimension' is that which is conceived of as between two limits.
>
> The first dimension is called 'line,' for 'line' is that which is extended in one direction. Two dimensions are called 'surface,' for a 'surface' is that which is extended in two directions. Three dimensions are called 'solid,' for a 'solid' is that which is extended in three directions. . . .
>
> . . .
>
> The point, then, is the beginning of dimension, but not itself a dimension, and likewise the beginning of a line, but not itself a line; the line is the beginning of surface, but not surface; and the beginning of the two-dimensional, but not itself extended in two directions. Naturally, too, surface is the beginning' of body, but not itself body, and likewise the beginning of the three-dimensional, but not itself extended in three directions.
>
> Exactly the same in numbers, unity is the beginning of all number that advances unit by unit in one direction; linear number is the beginning of plane number, which spreads out like a plane in one more dimension, and plane number is the beginning of solid number, which possesses a depth 3 in the third dimension, besides the original ones.[Ch2, 3]

In the second passage, Nicomachus describes a link between the implicit dimensions and the Pythagorean figurate numbers, the latter being a part of Greek arithmetic.

The primitive theory of the intermediate stage might be described as static. There is no indication of how generation is accomplished in the hierarchical sequence, point, line, surface, solid. Yet the static concept of limit or boundary is a basis of explanation. In this explanation one has a unifying static theory of the four geometrical elements together with the notion of limit or boundary.

[3] I wish to thank Brendon Lasell of St. John's College, Annapolis, Maryland, for pointing out this translation issue to me. I am leaving the given translation as is.

Then in what seems to be a later Pythagorean theory it is possible to find a more dynamic explanation of generation, for the key concept is motion. This later theory is succinctly described in two passages, one from Aristotle (*De anima* (*On the Soul*), I.4, 409a4; quoted in [Kirk & Raven, 1957, page 254]):

> For they say that the movement of a line creates a plane and that of the point a line; and likewise the movements of units will be lines. For the point is a unit having position.[Ch2, 4]

and one from Sextus Empiricus (ca.160–ca.210 CE)) [Sextus Empiricus, 1936, pp. 346–349]:

> But some assert that the body [solid] is constructed from one [a single] point; for this point when it has flowed produces the line, and the line when it has flowed makes the plane, and this when it has moved towards depth generates the body which has three dimensions. But this view of the (later) Pythagoreans differs from that of the earlier ones. For these latter formed the numbers from two principles, the One and the Indefinite Dyad, and then, from the numbers, the points and the lines and both the plane and the solid forms; but the former build up all of them from a single point. For from this the line is produced, and from the line the plane, and from this the body.[Ch2, 5]

In this unified dynamic motion theory of generation the basic geometrical sequence, point, line segment, triangle, tetrahedron, is replaced by the sequence, point, line segment, square, cube, generated by motion, successively in that order. This more advanced theory is often called the fluxion theory. The four geometrical elements are unified in a theory using motion. It can be viewed as having had a significant impact on later mathematicians, such as Newton and Riemann (see Sections 3.4 and 5.7, respectively).

2.3 Aristotle

I shall next consider Aristotle and some texts of Aristotle that appear to relate to dimension. Throughout the extant Aristotelian texts there are many mathematical examples supporting Aristotle's philosophical investigations and analyses, and in a number of places he seems to touch on dimension, doing so implicitly. He gives a direct 'theory of dimension' based on the notions of divisibility and continuity—linked concepts in his system of thought—in *On the Heavens* (*De caelo*) (I.l, 268a4-13, 268a20-268b5) and *Metaphysics* (V.6, 1016b23-31; V.13, 1020a7-14) (see [Heath, 1949, pp. 159–160, 206-207]). Part of the passage in *On the Heavens* (*De caelo*) is of special interest [Aristotle, 1984a, page 447] (cf. the slightly different translation of [Heath, 1949, page 159]):

> For of things constituted by nature some are bodies and magnitudes, some possess body and magnitude, and some are principles of things which possess these. Now a continuum is that which is divisible into parts always capable of

> subdivision, and a body is that which is every way divisible. A magnitude if divisible one way is a line, if two ways a surface, and if three a body. Beyond these there is no other magnitude, because the three dimensions are all that there are, and that which is divisible in three directions is divisible in all. For, as the Pythagoreans say, the universe and all that is in it is determined by the number three, since beginning and middle and end give the number of the universe, and the number they give is the triad.[Ch2, 6]

In this standard Oxford English translation the translator, J.L. Stocks, uses the word 'dimension,' though, strictly speaking, such a specific word is not in the Greek text. However, the concept of dimension would seem to be meant. In his Loeb translation Guthrie does not use the word 'dimension.' Again, the concept of dimension may be meant, though the specific concept given is divisibility [Aristotle, 1939, page 5]:

> For the sum of physically constituted entities consists of (a) bodies and magnitudes, (b) beings possessed of body and magnitude, (c) the principles or causes of these beings.
>
> The continuous may be defined as that which is divisible into parts which are themselves divisible to infinity, body as that which is divisible in all ways. Magnitude divisible in one direction is a line, in two directions a surface, in three directions a body. There is no magnitude not included in these; for three are all, and "in three ways" is the same as "in all ways." It is just as the Pythagoreans say, the whole world and all things in it are summed up in the number three.

More importantly, Aristotle explicitly adopts in this text a Pythagorean metaphysical argument for the 3-dimensionality of physical bodies. The fact that Aristotle relies on Pythagorean ideas suggests that these earlier philosopher-mathematicians likely took an interest in this special problem of dimension concerning the physical world. According to Simplicius in a commentary on Aristotle's work, Claudius Ptolemy later wrote an entire book *On Dimension* in which he put forward a better argument for the 3-dimensionality of bodies and the universe [Thomas, 1941, pp. 410–413]. Unfortunately, Ptolemy's book is not extant.

2.4 Euclid

It is important to examine the work of Euclid in the discussion of ancient Greek ideas related to dimension. In Euclid's *Elements* informal ideas relating to dimension are expressed through definitions. The definitions are among the first principles, the basics, of mathematical science. Other first principles include postulates and common notions or axioms. These are followed by a deductive structure of propositions, constructions, and theorems. The following definitions concerning the basic objects of geometry are found in Euclid's *Elements* [Heath, 1956, Vol. 1, page 153; Vol. 3, page 260] (original edition [Heath, 1926]; cf. a more recent translation by Fitzpatrick [Fitzpatrick, 2008]):

Book I.

Definitions.

1. A **point** is that which has no part.
2. A **line** is breadthless length.
3. The extremities of a line are points.
5. A **surface** is that which has length and breadth only.
6. The extremities [boundaries] of a surface are lines.

Book XI.

Definitions.

1. A **solid** is that which has length, breadth, and depth.
2. An extremity of a solid is a surface.[Ch2, 7]

It is possible to distinguish two rudimentary implicit theories of dimension adumbrated by these definitions: a direct theory given by definitions I.l, I.2, I.5, and XI.l and an indirect theory hinted at by definitions I.3, I.6, and XI.2. Following his usual way of inclusive compromise, Euclid presents both theories, although other sources separate them. The first theory offers a direct link between the basic geometrical figures and finite named dimensions. The definitions in effect nominate and enumerate the dimensions of the various figures. Points have no dimension ('no part'); while lines have one ('length'); surfaces, two ('length and breadth'); and solids, three ('length, breadth, and depth'). Euclid only implicitly touches on the broader concept of dimension. Neither does he provide any explanation of the concept, nor does he use any general term for it. Hence, these definitions hardly add up to a complete, well-defined theory of dimension. Rather, the general dimension concept is implicit.

Euclid's indirect theory, comprised in his subsidiary definitions 1.3, I.6, and XI.2 and based on the concept of extremity, has an affinity with the modern recursive definition of dimension using the concept of boundary. Euclid's subsidiary definitions suggest a progression from points to lines to surfaces to solids, or more accurately, the reverse progression. In modern dimension theory a set or space is n-dimensional according to a recursive definition, whereby n is the least integer for which every point has arbitrarily small neighborhoods with boundaries at most $(n - 1)$-dimensional (and the empty set is assigned dimension -1). Creators of the modern theory, such as Poincaré and Menger, have stressed the connection between the modern and Euclidean definitions, but one should not read too much of the modern theory into the ancient texts. The ancient mathematicians did not operate with genuine recursive definitions, even if their progressions have a semblance of recursion.

According to Heath [Heath, 1956, vol. 1, pages 155–156] the subsidiary definitions are older than Euclid's main ones, and he points to a passage in Aristotle (*Topics*, VI.4, 141b20) (see [Heath, 2016, pp. 85–86]) in which the philosopher speaks of the former as the definitions. Moreover, Aristotle provides a supplement to them (*Metaphysics*, XI.2, 1060b12-17) [Heath, 1949, page 224]:

> If we suppose lines or what immediately follows them (I mean the primary surfaces) to be principles, these are at all events not separable substances but are sections and divisions, the one of surfaces, the other of bodies (as points are of lines); they are also extremities or limits of the same things; but all of

> them subsist [originate] in other things, and no one of them is separable [none of them have an independent existence].[Ch2, 8]

Thus the various geometrical objects can be sections and divisions as well as extremities of the next higher ones in the hierarchy. Aristotle is, in fact, critical of all the definitions dependent upon the concepts of extremity and division (*Topics*, VI.4, 141b21) [Heath, 1949, pp. 85–86]:

> All these definitions explain the prior by means of the posterior, for they say that a point is an extremity of a line, a line of a plane, and a plane of a solid.[Ch2, 9]

To Aristotle's mind the progression from solid down to point, from posterior to prior in the older definitions, is mistaken, and so these are unscientific. In spite of Aristotle's criticism of the logic of the extremity definitions, these are the most interesting from the modern dimension-theoretic viewpoint (viewed with lots of hindsight, of course). The downwards progression from solid to point is what makes them seem recursive to a modern reader with a mathematical background. Heath remarks [Heath, 1956, vol. 1, 155–156] that Plato may well have been an object of Aristotle's criticism. Going beyond this criticism, I think that it may be possible to look for the origins of the Euclidean extremity definitions. It seems plausible that ultimately these derive from the first Pythagorean theory of generation. In that theory the essential concept was limit or extremity and, in particular, the line segment was viewed as dependent on its two endpoints. I think that a reasonable conjecture is that there is an indirect historical link between the very early Pythagorean theory and the extremity definitions. It would certainly be interesting to trace this link further through the intermediate generations of philosophers and mathematicians. It also seems plausible that just as the early Pythagoreans sought to explain the generation of physical things through numbers then points, lines, surfaces, and solids, the later extremity definitions constitute an attempt to explain the connections among the geometrical figures and to explain in a very rudimentary way a low finite dimension concept. Thus, the origins of an ancient theory resembling a simple dimension theory appear to lie in a cosmogonical theory aimed at explaining the beginnings of physical things.[4]

[4] One might assess Euclid's two theories from the viewpoint of modern logic as simple relational theories. The basic elements are the fundamental geometrical objects, point, line, surface, and solid. The relations are the pairwise relations among the elements and the entire hierarchy from point to solid.

2.5 Diophantus, Greek Algebra, Species of Numbers

In later Greek mathematics the *Arithmetica* of Diophantus of Alexandria (ca. 200–285 CE)[5] is an important work dealing with the algebra of numbers. It is a text showing how to find solutions for numerical problems, or as one would now say solving algebraic equations. The calculations concern numbers, numbers being collections of units, thus, positive integers, and fractions of positive integers. In modern terms, Diophantus deals with the solution of equations of positive integers and fractions of positive integers.

In 1910 Thomas Heath provided a translation, a lengthy paraphrase of Diophantus' work [Heath, 1910], and very recently Jean Christianidis and Jeffrey Oaks provided a full translation of Diophantus' work and included extensive commentary [Christianidis & Oaks, 2023]. There is also a classic study of Greek algebra by Jacob Klein [Klein, 1968].

Near the beginning of *Arithmetica* Diophantus includes definitions of kinds or species of numbers (cf. the discussion in [Christianidis & Oaks, 2023, pp. 26–79]). The names used are suggestive of higher dimensions [Heath, 1910, page 129] [Christianidis & Oaks, 2023, pp. 275–276] [Thomas, 1941, pp. 518–521]:

> As you know, in addition to these things, that all numbers are made up of some multitude of units, it being clear that their formulation has no limit. Among them are—
>
> *squares* [δύναμις or τετραγώνων], which are formed when any number is multiplied by itself, the number itself is called the *side of the square*;
>
> *cubes* [κύβων], which are formed when squares are multiplied by their sides;
>
> *square-squares* [δυναμοδυνάμεων], which are formed when squares are multiplied by themselves;
>
> *square-cubes* [δυναμοκύβων], which are formed when squares are multiplied by the cubes formed from the same side;
>
> *cube-cubes* [κυβοκύβων], which are formed when cubes are multiplied by themselves; and it is from the addition, subtraction, or multiplication of these numbers or from the ratio which they bear one to another or to their own sides that most arithmetical problems are formed. . . .[Ch2, 10]

Beyond these names there also is notation for positive fractions [Thomas, 1941, pp. 522–523].

While these names for numbers are suggestive of 'higher-dimensional' geometric objects, a kind of 'higher-dimensional geometric algebra,' the consensus is that this is not the case. These terms are merely conveniences for discussing numbers of

[5] Information on Diophantus and his work can be found in [Christianidis & Oaks, 2023, pp. 3–25], `https://mathshistory.st-andrews.ac.uk/Biographies/Diophantus/`, `https://en.wikipedia.org/wiki/Diophantus` and `https://www.encyclopedia.com/people/science-and-technology/mathematics-biographies/diophantus-alexandria#2830901182`.

higher powers, a mere *façon de parler*. In some cases there are alternatives without any geometrical connotation.

2.6 Pappus

In a later and quite different context or problem situation, Pappus (c. 300 CE) says in connection with the famous locus problem of five or six lines [Thomas, 1941, pp. 600–603]:

> . . . since no figure can be contained in more than three dimensions. It is true that some recent writers have agreed among themselves to use such expressions, but they have no clear meaning when they multiply the rectangle contained by these straight lines with the square on that or the rectangle contained by those. They might, however, have expressed such matters by means of the composition of ratios and have given a general proof both for the aforesaid propositions and for further propositions after this manner. . . . [Ch2, 11]

In this context Pappus' advice is negative: stick to tradition and avoid the impossible!

2.7 Place of Definitions in Ancient Greek Mathematics

At this point I would like to consider the subject matter in a broader context by examining the nature of definitions as part of the first principles of ancient Greek mathematics and science as well as philosophy. For this examination the focus is on Aristotle, who philosophically analyzes first principles in general in the *Posterior Analytics*, and Euclid, who uses definitions, postulates, and common notions or axioms as first principles as a basis for the mathematics in *The Elements*.

Aristotle's *Posterior Analytics* [Aristotle, 1984a, pp. 114–166] was written some decades before Euclid's *Elements*. Aristotle died in 322 BCE, and *The Elements* dates from around 300 BCE when Euclid lived and taught as a mathematician in Alexandria. In his work Aristotle considers the general principles of deductive science, his theory of demonstration or demonstrative science. Many of his examples are from mathematics. In Book I chapter 2, 71b 10 - 72b 5 [Aristotle, 1984a, pp. 115–116] and chapter 10, 76a 31 - 77a 4 [Aristotle, 1984a, pp. 124–125], he sets out the basics of a logico-deductive system. He includes posits, axioms, and definitions among the fundamentals. Thus in Book I chapter 2 he states (72a 15 - 24) [Aristotle, 1984a, page 116]:

> An immediate deductive principle I call a posit if one cannot prove it but it is not necessary for anyone who is to learn anything to grasp it; and one which it is necessary for anyone who is going to learn anything whatever to grasp, I call an axiom (for there are some such things); for we are accustomed to use

> this name especially of such things. A posit which assumes either of the parts of a contradiction—i.e., I mean, that something is or that something is not—I call a supposition; one without this, a definition. For a definition is a posit (for the arithmetician posits that a unit is what is quantitatively indivisible) but not a supposition (for what a unit is and that a unit is are not the same).[Ch2, 12]

Aristotle elaborates these matters in Book I, chapter 10 [Aristotle, 1984a, pp. 124–125].

Aristotle's assertions about demonstrative science and its constituent principles and his choice of mathematical examples as illustrations suggest that axiomatic systems of mathematics like *The Elements* existed before the time of Aristotle's writing of the *Posterior Analytics*. The question then arises: Was Euclid in *The Elements* following the methodology of the deductive sciences laid down by Aristotle? Euclid does begin *The Elements* with a list of definitions, five postulates, and five common notions or axioms; Euclid also includes more definitions in later books of the work. These basics appear to be similar to the basics of Aristotle's theory of demonstrative science. It would appear that Aristotle captured the basic methodology of mathematical practice of his day, and this methodological material may have influenced Euclid some time later.

In the introductory material to his classic, early twentieth-century translation of *The Elements* Heath discerns [Heath, 1956, vol. 1, pp. 116–124] a connection between Aristotle's notion of demonstrative science and what Euclid puts into practice in *The Elements*. Heath puts forward the evidence for this connection over several pages. However, more recent scholars have called into question this direct connection by making detailed analyses of Euclid's text and mathematical practice and offering differing results and conjectural interpretations.

Euclid's work has many definitions. There are 131 definitions (or 132, depending on the acceptance of an addition in Book VII) spread throughout the books of *The Elements*. They appear to serve broader purposes than those that Aristotle appears to envisage. In his thorough study of the Euclidean *Elements*, Ian Mueller [Mueller, 1969] [Mueller, 1981] notes that the many definitions in *The Elements* serve various functions. Mueller states [Mueller, 1969, page 294]:

> The definitions should be looked at as attempts to make clear the meanings of the terms to be used before argumentation begins, that is, to make clear the nature of the objects to be studied. That the most fundamental definitions (e.g., of point, line, straight line) succeed only with persons who already have some idea what the objects in question are does not really matter if these definitions are taken to represent preliminary agreements among people of presumably normal intelligence.

Some definitions are used to inform the reader broadly of the material under study; some are applied in the demonstrations and constructions; some add extra information; some are really not used at all in the *The Elements*.

On the matter of broader Euclidean mathematical derivations Mueller concludes [Mueller, 1969, page 295], 'that a Euclidean derivation is an experiment performed on idealised physical objects, the experiment being limited by preliminary agreements (first principles) concerning the nature of the objects, some of their properties, and the operations that may be performed on them.' The first principles include the definitions as well as the postulates and common notions or axioms. In his fuller analysis of *The Elements*, Mueller gives a further explanation of definitions [Mueller, 1981, 38-40]. There he contrasts Euclid's use with the much more modern, quite different use of definitions as embodied, for example, in Hilbert's *Grundlagen der Geometrie* (*Foundations of Geometry*) (1899) [Hilbert, 1899].

Considering the specifics of Greek mathematical language, Reviel Netz [Netz, 2003, pp. 89–103] analyzes the role of definitions in Greek mathematics. Having derived some of his ideas on definition from Fowler [Fowler, 1987], Netz states [Netz, 2003, page 94]:

> I trusted Heiberg and assumed that Greek mathematical texts often started with neatly numbered sequences of definitions. Seeing a specimen first page of a manuscript of the *Elements*, and glancing at Heiberg's apparatus, should dispel this myth. Numbers were not part of the original text.
>
> That numbers are absent from the original is not just an accident, the absence of a tool we find useful but the Greeks did not require. The absence signifies a different approach to definitions. The text of the definitions appears as a continuous piece of prose, not as a discrete juxtaposition of so many definitions. ... Mathematical texts start, most commonly, with some piece of prose preceding the sequence of proved results. Often, this is developed into a full 'introduction,' usually in the form of a letter (prime examples: Archimedes or Apollonius). Elsewhere, the prose is very terse, and supplies no more than some reflexions on the mathematical objects (prime example, Euclid).[6]

In his recent study of 'Aristotle and Mathematics' Henry Mendell [Mendell, 2019, page 42] concludes that, 'this correspondence between Aristotle's conception of first principles and Euclid's in *Elements* I is tenuous at best.' He further states [Mendell, 2019, pp. 42–43]:[7]

> A more fundamental distinction between Aristotle's treatment of first principles and those found in Greek mathematics is that Aristotle seems to think that each first principle has both a logical and an explanatory role in a treatise. Yet it is typical, especially in treatises [of mathematics] which are introductory to a topic, to have principles which serve a logical and explanatory role, but also to have principles whose only explicit role is pedagogical. For they serve no obvious role in the demonstrations. Such might be the definitions of point and line in *Elements* I. Hence, if there is a relation between Aristotle's conception of first principles and those of the mathematicians, Aristotle provides an ideal framework based on contemporary mathematical practice and which may or may not have been noticed by authors such as Euclid.

---

[6] I have not followed what Netz states in this text in my own presentation of Euclid in Section 2.4, but rather I have followed the earlier tradition from Heiberg using numbered definitions for convenience. By doing so, I do not think that I have invalidated my interpretation in Section 2.4.

[7] Compare also the earlier work of Wilbur Knorr [Knorr, 1980, pp. 164–168].

Aristotle's rendition of a logico-deductive or demonstrative science may have some similarities to modern thinking on an axiomatic theory, but it is definitely not the same as modern treatments. Also, it is necessary to look at *The Elements* of Euclid as quite different in approach from the approach of modern mathematics to axiomatic geometry as embodied in, for example, Hilbert's *Grundlagen der Geometrie* (*Foundations of Geometry*) [Hilbert, 1899]. The *Grundlagen der Geometrie* (*Foundations of Geometry*) offers a purely logical rendition of the subject. It is a refinement of extensive logical inquiries into the subject that occurred toward the end of the nineteenth century. On this matter see especially Mueller [Mueller, 1969] [Mueller, 1981] (and more generally cf. [Torretti, 2019] and [Gray & Ferreirós, 2021]).

2.8 Concluding Remarks

From the brief history and analysis just given it should be clear that several Greek philosophers and mathematicians had specific interests in problems implicitly connected with the dimension concept. Their efforts on these problems spanned centuries in ancient times. Their resulting treatments were almost exclusively confined to considering the basic objects of geometry, the point, line, surface, and solid, that is, the low-dimensional geometrical objects, and their mutual relations. Overall, their efforts yielded a variety of informal theories implicitly relating to dimension.

Chapter 3
Dimension from the Middle Ages to the End of the Eighteenth Century

3.1 Introduction

While this chapter covers an extensive period from the fourteenth century to the end of the eighteenth century, it is quite short. The reason for its shortness is that during these centuries I found only a few instances of new thinking on dimension and a few special references to ways of approaching problems of dimension.

3.2 Medieval and Early Modern European Concepts of Dimension

I shall first consider the important fourteenth-century French philosopher Nicole Oresme (1323?–1382) [Kirschner, 2021]. Avoidance of the impossible is characteristic of the prehistory of higher-dimensional spaces, that is, avoiding those spaces having more than three dimensions. In the fourteenth century when Oresme proposed a theory dealing with the quantitative measure or graphical representation of the intensities of qualities in things in his *Tractatus de configurationibus qualitatum et motuum* (*Treatise on the qualities and motions of configurations*), probably composed in the 1350s, he expressly denied the need to consider a 4-dimensional space even though his ideas led him in that direction. In chapter 4 of part I of this magnificent work Oresme introduced graphical representations of intensities of qualities in various geometrical subjects [Oresme, 1968, pp. 172–177]. For a point subject a line segment will represent a quantitative measure of a quality of that subject. A bounded surface will represent the measure of a quality in a linear subject, while a bounded solid will do for a planar subject. In the case of a solid subject Oresme suggests that it be broken into an infinite sequence of surfaces, each with a solid 'graph,' such that the graphs interpenetrate or are in mathematical superposition. However, he asserts categorically that this representation does not take place in the fourth dimension, since that does not exist and cannot be imagined. Notwithstanding his assertion,

D. M. Johnson, *Theories of Dimension from Ancient to Modern Times*, Sources and Studies in the History of Mathematics and Physical Sciences, https://doi.org/10.1007/978-3-032-02682-8_3

Oresme in effect has described a 3-dimensional image of a 4-dimensional solid. In all probability this was the first time that anyone had done so.

Following the lead of Diophantus' algebra, many Arab and early Western algebraists referred to supersolid magnitudes, although they did not intend to introduce spaces of dimensions greater than three. A noteworthy remark related to this tradition comes in Michael Stifel's (ca. 1487–1567) commentary to Christoff Rudolff's *Coss* (*Algebra*) [Rudolff & Stifel, 1554, folio 9]. Stifel points out that whereas in geometry we cannot progress beyond the 3-dimensional cube or solid, in algebra we can indeed have a progression (through multiplication) that goes beyond the cube to a 'solid line' ('corporliche lini,' which is 4-dimensional), a 'solid surface' ('corporliche superficies,' 5-dimensional), and so forth. In this possibility lies the great advantage of algebra over geometry. Pure algebra can break the bonds of the ancient geometrical formulation of the subject.

3.3 John Wallis and Henry More

A century later the Englishman John Wallis (1616–1703) inveighed against the mixing of geometrical terms or ideas with algebra in his important *Treatise of Algebra* [Wallis, 1685, page 126]:

> For whereas Nature, in propriety of Speech, doth not admit of more than *Three* (Local) Dimensions, (Length, Breadth and, Thickness, in Lines, Surfaces and Solids;) it may justly seem very improper, to talk of a Solid (of three Dimensions) drawn into a Fourth, Fifth, Sixth, or further Dimension.
>
> A Line drawn into a Line, shall make a Plane or Surface; this drawn into a Line, shall make a Solid; But if this Solid be drawn into a Line, or this Plane into a Plane, what shall it make? a *Plano-plane*? That is a Monster in Nature, and less possible than a *Chimera* or *Centaure*. For Length, Breadth and Thickness, take up the whole of *Space*. Nor can our Fansie imagine how there should be a Fourth Local Dimension beyond these Three.

Wallis' forceful words carry a message very similar to Pappus' of many centuries before (see Section 2.6). Notwithstanding Wallis' rhetoric and 'common sense,' a few mathematicians in the seventeenth century dared to conceive of the geometrical fourth dimension. No less than Descartes and Pascal were among their number, although they only flirted with the concept (see Wieleitner's paper [Wieleitner, 1925] for references).

In quite another sphere the Cambridge Platonist Henry More (1614–1687) defined a three-dimensional absolute space and included a fourth dimension for spirits ('Spissitudo essentialis'). He describes this idea in his *Enchiridion Metaphysicum* (*Manual of Metaphysics*) [More, 1671, page 384] (cf. [Cajori, 1926, pp. 399–401] [Giannetto, 2023] [Henry, 2023, pp. 35–44]). In the nineteenth century J.K.F. Zollner (1834–1882) held a view similar to More that led to some amusing consequences.

3.4 Isaac Newton

With regard to general approaches to dimension, the ancient Greek extremity theory of dimension and the motion theory of dimension (or 'fluxion theory' to use Newton's later term for the latter) have been very influential over the centuries (see Chapter 2). Thus, not surprisingly, Isaac Newton (1642–1727) called upon the ancient motion theory as a support for his fluxional principles and as one possible foundation for his version of the calculus. He refers directly to the theory in his well-known publication in Latin of 1704 *Tractatus de quadratura curvarum* (*Tract on the quadrature of curves*), which was published as an appendix to the *Opticks* (*Optics*) [Newton, 1704]. The *Tractatus* (*Tract*) derives from an earlier manuscript of 1693 (see `https://cudl.lib.cam.ac.uk/view/MS-ADD-03962/1` to view this manuscript and see [Whiteside, 1981, vol. 3, pp. 373–385; vol. 7, pp. 24–182; vol. 8, pp. 92–167] for Newton's development of the work). Newton begins the work by echoing ancient ideas and relying on intuitions about motion [Newton, 1704, pp. 165–166] (English translation, [Wilkins, 2002, page 1]; also [Struik, 1969, pp. 303–304]):

> I don't here consider Mathematical Quantities as composed of Parts *extreamly small*, but as *generated by a continual motion*. Lines are described, and by describing are generated, not by any apposition of Parts, but by a continual motion of Points. Surfaces are generated by the motion of Lines, Solids by the motion of Surfaces, Angles by the Rotation of their Legs, Time by a continual flux, and so in the rest. These *Geneses* are founded upon Nature, and are every Day seen in the motion of Bodies. And after this manner the Ancients by carrying moveable right Lines along immoveable ones in a Normal Position or Situation, have taught us the Geneses of Rectangles.
>
> Therefore considering that Quantities, encreasing in equal times, and generated by this encreasing, are greater or less, according as their Velocity by which they encrease, and are generated, is greater or less; I endeavoured after a Method of determining the Quantities from the Velocities of their Motions or Increments, by which they are generated; and by calling the Velocities of the Motions, or of the Augments, by the Name of *Fluxions*, and the generated Quantities *Fluents*, I (in the years 1665 and 1666) did, by degrees, light upon the Method of *Fluxions*, which I here make use of in the *Quadrature of Curves*.[Ch3, 1]

In this way Newton references the 'Ancients' as an intellectual support to his ideas about the calculus. At this place Newton is dealing with the problem of giving some foundations to a version of his calculus—quite a different problem than the ancients considered. In effect, Newton is seeking to introduce small or infinitesimal changes or motions as a support for his calculus. His method of fluxions was only one way he dealt with foundations for his calculus.

3.5 Eighteenth-Century Concepts of Dimension

Around the middle of the eighteenth century a serious view of the possibility of higher-dimensional spaces began to take shape. Immanuel Kant (1724–1804) touched upon the possibility of a science of hyperspaces in his very first publication, *Gedanken von der wahren Schätzung der lebendigen Kräfte und Beurtheilung der Beweise derer sich Herr von Leibnitz und andere Mechaniker in dieser Streitsache bedienet haben, nebst einigen vorhergehenden Betrachtungen welche die Kraft der Körper überhaupt betreffen* (*Thoughts on the true estimation of living forces* [*vis viva*] *and the assessment of the demonstrations that Leibniz and other scholars of mechanics have made use of in this controversial subject, together with some prefatory considerations pertaining to the force of bodies in general*) [Kant, 1746] [Kant, 2012b]. In this work dealing mainly with the well-known (at the time) question of *vis viva* Kant examines the cosmological problem of the 3-dimensionality of physical space, but he feels unable to provide an absolutely convincing explanation for this fact of nature (or of the universe). Nonetheless, he conjectures that it is probably explained by the inverse-square law governing the forces between substances (masses) in the universe. However, in another possible world this special law of action could be different, so that to cover all possible worlds one would need a general science of spaces of diverse dimensions [Kant, 1746, page 13] [Kant, 2012b, page 28]:

> A science of all these possible kinds of space would undoubtedly be the highest geometry that a finite understanding could undertake. The impossibility we notice in ourselves of representing to ourselves a space of more than three dimensions seems to me to stem from the circumstance that our soul likewise receives impressions from without according to the inverse-square law relation of distances, and because its nature is itself constituted so as not only to be it thus affected, but also to act external to itself in this way.[Ch3, 2]

Kant had a continuing interest in philosophical problems about space and geometry throughout most of his intellectual life. These interests were prominent in both his precritical and critical periods. In particular, he had further thoughts about conceiving a science of n-spaces. In his precritical essay 'Von dem ersten Grunde des Unterschiedes der Gegenden im Raume' ('Concerning the ultimate ground of the differentiation of directions in space') (1768) [Kant, 1768] [Kant, 1992a] , Kant had the penetrating insight that in our 3-dimensional world there exist bodies that are similar and have the same distances among all their respective parts, and yet they cannot be made directly congruent by superposition on the same space. The left and right hands give an immediate example of this paradoxical phenomenon. They are mirror images of one another, but they cannot be superimposed by simple translation and rotation. It is now known that in 4-space this phenomenon will no longer obtain—a geometrical fact that Kant probably could not know. Finally, one should note that for Kant geometry was primarily a science of space as we know it. Although he had some conception of other, more abstract geometries, as clearly evidenced by his prediction of a science of n-spaces, he still mainly thought of geometry as linked to physical space.

Late in the eighteenth century the two Greek theories of dimension, the extremity theory and the motion theory, are given an airing in the great *Encyclopédie* (*Encyclopedia*) of Denis Diderot (1713–1784), Jean Le Rond d'Alembert (1717–1783), and others (published 1751–1766; with supplements by others until 1780) [Diderot et al., 1765]. In his article 'POINT' the main author, d'Alembert, gives most prominence to the extremity theory, because he espoused an abstractionist philosophy of geometry. Only 3-dimensional solids really exist; points, lines, and surfaces merely exist by abstraction and are the boundaries of their respective higher-dimensional figures. However, he quite naturally includes a statement of the motion theory (or fluxion theory à la Newton) [Diderot et al., 1765, vol. XII, page 871]:

> If one pictures a *point* as flowing, it will trace a line; and a line that flowed would generate a surface, and so forth. This way of looking at the generation of dimensions or the properties of bodies appears to be the primary foundation of modern geometry, i.e., of the analytical geometry, which uses the differential and integral calculus.... [Ch3, 3]

For the entry 'DIMENSION' the main author, d'Alembert, enumerates merely the names of the three dimensions with an indication of the progression from line to surface to solid [Diderot et al., 1765, vol. IV, page 1009]:

> DIMENSION is the extension of body considered insofar as it is measurable, or susceptible of measurement....
>
> Thus, as we conceive that bodies are extended in *length*, *breadth*, and *depth* or *thickness*, we also conceive the three *dimensions* in the matter; length all alone is called *line*; length combined with breadth has the name of *surface*; finally, the length, breadth, and depth or thickness produces that named a *solid*. ... [Ch3, 4]

Also, Jean d'Alembert inserted a concrete suggestion of a space of more than three dimensions into the 'DIMENSION' article [Diderot et al., 1765, vol. IV, page 1010]:

> I have said... that it is not possible to conceive of more than three *dimensions*. An intelligent man of my acquaintance believes nevertheless that one can regard time as a fourth *dimension*, and that the product of time by a solid in this way will be a product of four *dimensions*; this idea can be disputed, but it seems to me to have some merit even though this may only be that of novelty. [Ch3, 5]

Joseph Louis Lagrange (1736–1813) later followed in the spirit of this suggestion, when he asserted in his *Theorie des fonctions analytiques* (*Theory of analytic functions*) [Lagrange, 1797, page 223] that mechanics could be taken to be a geometry of four dimensions, employing x, y, z as coordinates for space and t for the fourth dimension of time. Thus, by the end of the eighteenth century some well-known philosophers and mathematicians had pronounced on the desirability of considering hyperspaces, although no one had yet fashioned a genuine science of such 'unimaginable' geometrical objects.

3.6 Concluding Remark

At this point I end my broad survey of ideas on dimension from the middle ages to the end of the eighteenth century. Clearly, the problems that mathematicians and mathematical physicists considered and the problem situations they found themselves in varied considerably over the centuries. In these years the views, treatments, and results are very scattered. There was no unifying problem or set of problems to bring together the results. In the next chapter I shall provide a concentrated analysis of Bolzano's ideas on dimension and the fundamentals of geometry.

Chapter 4
Bolzano's Investigations into Geometry and Dimension

4.1 Introduction

In this chapter I shall examine in detail the mathematical investigations of Bernard Bolzano (1781–1848) into geometrical concepts relating to dimension—investigations that spanned his entire career. They were concentrated around his quest for what he deemed to be better and more precise or 'rigorous' definitions of the geometrical concepts of line, surface, solid, and extension.[1]

It is surprising to discover that many years before the main development of dimension theory into a fully fledged mathematical theory in the late nineteenth and twentieth centuries Bernard Bolzano examined a cluster of fundamental geometrical

[1] This chapter is a much revised version of my paper [Johnson, 1977]. Many new sources of information on Bolzano's life and work have been published since my paper appeared in 1977, and I have availed myself of a great many of them. The following are the main sources I have consulted. First, there is the magnificent *Bernard Bolzano-Gesamtausgabe* [Bolzano, 1969], which is an ongoing project to publish all of Bolzano's works, papers, and documents in his *Nachlass*. Many volumes have appeared, and many more are in preparation or planned. Bolzano himself wrote an autobiography [Bolzano, 1836]. There are now two main biographies of Bolzano of recent date [Winter, 1969] and [Rusnock & Šebestík, 2019]. Winter's biography is a classic, which I used in my earlier work. The work by Rusnock and Šebestík is an excellent up-to-date source of biographical information and philosophical interpretation. It is the most extensive to date. There are several main works covering Bolzano's logic and mathematics, including: [Berg, 1965], [Folta, 1966], [Folta, 1981], [Simon, 1981], [Šebestík, 1992], [Rusnock, 2000], [Behboud, 2000], [Bolzano, 2004b], [Rusnock & Šebestík, 2013], [Šebestík, 2017], [Morscher, 2018]. Four works have appeared covering Bolzano's geometrical and 'proto-topological' investigations: [Simon, 1981], [Šebestík, 1992, pp. 33–72], [Katětov & Simon, 1997], and [Rusnock, 2020]. Unfortunately, I have not been able to access an additional one, [Kraus, 2014]. English translations of Bolzano's main mathematical works by Steve Russ have appeared, collected in the wonderfully extensive [Bolzano, 2004a]. I have largely relied on my own translations, while consulting and comparing Russ' translations. Bolzano's *magnum opus*, his *Wissenschaftslehre* (*Theory of Science*) [Bolzano, 1837], has been translated into English (*Theory of Science*) in part in [Bolzano, 1972] and in its entirety in [Bolzano, 2014a]. There are many other articles and works related to Bolzano to be found in the bibliography.

I am grateful to Professor Rusnock for making comments on an earlier and a later version of this chapter. I found his comments most useful. They helped me clarify and correct my thoughts on Bolzano and his philosophical and mathematical investigations.

D. M. Johnson, *Theories of Dimension from Ancient to Modern Times*, Sources and Studies in the History of Mathematics and Physical Sciences, https://doi.org/10.1007/978-3-032-02682-8_4

problems that led him to formulate a theory of geometry, which, in the light of the later development of topology, looks very much like dimension theory, at least its first steps. It is well known to historians of mathematics that the great Bohemian priest, philosopher, logician, and mathematician from Prague prefigured much that is central to modern pure mathematics, especially in analysis and set theory. But in topology one can also count Bolzano as a precursor of more recent developments. Through a critical examination of the basic principles of Euclidean geometry, Bolzano was led to propose new definitions of key elementary geometrical concepts. In developing these definitions, he took what appears from a modern perspective as the point of view of set theory and point-set topology. These definitions have several features in common with the definitions of the dimension concept given later by Poincaré, Brouwer, Urysohn, and especially Menger.

Unfortunately, Bolzano's geometrical research suffered the same tragedy that befell much of his other mathematical and philosophical work: it did not receive due recognition either from his contemporaries (apart from a few friends or colleagues who were interested in philosophy and mathematics) or the immediately following generation of mathematicians. Yet in spite of this lack of influence on the main stream of topology and dimension theory, it is certainly worthwhile examining Bolzano's highly original way of dealing with geometrical problems. His investigations constitute a fascinating prelude, regrettably unnoticed by later mathematicians, to dimension theory. In the following I shall discuss the full details of Bolzano's contributions to 'topological' geometry.

4.2 Bolzano's Fundamental Geometrical Problem

What was the problem that led Bolzano to see geometry in a topological light and to formulate a rudimentary dimension theory? It was simply a search for exact definitions. In brief, Bolzano's problem was to define in a precise way the geometrical concepts of line (or curve), surface, and solid, and more generally to define geometrical extension or continuum. This fairly straightforward problem of definitions, conceived within the wider limits of Euclidean geometry, led Bolzano to break from the bonds of traditional geometry and to explore or 'invent' the entirely new domain of point-set theory, something like the beginnings of point-set topology. This special problem of definitions was certainly not the only geometrical problem that captured Bolzano's interest, but it was a prime underlying theme of his geometrical thought throughout his creative life.

Bolzano's interest in his definition problem began very early—by his own account at the age of 16—and continued throughout his life. His research on the problem neatly divides into two periods. His very first publication, which appeared in 1804 when he was 23, the tract *Betrachtungen über einige Gegenstände der Elementargeometrie* (*Considerations concerning some Objects of Elementary Geometry*) [Bolzano, 1804] (English translation in [Bolzano, 2004a, pp. 24–81]), contains hints about the problem, while his tract of 1817 *Die drey Probleme der Rectification, der Complanation und der Cubirung* (*The Three Problems of Rectification, Complanation, and Cubiture*) [Bolzano, 1817a] (English translation in [Bolzano, 2004a, pp. 278–344]) contains his tentative solution. In the latter he set down his first quasi-topological definitions of lines, surfaces, and solids, given according to set-theoretical and dimensional properties. The work of 1817 marks the end of the first period of Bolzano's investigations on this subject.

Later, in the 1830s and 1840s, Bolzano renewed his interest in his geometrical problem. This work forms a second period of geometrico-topological investigation, in which he revised and improved his earlier theory. The two main papers of this period are 'Uiber Haltung, Richtung, Krümmung und Schnörkelung bei Linien sowohl als Flächen sammt einigen verwandten Begriffen' ('Concerning the Disposition [Situation, Position, or Attitude], Direction, Curvature, and Torsion of Lines as well as Surfaces together with some Related Concepts') and 'Geometrische Begriffe, die Jeder kennt und nicht kennt. Versuch einer Erhebung derselben ins deutliche Bewußtseyn' ('Geometrical Concepts that Everyone Knows and yet does Not Know. An Attempt to Raise Them to Distinct Awareness'). Both of these were written around 1843–44, but were not published in his lifetime. The first paper 'Uiber Haltung' ('Concerning the Disposition') was published in 1948 [Bolzano, 1948b] and is planned to be published again in the *Bolzano-Gesamtausgabe* in [Bolzano, 20xxc]. The second was first published in [Bolzano, 2001a], the manuscript being [Bolzano, 0000b]. However, Bolzano included a short summary of his second 'dimension theory' in the *Paradoxien des Unendlichen* (*Paradoxes of the Infinite*), which appeared posthumously in 1851 [Bolzano, 1851](also [Bolzano, 2012] and in English translation in [Bolzano, 2014b] and [Bolzano, 2004a, pp. 590–678]).

To see why Bolzano took up his special geometrical problem, it is necessary to examine the origins of his interest in mathematics and the problem situation in traditional geometry around 1800 from which his work emerged.

The years from 1796 to 1804 were crucial in shaping Bolzano's intellectual development [Winter, 1969, pp. 21–32] [Rusnock & Šebestík, 2019, pp. 21–35]. As a young student he became intrigued with both mathematics and philosophy. From the start Bolzano looked upon mathematics with the eyes of a philosopher and foundationalist. In this respect he resembles Descartes, who saw a firm methodological link between mathematics and philosophy. In his autobiography Bolzano states [Bolzano, 1836, page 19]:

> My special *pleasure in mathematics* therefore rests only in its purely *speculative part*, or, in other words, I value in it only that which is at the same time *philosophy*.[Ch4, 1]

Throughout his career Bolzano was more deeply concerned with examining the logical foundations of mathematics than with creating new theories, deriving new results, or finding new applications. He was more a philosophical mathematician, a foundationalist, than a 'creative' mathematician. This is not to say that his insights into mathematical foundations were not extremely original. They certainly were. He undoubtedly broke new ground in various parts of mathematics and logic. Yet, to appraise Bolzano's contribution to mathematical thought, one must keep in mind the philosophical motivation for his work, a desire to have clear logical foundations.

One of the earliest influences on Bolzano's philosophical thought was Alexander Gottlieb Baumgarten's (1714–1762) widely available and much read and studied handbook *Metaphysica* (*Metaphysics*) (1779) [Baumgarten, 1779], which went through many editions. Baumgarten's textbook is largely a summary of the Leibniz–Wolff philosophy. Leibnizian ideas proved to have an enormous influence on Bolzano's philosophy of logic and his philosophy generally. Bolzano came to Leibniz first through Baumgarten. Bolzano attests to this influence in an autographical piece (a manuscript perhaps written in the 1830s), which Winter quotes in his biography of Bolzano [Winter, 1969, page 23]:[2]

> One of the first philosophical books, which Bolzano read at the age of about 16, was Baumgarten's *Metaphysica*, in which he believed he saw with full clarity that many definitions were wrong, several proofs, e.g., that of the principle of sufficient reason, assumed what they set out to prove, and several entirely incorrect assertions occurred (e.g., that of the construction of the line, surface, and solid from a finite set of points).[Ch4, 2]

Bolzano's last criticism is important for the origins of his geometrical ideas. Apparently it was through this book of philosophy that Bolzano came to his geometrical problem. In *Metaphysica* (*Metaphysics*) Baumgarten devotes several pages [Baumgarten, 1779, pp. 84–89] to definitions of geometrical concepts like point, line, and continuum. Points are the simple elements of geometry. A series of points with interposed points making up a continuum is, according to Baumgarten, a line. Similarly, a surface is a continuous series of lines, and a solid is a series of surfaces. Bolzano objected to these definitions, perhaps because they really are very uninformative. According to Bolzano, Baumgarten operates with only a finite set of points. Nevertheless, one finds in these definitions a primitive point-set-theoretic approach to the basic geometrical figures, which Bolzano also used later in his own definitions.

Bolzano's main teachers in mathematics at Prague were Stanislaus Vydra (1741–1804) and F.J. Gerstner (1756–1832). The latter took a particular interest in Bolzano's early career. However, it was a book that had the greatest effect on the formation of Bolzano's early mathematical attitudes: Abraham Gotthelf Kästner's (1719–1800) large compendium of mathematics, *Anfangsgründe der Arithmetik, Geometrie, ebenen und sphërischen Trigonometrie und Perspectiv* (*Foundations of Arithmetic, Geometry, Plane and Spherical Trigonometry and Perspective*) [Kästner, 1792] (fifth edition) [Kästner, 1800] (sixth edition).[3] This work, which several generations of students read as a basic text, went through six editions from 1758 to 1800 (cf. [Folta, 1966], [Folta, 1981], [Winter, 1969, pp. 21–22]). Bolzano learned much of his

[2] See Winter's full quotation from Bolzano's writing for more on Bolzano's philosophical background [Winter, 1969, pp. 22–23]

[3] In my earlier analysis of Bolzano's work [Johnson, 1977] I referred to the sixth edition of Kästner [Kästner, 1800]. However, Steve Russ in [Bolzano, 2004a, page 15] writes that Bolzano himself owned and annotated the fifth edition of Kästner [Kästner, 1792]. So I shall now refer to the fifth edition as well as the sixth.

Euclidean geometry from the geometrical sections of Kästner's textbook [Kästner, 1792, pp. 177–604] [Kästner, 1800, pp. 176–614]. In these pages Kästner does more than merely restate Euclid's results; he also restructures the logic of Euclid's proofs, so much so that there are passages that strike one as prefiguring ideas of Pasch and Hilbert (cf. [Goe, 1964]). Kästner's improved logic for conceiving and structuring Euclidean geometry is what impressed the young Bolzano [Bolzano, 1836, page 19]:

> Kästner proved there what one usually passes over entirely, because everyone knows it already; i.e., he sought to bring home clearly to the reader the basis on which each of his judgments rests: And that was what I liked best.[Ch4, 3]

Like Euclid, Kästner begins his treatment of geometry with definitions. The first few of these together with the definitions of Baumgarten must have been what originally inspired Bolzano to take up the problem of defining precisely the concepts of line, surface, solid, and continuum. Interestingly, Kästner himself starts off by attempting to define continuum or connected extension, a concept not defined in Euclid as such. Kästner's definition reads as follows [Kästner, 1792, page 177] [Kästner, 1800, page 176]:

> A *continuous* quantity (continuum) is something whose parts are so connected that where one stops another begins immediately, and between the end of the one and the beginning of the other there is nothing which does not belong to this quantity.[Ch4, 4]

Of course, this early attempt to define the continuity of a continuum or continuously connected set does not come up to the later standards of, say, Dedekind's set-theoretic definition involving the notion of cut [Dedekind, 1872], but it merits attention. Kästner's definition was partially motivated by the old criticism of Euclid's proof of the very first proposition of Book I of the *Elements*. Euclid, when constructing an equilateral triangle on a given line segment, assumes without justification that the two circles used in the construction will intersect in some point. Before Kästner, Christian Wolff (1679–1754), for example, knew about this criticism and sought to correct it in his textbook [Wolff, 1743, page 147]. To fill up this gap in Euclid's reasoning, Kästner laid down his definition and added a special axiom [Kästner, 1792, page 190, *Grundsatz* 7] [Kästner, 1800, page 189, *Grundsatz* 7] to his axiom system. Later this whole problem of continuity, implicit in Euclid's very first proof, impelled Bolzano to look more closely at the 'topology' (broadly speaking) of the situation and to come up with a surprising result, as I shall describe below.

For Kästner a geometric extension (*körperliche Ausdehnung*, *geometrische Ausdehnung*) is a space that a continuous quantity fills entirely. On this basis it is possible to define solid, surface, and line [Kästner, 1792, page 178] [Kästner, 1800, page 177]:

> *The solid extension, a geometrical solid* (solidum; corpus) is an extension which within its boundaries is surrounded everywhere on all sides. The extension of a solid at its boundaries is called *a surface* (superficies) and the extension of a surface at its boundaries, *a line* (linea).[Ch4, 5]

To complete this sequence, Kästner adds that the point is the boundary of the line and, hence, of all extensions. Thus, starting with solids (3-dimensional extensions), one has that the boundaries of these are surfaces, the boundaries of surfaces are lines, and the boundaries of lines are points. A few years later Bolzano came to criticize this progression in Kästner's definitions (see below).

Contrary to Baumgarten, Kästner argues that a line is *not* merely a set of juxtaposed points ('eine Menge von aneinandergesetzten Puncten macht keine Linie aus' ('a set of points placed together does not make a line') [Kästner, 1792, page 179] [Kästner, 1800, page 178]), and, similarly, a surface is not a set of lines, nor a solid, a set of surfaces. Now even with the boundary definition it is possible to imagine points placed anywhere on a line, i.e., the boundaries need not be conceived as occurring just at the ends. In fact, points are everywhere on a line, so that one can even think of a line as generated by the motion of a single point. But for Kästner these conceptions do not imply that a line is just a set of points. Indeed, in the first paper in which he set out his boundary definitions, Kästner [Kästner, 1746] tried to show that it is contradictory to regard a line as a set of points. If a line were made up of a set of points, then any point would have an immediate neighboring point. But these neighboring points, although distinct, would have no distance separating themselves, so that distances measured from any other point of the line to these two points would be the same. That is to say, the two points are really the same. A contradiction! Such metaphysical arguments suggest something akin to Zeno's paradoxes.

Kästner thought that he could trace his definitions of surface, line, and point as boundary elements back to a passage in the writings of William of Ockham [Kästner, 1782]. Going back even further, there is a suggestion of the boundary definitions in Euclid's *Elements* (Book I, Defs. 3, 6; Book XI, Def. 2) (see Section 2.4). However, the progression of Euclid's definitions is not uniform in going from solids down to points as in Kästner's progression.

Undoubtedly, the most original part of Kästner's treatment of elementary geometry is his discussion of the problem of parallels (cf. [Folta, 1966], [Goe, 1980], [Folta, 1981]). While he did not solve the logical problem of the role of Euclid's parallel postulate, i.e., he did not genuinely understand its independence, he did realize there was a logical difficulty surrounding the postulate. Kästner's textbook was read by the great early nineteenth century founders of non-Euclidean geometry and as a result it probably had some influence. Kästner's treatment of parallels particularly inspired Bolzano to examine the problem in his very first publication.

From what is known of Bolzano's early studies in mathematics, one can see that he must have derived his problem of geometrical definitions mainly from Baumgarten and Kästner. Bolzano's annotated copy of the fifth edition of Kästner's book [Kästner, 1792] still exists (cf. [Bolzano, 2004a, page 15]). Indeed, at that time there were others interested in the basic definitions of geometrical objects. For example, Karl Christian von Langsdorf (1757–1834) in his *Anfangsgründe* (*Foundations*) [Langsdorf, 1802], a rival to Kästner's text, defines lines, etc., both according to the boundary definitions and as sets of points lying next to one another. Thinking of the old theory of indivisibles of Cavalieri, he believes the latter definitions can provide better foundations for analysis [Langsdorf, 1802, Vorrede]. Johann Schultz (1739–

1805), a friend of Immanuel Kant (1724–1804) and Professor of Mathematics at Königsberg, in his defense of the philosopher's *Kritik der reinen Vernunft* (*Critique of Pure Reason*) [Kant, 1781] [Kant, 1787] [Kant, 1998] takes the boundary definitions as the correct ones and sees in them a justification for Kant's doctrine that our idea of space is a pure *a priori* form of our outer sensibility [Schultz, 1789, pp. 55–62]. Thus, around 1800 Bolzano's definition problem was a fundamental problem for several mathematicians. It seemed to go right to the heart of the foundations of geometry. It is no wonder Bolzano devoted his energies to it.

4.3 Bolzano's *Betrachtungen* (*Considerations*) of 1804

Betrachtungen über einige Gegenstände der Elementargeometrie (*Considerations concerning some Objects of Elementary Geometry*) [Bolzano, 1804] was Bolzano's first published work. He was only 23 when it appeared. He dedicated it to his teacher Professor Stanislaus Vydra, who had just retired because of increasing blindness and who died later in that year. Of the *Betrachtungen* (*Considerations*) Bolzano said it is no '*œuvre achevée*' ('finished work'), but only a trial (*Probe*), a mere test piece for his youthful investigations into geometry [Bolzano, 1804, Einleitende Vorrede, page viii]. Nevertheless, this little tract contains the seeds of many of Bolzano's mature mathematical ideas. Moreover, he retained a lifelong interest in the subjects dealt with in it, and just a few years before his death he wanted to bring out a new edition.

Bolzano's *Betrachtungen* (*Considerations*) is heavily philosophical. Indeed from the point of view of his intellectual development, the philosophical ideas contained in this work are more interesting than his mathematical theories. In the 'Vorrede' Bolzano stresses the theoretical role of mathematics and its 'usefulness' in exercising and sharpening the mind. Rigor in pure mathematics is uppermost in his thoughts. He opens with two methodological rules concerning proof, rules which always remained a part of his mathematical methodology. In the first rule Bolzano says that no sort of evidence for a statement will relieve him of the obligation to prove it, unless he can clearly see exactly why no proof is necessary. One must strive to derive all the truths of mathematics from their ultimate grounds and to obtain the greatest possible clarity and order among all the concepts of science. Thus, Bolzano's Euclidean methodology is thoroughgoing and complete. He even sees this strict logical ordering of concepts and truths as a way to discovering new truths in mathematics.

In a second methodological rule Bolzano says that he will not be satisfied with a proof if it cannot be carried out using just those concepts contained in the thesis to be proved. If a proof uses other accidental, foreign (fremdartig) concepts as well, then it cannot be satisfactory. For example, he regards it as a mistake in geometry to use the plane when considering the theorems about angles, straight lines, and triangles. The plane is foreign to these ideas. This application of his methodological rule, in fact, accounts for the peculiar treatment of geometry given in the *Betrachtungen* (*Considerations*).

Another concept that Bolzano considers foreign to the whole of geometry is the concept of motion. He gives two special reasons why this concept should be banned from geometrical investigations. First, he thinks that the idea of motion involves the notion of a moving object, which is different from the containing space. But the concept of a movable object in space is, even according to Kant (a philosopher with whom Bolzano often disagreed), an empirical concept and foreign to 'pure' geometry. Second, Bolzano claims the theory of motion actually assumes the theory of space, i.e., geometry, because in order to show the possibility of a certain motion used to prove some geometrical theorem, one must use the geometrical theorem itself for the proof. Hence, there is a *petitio principii*. One must deal with pure geometry first, before one can consider any motion. Starting from these philosophical arguments, Bolzano proceeds to investigate pure geometry. This purging of motion from geometry is relevant to Bolzano's dimension-theoretic definitions of line, surface, and solid. For instead of taking a line as the path of a moving point, as for example Kästner was willing to do, Bolzano attempts to define the concept of line independently of any idea of motion whatsoever.

Bolzano's central aim in the *Betrachtungen* (*Considerations*) is to prove elementary theorems about triangles and parallel lines solely on the basis of an assumed theory of the straight line. His objective, following his second methodological principle, is not to use the concept of plane in these derivations. Ultimately, he tries to justify the Euclidean parallel postulate, to prove that it and no other is true. Thus, Bolzano stays within the confines of Euclidean geometry, and subsequently he never ventured into any non-Euclidean geometrical investigations. Hence, the main part of his book is contained in the first part entitled, 'Versuch die ersten Lehrsätze von Dreyecken und Parallellinien mit Voraussetzung der Lehre von der geraden Linie zu beweisen' ('Attempt to Prove the First Theorems Concerning Triangles and Parallel Lines Assuming the Theory of the Straight Line').

In the second part, 'Gedanken in Betreff einer künftig aufzustellenden Theorie der geraden Linie' ('Thoughts Concerning a Prospective Theory of the Straight Line'), Bolzano outlines a theory of the straight line, upon which the first part of his tract is based. This second part is more of a sketch. Bolzano starts with the concepts of direction and distance. By doing this, he seems to be using a primitive notion of metric space or vector or inner product space (to use modern terms). Bolzano gives the following definitions [Bolzano, 1804, pp. 48–49] [Bolzano, 2004a, pp. 69–70]:

> I. That which belongs to the point b in relation to [a point] a, such that it is *independent* of the *determined point a* (which is precisely this and not another); hence, which could equally be in relation to *another* point, e. g., α: i*s called the distance of the point b from a*.
> II. That which belongs to the point b in relation to [a point] a such that it is *dependent just on the determined point a*; from which is separated what belongs already to the concept of distance, i.e., what we can associate with the point b also with respect to some other point: is called *the direction in which b lies to a*.[Ch4, 6]

Bolzano's 'philosophical' definitions may not be models of clarity, but his intentions are quite clear. The concepts of distance and direction always remained fundamental to Bolzano's geometrical research. On this basis Bolzano offers a definition of straight line segment [Bolzano, 1804, page 57] [Bolzano, 2004a, page 76]:

> Definition. An object which contains all and only those points which lie between the two points a and b, is called a *straight line between a and b*.[Ch4, 7]

In this definition the word 'between' ('zwischen') is a technical term, previously defined. This conceptual basis is a significant part of the background to Bolzano's later development of a theory of points, lines, surfaces, and solids. From a more modern mathematical viewpoint Bolzano in this instance is touching on some significant geometrical concepts, though not, of course, with the clarity of the later axiomatic work of Pasch, Hilbert, or Huntington.

In two sections of the second part of the *Betrachtungen* (*Considerations*) [Bolzano, 1804, pp. 46–48] Bolzano remarks on the appropriateness of certain definitions in geometry. These remarks have a bearing on his later theory of geometrical objects. He criticizes the boundary definitions of solid, surface, line, and point. Although Bolzano blames William of Ockham for these 'wrong' definitions, in effect, he is criticizing Kästner, from whose text Bolzano had learned the definitions and who had ascribed them to Ockham. Bolzano's criticism amounts to this. Obviously, one can think of a surface or a line or a point without also referring to some solid which they bound. So the Ockham–Kästner definitions are backwards. In contrast, Bolzano thinks it would be better if one constructed definitions so that the concept of line is based on the concept of point, the concept of surface on that of the line, etc. Bolzano cites Langsdorf [Langsdorf, 1802] in support of his criticism, but the criticism is of an older date. One can find it in Aristotle (cf. Section 2.4, referencing Aristotle *Topics*, VI.4, 141b21 [Heath, 1949, pp. 85–86], and also Section 2.3).

Overall, it is possible to see in Bolzano's *Betrachtungen* (*Considerations*) the seeds of many of his later geometrical ideas and some of the principles of his mathematical philosophy. By beginning his definitions with the concept of point as a basis Bolzano was implicitly pushing toward a simple point-set theory! The book was well received in Prague, including by his teacher Gerstner, and was a good recommendation for Bolzano's obtaining the professorship in mathematics, which Vydra left upon his retirement in 1804. However, Bolzano's friend Jandera was given the position. The story of how Jandera was chosen for the position over Bolzano is told in exquisite detail in [Guillén & Crippa, 2021] (cf. the older account in [Winter, 1969, pp. 34–40]). The two candidates had to submit to both a written and an oral examination. Bolzano was indeed able to demonstrate considerable knowledge of the mathematics of the time.[4] Given that he was passed over, Bolzano applied for a new professorship of religious instruction or religious doctrine (Religionslehre) at Prague, which he was given at first provisionally in 1805 and then, after some difficulties, permanently on 23 September 1806.[5]

---

[4] For example, see [Rusnock, 1997] for a detailed study of Bolzano's knowledge of the work of Lagrange.

[5] For details see [Guillén & Crippa, 2021] and especially [Pavlíková, 1985].

4.4 Bolzano's Philosophy of Definitions and Concepts

Throughout his career Bolzano's philosophical approach to mathematical foundations strongly influenced his mathematical research and yielded remarkably deep results. His search for what he viewed as the fundamental ideas, concepts, and theories of the science drove his mathematical work. After he published what might be called a piece of 'philosophical mathematics' as his first work, it was very much in line with his thinking that he should choose to outline his mathematical philosophy in his second publication, which appeared in 1810: *Beyträge zu einer begründeteren Darstellung der Mathematik. Erste Lieferung* (*Contributions to a Better-Founded Presentation of Mathematics. First Issue*) [Bolzano, 1810] [Bolzano, 2004a, pp. 82–137].[6] There exist in manuscript two versions of a planned sequel to the *Beyträge* (*Contributions*) [Bolzano, 1969, II, A, 5]. The *Beyträge* (*Contributions*) contains a number of suggestive ideas that Bolzano later modified and developed in his *magnum opus* on logic, philosophy, and philosophy of science, *Wissenschaftslehre* (*Theory of Science*) [Bolzano, 1837] (partially translated in English in [Bolzano, 1972] and in its entirety in [Bolzano, 2014a]). After completing and readying this major work for publication around 1830, Bolzano planned another major work, *Größenlehre* (*Theory of Quantities*), a treatise in which he intended to offer a complete foundation for all of contemporary mathematics. Bolzano was not able to complete this work, but he did write several hundred pages of it (cf. [Bolzano, 1969, II, A, 7, 8, 9, 10,1, 10,2]). An essay, 'Von der mathematischen Lehrart' ('On the Mathematical Method'), is part of the introductory matter of *Größenlehre* (*Theory of Quantities*), providing a summary of his philosophical thinking about mathematics at the time [Bolzano, 1975] [Bolzano, 1981] (in English [Bolzano, 2004b]).

At this point I shall examine the philosophical ideas most relevant to Bolzano's definition problem of geometry, as contained in the *Beyträge* (*Contributions*), *Wissenschaftslehre* (*Theory of Science*), 'Von der mathematischen Lehrart' ('On the Mathematical Method'), and a related, apparently later manuscript, 'Versuch einer Erklärung der Begriffe von Linie, Fläche und Körper' ('Attempt for a Definition of the Concept of Line, of Surface, and of Solid') [Bolzano, 0000d], [Bolzano, 2001c]. The last summarizes his philosophy in direct relation to his geometrical problem. Rusnock and Šebestík have carefully traced changes in Bolzano's philosophical view of definitions and, more broadly, his philosophy of logic and mathematics that occurred over the span of time over which he composed these works, especially noting changes from the early 1810 work to the later works [Rusnock & Šebestík, 2013] [Rusnock, 2020]. I am very much indebted to this work and also private communications from Professor Rusnock.

Bolzano opens his *Beyträge* (*Contributions*) of 1810 by stating that, while mathematics among all the sciences stands closest to perfection, it nevertheless has several gaps and imperfections—even among its most elementary concepts and theories. He mentions gaps in the theory of negative, irrational, and imaginary numbers, in

[6] An article by Paul Rusnock and Jan Šebestík [Rusnock & Šebestík, 2013] gives an excellent retrospective on the work from a broad philosophical point of view.

higher algebra, and in the differential and integral calculus. Essentially he is identifying gaps and problems that he considered open and needing filling or rigorous solutions.[7] One of his identified gaps is in geometry [Bolzano, 1810, Vorrede, page vi] [Bolzano, 2004a, page 87]:

> At the present time there is still lacking a precise definition of the important concepts: *line, surface, solid*.[Ch4, 8]

This marks the first time Bolzano gave a clear statement of his geometrical problem in print. In this philosophical essay he does not attempt to solve the problem, but the philosophy of definitions he puts forward is directly relevant to it.

The philosophy of mathematics of the *Beyträge* (*Contributions*) is characteristically 'Euclidean' in a broad sense. In comparison consider the Euclidean geometrical philosophy expressed in Pascal's 'De l'esprit géométrique et de l'art de persuader' ('Of the geometrical spirit and the art of persuasion') [Pascal, 1871].[8] In this work Pascal is concerned with the synthetic part of geometry and mathematics (and to some extend broader science) based on primitive terms, definitions of other terms, principles, and proved or demonstrated truths [`https://en.wikisource.org/wiki/`), 'Blaise Pascal/Of the Geometrical Spirit', page 431] [Pascal, 1871, page 165]:

> For there is one [order], and it is that of geometry, which is in truth inferior in that it is less convincing, but not in that it is less certain. It does not define every thing and does not prove every thing, and it is in this that it is inferior; but it assumes nothing but things clear and constant by the natural light, and this is why it is perfectly true, nature sustaining it in default of discourse. This order, the most perfect of any among men, consists not at all in defining every thing or in demonstrating every thing, nor in defining nothing or in demonstrating nothing, but in adhering to this middle course of not defining things clear and understood by all mankind, and of defining the rest; of not proving all the things known to mankind, and of proving all the rest. Against this order those sin alike who undertake to define everything and to prove every thing, and who neglect to do it in those things which are not evident of themselves.[Ch4, 9]

Pascal realizes that in the order of geometry as ordinarily understood not every term can be defined nor every truth proved. He recognizes the need for primitive terms from which other terms can be defined and for true fundamental principles from which other truths can be demonstrated. Rather idealistically, he also countenances the case of a 'perfect order' in which every term is defined and every truth proven, while realizing that this order not possible for humankind. Hence, the order of geometry or geometrical order is the 'second best' order. The perfect order would appear to involve an infinite regress as commonly understood (cf. [Lakatos, 1978c]).

---

[7] The history of mathematics shows that he and other mathematicians grappled with these over many years.

[8] See Wikisource for the work in French (`https://fr.wikisource.org/wiki/`) under 'Blaise Pascal' then 'De l'esprit géométrique et de l'art de persuader' and in English translation (`https://en.wikisource.org/wiki/` under 'Blaise Pascal/Of the Geometrical Spirit'.

In the lengthy second part of the *Beyträge* (*Contributions*) [Bolzano, 1810, pp. 38–134] Bolzano's philosophical analysis broadly follows the Euclidean mathematical framework and synthetic method, as described in Pascal's geometrical order. However, there are significant differences, which clearly distinguish Bolzano's special viewpoint from Pascal's. Pascal method is directed at *persuading the reader* of the truths of geometry, giving an order among the judgments of geometry so that the reader can gain conviction about them.[9] Bolzano seeks to outline a better-grounded presentation or description of mathematics in line with his own philosophical outlook.[10] A hallmark of Bolzano's philosophy is his view, already contained in the early *Beyträge* (*Contributions*), that mathematical and scientific truth is *objective*. In mathematics and science one is primarily concerned with objective connections among judgments (the classical term, or to use a later Bolzano term, propositions) based on objective concepts with objective definitions and not so much with subjective convictions (though they may be there too). Hence, proof in mathematics is not aimed mainly at persuading the reader or increasing his conviction; rather it is directed towards revealing objective connections among judgments (propositions), in particular, among objective truths. Thus, one finds Bolzano writing the following strong statement in 1810 [Bolzano, 1810, II §.2, pp. 39–40] (English translation in [Bolzano, 2004a, page 103]):

> In the domain of truth, i.e., in the totality of all true judgments, a certain *objective* connection prevails, which is independent of our accidental *subjective recognition* of the same independent *connection*, and consequently, some of the judgments are the grounds for others which follow from them. To present this objective connection of judgments, i.e., to choose a set of judgments and to order them so that each one that is a consequence is so represented and conversely, seems to me to be the proper *aim* that we should pursue in a scientific exposition.[Ch4, 10]

This strict objective and deductive view of scientific knowledge remained a part of Bolzano's view of the fundamentals of logic and mathematics throughout his career.[11]

Putting this view into practice is an ideal for mathematical exposition. However, in the *Beyträge* (*Contributions*) he does mention 'strength of conviction' ('Festtigkeit der Ueberzeugung') [Bolzano, 1810, II §.2, page 40] [Bolzano, 2004a, page 103].

---

[9] I should add that Bolzano is critical of previous philosophical analyses of mathematical method. He is especially critical of Immanuel Kant's analysis of mathematical method as portrayed in his theory of construction of concepts through intuitions [Kant, 1781] [Kant, 1787].

[10] Paul Rusnock in a private communication to me points to an important disagreement between Pascal and Bolzano: whereas Pascal says that proofs should begin with that which is clear by the natural light, Bolzano thinks that objective proofs or groundings should begin with objective axioms, which need not be evidently true and may even seem 'suspect or doubtful' (see especially *Beyträge* (*Contributions*) [Bolzano, 1810, II, §.10 and §.21 note, pp. 59–61, 92–95] [Bolzano, 2004a, II, §.10 and §.21 note, pp. 109, 118–119]).

[11] His later researches in logic led him to a theory of logical consequence, exposited in the *Wissenschaftslehre* (*Theory of Science*), which is similar to Alfred Tarski's theory of a hundred years later.

Later in his *Wissenschaftslehre* (*Theory of Science*) he considers 'certifications' ('Gewissmachungen'), which are not fully grounded in objective proof, whereas 'groundings' ('Begründungen') are so grounded in full objective proof. Nonetheless, certifications do aim at certainty [Bolzano, 1837, Vol. IV, §.525, page 261] [Bolzano, 2014a, Vol. IV, §.525, page 180].

Bolzano's geometrical problem is fundamentally a search for appropriate objective definitions. But what precisely are definitions according to Bolzano? In the *Betrachtungen* (*Considerations*) of 1804 one finds the following principle about the nature of definitions [Bolzano, 1804, page 46]:

> ... a genuine definition must contain only such marks of the concept to be defined that reveal its *essence* and without which it cannot be conceived at all ....[Ch4, 11]

At this point, Bolzano clearly took an essentialist position with respect to definitions. However, his logical characterization of a definition is different in the *Beyträge* (*Contributions*) of 1810 [Bolzano, 1810, pp. 42–43; cf. pp. 42–59]:

> Logicians understand by a *definition* (definitio [in Latin]) in the most proper sense of this word the *indication of the proximate* (two or more) *parts of which a given concept is composed.*[Ch4, 12]

Bolzano greatly expanded and deepened his exploration of logic and the logic of the sciences with the result being his monumental four-volume *Wissenschaftslehre* (*Theory of Science*). After the *Beyträge* (*Contributions*) of 1810 but before embarking on his logical studies, Bolzano developed the results for his three highly significant papers in analysis and geometry of 1816 and 1817 [Bolzano, 1816] [Bolzano, 1817b] [Bolzano, 1817a]. Then in the 1820s he devoted his primary efforts to his logical studies and completed the *Wissenschaftslehre* (*Theory of Science*) around 1830. Publication was delayed until 1837 due to restrictions placed on him after his dismissal and trial.

It is necessary to examine Bolzano's notion of definition and his broader views about definitions and their uses in mathematics and science as given in the *Wissenschaftslehre* (*Theory of Science*). In this work his attitude to treating definitions changes significantly.[12] Overall in the course of the work, he takes a more pragmatic view. In the main section on definition Bolzano writes [Bolzano, 1837, Vol. IV, §.554, pp. 330–331] [Bolzano, 2014a, Vol. IV, §.554, page 230]:

> The first sort of consideration of mere ideas and propositions that I shall say more about are the *definitions*. By this I mean ... nothing other than propositions that specify whether a certain idea or proposition is simple or composed of parts, and in the latter case, what these parts are and how they are combined.[Ch4, 13]

---

[12] The main discussion of definitions and related matters in the *Wissenschaftslehre* (*Theory of Science*) occurs in the following: primary section [Bolzano, 1837, Vol. IV, §.554, pp. 330–331] [Bolzano, 2014a, Vol. IV, §.554, pp. 230–231], along with [Bolzano, 1837, Vol. III, §§.350–351, pp. 397–405; Vol. IV, §§.554–559, pp. 330–350; §.668, pp. 542–551] [Bolzano, 2014a, Vol. III, §§.350–351, pp. 264–269; Vol. IV, §§.554–559, pp. 230–244; §.668, pp. 380–386].

So definitions in a strict sense are reports (objective propositions) of analyses of ideas (concepts) or propositions. A definition tells us either that a certain idea (concept) or proposition is simple or that it is complex and is composed of certain other ideas joined together in a particular way. In this case 'simple' ('einfach') is a technical term, meaning 'without parts.' Thus, a definition is either true or false, depending on whether it gives a correct report of an analysis of an idea or proposition or not. A definition has an objective truth value.

Bolzano's strict theory of definitions as conceptual analysis takes us into his world of propositions in themselves and ideas or concepts in themselves, a characteristic part of his later philosophy. There are indications of the theory of objective propositions and ideas already present in the early *Beyträge* (*Contributions*), but these matters are only fully explored in the *Wissenschaftslehre* (*Theory of Science*) [Bolzano, 1837, *Erster Theil. Fundamentallehre*; *Zweiter Theil. Elementarlehre*, Volumes 1 and 2] [Bolzano, 2014a, *Book One: Theory of Fundamentals*; *Book Two: Theory of Elements*, Volumes 1 and 2]. According to the latter work, both propositions and ideas in themselves are mind-independent [Bolzano, 1837, Vol. I, §§.19, 48, pp. 76–80, 215–218] [Bolzano, 1972, pp. 20–23, 61–62] [Bolzano, 2014a, Vol. I, §§.19, 48, pp. 58–61, 157–159]. They are to be differentiated from written, spoken, or mental propositions and ideas. An objective idea or proposition is independent of whether anyone has ever thought of it or expressed it. They are the objective contents or meanings of expressed or conceived ideas and propositions. As such, they are abstract entities, independent of thinking subjects.

Bolzano emphasizes that, strictly speaking, they do not exist and they do not have reality. Only subjective ideas and propositions really exist. The objective counterparts to subjective propositions and ideas only 'subsist' in what one might call a world of meanings. Bolzano's objective entities are close to being in something like Popper's 'world three' [Popper, 1972, pp. 106–190]. A proposition in itself asserts that something is or is not the case, so it can be true or false, whereas an idea in itself is only a part of a proposition and does not have a truth value. Propositions and ideas in themselves provide a central core element of Bolzano's philosophical system. Subjective ideas and propositions, spoken, written, or only thought, have a different kind of significance. They are, in fact, the manifestations in the mind of the objective entities. Bolzano thinks that definitions are one way to get at the objective world via the subjective mental world.

In the *Beyträge* (*Contributions*) there is not much indication of whether it is difficult to find a definition, but in the *Wissenschaftslehre* (*Theory of Science*) [Bolzano, 1837, Vol. III, §§.350-351, pp. 397–405] [Bolzano, 2014a, Vol. 3, pp. 264–269] Bolzano emphasizes that the task is normally quite difficult, and in many cases the exercise may not even be worth the effort.

In contrast to his narrow notion of definitions as analyses of objective ideas (concepts) or propositions, Bolzano takes a broader view of definitions and their value in the presentation of sciences. Thus in a passage in the *Wissenschaftslehre* (*Theory of Science*) following the one quoted above, he says [Bolzano, 1837, Vol. IV, §.554, page 331] [Bolzano, 2014a, Vol. IV, §.554, page 230]:

> We can go too far . . . in striving for distinctness in our own reflection: this is all the more certainly the case when communicating our thoughts to others, where we may lose ourselves in definitions and analyses, a blameworthy excess. . . . Thus in no science, even one mainly concerned with providing exercise in thinking, shall we undertake to give complete definitions, extending right down to the ultimate, simple parts, of every concept and proposition, nor even of all the essential theses.[Ch4, 14]

He adds that such analyses would be quite out of place in a science that has practical applications to life, such as medicine or practical wisdom. Hence, in the *Wissenschaftslehre* (*Theory of Science*) he provides some practical considerations for finding definitions or broader information about their related concepts [Bolzano, 1837, Vol. IV, §§.555-558, pp. 332–339] [Bolzano, 2014a, Vol. IV, §§.555–558, pp. 231–236].

Bolzano asks which definitions require a special proof of correctness. Some definitions are synthetic (synthetische), whereby one proposes or posits a sign or word to designate or express the idea arising from a combination of some set of ideas. Essentially, this would be a nominal definition for the proposed sign or word. Such definitions do not need proofs of correctness. In contrast, definitions of concepts that are given to us already are analytic (analytische), and their analyses may need proofs (or at least indications) of correctness [Bolzano, 1837, Vol. IV, §.555, pp. 332–333] [Bolzano, 2014a, Vol. IV, §.555, pp. 231–232].

Bolzano gives three heuristic rules for considering proofs (or just indications) of correctness of a definition [Bolzano, 1837, Vol. IV, §§.556–558, pp. 333–339] [Bolzano, 2014a, Vol. IV, §§.556–558, pp. 232–236], but I shall not go into the details. One noteworthy point is that Bolzano suggests in some cases comparing the extensions or contents of concepts to understand distinctions among them is useful. Essentially, he suggests considering various items that fit or do not fit under a pair (or set) of concepts to see if the concepts differ or coincide [Bolzano, 1837, Vol. IV, §.556, pp. 333–335] [Bolzano, 2014a, Vol. IV, §.556, pp. 232–233]. He uses this technique of extensional comparison in his later geometrical work (see below).

I need to consider an additional topic in the *Wissenschaftslehre* (*Theory of Science*) related to definitions: Bolzano's notion of 'Verständigung,' usually translated as 'explication' [Bolzano, 1837, Vol. IV, §.668, pp. 542–551] [Bolzano, 2014a, Vol. IV, §.668, pp. 380–386]. Bolzano's concern in this article of the *Wissenschaftslehre* (*Theory of Science*) is how the readers of a scientific book will come to understand the signs and concepts being used in the book. The understanding of readers is paramount. He describes a procedure of familiarization in terms of using known signs and concepts in [Bolzano, 1837, Vol. IV, §.668, pp. 543–544] [Bolzano, 2014a, Vol. IV, §.668, page 381]:

> It is easy to see that our procedure can only consist in using the signs which are already known to our readers to form certain propositions about the sign whose meaning we wish them to learn. Such propositions, however they may be constituted, I shall call *explications*. Accordingly, I use this term to cover all of the propositions occurring in a treatise which are put forward with the

> intention of making the meaning in which we use a certain sign known to our readers. If such a proposition serves its purpose, i.e., if it permits our readers to recognize the idea we want them to connect with our sign so well that it cannot be confused with any other, one may call it a *determination* of this meaning or a *determination of a concept*. Others ... also call such determinations *definitions*, either definitions of *signs* or definitions of their *meaning* or their *concept*, *definitions of concepts*.[Ch4, 15]

In the same article Bolzano drills down to a special procedure of concept determination and analysis to arrive at *definitions in the strict scientific sense* [Bolzano, 1837, Vol. IV, §.668, pp. 546–547] [Bolzano, 2014a, Vol. IV, §.668, page 383]:

> Here we will do well first to consider whether our idea is *complex* and, in particular, whether it can be decomposed into parts which can be represented by signs already known to our readers. If this case occurs, and we succeed in discovering the constituents of our idea, then we choose a very secure means of explaining our meaning if we set out the signs which express these constituents in the very sequence in which the ideas they designate must follow one another in our mind in order to produce a complex idea, and then observe that the idea which the reader obtains through the consideration of these signs is the very one we connect with our new sign. By this means, the reader not only learns the concept that we connect with our sign, but also knows, and indeed clearly, the constituents (at least the proximal ones) of which it is composed, i.e., their idea of the concept becomes *distinct*. ... Since this cannot be accomplished using the procedures ... above, it is reasonable to give the present procedure a special name. It may be called a *determination of a concept by its parts*, an *analysis*, or a definition in the *strict* or rigorously scientific sense. (cf. §.555.) One may find a great many examples of such definitions in treatises of mathematics.[Ch4, 16]

This passage expresses the strongest scientific sense of 'definition' for Bolzano, the analysis of complex ideas and concepts into simpler ones coupled with laying out the structure of their connection. In the *Wissenschaftslehre* (*Theory of Science*) Bolzano recognizes the difficulty of following this procedure, but he does not abandon it entirely. Yet, Bolzano's emphasis is on bringing the readers of a scientific book to an understanding of its ideas, concepts, and propositions by relying on better known concepts and propositions.

I shall next consider the treatment of definitions and related matters in 'Von der mathematischen Lehrart' ('On the Mathematical Method'), which is an introductory methodological part of the *Größenlehre* (*Theory of Quantities*) [Bolzano, 1975] [Bolzano, 1981] [Bolzano, 2004b]. This short piece is a compact summary of the methods that are given more extensive treatment in the *Wissenschaftslehre* (*Theory of Science*). This is Bolzano's exposition of method for mathematical works, and mathematical examples occur frequently in it. The examples are a significant part of his exposition. The entire exposition is intended to help the prospective writer of a mathematical text determine what might be or should be included as methodological content and logical structure to aid the future reader's understanding of the subject of the text. It clearly was written after the completion of the *Wissenschaftslehre* (*Theory of Science*), but an exact date for it is not known.

In the first part of the exposition Bolzano summarizes the elements concerning objective propositions and objective ideas or concepts. He describes simple and complex ideas. He gives some emphasis to the *extension* (*Umfang*) of an idea: the collection of all the objects falling under it.

In §.9 Bolzano explains the nature of 'explications' along with the narrower nature of definitions. The emphasis is on teaching the beginner to enable understanding of the subject of a mathematical work. Explications are taken to be definitions in a broad sense. However, he narrows his subject by wanting [Bolzano, 1975, page 69] [Bolzano, 1981, page 47] [Bolzano, 2004b, page 59]:

> . . . to save the word *definition* for only a special kind of these [explications]: those, namely, that acquaint us with the meaning of a sign by indicating to us the components of which the designated concept is composed.[Ch4, 17]

From the viewpoint of Bolzano's problem of defining line, surface, and solid, and implicitly dimension §§.10,11 are of interest. In §.10 Bolzano asks the rhetorical question: 'Which concepts and propositions should be rendered as distinct as possible?'[13] [Bolzano, 1975, page 69-70] [Bolzano, 1981, pp. 48–49] [Bolzano, 2004b, pp. 59–60]. The terms 'distinct' and 'distinctness' ('deutlich' and 'Deutlichkeit') are technical terms (also treated earlier in the *Wissenschaftslehre* (*Theory of Science*) [Bolzano, 1837, Vol. III, §.281, pp. 39–45] [Bolzano, 2014a, Vol. III, §.281, pp. 25–29 383]).[14] Distinctness answers the question of which ideas or propositions are simple or complex, and for complex ideas of what simple ideas they are composed. Thus, distinctness refers to the process of analyzing complex propositions or complex ideas or concepts down to simpler or even simplest ones. Degrees of distinctness relate to the levels of analysis achieved. Bolzano does not advocate that all propositions or ideas be brought to the same degree of distinctness. However, he sees an advantage in the analysis of concepts and propositions *proper* to a given science, because such analysis allows us to recognize the truth of other propositions or their grounds.

In §.11 Bolzano explores how propositions and ideas may be rendered as distinct as possible. I need to mention only a few highlights. He recognizes that one learns about concepts from experiences, perhaps learning about them from childhood. However, that acquired knowledge of concepts does not necessarily enable one to determine whether they are simple or not. It is actually quite difficult to determine if a concept is simple or complex.

In the course of his discussion, Bolzano questions whether he has given the correct conception of the expression '*extended spatial object*' ('*ein ausgedehntes Raumding*') when he defines it 'as a spatial object of a kind such that "every one of whose points, at every distance no matter how small, has certain neighbors." [and

[13] 'Welche Begriffe und Sätze zur möglich grössten Deutlichkeit erhoben werden sollen?'

[14] Historically, the terms 'clear' and 'distinct' [perceptions] derive from Descartes' epistemology.

he continues] – But if we are in a position to derive from the concept all the known properties of extended spatial objects, then it will be shown that our concept, if not identical with the customary one, is at least equivalent to it, and one will have cause to be satisfied with our definition.'[15] I note that in this example Bolzano brings up a kind of point-set-theoretic point of view.

Bolzano is critical of all previous treatises of mathematics that claim to be rigorously scientific, because they do not define the first and most important concepts, or, if they do give a kind of definition, they do not prove its correctness. In this context he mentions the basic concepts of time, space, line, surface, and solid. He asserts that one should begin a treatise of the general *theory of quantities* (the subject of the *Größenlehre* (*Theory of Quantities*)), if it is to be called rigorously scientific, with a definition of the concept of *quantity* and a treatise of the *science of space* with a definition of the concept of *space*.

In mathematics, he thinks that the emphasis should not only be on the clarity of concepts but also on the distinctness of concepts in the rigorous sense of conceptual analysis. Clarity of concepts is the basis for understanding them; distinctness provides a deeper analysis of the concepts and their relationships to others. According to Bolzano, only mathematics above the other sciences has so many concepts that are not only clear but also have been raised to a greater level of distinctness (see footnote 14).

For Bolzano, it is preferable that presentations of mathematics, as those that may be given in treatises, begin by laying out the clear or clearly understood concepts together with their important concepts raised in a suitable degree of distinctness. However, he appears to take a pragmatic view, since conceptual analysis is difficult and the identification of simple and complex concepts and propositions is hard. He is expressing an ideal that he would like to follow, if possible.

To conclude my summary and analysis of Bolzano's philosophy of definitions and concepts, I would like to point to a methodological remark Bolzano makes in the apparently later fragmentary work: 'Versuch einer Erklärung der Begriffe von Linie, Fläche und Körper' ('Attempt for a Definition of the Concepts of Line, Surface, and Solid') [Bolzano, 0000d] [Bolzano, 2001c].[16] The remark suggests an even more pragmatic view of definitions in mathematics [Bolzano, 0000d, sheet 12r] [Bolzano, 2001c, page 131]:

> It is a good thing that the question upon whose correct answer so much depends is not "of which parts does a certain idea, as *at present* found in our consciousness, or even as it will be in the *future*, consist?" but rather "which parts must we ascribe to it so that it will best serve the ends for which we wish to use it?" [Ch4, 18]

[15] 'als ein Raumding von solcher Art erklären, „dessen jeder Punkt für jede noch so kleine Entfernung gewisse Nachbarn hat"; – wenn wir imstande sind, alle Beschaffenheiten, die man von ausgedehnten Raumdingen kennt, aus diesem Begriffe herzuleiten; so wird erwiesen sein, daß unser Begriff wo nicht der nämliche mit dem gewöhnlichen, doch ein ihm gleichgeltender sei, und man wird Ursache haben, mit unserer Erklärung zufrieden zu sein.'

[16] Cf. [Rusnock, 2020, pp. 5–6].

Rather than focusing on the analysis of concepts in a definition, Bolzano clearly is emphasizing the usefulness of definitions in presenting a science. Bolzano makes a similar point about definitions in a more extended way in a work on aesthetics of 1843 [Bolzano, 1843a, §.1, no. 4, page 12] [Bolzano, 1845a, §.1, no. 4, page 12] [Bolzano, 1989a, §.1, no. 4, page 104]:[17]

> ... in most cases when we put forward a concept *in a scientific context*, it matters little whether the definition we propose for it actually presents in all its parts the concept we have hitherto connected with the word chosen to designate it, provided only that it is an appropriate concept, one that merits consideration at the place where we present it, and also to serve as the basis of the theses we are to present concerning its object.[Ch4, 19]

Seemingly, Bolzano realized in the later part of his career the immense difficulty (if not the impossibility) of reducing complex concepts to simple ones and their connections, so he apparently came to settle on the criterion of usefulness of definition in a well-founded science.[18]

Considering the whole span of Bolzano's views on definitions in science and mathematics, he may have developed in the later part of his career some ambivalence about whether to pursue strict definitions as conceptual analysis or to take a more pragmatic view in building mathematical structures for understanding and for presentation.

4.5 First Solution: The Geometrical Definitions of 1817

During the same year in which he brought out his well-known *Rein analytischer Beweis* (*Purely Analytical Proof*) [Bolzano, 1817b] [Bolzano, 2004a, pp. 250–277], 1817, Bolzano published another significant work: *Die drey Probleme der Rectification, der Complanation und der Cubirung, ohne Betrachtung des unendlich Kleinen, ohne die Annahmen des Archimedes, und ohne irgend eine nicht streng erweisliche Voraussetzung gelöst; zugleich als Probe einer gänzlichen Umstaltung der Raumwissenschaft, allen Mathematikern zur Prüfung vorgelegt* (*The Three Problems of Rectification, Complanation, and Cubature, solved without consideration of the infinitely small, without the hypotheses of Archimedes, and without any assumption that is not strictly provable; at the same time being presented for the scrutiny of all mathematicians as a sample of a complete reorganization of the science of space*) [Bolzano, 1817a]. In this work Bolzano seeks rigorous proofs—proofs that achieve the standards laid down in the *Beyträge* (*Contributions*)—for the usual integral formulae for the length of a curve, the area of a surface, and the volume of a solid. All previous demonstrations are unacceptable to Bolzano, and it is evident from

[17] As Rusnock has pointed out [Rusnock, 2020, pp. 5–6].

[18] I shall return to the 'Versuch' ('Attempt') document in Section 4.6 for a fuller analysis of its mathematics.

his citations that he has surveyed a vast number of these. In particular, he censures those who use infinitesimals. Such criticism was an integral part of Bolzano's early arithmetization program. Also, he finds fault with those who base their arguments on the so-called axiom of Archimedes, viz., the assumption that of two curved lines with the same chord the one enclosed within the chord and the outer curve must be the shorter and a similar assumption for surfaces. Bolzano's extensive critical remarks in the 'Vorrede' demonstrate his desire to stay within the bounds of the rigorous program of his *Beyträge* (*Contributions*). However, judged by later standards of rigor [Stolz, 1881, pp. 267–268], his analytic results fall a little short.

Embedded in this work of analysis, one finds Bolzano's first published solution to his fundamental geometrical problem, a group of modern-sounding 'topological' definitions of the basic objects of geometry [Bolzano, 1817a, §.11-15, pp. 20–30; §.35–42, pp. 51–54; §.52, page 66] [Bolzano, 2004a, pp. 301–309, 322-326, 334]. Bolzano's definitions show the high standard of rigor that he sought in the foundations of geometry and, inasmuch as they appear within a book on mathematical analysis, they suggest that he was searching for some kind of 'geometrico-topological' basis for the entire edifice of analysis. Bolzano's whole approach to his geometrical problem strikes one as quite similar to Karl Menger's intuitive approach to dimension during the early 1920s.

Contrary to his own early criticism of Baumgarten (see above), Bolzano begins his analysis of geometrical concepts by assuming that any geometrical figure is a system or collection of points [Bolzano, 1817a, page 20 footnote] [Bolzano, 2004a, page 301 footnote]:

> *Spatial object* is in general the name for any *system* (any collection) *of points* (it may consist of a finite or an infinite multitude).[Ch4, 20]

Thus Bolzano immediately puts his geometry on an intuitive point-set-theoretic basis. The main objects of geometry, lines, surfaces, and solids, are not *merely* sets of points, for the sets of points are connected in certain ways. Bolzano sees the objective of his definitions as a way to explain how points are composed together to make lines, surfaces, and solids. For example, the definition of *line* reads as follows [Bolzano, 1817a, page 20] [Bolzano, 2004a, page 301]:[19]

> A spatial object with the property that every point of it has at least *one* and at most only a *finite* set of points as neighbors, corresponding to each distance smaller than a certain given distance, is called a *line in general* (Fig. 1–7.).[Ch4, 21]

By a 'neighbor' of some point of a spatial object with respect to a given distance, Bolzano really means a point of the intersection of the spatial object with the surface of a sphere with radius equal to the given distance. Thus, Bolzano uses the boundaries of spherical neighborhoods to determine the neighbors, although he does not state this explicitly. This interpretation of his neighbor concept is only revealed by the

[19] See Figure 4.1 for a copy of Bolzano's sheet of figures, Figs. 1–21, [Bolzano, 1817a, foldout after page 80]. These figures nicely illustrate Bolzano's line of geometrical thinking as he works through his definitions of the basic entities of geometry.

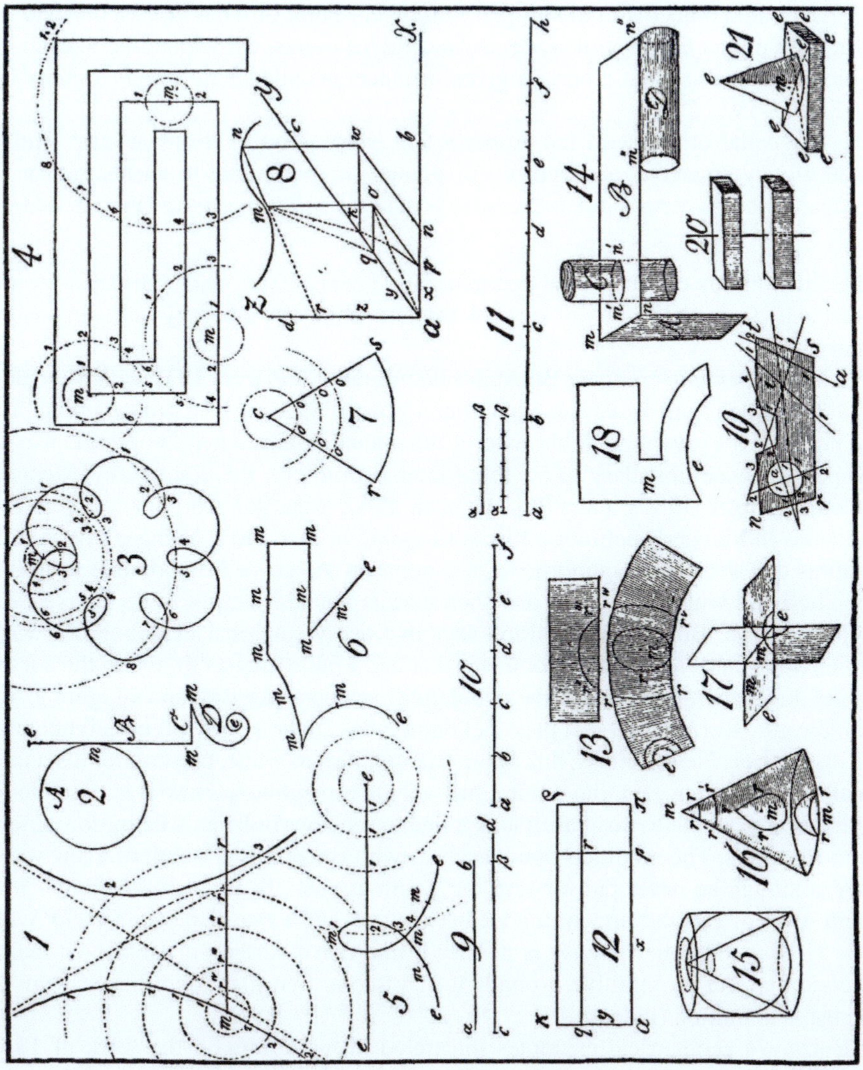

Fig. 4.1 Sheet of Figures from Bolzano's book *Die drey Probleme* (*The Three Problems*) [Bolzano, 1817a, foldout after page 80].

way he applies it and through the diagrams he gives. In modern terms, he defines a line or curve as a point set with the property that the boundaries of arbitrarily small spherical neighborhoods have finite intersections with the point set. Continuing with Bolzano's main series of definitions, one has [Bolzano, 1817a, pp. 51, 66] [Bolzano, 2004a, pp. 322, 334]:

> A spatial object with the property that every point of it has at least *one* and at most only a *finite* set of *separate lines* full of points, corresponding to each distance smaller than a certain given distance, is called a *surface in general* (Fig. 13-19.).[Ch4, 22]
>
> A spatial object with the property that every point of it has at least *one* absolutely connected surface full of points, corresponding to each distance smaller than a certain given distance, is called a *solid in general* (Fig. 20 and 21.).[Ch4, 23]

These definitions of the typical geometrical objects of 1, 2, and 3 dimensions are 'prescient' from the point of view of modern point-set topology and dimension theory.

It is interesting to compare Bolzano's definition of line from 1817 with Menger's definition of a regular curve as a continuum from 1925, each of whose points has the property that every neighborhood of the point contains a neighborhood, also of the point, whose boundary has a finite intersection with the continuum [Menger, 1926a] [Menger, 1932a, page 96] [Menger, 1967, page 96]. To be sure, there are subtleties in Menger's definition that are missing in Bolzano's. Menger's definition requires that for each neighborhood of a point on the curve it is possible to find a neighborhood within the given one such that its boundary has a finite intersection with the curve. Bolzano's definition says that every spherical neighborhood with sufficiently small radius has a boundary that has a finite intersection with the curve. To see the distinction between the two definitions, consider Bolzano's Figure 7. As he indicates [Bolzano, 1817a, page 22] point c has a whole line full of neighbors at the distance cr. Nevertheless, this figure still qualifies as a line, because for distances smaller than cr, e.g., co, the point c has only two neighbors. Now it is possible to modify the example to disqualify it as a line to see how Bolzano's definition differs from Menger's. The modified figure accords with Bolzano's thinking of a few years later, although he never put forward the figure as such. In Bolzano's Figure 7 just insert a set of circular arcs around c between cr and cs at distances $(\frac{1}{2})^n\overline{cr}$ with $n = 1, \ldots \infty$. This new figure is neither a line nor a surface in Bolzano's realm of concepts. Yet on intuitive grounds it may seem 'line-like,' and, hence, may fit Menger's definition (it does).

Bolzano's geometrical examples illustrated in his figures in the work of 1817 [Bolzano, 1817a, foldout after page 80] (see Figure 4.1) exhibit some of the subtleties of his definitions for varieties of lines, surfaces, and solids. Further analysis given below shows that Bolzano changed his definitions over time. Such changes provide evidence of one of his modes of thought: confronting examples with alternative examples and counterexamples to refine definitions and theorems. Overall, the collection of his geometrical examples accumulated over the years of mathematical work on defining the basics of geometry provides us with broader evidence of his example-counterexample line of thinking. It seems to me to be a natural line of thinking for Bolzano in his quest for logically precise results. To summarize, Bolzano's approach seems to follow a definitions–refutations path guided through examples and counterexamples to achieve greater precision. It is analogous to Lakatos' philosophy of mathematics of proofs and refutations [Lakatos, 1964] [Lakatos, 2015].

Looking forward to later developments, mathematical thinking in point-set theory and point-set-theoretic topology very much follows the line of examples confronted with counterexamples to sharpen definitions and theorems and their proofs. Such thinking is an essential feature of the field.

Bolzano sometimes looks upon a group of unconnected curves as a single line. His Figure 1, a hyperbola, is such a case. One may conceive of the hyperbola as 'connected' by a single equation, although it is separated into two branches as a graph. Nevertheless, Bolzano regards connectedness, in the sense of being composed of one piece, as an important property for lines and defines it as follows [Bolzano, 1817a, pp. 20–21] [Bolzano, 2004a, page 302]:

> A spatial object is called an *absolutely connected line*, if every part which can be considered as a line according to the given definition has at least *one point* in common with the remaining parts, which must themselves also be considered as lines (Fig. 2-7.).[Ch4, 24]

In order to make sense of Bolzano's definition of connectedness for lines, it must be assumed that he thinks that a line segment or an arc always includes its endpoints. If one does not make this assumption, then his definition of connectedness falls to pieces. Even the open segment $(0, 1)$ would be disconnected. In fact, Bolzano does define a line with its endpoints [Bolzano, 1817a, page 21] [Bolzano, 2004a, page 303]:

> A line in which there are points that have only *single* neighbors, corresponding to each distance smaller than a certain distance, is called a *bounded line* [or more accurately, a line with endpoints] (Fig. 2, 5, 6.).
>
> Those points in it (e.g., e, e) are called *boundary-* (or *end-*) *points*; the rest (e. g., m, m), *interior points*.[Ch4, 25]

Yet in the opening definition of line (given above) Bolzano does not explicitly require that segments contain their endpoints; only his figures suggest that this is a requirement. During the 1840s when he made a second attempt to define the concept of line, he definitely did not wish to require that segments or arcs necessarily include their endpoints.

Bolzano also defines connected surfaces and solids, relying on the tacit assumption that nonclosed surfaces include their perimeters and solids include their boundary surfaces [Bolzano, 1817a, pp. 51–52, 66] [Bolzano, 2004a, pp. 322, 334]:

> A spatial object is called a single absolutely *connected surface*, if every part which can be considered as a surface according to the given definition has at least *one line* in common with the remaining parts, which must themselves also be considered as *connected surfaces*. (Fig. 14-19.)[Ch4, 26]
>
> A spatial object is called a *single absolutely connected solid*, if every part which can be considered as a solid according to the given definition has at least *one surface* in common with the remaining parts, which must themselves also be considered as solids (Fig. 21.).[Ch4, 27]

A further implication of Bolzano's definitions of lines, surfaces, and solids, and their connected cases is that the lines, surfaces, and solids must be homogeneously 1-, 2-, and 3-dimensional. For example, a square surface with a line segment sticking out from it or a figure consisting of two disjoint circular areas joined together by a line cannot be a connected surface according to Bolzano's definitions. Whether Bolzano fully understood this implication is hard to tell, but he probably was aware of it to some extent.

Among his several further definitions Bolzano lays down defining conditions for simple, closed, and simple closed lines as well as surfaces. His proto-topological method is eminently suitable for this task. For example, simple closed lines are lines made up of points having exactly two neighbors for sufficiently small distances and such that the distances between pairs of points on the line never exceed some specified maximum [Bolzano, 1817a, page 21] (see Fig. 3, 4, 7, given in Figure 4.1). This definition is all the more interesting, when one finds that some years later Bolzano thought of stating the Jordan curve theorem for simple closed curves and asserting that the 'theorem' needed a proof!

Bolzano's 'topological' theory of geometrical figures as given expression in the twenty definitions of his work of 1817 is certainly general, in fact, so general that it is of little direct use to him in the analytic part of the work. He admits this himself [Bolzano, 1817a, page 21] [Bolzano, 2004a, pp. 304–305]:

> One does not have to keep these definitions in mind in order to be able to understand and judge the present theory [of analysis]. Nevertheless, it seemed appropriate to put them here, even if only to give a glimpse of the structure of the entire *domain of geometry*, upon which I have worked for several years, but about which I have communicated very few examples to the public before now.[Ch4, 28]

In his rectification theory Bolzano confines his attention to 'determinate' or 'determinable spatial objects,' roughly speaking, those figures that can be specified by single equations or sets of equations. So in *Die drey Probleme* (*The Three Problems*) the general topological theory quickly drops away from the exposition.

One should see his definitions of the basic objects of geometry in the 1817 work as noteworthy and insightful, even though there are no theorems or proofs that rely on his definitions. More can be found in his second period of research on his geometrical problem, which I consider next.

4.6 Second Solution: Bolzano's Later Program in Geometry

Two years after Bolzano published his highly significant mathematical tracts of 1816 and 1817, he lost his position as Professor of Religious Instruction at Prague. Emperor Franz I signed the final decree on 24 December 1819. It was almost inevitable that Bolzano's liberal Enlightenment views on moral and political matters should come into conflict with those of the civil and ecclesiastical authorities during this time of the 'Austrian-Catholic Restoration' [Winter, 1969, pp. 52–71] [Rusnock & Šebestík, 2019, pp. 50–63]. With the enforced end of his teaching career, Bolzano was able to devote much of his energy to a study of logic, logic as a foundation for science. His intellectual work was enabled by his friends, particularly, the Hoffmans, Josef and especially his wife Anna [Winter, 1969, pp. 72–90] [Rusnock & Šebestík, 2019, pp. 64–77]. The product of his researches into logic and the philosophy of science, which took up most of his time during the decade from 1820 to 1830, was the *Wissenschaftslehre* (*Theory of Science*) [Bolzano, 1837] [Bolzano, 1972] [Bolzano, 2014a].

Around 1830, after the decade devoted to studies in logic, Bolzano turned his attention to mathematics, among other research topics. During the 1830s and 1840s, the last two decades of his life, Bolzano investigated a number of mathematical topics, wrote several short papers, and worked towards putting together his large-scale mathematical treatise, *Größenlehre* (*Theory of Quantities*)—writing many parts of this but not publishing any.[20] This is the time of his second period of interest in his problem of defining the basic entities of geometry. The *Wissenschaftslehre* (*Theory of Science*) stands as a broad methodological background to this second period.

There are five relevant documents belonging to this second period. The first document is 'Versuch einer Erklärung der Begriffe von Linie, Fläche und Körper' ('Attempt for a Definition of the Concepts of Line, Surface, and Solid') [Bolzano, 0000d] [Bolzano, 2001c], a short fragmentary work, touched on in Section 4.4, summarizing the logical programmatic elements of the *Wissenschaftslehre* (*Theory of Science*) and analyzing the concept of extension. The second document is 'Anti-Euklid' ('Anti-Euclid') [Bolzano, 0000a] [Bolzano, 1967] [Bolzano, 20xxa], which contains an unusual geometrical analysis, to be discussed below. The three other mathematical documents that directly seek to define the elements of geometry are: 'Uiber Haltung, Richtung, Krümmung und Schnörkelung bei Linie sowohl als Flächen sammt einigen verwandten Begriffen' ('Concerning the Disposition [Situation, Position, or Attitude], Direction, Curvature, and Torsion of Lines as well as of Surfaces Together with Some Related Concepts') [Bolzano, 1948b] [Bolzano, 20xxc], 'Geometrische Begriffe, die Jeder kennt und nicht kennt. Versuch einer Erhebung derselben ins deutliche Bewußtseyn' ('Geometrical Concepts, which Everyone Knows and Does Not Know. An Attempt to Raise Them to Distinct Awareness')[Bolzano, 0000b] [Bolzano, 2001a], and *Paradoxien des Unendlichen* (*Paradoxes of the Infinite*) (1851) [Bolzano, 1851] [Bolzano, 2012] [Bolzano, 20xxb]. I shall examine the first two documents in the remainder of this section and the other three in the next section.

First, I note that in the *Wissenschaftslehre* (*Theory of Science*) there are references to a basic 'theory of point collections or sets,' inasmuch as points are recognized as the basic elements of geometry. I have already mentioned Bolzano's argument for simple ideas in that work [Bolzano, 1837, Vol. I, §.61, pp. 263–265] [Bolzano, 2014a, Vol. I, §.61, pp. 191–192]. Lines, surfaces, and solids are composed of infinitely many

[20] See [Bolzano, 1969, II, A, 7, 8, 9, 10,1, 10,2], containing material from his *Nachlass*.

points. There is an elaboration of this view later in the *Wissenschaftslehre* (*Theory of Science*) [Bolzano, 1837, Vol. III, §.317, 7, pp. 250–255] [Bolzano, 2014a, Vol. III, §.317, 7, pp. 168–172]. In particular, Bolzano states [Bolzano, 1837, Vol. III, §.317, 7, page 252] [Bolzano, 2014a, Vol. III, §.317, 7, page 169]:

> ... we should not hesitate to admit that every extended space, whether line, surface, or solid, also consists of simple parts, namely, points; it is nothing but a certain collection of such points.[Ch4, 29]

Thus, from an examination of the *Wissenschaftslehre* (*Theory of Science*) one sees Bolzano founding geometry on a simple theory of collections (structured sets) of points. This foundation becomes increasingly evident in his later mathematical work.

In Section 4.4 I referred to the fragmentary 'Versuch' ('Attempt') document because of its more pragmatic statement about definitions and the need for them to be useful in scientific presentations [Bolzano, 0000d, sheet 12r] [Bolzano, 2001c, page 131]. The statement is similar to one in Bolzano's 1843 publication in aesthetics [Bolzano, 1843a, §.1, no. 4, page 12] [Bolzano, 1989a, §.1, no. 4, page 104]. The fragment is not dated, but it was likely written between the time Bolzano completed writing of the *Wissenschaftslehre* (*Theory of Science*), around 1830 or perhaps even earlier, and the later work on aesthetics of 1843.[21]

More than half of the incomplete 'Versuch' ('Attempt') is taken up with a summary of Bolzano's theory of ideas or concepts and definitions, a theory which finds much fuller expression in the *Wissenschaftslehre* (*Theory of Science*). Then in the second part of the 'Versuch' ('Attempt') Bolzano makes a fresh attack on his geometrical definition problem [Bolzano, 0000d, sheets 18r-32v] [Bolzano, 2001c, pp. 138–150]. Since lines, surfaces, and solids are various kinds of spatial extension, he first proceeds to analyze the concept of extension (Ausdehnung), never returning in this unfinished paper to a specific analysis of the concepts of line, surface, and solid.

What is an extension? In the first place, Bolzano rejects the idea that an extension must be continuous (stetig) in the sense of connected, because he does not want to exclude curves which fall into two or more parts. For example, he wishes to include the curve defined by the equation $y = \sqrt{(1-x)(2-x)(3-x)}$, which falls into two separate pieces, as a single extension.[22] This view is the same as the view in the tract of 1817. However, in the second place, he does not want to regard finite sets of points as proper extensions. Extensions are a certain type of 'whole' or 'totality' ('Ganze', 'Inbegriff'), but they are not merely sets ('Mengen') with no form or order. They are somehow united or connnected (zusammengesetzt) into a whole. The problem is how to explain the connection of the parts in a geometrical extension. Why can one not regard a set of, say, 100 points as an extension? Bolzano's answer is that the parts, i.e., the points, are not connected (verbunden), but are isolated; they stand alone (vereinzelt oder isolirt stehen). Corresponding to each point in a finite system of points there is a smallest distance for which the point has a neighbor, while for all

[21] Cf. [Rusnock, 2020, page 11, footnote 45].

[22] As given, Bolzano may mean the positive value of the square-root expression, or he may really mean $y^2 = (1-x)(2-x)(3-x)$, i.e., the square of Bolzano's equation, allowing both positive and negative values for y.

smaller distances the point has no neighbors. Consequently, in order to have points joined together into an extension the opposite must be the case, viz., for none of the points can there be a smallest distance such that for smaller distances the point has no neighbors. Is this condition sufficient to guarantee that a collection of points form an extension? Bolzano answers 'no.' Consider the following example. Take two points a and b and insert a point c half way between them; then insert points half way between a and c, and c and b; and so on. Each point in this infinite system has neighbors for certain small distances, but not for all arbitrarily small distances. For example, a does not have neighbors at the distances $\frac{1}{3}\overline{ab}$, $\frac{1}{5}\overline{ab}$, etc. Thus, Bolzano ends up with the following definition of extension [Bolzano, 0000d, sheet 22v] [Bolzano, 2001c, page 143]:

> Every totality of points will be hereafter called an *extended spatial object*, in case each point is connected with the others, i.e., it has neighbors in this spatial object that are as close as you please; or, in other words, for every small distance there are points in the spatial object at this distance from it.[Ch4, 30]

This definition of an extended spatial object is the one that Bolzano finally settled on for his last attack on his geometrical problem. Although the terms may be different in the papers of the 1840s (*ausgedehntes, continuirliches Raumding*; *Kontinuum*), the definition remains the same. In the rest of the incomplete 'Versuch' ('Attempt') Bolzano tries to justify his definition of extension or continuum. He attempts to show that it is not too narrow and not too wide, but that it is the *right* definition [Bolzano, 0000d, sheets 22v-32v] [Bolzano, 2001c, pp. 143–150]. Bolzano's desire to justify his definition as uniquely correct comes to the fore. His deliberations show the close link between his philosophical view and his mathematics.

At this stage in his research one sees Bolzano heading for a new conception of geometrical objects based on the concept of isolated point. This concept is implicit in the 'Versuch' ('Attempt'), and in his paper of 1843–44, 'Uiber Haltung' ('Concerning the Disposition'), there is an explicit definition [Bolzano, 1948b, page 143] [Bolzano, 20xxc]:

> If the point i so lies in a spatial object such that there is no distance, however small, for which it can be asserted that *for this distance and for all smaller distances* i possesses one or several neighbors, then I say that i stands *isolated* or *alone* in the spatial object.[Ch4, 31]

Using this definition, one may describe a Bolzano extension or Bolzano continuum as a set with no isolated points.

Bolzano's concept of isolated point is quite different from that in present-day topology, which has come down to us from Cantor. Rather Bolzano's concept of isolated point resembles the notion of a point at which a set is 0-dimensional in the sense of Urysohn–Menger dimension theory (cf. [Hurewicz & Wallman, 1948, page 10]):

A separable metric space (in particular, a Euclidean space) has *dimension* 0 *at a point* p, if p has arbitrarily small neighborhoods with empty boundaries, i.e., if for each neighborhood U of p there exists a neighborhood V of p such that $V \subset U$ and $bdry(V) = \phi$.

It is not difficult to see that a Bolzano-isolated point in a set must also be a point at which the set is 0-dimensional in the sense of Urysohn and Menger. However, the converse is not true, since Bolzano restricts his attention to spherical neighborhoods.

In his works of the 1840s Bolzano provides some interesting examples of isolated points. For example, in 'Uiber Haltung' ('Concerning the Disposition') [Bolzano, 1948b, page 144] [Bolzano, 20xxc] Bolzano describes an infinite set, all of whose points are isolated: the set of points x of the form

$$x = \frac{m}{2^n},$$

where m, n are integers. In the *Paradoxien des Unendlichen* (*Paradoxes of the Infinite*) [Bolzano, 1851, §.41, page 82] [Bolzano, 20xxb] [Bolzano, 2004a, page 648] one finds the following example. Consider line segment az. Let point b be the midpoint between a and z; c, the midpoint between b and z; d, the midpoint between c and z; and so forth. If one takes the set consisting of the segment az minus the midpoints but including z, then z is an isolated point and so the set cannot be a Bolzano continuum. However, if one removes z from the set, one is left with a set with no isolated points and, hence, a Bolzano continuum. Clearly, Bolzano is well aware of the subtle distinction between his continua and noncontinua.

Some decades after Bolzano's death Georg Cantor felt obliged to criticize Bolzano's notion of continuum. Cantor charged Bolzano with focusing on just one property of a continuum and provided sets purporting to be counterexamples satisfying the property that do not appear intuitively to be continua [Cantor, 1883b, page 576] [Cantor, 1932, page 194]:

> The Bolzano definition of continuum (*Paradoxien* §38) [*Paradoxes of the Infinite* §38 [Bolzano, 1851]] is certainly not correct; it exclusively expresses just *one* property of a continuum, which, however, is also fulfilled by sets that result from [Euclidean n-space] G_n by imagining any "isolated" point set removed (*Math. Ann.* vol. 21, page 51). Likewise it is fulfilled by sets consisting of several separated continua. Clearly, in such cases we have no continuum, although according to Bolzano this would be the case.)[Ch4, 32]

Cantor's criticism is misplaced, at least in part. On the one hand, Cantor seems to confuse his own notion of isolated point (the modern one) with Bolzano's. It is by no means true that any set of Euclidean n-space G_n (Cantor's symbol) with some set of 'Cantor-isolated' points removed is a Bolzano continuum. The straight line minus all the points $(\frac{1}{2})^n$ and $-(\frac{1}{2})^n$ is a set such as Cantor describes; yet, because 0 is Bolzano-isolated in it, it is *not* a Bolzano-continuum. On the other hand, Cantor is right when he says that a Bolzano-continuum may be composed of several pieces. Bolzano fully recognized this possibility; he just did not always require his continua to be connected. Clearly, Bolzano's and Cantor's ideas diverge. Cantor's ideas are the antecedents of the modern notion of a connected continuum, whereas Bolzano's are the forerunners of the concept of a set of positive dimension.

I shall now turn to look briefly at Bolzano's paper, 'Anti-Euklid' ('Anti-Euclid') [Bolzano, 0000a] [Bolzano, 1967] [Bolzano, 20xxa]. This short sketch, covering just a few sheets, offers a vigorous critique of Euclid's presentation of geometry, thereby suggesting an improved version of traditional geometry. Bolzano criticizes the whole basis for the proofs of Euclidean geometry. Constructively, he seeks a 'learned' or rigorous presentation (eine gelehrte Darstellung) of Euclidean geometry, given according to the principles of science laid down in the *Wissenschaftslehre* (*Theory of Science*). In a rigorous development of geometry Bolzano requires that one proceed from simplest concepts to more complex ones through scientific definitions and from simplest to more complex truths through objective proofs.

Given the pragmatic view of definitions expressed in 'Versuch' ('Attempt') and the stricter view expressed in 'Anti-Euklid' ('Anti-Euclid'), it could be the latter should be dated earlier and the former closer to 1843 when the work on aesthetics was published. The manuscripts do not make the situation clear.

Bolzano cites many examples of concepts that, according to his view, lack rigorous definitions in traditional presentations of geometry. He holds that even the concept of space itself has no definition. Not surprisingly, Bolzano demands definitions for the well-known simpler concepts of geometry: extension, line, surface, solid, as well as distance and direction. This demand for rigorous definitions is just a restatement of his old long-standing problem. However, he also goes on to demand definitions for a whole host of other concepts, thereby vastly extending the scope of his geometrical problem. For example, he wants definitions for the two sides of each point on a line, of each line on a surface, of each surface in a solid, as well as definitions for closed lines and surfaces and when a point lies on the interior or the exterior of a closed line drawn on a surface. It is most interesting that Bolzano should want to have a definition for this last seemingly 'topological' situation.

In a similar way Bolzano criticizes previous expositions of geometry for giving inadequate proofs or no proofs at all for several significant propositions. Undoubtedly, the most significant theorem that Bolzano cites as lacking a proof in the geometry textbooks comes at the end of his manuscript [Bolzano, 0000a] (not included in [Bolzano, 1967]; though it should be included in [Bolzano, 20xxa]):

> *Theorem*: If a closed line lies in a plane and if by means of a connected line one joins a point of the plane which is enclosed within the closed line with a point of the plane which is not enclosed within it, then the connected line must cut the closed line.[Ch4, 33]

In this assertion one has Bolzano's 'forward-looking' theorem akin to the celebrated Jordan curve theorem. It may seem almost incredible to find it as part of Bolzano's critical program for reorganizing elementary geometry. Yet the motivation for his statement of the theorem is straightforward to discern. It is the old criticism of Euclid's first existence proof concerning the construction of an equilateral triangle on a given line segment, which Wolff and Kästner knew about before Bolzano. However, in contrast, Bolzano's insight into the problem is deeper than the insights of his predecessors, for he sought to base this and other elementary geometrical theorems on a rigorous foundation of something like point-set theory or point-set topology. This proposed theorem is also related by analogy to the earlier Intermediate Value Theorem that Bolzano clearly stated and proved in *Rein analytischer Beweis* (*Purely Analytical Proof*) [Bolzano, 1817b]. Consequently, in the mid-1840s, unfortunately when his strength was fading, Bolzano started to carry out the ambitious program sketched in 'Anti-Euklid' ('Anti-Euclid').

4.7 Second Solution: Continuing Geometrical Investigations of the 1840s

I shall now consider the last three documents of the five listed at the beginning of Section 4.6. The first two papers, 'Uiber Haltung, Richtung, Krümmung und Schnörkelung bei Linie sowohl als Flächen sammt einigen verwandten Begriffen' ('Concerning the Disposition [Situation, Position, or Attitude], Direction, Curvature, and Torsion of Lines as well as of Surfaces Together with Some Related Concepts') [Bolzano, 1948b] [Bolzano, 20xxc] and 'Geometrische Begriffe, die Jeder kennt und nicht kennt. Versuch einer Erhebung derselben ins deutliche Bewußtseyn' ('Geometrical Concepts, which Everyone Knows and Does Not Know. An Attempt to Raise Them to Distinct Awareness') [Bolzano, 0000b] [Bolzano, 2001a] were written during 1843–44, but were left incomplete by Bolzano. They are the result of a burst of intellectual activity, even though he was not well at the time. They are two of his final attempts to put his mathematical ideas down on paper, in order that others might carry out his program.[23] The *Paradoxien des Unendlichen* (*Paradoxes of the Infinite*) contains, *inter alia*, a summary of his last geometrical ideas, in particular, the ideas relating to dimension.

In beginning an analysis of Bolzano's final investigations into basic geometrical problems as he exposited in the late works of the 1840s, I first note that the definition problem for lines, surfaces, and solids remains fundamental. The two papers written during 1843–44, 'Uiber Haltung' ('Concerning the Disposition') and 'Geometrische Begriffe' ('Geometrical Concepts'), are the most significant. *Paradoxien des Unendlichen* (*Paradoxes of the Infinite*) [Bolzano, 1851, §§.38, 40–41; pp. 71–75, 78–83] only has a very brief summary of his last results on his chosen range of geometrical problems. Unfortunately, these final investigations are incomplete. Bolzano left neither of the two main papers in anything like a finished state. In the 'Vorwort' ('Forward') to 'Uiber Haltung' ('Concerning the Disposition') [Bolzano, 1948b, page 142] [Bolzano, 20xxc] he freely admits that this work must be considered as an 'incomplete attempt' ('unvollständiger Versuch'). He even says that he has not carried out the calculations required for a complete solution of the problems

[23] At the same time, 1843, Bolzano produced his publications on aesthetics [Bolzano, 1843a] [Bolzano, 1989a] and on the three dimensionality of space [Bolzano, 1843b] [Bolzano, 1948c] [Bolzano, 1989b].

under examination. He says he lacks the necessary strength, but hopes that someone else will finish the task. In his last years Bolzano was not well, though his interest in his research remained undiminished.

In the paper 'Uiber Haltung' ('Concerning the Disposition') [Bolzano, 1948b] [Bolzano, 20xxc] Bolzano is especially concerned with the problem of curvature, particularly the curvature of skew or twisted (nonplanar) curves in 3-dimensional space. He is not satisfied with previous efforts to characterize the 'double curvature' of skew curves. No arc of such curves, however small, is planar; hence, the usual circle of curvature fails to characterize the full idea of curvature for these curves, just as the straight line tangent to a planar curve fails to embrace the curvature in this instance. Bolzano also rejects an old suggestion of Newton to use a parabola touching the skew curve, for that alternative is no better than using the circle. What Bolzano seems to be trying to find is a measure of torsion for spatial curves (and surfaces) (as in modern differential geometry). He employs a special word, 'Schnörkelung,' for what might be considered torsion. His partial 'solution' for curves is to use osculating cylindrical helices as a way to measure the spatial curvature of skew curves, but he realizes that this approach does not provide a complete solution either [Bolzano, 1948b, pp. 177–181].

Curvature is not the only problem that Bolzano intended to deal with in the 'Uiber Haltung' ('Concerning the Disposition'). He planned to examine a whole class of neighborhood properties of both curves and surfaces. In fact, he coined the term 'Haltung' as a general name for these neighborhood properties (perhaps something like 'topological disposition' or 'situation'). In particular, 'Haltung' is supposed to cover the concepts of direction, curvature, and rate of change of curvature ('Schnörkelung,' another term invented by Bolzano for this particular situation). Bolzano only wrote down a partial analysis of the 'Haltung' of lines, mainly investigating direction and curvature. A treatment of the 'Haltung' of surfaces is lacking. On the whole, the specific results on curvature and 'Haltung' that Bolzano committed to paper are relatively meagre. However, they are suggestive of his apparently broader insights concerning the 'geometrico-topological' situation.

Undoubtedly, the most significant feature of the 'Uiber Haltung' ('Concerning the Disposition') paper is the very general proto-topological viewpoint that Bolzano adopts throughout. It is clear that he intended to weave his topological fundamentals much more solidly into the fabric of his geometrical work. The concepts of distance and neighbor form the foundation of his ideas. Using this basis, Bolzano defines an isolated point of a set as hinted at in his earlier 'Versuch' ('Attempt') [Bolzano, 0000d] [Bolzano, 2001c] (discussed in Section 4.6) as a point for which there is no distance, however small, such that the point has one or more neighbors in the set for this distance and all smaller distances. A continuum or extension is a set with no isolated points, whereas a discontinuum ('discontinuirliches Raumding') is a set all of whose points stand isolated [Bolzano, 1948b, pp. 143–144].

Now Bolzano is able to define the basic objects of elementary geometry in a new way, bringing ideas of dimension to the fore [Bolzano, 1948b, pp. 144–145]:

> I call an extension whose every point has only so many neighbors for each sufficiently small distance that the set of these, considered in itself for each of the distances, still does not represent an extension, a *spatial object of a single or simple extension* or a *line*. I call a spatial object whose every point has so many neighbors for each sufficiently small distance that the set of these, considered in itself for each of the distances, still represents a spatial object of simple extension, a *spatial object of twofold or double extension* or a *surface*. Finally, I call a spatial object whose every point has so many neighbors for each sufficiently small distance that the set of these, considered in itself for each of the distances, already represents a spatial object of double extension, a *spatial object of threefold extension* or a *solid*.[Ch4, 34]

These revised definitions broaden the ones Bolzano put forward in 1817. They provide the makings of an inductive definition of geometrical objects of dimensions 0, 1, 2, and 3. The '0-dimensional' basis of isolated point sets is far richer than his older basis of finite sets used to define lines. Bolzano proudly claims [Bolzano, 1948b, page 145] that the new definitions are essentially better than his former ones. The important question then arises: Why the change?

There are probably a number of reasons that made Bolzano modify his earlier definitions. His discovery of the idea of isolated point and more generally his willingness to deal with infinite sets in his later work almost certainly influenced the change in his thinking. However, there is a specific example in the 'Uiber Haltung' ('Concerning the Disposition') [Bolzano, 1948b, page 183, fig. 9] (see Figure 4.2 for a copy of Bolzano's sheet of figures from 'Uiber Haltung' ('Concerning the Disposition') [Bolzano, 1948b, page 183]) that suggests to me why he altered his theory. The example consists of an infinite set of circles $MA, MA', \ldots$, with increasing diameters, but all tangent to a straight line (RMS) at a single point M. If one considers this figure as a single line, then the point M has infinitely many neighbors for each small distance, but these do not form a Bolzano-continuum. Consequently, one has a line in Bolzano's new sense, but not in the sense of the work of 1817. Perhaps an example like this motivated Bolzano to rework his older theory. Certainly in his later research Bolzano developed a facility for dealing with infinite sets and for seeing their important role in geometry and analysis. Hence, he was able to handle a wider variety of geometrical objects, considered as sets of points.[24]

In his work 'Uiber Haltung' ('Concerning the Disposition') [Bolzano, 1948b, page 146] Bolzano defines the special class of 'determinable' spatial objects in a way similar to that in *Die drey Probleme* (*The Three Problems*). An object of geometry is determined with respect to a right-angled coordinate system and a measure of distance. In particular, lines are determined through one or more pairs of equations of the form

$$f(x, y, z) = 0, \qquad \phi(x, y, z) = 0,$$

[24] In a short paper Petr Simon [Simon, 1981] provides a modernized dimension-theoretic explanation of Bolzano's analyses of the concept of line. He gives some differentiating geometrical examples that illustrate some of the distinctions between Bolzano's two concepts of line. See also [Šebestík, 1992, pp. 61–63], as well as [Katětov & Simon, 1997].

where f and ϕ are functions that are continuous and have derivatives except possibly at an isolated set (finite or infinite) of values x, y, z. The cases of surfaces and solids are similar.

Unlike his treatment in *Die drey Probleme* (*The Three Problems*), Bolzano does not restrict his attention to determinable geometrical objects in the 'Uiber Haltung' ('Concerning the Disposition') [Bolzano, 1948b] [Bolzano, 20xxc]. Instead he wishes to consider lines, surfaces, and solids in full generality (where possible), so that his concept of topological neighbor becomes a more genuine foundation for his geometrical investigations. Thus, when it comes to indicating the scope of 'Haltung,' he offers a very general description [Bolzano, 1948b, page 150]:

> We are therefore permitted to attribute to *every extension which has no solid part*, a *property* at each of its points, which we think of as the *proximate cause* why this spatial object has as neighbors, for each of its points at every sufficiently small distance, *only those neighbors and, otherwise, no others from the entire stock of points which have this distance from the point*. Allow me to call this property of a spatial object at each of its points its *situation* [*Haltung*] *at this point*.[Ch4, 35]

As mentioned above, Bolzano unfortunately was not able to carry out a very complete analysis of the general 'Haltung' of lines, nor any analysis of surfaces. Primarily he treats the 'Haltung' of the straight line, the circle, and the helix. So at the end of the 'Uiber Haltung' ('Concerning the Disposition') Bolzano's geometrico-topological program still needs further development.

To some extent the topological investigations of 'Uiber Haltung' ('Concerning the Disposition') are continued in the fragmentary paper 'Geometrische Begriffe' ('Geometrical Concepts')[Bolzano, 0000b] [Bolzano, 2001a], which fills in more details. The paper aims at clarifying some of the ordinary concepts of geometry that we, whether mathematicians or non-mathematicians, know and yet do not really grasp with complete clarity. It treats lines and surfaces in full generality according to the definitions of 'Uiber Haltung' ('Concerning the Disposition'), and it examines special types of these. 'Geometrische Begriffe' ('Geometrical Concepts') encompasses a long sequence of definitions about lines and surfaces. Bolzano defines several geometrical concepts in two or three distinct ways. A later definition appears to improve on an earlier one. Bolzano defines in rapid succession special kinds of lines and surfaces, simple, bounded, closed, simple closed, and so forth, and then takes up the problem of betweenness for various types of geometrical figures. With respect to the latter problem he considers, for example, the cases when one point is between two others on a line and when a point is said to be enclosed within a closed curve. Interestingly, Bolzano includes some very intriguing, quite insightful geometrical figures in the text and in the margins of the manuscript to illustrate his quest for improved definitions though differentiating examples (see Figures 4.3 and 4.4 for copies of Bolzano's figures collected from the margins of 'Geometrische Begriffe' ('Geometrical Concepts') [Bolzano, 0000b] [Bolzano, 2001a, pp. 168–169]). The lengthy series of definitions in the 'Geometrische Begriffe' ('Geometrical Concepts') emphasizes Bolzano's strong inclination to examine mathematical matters

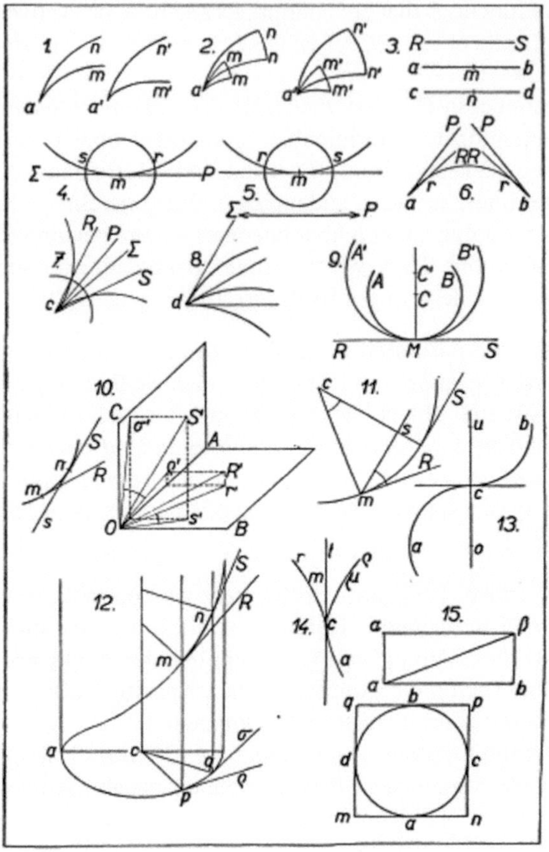

Fig. 4.2 Sheet of Figures from Bolzano's 'Uiber Haltung' ('Concerning the Disposition') [Bolzano, 1948b, page 183].

using conceptual analysis as a tool to strive for clarity and distinctness and as a method to lay out diverse cases. An important job of the mathematician is to seek good, i.e., precise, definitions for basic mathematical concepts in order to bring them into 'clear consciousness' or clear understanding and to understand their distinctness and interrelationships. Correct definitions are essential to the proper development of a mathematical science. In 'Geometrische Begriffe' ('Geometrical Concepts') Bolzano lays down definitions for many of the geometrical concepts that he suggested needed defining in 'Anti-Euklid' ('Anti-Euclid')[Bolzano, 0000a] [Bolzano, 1967] [Bolzano, 20xxa]. 'Geometrische Begriffe' ('Geometrical Concepts') appears as an important *experimental work*, informally testing geometrical definitions, ideas, and concepts with illustrative diagrams.

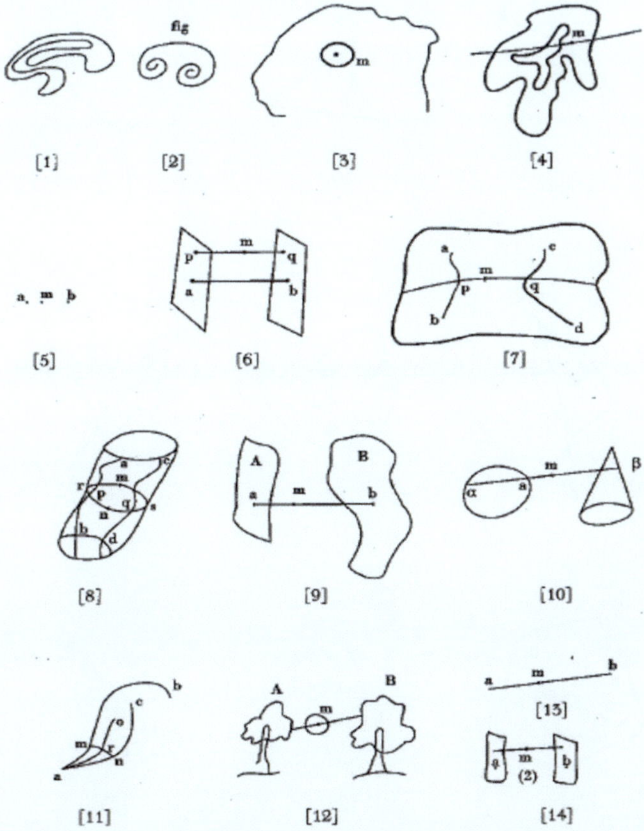

Fig. 4.3 Sheet 1 of Figures from Bolzano's 'Geometrische Begriffe' ('Geometrical Concepts') [Bolzano, 2001a, page 168]; by permission of the publisher of the *Bernard Bolzano: Gesamtausgabe*, frommann-holzboog.

As an example of the investigations of 'Geometrische Begriffe' ('Geometrical Concepts') I shall consider the definitions for closed and simple closed lines. Recall that in *Die drey Probleme* (*The Three Problems*) Bolzano tacitly assumed that lines that are not closed, such as segments and arcs, always include their endpoints, just as surfaces that are not closed always include their boundary curves. In 'Geometrische Begriffe' ('Geometrical Concepts') he comes to realize this tacit assumption. At the beginning of 'Geometrische Begriffe' ('Geometrical Concepts') he gives the following definition for a simple closed curve [Bolzano, 0000b, sheet 2r] [Bolzano, 2001a, page 157]:

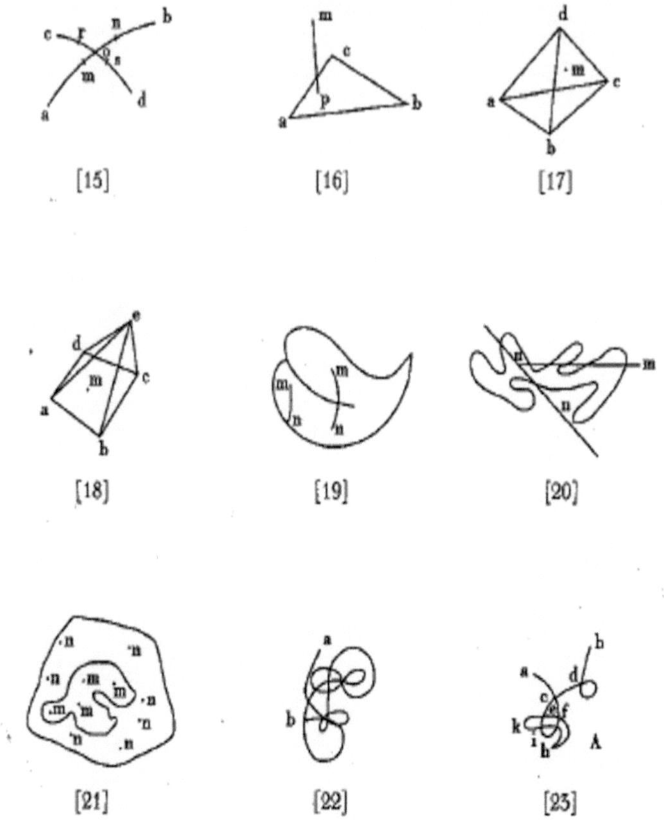

Fig. 4.4 Sheet 2 of Figures from Bolzano's 'Geometrische Begriffe' ('Geometrical Concepts') [Bolzano, 2001a, page 169]; by permission of the publisher of the *Bernard Bolzano: Gesamtausgabe*, frommann-holzboog.

> §.2. A *simple closed line* is one for which every point has two neighbors, but no point possesses neighbors at a distance greater than a given distance. ***And as false see fig.***[25] [Ch4, 36]

Under this definition an open line segment such as (0, 1) or an open arc qualifies as a closed line, which is likely not Bolzano's intention. Perhaps that observation is the critical meaning Bolzano intended with his words in Latin, which I translated and put in the original boldface in the quotation above. In any case, the definition will appear unsatisfactory to a careful modern reader. Apparently in order to improve

[25] See figure [1], a simple closed line/curve, and figure [2], an open line/cure with 'fig' above, in Figure 4.3.

matters, Bolzano offers a definition of closed line in an addendum at the very end of the manuscript [Bolzano, 0000b, sheet 9r] [Bolzano, 2001a, page 166]:

> §. A line is *called closed*, if (1) no point of it has a distance from any other greater than a given **E**, (2) every point has two neighbors for every sufficiently small distance, and finally (3) no single point or set of several points, which does not in itself form a line, can be added to it without the new totality ceasing to be a line and having two neighbors for each sufficiently small distance.[Ch4, 37]

This definition overcomes the difficulty about open segments and arcs. Nevertheless, clause (2) effectively requires the curve to be simple. At another place [Bolzano, 0000b, sheet 8r] [Bolzano, 2001a, page 164] Bolzano seems to offer his final attempt at defining closed and simple closed curves. For an arbitrary closed line (with possible multiple points) it is only necessary to require an even number of neighbors in clauses (2) and (3); exactly two neighbors are required for simple closed lines. These 'final' definitions are quite good and are typical of Bolzano's late geometrico-topological and distinctively experimental thinking.

A pair of disjoint figure-eights can be taken as one closed curve under Bolzano's 'final' definition, but, of course, they are not connected. To sharpen his definition Bolzano gives a condition to describe when a closed curve is all in one piece [Bolzano, 0000b, sheet 8r] [Bolzano, 2001a, page 164]:

> §. A *line* is called a *single* closed line, if no part of it, which by itself already represents a closed line, can be omitted without the remainder ceasing to be a closed line in itself.
> §. In a similar way a *single closed surface*.[Ch4, 38]

In this case there is some improvement on the definitions of connectedness of 1817. Nonetheless, these definitions are very restricted in scope, since they only apply to closed curves and surfaces.

In the mass of definitions of 'Geometrische Begriffe' ('Geometrical Concepts') Bolzano states virtually only one 'theorem' [Bolzano, 0000b, sheet 8r] [Bolzano, 2001a, page 165]:

> §. Every simple closed line which lies on a surface separates the latter into two parts, which are distinguished by the fact that all points of the surface, which lie outside the line [which are not on the line], lie either on one or the opposite side of the line.[26] [Ch4, 39]

Unfortunately, if one takes 'surface' to mean any figure which falls under Bolzano's very general definition, then this new version of the closed curve theorem is simply false. For example, it fails for certain nonbounding closed curves on the torus. Bolzano's diagram accompanying his assertion shows that he was really thinking of quite simple types of surfaces.[27] Yet his surface definition takes in much more. Consequently, his version in 'Anti-Euklid' ('Anti-Euclid') of the closed curve theorem

[26] See figure 21 in Figure 4.4.

[27] See figure 21 in Figure 4.4

for the plane—akin to the celebrated Jordan curve theorem—is preferable to the one just quoted. To be fair to Bolzano, in other places in his text he does contemplate more complex situations, nicely illustrating them in his figures 20, 22, and 23 in Figure 4.4.

Bolzano does not include in 'Geometrische Begriffe' ('Geometrical Concepts') any proof of either version of the closed curve theorem. However, before asserting the version above, he carefully defines the main concepts of the theorem: the concepts of simple closed line (as given above) and of points lying on the same or opposite sides of a line on a surface [Bolzano, 0000b, sheet 8r] [Bolzano, 2001a, page 164]:

> §. *Two points on a given surface are said to be (1) on the same side or (2) on opposite sides of a line lying on the surface*, according to whether every plane passing through them (at most with the exception of one or several isolated cases) has either (1) none or 2, 4, . . . or in general *an even number* of intersection points or (2) one or 3, . . . or in general an *odd number* of intersection points with the line which lie in the piece of the line of intersection of the plane with the surface that has those points as endpoints.[Ch4, 40]

Perhaps Bolzano thought that his definitions cleared the way for an easy proof of the conjectured theorem or that it was a direct consequence of the definitions. These thoughts would appear to be in line with his philosophy. In any case, he did not write down a proof, and so one cannot be sure of his attitude towards the nature of a proof.

Bolzano's bare statement, with no supporting proof, of the closed curve theorem is the highpoint of 'Geometrische Begriffe.' Beyond this paper he produced no new results. Hence, one is left with the interesting beginnings of what might have been a fruitful program in set-theoretic topology, but which Bolzano did not develop any further. It is clear that Bolzano had many exciting 'topological' ideas, which could have been the framework for a detailed geometrico-topological theory as one might expect one today. However, in his last years he could not summon up the energy to work out the details.

4.8 Bolzano's Attempted Objective Foundation for Three Dimensions of Space

On 7 December 1842 Bolzano presented a philosophical paper to the Königliche böhmische Gesellschaft der Wissenschaften (Royal Bohemian Society of Sciences) on a topic of dimension. The discussion in the paper occurs in a different conceptual context and is of quite a different character than his other work on dimension that I have examined it in the foregoing sections. The paper was soon published in 1843 as a separatum, *Versuch einer objectiven Begründung der Lehre von den drei Dimensionen des Raumes* (*Attempted Objective Foundation for the Doctrine [Theory] of the Three Dimensions of Space*) [Bolzano, 1843b] [Bolzano, 1845b] [Bolzano, 1948c] [Bolzano, 1989b].

The topic of this work is the three-dimensionality of space. The work contains his 'attempted' objective foundation, justification, or proof that space has three dimensions. The general philosophical reference Bolzano provides for an objective foundation is to the *Wissenshaftslehre* (*Theory of Science*), volume 4, §.525. 'Explaining the objective ground of truth' [Bolzano, 1837, §.525, pp. 261–263] [Bolzano, 2014a, §.525, pp. 180–181]. Bolzano attempts to lay out objective conceptual grounds or an objective basis or detailed objective proof for the three-dimensionality of space. The fact that the basis is conceptual would appear to remove it from the empirical realm for understanding the dimensionality of space. In this section I shall only give a short summary of Bolzano's work.

Bolzano first develops a conceptual framework in which he elaborates the basic characteristics of both time and space according to his objectivist philosophical views. Bolzano sees time and space as closely related, just as many other philosophers of that time (e.g., Kant) did. The concept of time is simpler than that of space. The concept of space actually contains the concept of time as a component [Bolzano, 1843b, §.1, page 5] [Bolzano, 1845b, §.1, page 205] [Bolzano, 1948c, §.1, pp. 52–53] [Bolzano, 1989b, §.1, pp. 225–226]. He provides a conceptual analysis of each with the aim of enumerating their properties, showing their close relationship, and seeking to bring out characteristics relevant to the three-dimensionality of space.

He first analyzes the notion of time, offering both negative and positive characteristics [Bolzano, 1843b, §§.2,3, pp. 6–10] [Bolzano, 1845b, §§.2,3, pp. 206–210] [Bolzano, 1948c, §§.2,3, pp. 53–58] [Bolzano, 1989b, §§.2,3, pp. 226–231]. For time there is a notion of distance between time instances and a notion of duration or time interval. These temporal notions have spatial implications.

He then analyzes the notion of space in a parallel way to time with both negative and positive characteristics [Bolzano, 1843b, §§.4,5, pp. 10–12] [Bolzano, 1845b, §§.4,5, pp. 210–212] [Bolzano, 1948c, §§.4,5, pp. 58–61] [Bolzano, 1989b, §§.4,5, pp. 231–234]. He develops dimensional characteristics, in part deriving them from dimensional characteristics of time. I must confess that I find his discussion of time and space coupled with dimension rather opaque.

However, his framework, including the embedded material on three-dimensionality, enables him to assert his principal theorem (Lehrsatz) asserting the three-dimensionality of space [Bolzano, 1843b, §.6, page 13] [Bolzano, 1845b, §.6, page 213] [Bolzano, 1948c, §.6, page 62] [Bolzano, 1989b, §.6, page 235]:

> *There are systems of four points, such that no one of the four as such is determined by its relationship to the other three, insofar as it can be comprehended purely conceptually. However, if such a system of four points is given, then every other point and every set of points* (i.e., every spatial thing) *can be determined by mere concepts that express its relationship to those four points.*[Ch4, 41]

A proof follows the statement of the theorem [Bolzano, 1843b, §§.6,7, pp. 13–15] [Bolzano, 1845b, §§.6,7, pp. 213–215] [Bolzano, 1948c, §§.6,7, pp. 62–65] [Bolzano, 1989b, §§.6,7, pp. 235–238].

Bolzano's statement expressing the three-dimensionality of space is, I think, quite clear, at least as it might be interpreted from a modern point of view. While four points in general position in space (i.e., no three are collinear) are linearly independent, whereby no one of the four can be determined by a relation with the other three; any additional point or additional set of points is dependent on these four or can always be determined by a relation to these four. In terms of modern vector space theory Bolzano's theorem is effectively about linear independence and dependence. Vectors drawn to the first four points are linearly independent of one another, while a vector drawn to an additional point will be linearly dependent on the first four. The first part of Bolzano's theorem statement asserts that (our) space is *at least* three-dimensional, while the second part asserts that (our) space is *at most* three-dimensional. Bolzano's proof itself brings in much of the conceptual material presented in the earlier sections, so the modern mathematical interpretation does not do justice to Bolzano's proof scheme.

Bolzano's full argument for the three-dimensionality of space appears labored from a modern vantage point, given its assumed conceptual basis. However, that does not necessarily make it wrong.

Much later Hermann Weyl in his well-known book on philosophy of mathematics and science indicated that he did not rate Bolzano's argument highly (using the 'absurd' in the context) [Weyl, 1949, page 137]. I think that Bolzano's argument rests very much on what he includes in the middle sections on dimensionality and then brings to bear in the final section of his paper [Bolzano, 1843b, §.7, page 15] [Bolzano, 1845b, §.7, page 215] [Bolzano, 1948c, §.7, pp. 64–65] [Bolzano, 1989b, §.7, pp. 237–238]. In the first part of this section he uses an orthogonality construction to yield four points a, b, c, d that are in general position, so that his notional space is at least three-dimensional. Then his final statement, indeed his last sentence, is crucial to showing that his notional space is not more that three-dimensional [Bolzano, 1843b, §.7, page 15] [Bolzano, 1845b, §.7, page 215] [Bolzano, 1948c, §.7, page 65] [Bolzano, 1989b, §.7, page 238]:

> ... Finally, if the four points a, b, c, d have the specified property [of being independent or in general position], then there is *no fifth one*, that could be fully determined [as independent] by its conceptual relationship to the others [other four points]: therefore, it follows that every other prescribed direction emerging from a, which stands perpendicular [or orthogonal] to two of the others [other points] just found, is either coincident or opposite in direction to the third, and that is exactly what is stated in the theorem of the three dimensions of space.[Ch4, 42]

In my opinion Bolzano's final statement is not an entirely cogent completion of the argument. However, taking a more sympathetic and positive view, and using all the framework of Bolzano's conceptual world, Rusnock and Šebestík give an extended and very valuable exposition of Bolzano's argument, which one should consider [Rusnock & Šebestík, 2019, pp. 458–469].

4.9 Concluding Remarks on Bolzano's Geometrical Investigations

Bolzano's interest in his fundamental geometrical problem of defining the concepts of line, surface, solid, and extension or continuum spanned his entire creative lifetime, from his earliest thoughts on mathematics and first published works to his last research papers, notes, and sketches. Having examined Bolzano's attempted solutions to his problem in detail, I shall in closing this chapter try to put his work in some perspective.

Fundamental to Bolzano's outlook on research in mathematics is his view that mathematics stands in close relation to philosophy. His autobiographical statements reveal that he preferred 'philosophical mathematics,' and his published work shows his deep concern with logical and foundational issues in mathematics. One is tempted to call Bolzano a 'philosophical mathematician.' In this respect he reminds one of Descartes, whose mathematical work was closely linked to his philosophy and his published and well-known method of discovery.

In the particular case of his basic geometrical problem Bolzano's philosophy of definitions and concepts had a strong influence, seemingly guiding his quest for a solution. Yet, as described in Section 4.4 Bolzano's adherence to his strict theory of definitions as the analysis of concepts down to simples and their connections accompanied with proofs apparently softened over his research career. He realized that such analyses in many cases could be very difficult to bring to fruition.

While persisting in taking a definitional view of his geometrical problem over his career, Bolzano was able to explore broadly a variety of definitions and results—many very insightful. He proposed definitions, examined examples and counterexamples with respect to the definitions, and then changed his definitions to make improvements. Over time, he followed a pattern of offering tentative definitions, sharpening them with examples and counterexamples, and then developing improved definitions, while taking a very self-critical attitude to this work. This progressive Popperian model is analogous to Lakatos' pattern of proofs and refutations [Lakatos, 1976] [Lakatos, 2015]. This is a pattern of mathematical thinking that one finds frequently evident in topology and some other parts of mathematics.

Bolzano is perhaps best seen as an early enthusiastic explorer in a new territory of mathematics, who remained grounded in his unique logical view of foundations. Although he had several forward-looking ideas, he was not able to conquer the whole of the new domain suggested by them. His viewpoint and results were both foundational and modestly expansive. Perforce his conquests were of necessity limited. Like Leibniz when he was trying to open up his new domain of 'analysis situs,' Bolzano could not provide those following him with a detailed map of the new land, but only a few impressions of its terrain.

Chapter 5
Nineteenth-Century Theories of Higher-Dimensional Spaces

5.1 Introduction

In this chapter I shall examine the development of theories of higher-dimensional spaces, that is, spaces beyond 3 dimensions to *n* or infinite dimensions, that occurred during the nineteenth century. Early in the century there were mathematical forays into the world of higher-dimensional spaces, with Gauss being a principal explorer. The main development of such theories then took off in the 1840s, 1850s, and 1860s with the publication in rapid succession of works by Grassmann, Cayley, Riemann, and several others. The theories in these works clearly demonstrated that geometry and its related spaces ought to be separated from the usual physical space of human perception or understanding. Moreover, they forced a radical change in the philosophy of geometry.[1] These changes in geometry and its accompanying philosophy are to an extent part of the trend of arithmetization of the nineteenth century. They form a prime background to Cantor's paradoxical discovery about dimension. In Chapters 2 and 3 I have already called attention to some scattered developments of 'higher-dimensional' spaces and some warnings against the existence of those spaces.

[1] For example, cf. [Torretti, 2019].

D. M. Johnson, *Theories of Dimension from Ancient to Modern Times*, Sources and Studies in the History of Mathematics and Physical Sciences, https://doi.org/10.1007/978-3-032-02682-8_5

5.2 Jean-Robert Argand

The amateur mathematician Jean-Robert Argand (1768–1822)[2] is well known for his geometric representation of complex numbers in the plane,[3] as described in his 'Essai sur une manière de représenter les quantités imaginaires dans les constructions géométriques' ('Essay on a method of representing imaginary quantities by geometric constructions'), privately published in 1806 [Argand, 1806]. A modified and shorter version of the 'Essai' was then published in Gergonne's *Annales de Mathématiques* in 1813 [Argand, 1813] (cf. [Argand, 1874], which contains both versions of the 'Essai,' and the English translation [Argand, 1881]).[4] In the 1813 version he makes a brief remark suggesting the possibility of spatial extensions beyond three to higher dimensions, derived from his algebraic analysis [Argand, 1813, page 146]:

> . . . we. . . observe only in closing that the expressions a, a_b, a_{b_c}, which designate lines considered in reference to one, two, and three dimensions, are only the first terms of a series that can be indefinitely extended.[Ch5, 1]

Argand never produced anything on this subject beyond this brief remark.

5.3 Carl Gustav Jacob Jacobi and George Green

Both Carl Gustav Jacob Jacobi (1804–1851)[5] and George Green (1793–1841)[6] used n-dimensional generalizations in the course of their analytical researches. Both found higher-dimensional spaces very convenient: Jacobi in a work of 1834 [Jacobi, 1834] [Jacobi, 1881, Volume 3, pp. 191–268] and Green in a work of 1835 [Green, 1835]. Jacobi uses them quite naturally from the beginning of his paper. Green, after

---

[2] For biographical material on Argand see
`https://mathshistory.st-andrews.ac.uk/Biographies/Argand/` and
`https://www.encyclopedia.com/science/dictionaries-thesauruses-pictures-and-press-releases/argand-jean-robert`.

[3] On the topic of representing complex numbers geometrically cf. [Lützen, 2001] and, in particular [Schubring, 2001].

[4] Gergonne's 1813 volume of *Annales de Mathématiques* captures a scientific interchange between Argand and Professor J.F. Francais on the issues of representing complex numbers. The exchange is fully presented in the later edition by J. Hoüel [Argand, 1874] and its English translation [Argand, 1881]. In 1814 Argand also gave a proof of the fundamental theorem of algebra [Argand, 1814].

[5] For biographical information on Jacobi cf.
`https://mathshistory.st-andrews.ac.uk/Biographies/Jacobi/` and
`https://en.wikipedia.org/wiki/Carl_Gustav_Jacob_Jacobi`.

[6] For biographical information on Green cf.
`https://mathshistory.st-andrews.ac.uk/Biographies/Green/`,
`https://www.nottingham.ac.uk/physics/about/history/george-green.aspx`, and
`https://en.wikipedia.org/wiki/George_Green_(mathematician)`.

reducing the physical problem of the attraction of ellipsoids with variable densities to mathematical analysis, says [Green, 1835] [Green, 1871, page 188]:

> The original problem being thus brought completely within the pale of analysis, is no longer confined as it were to the three dimensions of space.

After that point he develops his theorems of analysis quite readily in a space of arbitrary finite dimension.

5.4 Carl Friedrich Gauss

Carl Friedrich Gauss (1777–1855)[7] was the most significant mathematician to consider the geometry of hyperspaces during the first half of the nineteenth century. However, he published little on the subject, and even his notes and letters, which only came to light after his death, probably do not represent the full scope of his thoughts on this matter. Yet, the short remarks and notes that are available indicate a lifelong interest in multi-dimensional spaces. His interest was related to many of his other investigations in mathematics. Gauss undoubtedly thought often and deeply about the abstract foundations of geometry. Gauss' notes on hyperspaces are indicators of something of broader significance. One might look upon them as pointing to a maturing development of the concept and theory of multi-dimensional spaces. Such spaces fit well with the developments of other domains of mathematics.[8]

The earliest record of Gauss' knowledge of higher spaces comes from 1816 in a letter of Friedrich Ludwig Wachter to Gauss of 12 December 1816 [Stäckel, 1901, pp. 61–69] (in full) [Gauss, 1917, pp. 481–482] (in part). This letter refers to a conversation that Wachter had with Gauss during April of that year. It demonstrates Gauss' (as well as Wachter's) early understanding of multi-dimensional analytic geometry. Wachter specifically mentions infinite-dimensional spaces as well as finite-dimensional ones, so he and Gauss were seemingly well advanced in the theory.

Gauss' most important references to the problem of defining 1-, 2-, and n-dimensional manifolds (Mannigfaltigkeiten) are to be found in a work published in 1832 on the theory of biquadratic residues: 'Theoria residuorum biquadraticorum. Commentatio secunda' ('Theory of biquadratic residues. Second commentary') [Gauss, 1832] [Gauss, 1863, pp. 93–148] and in his presentation ('Anzeige') of that work to the Königliche Societät der Wissenschaften zu Göttingen (Royal Society of Sciences at Göttingen), given the previous year on 15 April 1831 [Gauss, 1831] [Gauss, 1863, pp. 169–178]. The reference to higher-dimensional developments consists of just a single remark. One of Gauss' concerns in these works is the 'metaphysics' or foundations of the complex number system. Although he describes the

[7] There are numerous sources of information about Gauss' life and work. One may start with `https://mathshistory.st-andrews.ac.uk/Biographies/Gauss/` and `https://en.wikipedia.org/wiki/Carl_Friedrich_Gauss` for references.

[8] Cf. [Ferreirós, 2007, pp. 43–45] for a short summary of Gauss' work on manifolds and hyperspaces in relation to Riemann.

usual planar representation of complex numbers, feeling even as late as 1831 that this number system needed a kind of justification by means of a geometrical model [Gauss, 1863, pp. 109–110, 174], in his companion presentation he puts more emphasis on an abstract approach to the 'true metaphysics of imaginary quantities' [Gauss, 1863, pp. 175–178]. He only treats the complex ('Gaussian') integers in detail, but presumably an extension to all complex numbers can be made easily. The kernel of Gauss' abstract theory arises from his observation that numeration consists in a set of relations. If one has a linear sequence of objects, it is the relations among the objects in the sequence and their inverse relations that are important. From these it is possible to pick an origin, derive the inverse units +1 and −1, and then one obtains the usual integers. If one has a sequence of sequences or, in other words, a manifold of two dimensions, one must examine the relations between one sequence and an immediately neighboring sequence and thence derive the inverse units $+i$ and $-i$ and then all the complex integers. In this way it is possible to abstract the complex integers from a twofold variety or sequence of sequences.

Thus, for Gauss, number systems in the most proper mathematical sense of that term are about relations, their comparison and 'numbering,' and not about the objects numbered. However, to see these relations intuitively one needs to use geometrical models. The plane, a well-understood manifold of two dimensions, serves this purpose for the complex number system. Gauss regarded his theory of complex quantities as a way of clearing up 'a mysterious obscurity' ('eine geheimnißvolle Dunkelheit') falsely attributed to the imaginary numbers. His abstract theory is a genuine ancestor of the rigorous logical theories of numbers proposed during the latter half of the nineteenth century.

At the very end of the 1831 presentation Gauss [Gauss, 1831] [Gauss, 1863, page 178] mentions a manifold of more than two dimensions ('eine Mannigfaltigkeit von mehr als zwei Dimensionen') in relation to his algebraic work, but only promises to develop the idea in a future work. Thus, it appears that his specific theory for the complex integers and 2-dimensional manifolds is related to a more general theory of manifolds that he held. Such a general theory would cover hypernumber systems and n-dimensional manifolds and would constitute a branch of abstract geometry. There is a brief indication of the more general scope of such a theory in a letter of Gauss to Hansen of 11 December 1825 [Gauss, 1929, page 8]:

> During this autumn I have concerned myself very much with the general consideration of curved surfaces, which leads into an immense field. . . . Those investigations penetrate deeply into many others, I may even say into the metaphysics of the theory of space, and only with difficulty can I tear myself away from such results arising therefrom, as, for example, the true metaphysics of negative and imaginary quantities. The true meaning of $\sqrt{-1}$ stands very vividly before my mind, but it will be very difficult to put it into words, which can only give but a vague fleeting image.[Ch5, 2]

It is clear that Gauss' ideas on curved surfaces, negative and imaginary quantities, and manifolds and space were all linked. His referencing curved surfaces must be to his celebrated work on the differential geometry of curved surfaces [Gauss,

1828] [Gauss, 1900a] [Gauss, 2005] (cf. the very useful summary in [Laugwitz, 2008, pp. 226–230] and [Reich, 1973] for an extended treatment of the history of differential geometry).

During the last years of his life Gauss explicitly considered general n-dimensional manifolds in his lectures, 'Über die Methode der kleinsten Quadrate' ('Concerning the method of least squares'), delivered during the winter semester of 1850–51. For the content of these there are the records of his student August Ritter (1826–1908) [Gauss, 1917, pp. 469–482] (cf. Richard Dedekind's work, who also reported on these lectures [Dedekind, 1931, pp. 293–306]). From these records it is easy to see why Gauss chose abstract manifolds or analytic spaces of n dimensions as the natural vehicle for his investigations into the method of least squares. Such spaces involve a generalized analytic geometry based on n variable coordinates and the usual Euclidean metric, which can be applied to the solution of the minimum problem.

Gauss' ideas on multi-dimensional spaces readily leads to his philosophical thoughts on geometry. His presentation ('Anzeige') of 1831 and his letter to Hansen of 1825 give definite signs of his abstract view of the subject. There are further indications of this view from the later part of his life. In his Jubiläumsschrift (Jubilee Publication) of 1849, 'Beiträge zur Theorie der algebraischen Gleichungen' ('Contributions to the theory of algebraic equations') [Gauss, 1849a] [Gauss, 1849b] [Gauss, 1866, pp. 71–102], containing an important revision of his very first attempt to prove the fundamental theorem of algebra as given in his Inaugural Dissertation of 1799, he excuses himself for a reliance on geometrical reasoning [Gauss, 1866, page 79]:[9]

> I shall represent the proof in clothing borrowed from the geometry of position, because this yields the greatest lucidity and simplicity. However, the real content of the entire argument belongs fundamentally to a higher domain of the general abstract theory of quantity, independent of spatial things, whose subject is the combinations of quantities connected according to continuity; a domain which at this time is still little cultivated, and in which one cannot express oneself without a language based on spatial images.[Ch5, 3]

The view of an abstract geometry divorced from spatial intuition expressed in this passage must be related to Gauss' longstanding desire for a development of 'geometria situs' (see Paul Stäckel's extended work on Gauss as a geometer [Stäckel, 1917, pp. 46–49]). At the end of his life he persisted in his hope for a flowering of this subject, as Walter Sartorius von Waltershausen reported from conversations that he had with Gauss during his last years, 1847–1855 [Walterhausen, 1856, page 88]:

> . . . yet he held an extraordinary hope for the development of geometria situs, in which there might be extensive but entirely uncultivated fields that could hardly be controlled by our present methods.[Ch5, 4]

---

[9] Cf. [Ostrowski, 1920] for a modern analysis of these two proofs. Cf. also [Schlesinger, 1912].

Gauss thus connected abstract geometry with geometria situs, or topology, as it would be called now. He published virtually nothing on this subject, nor did he even write much down as notes on it. Yet he had a vision of its importance. It remained for others to transform this vision into a reality.[10]

Gauss thought that it was necessary to relate abstract geometry to the concrete geometry of physical space (the latter being applied geometry). Abstract geometry deals with n dimensional manifolds in general, while spatial geometry examines a special 3-dimensional manifold for its subject. He thought that knowledge of this special manifold is not entirely *a priori* (as evidenced in letters to Bessel [Gauss, 1900b, pp. 200–201]). Consequently, he did not subscribe to Kant's view of space as merely a form of our outer intuition and explicitly said so on a number of occasions [Gauss, 1863, page 177] [Gauss, 1900b, page 224]. He believed that space is real and our knowledge of space is at least partly empirical. One could even prove this fact from the writings of Kant himself.[11]

Over the course of his mathematical career, Gauss touched upon 2- and multi-dimensional manifolds with respect to several domains of mathematics, as my survey of his work has shown. He found that his basic geometrical concept of manifold could support several mathematical domains, while holding out the promise of developing the concept more abstractly.

Gauss' general attitude towards geometry marks the beginning of a new philosophical view of the subject that became established during the nineteenth century. For him geometry in large part became freed from physical space. On one level it became a branch of 'pure mathematics.' One may rely on spatial intuition and images, but in the final analysis these can be eliminated when dealing with abstract geometrical theories. While Gauss cannot be said to have completely envisaged the purely logical, abstract treatment of geometry, so masterfully advocated at the end of the century by the Italian school and Hilbert, he did see the subject as encompassing much more than the metrical relations of physical space. His notion of n-dimensional manifolds as well as his theory of non-Euclidean geometry were a part of this view. Gauss' philosophical views about the nature of geometry were soon to influence the young Riemann (see Section 5.7).

5.5 Other Mathematical Developments

A host of works published during the 1840s and 1850s dealing with hyperspaces mark the beginnings of the mature development of multi-dimensional geometry. A wide variety of algebraic, analytic, and geometrical problems spurred mathematicians to consider such spaces.[12] The list of mathematicians who started to grapple with the

[10] For comments on Gauss' work in 'geometria situs' or 'topology' and the foundations of the complex number system see [Stäckel, 1917, pp. 46–68], [Fraenkel, 1920], [Schlesinger, 1912, pp. 53–57, 202–210]. Also see [Gauss, 1917, pp. 106–107, 396–397, 407–412, 436–437].

[11] Cf. [Gauss, 1863, page 177, footnote] and [Schlesinger, 1912, pp. 177–181].

[12] Cf. works by C. Segre [Segre, 1904] and [Segre, 1912].

problems of hyperspace geometry at this time includes Arthur Cayley (1821–1895), Augustin-Louis Cauchy (1789–1857), James Joseph Sylvester (1814–1897), Ludwig Schläfli (1814–1895), Julius Plücker (1801–1868), Hermann Grassmann (1809–1877), and Bernhard Riemann (1826–1866). Most of these mathematicians simply introduced hyperspaces through n-fold systems of coordinates and then investigated the metric, affine, and projective geometry of the resulting coordinate spaces using analytic techniques. In his first paper on the subject, 'Chapters in the Analytical Geometry of (n) Dimensions' [Cayley, 1844] (read in 1843), the young Cayley proceeds in this way. In this paper as elsewhere he is not concerned with giving an explanation of the possible 'metaphysics' of higher space. Cauchy's very short 'Mémoire sur les lieux analytiques' ('Memoir on analytical positions/spaces') [Cauchy, 1847], in a similar vein to Cayley's first paper, merely defines some geometrical terms for n-dimensional analysis. Likely following the lead of Cayley, Sylvester began to work with n-dimensional spaces in 1850 [Sylvester, 1850] [Sylvester, 1851a] [Sylvester, 1851b]. In one of his early papers he makes the following apposite methodological remark justifying the geometry of n-spaces [Sylvester, 1851b, page 120] [Sylvester, 1912, Volume 1, page 219, footnote]:

> If the impressions of outward objects came only through the sight, and there were no sense of touch or resistance, would not space of three dimensions have been physically inconceivable? The geometry of three dimensions in ordinary parlance would then have been called transcendental. But in very truth the distinction is vain and futile. Geometry, to be properly understood, must be studied under a universal point of view; every (even the most elementary) proposition must be regarded as a fact, and but as a single specimen of an infinite series of homologous facts.
>
> In this way only (discarding as but the transient outward form of a limited portion of an infinite system of ideas, all notion of extension as essential to the conception of geometry, however useful as a suggestive element) we may hope to see accomplished an organic and vital development of the science.

George Boole (1815–1864) subscribed to much the same view as Sylvester, as shown in a brief note in his *Laws of Thought* [Boole, 1854, page 136].

Ludwig Schläfli, proceeding along the same methodological path as Cayley, Sylvester, and Cauchy, wrote one of the most extensive accounts of the metrical geometry of n coordinate spaces during the years 1850–1852: *Theorie der vielfachen Kontinuität* (*Theory of multifold Continuity*) [Schläfli, 1901]. However, he ran into difficulties with the publication of this immense work, so that it was not published in its entirety until after his death [Schläfli, 1950, pp. 388–390]. The excerpts that did appear soon after its composition in the years 1855–60 unfortunately did not attract much attention [Schläfli, 1953, pp. 164–190, 219–279].

Julius Plücker's approach to introducing hyperspaces was quite different from the approaches of Cayley, Cauchy, Sylvester, and Schläfli. In a remark in his *System der Geometrie des Raumes in neuer analytischer Behandlungsweise, . . .* (*System of Geometry of Space in a New Analytical Treatment, . . .*) [Plücker, 1846, pp. 322–323] he shows how to create a 4-dimensional geometry in ordinary 3-dimensional space through a system of line coordinates. Straight lines in ordinary space defined by four parameters become the elements of a hyperspace. This innovative and distinctly geometrical way of introducing higher spaces through basic elements other than points gave Plücker an entire program of research, which culminated in his posthumously-published *Neue Geometrie des Raumes gegründet auf die Betrachtung der geraden Linie als Raumelement* (*New Geometry of Space Based on Considering the Straight Line as the Spatial Element*) [Plücker, 1868] [Plücker, 1869].

5.6 Hermann Grassmann

At this point I shall consider the work of Hermann Grassmann (1809–1877) related to the nineteenth-century development of ideas of higher-dimensional spaces. His is an interesting case for this development because of his own life story[13] and because of the limitations of his career trajectory in the mathematical world of his day. In effect, his influence on mathematical developments was considerably delayed, though he enjoyed recognition during the latter part of his career.

Grassmann was born in 1809 in Stettin, which was then part of Pomerainia in Prussia. (Stettin is now Szczecin, Poland.) He spent most of his life and professional career in Stettin. For his university education he attended the University of Berlin (1827–1830), where he studied theology, philosophy, linguistics, and literature. He did not formally study mathematics at the university, though about this time he developed a strong interest in mathematics, He began exploring the subject through independent study by reading mathematics texts by his father and by others. His father, Justus Grassmann, was strongly interested in mathematics, science, and scientific subjects. He was a teacher and the professor of mathematics at the Stettin Gymnasium and wrote several texts on various parts of mathematics. Justus' interests and mathematical work had a definite impact on his son.

Following his father in his choice of career, Hermann Grassmann became a school teacher after obtaining certification. He taught mathematics along with other subjects. In an effort aimed at obtaining a university position, Grassmann embarked on a line of mathematical research and assiduously pursued it in spite of his heavy teaching load. The result was a solid work in applied mathematics of considerable length: *Theorie der Ebbe und Flut* (*Theory of Tides and Floods*) [Grassmann, 1911, 1 Theil, pp. 1–238]. The work, completed in 1840, explores the problem of tides in a

[13] There are three substantial biographies of Grassman: one by Victor Schlegel [Schlegel, 1878], an early sympathetic work by someone who followed and supported his work published a year after Grassmann's death; one by Friedrich Engel [Engel, 1911], a work that was included as part of Grassmann's collected works by the principal editor of the collected works; and one by Hans Joachim Petsche [Petsche, 2009], a recent work taking into account up-to-date research and providing additional details. David Rowe provides a broader picture of Grassmann in the context of German culture and mathematics, including some information about the appreciation of his mathematical work around the time of his death and some years after [Rowe, 2010]. In addition, Petsche and others have assembled a book of background material on Grassmann and his family [Petsche et al., 2010]. Also cf. `https://mathshistory.st-andrews.ac.uk/Biographies/Grassmann/`.

way that Grassmann felt improved on the work of Laplace in the *Traité de mécanique céleste* (*Treatise on celestial mechanics*). In the work he draws on ideas from his father's texts, and so he begins his own development of ideas of vector analysis (see detailed analyses in [Crowe, 1967, pp. 55–63] and [Petsche, 2009, pp. 171–175]).

The work on the theory of tides, a *Prüfungsarbeit* (*Trial Work*) intended to demonstrate his research competence in the field and not intended for wide circulation, laid some of the groundwork for his subsequent contributions to vector analysis. This work motivated further development of his mathematical ideas, resulting in his classic book: *Die lineale Ausdehnungslehre, ein neuer Zweig der Mathematik* (*The Theory of Linear Extension, a New Branch of Mathematics*), published in 1844 [Grassmann, 1844] [Grassmann, 1896, 1 Theil, pp. 1–319] [Grassmann, 1995, pp. 9–278].

Grassmann regarded his *lineale Ausdehnungslehre* (*Theory of Linear Extension*) '. . . as the exposition of a new mathematical discipline . . . ' ('. . . als Bearbeitung einer neuen mathematischen Disciplin . . . ') [Grassmann, 1995, page 9] [Grassmann, 1844, Vorrede, page v], and he felt that the book itself would provide ample justification for his claim for its being a new discipline. However, in expositing the new discipline, Grassmann in the early parts of the book cast his ideas in a more philosophical way, and that form put off many contemporary readers of mathematics.

In the Vorrede (Forward) he gives a rational reconstruction of how he arrived at his new mathematical discipline. He says that negatives of displacements in geometry made him realize the importance of direction. In effect, it is possible to regard displacements as vectors, and that the addition of vectors makes sense when taking into account direction. Similarly, multiplication of displacements makes sense and gives a parallelogram area. For this insight he credits two texts by his father, *Raumlehre* (*Space Theory*) and *Trigonometrie* (*Trigonometry*).

He then refers to his work on the theory of tides as providing an application of his basic ideas and one that transformed the research and made the calculations much simpler. For him the new analysis provided a natural way to apply mathematics to natural phenomena (thus, applied mathematics) and explore geometry. He found too that his new vectorial methods naturally applied to finding centers of gravity, such as exposited in Möbius' *Barycentrishe Kalkül* (*Barycentric Calculus*) [Möbius, 1827].

With this background driving his new vectorial approach and its immediate successes, Grassmann feels able to state in his Vorrede [Grassmann, 1995, pp. 11–12] [Grassmann, 1844, Vorrede, pp. IX-X]:

> Thus when I proceeded to work out my results consistently and from the beginning, being careful to appeal to no principle proven in any other branch of mathematics, I found that the analysis I had discovered touched not only on the subject of geometry, as it had seemed to me at the start. Rather, I soon realized that I had come upon the domain of a new science, of which geometry itself is only a special application. For a long time it had been evident to me that geometry can in no way be regarded as a branch of mathematics like arithmetic or combination theory; instead, geometry relates to something already given in nature, namely space. I had also realized that there must be a branch of mathematics that yields in a purely abstract way laws similar to those that in

> geometry seem bound to space. The possibility of constructing such a purely abstract branch of mathematics was provided by the new analysis; indeed this analysis, developed without the assumption of any principles established outside its own domain, and proceeding purely by abstraction, was itself this science. The essential advantages of this approach are, in regard to form, that all principles expressing perceptions of space are now entirely omitted, and consequently the origins of this science are as immediate as those of arithmetic; and in regard to content, that the limitations to three dimensions is absent. Only thus do the laws come to light in their full clarity and generality, and are their essential interrelationships revealed; and many regularities, which for three dimensions either do not appear at all, or at best obscurely, present themselves perfectly clearly with this generalization.[Ch5, 5]

In this way Grassmann asserts that on the basis of his broad general approach he has no problem breaking through the usual, accepted limitation to three dimensions. Higher-dimensional spaces are quite appropriate in his new mathematical discipline.

To orient his new discipline in the realm of mathematics, Grassmann devotes his Introduction (Einleitung) [Grassmann, 1844, pp. XIX–XXXII] [Grassmann, 1995, pp. 23–32] to a philosophical treatment of the nature of the whole mathematical domain and the placement of the new Extension Theory within it. Thus, the Introduction has subsections: Derivation of the Concept of Pure Mathematics, Derivation of the Concept of Extension Theory, Exposition of the Concept of Extension Theory, Form of Presentation (Ableitung des Begriffs der reinen Mathematik, Ableitung des Begriffs der Ausdehnungslehre, Darlegung des Begriffs der Ausdehnungslehre, Form der Darstellung). After the Introduction Grassmann offers a very abstract Survey of the General Theory of Forms (Uebersicht der allgemeinen Formenlehe), dealing with the concepts of equality and conjunction and the combinability of factors according to abstract laws or rules [Grassmann, 1844, pp. 1–14] [Grassmann, 1995, pp. 33–43].

The Introduction is heavily philosophical, but it is not entirely out of keeping with some presentations in nineteenth-century mathematics. Broadly speaking, I think that it could be regarded as in the same category as something that Bolzano might have presented in a philosophical context (see Chapter 4).[14] To a modern reader the material of the immediately following Survey looks like a piece of abstract algebra. For the mid-nineteenth century it was a little ahead of its time.

After the initial philosophical argument of the Introduction and the opening abstract Survey, Grassmann finally begins his treatment of vector algebra and analysis—the main content of the book. He introduces the notion of a 1-dimensional vectorial displacement in a rather philosophical manner [Grassmann, 1995, pp. 46–47] [Grassmann, 1844, §§13, 14, pp. 17–18]:

> We call the transition of the generating element from one state to another an evolution; and this abstract evolution of the generating element thus corresponds to the evolution in position or motion of a point in geometry.

[14] Petsche devotes considerable space in his book to Grassmann's philosophical views [Petsche, 2009, Chapter 4, pp. 221–255]. Grassmann's university course concentrated on philosophy.

> ...
>
> "By an extensive structure of first order we mean the collection of elements into which a generating element is transformed by a continuous change."
>
> ...
>
> We call such an evolution, in which there is generated from one element of a continuous form its next adjacent element, a fundamental evolution, and can therefore say: "The *elementary extensive structure* is that which results from the continuous action of the same fundamental evolution."
>
> ...
>
> ... we call the elementary extensive structure of the first order an *extensive magnitude* or an *extension of the first order*, or a *displacement*.[Ch5, 6]

A little later Grassmann offers his main statement about systems of higher order [Grassmann, 1995, pp. 50–51] [Grassmann, 1844, §16, pp. 21–22]:

> In order to proceed to the conjunction of dissimilar displacements, I consider next two dissimilar fundamental evolutions, and let the first fundamental evolution (or its opposite) arbitrarily transplant an element, and then let the element thus generated likewise he arbitrarily transplanted by the second method of evolution. In this way I can therefore generate an infinite set of new elements from a single element, and the collection of elements so generated I call a *system of second order*. I then assume a third fundamental evolution, which does not lead from that same initial element again to one of the elements of this system of second order, and which I therefore designate as *independent* of these first two, and let this third evolution (or its opposite) arbitrarily transplant an arbitrary element of that system or second order, the collection of elements so generated forming a system of third order. Since this method of generation has no limitation on its concept, I can in this way proceed to systems of arbitrary high order.
>
> ...
>
> ... Thus the plane clearly corresponds to the systems of second order, and the plane is thus regarded as generated when all the points of a straight line are moved in a new direction (or its opposite) not included in it; the collection of points so generated thus forms precisely an infinite plane. A plane thus appears as a collection of parallels, all of which intersect a given line, and clearly, since parallels do not intersect, and since in addition the original line does not intersect itself, all the points so generated differ from one another, and consequently the analogy is perfect. Likewise one proceeds to the whole of infinite space as a system of third order by moving the points of a plane in a new direction (or its opposite) not lying in the plane. Further geometry cannot progress, but the abstract science knows no limitation.[Ch5, 7]

From this point in the book Grassmann proceeds to develop his full theory of vectorial analysis. It is a new mathematical theory with new abstractions, new operations, new forms of calculation, and new results. Grassmann is comfortable in working freely with m-dimensional spaces [Grassmann, 1995, page 60] [Grassmann, 1844, §20, page 32]:

> "Every system of mth-order can be regarded as generated by those same m independent methods of evolution from any arbitrary element; that is, all other elements can be generated from a single such element by those methods of evolution."[Ch5, 8]

Grassmann includes a wide variety of applications: to geometry, offering new foundations for geometry, to algebra, giving results for the theory of equations and theorems on algebraic curves, and to physics, such as to Newtonian mechanics and crystal structures. The book holds an impressive array of new mathematical ideas (cf. analyses of the mathematics in [Crowe, 1967, pp.63–77] and [Petsche, 2009, pp. 175–192]).

The mathematical community hardly took notice of the book in the years immediately after its publication in 1844. Grassmann did send copies to various prominent mathematicians, including Gauss and Möbius, but those receiving the copies hardly recognized its value. The lack of early recognition was likely due to its philosophical presentation style as well as to the novelty of its ideas. Friedrich Engel, editor of Grassmann's collected works and a principal biographer of Grassmaan sadly reports the reception of the work [Engel, 1911, page 97]:

> Thus *Grassmann* experienced what must be the most painful experience for the author of a new work: his book nowhere received attention; the public was completely silent about it; there was no one who discussed it or even publicly found fault with it. The mathematicians to whom he had sent the work expressed themselves as not unfriendly to it, to some extent even as benevolent to it, but no one really studied it.[Ch5, 9]

The book did not sell well, with the result that the publisher shredded some remaining copies of the book for waste paper, as Grassmann himself related much later in 1877 [Grassmann, 1896, page 18] [Grassmann, 1995, page 19]. Crowe provides a detailed summary of its poor reception in the early years after its publication [Crowe, 1967, pp. 77–89].

In the years just after the 1844 publication Grassmann wrote two additional works presenting some of his ideas. In 1845 he wrote a kind of review of his work: 'Kurze Übersicht über das Wesen der Ausdehnungslehre' ('Short Overview of the Essence of the Extension Theory') [Grassmann, 1845], though this had little impact. Then in response to a suggestion by Möbius, he wrote a longer work, *Die Geometrische Analysis* (*The Geometrical Analysis*), as a contribution to a prize contest. He was awarded the prize, and the work was published along with an appendix by Möbius in 1847 [Grassmann & Möbius, 1847] [Grassmann, 1896, 1 Theil, pp. 321–399] [Grassmann, 1995, pp. 313–414]. The work expands on certain aspects of the original work of 1844.

By the 1850s some recognition of Grassmann's 1844 work began to emerge. Sir William Rowan Hamilton (1805–1865) gave an appreciation of the work in relation to his own in his *Lectures on Quaterions*, published in 1853 [Hamilton, 1853, page 62, footnote]. Others who around this time came to recognize the value of Grassmann's work included Möbius, Comte de Saint-Venant (1797–1886), Luigi Cremona (1830–1903), and Giusto Bellavitis (1803–1880). Crowe provides the details of this delayed early recognition [Crowe, 1967, pp. 81–89].

Grassmann's further development of his theory led to a new book, new in mathematical style with much new content, published in 1862, *Die Ausdehnungslehre* (*The Extension Theory*) [Grassmann, 1862] [Grassmann, 1896, 2 Theil, pp. 1–383] [Grassmann, 2000]. This work is written in a standard mathematical style, recognizable by a modern reader of mathematics. The move to domains of order n is straightforward, as Grassmann presents in the following definition, which occurs near the beginning of the book [Grassmann, 2000, page 8] [Grassmann, 1862, 1.1, §2, page 6] [Grassmann, 1896, 2 Theil, page 16]:

> *Definition*. The collection of magnitudes that are numerically derivable from a series of magnitudes $a_1, a_2, \ldots, a_n$ I call the *domain* derivable from these magnitudes (the domain of the magnitudes $a_1, \ldots, a_n$) and in particular a domain of nth *order* if those magnitudes are of first order (that is, numerically derivable from n original units), and if the domain is not derivable from less than n such magnitudes. A domain that includes no magnitude but zero is called a domain of *zeroth* order.[Ch5, 10]

Essentially, this is a definition of an n-dimensional vector space (domain), including the condition of linear independence of the defining 1-dimensional basis vectors (magnitudes that include direction). A null space is defined as a special case. I shall not summarize the book's content further, as it would be beyond the scope of this book. See [Crowe, 1967, pp. 89–90] and [Petsche, 2009, pp. 196–198] for short summaries of the content.

Toward the end of his career Grassmann received some well-deserved, though limited recognition for his mathematical work (cf. [Crowe, 1967, pp. 89–96]). As well as Hamilton, Hermann Hankel (1839–1873), Alfred Clebsch (1833–1872), Victor Schlegel (1843–1905), and William Kingdom Clifford (1845–1879) took note of Grassmann's contributions.

Grassmann is now recognized as a visionary, ahead of his time. Crowe views him as a revolutionary [Crowe, 1967, pp. 94–96]. His work, starting with the basis of linear or vector spaces and vector analysis, is an important foundation for modern mathematical theories, such as geometrical algebra and exterior algebra. He easily moved into thinking in abstract higher-dimensional terms. After the slow recognition of his work late in his career, Grassmann's mathematical ideas became enormously influential. I have only scratched the surface of Grassmann's world by touching on his fundamentals of n-dimensional spaces.[15]

[15] There are numerous recent appraisals of Grassmann's mathematical work. In addition to [Crowe, 1967] and [Petsche, 2009, Chapter 3, pp. 165–220], these include: [Dieudonné, 1979], [Fearnley-Sander, 1979], [Fearnley-Sander, 1982], [Otte, 1989]. [Hestenes, 1996], [Hestenes, 2011], [Schubring, 1996] (an entire collection of essays).

5.7 Bernhard Riemann

I shall now examine the work of Bernhard Riemann (1826–1866) on higher-dimensional spaces and to a limited extent the associated geometry of these spaces. Riemann, one of the greatest mathematicians of all time, possessed an extraordinary mathematical imagination, which yielded many fruitful ideas, theories, and results. He is a pivotal figure in the nineteenth-century work on higher-dimensional spaces and related areas of mathematics and mathematical physics. His philosophical views concerning the foundations of hyperspace geometry exemplify some of the essential changes in mathematical thinking concerning space, geometry, and dimension that were taking place around the middle of the nineteenth century. His student friend and mathematical colleague, Richard Dedekind (1831–1916), wrote a biographical sketch of Riemann's life and work after his death, which is a standard source [Riemann, 1892, pp. 539–558] [Riemann, 1990, pp. 571–590] (in English translation [Riemann, 2004, pp. 518–533]). Detlef Laugwitz has written an extensive, in-depth analytical book covering the man and his mathematics [Laugwitz, 2008] (originally in German [Laugwitz, 1996]).[16] José Ferreirós' has contributed a detailed analysis of Riemann's work on the concept of manifold in an important paper [Ferreirós, 2006] and in his book on the history of set theory [Ferreirós, 2007, Chapter II, pp. 39–80]. Also, Jeremy Gray has written a short paper covering aspects of Riemann's mathematics [Gray, 2017] and a charming survey of Riemann's mathematics [Gray, 2019].

Bernhard Riemann's principal statement concerning the domain of higher-dimensional geometry comes in his celebrated Habilitationsvortrag (habilitation lecture) to become a *Privatdozent* or unpaid lecturer in the Georg-August University at Göttingen), 'Über die Hypothesen, welche der Geometrie zu Grunde liegen' ('On the Hypotheses that Lie at the Foundation of Geometry'). Riemann delivered it on 10 June 1854 to the Philosophical Faculty, with Gauss among others attending, but it was not published until 1868 after Riemann's death [Riemann, 1868]. As a presentation to a broad faculty audience, Riemann could not elaborate the details in the course of delivery, though the lecture is replete with profound ideas, boldly sketched, as subsequent analyses have shown. These include an intuitive, genetic description of n-dimensional manifolds, a generalization to n dimensions of Gaussian curvature, foundations for 'Riemannian' geometry, and a description of the elliptic variety of non-Euclidean geometry, along with with some discussion of the relationship between pure geometry and physical space. Today mathematicians clearly recognize in Riemann's short work the seeds of a tradition that led to the important mathematics behind Einstein's theory of relativity. Riemann himself left the work aside after giving the lecture. Hence, it only could begin to become influential through the working out of its ideas by others after its publication in 1868 when Riemann had already died.

For the present discussion I shall only be concerned with the first section of the lecture. It is in this section that Riemann offers his genetic notion of an n-dimensional manifold. When the lecture was first published in 1868, philosophers

[16] See also [Laugwitz & Laugwitz, 1995] as well as [Narasimhan, 2010].

took most interest in this section, because the treatment of the manifold concept in it is largely 'philosophical.' In fact, the philosophers quickly offered objections to Riemann's unclear methods and presentation. Yet there is a key point in the first section that nearly all the early readers of the published lecture probably could not fully grasp. Riemann primarily aimed at defining n-dimensional manifolds as *topological* objects. His fundamental concern was topology, or analysis situs, as he would say, à la Leibniz. With topology in its infancy, it is not surprising that early readers did not understand Riemann's aim. Nonetheless, it is worth while examining this very early statement of a topological program, especially in order to understand some of the difficulties faced by a great mathematician when presenting his original ideas. Below I shall summarize some other significant examples of topological thinking in Riemann's work, some of which formed a background to the Habilitationsvortrag (habilitation lecture).

Dedekind told the story of how Riemann came to lecture on the foundations of geometry for his habilitation to *Privatdozent* in some detail in his biography [Riemann, 1892, pp. 547–549] [Riemann, 1953, pp. 547–549] [Riemann, 1990, pp. 579–581] [Riemann, 2004, pp. 523–526]. By the end of 1853 Riemann had finished and submitted his habilitation thesis [Habilitationsschrift], 'Über die Darstellbarkeit einer Function durch eine trigonometrische Reihe' ('Concerning the representation of a function by a trigonometric series'). The habilitatiion lecture [Habilitationsvortrag] still had to be arranged. Concerning the prospects for this he wrote to his brother Wilhelm on 28 December 1853 [Riemann, 1953, page 547] [Riemann, 1990, page 579] [Riemann, 2004, pp. 523–524]:

> My work is now progressing tolerably well; at the beginning of December I submitted my habilitation thesis [Habilitationsschrift] and besides had to propose three subjects for the trial lecture, from which the faculty could then choose one. I had the first two complete and hoped that they would choose one of them; Gauss, however, chose the third and so now I am in something of a quandary, since I must still work this one out.[Ch5, 11]

Gauss passed over the first two of the suggested themes, one on the history of the representation of functions by trigonometric series, the second on the solution of two equations of the second degree in two unknowns, in order that he might hear something on the foundations of geometry, a subject close to his own interests. Laugwitz claims, and I think rightly, that it was not that surprising that Gauss chose the third topic [Laugwitz, 2008, pp. 219–220]. Moreover, Riemann had some months to prepare for his lecture in the following year. As it turned out it was given in June. During the first part of 1854 Riemann was ill much of the time; he overworked, mainly on other topics of physics, and suffered from the inclement weather. It was not until a fortnight after Easter that he began to prepare his trial lecture in earnest, though it nevertheless was complete by Whitsun.[17] The preparation time was certainly remarkably short, when one considers the stature of the final product, although surely he had his principal ideas in mind already earlier. Gauss himself

[17] Whitsunday or Pentecost was 4 June 1854.

wanted to examine the young geometer, but he was in poor health. Many even thought his death was imminent. So he asked that Riemann's lecture be postponed until his own health might improve, and Riemann accepted the inevitable delay. But then without warning Gauss decided to let Riemann give the lecture, as the latter told his brother in a letter of 26 June 1854 [Riemann, 1953, page 548] [Riemann, 1990, page 580] [Riemann, 2004, pp. 524–525]:

> Then suddenly, after my repeated requests 'to take the yoke from my neck,' he decided at noon on the Friday after Whitsun to set the colloquium for the next day at half past ten and so on Saturday at one o'clock I was fortunately finished with it.[Ch5, 12]

After the lecture, which was given so suddenly on 10 June, Gauss highly praised Riemann's work to the physicist Wilhelm Weber on their way back to the faculty meeting. He expressed himself with unaccustomed excitement on the depth of Riemann's thought. In the eyes of the 'Prince of Mathematicians' the young mathematician was a success.

Riemann's lecture contains only an extended sketch of his geometrical ideas. At the time of delivery he was required to present his ideas to the entire Philosophical Faculty of Göttingen, so he needed to make them as clear as possible to nonmathematicians as well as mathematicians (Dedekind in [Riemann, 1953, page 549] [Riemann, 1990, page 581] [Riemann, 2004, page 525]). If he had had time in his brief life to revise and expand the lecture for publication in scientific proceedings, he surely would have expressed himself differently and more fully. As it was, his only further treatment of part (the more metrical part) of the material of the lecture was in his Paris Academy essay of 1861 [Riemann, 1953, pp. 391–404] [Riemann, 1990, pp. 423–436] [Riemann, 2004, pp. 365–381]. Thus, his friend Dedekind almost certainly published a work that Riemann himself would have preferred to put into a better and more polished form. It was, of course, appropriate for Dedekind to publish the original lecture, but one must be prepared for its brevity and perhaps infelicitous expressions.

The overall theme of Riemann's habilitation lecture [Riemann, 1868] (in editions of his collected works: [Riemann, 1953, pp.272–287] [Riemann, 1990, pp. 304–319] [Riemann, 2004, pp. 257–272]) is to determine logical relationships between geometry conceived, more or less, as a discipline of 'pure mathematics' and its application to physical space. Within the scope of this theme he examines three special problems. The first of these is the important one for the history being covered in this work [Riemann, 1953, page 272] [Riemann, 1990, page 304] [Riemann, 2004, page 257]:

> In the first place I have thus set myself the task of constructing the concept of a multiply-extended magnitude from the general concepts of magnitude.[Ch5, 13]

Riemann thought that a clarification of the notion of multiply-extended [i.e., multidimensional] magnitude was a crucial prerequisite for improving our understanding of the relationship of geometry to the concept of space. Contained in his discussion of n-fold magnitudes given in section I of his lecture is an informal theory of topological manifolds and dimension. He intended this theory to serve as a general framework ('Vorarbeit', 'preparation') for contributions to analysis situs. A statement of this intention is tucked away in a footnote at the end of the published version of the lecture [Riemann, 1953, page 286] [Riemann, 1990, page 318] [Riemann, 2004, page 271].[18]

When embarking on his discussion of the general notion of manifold, Riemann asks the indulgence of his audience, for, as he puts it, the difficulty lies more in the philosophical nature of the subject, and he is not practiced in such investigations [Riemann, 1953, page 273] [Riemann, 1990, page 305] [Riemann, 2004, page 258]. He claims besides that there was little background for him to rely upon—only a few remarks in papers of Gauss and some philosophical researches of Herbart.

He explicitly cites those papers of Gauss that he found useful: 'Theoria residuorum biquadraticorum. Commentatio secunda' ('Theory of biquadratic residues. Second commentary') [Gauss, 1832] [Gauss, 1863, pp. 93–148], the accompanying 'Selbstanzeige' ('self-presentation' or 'presentation by himself') [Gauss, 1831] [Gauss, 1863, pp. 169–178], and the Jubiläumsschrift (Jubilee Publication) of 1849, 'Beiträge zur Theorie der algebraischen Gleichungen' ('Contributions to the theory of algebraic equations') [Gauss, 1849a] [Gauss, 1849b] [Gauss, 1866, pp. 71–102] (see the discussion in Section 5.4 on Gauss). Thus Gauss' hints concerning the abstract and geometrical foundations of the complex number system, the definition of 2- and higher-dimensional manifolds, and the broader need to develop the subject of analysis situs were a direct stimulus for Riemann's ideas about manifolds. In his discussion Riemann goes so far as to borrow some expressions from Gauss.

In the case of the reference to Herbart's works in the lecture Riemann does not explicitly say which he found helpful. However, it is known from Riemann's notes on philosophy as published much later [Riemann, 1953, pp. 507–538] [Riemann, 1990, pp. 539–570] as well as Dedekind's biography of Riemann [Riemann, 1953, pp. 544–545] [Riemann, 1990, pp. 576–577] [Riemann, 2004, pp. 521–522] that the philosophical work of this German philosopher had a significant influence on the young mathematician. Dedekind reports that during three semesters at Göttingen from Easter 1849 through 1850 Riemann attended some philosophical lectures and during that time made a special study of Herbart's thought [Riemann, 1953, pp. 544–545] [Riemann, 1990, pp. 576–577] [Riemann, 2004, pp. 521–522]. In particular, in an important and often quoted fragment Riemann makes the following assertion about his relation to Herbart's thought [Riemann, 1953, page 508] [Riemann, 1990, page 540]:

> The author is a Herbartian in psychology and epistemology (methodology and eidolology), but for the most part he cannot subscribe to Herbart's natural philosophy and the related metaphysical disciplines (ontology and synechology).[Ch5, 14]

[18] Many apparently missed this statement of intention. For example, in his English translation Clifford (1873) fails to include it [Clifford, 1882, pp. 55–71].

Several philosophers and historians of mathematics have examined the likely influences of Herbart's thought, philosophy, and scientific work on Riemann. In his classic work from 1897, *An Essay on the Foundations of Geometry*, Bertrand Russell (1872–1970) examines Herbart's relation to Riemann's geometrical thinking [Russell, 1897, pp. 13–22, 62–70]. Much more recently several scholars have looked for Herbart's influence on Riemann and his mathematical thinking. These include Erhard Scholz [Scholz, 1982a] [Scholz, 1992], Luciano Boi [Boi, 1994], Umberto Bottazzini and Rossana Tazzioli [Bottazzini & Tazzioli, 1995], Eric Banks [Banks, 2005], and Dinçer Cevik [Çevik, ND].

Erhard Scholz' search for Herbart's influence on Riemann is especially significant. By carefully examining Riemann's notes on Herbart's philosophy and the excerpts Riemann himself made from Herbart's writings as found in his *Nachlass* as well as in his notes and references as published in Riemann's *Gesammelte mathematische Werke* (*Collected mathematical works*) in relation to Herbart's stated philosophical positions and published writings, Scholz has provided an extensive analysis of Herbart's influence on Riemann's formulation of the concept of manifold as well as several other aspects of Riemann's mathematics and scientific thinking.[19]

Johann Friedrich Herbart (1776–1841), born in Oldenburg on 4 May 1776, studied philosophy, law, literature, and mathematics at the university at Jena, 1794–1796.[20] There the philosopher Johann Fichte (1762–1814) exerted a strong influence on the young Herbart, through Fichte's critical philosophy of the Self. This influence led Herbart to his own critical studies of the self, ego, and self-consciousness. A little later Herbart studied at Göttingen from 1801 to 1809, where he completed his doctorate and his habilitation. Herbart was appointed to the chair formerly occupied by Kant at Königsberg in 1809. Years later he became professor at Göttingen from 1833 to his death in 1841. Early on in his career Herbart also developed broader interests and research in philosophy of education, pedigogy, and psychology. Herbart was strongly interested in mathematics throughout his career. In his writings on psychology there is a considerable amount of mathematics used to analyze aspects of the field. In a sense he treated philosophy and science on an equal footing.

Herbart systematically divided metaphysics into four parts:

1. Methodology (Methodologie) — study of the theory of method
2. Ontology (Ontologie) — study of the nature of categories of things in the world, in reality
3. Synechology (Synechologie) — theory of serial forms and spatial things
4. Eidolology (Eidelologie) — theory of the self, philosophy of mind

In addition to metaphysics Herbart investigated:

5. Psychology (Psychologie) — scientific study of mind; closely related to the philosophy of self and mind
6. Natural Philosopy (Naturphilosophie) — philosophy of nature; physics

[19] Cf. also Scholz' earlier book [Scholz, 1980] and later extensive survey [Scholz, 1999].

[20] See [Kim et al., 2024] and [Scholz, 1982a, pp. 429–431].

The Riemann quotation given above indicates he followed Herbart in 5, 1, and 4, but not in 6, 2, or 3.

From his careful sifting through the sheets of Riemann's *Nachlass* that contain excerpts from and references to Herbart, Scholz has produced a table listing the excerpts organized according to Herbart's four parts of metaphysics with the sources from his published works [Scholz, 1982a, Table 1, page 416]. Under methodology, ontology, and synechology the main sources are Herbart's works on psychology, *Psychologie als Wissenschaft* (*Psycholoogy as Science*), [Herbart, 1824] [Herbart, 1825] and metaphysics, *Allgemeine Metaphysik* (*General Metaphysics*, Historical-Critical Part), [Herbart, 1828].[21] Under eidolology the quotations from Herbart come from his critical work, mostly from early in his career, on the philosophy of the self. It is interesting that one quotation under eidolology comes from a short work entitled, *Ueber die Möglichkeit und Nothwendigkeit, Mathematik auf Psychologie anzuwenden* (*On the Possibility and Necessity of Applying Mathematics to Psychology*) [Herbart, 1822]. As already indicated, Herbart applied a considerable amount of mathematics in his studies in psychology.[22] The use of mathematics itself may have attracted Riemann in his reading of Herbart. Riemann's excerpting of Herbart's works suggests to me a keen student's exercise in grappling with the material. Perhaps these excerpts were made around the time of Riemann's study of philosophy, from Easter 1849 through 1850? It is not possible to know for sure.

Scholz draws several important conclusions from his analysis of the excerpts Riemann chose from Herbart in relation to the main elements of Herbart's philosophy and more generally the philosophy of that period.[23] For this work I need only consider those that relate to Riemann's subsequent treatment of n-dimensional manifolds.

Herbart did not subscribe to Kant's view of space as a pure (a priori) form of sensible intuition.[24] Rather Herbart took the view that spatial concepts originate in our sense experiences, and that we develop spatial concepts from these sense experiences. He was clearly an empiricist. Our movements in the neighborhood or region about us relative to our perceptions give us the basis for generating an ordered series and even multiply ordered series and thus a concept of space or spaces. Riemann too was an empiricist with regard to generating or deriving concepts of space.

Thus, Herbart put forward a psychological theory for generating or constructing space-like concepts or spaces [Herbart, 1824, pp. 374–376] [Herbart, 1912, Volume 5, pp. 424–425] [Herbart, 1825, §§110–114, pp. 123–152] [Herbart, 1912, Volume 6, pp. 88–106] [Herbart, 1825, pp. 297–300] [Herbart, 1912, Volume 6, pp. 192–193]. Among our perceptions or presentations (Vorstellungen) in our mind we have notions of directions, right and left, up and down, forwards and backwards, and each

---

[21] Cf. the Systematic Part [Herbart, 1829].

[22] Cf. [Kim et al., 2024, pp. 7–54] for a recent discussion of Herbart's psychology.

[23] I find Scholz' analysis of the relation of Herbart to Riemann on the philosophy of change and the Real and the process of acquiring knowledge particularly perceptive [Scholz, 1982a, pp. 418–421].

[24] Herbart offers a critique of Kant's view of space in [Herbart, 1824, pp. 378–383] [Herbart, 1912, Volume 5, pp. 427–429]. Note that Kant viewed time as the other pure (a priori) form of sensible intuition.

of these directions yields or has an associated continuous serial form. The serial form is produced when a specific set of perceptions or presentations undergoes a 'graded fusion' ('abgestufte Verschmelzung'), thereby ordering them as a serial (linear) space [Herbart, 1825, pp. 297–300] [Herbart, 1912, Volume 6, pp. 192–193].

This psychological theory then led Herbart to his extensive metaphysical theory of continuous serial forms or Synechology [Herbart, 1829, pp. 183–339] [Herbart, 1912, Volume 8, pp. 110–197]. Herbart also thought of other space-like serial forms such as the line or ordered sequence of sounds or the mixing of colors as in the triangle of colors with primary red, yellow, and blue.[25]

As he clearly stated in the quotation given above, Riemann did not subscribe to Herbart's Synechology, though in his philosophical studies he must have grappled with the metaphysical theory as well as the psychological one, in order to be able to discard at least the metaphysical one.[26] He must have taken some interest in Herbart's psychological theory for constructing spaces.

Scholz makes a very interesting point. Whereas Herbart saw continuous spatial forms and magnitudes ubiquitous in ordinary life, Riemann saw these as rare, except in mathematics.

Apart from the background researches of Gauss and Herbart, Riemann's own research work in mathematics, especially in the theory of functions of a complex variable and the theory of algebraic functions, provided a strong motivation for him to put forward in the *Habilitationsvortrag* (habilitation lecture) a potential framework for contributions to analysis situs through a definition of manifold. His investigations in complex function theory as presented in his *Inauguraldisseration* (inaugural dissertation) of 1851, 'Grundlagen für eine allgemeine Theorie der Functionen einer veränderlichen complexen Grösse' ('Foundations for a general theory of functions of one complex variable'), led him to consider the topology of surfaces, i.e., 2-dimensional manifolds [Riemann, 1851] (and in [Riemann, 1953, pp. 3–48] [Riemann, 1990, pp. 35–80] [Riemann, 2004, pp. 1–41]). He introduced this surface topology in sections 5 and 6, including the notions of simply and multiply connected surfaces [Riemann, 1953, pp. 7–12] [Riemann, 1990, pp. 38–44] [Riemann, 2004, pp. 4–10]. Then he proceeded to develop his analytical theory of Riemann surfaces and apply these new ideas in the rest of the work to the analysis of complex functions.[27] Later it then was natural for him to contemplate a study of the general case. Thus, in his *Habilitationsvortrag* (habilitation lecture) of 1854 he briefly refers to his own special way of dealing with multivalued analytic functions by the introduction of Riemann surfaces to persuade his listeners that a topological treatment of manifolds is important and even necessary for an understanding and further development of some parts of mathematics.

In looking for the background to Riemann's *Habilitationsvortrag* (habilitation lecture), Erhard Scholz carefully examined Riemann's *Nachlass* for evidence of what Riemann was researching around this time [Scholz, 1982b]. Scholz has conjectured

[25] It would seem that the triangle of colors is not linear, but 2-dimensional.

[26] Authors such as Russell and Torretti differ on this point.

[27] He also therein provided a statement of the 'Riemann mapping theorem' and what later turned out to be an inadequate proof thereof.

that some of Riemann's notes on the concept of manifold and the foundations of geometry found in the *Nachlass* come from the period 1852–53 just before Riemann prepared and gave his lecture. Scholz has also conjectured that the 'Fragment aus der Analysis Situs' ('Fragment on Analysis Situs'), which Weber had published as part of Riemann's *Gesammelte Werke* (*Collected works*), comes from this period [Riemann, 1953, pp. 479–482] [Riemann, 1990, pp. 511–514] [Riemann, 2004, pp. 475–477]. The 'Fragment' relates to and expands on his theory of surfaces. The material is of a combinatorial character, material that others developed later into combinatorial or algebraic topology. Given the interrelated documentary evidence, Scholz's conjectures appear plausible.

In his magnificent series of papers 'Theorie der Abel'schen Functionen' ('Theory of Abelian Functions') [Riemann, 1857b], published in the *Journal für die reine und angewandte Mathematik* (*Journal for pure and applied Mathematics*) and also issued as a separate booklet [Riemann, 1857a] ([Riemann, 1953, pp. 88–144)] [Riemann, 1990, pp. 120–176] [Riemann, 2004, pp. 79–134]), appearing three years after delivering the lecture, Riemann provides a reasonably clear statement of his conception of topology [Riemann, 1953, page 91)] [Riemann, 1990, page 123] [Riemann, 2004, page 82]:

> In the investigation of functions which arise from the integration of total differentials several theorems belonging to analysis situs are almost indispensable. With this name, used by Leibniz, though perhaps not entirely with the same meaning, one may be permitted to denote a part of the theory of continuous quantities which considers such quantities not as existing independently of their position or measurable by one another, but, on the contrary, which investigates only their local and regional properties entirely divorced from measure-relations.[Ch5, 15]

With the background to Riemann's *Habilitationsvortrag* (habilitation lecture) just summarized, I shall proceed to examining the details of the first section of his lecture, seeking to give a sympathetic reading to some of his obscure expressions.

In the lecture when attempting to construct the concept of a multiply extended magnitude (mehrfach ausgedehnte Grösse) from general concepts of magnitude, Riemann has much to say about the concept of manifold (Mannigfaltigkeit). A close reading of the text shows that he regards the manifold concept as more general than the concept of magnitude. Most of the early expositors and critics of his work did not recognize this important point, probably because it is not declared explicitly. Julius Jacobson [Jacobson, 1883, pp. 318–320, 325] was an exception. From the background I have examined, it seems clear that Riemann was seeking to define a notion something akin to the concept of the modern topological manifold or even the concept of general topological space. However, Riemann had limited tools with which to work. Not surprisingly, there remains an implicit link between his general manifold concept and the narrower domain of algebraic functions. One cannot really expect fully generality at this point in his development of the mathematics.

Riemann speaks of discrete and continuous manifolds with elements and points as corresponding specializations (Bestimmungsweisen) [Riemann, 1953, pp. 273–274)] [Riemann, 1990, pp. 305–306] [Riemann, 2004, pp. 258–259]. In essence the discrete manifolds are quantified by the integers. However, it is the continuous manifolds that are of prime interest to him. Riemann suggests that in most examples manifolds are measured quantities, i.e., magnitudes. In this case measure consists in the superposition of the magnitudes to be compared. However, there are important examples when this method does not apply. He says [Riemann, 1953, page 274)] [Riemann, 1990, page 306] [Riemann, 2004, page 259]:

> In the absence of this, two magnitudes can only be compared when one is a part of the other; in which case also we can only determine the more or less and not the how much: The researches which can in this case be instituted about them form a general part of the science of magnitude in which magnitudes are regarded not as existing independently of position and not as expressible in terms of a unit, but as regions in a manifold.)[Ch5, 16]

This is the main passage in which Riemann appears to point to a nascent science of analysis situs. It is very similar to the one quoted above from his work of 1857 on Abelian functions. It is clear that he is referring to topological manifolds, because he immediately cites the cases of multi-valued analytic functions and problems related to complex analysis. 'Riemann surfaces' are uppermost in his mind. Although he uses the term 'Grösse,' it is apparent that he does not mean 'magnitude' in the usual measured or metrical sense, but rather in the sense of general and not necessarily metrical extension. Perhaps he could have used Grassmann's term 'Ausdehnung,' but given it a broader connotation. However, he most probably did not know Grassmann's work.

Riemann chooses to examine only two aspects of his general topological manifolds. He offers a construction method for producing multi-dimensional manifolds and a reduction method for determining points in such manifolds by sets of quantities, i.e., real number coordinates.

Riemann's construction method is akin to the ancient motion theory (more or less fluxion theory à la Newton) extended to n dimensions [Riemann, 1953, page 275)] [Riemann, 1990, page 307] [Riemann, 2004, pp. 259–260]. His description has something of the form of an inductive definition. The true character of a simply extended or 1-dimensional manifold (eine einfach ausgedehnte Mannigfaltigkeit) is that continuous progress (motion; stetiger Fortgang) is only possible in two directions, forwards and backwards (terms borrowed from Gauss). If one supposes that an entire simply extended manifold passes over into a series of such manifolds in point-to-point correspondence, then one obtains a doubly extended (2-dimensional) manifold. This construction is exactly analogous to the ancient idea that a continuously moving line produces a surface. In general, one can continue the process to give an n-fold or n-dimensional manifold. Riemann concludes [Riemann, 1953, page 275)] [Riemann, 1990, page 307] [Riemann, 2004, pp. 259–260]:

> If instead of considering the concept as determinable, one considers its objects as a variable, then this construction can be described as a composition of a variable of $n + 1$ dimensions from a variable of n dimensions and a variable of one dimension.[Ch5, 17]

Riemann's language appears philosophical, possibly a little suggestive of Herbart's. However, Riemann is bold in moving to arbitrary dimensions. Herbart took 3 as the limit for the number of dimension.

In order to show how to reduce the positions in an n-dimensional manifold to the determination of n real quantities, Riemann reverses the construction process [Riemann, 1953, pp. 275–276)] [Riemann, 1990, pp. 307–308] [Riemann, 2004, page 260]. He assumes that there is a 1-dimensional manifold, which one can use as a device to determine the positions of points in a given n-dimensional manifold: a 1-dimensional measuring stick, so to speak. A point on the 1-manifold is taken as the origin; other points on it are assigned appropriate distances from the origin. Corresponding to any given point on the 1-manifold, one can assign a whole set of points in the n-manifold in such a way that different points on the 1-manifold correspond to distinct submanifolds of the n-manifold. Expressing this idea in the language of function theory, Riemann says that one can take a continuous function of positions within the n-manifold to values on the 1-manifold, assuming that the function is not constant over an entire (n-dimensional?) region of the manifold. Accordingly, every system of points for which the function has a constant value forms a continuous submanifold of fewer than n dimensions, in fact, of $n-1$ dimensions. As the function value on the 1-manifold changes, these $(n - 1)$-dimensional manifolds pass over continuously into one another. Hence, one may think of one of them as the original $(n - 1)$-manifold from which the others proceed, such that each point on the original manifold corresponds (in a one-one onto manner) to a distinct point on each image manifold. This correspondence is part of the construction process. Riemann casually mentions possible exceptional cases, i.e., singularities in the correspondence; he is apparently thinking of branch points on a Riemann surface. In fact, most of his ideas about the functional reduction of manifolds can be related to complex, or even algebraic functions.

The process just described shows that one can reduce the determination of a position in an n-manifold to the determination of a number and an $(n - 1)$-manifold. Continuing the process, one sees that a point in an n-manifold can normally be determined by n real numbers: $x_1, x_2, \ldots x_n$. However, Riemann asserts that sometimes n steps/numbers will not suffice to determine a point in a manifold. In this case infinite-dimensional manifolds arise, and he mentions function spaces as examples!

In brief, Riemann characterizes an n-dimensional manifold as a 1-parameter family of $(n-1)$-dimensional manifolds. His manifolds are Euclidean in their smallest parts and even globally Euclidean (apart from possible singularities which he decides to ignore in his lecture). He assumes differentiability in the later sections of the *Habilitationsvortrag* (habilitation lecture), so Riemannian manifolds are intended to be differentiable manifolds.

Riemann sought a manifold concept of great generality, yet his construction methods are informal. Questions remain. What is one to understand by 'continuity' or the implicit use of motion in his abstract characterization? One may suppose that these concepts are supposed to have the informal meanings current around 1854. Not surprisingly later mathematicians following the critical principles of the broad arithmetization program, especially as laid down by Weierstrass, found it necessary to probe more deeply into these points in Riemann's lecture.

From a later perspective there is a serious logical problem related to Riemann's approach. Ultimately he links his basic topological objects with numbers and coordinate systems. In other words, a Riemannian manifold as constructed always appears to be reducible to a 'number-manifold' ('Zahlenmannigfaltigkeit,' a term often used by Lie and Klein). Consequently, when trying to develop a framework for nonmetrical analysis situs, he seemingly is forced to fall back on concepts of measurement and ordinary analytic geometry. In this way he appears to complete a kind of logical circle. If the coordinates of the points in a manifold are given by the usual Euclidean or similar distance measure, then a specific metric is already assumed for the supposedly nonmetrical structure. How can he then go on to consider different possible measure-relations in the second section of his lecture? At the heart of this logical problem is a deep issue in the foundations of geometry of which Riemann's later followers and critics became acutely aware; for example, consider Felix Klein [Klein, 1898, especially pp. 587–588] [Klein, 1921, pp. 384–401, especially pp. 388–389] and Bertrand Russell [Russell, 1897, pp.30–33]. Towards the end of the nineteenth century mathematicians discerned a directly related problem in the case of projective geometry, ostensibly nonmetrical. They wondered: Is it really permissible to use a coordinate system, normally defined by a metric, to study this geometry in a purely projective way? Klein pointed the way to a solution for pure projective geometry to avoid the vicious circle, through von Staudt's quadrilateral construction.[28] By means of this construction it is possible to avoid the usual metric. One can coordinatize the structures investigated in projective geometry in such a way that their points are consistently assigned numbers by convention without bringing in a metric. By analogy, what Riemannian manifolds need is a method for arbitrarily assigning coordinates consistent with the topologies of their structures. In his work Riemann gives no indication of such a method, although he most probably thought of his coordinates as arbitrary. They constitute parameterizations of the manifolds. Still, the solution by arbitrary coordinates is only a partial one. What one finds conspicuously lacking in Riemann's work is the notion of a topological mapping. For modern mathematicians topology is inseparable from topological mappings or homeomorphisms. Riemann never contemplated these as such in his program of analysis situs. He was much more fluid and informal in his treatment of manifolds—his way of proceeding is not surprising at this juncture in the development of mathematics.

In a broader context mathematicians working towards the end of the nineteenth century began to feel the need for a deeper analysis of the foundations of nonmetrical geometry, in particular, of topology. Felix Klein especially saw the logical difficulty

[28] See Wikipedia entries for 'Projective harmonic conjugate' and 'Complete quadrangle.'

in basing geometry on number-manifolds and pressed for a solution [Klein, 1898, especially pp. 587–588] [Klein, 1921, pp. 384–401, especially pp. 388–389]. This difficulty was a motivation for the following development of the concept of a topological space by David Hilbert [Hilbert, 1902b] [Hilbert, 1903], Hermann Weyl [Weyl, 1913], Felix Hausdorff [Hausdorff, 1914], and others.

In spite of any obscure points there might be in Riemann's *Habilitationsvortrag* (habilitation lecture), the work became quite significant in the development of geometrical thought after it was published in 1868. It was primarily Hermann von Helmholtz (1821–1894) who made mathematicians and philosophers aware of Riemann's brilliant ideas. Since about 1866 Helmholtz had himself been working on a philosophico-geometrical problem closely related to Riemann's problem, but derived from a different source. When reading accounts of Riemann's work in two obituaries by Ernst Schering, Helmholtz was surprised—and dismayed—to find that Riemann had previously explored fundamental questions about space and the geometrical axioms. He quickly obtained a copy of the *Habilitationsvortrag* (habilitation lecture) from Schering. Shortly thereafter he presented a paper to the Naturhistorisch-medicinischer Verein (Natural-History-Medical Union) at Heidelberg on 22 May 1868 [Helmholtz, 1868b] [Helmholtz, 1883, pp. 610–617], and he sent a further paper to the Königliche Gesellschaft der Wissenschaften (Royal Society of Sciences) at Göttingen, which was presented on the third of June. The second paper, 'Über die Thatsachen, die der Geometrie zum Grunde liegen' ('Concerning the facts that lie at the foundation of geometry') [Helmholtz, 1868a] [Helmholtz, 1883, pp. 618–639] is his most important on the subject.

In general terms Riemann and Helmholtz had attacked the same problem, the so-called Riemann–Helmholtz (afterwards Riemann–Helmholtz–Lie) space problem. However, a closer analysis reveals that Riemann had started from a far more general position, a position that in more modern times would be characterized as partly topological. Indeed, Helmholtz' much more widely-read and influential papers tended to conceal the deeper springs of Riemann's thought. Because of Helmholtz' greater influence virtually all nineteenth-century thinkers lost sight of Riemann's more general but informal topological basis. In his second paper of 1868 Helmholtz simply reduces Riemann's discussion of n-dimensional manifolds to a succinct assumption or axiom that declares that n real-valued continuous coordinates determine the points in such manifolds [Helmholtz, 1868a, pp. 197–198]. At the time anyone reading both Riemann and Helmholtz on manifolds would probably regard the latter's account as simpler and clearer.

5.8 Concluding Remarks on Developments of Higher-Dimensional Spaces

After the burst of activity on higher-dimensional geometry of the 1840s, 1850s, and 1860s, numerous mathematicians continued the research. They investigated hyperspaces both from an analytical and from a more purely geometrical direction. Most attention was given to n-spaces of the Euclidean linear type, 'homaloidal spaces,' like Grassmann's domains. Hence, during the final third of the nineteenth century hyperspaces became commonplace in mathematical works. Following Gauss and Riemann many employed the term 'Mannigfaltigkeit' ('manifold', and in French, 'variété'), but giving the term a diversity of meanings. In his Erlanger Programm Felix Klein (1849–1925) prominently uses the term to refer to the basic abstract spaces underlying his group-theoretic definitions of geometries [Klein, 1872, §10, Note IV]. Mappings or transformations are, of course, a feature of Klein's approach to geometry. As noted, those were lacking in Riemann's analysis situs. Karl Weierstrass (1815–1897) commonly used the term 'Mannigfaltigkeit' in his lectures on multi-variable analysis [Weierstrass, 1927, pp. 55–60] (cf. Pincherle's summary of analytic function theory [Pincherle & Weierstrass, 1880]). He treated his n-fold manifolds as arithmetical continua (Euclidean spaces) founded on a logical theory of the real numbers, and he introduced basic analytical concepts such as limit point, continuum, and interior in order to develop analysis rigorously. In this work lies the origin of general or analytic topology. Cantor, taking his cue from Weierstrass, one of his teachers in Berlin, then employed the term 'Mannigfaltigkeit' and the related topological concepts. However, in several of his important contributions the term merely means an arbitrary set of numbers or points; it can even mean just an abstract set (Menge). It was only at the very end of the nineteenth century and then in the first years of the twentieth that a few mathematicians started to propose rigorous definitions of the topological concept of manifold. Walter von Dyck (1856–1934), Henri Poincaré (1854–1912), Hermann Weyl (1885–1955), and L.E.J. Brouwer (1881–1966) were the principal drivers of this move to formalization. Nevertheless, there is some historical justice in linking Riemann's earlier basic ideas with the more elaborated formal definitions of these later mathematicians.

The publications of Riemann and especially Helmholtz initiated a wide-ranging philosophical discussion of the nature of space and the foundations of geometry including its non-Euclidean and multi-dimensional branches (cf. [Torretti, 2019], [Gray & Ferreirós, 2021], and also [Gray, 2011]). In the 1870s and 1880s the Germans were the most active participants in this discussion, while in the 1890s the French took the leading role with Bertrand Russell as a more 'philosophical' English contributor. From the beginning philosophers strongly criticized Riemann's entire way of proceeding and implicit philosophy, especially his starting point: coordinatized manifolds. The whole gamut of philosophical positions from empiricism to idealism, including Kantianism, emerged in the discussion. Helmholtz was the principal empiricist; his student, Benno Erdmann, expanded upon his philosophy at great length in the book *Die Axiome der Geometrie* (*The axioms of geometry*) [Erdmann, 1877]. Apart perhaps from a certain amount of almost inevitable polemicizing, on the whole the standard of argument in this discussion was high.

The rapid developments in geometry forced philosophers to see the science in a new light. It became clear that many significant branches of geometry had to be separated from the notion of space. Physical space could no longer be the ultimate foundation for the entire discipline. Philosophers and mathematicians thus spoke of 'metageometry' and 'géométrie générale' ('general geometry'). Geometry presented a more abstract face. To be sure, the geometry of our space still remained uppermost in the minds of many philosophers. If Erdmann or Russell discussed the axioms of geometry, in the final analysis they were concerned with those assumptions that characterize the 3-fold manifold that is our space. The more radical split between pure geometry as an uninterpreted deductive system and applied geometry as a way of modeling physical space did not become fully understood until after the later work of Pasch and the Italian school and the publication of Hilbert's *Grundlagen der Geometrie* (*Foundations of Geometry*) [Hilbert, 1899].

The approach to higher spaces through coordinate manifolds and analytical methods that geometers of the mid-nineteenth century took up presents a definite historical indicator of the developing cleavage between geometry and theories of space. Abstract spaces took on a fundamental significance; 'space' became arithmetized. This development in geometry may be viewed as part of the trend of nineteenth-century arithmetization (cf. Felix Klein's paper [Klein, 1896] [Klein, 1922, pp. 232–240]). The application of the new rigorous theories of real numbers (proposed around 1870) then brought about a synthesis between the programs of arithmetization in analysis and in geometry. Geometrico-mechanical continuity based on the intuitive idea of motion was thereby replaced by analytically-defined continuity. With Weierstrass and Cantor multi-dimensional coordinate manifolds became fully arithmetized.

Nearer the end of the century, however, a trend contrary to the analytic one developed. Several mathematicians argued that geometry ought to be constructed as a pure autonomous science and not on the basis of number manifolds. This is the axiomatic-deductive trend. From it arose the criticism of Riemann, Helmholtz, and Lie for basing geometry on Zahlenmannigfaltigkeiten (number manifolds), because ultimately such a use appears to complete a vicious circle. Hence, mathematicians initiated a deeper analysis of the foundations of geometry, which had important consequences for the subsequent development of a genuinely nonmetrical topology. Riemann (and others) had sought to develop this subject, but without full success. Only in the twentieth century could this subject grow according to uniform principles.

To conclude this chapter let me turn specifically to the impact of the proposal and growth of analytical theories of hyperspaces and n-dimensional manifolds on ideas of dimension itself. Without doubt the growth of these theories made the dimension concept more easily comprehensible. Implicit in them is what one might appropriately call the coordinate idea of dimension. This is the simple, naive idea that the dimension of a space is merely the apparently unique number of coordinates needed to determine a point in that space. Strictly speaking, neither Grassmann, nor Riemann, nor Helmholtz ever espoused this theory of dimension, for they associated continuity and even differentiability with their abstract spaces. Yet it must be admitted that they were not always clear about their continuity and differentiability assumptions. Grassmann and Riemann really thought of continuity in terms of the ancient motion theory or the fluxion theory à la Newton. However, when Georg Cantor probed deeper into the coordinate idea of dimension and discovered his paradoxical result showing that the simple idea is untenable, he forced mathematicians to take a fresh look at dimension. Cantor's work could only have taken place in the context of the development of arithmetized hyperspaces. His discovery, the subject of the next chapter, is inconceivable without the prior growth of such ideas.

Chapter 6
Cantor and Dedekind: Discovery of a Counterintuitive Example concerning Dimension

6.1 Introduction

In 1877 Georg Cantor (1845–1918)[1] discovered to his own amazement that the points of a unit line segment can be put into one-one correspondence with the points of a unit square or more generally with the points of a ρ-dimensional cube. This counterintuitive result at first sight called into question the concept of dimension. Was it well-defined or even meaningful in mathematics? [2] Cantor's discovery marks the origin of the most important modern mathematical problem of dimension, the problem of its invariance. Is dimension an invariant with respect to any class of mappings? This problem spurred mathematicians to search for an appropriate invariance proof for dimension, a search that lasted for over three decades. Cantor's discovery and the immediately linked invariance problem together have been the most important impetus for the growth of modern dimension theory. In this chapter I shall examine how Cantor arrived at his surprising result and consider its immediate implications. Cantor's interactions with Richard Dedekind (1831–1916)[3] around this time played a significant part in this mathematical development; they discussed the issues in detailed mathematical correspondence.

Several historians of mathematics have examined the work of Georg Cantor and Richard Dedekind that relates to Cantor's discovery of the counterintuitive correspondence and its implications for the dimension concept and the invariance problem. These include Meschkowski [Meschkowski, 1967, pp. 31–41], Dugac [Dugac, 1976b, Chapter XI, pp. 115–130], Dauben [Dauben, 1979, Chapter 3, pp. 47–76]

---

[1] For a full-length biography of Cantor see especially [Dauben, 1979]. Cf. the earlier [Fraenkel, 1930], abridged in [Cantor, 1932, pp. 452–483], as well as [Meschkowski, 1967] and [Purkert & Ilgauds, 2013].

[2] Perhaps the concept could even be questioned from a common intuitive point of view?

[3] Compare also Ferreirós' careful historical analysis of Dedekind's approach to set theory through his work in algebra [Ferreirós, 2007, Chapter III, pp. 81–116]. For biographical information on Dedekind see `https://mathshistory.st-andrews.ac.uk/Biographies/Dedekind/`, `https://mathshistory.st-andrews.ac.uk/Strick/dedekind.pdf`, and `https://en.wikipedia.org/wiki/Richard_Dedekind`.

D. M. Johnson, *Theories of Dimension from Ancient to Modern Times*, Sources and Studies in the History of Mathematics and Physical Sciences, https://doi.org/10.1007/978-3-032-02682-8_6

(cf. [Dauben, 1974]), Purkert and Ilgauds [Purkert & Ilgauds, 2013, pp. 48–53], and above all Ferreirós [Ferreirós, 1993] [Ferreirós, 2007, Chapter VI, pp. 171–201], [Ferreirós, 2024]. Ferreirós provides a detailed, nuanced treatment of the history that takes account of virtually all previous work related to the topic. As indicated, the correspondence between Cantor and Dedekind figures prominently in this history: letters, cards, and Dedekind's related notes from the years 1873 to 1879, [Cantor & Dedekind, 1937] supplemented with [Dugac, 1976b, pp. 223–262] (English translation [Ewald, 1996, Volume II, pp. 843–871]).[4] For sake of comprehensiveness, I feel that I must include this part of the history, while relying on the excellent work others have published. When writing my original article covering this topic, many of these sources were not then available. It is gratifying to see how much work has been done on this area of history.

6.2 First Interactions between Cantor and Dedekind and Early Results in Set Theory

Cantor's work on set theory arose from his investigations into the uniqueness of representing a function by a trigonometric series (for details see [Dauben, 1971]). In 1874 he published his first purely set-theoretic paper, 'Über eine Eigenschaft des Inbegriiffes aller reellen algebraischen Zahlen' ('Concerning a property of the set of all real algebraic numbers') [Cantor, 1874] [Cantor, 1932, pp. 115–118].[5] In the paper he shows that the set of real algebraic numbers can be conceived, indeed ordered as an infinite sequence

$$\omega_1, \omega_2, \ldots, \omega_\nu, \ldots$$

(i.e., the set is countable, abzählbar, to use Cantor's later term). This is the property to which he refers in the paper's title. More importantly, Cantor proves a strong contrasting result, namely, the set of all real numbers cannot be given in the form of an infinite sequence (i.e., the set is uncountable, unabzählbar).[6] In fact, he proves the latter result in the stronger form: for every sequence of real numbers and every interval of the real line a number can be determined in the interval that is not in the given sequence. From a modern standpoint this result is the more important one. Cantor himself saw through the medium of these two results a clear distinction between two types of infinite sets of numbers on the real line, which he had questioned

[4] Cf. [Grattan-Guiness, 1974]. This paper contains important historical analysis and commentary on the rediscovery of the Cantor–Dedekind correspondence.

Many letters from and to Cantor have been published. See [Cantor, 1932, pp. 443–451] [Cantor & Dedekind, 1937] [Cavaillès, 1962, pp. 177–251] [Dugac, 1976b, pp. 223–262] [Lipschitz, 1986, pp. 28–37] [Meschkowski & Nilson, 1991].

[5] See also the Wikipedia article 'Cantor's first set theory article' `https://en.wikipedia.org/wiki/Cantor%27s_first_set_theory_article`.

[6] Cantor's term 'abzählbar' first appears in [Cantor, 1882] [Cantor, 1932, pp. 149–157].

earlier. It was then perfectly natural for him to wonder whether there are other types of infinite sets, say, for the plane or for higher-dimensional spaces.

There is a surprising back story to these results, which Ferreirós tells in fine detail [Ferreirós, 1993] [Ferreirós, 2007, Chapter VI, pp. 171–201]. The story emerges out of the correspondence between Cantor and Dedekind, on which Ferreirós provides a careful gloss with supporting evidence. Dedekind was the senior and more mature mathematician of the two, having a well-developed interpretation of the fundamentals of mathematics and substantial published works to his credit [Ferreirós, 1993, pp. 346–347] [Ferreirós, 2007, pp. 172–175], and having published in 1872 his well-known *Stetigkeit und irrationale Zahlen* (*Continuity and irrational Numbers*) [Dedekind, 1872], in which he introduced 'Dedekind cuts' as a theory of real numbers.[7] Cantor had met Dedekind by chance in Gersau during a trip to Switzerland in 1872 [Fraenkel, 1930, page 196]. On occasions they exchanged mathematical papers by post, and intermittently they corresponded. They discussed the pair of results of Cantor's paper before its publication in 1874 in letters dated in November and December 1873. Letters from Cantor to Dedekind plus Dedekind's notes on this part of their correspondence exist [Cantor & Dedekind, 1937, pp. 12–20]. The correspondence reveals that Cantor's prime motivating problem was whether the set of all real numbers could be enumerated by the positive integers. Thus, the second result of the paper should be the more important for him, even though it is not mentioned in the title. In the correspondence Dedekind says that he could not solve that problem. However, as part of the correspondence he provides Cantor with a proof of the countability of the set of algebraic numbers, including a bit of technical apparatus called the 'height' ('Höhe') of an algebraic equation.[8] It is not entirely clear from the evidence whether Cantor himself might have had a similar or different proof. However, Dedekind complains in his notes that Cantor's published paper contains Dedekind's proof nearly word for word, including the use of the auxiliary height of an algebraic equation. The case of plagiarism may not be entirely clear—Cantor may have had a very similar proof, though there is scant evidence for that. Surely Cantor slighted Dedekind in not giving the latter any credit at all in the published paper. Notwithstanding this slight, Dedekind is complimentary about Cantor's result concerning the set of all real numbers, having provided help in sharpening the result. Indeed, his improvement went straight into Cantor's paper—also without credit.[9] Cantor did show his results in December 1873 to Karl Weierstrass (1815–1897), one of Cantor's celebrated professors in Berlin, who had some appreciation of it. Cantor's pair of results shows the existence of transcendental numbers, those numbers among

---

[7] Cantor had put forward an alternative theory of real numbers in 1872 [Cantor, 1872] [Cantor, 1932, pp. 92–102].

[8] The height, N, of an algebraic equation of degree n, $a_0\omega^n + a_1\omega^{n-1} + \cdots + a_n = 0$, is $N = n - 1 + [a_0] + [a_1] + \cdots + [a_n]$. For each finite height there are only a finite number of algebraic number solutions of equations of that height.

[9] Perhaps the junior mathematician Cantor thought that the mentoring Dedekind gave him was akin to a senior professor giving advice to a 'student' and did not need to be given credit. If this was Cantor's view, it seems naive. Note that Cantor wrote to Dedekind in 1874, and apparently Dedekind did not respond.

the reals that are not algebraic. Ferreirós additionally discusses other professional overtones and disagreements among a broader set of mathematicians of the time, which I need not go into at this point. Suffice it to say that the relationship just between Cantor and Dedekind was not an entirely smooth one over the years.

6.3 A Problem of Geometry and Sets

Having investigated the power or cardinality of two important linear sets, the set of real algebraic numbers and the set of all real numbers, Cantor turned to examining higher-dimensional sets.[10] In a letter to Dedekind dated 5 January 1874 he posed a new research question [Cantor & Dedekind, 1937, page 20]:

> Can a surface (perhaps a square including its boundary) be put into one-one correspondence with a line (perhaps a straight line segment including its endpoints), so that to each point of the surface there corresponds a point of the line and inversely to each point of the line there corresponds a point of the surface? [Ch6, 1]

This problem is fundamental to the modern development of dimension theory. From the start Cantor was convinced of the importance and difficulty of his research question. He realized that some, indeed most, would regard a negative answer to it as so obvious that a proof would hardly be necessary. When he discussed it with a friend in Berlin some time during the first part of 1874, the friend explained that the matter was absurd, so to speak [Cantor & Dedekind, 1937, page 21]:

> ... since it is obvious that two independent variables cannot be reduced to one.[Ch6, 2]

In relating this encounter in a letter of 18 May 1874, the young Cantor sought Dedekind's reassurance that he was not chasing after a delusion! As the subsequent history shows, Cantor was not deceived in considering this problem. His own investigations and those of other mathematicians show the problem's value. Even in posing his research question Cantor introduced something new and important into thinking about dimension. *He related mappings and correspondences to considerations of the dimensions of figures and spaces.* For Cantor this new way of thinking was natural, because he was already interested in cardinality. In this way he was led to a novel treatment of the dimension concept with surprising consequences.

There is no record of Dedekind's initial reaction to Cantor's research problem. Dedekind may have still felt some annoyance at Cantor's treatment of him, although the record seems to show that Cantor did not recognize any difficulties in their relationship. In fact, there is no written record of exactly how Cantor dealt with it from May 1874 until April 1877. No letters or documents concerning the matter

---

10 With different emphases Dauben [Dauben, 1979, pp. 54–66] covers Cantor's work on mappings between manifolds of different dimensions that I am treating in this work in Sections 6.3–6.7.

have come to light covering this period. Perhaps nothing significant on the subject was written. It is likely that he only worked intermittently on his research problem from mid-1874 to mid-1877, and indeed without success. However, he persisted in regarding it as important. For example, when he attended the *Gaussjubiläum* (Gauss Jubilee) in Göttingen on 30 April 1877, the one-hundredth anniversary of Gauss' birth, he told various colleagues, among them Heinrich Weber and Rudolph Lipschitz, of his problem, which he felt was fundamental to geometry. Again, most of these colleagues thought the answer was obvious; a one-one correspondence between geometrical figures of differing dimensions is not possible. Still, Cantor felt that a *proof* was needed.

6.4 Cantor's Early Efforts Investigating the Problem of Dimension

Some time after the Gauss Jubilee in April 1877 Cantor changed his approach. He switched from trying to show the impossibility of a one-one correspondence between the points of figures of differing dimensions to showing that such a correspondence is indeed possible. In a letter to Dedekind of 20 June 1877 he offers a positive solution by specific construction of the correspondence! In this letter he first states his solution in geometrical terms, namely [Cantor & Dedekind, 1937, page 25]:

> ... that surfaces, solids, even continuous figures of ρ dimensions can be put into one-one correspondence with continuous lines, thus figures of only *one* dimension; therefore, that surfaces, solids, even figures of ρ dimensions have the same *power* as curves ....[Ch6, 3]

Immediately Cantor saw his result on the equal power or cardinality of sets of various dimensions as a criticism of some assumptions that geometers of his time commonly held about dimension. Contemporary geometers and mathematicians spoke of simply infinite, twofold, threefold, ... ρ-fold infinite figures. They informally regarded the infinity of points of a surface as the square of the infinity of points of a line or the infinity of points of a solid as the cube of the infinite set of points of the line. Such thoughts were prevalent at the time. Essentially, Cantor changed the situation entirely. He injected his new ideas about mappings and correspondences into thinking about dimension. Consequently, he regarded his proposed proof of equal power as an attack on the foundations of geometry, on the very concept of dimension, which he felt geometers were using uncritically.

In the letter of 20 June Cantor presents a demonstration that is arithmetical, not geometrical, in character. First of all, he claims that all (connected) figures of the same dimension number can be mapped analytically onto one another. By 'figures' he seems to mean just a variety of relatively simple geometrical ones, for otherwise the claim would amount to a very broad generalization of the 2-dimensional Riemann mapping theorem. Surely he does not intend to make such a generalization, for he must have been well aware of the virtual impossibility of proving such an extreme—and ultimately false—generalization.

Cantor is really interested in figures as sets of points and of different dimensions in the usual geometrical sense of dimension. He seeks to show that systems of values $(x_1, x_2, \ldots, x_\rho)$, with each x_i, $0 \leq x_i \leq 1$, can be put into one-one correspondence with values of a variable y, $0 \leq y \leq 1$. In other words, he aims to show that the points of a ρ-dimensional cube with all sides of length 1 can be put into one-one correspondence with the points of a unit line segment. He tells Dedekind that he himself changed his position on the question of the possibility of the correspondence from negative to positive [Cantor & Dedekind, 1937, page 26]. Cantor's technique of proof rests on his assertion that every number x, $0 \leq x \leq 1$, can be represented uniquely in the form of an *infinite* decimal expansion:

$$x = \alpha_1 \frac{1}{10} + \alpha_2 \frac{1}{10^2} + \cdots + \alpha_\nu \frac{1}{10^\nu} + \cdots,$$

where the α's are the digits, 0, …, 9 (with infinitely many being non-zero). Using simple infinite decimal expansions instead of writing Cantor's equivalent decimal series, one can write ρ decimal values for each point of the ρ-cube:

$$\begin{aligned} x_1 &= 0.\alpha_{1,1}\alpha_{1,2} \ldots \alpha_{1,\nu} \ldots, \\ x_2 &= 0.\alpha_{2,1}\alpha_{2,2} \ldots \alpha_{2,\nu} \ldots, \\ &\ldots\ldots\ldots\ldots\ldots\ldots\ldots\ldots \\ x_\rho &= 0.\alpha_{\rho,1}\alpha_{\rho,2} \ldots \alpha_{\rho,\nu} \ldots, \end{aligned}$$

and a single decimal for each point or value of y:

$$y = 0.\beta_1\beta_2 \ldots \beta_\nu \ldots .$$

Now the x's and the y can be put into correspondence according to the following equations:

$$\begin{aligned} \beta_{(n-1)\rho+1} &= \alpha_{1,n}; \qquad \beta_{(n-1)\rho+2} = \alpha_{2,n}; \ldots \\ \beta_{(n-1)\rho+\sigma} &= \alpha_{\sigma,n}; \ldots \qquad \beta_{(n-1)\rho+\rho} = \alpha_{\rho,n}. \end{aligned}$$

The equations define a 'lacing up' of ρ decimal digits of the x's in succession, decimal place by decimal place, to yield a single y number:

$$y = 0.\alpha_{1,1}\alpha_{2,1} \ldots \alpha_{\rho,1}\alpha_{1,2}\alpha_{2,2} \ldots \alpha_{\rho,2}\alpha_{13}\alpha_{2,3} \ldots \alpha_{\rho,3} \ldots .$$

In the opposite direction, the digits of a y number can be systematically 'unlaced' in sequence to give the x numbers.

6.5 Dedekind's Objection to Cantor's First Attempted Proof of a Counterintuitive Result

Dedekind quickly responded in a letter of 22 June with an objection to Cantor's proof. First, Dedekind assumed that Cantor, by requiring *infinite* decimal expansions, wished to exclude decimal numbers with trailing 0's. Of course, these can always be replaced with trailing 9's (with the single exception of the number 0 itself). For example, the finite decimal

$$0.3000\ldots = 0.3$$

can be replaced by the infinite decimal

$$0.2999\ldots .$$

With this assumption Dedekind's objection amounts to the following. Take the simplest case when one laces up the digits of two real numbers, x and y:

$$x = 0.\alpha_1\alpha_2\ldots\alpha_\nu\ldots,$$
$$y = 0.\beta_1\beta_2\ldots\beta_\nu\ldots$$

to derive a third number z:

$$z = 0.\gamma_1\gamma_2\ldots\gamma_\nu\ldots,$$

where

$$\gamma_1 = \alpha_1, \gamma_2 = \beta_1, \gamma_3 = \alpha_2, \gamma_4 = \beta_2, \ldots, \gamma_{2\nu-1} = \alpha_\nu, \gamma_{2\nu} = \beta_\nu, \ldots .$$

Now if one takes a specific z number:

$$0.478310507090\alpha_7 0\alpha_8 0\alpha_9 0\ldots\alpha_\nu 0\ldots,$$

then the unlacing process will yield a y number,

$$y = 0.730000\ldots,$$

which does not have the required infinite decimal form. Moreover, the infinite decimal form for this y number does not correspond to the specified z number. In general, there are infinitely many of these z numbers that yield inadmissible x or y decimal numbers. Dedekind concluded this letter with the sentence [Cantor & Dedekind, 1937, page 28]:

> I do not know whether my objection is of essential significance for your idea; however, I did not want to hold it back from you.[Ch6, 4]

6.6 Cantor's Response: An Elaborate Proof

Cantor's reply was swift. In a card postmarked 23.6.77, the day after Dedekind wrote his objection, he accepted the criticism as a criticism of the proof but not of the theorem itself. Cantor thought his result could be salvaged. His immediate reaction was to claim that he had 'proved' more than he intended. As given, his lacing process brings a system of several real variables $x_1, x_2, \ldots, x_\rho$, with $0 \leq x_i \leq 1$, into correspondence with a single variable y, with $0 \leq y \leq 1$. Now the variable y does not take on all values in the unit interval, but only certain values y' that exclude those suggested in Dedekind's objection, viz., those with certain alternating patterns of 0's. In his postcard Cantor suggests that to fix up his proof all that is needed is to show how to bring the y' into one-one correspondence with another variable t which varies over *all* real values between 0 and 1.

As it turned out, Cantor was not immediately able to repair his proof as suggested in his postcard.[11] Instead in a letter of just two days later, 25 June 1877, he presented Dedekind with a completely new proof, which, though overcoming Dedekind's objection, is not nearly as simple as the earlier proposed proof. He must have been anxious to see his research problem through to an acceptable, provable conclusion. In the letter he states his theorem as follows [Cantor & Dedekind, 1937, page 29]:[12]

> (A) "A continuous manifold extended in ρ dimensions can be put into one-one correspondence with a continuous manifold of one dimension, or (what is only the same theorem in a different guise) the points (elements) of a manifold extended in ρ dimensions can be determined by one real coordinate t, so that to every real value of t in the [closed] interval $(0 \ldots 1)$ there corresponds a point of the manifold and also conversely to every point of the manifold there is a definite value of t in the [closed] interval $(0 \ldots 1)$." [Ch6, 5]

To prove the theorem Cantor begins by using the fact that every irrational number e between 0 and 1 can be represented by a unique infinite continued fraction:

$$e = \cfrac{1}{\alpha_1 + \cfrac{1}{\alpha_2 + \ddots + \cfrac{1}{\alpha_\nu + \ddots}}} = (\alpha_1, \alpha_2, \ldots, \alpha_\nu, \ldots),$$

[11] It turns out that it is quite easy to fix up Cantor's first proof. One merely blocks 0's with nonzero digits to eliminate the difficulty which Dedekind pointed out. Thus one blocks the example of Dedekind as follows: 0.47831 | 05 | 07 | 09 .... One then treats blocks as single digits for lacing and unlacing.

[12] Cantor states theorem (A) a little imprecisely. By ρ-dimensional manifold he really must mean a ρ-dimensional cube. Also, the first ρ in the edited German text is given as an e—clearly an error.

where the α's are positive integers. Since there is no difficulty with 0 (since it is not irrational), it is now possible to salvage the interlacing argument. Thus if there are ρ irrational numbers between 0 and 1, represented by continued fractions:

$$e_1 = (\alpha_{1,1}, \alpha_{1,2}, \ldots, \alpha_{1,\nu}, \ldots),$$
$$e_2 = (\alpha_{2,1}, \alpha_{2,2}, \ldots, \alpha_{2,\nu}, \ldots),$$
$$\cdots\cdots\cdots\cdots\cdots\cdots\cdots\cdots\cdots$$
$$e_\rho = (\alpha_{\rho,1}, \alpha_{\rho,2}, \ldots, \alpha_{\rho,\nu}, \ldots),$$

it is possible to lace up these to derive another irrational number $\delta, 0 < \delta < 1$:

$$\delta = (\beta_1, \beta_2, \ldots, \beta_\nu, \ldots)$$

according to the equations:

$$\beta_{(n-1)\rho+1} = \alpha_{1,n}, \ \ldots, \ \beta_{(n-1)\rho+\sigma} = \alpha_{\sigma,n}, \ \ldots, \ \beta_{n\rho} = \alpha_{\rho,n}.$$

Consequently, ρ irrational numbers can be laced up to make one irrational number and, conversely, one irrational number can be unlaced to give ρ irrationals, yielding a one-one correspondence.

The problem of giving a correspondence between points of the closed unit segment and points of the closed ρ-cube is now driven back to the problem of showing that there is a one-one correspondence between the set of irrational numbers in the unit interval and the set of all numbers in that interval, i.e., the problem of proving the supporting theorem [Cantor & Dedekind, 1937, page 30]:

> (B) "A variable number e that can take all *irrational* number values of the [closed] interval $(0 \ldots 1)$ can be put in *one-one* correspondence with a number x that takes *all* values of this interval without exception." [Ch6, 6]

Not having well-developed set-theoretic tools to hand, Cantor initially gives an rather complicated proof in his letters to Dedekind of 25 and 29 June 1877. The proof reveals how he first grappled with the problems of the cardinality of infinite sets.

To demonstrate (B) Cantor begins by considering all the rational numbers in the closed interval[13] [0, 1] given in the form of a sequence:

$$r_1, r_2, \ldots, r_\nu, \ldots$$

and an arbitrarily chosen infinite sequence of irrational numbers ε_ν from the interval $[0, 1]$ that obey the conditions that $\varepsilon_\nu < \varepsilon_{\nu+1}$ (monotonically increasing) and $\lim \varepsilon_\nu = 1$. The sequence of rational numbers and the chosen sequence of irrational numbers can be put into one-one correspondence in the obvious way: r_ν corresponds to ε_ν. On the basis of this correspondence immediately there is a correspondence

[13] In my explanatory text it is convenient to use the modern terminology and symbols, (x, y), $[x, y]$, $(x, y]$, $[x, y)$, for open, closed, and half-open intervals.

between the real numbers of the closed unit interval minus the rationals (i.e., the irrationals e of the interval) and the numbers of $[0, 1]$ minus the chosen irrationals ε_ν and so theorem (B) is reduced to [Cantor & Dedekind, 1937, page 31]:

> (C) “A number f that can take all values of the [closed] interval $(0 \ldots 1)$ with the exception of certain ε_ν, which are bound by the conditions $\varepsilon_\nu < \varepsilon_{\nu+1}$ and $\lim \varepsilon_\nu = 1$, can be put into one-one correspondence with a continuous variable x that takes all values of the [closed] interval $(0 \ldots 1)$ without exception.” Ch6, 7

Reducing theorem (B) to (C) simplifies the task of finding a correspondence. The irrational numbers ε_ν just break up the unit interval $[0, 1]$ into infinitely many subintervals. The first of these is half-open and the remaining are open intervals; additionally, there is the single point 1:

$$[0, \varepsilon_1), (\varepsilon_1, \varepsilon_2), \ldots, (\varepsilon_\nu, \varepsilon_{\nu+1}), \ldots, 1.$$

These can be lined up with the complete closed interval $[0, 1]$. Thus towards the end of his letter of 25 June Cantor says the proof of theorem (C) is to be carried out by successive applications of [Cantor & Dedekind, 1937, page 32]:

> (D) “A number y which can take all values of the interval $(0 \ldots 1)$ with the single exception of the value 0 can be put into one-one correspondence with a number x which takes all values of the interval $(0 \ldots 1)$ without exception.” Ch6, 8

Hence, the aim is to show that the half-open interval $(0, 1]$ can be put in one-one correspondence with the closed interval $[0, 1]$. Cantor's proof of (D) proceeds by means of a complex diagram defining a curve and mapping [Cantor & Dedekind, 1937, page 32] (reproduced as Figure 6.1). The curve consists of the parallel segments $\overline{ab}, \overline{a'b'}, \overline{a''b''}, \ldots$, and the point c, but excluding the endpoints $b, b', b'', \ldots$. The lengths are defined as:

$$\overline{op} = \overline{pc} = 1;\ \overline{ob} = \tfrac{1}{2};\ \overline{bb_1} = \tfrac{1}{4};\ \overline{b_1b_2} = \tfrac{1}{8};\ \overline{b_2b_3} = \tfrac{1}{16}; \ldots$$
$$\overline{oa} = \tfrac{1}{2};\ \overline{a'd'} = \tfrac{1}{4};\ \overline{a''d''} = \tfrac{1}{8};\ \overline{a'''d'''} = \tfrac{1}{16} \ldots.$$

The segments of the x-axis are mapped or ‘flipped’ onto the y-axis using the curve to define the ‘segment-by-segment reflective’ mapping. Clearly, the curve maps the closed unit interval $[0, 1]$ on the x-axis onto the half-open unit interval $(0, 1]$ on the y-axis.

The curve and theorem (D) do not quite complete the proof of (C), because, first, (D) needs to be generalized to intervals with arbitrary endpoints and, second, it is necessary to show the related result that an open interval can be put into correspondence with a closed interval. Cantor soon realized the need to fill out his proof of (C) and sent a completion to Dedekind in a letter of 29 June [Cantor & Dedekind, 1937, pp. 34–37]. The completion of (C) is accomplished as follows. First, it is possible to generalize theorem (D) to intervals with arbitrary endpoints in a straightforward manner by using linear substitutions, yielding theorem (F) [Cantor & Dedekind, 1937, page 35]:

> (F) "A variable number z that can take all values of an interval $(\alpha \ldots \beta)$ with the exception of α can be put in one-one correspondence with a variable number u that can take all values of the interval $(\alpha \ldots \beta)$ without exception." [Ch6, 9]

Second, it is possible to prove the following theorem with the help of (F) by aligning appropriate intervals [Cantor & Dedekind, 1937, page 35]:

> (G) "A variable number w that takes all values of the interval $(\alpha \ldots \beta)$ with the exception of the two endpoints α, β can be put into one-one correspondence with a variable number u that takes all values of the interval $(\alpha \ldots \beta)$." [Ch6, 10]

Cantor also needs an auxiliary theorem (E) stating that for two collections (in Cantor's notation), $a \equiv (a', a'', \ldots)$ and $b \equiv (b', b'', \ldots)$, of pairwise disjoint sets that can be lined up and put in one-one correspondences set by set across the collections, the two unions, one for each of the collections, a and b, can be put into one-one correspondence [Cantor & Dedekind, 1937, page 35]. Finally, one can conclude that it is possible to have a one-one correspondence between all the intervals

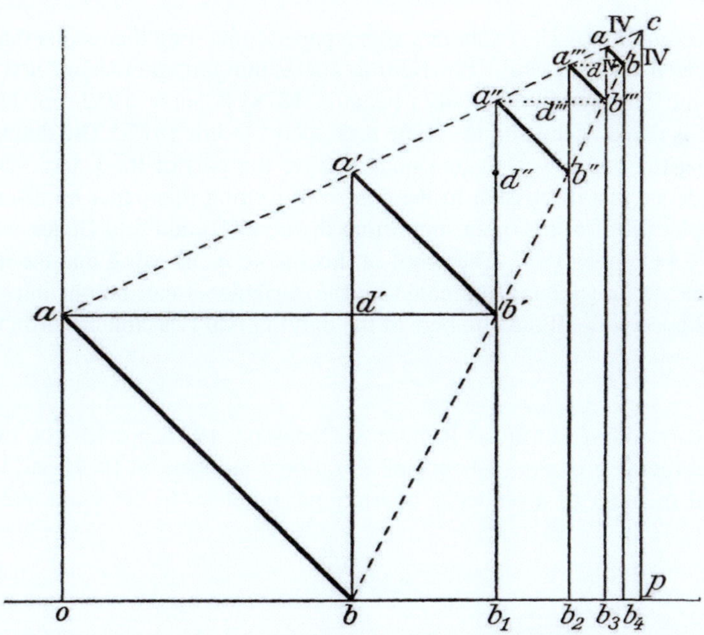

Fig. 6.1 Diagram from Cantor's Letter to Dedekind, 25 June 1877, [Cantor & Dedekind, 1937, page 32].

$$[0, \varepsilon_1), (\varepsilon_1, \varepsilon_2), \ldots, (\varepsilon_\nu, \varepsilon_{\nu+1}), \ldots, 1.$$

and the entire closed interval $[0, 1]$ by using theorem (G) to give correspondences between open and closed intervals $(\varepsilon_1, \varepsilon_2)$ and $[\varepsilon_1, \varepsilon_2], \ldots, (\varepsilon_{2\nu-1}, \varepsilon_{2\nu})$ and $[\varepsilon_{2\nu-1}, \varepsilon_{2\nu}], \ldots$. This completes Cantor's lengthy and rather labored proof.

The summary above shows that Cantor expended some effort to complete his proof. In a letter to Dedekind of 29 June 1877, he expresses some surprise about his counterintuitive result, exclaiming [Cantor & Dedekind, 1937, page 34]:

> . . . I see it, but I don't believe it.[Ch6, 11]

In his response to Cantor of 2 July 1877, Dedekind says that he cannot find any gaps in Cantor's proof and believes that the result is correct [Cantor & Dedekind, 1937, page 37]:

> I have checked your proof once again, and I can find no gaps; I certainly believe that your interesting theorem is correct, and I congratulate you for it.[Ch6, 12]

6.7 Cantor's Paper, 'Ein Beitrag zur Mannigfaltigkeitslehre' ('A Contribution to the Theory of Manifolds')

During the summer of 1877 Cantor wrote a paper containing the results that he had communicated to Dedekind, 'Ein Beitrag zur Mannigfaltigkeitslehre' ('A Contribution to the Theory of Manifolds') [Cantor, 1878] [Cantor, 1932, pp. 119–133]. The paper is dated at the end as 'Halle a. S., den 11. Juli 1877.' The dating would suggest that the paper was written quickly after the part of the Cantor–Dedekind correspondence just discussed. In the course of writing this paper he discovered a much simpler proof of his main supporting theorem (B) and told Dedekind of it in a letter of 23 October 1877. The proof of this theorem, asserting that the irrational numbers in the unit interval (denoted by the variable e) can be put into one-one correspondence with all the numbers of the unit interval (variable x), or in Cantor's notation:

$$e \sim x,$$

is simply carried out as follows [Cantor & Dedekind, 1937, pp. 40–41]. Let ϕ_ν be the general member of the sequence of all rational numbers in $[0, 1]$ and let η_ν be the general member of a sequence of irrational numbers in the same interval, for example,

$$\eta_\nu = \frac{\sqrt{2}}{2^\nu}.$$

Also, let h represent all the other values of the interval $[0, 1]$, except those represented by ϕ_ν and η_ν. Then:

$$x \equiv \{h, \eta_\nu, \phi_\nu\},$$

$$e \equiv \{h, \eta_\nu\} \equiv \{h, \eta_{2\nu-1}, \eta_{2\nu}\},$$

But

$$h \sim h; \quad \eta_\nu \sim \eta_{2\nu-1}; \quad \phi_\nu \sim \eta_{2\nu}.$$

Hence, it follows that

$$x \sim e.$$

What an elegant proof! It obviates the need for the machinery of the initial proof. However, in the extended paper of some 17 pages, he includes both the extensive proof and the short elegant one. Cantor must have thought that the supporting lemmas had value in their own right. In the paper he also includes some statements and propositions about the 'power' ('Mächtigkeit') or cardinality of sets.

6.8 Discussion between Cantor and Dedekind on the Counterintuitive Result on Dimension

Having discovered the counterintuitive result that 1- and ρ-dimensional figures can be put into one-one correspondence, Cantor was quick to draw some mathematical and philosophical consequences of his unexpected find. His correspondence with Dedekind reveals some first reactions of the two mathematicians to Cantor's result. In the last paragraphs of his letter of 25 June 1877 [Cantor & Dedekind, 1937, pp. 33–34] Cantor remarks about the implications of his discovery. He says that he has followed with interest the efforts of Gauss, Riemann, Helmholtz, and others directed towards understanding the foundations of geometry, but his result has now made him doubt the validity of their work [Cantor & Dedekind, 1937, page 33]:

> It strikes me that all investigations taken up in this field begin *for their part* from an unproved assumption, which does not appear to me to be obvious, but rather seems to need a proof. I mean the assumption that a ρ-fold extended continuous manifold requires ρ independent real coordinates for the determination of its elements and that this number of coordinates can be neither increased nor decreased for one and the same manifold.[Ch6, 13]

At first he had thought that this assumption of geometers was correct, but he had differed from them in that he had thought it required a rigorous proof and so persisted in trying to find one. However, he was completely surprised when he finally found a counterexample: the assumption is false. A ρ-fold manifold can be 'coordinatized' by a single coordinate. He ascribed this strange result to the 'wonderful power in the usual real and irrational numbers' ('wunderbare Kraft in dem gewöhnlichen reellen rationalen und irrationalen Zahlen'). Moreover, he recognized that his result can be easily extended from ρ-dimensional manifolds to infinite-dimensional manifolds, assuming that their infinitely many dimensions have the form of a simple infinite sequence (i.e., that the dimension is countably infinite).

Cantor's letter ends with the following paragraph [Cantor & Dedekind, 1937, page 34]:

> *Now it seems to me* that all philosophical or mathematical deductions that make use of this mistaken assumption are inadmissible. Rather the distinction which exists between figures of *different* dimension numbers must be sought in entirely different aspects than in the number of independent coordinates, which is normally held to be characteristic.[Ch6, 14]

Thus, he saw his result as a direct and devastating blow to the 'coordinate concept of dimension.' Another dimension concept is needed.

Dedekind responded to Cantor's observations as well as to his proof, first in a postcard (now lost) and then more fully in a letter of the second of July. Dedekind had gone over the second proof and reported that he could find no mistakes. However, he was not able to agree with Cantor's reading of the consequences of his theorem. In spite of the theorem or rather because of considerations occasioned by it, he still affirmed his conviction [Cantor & Dedekind, 1937, page 37]:

> . . . that the dimension number of a continuous manifold is now as before the first and most important invariant of a manifold . . . . [Ch6, 15]

In the light of Cantor's counterintuitive result the constancy or invariance of the dimension number certainly required proof, and as long as this proof was lacking dimensional invariance was in doubt. But Dedekind was convinced that a proof is possible.

In his letter of 2 July Dedekind goes on to give an insightful explanation of the mathematical situation surrounding dimension and Cantor's result. Giving the benefit of the doubt to previous writers on this subject and thereby possibly deferring to his friend and mathematical colleague Riemann, now deceased, he says that these writers clearly made the implicit but quite natural assumption that, when giving a new coordinate system to the points of a continuous manifold, it is assumed that the new coordinates are *continuous* functions of the old coordinates, so that what is continuously connected according to the first coordinate system is also continuously connected according to the second. In consequence, Dedekind arrives at the following theorem [Cantor & Dedekind, 1937, page 38]:

> "If one succeeds in setting up a one-one and complete correspondence between the points of a continuous manifold A of a dimensions on the one hand and the points of a continuous manifold B of b dimensions on the other, then this *correspondence itself* must necessarily be *discontinuous throughout* if a and b are *unequal*. "[Ch6, 16]

Dedekind generously defends previous writers as having made 'the tacit, completely natural presupposition' ('die stillschweigende, ganz naturgemässe Voraussetzung') concerning continuity and connectedness. Dedekind can state this with some authority, for he himself had thought long and hard about matters of continuity. Cantor's discovery does bring to the fore a new understanding of the situation surrounding dimension and basic assumptions of geometry. The implicit presupposition should be examined further. Thus, Cantor's result drove Dedekind to conjecture a new theorem, which effectively is a reasonable statement of the invariance of dimension theorem. This is a very apposite conjecture for Dedekind to make. The conjectured theorem will become the central issue in the subsequent development of dimension theory.

Additionally, Dedekind offers a specific explanation of Cantor's correspondence result relative to Cantor's first mechanism of proof using decimal expansions and his second mechanism of proof using continued fractions with extra machinery, all in relation to Dedekind's conjectured invariance theorem. It turns out—with the advantage of hindsight given by later developments of set theory and topology—that his explanation is not entirely correct. Concerning Cantor's first attempt at defining the correspondence using the mechanism of decimal expansions, Dedekind claims somewhat hesitantly that if it had been possible to complete the direct correspondence between the ρ-dimensional cube and *all* the points of the unit segment, then the correspondence would need to be continuous [Cantor & Dedekind, 1937, page 38]. However, Dedekind realizes that his counterexample to Cantor's initial attempted proof apparently shows that the completion is not really possible. In this case Dedekind essentially is on the right track.[14] Unfortunately, he is not correct in his explanation of Cantor's second way of setting up a correspondence between the ρ-dimensional cube and the points of the unit segment, starting with the mechanism of infinite continued fractions for the irrational numbers. Dedekind says that this initial part of the correspondence is continuous [Cantor & Dedekind, 1937, page 38]:

> . . . it seems to me that in your present proof the *initial* correspondence between the points of the ρ-interval (whose coordinates are all irrational) and the points of unit interval (also with irrational coordinates) is in a certain sense (smallness of the changes) as continuous as possible . . . [Ch6, 17]

This assertion is not correct. From a modern standpoint one would naturally employ the usual Euclidean space topology, subsetting over the irrationals for both the ρ-interval and the unit interval to derive two topologies. If there were a homeomorphism between the two spaces with these derived topologies and not just a one-one correspondence, then in essence by extension one would contradict the dimension invariance theorem. Of course, this way of thinking is all with hindsight. Still, one should not be overly critical of Dedekind. Dedekind goes on to claim that extending the correspondence to include the rationals makes it thoroughly discontinuous. As Dedekind says in rather colorful language [Cantor & Dedekind, 1937, page 38]:

> . . . you are compelled to admit a frightful, dizzying discontinuity emerge in the correspondence, which dissolves everything into atoms . . . . [Ch6, 18]

[14] For example, if one considers the correspondence of the unit segment with the square and consider on the line the point $0.5 = 0.4999\ldots$ and any point close to 0.5 but slightly greater in value, then in the correspondence these points are unlaced to give points in the square which are far apart. So the partial correspondence is not continuous in the sense in which Dedekind thought of it, namely, in the sense that nearby points are mapped to nearby points.

At the end of his letter Dedekind warns the younger Cantor not to polemicize openly against the original 'article of faith' ('Glaubenartkel') of manifold theory unless Cantor examines the conjectured invariance theorem. Dedekind did not want to read Cantor's counterintuitive discovery in the same way as Cantor himself did at first. Dedekind wanted to save the dimension concept by means of an invariance proof and not to criticize overly any deficiencies of previous geometers in handling the concept. His topological understanding of the geometrical situation is acute, even with the error noted. His idea of dimensional invariance based on continuity is certainly insightful.

Cantor swiftly responded to Dedekind's letter two days later on July 4. In his reply he immediately concedes Dedekind's points over the interpretation of the correspondence result. He says that he overstated his case and did not intend to criticize the concept of ρ-fold extended continuous manifold as such, but only to clarify it. He did not intend to deny the 'definiteness of the dimension number' ('Bestimmtheit der Dimensionenzahl'), but only the invariance of the number of coordinates, which some authors seemed to accept too easily as the same as the dimension number. So to Cantor's mind coordinate dimension is not viable in mathematics. It must be distinguished from some other notion of dimension, not yet made explicit.

Cantor readily accepted Dedekind's conjectured invariance theorem as a possible way out of the difficulty over dimension, but, of course, he wanted a proof. In his letter he immediately points to some apparent difficulties likely to be encountered in any attempt to demonstrate the conjectured theorem. In the first place, he foresees a problem in defining the concept of continuous correspondence in full generality; yet all hangs on such a definition. In the second place, he recognizes that there may be some difficulties with the approach to dimension number through an invariance proof if one takes as in scope figures that are not continuous throughout (i.e., in modern terms, not connected), but that may reasonably be assigned a dimension number. This latter point is perceptive. Not only does Cantor think an invariance proof is required, but also he hints at the need for an independent definition of dimension to cover arbitrary figures (or, as one would say now, arbitrary point sets). At this stage in their critical discussion of the dimension problem one can see that Cantor and Dedekind achieved a very high level of understanding and insight into this situation.

Cantor's important paper, 'Ein Beitrag zur Mannigfaltigkeitslehre' ('A Contribution to the Theory of Manifolds') [Cantor, 1878] [Cantor, 1932, pp. 119–133], written during the early summer of 1877 and published in the following year, is an exposition of his discoveries about correspondence and dimension and incorporates all his results largely in the form in which he had related them to Dedekind. However, certain general points are emphasized in the paper. From the beginning he uses the term 'Mannigfaltigkeit' ('manifold') to mean virtually any 'well-defined' set of elements; thereby putting the paper nearly in the realm of general set theory. He devotes the first few paragraphs to a delineation of his idea of power (Mächtigkeit) based on one-one correspondence, giving definitions of equal power (gleiche Mächtigkeit; or äquivalent), lesser and greater power, and providing several significant mathematical examples. He emphasizes the sharp distinction between the power of finite sets and the power of infinite sets. The concept of power for finite sets is equivalent to that of positive whole number, so that two finite sets are of equal power if and only if they have the same number of elements and a proper subset (Bestandteil) of a finite set always has smaller power than the set itself. However, the situation is fundamentally different for infinite sets. An infinite set can be put into one-one correspondence with certain of its infinite subsets, and Cantor gives the simple example of the sequence of positive integers and the sequence of positive even integers, which can be put into one-one correspondence and which are, therefore, of equal power.

Although in his paper Cantor briefly deals with sets of the power of the positive integers, sets of the smallest infinite power (i.e., countable sets), he, of course, concentrates on the power of 'so-called continuous n-fold manifolds' (i.e., the power of the continuum). Near the beginning of the paper he mentions Riemann, Helmholtz, and others[15] who first dealt with manifolds. However, after the discussion in the correspondence with Dedekind his criticism of these writers is toned down. He only points to the 'tacit' assumption of most of these that the basic correspondence of the elements of an n-dimensional manifold with n real coordinates must be continuous and promises in a footnote to return to this matter on another occasion. In fact, a year later he did consider the matter, the fundamental problem of dimension. Then he launches into his demonstration that the elements of an n-fold extended continuous manifold can be determined by a single coordinate or, more specifically, that the points of an n-dimensional cube can be put into one-one correspondence with the points of a unit line segment. Thus, all continuous manifolds have the same power. His proof is essentially as he communicated it to Dedekind. However, one very important new thing is added at the end of the paper: a mention—the first by Cantor—of the celebrated continuum hypothesis! At this point he formulates it by saying that the infinite linear manifolds (i.e., sets of real numbers) divide into two classes according to whether they can be brought into the form of the sequence of the positive integers or into the form of the real numbers in the unit interval. In other words, there are only two powers of infinite linear sets. In the light of his recent discoveries it is a perfectly natural way to state the continuum hypothesis. Thus far he had found only two kinds of infinite sets even among the n-dimensional and infinite-dimensional manifolds.[16]

Cantor submitted his paper to *Crelles Journal*, the *Journal für die reine und angewandte Mathematik* (*Journal for pure and applied Mathematics*), on 11 July 1877. It is well known that he felt frustrated by delays with its publication [Fraenkel, 1930, page 198] [Grattan-Guiness, 1974, pp. 111–113]. Borchardt, the managing editor of the journal, seemed to impede its progress, as he learned from his old friend Emil Karl Lampe, who dealt with printer's proofs for the journal. Borchardt put other papers that were submitted later before Cantor's in the order for publication. After three months he complained to Dedekind and sought his advice about publishing the work as a separate pamphlet. Dedekind thought this was not a wise course, when he considered his own difficulties with publishing his pamphlet *Stetigkeit und irrationale*

[15] He mentions J. Rosanes, O. Liebmann, and B. Erdmann.

[16] In the paper he implies that he has a proof of the hypothesis. However, the later history reveals that he struggled for many years to find a proof and in the end failed to find one which satisfied him.

Zahlen (*Continuity and Irrational Numbers*) [Dedekind, 1872]. Finally in November, Cantor heard that his 'Beitrag' ('Contribution') paper would be published.

In his biography of Cantor Fraenkel [Fraenkel, 1930, page 198] suggests that Leopold Kronecker (1823–1891) may have had a hand in the delay over publication in *Crelles Journal*. Others have followed up on this possibility [Grattan-Guiness, 1974, pp. 111–113] [Dauben, 1979, pp. 66–70]. However, Ferreirós casts some doubt on the possibility of Kronecker's interference over this publication [Ferreirós, 1993, pp. 351–352] [Ferreirós, 2007, pp. 197–201]. The evidence is not entirely clear. Whatever the case may be, Cantor's paper certainly offered a surprising and counterintuitive result in the emerging field of set theory. The editors of the journal may have had some lingering doubts about the validity of Cantor's (admittedly long) proof of the result. If they were not able to detect a specific error in the proof, they still may have found the result unsatisfactory, contrary to accepted mathematical intuition of the time.

6.9 Concluding Remarks

To conclude this chapter I shall sum up the achievements concerning dimension that Cantor and Dedekind made up through 1878. Prior to their work Riemann and Helmholtz, among others, had put forward informal or semi-formal theories of continuous manifolds or n-dimensional geometrical objects; each included an implicit notion, definition, or basic theory of dimension. Riemann and Helmholtz each intended his theory of manifolds and dimension as a general framework for investigations into geometry.

A few years after the appearance of the publications of Riemann, Helmholtz, and other nineteenth-century mathematicians with an early interest in multi-dimensional geometry Cantor approached the basic concept of dimension from a position outside of geometry, from the viewpoint of his work on one-one correspondences and power, hence, from the viewpoint of the set theory he was in the midst of creating. Consequently, he brought a new set-theoretic approach to problems of geometry. After his initial stumbling over a proof, with Dedekind pointing out the flaw, he was able to demonstrate that a ρ-dimensional cube can be put into one-one correspondence with a unit line segment—a surprising and counterintuitive result. Apparently the editors of *Crelles Journal* found this result more than a little disconcerting. Yet Cantor took the bull by the horns, so to speak. Initially, he read into his result a strong criticism of the dimension concept implicit in the work of Riemann, Helmholtz, and others: the concept of dimension based on the number of coordinates needed to determine a point.

Dedekind, while accepting Cantor's counterintuitive result whole-heartedly, did not read the same message in it. He immediately saw a way out of the difficulty through continuity and gave credit to the older geometers for this means of escape from the consequences of Cantor's result. It is certainly fair to say that Riemann and Helmholtz included continuity (and differentiability) naturally in their concept of dimension. Nonetheless, Cantor's discovery put an entirely new light on the coordinate theory of manifolds and dimension. A fresh examination of the informal concept of dimension was now needed. Dedekind even realized this: he quickly saw that a proof of some kind of theorem about dimensional invariance incorporating the idea of continuity was required. Hence, he came to state very clearly the crucial problem of dimension, which the counterintuitive correspondence result forced upon mathematicians. At once Cantor accepted that Dedekind's reading of the problem situation was better than his own initial one. However, he also recognized possible difficulties lurking in the background of the proposed dimensional invariance theorem. In fact, the subsequent history of the search for a proof of invariance of dimension has shown that the theorem was far easier to state than to prove. Nevertheless, with hindsight one can see that Cantor and Dedekind pinpointed the sensitive spot in the problematic notion of dimension.

The critical discussion in which Cantor and Dedekind engaged concerning the dimension problem was unquestionably of a high standard. They discerned the problems involved in the situation and immediately sought to explain the difficulties. Their explanation of the counterintuitive result was very perceptive. In this way Dedekind was led to formulate succinctly the next problem to be tackled after Cantor's important discovery. The Cantor–Dedekind discussion appears as paradigmatic of a good critical discussion of a mathematical or scientific problem.

Chapter 7
Early Efforts to Prove Invariance of Dimension, 1878–1879

7.1 Introduction

The publication of Cantor's 'Beitrag' ('Contribution') towards the beginning of 1878 immediately caused a flurry of mathematical activity. The objective at this point was to save the concept of dimension. The counterintuitive correspondence between geometrical objects of different dimensions had to be explained. During the months July to October 1878, when Cantor's paper had barely left the presses, five mathematicians attempted to demonstrate the invariance of dimension through a consideration of continuity—just as Dedekind had suggested in his letters to Cantor. These were Jacob Lüroth, Johannes Thomae, Enno Jürgens, Eugen Netto, as well as Cantor himself. Then during 1879 both Jürgens and Cantor came back to consider the problem again. Jürgens published a much fuller account of his solution in a separatum, and Cantor gave an entirely new and general approach to dimensional invariance after discussing it in correspondence with Dedekind. These early efforts towards showing dimensional invariance of 1878–79 were only partially successful. Lüroth and Jürgens, and also Cantor in his first work of 1878 only aimed at demonstrating invariance for low dimension numbers. The proofs, though interesting, are rather complex. Thomae, Netto, and Cantor all tried to prove dimensional invariance generally, but subsequent criticism has revealed flaws in their proofs. Without doubt the greatest difficulty, which all these mathematicians came up against, was the primitive state of topology. Topology, at least the part that is now known to be most relevant dealing with the continuity and connectedness of sets or spaces, was hardly developed at the time. The proofs from this period mainly use methods of real analysis and simple geometry, since the topological tools were not yet available. These tools were only created in the later decades of the nineteenth century and the first decades of the twentieth (see my later chapters). One cannot really blame the mathematicians of this period for their lack of full success. Yet it is worth while examining this early work to see how these mathematicians

D. M. Johnson, *Theories of Dimension from Ancient to Modern Times*, Sources and Studies in the History of Mathematics and Physical Sciences, https://doi.org/10.1007/978-3-032-02682-8_7

struggled with a difficult problem. In the history of mathematics 'failures' or really inadequacies of proof can be interesting.[1]

7.2 Jacob Lüroth

Jacob Lüroth (1844–1910)[2], a young professor at the Technische Hochschule in Karlsruhe, was the first off the mark in 1878 to make a public contribution towards solving the dimensional invariance problem. He presented a paper to the Physikalisch-medizinische Sozietät (Physical-Medical Society) in Erlangen as early as 8 July 1878 [Lüroth, 1878]. In his brief note there are proofs of dimensional invariance for two special cases only. Later in his career he returned to the problem and extended his results. These later contributions date from 1899 and 1906 [Lüroth, 1899] [Lüroth, 1906]. In all of his papers Lüroth's procedure is to assume that there is a one-one (injective) continuous correspondence between coordinate or number manifolds M_m and M_n, with $n < m$, and then to show that this assumption leads to a contradiction by deducing that the correspondence cannot be one-one. In his note of 1878 he does this for the cases $n = 1$ and $n = 2$, while in his note of 1899 he extends his results to $n = 3$. Finally in his long paper of 1906, which appeared in *Mathematische Annalen*, he collected all his results and presented them in a more complete form [Lüroth, 1906].

The following summarizes some of the details of Lüroth's first note 'Ueber gegenseitig eindeutige und stetige Abbildung von Mannigfaltigkeiten verschiedener Dimensionen aufeinander' ('Concerning one-one and continuous mappings of manifolds of differing dimensions onto one another in both directions') [Lüroth, 1878]. Letting $x_1, x_2, \ldots, x_m$ denote the coordinates of points x of the m-dimensional manifold M_m and $y_1, y_2, \ldots, y_n$ denote the coordinates of points y of the n-dimensional manifold M_n, with $n < m$, assume that there is a one-one correspondence between points x and y and that the correspondence is continuous. Think of each y coordinate, y_i, as a function of the coordinates of x.

The proof for the case $n = 1$ and $m \geq 2$ is easy. First, let the coordinates $x_3, \ldots, x_m$ in M_m all be held constant, so that one considers only a plane section of M_m. Moreover, let $x_1 = r \cos \theta$ and $x_2 = r \sin \theta$, where r is small, so that one just considers a little circle in this plane section. Let A and B be any two points on this circle, for which y_1 has the values a and b, respectively. Since y_1 is a continuous function of points on the circle, y_1 must take the value $\frac{1}{2}(a + b)$ at least once on each of the two arcs from A to B according to the intermediate value theorem and, hence, it must take this value at least twice on the circle. But this result contradicts the supposition that the function y_1 is one-one (or injective). According to Lüroth, Cantor discovered this same simple proof and told him of this fact in a letter [Lüroth, 1878, page 191].

---

[1] The material covered in this chapter has also been dealt with by Dauben in [Dauben, 1975].

[2] For biographical material on Lüroth cf. [Brill & Noether, 1911] and [Voss, 1911].

The mathematicians trying to prove dimensional invariance at this time commonly used the intermediate value theorem as an important element in their proofs. It is possible to formulate the theorem (a theorem that goes back at least to Bolzano [Bolzano, 1817b] and Cauchy [Cauchy, 1821, pp. 460, 463]) in the following way:

> Let $y = f(x)$ be a continuous real function defined on a closed interval of $\mathbb{R}$ (the real numbers). If the function takes two values $f(x_1) \neq f(x_2)$ at two points x_1, x_2, with $x_1 < x_2$, in the interval, then for any number c between $f(x_1)$ and $f(x_2)$ $(f(x_1) < c < f(x_2))$ there exists x_3, such that $x_1 < x_3 < x_2$ and $f(x_3) = c$.

In the argument given above Lüroth used the theorem in a more general version for a continuous function defined on the arcs of a circle. Others employed versions of the theorem covering continuous functions defined on curves or arcs of curves. In general, the mathematicians of this time recognized—perhaps somewhat vaguely—the topological significance of this theorem. One now sees the intermediate value theorem in a topological context as, for example, Munkres states it in the following way [Munkres, 2000, page 154]:

> **Theorem . . . Intermediate value theorem.** Let $f : X \to Y$ be a continuous map, where *X* is a connected space and *Y* is an ordered set in the order topology. If *a* and *b* are two points of *X* and if *r* is a point of *Y* lying between $f(a)$ and $f(b)$, then there exists a point *c* of *X* such that $f(c) = r$.

Mathematicians around 1880 did not possess anything as general as this theorem. However, they chose the special case, the intermediate value theorem, as the best tool available for attacking the problem of dimensional invariance.[3]

Returning to Lüroth's note of 1878, there is a more elaborate proof for the case $n = 2$ and $m \geq 3$, and he devotes the rest of his note to it [Lüroth, 1878, pp. 191–195]. The bare bones of the proof are as follows. First of all, consider a 3-dimensional section of M_m by holding the coordinates $x_4, x_5, \ldots, x_m$ constant. Consider two points of this section, A and B, for which y_1 has different values, a_1 and b_1, respectively, and assuming $a_1 > b_1$. Lüroth considers the sphere with diameter AB and thinks of it as a globe representing the surface of the earth with A as the North pole and B as the South pole. Under this representation one regards y_1 and y_2 as functions of position on this globe. The objective of the proof is to show that there are two points on the globe for which y_1, y_2 take the same pair of values. Hence, the correspondence is not one-one (injective).

Because of the continuity of the function y_1 there exist circles around the poles A and B, such that the variation of the function is less than $\delta_1 = \frac{1}{4}(a_1 - b_l)$ within these circles; that is, in the circle around A $y_1 > a_1 - \delta_1$ and in the circle around B $y_1 < b_1 + \delta_1$. Thinking of the sphere as a geographical globe, it is possible to give it lines of latitude and longitude so as to divide it up into $2q^2$ surface pieces.[4]

---

[3] Heinz Hopf wrote an excellent paper [Hopf, 1953] on the deeper topological significance of the intermediate value theorem.

[4] To make this subdivision of the globe one needs $q - 1$ parallel circles of latitude (rather than q circles as given in [Lüroth, 1878]) and $2q$ equidistant meridians of longitude.

If one takes q equal to some minimal p, then the two sets of triangular pieces close to the North and South poles, covering two circular regions, respectively, fall within the circles around the poles. Beginning from the circular region covering the North pole, one can construct a surface over the pole for which $y_1 \geq c_1 = \frac{1}{2}(a_1 + b_1)$ by adding little square-like geographical regions with this property, so long as they are connected on their sides with regions already added. The focus is on the southernmost boundary of this North-pole-covering surface: it is a connected path, having no self-intersections and consisting of pieces of meridians and parallel circles, and it goes around the North pole, touching every meridian as it passes.

One can carry out the construction just described for every q of the form $p, 2p, 4p, \ldots, 2^i p, \ldots$, thereby deriving North-pole-covering surfaces $R_0, R_1, R_2, \ldots$ with southernmost boundary curves $G_0, G_1, G_2, \ldots$. Since each surface R_i includes its predecessors, the boundary curves G_i in general approach the south or at least do not go north. Hence, one can always find a boundary curve such that the points of it and all succeeding curves in the sequence are less than a prescribed small distance from points for which $y_1 < c_1$.

Similarly, one can construct connected South-pole-covering surfaces and boundary curves F_i around the South pole for every q with form $2^i p$. In constructing these surfaces covering the South pole one joins little square regions on the sides of previously added regions under the conditions that none of these touch any of the curves G_k and that each of the resulting boundary curves F_i does not intersect itself. With increasing index the curves F_i proceed northwards, or at least not southwards, and their points come as close as one pleases to points of surfaces R_k for sufficiently large k. Yet the curves F_i and G_k do not intersect, so that between each pair of these there is a ring or 'stream' around the globe whose 'banks' do not cut one another.

If for each F_i one considers its northernmost point lying on the 0° meridian, then with increasing i these points proceed north and approach a limit position C. There is a similar limit position D on the 180° meridian. Lüroth claims without proof that $y_1 = c_1$ for these points. Hence, there are points on 'opposite sides' of the globe with the same y_1 value c_1. Supposing that the function y_2 has value a_2 at C and value b_2 at D and setting $c_2 = \frac{1}{2}(a_2 + b_2)$, Lüroth gives in the remainder of his paper a lengthy proof that there are two distinct points within the rings with the same y_1, y_2 values, c_1, c_2. Consequently, he concludes [Lüroth, 1878, page 195]:

> *... if on a sphere two single-valued and continuous functions are given, then there are always values which are taken simultaneously by both functions in two points of the sphere ....* [Ch7, 1]

Hence, one must conclude that a one-one continuous correspondence between a manifold $M_m, m > 2$, and a manifold M_2 is not possible.

Lüroth's proof with its strong geometrical flavor is appealing. The tools of real analysis are used carefully to prove the existence of the two different points on the sphere with the same pair of values.[5]

---

[5] By later standards of rigor some aspects of the proof are probably open to criticism.

When Lüroth returned to the problem of dimensional invariance in his note of 1899 with a proof for the case $n = 3$ and $m \geq 4$ [Lüroth, 1899], he generalized the basic idea of his earlier note. Additionally, in the new case he employed Gauss' integral for the linking of two curves in space [Gauss, 1867, page 605], although he did not use this important topological idea in any exceptional way.[6] In general he found it extremely difficult to extend his work to higher-dimensional cases and as a result his 1889 proof is much more complex.

In a sense Lüroth's *Mathematische Annalen* paper of 1906 [Lüroth, 1906], in which he set out proofs in full for all the cases he had considered earlier, is an admission of defeat. He was convinced that his basic idea for a proof could be extended to a general demonstration of dimensional invariance, but he could not see how to make the extension. His 'geometrical intuition' did not afford him much insight into the general case.

7.3 Johannes Thomae

At virtually the same time as Lüroth was first proving special cases of dimensional invariance, Johannes Thomae (1840–1921),[7] a prolific writer of mathematical books and papers and then *ordentlicher Professor* (*ordinary Professor*) at Freiburg, tried to prove the invariance in full generality. He presented a note to the Königliche Gesellschaft der Wissenschaften (Royal Society of Sciences) at Göttingen in August 1878 [Thomae, 1878], just a month after Lüroth's first note. His general proof is based on an assumption from analysis situs 'against whose general validity there can be no real objections'[8] [Thomae, 1878, 466–467]:

> A connected continuous manifold M_n of n dimensions cannot be divided into separate pieces by one or several manifolds of $n-2$ or fewer dimensions $(M_\nu, M'_{\nu'}, M''_{\nu''}, \ldots; \nu, \nu', \nu'' \cdots \leq n-2)$.
>
> However, it must be assumed that the number of manifolds $M_\nu, M'_{\nu'}, M''_{\nu''}, \ldots$ in each small region of a continuous manifold of $n-1$ dimensions is not countably infinite.[Ch7, 2]

Briefly, Thomae's proof runs as follows. One supposes that there is a one-one (injective) continuous correspondence between manifolds M_m (with points x having coordinates $x_1, x_2, \ldots, x_m$) and M_n (with points y having coordinates $y_1, y_2, \ldots, y_n$), $m > n$. One considers just one coordinate y_1 as a continuous function of $x_1, x_2, \ldots, x_m$ for a bounded region of M_m. If A and B are points of this region

[6] Gauss' linking integral leads to the concept of linking number, a numerical invariant that describes the linking of two closed curves in three-dimensional space. Intuitively, the linking number represents the number of times that each curve winds around the other. Brouwer used the 'looping coefficient.' These notions have an important place in modern topology.

[7] For biographical material on Thomae cf. [Liebmann, 1921].

[8] In the original German: 'deren allgemeiner Giltigkeit keine erheblichen Bedenken entgegen stehen dürften.'

giving maximum and minimum values a and b to y_1, then, by a generalization of the intermediate value theorem, which Thomae assumes as proved, on every curve connecting A and B there is a point for which $y_1 = c$, c being an intermediate value between a and b. In view of his main topological assumption he concludes that the infinite set of points for which $y_1 = c$ must form a continuous (i.e., connected) manifold of $n-1$ dimensions in at least one place. Hence, a continuous function y_1 of a manifold M_m, of m dimensions takes a certain value along a continuous manifold M_{m-1} of $m-1$ dimensions. Similarly a continuous function y_2 of a manifold of $m-1$ dimensions takes some fixed value along a manifold M_{m-2} of $m-2$ dimensions, and so on. In the end the n functions $y_1, y_2, \ldots, y_n$ of points in some region of M_m take at least one system of values in a continuous manifold of $m-n$ dimensions. Thus, there is a fixed y in M_n corresponding to infinitely many x points in M_m, and as a result the correspondence is certainly not one-one.

Thomae must have derived his fundamental topological assumption from a fragment on analysis situs by Riemann [Riemann, 1953, page 481] [Riemann, 1990, page 513] [Riemann, 2004, page 477] (cf. [Cantor & Dedekind, 1937, page 44]). Riemann's separation property for dimension is indeed interesting. Something like it figures in the later history of dimension theory beginning with Poincaré and Brouwer.[9] However, Thomae's contemporaries were quick to point out that the separation property was no less in need of proof than the invariance of dimension. At a session of the 51st Versammlung Deutscher Naturforscher und Aerzte (Meeting of German Natural Scientific Researchers and Doctors) held in Cassel in September 1878, Lüroth rightly criticized Thomae's 'proof' of dimensional invariance, declaring that the separation property and invariance were on the same footing with respect to their importance and difficulty of proof (see Lüroth's statement on this matter recorded in [Jürgens, 1878, pp. 139–140]). Thomae's 'proof' was a non-starter![10] Still, in defense of Thomae it is significant that he chose to put the problem of dimensional invariance squarely in the domain of analysis situs.

7.4 Enno Jürgens

At this same session (51st) of the Versammlung Deutscher Naturforscher und Aerzte (Meeting of German Natural Scientific Researchers and Doctors), held on 13 September 1878, Enno Jürgens (1849–1907)[11] sketched an alternative proof for the 2-dimensional case of invariance [Jürgens, 1878]. A year later he published a full version of his proof. Jürgens was a minor figure in nineteenth-century mathematics. During 1873 and 1874 he had studied at Berlin, where he developed an interest in

[9] A version of it is provable in modern dimension theories; e.g., cf. [Hurewicz & Wallman, 1948, page 48].

[10] Lüroth also criticized Thomae's assertion that the point set for which $y_1 = c$ must form a *connected* $(n-1)$-dimensional manifold in at least one place. Cf. also [Cantor & Dedekind, 1937, page 44].

[11] On Jürgens' life cf. [Krause, 1908].

function theory through Weierstrass. The latter advised him to go to Halle University for his Habilitation, where he remained from 1875 until 1883. Halle was, of course, Cantor's university, so it is not surprising that Jürgens became interested in the problem of dimensional invariance. The end product of his investigations was a proof whose method is especially interesting in the light of later developments. Unfortunately, he chose to publish the full version of his alternative proof in a separatum, *Allgemeine Sätze über Systeme von zwei eindeutigen und stetigen reellen Functionen von zwei reellen Veränderlichen* (*General Theorems concerning Systems of two single-valued and continuous real Functions of two real Variables*) [Jürgens, 1879]. Due to its separate publication it probably did not receive wide circulation.

The most significant feature of Jürgens' proof of the 2-dimensional case of dimensional invariance is the fact that he laid its foundation on a fairly rigorous proof of the invariance of domain for the plane. He was the first mathematician to discern the close connection between invariance of domain and invariance of dimension—an important observation that gave him a better insight into the problem situation than his contemporaries possessed. Jürgens' planar domain invariance theorem is a little more general than usual [Jürgens, 1879, page 3]:

> *If two independent real variables x_1 and x_2, conceived as rectangular coordinates in the plane, range over all points in the interior of and on a circle, and if two other real variables y_1 and y_2 are single-valued and continuous functions of them, whereby the same value pair y_1, y_2 belongs to a finite number of value pairs x_1, x_2, then, when the variables y_1 and y_2 are thought of as rectangular coordinates in a second plane, the part of this latter plane formed by the points y_1, y_2 contains a twofold extended piece of the plane, say the entire surface of a circle.*[Ch7, 3]

The usual domain invariance theorem is a corollary deducible from this theorem.

Jürgens' demonstration of the theorem is complicated but, I think, adequate. The proof is based on Weierstrassian methods of real analysis. Because of the importance of domain invariance in the history of dimension theory and topology generally, it is worth while taking a closer look at the proof [Jürgens, 1879, pp. 3–9].

Let a_1, a_2 denote the center of the given circular region in the x_1, x_2 plane and let A_1, A_2 be the corresponding y_1, y_2 point. Because the point A_1, A_2 corresponds to only a finite number of x_1, x_2 points, there must exist a circle, say with equation

$$(x_1 - a_1)^2 + (x_2 - a_2)^2 = r^2, \tag{7.1}$$

such that none of the points within the circle is mapped to the point A_1, A_2. Let ρ be the minimum value that the expression

$$+\sqrt{(y_1 - A_1)^2 + (y_2 - A_2)^2} \tag{7.2}$$

takes for points on circle (7.1). If R is a positive number $< \rho$, then every radius emanating from a_1, a_2 to a point on the circumference of circle (7.1) has a point that is mapped by y_1, y_2 to a point that is at a distance R from A_1, A_2, a fact guaranteed by the intermediate value theorem. Thus one finds that on the circle

$$(\eta_1 - A_1)^2 + (\eta_2 - A_2)^2 = R^2 \tag{7.3}$$

(where η_1, η_2 are the coordinates for the second plane) there are infinitely many points that are images of points of the surface within circle (7.1). Jürgens' first objective is to show that these y_1, y_2 points fill up an entire connected arc of circle (7.3).

In order to achieve this objective he borrows Lüroth's method of cartographical subdivision and divides the surface of circle (7.1) into $4q^2$ little elements.[12] One is only interested in those elements all of whose points are mapped to y_1, y_2 points that lie within circle (7.3). Provided q is sufficiently large, all the little elements around a_1, a_2 will be of this type. As Lüroth had done [Lüroth, 1878], one constructs a surface within circle (7.1) by successively adding elements of the appropriate type to those around a_1, a_2, provided they are adjacent on a side to previously added elements. The result is a surface F all of whose points are mapped to y_1, y_2 points within circle (7.3). Jürgen is concerned with the outermost boundary C of F; thus, he takes the trouble to prove that the curve C traces a continuous circuit around the point a_1, a_2. C has the property that its points are mapped to points y_1, y_2 that are less than distance R from A_1, A_2, but within a well-defined small distance of its points[13] there are points with images at a distance $\geq R$ from A_1, A_2.

For an appropriate sequence of increasing q's one derives a sequence of curves $C_1, C_2, \ldots$, such that each curve encloses its predecessors unless they coincide. Any limit point of the curves C_i (i.e., a point x_1, x_2 that comes arbitrarily close to points of the curves C_i) falls within circle (7.1) and for it the expression (7.2) has the value R. The objective is to show that the y_1, y_2 images of these limit points form a whole connected arc on circle (7.3). Jürgens achieves this first objective by giving a rigorous limit argument based on the y_1, y_2 images of the curves C_i and the intermediate value theorem.

In the first part of his proof he only required that the radius R of circle (7.3) be less than ρ. Consequently, he really showed that the image points y_1, y_2 fill up a connected arc on every circle with center A_1, A_2 and radius $< \rho$. Jürgens' second objective is to show that these arcs fill up an entire piece of surface. This he accomplishes through a modified limit argument, hence, completing his demonstration of the invariance theorem.

Jürgens' argument, though of a high standard of rigor, is not very attractive. Its complicated applications of methods of Weierstrassian analysis do not reveal in a clear way why the theorem should work. Moreover, it is difficult to imagine extending the argument to higher dimensions. From the present vantage point of several more generations of mathematical experience, the proof clearly lacks topological insight. Of course, one cannot blame Jürgen for not using mathematical tools that did not exist at the time. One must give him credit for his skillful handling of the tools

[12] Jürgens uses $q - 1$ equidistant concentric circles and $4q$ radii at equal angles to obtain this subdivision.

[13] distance $(r/q)(\pi/2 + 1)$.

he did have at his disposal. A significant feature of his proof is its dependence upon the intermediate value theorem, a theorem with topological import concerning connectedness. Lüroth and Thomae also employed this theorem in their arguments. Thus, all these mathematicians recognized an importance in this quasi-topological result.

Jürgens expresses his result in an alternative way in his separatum [Jürgens, 1879, page 9]: the system of image points y_1, y_2 of the circular region in the first plane must contain an interior point in the second plane. However, this does not mean that interior points of the region in the first plane are necessarily mapped to interior points in the second. A simple example—a circle folded along one of its diameters—shows that this need not be the case. In this example some interior points become boundary points. Yet Jürgens' theorem does guarantee that there are interior points arbitrarily close to images of interior points. Motivated by these considerations, he strengthens the conditions of his theorem in order to prove the more usual domain invariance theorem for the plane as a corollary [Jürgens, 1879, page 11]:

> *If two single-valued and continuous real functions of two real variables do not take the same pair of values more than once, then to an interior point of the domain of the independent variables there corresponds an interior point in the domain of the dependent variables.*[Ch7, 4]

In other words, a one-one (injective) continuous mapping of a region in one plane to another takes interior points to interior points. Moreover, as Jürgens adds, the inverse mapping must be continuous [Jürgens, 1879, page 11]: so in modern terms one has a topological (i.e., one-one bicontinuous onto) mapping.

It is an easy step to deduce the invariance of dimension theorem for $n = 2$ from the planar domain invariance theorem [Jürgens, 1879, pp. 17–18]. Jürgens expresses the former theorem in the following way [Jürgens, 1879, page 17]:

> *No part of threefold extended space that contains an entire sphere can be mapped onto any part of the plane in a single-valued and continuous way.*[Ch7, 5]

For the proof suppose that one can map a sphere continuously onto a part of the plane and assume that the mapping is one-one (injective).[14] Then an equatorial planar section of the sphere will be mapped onto a part of the plane in such a way that interior points are mapped to interior points. Hence, the center of the sphere, a_1, a_2, a_3, which is an interior point in the equatorial plane, will be mapped to an interior point, B_1, B_2, in the image plane. But then points not in the equatorial plane but close to a_1, a_2, a_3 cannot be mapped to points close to B_1, B_2—contrary to the assumption of continuity. Thus, from this contradiction the invariance of dimension for dimension two follows.[15]

---

[14] The assumption that the mapping is one-one is a slight simplication of Jürgens' argument. He only assumes that the mapping is of the type described in his first theorem, viz., that the mapping is such that each image point corresponds to at most finitely many preimage points.

[15] Jürgens' separatum [Jürgens, 1879, pp. 13–16] contains another interesting application of his main theorem: a proof of the fundamental theorem of algebra. The connection between this theorem and domain invariance is now well known.

7.5 Eugen Netto

Following close on the heels of the contributions of Lüroth, Thomae, and Jürgens, Eugen Netto (1846–1919)[16] completed a general proof of dimensional invariance in October 1878, which was published in *Crelles Journal* at the very end of the year. Netto is best remembered for his contributions to the development of group theory; his textbook of 1882 *Substitutionentheorie und ihre Anwendung auf die Algebra* (*Substitution Theory and Its Application to Algebra*) [Netto, 1882] (in English [Netto, 1892]) is a minor classic expositing the subject. From 1866 until 1870 he had studied at Berlin under the great masters Kummer, Kronecker, and Weierstrass. Weierstrass had been the chief examiner for his dissertation. During the 1870s he was a teacher at the Friedrich-Werder'sches Gymnasium in Berlin. While there he wrote his paper on dimensional invariance, virtually his only contribution to *Mannigfaltigkeitslehre* (*Manifold Theory*).

Netto's 'Beitrag zur Mannigfaltigkeitslehre' ('Contribution to manifold theory'), setting out his attempt to prove dimensional invariance in general, is an important paper for its time [Netto, 1879]. Even though the attempt does not succeed, it reveals interesting facets. Netto put the problem firmly in the realm of topology and in a much better way than Thomae had. Netto's topological concepts are largely borrowed from Weierstrass, who used them in his work in function theory.

Netto's proof is inductive. He starts by examining the three simplest cases [Netto, 1879, pp. 264–265]. First, a one-one correspondence between a 1-dimensional manifold, a line, and a 0-dimensional manifold, which he takes to be a single point, is obviously not possible. So clearly a one-one correspondence between a manifold M_n $(n \geq 1)$ and M_0 is impossible. Second, a correspondence between manifolds M_n $(n \geq 2)$ and M_1 is not possible either. Suppose one can map a 2-dimensional manifold M_2 one-one continuously onto a simple manifold M_1. Consider a simple closed line $\mathfrak{A}$ in M_2 and its image A in M_1. Since A cannot be a point, Netto asserts it must be a certain piece of the line M_1, although it cannot be the entire line. Consider an interior point of A. In order to pass from this point to a point not belonging to A, one must pass over one of the boundary points (Grenzpunkte) of A, say α_1. Let α_1 in M_1 correspond to a_1 in M_2. Now in M_2 in order to pass from points of $\mathfrak{A}$ to points not belonging to this figure it is by no means necessary to pass through a_1. But given the continuous correspondence, it should be necessary. Hence, the supposed correspondence between M_1 and M_2 is impossible, and also one between M_1 and M_n $(n > 2)$. Third, consider the possibility of mapping a continuous threefold manifold M_3 one-one continuously onto a continuous twofold manifold M_2. Choose a surface $\mathfrak{A}$ in M_3, which will be mapped to a surface A in M_2. Surface A will be bounded by one or more curves $\alpha_1, \alpha_2, \ldots$, such that it is impossible to go from a point of A to a point not in A without passing through one of these curves. Yet the corresponding curves $a_1, a_2, \ldots$ in $\mathfrak{A}$ do not have this bounding property. It is always possible to go from a point of a surface in space immediately to a point not on the surface. Hence, a correspondence between M_2 and $M_n (n \geq 3)$ is not possible.

[16] On Netto's life cf. [Biermann, 1990].

The modern topologist will likely find the proofs above unsatisfactory.[17] They are not so much proofs, but more in the way of useful intuitive observations. Netto accepts ideas and facts related to the Jordan curve theorem (1887) and its analogues for other dimensions, which he does not develop as a basis for his ideas; those underlying topological notions came later. In his defense one should recognize his good intuitive understanding of the topological situation, although he expresses no need to prove his most significant assumptions about separation. Topology was indeed in a very primitive state when Netto wrote his paper. He could only indicate the topological problem situation without really proving anything as one might expect today.

In his paper Netto proceeds to the inductive step of his general invariance proof, trying to demonstrate the impossibility of a one-one continuous mapping from an n-dimensional manifold M_n onto an $(n-1)$-dimensional manifold M_{n-1}. His demonstration continues the line of thought of his first three cases, which he connects with the following principle [Netto, 1879, page 265]:

> "In a manifold of νth degree [i.e., dimension] every figure of νth degree is bounded by another of lesser degree; in a manifold of $(\nu+1)^{st}$ degree every figure of νth degree coincides with its boundary." [Ch7, 6]

In tackling the general case he more clearly recognizes the need to define rigorously the geometrico-topological concepts to be used. Hence, he adopts Weierstrassian neighborhood/proximity topology, although he does not credit Weierstrass explicitly.[18] These topological ideas were the best tools that Netto had available. His use of them probably constitutes the most important feature of his entire paper. Thus, he gives reasonable definitions of interior point (ein Punkt im Innern) and boundary point (ein Punkt auf der Grenze) and takes connectedness in the Weierstrassian sense that a figure will be connected if every pair of points of it can be joined by a line belonging entirely to the figure (path connectedness). Unfortunately, these basic tools from general topology were not sufficient for Netto to construct a cogent inductive step for his proof.

Jürgens was very quick to detect an important error in Netto's general invariance proof. He included a very astute criticism of the proof in his separatum of 1879 [Jürgens, 1879, page 17], which appeared a few months after Netto's paper. Having *proved* the 2-dimensional domain invariance theorem himself, Jürgens realized quite correctly that Netto merely *assumed* the general invariance of domain theorem in his proof. For example, in his third special case he took it for granted that the surface in M_3 will be mapped to a surface in M_2 *that possesses interior points*. Yet only a proof of something like Jürgens' planar domain invariance theorem will guarantee

[17] However, Sagan thinks that Nette provided a correct proof [Sagan, 2012, pp. 5–6].

[18] Cf. Weierstrass' works in selected places such as the following: [Weierstrass, 1894, pp. 70–71, 83–84] [Weierstrass, 1880, page 721] = [Weierstrass, 1895, page 203] [Weierstrass, 1927, pp. 56—58]; also [Pincherle & Weierstrass, 1880, pp. 234–237]. Weierstrass certainly used such ideas in his lectures around 1879. However, in the case of connectedness Netto's definition appears to be slightly more general than that of Weierstrass.

that this is the case. Consequently, Jürgens was fully justified in further condemning Netto's proof [Jürgens, 1879, page 17]:

> ... the induction step is then only correct if the impossibility of a single-valued and continuous mapping [of the type Jürgens had considered] of a continuous manifold of the nth dimension onto a discontinuous part of a second continuous manifold of the nth dimension, i.e., onto a part without interior points, is also proved.[Ch7, 7]

Jürgens, equipped with a demonstration of a special domain invariance theorem, had a good insight into the close connection between any attempt to prove dimensional invariance in general and the invariance of domain.[19]

7.6 Georg Cantor and Richard Dedekind

During the time that Lüroth, Thomae, Jürgens, and Netto were busily trying to prove dimensional invariance in the last months of 1878, Cantor did not remain inactive over the problem to which his own discovery of the counterintuitive result had given birth. Several times in his correspondence he gently urged Dedekind to prove the conjecture on the invariance of dimension that Dedekind himself had put forward. However, Dedekind did not seem to have the time or inclination to think about a proof ([Cantor & Dedekind, 1937, pp. 37, 40, 41, 42], [Grattan-Guiness, 1974, page 111]). Cantor also kept abreast of the publications on the subject, citing all that had appeared thus far in a letter to Dedekind of 29 December 1878 [Cantor & Dedekind, 1937, page 43]. He was not entirely satisfied with the published proofs [Cantor & Dedekind, 1937, page 43]:

> ... it seems to me, however, that the situation is still not entirely resolved.[Ch7, 8]

Netto's proof interested him most, undoubtedly because it was the only reasonably good attempt at a general proof. Nevertheless, it did not make him feel that the problem was solved, as he admits in a letter to Dedekind of 5 January 1879 [Grattan-Guiness, 1974, page 114]:

> ... No matter how commendable this penetrating attempt at a proof appears to me, I still cannot banish certain doubts about it and I fear that it is *only* an attempt, which nonetheless will certainly contribute to clarifying the situation.[Ch7, 9]

In view of these lingering doubts he presented his own proof of dimensional invariance to Dedekind in a letter dated 17 January 1879.

---

[19] Jürgens' paper [Jürgens, 1899, pp. 52–53] also contains this criticism.

In his letter Cantor expresses the theorem to be proved in the following terms [Cantor & Dedekind, 1937, page 45]:

> "A continuous [i.e., connected] M_μ and a continuous M_ν, in case $\mu < \nu$, cannot be put into *continuous* correspondence with one another such that to each element of M_μ there belongs *a single* element of M_ν and to each element of M_ν there belongs *one or more* elements of M_μ." Ch7, 10

He claims that he already had a proof of this for over a year, but previously had serious doubts about its validity, since it depends upon a multi-valued correspondence. Only now has he resolved the doubts in his own mind.[20]

Cantor's proof is complex [Cantor & Dedekind, 1937, pp. 44–46]. It proceeds by induction on ν. According to him the theorem is obvious for $\nu = 1$, since, like Netto, he thinks of M_0 as a single point. For the induction step one assumes the case $\nu = n - 1$ and seeks to deduce the case $\nu = n$. To this end, one supposes that there can be a continuous many-one correspondence as described between M_μ and M_n, $\mu < n$. Cantor's objective is to show that this supposition leads to a clash with the intermediate value theorem. As has already been seen, the intermediate value theorem figures in nearly all the dimensional invariance proofs from the years 1878–1879. Cantor's intention is to use the theorem in a modified form, as the 'Jordan–Brouwer separation theorem' for the simple case of the hypersphere. Taking a distance function from the center of the hypersphere relative to its surface, so that points within have a negative distance, while points outside have a positive distance, and points on the sphere have zero distance, one derives the result that a connected figure with points on both the inside and outside must intersect the sphere. This result follows, since a continuous function with a positive and a negative value must have a zero in between.

For the main body of Cantor's proof let a and b be two interior points of M_μ with A and B the corresponding points of M_n. In M_n construct an $(n-1)$-dimensional sphere K_{n-1} around A sufficiently small to exclude B, while in M_μ construct a sphere $K_{\mu-1}$ around a sufficiently small so that it excludes b and so that its image in M_n, a figure $G_{\mu-1}$ (which is $(\mu-1)$-dimensional by the inductive hypothesis), lies entirely within K_{n-1}. The latter condition can be met according to Cantor because of the continuity of the correspondence. Letting ζ be the point (of $G_{\mu-1}$) corresponding to an arbitrary point z of $K_{\mu-1}$, draw a straight line segment from A to ζ and extend it so that it intersects the sphere K_{n-1} in a determined point Z. Carrying out this construction with respect to all points z of $K_{\mu-1}$ furnishes a continuous correspondence between points z of $K_{\mu-1}$ and certain points Z of K_{n-1}. The points Z cannot cover all of K_{n-1}, for otherwise the inductive hypothesis would be contradicted. Hence, there must be a point P of K_{n-1} that is not a Z point. If one draws segment AP, it does not touch the figure $G_{\mu-1}$. Moreover, if one connects P to the point B lying outside K_{n-1} by a curve, then one obtains a connected line APB, which has no point in common with

[20] Cantor's invariance 'theorem' as stated above can be falsified by a counterexample: Peano's space-filling curve. However, Peano's example was not published until eleven years later and by then the problem situation in topology and dimension theory had changed considerably. Cf. Chapter 8.

$G_{\mu-1}$. But this line corresponds to one or more curves in M_μ that run continuously from a to b, and, according to Cantor's intermediate value or separation theorem, these curves must cut the sphere $K_{\mu-1}$. However, it is impossible for the curves to cut $K_{\mu-1}$, since they correspond to line APB, which does not cut $G_{\mu-1}$. Thus by *reductio ad absurdum* the result follows.

Two days after Cantor wrote out his proof (i.e., 19 January) Dedekind responded with two specific objections [Cantor & Dedekind, 1937, pp. 46–48]. First, it is conceivable that a point ζ of $G_{\mu-1}$ corresponding to a point z of $K_{\mu-1}$ is A, since several points of M_μ can correspond to one and the same point of M_n. In this case the image Z would be indeterminate. Dedekind immediately saw a way out of this difficulty if one assumes that the number of points a' in M_μ that correspond to point A in M_n is *finite*. Under this assumption one can make the sphere $K_{\mu-1}$ sufficiently small to exclude the other points a'. However, if the number of corresponding points can be *infinite*, then he could see no exit from the difficulty, and, moreover, he then discerned a second difficulty. According to Cantor the line APB corresponds to one or more continuous lines in M_μ going from a to b. But if B can correspond to infinitely many points b', then he was doubtful about the existence of a particular line from a to b that has the image APB. Hence, what concerned Dedekind most was the possibility of an infinite many-one correspondence in Cantor's proof.

Beyond his specific criticisms Dedekind had some interesting programmatic ideas about *Mannigfaltigkeitslehre* (*Manifold theory*) or *Gebietslehre* (*Domain theory*) (his preferred term). In his letter of 19 January 1879 he calls for a more rigorous development of the subject, saying [Cantor & Dedekind, 1937, page 47]:

> ... It would be very meritorious if this entire "domain theory" were developed ab ovo without bringing in geometrical intuition, and in doing that, e.g., the concept of a line proceeding continuously from the point a to the point b within the domain G would have to be defined quite definitely and clearly.[Ch7, 11]

On the face of it, this is a proposal for a program of topology based on arithmetical and analytical principles. In a sense Dedekind's remark foreshadows the subsequent development of point-set topology, which Cantor in large part and Dedekind to some extent initiated (see Chapter 9).

In his letter Dedekind praises Netto's paper, saying Netto's definitions form a good kernel for further developments. At the same time he reveals to Cantor some of his own attempts to elaborate the basic concepts of the theory of domains. He seems to be referring to a little paper, 'Allgemeine Sätze über Räume' ('General theorems concerning spaces'), probably written in the early 1870s but not published until 1931 [Dedekind, 1931, pp. 353–355]. This paper is brief, containing just a few theorems on open sets (Körper) and their boundaries (Begrenzungen). In his letter Dedekind tells Cantor that his motivation for writing such a paper was to shore up the Dirichlet principle for a contemplated edition of Dirichlet's lectures on potential theory. He thought that an analytic theory of domains could be used to avoid

Weierstrauss' devastating critique (referenced without publication in [Weierstrass, 1870], then published in [Weierstrass, 1895, pp. 49–54]).[21]

Cantor partly anticipated the objections that Dedekind included in his letter. This is evidenced by a postcard of 20 January [Cantor & Dedekind, 1937, page 48]. There Cantor suggests that it would be better to think of the correspondence as proceeding from points A and B of M_n to points of M_μ. If we consider just one of the points a corresponding to A and yet all of the points $b, b', b'', \ldots$ corresponding to B, then the sphere $K_{\mu-1}$ around a can be made sufficiently small so that *all points* $b, b', b'', \ldots,$ will fall outside $K_{\mu-1}$.[22] According to Cantor this is possible because of the continuity of the correspondence that prevents the points $b, b', b'', \ldots$ from coming 'infinitely close' to a.[23] Now there can be no doubt that one of the curves corresponding to APB goes from a to at least one of the points $b, b', b'', \ldots,$ thus cutting the sphere $K_{\mu-1}$. Hence, Dedekind's second objection is overcome.

When Cantor heard from Dedekind, he was not immediately sure how to overcome the first objection [Cantor & Dedekind, 1937, page 49]. Yet he was certain that he did not want to limit his theorem to finite many-one correspondences with the consequent loss of generality. Then a little while later he found a way around the first objection. This discovery led him to write out a fuller version of his proof in a paper which was presented to the Königliche Gesellschaft der Wissenschaften (Royal Society of Sciences) at Göttingen on 12 February 1879. Having only recently been elected a Corresponding Member of the society, he felt the society's *Nachrichten* (*Proceedings*) were a good medium for publication. In this paper, 'Über einen Satz aus der Theorie der stetigen Mannigfaltigkeiten' ('Concerning a Theorem from the Theory of Continuous Manifolds') [Cantor, 1879a], his theorem appears in a slightly more general form [Cantor, 1879a, page 131] [Cantor, 1932, page 136]:

> (III) "If one has a correspondence between two continuous domains M_μ and M_ν, such that to each point z of M_μ there corresponds *at most* one point Z of M_ν and to each point Z of M_ν there corresponds *at least* one point z of M_μ, and further if this relation is continuous, such that infinitely small changes of Z correspond to infinitely small changes of z and also, conversely, infinitely small changes of z correspond to infinitely small changes of Z, then $\mu \geq \nu$."
> Ch7, 12

The fact that in this version of the theorem Cantor just assumes that the many-one correspondence from M_μ to M_ν is partial does not affect the proof in any way.[24] His published proof is in essence the same as the one he communicated to Dedekind by letter. It differs only by including attempts to overcome Dedekind's two objections.

---

[21] In the end someone else edited Dirichlet's lectures, viz., P. Grube ([Dirichlet, 1876] [Dirichlet, 1887]). However, Dedekind continued working out his ideas on analytic topology in another paper of about 1892, published much later in his *Gesammelte mathematische Werke* (*Collected mathematical Works*), volume 2 [Dedekind, 1931, pp. 356–370].

[22] One still must insure that $G_{\mu-1}$ corresponding to $K_{\mu-1}$ falls within K_{n-1}.

[23] That is, the continuity prevent a being a limit point of the set of points $b, b', b'', \ldots$.

[24] Note that in this version of the theorem he has now spelled out the requirement that his correspondence be continuous in both directions, i.e., that it be bicontinuous.

The second objection is dealt with in the way suggested above. With regard to the first, he became convinced that he could *ignore* the possibility that points ζ may coincide with A (inasmuch as some of the a's associated with A may be on the sphere $K_{\mu-1}$). One still gets a continuous correspondence between those points of $K_{\mu-1}$ that have corresponding points in M_n *and yet are not mapped to* A and certain points Z' on K_{n-1}. One can still form the line APB in M_n. It corresponds to a continuous domain N in M_μ that runs from a to one or more of the b's associated with B and that has no point in common with $K_{\mu-1}$. Indeed N includes no a points even if some of these happen to be on $K_{\mu-1}$, because the line APB does not return to A after leaving A. Hence, when it came to publication Cantor had no qualms about the validity of his theorem and proof.

In spite of Cantor's convictions, his final proof is still open to criticism. It is flawed. Dedekind put his finger on a sensitive place in the proof when he drew attention to the possibility of an infinite many-one correspondence, especially in his second objection. In this case he wondered whether there would always be a curve leading from a to b corresponding to APB. However, when Cantor's paper appeared no one redeveloped or continued the line of criticism that Dedekind had put to Cantor in their private correspondence. It seems the paper of 1879 attracted little attention. It was not until 1899 that Jürgens published a direct criticism of Cantor's proof in a review of the problem situation concerning dimensional invariance [Jürgens, 1899, page 53]. He refuted the intermediate value theorem upon which the proof is based by providing the following counterexample. Consider the mapping from points x in the open interval $(1, 2)$ to all irrational numbers y for which $\frac{3}{8}x < y < \frac{5}{8}x$. In an intuitive sense this function is continuous, although as Jürgens pointed out there are difficulties with defining continuity for multi-valued functions.[25] Yet the intermediate value theorem does not hold, because y does not take the rational values. Furthermore, if one considers the images of the interval $(1, 2)$, there is no continuous path stretching from an image point of an x close to 1 to an image point of an x close to 2. However, Cantor required a similar path in his proof, since he relied on the existence of a continuous path from a to one of the b's corresponding to the line APB. Jürgens' counterexample shows that such a path need not exist. So Cantor's proof fails.

7.7 Concluding Remarks

From the present day vantage point the attempted proofs of dimensional invariance put forward during 1878 and 1879 do not appear satisfactory. Having the great benefit of hindsight, one can see that the mathematicians of the period were struggling to subdue a difficult problem with inadequate means. They did not have fully developed topological ideas and methods available to them. Their best means of attack derived from analysis. In fact, with only these means they handled their problem

[25] Cantor did not reckon with these difficulties explicitly.

very skillfully. It must be recognized that the standard of rigor and critical argument among them was high, given the constrained nature of their methods. Of course, some topological ideas were implicit in the Weierstrassian analysis of their day. For example, there were reasonable definitions of interior point, boundary point, and connectedness. They could and did use the intermediate value theorem together with its implications with some topological understanding. One can virtually see that the mathematicians developing the proofs around 1878–1879 frequently felt the need to go beyond the slender means at their disposal. The general proofs of Thomae and Netto give evidence of this fact. However, their contemporaries quickly sensed the lack of rigor in these proofs. They soon realized that much more was required to make them cogent.

In my judgment the best proof of all from this period is Jürgens'. His linking of a demonstration of the planar domain invariance theorem with dimensional invariance gave him a significant insight into the problem situation. However, not being a first-class mathematician, he was not able to exploit fully his insight to go on to develop more general results. He was no Brouwer. Perhaps if he had chosen a better way of publishing his research, his work might have received more attention. Then others might have taken it up and developed it further.

It is clear that around 1880 adequate topological tools in mathematics were badly needed. The attempts to prove invariance provided some motivation for these to be fashioned. Other parts of mathematics cried out for their development. Dedekind's letter to Cantor of 19 January 1879 (quoted in Section 7.6) expresses a definite need for a full program of analytic or point-set topology. Hence, before considering twentieth-century efforts towards solving the dimensional invariance problem, it is necessary for me to take account of the earlier development of point-set topology in the next Chapter 8.

Chapter 8
Early Development of Point-Set Theory and Point-Set Topology

8.1 Introduction

By the 1880s most mathematicians thought that Cantor's paradox about dimension had been dealt with or explained away by either Netto's or Cantor's proof of dimensional invariance. Jürgens was virtually the only one to voice dissent, but his acute criticism of Netto's proof went largely unnoticed. The consensus apparently was that the invariance problem for dimension had been solved. For example, Felix Klein, when drawing attention to Cantor's paradoxical result in his important little tract of 1882, *Ueber Riemanns Theorie der algebraischen Funktionen und ihrer Integrale* (*On Riemann's Theory of Algebraic Functions and Their Integrals*) [Klein, 1882, page 26] [Klein, 1923, page 527], appears not to be overly concerned about the difficulties that it presented.[1] Walter von Dyck (1856–1934) in his first 'Beiträge zur Analysis situs' ('Contributions to Analysis Situs'), published in 1888, pointed to the result of Cantor, but then he immediately held up Lüroth's special proof and Cantor's general proof of invariance as guarantees for the soundness of the dimension concept [Dyck, 1888, pp. 457–458]. In a short paper published in 1887 Gino Loria (1862–1939) described the situation concerning dimension and its invariance for an Italian readership. In the paper he summarizes Cantor's correspondence result between the unit interval and the unit cube of n dimensions and then repeats Netto's general invariance proof [Loria, 1887]. Hence, among mathematicians the generally held opinion during the last two decades of the nineteenth century was that the invariance of dimension had been proved. This opinion is echoed in Arthur Schoenflies' *Encyklopädie* article, 'Mengenlehre,' submitted in November 1898 [Schoenflies, 1898, page 187]. However, later mathematical developments show that the situation with regard to dimensional invariance of the late nineteenth century was not as satisfactory as it then seemed.

[1] Klein in this little work examines Riemann's function theory on the intuitive geometrical level. The basic correctness of the dimension concept and its fundamentals should be relevant in this case.

D. M. Johnson, *Theories of Dimension from Ancient to Modern Times*, Sources and Studies in the History of Mathematics and Physical Sciences, https://doi.org/10.1007/978-3-032-02682-8_8

It was a revolution that changed attitudes to the dimension problem—a revolution in topology brought about by the rise of the theory of sets, especially the theory of sets of points. Georg Cantor is a primary creator of point-set theory (Punktmannigfaltigkeitslehre), though others, especially Richard Dedekind, had significant roles in its initial development. While Cantor did not contribute much to topology directly, his development of point-set theory added an entirely new perspective to topological thinking. Today dimension theory is regarded primarily as a branch of set-theoretic or point-set topology, and so the development of point-set theoryx eventually had the most profound effect on the subject. In this chapter I shall go beyond the confines of the narrow history of dimension theory and take account of the rise of set theory and point-set theory.

8.2 Cantor's Development of Point-Set Theory

Cantor published his most important investigations into the theory of sets of points in a series of six papers entitled 'Ueber unendliche, lineare Punktmannichfaltigkeiten' ('On infinite, linear Point Manifolds [i.e., Sets]'), which appeared between 1879 and 1884 [Cantor, 1879b] [Cantor, 1880] [Cantor, 1882] [Cantor, 1883a] [Cantor, 1883b] (also published as a separate tract [Cantor, 1883c]) [Cantor, 1884b] [Cantor, 1932, pp. 139–246]. These brilliant papers constitute the 'quintessence of Cantor's lifework' as Ernst Zermelo, set theorist and editor of Cantor's collected papers, wrote [Cantor, 1932, page 246].[2] Much of the material concerns the founding of the theory of transfinite cardinals and ordinals. However, Cantor's main object of study is linear point-sets and point-sets in the n-dimensional arithmetic continuum (Euclidean n-space) and their properties. The resulting theory became the basis for point-set topology. The fundamental concept is that of limit point (Grenzpunkt; or in later terminology, Häufungspunkt, accumulation point): a point p is a limit point of a set P if every neighborhood (Umgebung) of the point, however small, contains points of the set [Cantor, 1882, pp. 113–114] [Cantor, 1932, page 149] (cf. the earlier paper [Cantor, 1872, page 129] [Cantor, 1932, pp. 97–98]). The fundamental theorem is the so-called Bolzano–Weierstrass theorem, used by Weierstrass in his lectures and known to Cantor from these, which states [Cantor, 1882, page 114] [Cantor, 1932, page 149] that every infinite set of points in a bounded region of n-space possesses at least one limit point.[3] From this fundamental concept and theorem flow all of Cantor's deep point-set-theoretic concepts. Among these are the notions of a derived set (abgeleitete Punktmenge, Ableitung) [Cantor, 1872, page 129] [Cantor, 1879b, page 1] [Cantor, 1932, pp. 98, 139–140], an everywhere dense set in a given interval (überall-dichte in einem gegebenen Intervalle Punktmenge) [Cantor, 1879b, page 3] [Cantor, 1932, pp. 140–141], an isolated point set (isolierte Punktmenge) [Cantor, 1883a, page 51] [Cantor, 1932, page 158], and a nowhere dense set (in keinem

[2] In German, 'die Quintessenz des Cantorschen Lebenswerkes'.

[3] Cf. Bolzano [Bolzano, 1817b] and Weierstrass [Weierstrass, 1927, page 56].

Intervalle überall-dichte Punktmenge) [Cantor, 1883a, page 55] [Cantor, 1932, page 161]. By introducing these notions Cantor opened up a whole new world of ideas for the emerging field of topology.

As an example of his procedure, consider how Cantor investigates the concept of continuum in the fifth in his series of six papers, which he also entitled separately as *Grundlagen einer allgemeinen Mannigfahigkeitslehre* (*Foundations of a General Theory of Manifolds*) [Cantor, 1883b] [Cantor, 1883c]. Cantor was very proud of his definition of continuum [Cantor, 1883b, §10, pp. 572–576] [Cantor, 1883c, §10] [Cantor, 1932, §10, pp. 190–194] as he felt it solved an ancient problem of both mathematical and philosophical importance.[4] In the *Grundlagen* (*Foundations*) he gives his solution within the scope of 'the n-dimensional planar arithmetic space G_n' ('der n-dimensionale ebene arithmetische Raum,' i.e., n-dimensional Euclidean space), consisting of the set of all value systems (points)

$$(x_1|x_2|\dots|x_n)$$

of real numbers x_i (according to the definition of Cantor) provided with the usual Euclidean metric or measure of distance between points. The question then is: When is a set of points of G_n a continuum or a continuous set?

In order to answer this question, Cantor first introduces the concept of a perfect set (perfekte Punktmenge): a perfect set is one which is identical with the set of all its limit points, i.e., its first derived set,

$$S \equiv S^{(1)}$$

and, hence, is identical with its derived sets of all orders. Very perceptively he realizes that a perfect set need not be dense in any interval and in an annotation in an endnote [Cantor, 1883b, page 590] [Cantor, 1932, page 207] he gives the classic example of a nowhere dense perfect set, the well-known Cantor ternary set or discontinuum: the set of all real numbers of the form

$$z = \frac{c_1}{3} + \frac{c_2}{3^2} + \dots + \frac{c_\nu}{3^\nu} + \cdots,$$

where the coefficients c_i can only take the values 0 and 2.[5] Cantor then introduces the concept of connectedness (zusammenhängende Punktmenge, connected point set). Previously Weierstrass had used the notion that a set is connected if every pair of points of it can be joined by a polygonal path belonging to the set.[6] However, Cantor wanted a concept within the domain of his new theory of sets. A set T is connected

---

[4] Compare my analysis of Cantor's criticism of Bolzano's notion of continuum in Section 4.6.

[5] There is an unusual background history to Cantor's ternary set. Unknown to Cantor, Henry John Stephen Smith (1826–1883) in 1875 had already defined and examined a ternary set as part of his analysis of the integration of discontinuous functions [Smith, 1875]. See [Fleron, 1994]. Cf. `https://en.wikipedia.org/wiki/Cantor_set` and `https://en.wikipedia.org/wiki/Smith-Volterra-Cantor_set`.

[6] Cf. [Weierstrass, 1895, page 203] [Weierstrass, 1927, page 57]. However, he only defined connectedness in conjunction with his concept of continuum or connected open set.

in Cantor's sense if for every pair of points t, t' of T and every given arbitrarily small number ε it is possible to choose a finite number of points $t_1, t_2, \ldots, t_\nu$ in T such that the distances $\overline{tt_i}$, $\overline{t_1t_2}$, $\overline{t_2t_3}, \ldots,$ $\overline{t_\nu t'}$ are all less than ε. This form of metric connectedness, sometimes called the property of being well-chained, remained a part of point-set theory for a long time after Cantor, until Riesz, Lennes, and Hausdorff put forward the 'modern' concept of connectedness in the first years of the twentieth century.

Finally, Cantor asserts that a point set of G_n will be a continuum (Kontinuum) if it is both perfect and connected. In a endnote he adds [Cantor, 1883b, page 590] [Cantor, 1932, page 208] the definition of a semicontinuum as an imperfect connected set of the second number class ($\aleph_1$) with the property that every pair of points can be joined by a complete continuum. He boasts that his conception of continuum is superior to those of Bolzano and Dedekind, although his criticism of the Bolzano notion is founded on a misunderstanding of Bolzano's ideas.[7] In any case, Cantor's new set-theoretic idea of continuum was of major significance in the early development of analytic topology. It constituted a new insight into a traditional informal concept.

In the course of analyzing the continuum concept Cantor asserted a false proposition [Cantor, 1883b, page 575] [Cantor, 1932, page 193], which the Swedish mathematician Ivar Bendixson (1861–1935)[8] swiftly criticized.[9] The revision of the offending proposition led Cantor to define the extremely important notion of a closed set (abgeschlossene Menge) [Cantor, 1884b, page 470] [Cantor, 1932, page 226]: a closed set is one that contains all of its limit points. On the other hand, a dense-in-itself set (in sich dichte Menge) [Cantor, 1884b, page 471] [Cantor, 1932, page 228] is one which is contained in its set of limit points or its first derived set. A perfect set is then both closed and dense-in-itself.

When investigating the concept of continuum, Cantor noted that his notion was independent of the idea of dimension [Cantor, 1883b, page 590] [Cantor, 1932, pp. 207–208]. He promised to take up the problem of defining the dimension concept for continua, but he never did. Perhaps his other research or even his mental illness prevented him from doing so. Nevertheless, he had a clear interest in the definition problem for dimension motivated by his own paradoxical discovery.

---

[7] Cf. Section 4.6.

[8] For biographical material on Bendixon cf.
`https://mathshistory.st-andrews.ac.uk/Biographies/Bendixson/` and
`https://en.wikipedia.org/wiki/Ivar_Otto_Bendixson`.

[9] During 1883 Cantor and Bendixson had a substantial correspondence concerning the difficulty. Cf. [Bendixson, 1883].

8.3 Developing Point-Set Theory and Set Theory

Cantor's imaginative ideas had the most profound effect on the growth of mathematics. Analysts were the first to discern their usefulness. Point-set theory offered wonderful new instruments for a detailed study of the nature of functions, with the result that the growth of real and complex function theory was greatly accelerated in the years after the publication of Cantor's great papers. Applications of the Cantorian toolkit to the fundamental notions of geometry came a little later. Camille Jordan and Giuseppe Peano were among the first to take the Cantorian ideas into the domain of geometry, so their work became a vital background to the development of point-set topology.

8.4 Camille Jordan and the Jordan Curve Theorem

Among topologists Camille Jordan (1838–1922)[10] is best known for his celebrated theorem on closed curves in the plane. He first stated this result in a 'Note sur quelques points de la théorie des fonctions' ('Note on some points in the theory of fundtions') at the end of the first edition of his *Cours d'Analyse* (*Course of Analysis*) [Jordan, 1887, pp. 549–615]. In order to understand why Jordan wanted to state and prove this theorem, one must consider a few aspects of his important and very influential analysis text.[11]

The differences between the first and second editions of Jordan's *Cours d'Analyse* could hardly be greater.[12] In terms of the state of mathematics and standards of rigor of the 1880s and 1890s the editions are worlds apart. Jordan's attempt to prove the curve theorem is bound up with these differences. The first edition, published in three volumes between 1882 and 1887, is very much a *Cours d'Analyse de l'École Polytechnique* (*Course of analysis of the École Polytechnique*) [Jordan, 1882] [Jordan, 1883] [Jordan, 1887]. It is not a research treatise, but a textbook for students with an emphasis on methods. However, in the 'Preface' Jordan writes [Jordan, 1882, page v]:

> We have taken particular care in the establishing of fundamental theorems. There is none for which the demonstration has not been subordinated to certain restrictions. We have tried to bring into the discussion, at times delicate, all the precision and rigor compatible with an elementary education.[Ch8, 1]

[10] The standard biography of Jordan is [Lebesgue, 1926]. See also `https://www.encyclopedia.com/people/history/french-history-biographies/camille-jordan#2830902220` and `https://mathshistory.st-andrews.ac.uk/Biographies/Jordan/`.

[11] Cf. [Guggenheimer, 1977], which contains an analysis of some of the history of the theorem.

[12] On these significant differences cf. especially the thesis [Gispert-Chambaz, 1982].

In spite of this declaration of intended rigor[13] the first *Cours* (*Course*) is virtually pre-Weierstrassian on the fundamentals of analysis. Of course, Jordan was well aware of the new Weierstrassian ideas, which by the 1880s were having a powerful impact in France, but apparently he did not wish to expose his students to their full force. Nevertheless, in order to assuage his conscience about matters of rigor Jordan appended the 'Note sur quelques points de la théorie des fonctions' ('Note on some points in the theory of functions') at the end of the last volume of the first edition [Jordan, 1887, pp. 549–615]. In this place he revises proofs of the main text according to the highest standards of rigor then current. It is in this 'Note' that the closed curve theorem first appeared.

The second edition of the *Cours* (*Course*), published between 1893 and 1896, contrasts markedly with the first edition in both content and style [Jordan, 1893] [Jordan, 1894a] [Jordan, 1896]. In 1883 Jordan succeeded Liouville to the chair of mathematics at the Collège de France; he had been suppliant to Serret at the Collège since 1875. At this ancient institution (founded in 1530) he could lecture on his own research interests, and this fact is reflected in the second edition of his book. Perhaps it should really be entitled *Cours d'Analyse du Collège de France* (*Course of analysis of the Collège de France*), for it more accurately describes his analysis lectures at the Collège than at the École Polytechnique.[14] It is much more a research treatise; the standard of rigor is far higher. Jordan incorporates the material of the earlier 'Note' into the main text of the first volume and includes much new research material.[15]

The theorem on closed curves is a part of this radical transformation from the first to the second edition and is connected with proofs of Cauchy's integral theorem in complex analysis. In the first edition Jordan [Jordan, 1883, pp. 275–280] proves the Cauchy theorem in much the same way as Cauchy himself had done in 1825—by a variational calculus technique showing that the first variation of the integral is zero. This proof hardly comes up to standards being expected in the 1880s, when quite a few mathematicians were scrutinizing the theorem and subjecting it to further proof analysis, although the proof is probably sufficient for a student text.[16] However, in the 'Note' Jordan reexamines the Cauchy theorem, stating it in the following form [Jordan, 1887, page 605]:

> *Let C be a closed continuous line without multiple points, not containing any critical points of the function $f(z)$ on its interior. The integral $\int f(z)\,\mathrm{d}z$, taken along any closed and rectifiable line K, is zero.*[Ch8, 2]

[13] Jordan's declaration harks back to the opening of Cauchy's *Cours d'Analyse* (*Course of Analysis*) [Cauchy, 1821].

[14] The student editions of Jordan's analysis courses at the École Polytechnique reveal that he never treated analysis there as rigorously as in his published second edition. The second edition owes much to a course which he gave at the Collège de France during the academic year 1891–92 entitled, 'Principes du calcul infinitésimal'.

[15] Lebesgue in [Lebesgue, 1926] essentially makes the point of this paragraph.

[16] Probably the most important new proof of the Cauchy theorem to appear during the 1880s was Goursat's proof in [Goursat, 1884]. This led to his celebrated improvement of the theorem in [Goursat, 1900].

From the very statement of the theorem one can discern that Jordan had himself subjected the theorem and its proof to a detailed analysis. For example, simple closed lines or curves, which are assumed to have interiors, figure in the restatement of the old theorem. Moreover, in the 'Note' he provides his own definition of rectifiability [Jordan, 1887, pp. 594–598], which he immediately links with his earlier defined concept of a function of bounded variation [Jordan, 1881]. He also takes the trouble to prove the existence of the integral of a complex function [Jordan, 1887, pp. 603–605].

Thus, while previous mathematicians hardly saw the need to prove the 'obviously true' statement that a simple closed curve divides the plane into an inside region and an outside region, Jordan required a proof, and a lengthy one at that [Jordan, 1887, pp. 587–594], as a consequence of his analysis of the Cauchy theorem.[17] In the 'Note' he defines a curve as the sequence of points [i.e., graph] represented by the equations

$$x = f(t),\ y = \phi(t),$$

where f and ϕ are functions of the parameter t. If the functions are continuous, the curve is said to be continuous, and if they have a common period, the curve is said to be closed. If for certain distinct values of t within the period the values of x are the same and the values of y are the same, then the closed curve has multiple points. However, Jordan only considers continuous closed curves without multiple points, i.e., simple closed curves, and so states the theorem [Jordan, 1887, page 593]:

> ... Every continuous curve C divides the plane into two regions, the one exterior, the other interior; the latter cannot be reduced to zero, because it contains a circle of finite radius.[Ch8, 3]

Jordan's proof is based on polygons approximating the curve C from the inside and from the outside. He assumes the theorem for simple polygons, i.e., he assumes that a polygon without multiple points divides the plane into an interior and an exterior region. First, according to Jordan it is possible to construct a simple polygon P' that approximates C to any degree of accuracy. Then he offers a way to construct simple polygons S and S' from P' that are strictly inside or outside of C, respectively. These approximate C and contain it within an annular region. Indeed an entire sequence of interior polygons $S, S_1, \ldots$, each inside the next, and an entire sequence of exterior polygons $S', S_1', \ldots$, each outside the next, can be constructed. These approximate C with increasing accuracy and corresponding pairs form smaller and smaller annular regions which squeeze down to C. Hence, the theorem follows according to Jordan.

Later generations of mathematicians have come to regard the closed curve theorem as a somewhat difficult part of elementary topology. As a first attempt Jordan's proof is quite good, although it was not many years before various mathematicians began to point out what seemed to them to be flaws.[18] Arthur Schoenflies was the first critic,

[17] Bolzano was actually the first to state the Jordan curve theorem, and he did so many years before Jordan, but he gave no proof. Cf. Sections 4.6 and 4.7.

[18] Use of the the method of polygonal approximations has long been considered an inefficient way of proving the Jordan curve theorem. Thus, Oswald Veblen (1880–1960) initiated the direct method of proving the theorem [Veblen, 1905]; cf. Brouwer's paper [Brouwer, 1910b]. Early criticisms of

but initially he only criticized the proof for its complexity. He followed his criticism with a proof of a less general version of the theorem; indeed, his proof of this version is not completely adequate [Schoenflies, 1896]. Schoenflies was also the first to draw attention to the topological significance of the result. Jordan was interested in the theorem for its immediate applications in analysis. In fact, the reason why he employed polygonal approximations in the proof is to be found in his application of the result to the Cauchy integral theorem. For that theorem he wanted a polygonal path of integration on the interior of the simple closed curve [Jordan, 1887, page 605]. Still, while Jordan may not have considered fully the topological significance of the closed curve theorem, he should be credited with stating and seeking to prove something that has great topological importance.[19]

Mathematicians thinking about the dimensional invariance problem soon realized that the Jordan curve theorem could be useful to them, because they perceived that an n-dimensional generalization—the theorem that states that an $(n-1)$-dimensional sphere and its homeomorphic images divide an n-space into two regions—yields an n-dimensional domain invariance theorem and then the invariance of dimension. However, no one genuinely understood the difficulty of proving an n-dimensional Jordan separation theorem until Brouwer.

In the first volume of the second edition of his *Cours d'Analyse* (*Course of Analysis*) Jordan included a slightly revised proof of the curve theorem [Jordan, 1893, pp. 90–100]. This is where most mathematicians first came across the theorem. He also included a full section [Jordan, 1893, pp. 18–31] on Cantorian set theory, incorporating material from a research paper on the integral of the previous year [Jordan, 1892]. He based the set theory on n-spaces provided with an *écart* (distance measure).[20] Within the section on set theory there is a new definition of connectedness (d'un seul tenant) [Jordan, 1893, page 25]: a set is connected if it cannot be separated into several closed sets. This definition is equivalent to Cantor's inasmuch as Jordan only considered closed and bounded (i.e., compact) sets. In this special case it is also equivalent to the usual definition (due later to Riesz, Lennes, and Hausdorff). Jordan's overall contribution to point-set topology was substantial, although he only viewed the subject in the context of analysis. Numerous mathematicians read the second edition of the *Cours d'Analyse* (*Course of Analysis*) and through it learned about the set-theoretic foundations of modern analysis.

Jordan's proof are to be found in: [Schoenflies, 1896], [Veblen, 1905], [Ames, 1905b], [Schoenflies, 1906a], [Schoenflies, 1908b, page 169], [Schoenflies, 1925]. Cf. [Guggenheimer, 1977].

[19] Thomas Hales provides a modern analysis of Jordan's proof with an argument for its validity. See [Hales, 2007b] and [Hales, 2007a]. See also [Thomassen, 1992].

[20] Jordan's *écart* idea seems to have been an important source of inspiration for Fréchet when defining the concept of (E) class (metric space) in [Fréchet, 1906].

8.5 Giuseppe Peano and Space-Filling Curves

Giuseppe Peano (1858–1932)[21] was in many ways Jordan's rival in mathematics. Peano and Jordan attacked the problems of measure and integration independently of one another, and yet their results were remarkably close [Peano, 1887] [Jordan, 1892]. They proposed similar definitions of rectifiability, although Peano's is somewhat more general [Jordan, 1887, page 594] [Peano, 1890b]. It is interesting to find an early criticism by Peano of a result in the first edition of Jordan's *Cours*. The dispute over Jordan's mistake occupies several pages of the 1884 volume of the *Nouvelles Annales de Mathématiques*.[22] From the point of view of the development of topology both saw the importance of applying Cantorian set theory to intuitive geometrical ideas.

Peano's example of a space-filling curve, a curve which covers all the points of a square, is the most spectacular example he ever devised [Peano, 1890a].[23] His example (like many other similar ones) is really a surface-filling or surface-section-filling curve. The example coming out in 1890 completely upset the 'geometrical intuitions' of the mathematicians of the day. It is not known precisely how Peano came to devise such a counterintuitive curve, but one can see that much of his earlier work of the 1880s was directed towards a critical appraisal of commonly-held mathematical notions. The analysis text *Calcolo Differenziale e Principii di Calcolo Integrale* (*Differential Calculus and Principles of Integral Calculus*), written by Angelo Genocchi (1817–1889) with Peano's assistance [Genocchi, 1884], abounds with examples demonstrating the need to revise the fundamentals of the subject, which were then taken for granted. In his *Applicazioni Geometriche del Calcolo Infinitesimale* (*Geometrical Applications of Infinitesimal Calculus*) [Peano, 1887, pp. 152–259] Peano devotes a chapter to the study of geometrical magnitudes, giving definitions of the interior (inner) and exterior (outer) measure of linear sets, plane areas, and spatial volumes. One would expect that the curve bounding an area would contribute nothing to the measure of that area, i.e., that the curve itself could be squeezed within an arbitrarily small region. However, by taking a set-theoretic viewpoint, Peano realizes that the interior (inner) measure of area and the exterior (outer) measure of area of a planar region can be different. Possibly these ideas motivated Peano's investigation of planar curves defined by continuous functions.

In the short paper 'Sur une courbe, qui remplit toute une aire plane' ('On a curve that fills an entire planar area') [Peano, 1890a] Peano defines his curve in a purely analytic way without reference to geometrical considerations or diagrams. This procedure accords with his critical 'anti-geometrical' attitude to analysis. For the construction, consider x and y as continuous functions of the parameter t in the interval $[0, 1]$; as numbers between 0 and 1, x, y, and t are given ternary (base 3)

---

[21] H.C. Kennedy provides a short biography of Peano in [Peano, 1973, pp. 3–10]. See also [Kennedy, 2002], `https://mathshistory.st-andrews.ac.uk/Biographies/Peano/`, and further `https://en.wikipedia.org/wiki/Giuseppe_Peano`.

[22] Cf. *Nouvelles Annales de Mathématiques* Troisième Série. 3(1884), 45–47, 153–155, 252–256, 475–482.

[23] For an excellent mathematical treatment of space-filling curves see [Sagan, 2012].

representations.[24] The aim is to put the ternary fractions

$$t = 0.a_1a_2a_3\ldots,$$

representing points of the unit interval, into continuous correspondence with the ternary fractions

$$x = 0.b_1b_2b_3\ldots,$$
$$y = 0.c_1c_2c_3\ldots,$$

representing points in the unit square. To achieve this aim one lets $\boldsymbol{k}a$ (the 'complement of a') denote the digit $2 - a$, where a is one of the digits 0, 1, or 2. Thus

$$\boldsymbol{k}0 = 2, \ \ \boldsymbol{k}1 = 1, \ \ \boldsymbol{k}2 = 0.$$

$\boldsymbol{k}^n a$ is the result of applying the operation $\boldsymbol{k}$ n times to a, and it has the value a, if n is even, and $\boldsymbol{k}a$, if n is odd. The correspondence between t and (x, y) is given by the following equations:

$$\begin{aligned} b_1 &= a_1, & c_1 &= \boldsymbol{k}^{a_1}a_2,\\ b_2 &= \boldsymbol{k}^{a_2}a_3, & c_2 &= \boldsymbol{k}^{a_1+a_3}a_4,\\ b_3 &= \boldsymbol{k}^{a_2+a_4}a_5, & c_3 &= \boldsymbol{k}^{a_1+a_3+a_5}a_6,\\ &\ldots & &\ldots\\ b_n &= \boldsymbol{k}^{a_2+a_4+\cdots+a_{2n-2}}a_{2n-1}, & c_n &= \boldsymbol{k}^{a_1+a_3+\cdots+a_{2n-1}}a_{2n}. \end{aligned}$$

It is easy to see that every point of the square is covered and that the correspondence is continuous by the very way the ternary representation proceeds. In his paper Peano briefly indicates how the correspondence can be generalized to the cube. In fact, extending it to an n-cube is not difficult.

Although Peano described his curve analytically, soon after the publication of his paper, Hilbert laid out a geometrical construction method for defining a different space-filling curve as the limit of a sequence of polygonal paths traversing the square [Hilbert, 1891]. A few years later Schoenflies [Schoenflies, 1900, pp. 121–123] and the American mathematician E.H. Moore (1862–1932) [Moore, 1900] each presented Peano's curve itself as a limit of a sequence of polygonal paths traversing the square. Moore provided an extended mathematical analysis of Peano's curve [Moore, 1900, pp. 77–80].[25] The basic figures of the construction are given in Figure 8.1, taken from [Moore, 1900, page 77]. Consider the left diagram as the first step of an iterative construction, dividing the square into 9 subsquares and drawing a continuous line composed of diagonal line segments in the subsquares joined together

[24] Peano carefully distinguishes between a written numeral and the abstract number it represents, but to simplify my summary of his exposition I have not emphasized this distinction. Also in summarizing Peano's continuous correspondence, I have simplified his exposition.

[25] In his book Sagan provides a full explanation and proof to connect Peano's arithmetic definition of Peano's curve and the geometrical construction and depiction [Sagan, 2012, §3.4, pp. 36–40].

from the bottom left to the top right as given. Then consider the right diagram in Figure 8.1 as the second step, dividing each subsquare into 9 subsubsquares and drawing a continuous line composed of diagonal line segments through the newly divided square in similar ways through the 81 subsubsquares, moving from bottom left to top right. Continue the construction by dividing squares in steps ad infinitum, and the result is Peano's curve at the limit covering all the points of the original square. Peano's curve is continuous, though certainly not one-one (injective), with many points covered multiple times.

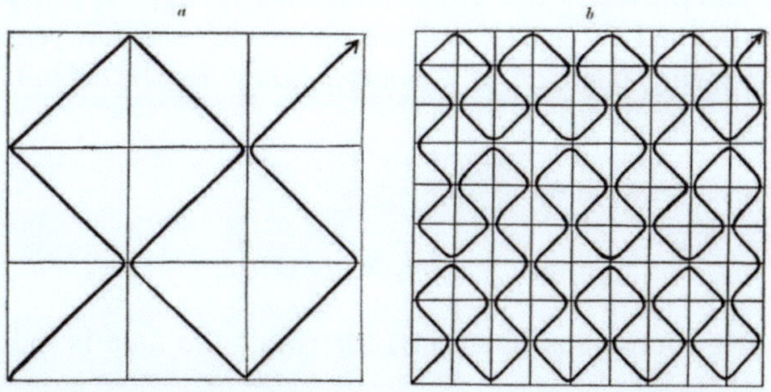

Fig. 8.1 Diagrams for the Two Initial Steps in Construction of Peano Curve, [Moore, 1900, page 77]

Peano was quite proud of his counterintuitive curve. He had one of the curves in the sequence put on the terrace of his home, in black tiles on white.[26]

Peano's curve was a second blow to the naive concept of dimension after Cantor's one-one correspondence, for it provides a continuous mapping between a 1-dimensional object and a 2-dimensional one. However, Peano himself was quick to explain away the difficulty. Because some numbers have two distinct ternary representations, certain points of the square are covered twice and certain others are covered four times by the curve. Thus, the Peano curve describes a continuous, but not a one-one (injective) correspondence. Peano interpreted this fact as a confirmation of Netto's proof of the invariance of dimension, which he cites in his paper. Hence, it may be noted that even Peano, usually a very perceptive critic, discerned no flaw in the earlier general invariance proofs.

Hilbert in his paper [Hilbert, 1891, page 461] notes that it is possible to devise space-filling curves that cover the points of the square at most three times. Much later Lebesgue found that it is not possible to give such curves with the points covered at most twice. He discerned in this fact a key to proving dimensional invariance![27]

Peano's 'curve' clearly demonstrates that the usual way of defining a curve through a pair of continuous functions of a parameter leads to counterintuitive results. Presumably no one wants to think of a square as a curve. It is not surprising that

[26] See Kennedy's statement in [Peano, 1973, page 7].

[27] See Chapter 12.

Peano's example immediately captured the attention of mathematicians, and soon some were proposing new definitions of the concept of curve in order to exclude the Peano counterintuitive cases. This work marks a starting place for the history of topological curve theory, a subject closely related to topological dimension theory. Jordan, however, never publicly recognized Peano's counterintuitive example. In the first edition of the *Cours* [Jordan, 1887, page 587], written before Peano's paper appeared, he had defined curves through continuous functions, and in subsequent editions he never bothered to change his old definition. Moreover, he had an interest in determining when a closed curve encloses a well-defined area in the sense of measure theory and proved in the 'Note' [Jordan, 1887, pp. 599–600] that the area will certainly be defined in the case of a rectifiable curve. A few years later in 1894 he asked the following question in *L'Intermédiaire des Mathématiciens* [Jordan, 1894b]:

> Can one describe a curve
>
> $$x = f(t), \ y = \phi(t)$$
>
> (f and ϕ being continuous functions) for which the area is indeterminate?
> Ch8, 4

Peano [Peano, 1896] responded to Jordan's question by pointing to his space-filling curve. If one takes any arc of the curve and connects the endpoints with an ordinary curve so as to give a closed curve, then one has a curve with an interior area strictly less than its exterior area. But Jordan never acknowledged Peano's response and did not alter his definition of continuous curve in the later third edition of the *Cours d'Analyse* (*Course of Analysis*) from 1909–1915.

Peano's spectacular example is but one in a long list of strange curves and point sets that run counter to 'naive geometrical intuition.' Set theory has been the main breeding ground for such examples. Two other striking entries in this list may be mentioned. In 1903 William Osgood [Osgood, 1903] and Henri Lebesgue [Lebesgue, 1903] (correction in [Lebesgue, 1905a]) each published an example of a Jordan curve (i.e., a simple closed curve) possessing a positive exterior measure, i.e., the curve itself has a measurable area. Each of these curves is even stranger than the one that Peano suggested in his response to Jordan's question, because each has no multiple points. In 1910 L.E.J. Brouwer published the first examples of indecomposable continua and curves that divide the plane into three or more regions but are the common boundaries of all the regions [Brouwer, 1910a]. One might say that in the years around 1900 set theory bred a plethora of topological monsters. An early catalogue of these is contained in the first book in English on set theory, William H. and Grace Chisholm Young's *Theory of Sets of Points* [Young & Young, 1906].[28] For mathematicians well trained in Cantorian methods the monsters became manageable, indeed friendly, and they guided the way to quite a few new results. Thus, pathological oddities were an important motivation to growth in set-theoretic topology.

[28] Cf. [Schoenflies, 1906b].

8.6 Adolf Hurwitz: Outline of a Program of Set-Theoretic Topology

Given the wealth of results from the papers of Cantor and many others, several mathematicians recognized the need for an entire program of exploration in point-set theory and set-theoretic topology. At the First International Congress of Mathematicians, held in Zürich from 9 to 11 August 1897, Adolf Hurwitz (1859–1919)[29] sketched such a plan of investigation [Hurwitz, 1898]. In his lecture reviewing progress in analytic function theory Hurwitz seeks to determine the precise domain of validity of the Cauchy integral theorem and in so doing points to the important work of Jordan [Hurwitz, 1898, pp. 101–104]. He asks the highly relevant general questions [Hurwitz, 1898, page 101]:

> . . . What is a simple closed line, what is a line, especially, a closed line in general, and are all or only certan closed lines admissible in the enunciation of the Cauchy theorem? [Ch8, 5]

In attempting to grapple with these questions he puts them into the general context of the topology of closed sets. In a way similar to Cantor's definition of cardinals and ordinals, one can assign closed point sets to classes: each class containing those sets that can be mapped one-one continuously onto one another. The sets in each class may be called 'equivalent' ('äquivalent'). Then Hurwitz asserts [Hurwitz, 1898, page 102]:

> This distribution of point sets into classes forms . . . the most general foundation of *analysis situs*. The task of analysis situs is to search for the invariants of the single classes of point sets. [Ch8, 6]

Hurwitz gives a specific example: a geometrico-set-theoretic definition of Jordan's idea of a simple closed curve, as a point set (in the plane) which belongs to the class of the boundary of a square. In essence, he states the Klein *Erlanger Programm* (*Erlanger Program*) (1872) for analysis situs in a sharper form by placing the central problem into the framework of point-set theory.

---

[29] For biographical information on Hurwitz cf.
`https://mathshistory.st-andrews.ac.uk/Biographies/Hurwitz/` and
`https://www.encyclopedia.com/science/dictionaries-thesauruses-pictures-and-press-releases/hurwitz-adolf`.

8.7 Arthur Schoenflies: Further Development of Point-Set Theory and Point-Set Topology

Hurwitz never contributed to the program outlined in his Zürich lecture. It was Arthur Schoenflies (1853–1928)[30] who first embarked upon such a program. At the turn of the century Schoenflies was a forceful advocate for Cantor's set theory and point-set theory. He wrote the article on 'Mengenlehre' ('Set Theory') for the *Encyklopädie der mathematischen Wissenschaften mit Einschluss ihrer Anwendungen* (*Encyclopedia of the Mathematical Sciences Including their Applications*) [Schoenflies, 1898] and composed the two part *Bericht* (*Report*) [Schoenflies, 1900] [Schoenflies, 1908b] on progress in the field in response to a commission by the Deutsche Mathematiker-Vereinigung (German Association of Mathematicians).[31] As noted above, he was the first mathematician to draw attention to Jordan's curve theorem by drawing attention to its topological significance and giving an alternative proof.

In a short early paper, 'Ueber einen Satz der Analysis Situs' ('Concerning a theorem of analysis situs') [Schoenflies, 1899], Schoenflies reproves the planar domain invariance theorem in the formulation [Schoenflies, 1899, page 282]:

> . . . that *the one-one continuous image of the surface of a square is again a simply connected piece of surface*, viz., that it consists of the totality of all points which belong to the interior of a *closed curve*.[Ch8, 7]

At the time when this paper was presented, 13 January 1900, Schoenflies was unaware of Jürgens' proof of 1878–79 of this same result. However, in contrast to Jürgens' way of proceeding Schoenflies relies heavily on Jordan's demonstration of the curve theorem.[32] At the end of his paper he deduces the special case of dimensional invariance from the planar domain invariance theorem [Schoenflies, 1899, pp. 289–290]. For this he uses a theorem of Cantor [Cantor, 1882, page 117] [Cantor, 1932, page 153] to the effect that a collection of closed planar surfaces (having interior points), for which no two have interior points in common, is at most countable. The objective is to show that a 3-dimensional or an n-dimensional cube cannot be mapped in a one-one continuous manner onto a planar surface. Suppose such a mapping is possible. Then by the domain invariance theorem parallel square slices of the cube will be mapped onto simply connected regions bounded by Jordan curves, with interior points corresponding to interior points. Because the mapping is one-one, the images of the square slices cannot overlap. However, as there are uncountably many square sections, one has a contradiction with Cantor's theorem.[33]

---

[30] For information on Schoenflies' life and work cf. https://mathshistory.st-andrews.ac.uk/Biographies/Schonflies/ and https://www.encyclopedia.com/science/dictionaries-thesauruses-pictures-and-press-releases/schoenflies-arthur-moritz.

[31] See also [Schoenflies & Hahn, 1913].

[32] Cf. two subsequent proofs by Osgood [Osgood, 1900] and Bernstain [Bernstein, 1900].

[33] Cf. a similar earlier proof of Milesi [Milesi, 1892]; criticized by Schoenflies [Schoenflies, 1908b, page 165].

In the last paragraph of his paper Schoenflies suggests that his method of proof for domain invariance is capable of generalization to arbitrary spaces. At the time he probably did not fully grasp the difficulties of such a generalization.

Around 1902 Schoenflies began an intense and rapid development of a program of set-theoretic topology. In that year he published a paper containing a statement and proof of a converse of Jordan's theorem [Schoenflies, 1902]. This little paper led to three long papers, 'Beiträge zur Theorie der Punktmengen' ('Contributions to the Theory of Point Sets') [Schoenflies, 1903] [Schoenflies, 1904a] [Schoenflies, 1906a] (cf. [Schoenflies, 1908a]), having as their central theme the topology of closed curves and simple closed curves in the plane. The opening paragraph of the first 'Beiträge' ('Contributions') paper sets the scene [Schoenflies, 1903, page 195]:

> As one of the most general problems of the theory of point sets one can point to the task of formulating and establishing the fundamental theorems of *analysis situs* set-theoretically and setting forth the relationships that exist between the *set-theoretico-geometric* and the *analytic* modes of expressing these concepts and theorems. The paradoxical results, as they occur, e.g., in the one-one mapping of continua and in the Peano curve, have completely destroyed the naive ideas of analysis situs. All the more we must demand that set theory provide a substitute and define the basic geometrical concepts in a way that gives back to them their natural content characteristic of analysis situs. Even if the much maligned intuition is no source of proof, it still seems to me that it is a goal of research to reconcile the content of geometrical definitions with the *content* of intuition — at least in the domain of analysis situs.[Ch8, 8]

When submitting his first 'Beiträge' ('Contributions') paper Schoenflies put the following methodological point in his accompanying letter to the editor and friend David Hilbert (letter of 28 April 1903):[34]

> With regard to what constitutes the content of analysis situs, to a large extent we still rely on intuition, which at the same time must serve as a basis of proof. Yet it may be necessary to strive for even greater exactness in this domain and derive its theorems from geometric and set-theoretic fundamentals. It is my keen wish to do this at last. I have felt like working on this more intensively and I hope that I shall succeed in time.[Ch8, 9]

Schoenflies was certainly aware of the difficulties of 'naive geometrical intuition' and sought to put topology on a sound, purely set-theoretic footing. However, his methodological remark in the letter to Hilbert would come to have an ironic ring after Brouwer delivered his devastating critique of Schoenflies' work.

In executing his program Schoenflies had most success with the converse to the Jordan curve theorem. For this he introduced the concept of accessibility (Erreichbarkeit), at first implicitly [Schoenflies, 1902] [Schoenflies, 1903], then explicitly [Schoenflies, 1904b] [Schoenflies, 1906a]. If M is a connected domain (open set)

[34] This letter is in the Hilbert *Nachlass* in the Niedersächsische Staats- und Universitätsbibliothek, Göttingen (signature: HILBERT 355, no. 15).

and t is one of its boundary points, then one says that t is accessible from M if for every point m of M there is a simple path (einfacher Weg) in M leading from m to t [Schoenflies, 1906a, pp. 296–297].[35] The example that motivates Schoenflies' definition is the curve $\sin(1/x)$, where $-1 \leq x \leq +1$, together with its limit points on the y-axis that lie between -1 and $+1$ (a variant of the topologist's sine curve). The points on the y-axis, $-1 < y < +1$, are not accessible from an open domain surrounding the curve. However, Schoenflies recognized that all the points of a Jordan curve, i.e., a homeomorphic image of a circle, are accessible from both the interior domain and the exterior domain and that this fact is the key to giving a converse to Jordan's theorem. In Schoenflies' terminology a closed curve is a bounded nowhere dense perfect connected point set that divides the plane into two connected domains having the point set as their common boundary [Schoenflies, 1904a, pp. 146–147] and a simple closed curve is a closed curve with all of its points accessible from each of the two domains [Schoenflies, 1903, page 217] [Schoenflies, 1906a, page 305]. Then the converse of the Jordan curve theorem states that every simple closed curve can be mapped in a one-one continuous manner onto the points of a circle (proofs in [Schoenflies, 1902] [Schoenflies, 1903] [Schoenflies, 1906a]). This result is the high point of Schoenflies' topological investigations.

Schoenflies gathered the results of his 'Beiträge' ('Contributions') series of papers and some others [Schoenflies, 1907a] [Schoenflies, 1907b] into the second part of his *Bericht* (*Report*) for the Deutsche Mathematiker-Vereinigung (German Association of Mathematicians): *Die Entwickelung der Lehre von den Punktmannigfaltigkeiten* (*The Development of the Theory of Point Manifolds/Sets*)[Schoenflies, 1908b]. In his eyes he had completed a beautiful chapter in topology, meeting the highest standards of set-theoretic rigor. Yet, it was scarcely a year after the appearance of this 'complete manual' of planar analysis situs that L.E.J. Brouwer delivered a resounding critique of the theory through a set of ingenious counterexamples [Brouwer, 1910a].[36] Nearly all of the theory of closed curves suffered destruction. Only the converse to the Jordan theorem based on accessibility escaped unharmed. It turned out that the topology of the plane was more subtle than Schoenflies had ever imagined.

In spite of Brouwer's critique Schoenflies should still be recognized as a pioneer. He strove for a high standard of rigor through his use of a fairly good set of Cantorian tools. He also had some good insights into the pathology of curves. However, many times he was careless and somewhat disorganized in his research. 'Chaotic' may not be an entirely unfair description for some of his books and papers. More importantly, one now knows that the backbone of his topological investigations, polygonal approximations, is a very inefficient supporting device, which can be mathematically misleading. Thus in the end the work of Schoenflies in topology was overshadowed historically by that of Brouwer and others researching in this domain.

---

[35] In this definition a simple path is just a polygonal path which does not cross itself. It may have infinitely many edges, in which case the only limit point of the vertices is an endpoint.

[36] Cf. also Denjoy's paper [Denjoy, 1910].

8.8 Concluding Remarks

To conclude this chapter I should like to describe the problem situation concerning dimensional invariance around the beginning of the twentieth century. As noted above, after the intensive work of 1878–79 most mathematicians forgot about the problem during the last two decades of the nineteenth century; they simply believed that Netto and/or Cantor had solved it. However, Jürgens had never been convinced by the earlier general proofs; so in 1899 he published his note on 'Der Begriff der n-fachen stetigen Mannigfaltigkeit' ('The concept of the n-fold continuous manifold') [Jürgens, 1899] in which he thoroughly criticized the old proofs and thereby reopened the discussion of the general problem. Jürgens was acutely aware of the difficulties in proving the invariance of dimension. As one of the major stumbling blocks to proof he discerned the difficulty of providing an adequate definition of continuous connection and small displacement (stetiger Zusammenhang; kleine Verschiebung) for manifolds, i.e., the property of nearness of points in manifolds. He saw the need for a better definition of the very notion of continuous manifold. This is a problem that Brouwer later attacked and solved (see Chapter 11). In a sense, only after this problem had been solved could one assign a dimension number to a manifold and then consider the invariance problem. Related to these questions is the problem of assigning dimension numbers to arbitrary subsets of a manifold. Jürgens realized that, apart from associating the numbers 1 through n to most subsets as their dimension numbers, there are certain subsets for which the notion of extension is inapplicable. He thought of these sets as consisting of infinite systems of separate points; one might call them 0-dimensional. By thus recognizing the existence of 0-dimensional sets, however vaguely, he was able to criticize the proof of Netto (see Chapter 7) and also clarify the general definition problem for dimension. Overall, in his short article Jürgens put the problems of dimension into sharper focus and so prepared the way for the work of the twentieth-century topologists. In particular, he emphasized the close relation between domain and dimensional invariance.

In the same year as Jürgens published his critical note Jacob Lüroth returned to the invariance problem and proved another special case [Lüroth, 1899]. Then a few years later he brought together all of his results, fully worked out, in a final contribution to the subject [Lüroth, 1906]. In the second part of his *Bericht* (*Report*) Schoenflies [Schoenflies, 1908b, pp. 164–168] drew attention to the open problem of dimensional invariance, which just a few years before he had considered closed. He suggested in the second part of the *Bericht* (*Report*) that the best way to tackle the problem would be through a general proof of domain invariance in n dimensions.

Solutions to the problems of dimensional and domain invariance were all the more urgent at this time, because some mathematicians were recognizing their relevance to certain 'higher' branches of mathematics. Robert Fricke (1861–1930),[37] the expert on automorphic functions at the beginning of the century and co-author with Klein

[37] For biographical material on Fricke see
https://mathshistory.st-andrews.ac.uk/Biographies/Fricke/ and
https://en.wikipedia.org/wiki/Robert_Fricke.

of the classic treatise on the subject [Fricke & Klein, 1897] [Fricke & Klein, 1912], underlined the pressing need to solve these problems: their solution would be critical to shoring up the continuity method of proof employed in the theory of automorphic functions and in uniformizing analytic functions.[38] Poincaré and Klein had first introduced the method in the 1880s, but by about 1900 Fricke became convinced that it required further justification. In a lecture delivered to the Third International Congress of Mathematicians held in Heidelberg during August 1904 Fricke declared that the development of the theory of automorphic functions had come to a virtual standstill over this difficult point [Fricke, 1905]. The method was used for proving some fundamental theorems on the existence of inverse functions to the automorphic functions (the so-called linear polymorphic functions). The leading idea of the method was to compare two manifolds of equal dimension, one of all the groups or fundamental domains of a given signature associated with a set of automorphic functions and the other of all the related Riemann surfaces of the same signature. The dimensions of the manifolds were determined by counting the parameters given in the signatures. Hence, the method partly rested on the simple coordinate or parameter idea of dimension. In the light of Cantor's paradoxical discovery Fricke concluded that the method was not completely rigorous. Domain or dimensional invariance was needed in some form. Although he did not expect that the continuity method would be invalidated over this point, he wanted to see a full topological foundation for the fundamental existence theorems. When he had struggled earlier to obtain partial proofs of the existence theorems [Fricke, 1904], he had been reassured by Schoenflies' proof of the planar domain invariance theorem that full rigor could be attained. Consequently, at Heidelberg he urged set theorists to bring the problems of domain and dimensional invariance to a satisfactory conclusion by proving the general n-dimensional cases.

Undoubtedly, at the start of the twentieth century the problem of the invariance of dimension was wide open. It virtually demanded solution, and mathematicians expected a positive solution. The rapid developments in set-theoretic topology seemed to give strong hints towards this end. The theory had become a very helpful tool for clarifying the basic notions and difficulties surrounding the problem. Partial solutions also seemed to open up the way to a full solution. Undoubtedly, the most promising route appeared to be through invariance of domain. Yet in spite of the hopeful expectations that mathematicians held at the time, the road to success turned out to be more difficult than imagined, as the following chapters will show.

---

[38] Cf. the details in Fricke [Fricke, 1913, pp. 445–452]. Also cf. Freudenthal's historical remarks in [Brouwer, 1976, pp. 572–573] and the remarks of Klein in [Klein, 1923, pp. 731–741].

Chapter 9
Poincaré's Theories of Dimension

9.1 Introduction

In the previous chapter I noted that several mathematicians around the beginning of the twentieth century recognized the urgent need to solve the problem of dimensional invariance. In a much broader direction during the last years of the nineteenth century and the first decade of the twentieth a number of prominent mathematicians put forward new ideas about dimension. Some of these mathematicians offered methods and ways for solving the invariance problem, while others presented novel ways of looking at the notion of dimension itself.[1] Among these mathematicians Henri Poincaré was the most important. I shall cover his work concerned with dimension in this chapter. In the next chapter I shall cover the work on dimension of three other significant mathematicians, Frigyes Riesz, René-Louis Baire, and Maurice Fréchet. The work of these mathematicians, especially that of Poincaré, formed an important background to Brouwer's breakthrough to solutions of the invariance problem and his analysis of the dimension concept itself, which will be discussed in Chapters 11 and 12.

[1] During the last decade of the nineteenth century Wilhelm Killing had shown some interest in problems of dimension, but he did not produce anything of significance for the subsequent development of dimension theory. In his long paper 'Ueber die Grundlagen der Geometrie' ('Concerning the foundations of geometry') he proposes a geometrical definition of dimension [Killing, 1892, pp. 134–137]. Unfortunately, the definition is far from clear and almost certainly incorrect [Torretti, 1978, pp. 183–184]. In his book on the foundations of geometry Killing discusses dimension, including the results of Cantor and Peano, but he does not make any progress on the problems of defining dimension in general or proving dimensional invariance [Killing, 1893, pp. 166–270]. However, his discussion of the 3-dimensionality of space is interesting. The 3-dimensionality is explained by means of an informal idea of cut [Killing, 1893, pp. 174–176]. The explanation is derived from traditional Euclidean geometry, and in it there is a suggestion of something like of Poincaré's later cut definition of dimension (1903) (1912) (discussed later in Section 9.4.)

D. M. Johnson, *Theories of Dimension from Ancient to Modern Times*, Sources and Studies in the History of Mathematics and Physical Sciences, https://doi.org/10.1007/978-3-032-02682-8_9

9.2 Poincaré's Initial Analysis of Ideas Concerning Space and Geometry

Among the new ideas about dimension of the last years of the nineteenth century and the first years of the twentieth the most interesting came from Henri Poincaré[2] (1854–1912).[3] Poincaré's concern with dimension did not arise from his mathematical work in topology. Rather, it developed from his philosophical investigations spanning many years into the origin and nature of our ideas of space and their relation to geometry and other parts of mathematics. To be specific, his philosophy of space and geometry grew out of an examination of three main problems: first, the problem of the genesis and development of our fundamental ideas concerning space; second, the problem of the relation of our ideas of space to geometry and other parts of mathematics, primarily continuous groups or Lie group theory and then later topology; and third, a subsidiary to the first two, the problem of the nature of the concept of dimension and why we might be led to assert that our space has three dimensions. Poincaré's extensive treatment of these problems dealing with the nature of space, the nature of the accompanying and supporting geometry, the dimensionality of our space, and the abstract concept of dimension starts with his work of the last years of the nineteenth century beginning in 1887 and extends to his death in 1912. He often examined and reexamined these topics in the spirit of philosophical inquiry and in response to criticism, with the result that he wrote several significant essays combining philosophy and mathematics, which were intended to appeal to a broad educated audience.

Poincaré constructed two theories of dimension. The first, developed in the 1890s, is based on continuous group theory or Lie group theory (or as one would say today more generally, topological group theory) and the second, developed in the first years of the twentieth century, is based more specifically on topology. The latter is the one that influenced the subsequent development of dimension theory. In this chapter I shall examine the origin and development of Poincaré's ideas that first led to his continuous group-theoretic dimension theory and that finally led to his topological dimension theory.[4]

---

[2] In the earlier version of this part of my work on Poincaré I thanked Prof. Dr. H. Freudenthal, Dr. J. Giedymin, and Prof. R. Torretti for kindly reading a draft of what I had written and offering me helpful comments and criticisms. I took account of these for that earlier version, and for this revised version I remain thankful for their comments. Regrettably, all three have died.

[3] There are several works covering Poincaré's life and work, including [Lebon, 1912] [Julia, 1955] [Bellivier, 1956] [Verhulst, 2012] [Gray, 2013] [Nordmann & Boutroux, 2021] [Heinzmann & Stump, 2024] [Murzi, 2024a]. The recent books by Gray and Verhulst are the most extensive. There are also retrospectives on his work: [Browder, 1983] and [Charpentier et al., 2010]

[4] Numerous authors in many works have commented upon and criticized Poincaré's philosophy of space and geometry. I have found the following the most helpful for my own discussion. Rougier [Rougier, 1920] provides a good descriptive account of Poincaré philosophy of geometry and the historical and mathematical background to it. Freudenthal [Freudenthal, 1960] contains some very useful insights into the historical development of Poincaré's geometrical ideas. Giedymin [Giedymin, 1977] (reprinted in [Giedymin, 1982, pp. 1–41]; cf. [Giedymin, 1991]) gives an excellent critical treatment of the historical origin and significance of Poincaré's geometrical conven-

Poincaré started publishing philosophical articles in the 1890s. However, in 1887, just a few years before these articles began to appear, he published a short mathematical paper 'Sur les hypothèses fondamentales de la Géométrie' ('On the fundamental hypotheses of geometry') [Poincaré, 1887] with some general philosophical remarks at the end. These concluding remarks are quite significant, for they form the germ of the philosophy of conventionalism, which he subsequently developed in his later philosophical articles.[5]

In the paper of 1887 Poincaré seeks to find the necessary and sufficient hypotheses for the foundations of the geometries of two dimensions or of the geometric plane, and in this search he adopts the group-theoretic approach to geometry of Helmholtz, Klein, and Lie. He follows the particular approach of Sophus Lie [Lie, 1880][6] as a method for finding all possible continuous (Lie) groups for the various geometries of a 2-dimensional manifold or geometric plane. In modern formal terms a Lie (continuous) group is a group that is also a differentiable manifold. Hence, in his concluding remarks he asserts [Poincaré, 1887, page 215]:

> . . . geometry is nothing but the study of a group . . . .[Ch9, 1]

This assertion provides a key to understanding both Poincaré's mathematical view and his philosophical view of geometry in 1887 and for many years after. For him group theory is the proper foundation for geometry. Hence, each of the most important geometries, those of Euclid, Lobachevskii (Bolyai, Gauss), and Riemann,[7] correspond to a particular continuous group. Topology is part of the continuous group theory.

In the paper of 1887 Poincaré poses the philosophical question: what is the nature of the hypotheses at the foundation of the various geometries? He asks [Poincaré, 1887, page 215]:

> Are they experimental facts, or analytic or synthetic judgments *a priori*?[Ch9, 2]

tionalism. Torretti [Torretti, 1978, pp. 320–358] offers a careful analysis of Poincaré's philosophy of geometry with many penetrating insights into the mathematical and philosophical details. [Torretti, 1984] includes a correction. Additionally, [Torretti, 1978] in its entirety is a very important reference for Chapter 5 of this work. Torretti's recent article [Torretti, 2019] is a summary of some important developments in nineteenth-century geometry. Two works treating Poincaré's philosophy of geometry are [Zahar, 1997, pp. Chapter 3] and [Soltani, 2020]. Zahar covers important aspects of Poincaré's philosophy of geometry, but these aspects are only tangentially related to the discussion in this work. Soltani's dissertation treats the historical and philosophical background to Poincaré's philosophy of geometry, the content of his philosophy, and also the views of other philosophers concerning his philosophy. The work contains some analysis of Poincaré's views of dimension and dimensionality.

[5] Giedymin [Giedymin, 1977] has stressed the importance of this early paper for the subsequent development of Poincaré's philosophy of geometry.

[6] Cf. Lie's earlier papers of 1874–1879 which are preliminary to [Lie, 1880] in [Lie, 1924, pp. 1–223].

[7] Poincaré normally uses the term 'geometry of Riemann' to refer to either spherical geometry, as in [Poincaré, 1887] [Poincaré, 1891], or elliptical geometry, as in [Poincaré, 1902a] [Poincaré, 1903a]. To be sure, he is well aware of the distinction between the two geometries of Riemann for spaces having constant positive curvature. He is also aware of the more general geometries of Riemann; see below and [Poincaré, 1887] [Poincaré, 1891].

His answer is negative to each of these alternatives. The hypotheses cannot be the results of experiments, because then geometry would be constantly open to revision and, hence, not an exact science. They cannot be synthetic or analytic judgments *a priori*, because, according to Poincaré, it would be impossible to have a geometry based on their negation.[8] Geometry is the study of continuous groups. One cannot say that Euclidean geometry is true and Lobachevskian geometry is false. The group for one is not incompatible with that of the other. However, it is appropriate to ask: Which geometry do we use for understanding physical phenomena? Poincaré's response is that we choose a geometric group from those possible in the same way that we choose a coordinate system from those available—*according to convenience*. We are led to choosing a geometry by considerations of simplicity and also through recognition of the apparent fact that in our world rigid motions of solids are possible. With regard to this latter point one sees that Poincaré excludes the more general geometries that Riemann contemplated in his *Habilitationsvortrag* (habilitation lecture), 'Über die Hypothesen, welche der Geometrie zu Grunde liegen' ('On the Hypotheses that Lie at the Foundation of Geometry'), and restricts himself to the viewpoint of Helmholtz and Lie on the Riemann–Helmholtz–Lie space problem.[9] As a result he permits only spaces with constant curvature in his foundational inquiry; he excludes those of variable curvature, although he certainly would accept them as mathematically possible.

In the paper of 1887 Poincaré offers his earliest statement of his philosophy of geometrical conventionalism. Though brief, it contains several of the essential features of his philosophy. Having rejected the extremes of empiricism and rationalism in geometry, he puts forward his conventionalist thesis as a middle course. There are various possible geometries; one chooses the one that is most convenient for describing spatial phenomena.[10]

From the 1890s until his death in 1912 Poincaré expanded upon his conventionalist position through a long series of essays. Most of his essays first appeared in the *Revue de Métaphysique et de Morale* (*Review of metaphysics and moral philosophy*), a newly-founded journal that above all sought a rapprochement between philosophy and contemporary science. Only later were his essays collected into his four well-known books: *La Science et l'Hypothése* (*Science and Hypothesis*) [Poincaré, 1902b], *La Valeur de la Science* (*The Value of Science*) [Poincaré, 1905a], *Science et Méthode* (*Science and Method*) [Poincaré, 1908b], and *Dernières Pensées* (*Last Essays*) [Poincaré, 1913a].[11]

---

[8] For a criticism of this position of Poincaré cf. [Torretti, 1978, pp. 329–330].

[9] Freudenthal [Freudenthal, 1960, pp. 17–25] regards this restriction as very important for understanding Poincaré's view of geometry. Cf. also [Freudenthal, 1955], [Freudenthal, 1964].

[10] Giedymin [Giedymin, 1977] has argued that Poincaré's paper [Poincaré, 1887] is really a critical response to the kind of empiricism advocated by Riemann in his *Habilitationsvortrag* (habilitation lecture), 'Über die Hypothesen, welche der Geometrie zu Grunde liegen' ('On the Hypotheses that Lie at the Foundation of Geometry') [Riemann, 1868] [Riemann, 1953, pp.272–287] [Riemann, 1990, pp. 304–319] [Riemann, 2004, pp. 257–272]).

[11] Freudenthal [Freudenthal, 1960, page 17] has stressed the need to be aware of the earlier dates of Poincaré's essays in order to be able to understand them in their proper historical context. The four books have been translated into English as *Science and Hypothesis* [Poincaré, 1905b], *The Value*

In the very first volume of the *Revue* (*Review*) Poincaré published an article entitled, 'Le Continu Mathématique' ('The mathematical continuum') [Poincaré, 1893]. In it he offers an explanation of the notions of physical continuum and mathematical continuum.[12] These continua are linear or 1-dimensional. For Poincaré the leading idea of our concept of physical continuum is captured in a set of relations deriving from simple observations about sensations and psycho-physiological experiments. For example, experimental results concerned with 'weighing' objects in our hands show that we cannot distinguish between a weight A of 10 grams and a weight B of 11 grams, nor between weight B and a weight C of 12 grams, but we can distinguish between weights A and C. Similar results obtain for other experiments dealing with sensations and our ability to order or measure them in a primitive way. According to Poincaré we can express these results by the relations $A = B$, $B = C$, and $A < C$. These are contradictory relations, which, nonetheless, form the rudimentary basis for the notion of physical continuum, the rough continuum of the senses. In order to eliminate such an intolerable contradiction we proceed to invent the mathematical continuum of the real line of analysis, the perfect continuum of reason, which may be described in terms of the integers, the rational numbers, and Dedekind cuts (for the full set of real numbers).[13] Torretti (in [Torretti, 1978, pp. 345–347]) has emphatically asserted that Poincaré is mistaken in his explanation of the origin of the concept of mathematical continuum through a 'contradiction.' Poincaré confuses his relation of indistinguishability in the weighing experiment with the relation of equality or identity of quantity by using the sign '='. Yet indistinguishability, which is not a transitive relation, is by no means the same as equality or identity, which is a transitive relation (and indeed an equivalence relation). Hence, his analysis of the origins of the concepts of physical continuum and mathematical continuum is put in doubt. Torretti's critique of Poincaré is based on a modern interpretation of the equality relation. From the viewpoint of the geometry of his day and later dimension theory it is also significant that in the original version of his paper there is no analysis of multi-dimensional continua, although such a natural extension of his ideas would seem to be very desirable.[14]

In a paper on the origins of our knowledge of space and its geometry, 'L'espace et la géométrie' ('Space and geometry') [Poincaré, 1895b], Poincaré offers for the first time a brief explanation of the dimensionality of our space partly based on the Lie theory of continuous groups [Poincaré, 1895b, pp. 640–641]. This paper immediately drew criticism from the young Louis Couturat (1868–1914), once a

---

of Science [Poincaré, 1913b], *Science and Method* [Poincaré, 1914], and *Last Essays* [Poincaré, 1963].

[12] This article was prompted by criticism of M. Mouret against an earlier article, [Poincaré, 1891]; cf. [Poincaré, 1892]. Frequently the comments and criticisms of others motivated Poincaré to write fresh articles defending and developing his philosophical position. Louis Couturat, Bertrand Russell, and Édouard LeRoy were among those who prompted him to expand upon his position.

[13] In [Poincaré, 1893] and the first printing of [Poincaré, 1902b] Poincaré attributes Dedekind cuts to Kronecker!

[14] He added an analysis of multi-dimensional continua to the version of the paper appearing in *Science et l'Hypothèse* (*Science and Hypothesis*) [Poincaré, 1902b, Chapitre II]. The brief analysis forms a prelude to his first paper defining dimension inductively with cuts [Poincaré, 1903b].

student of Poincaré at the École Normale Supérieure. Couturat [Couturat, 1896, pp. 659–662], who took a Kantian position opposed to Poincaré, bluntly accused him of arguing in a circle in his attempt to explain the dimension of space and the related notion of point. Naturally, Poincaré felt obliged to reply, at first briefly [Poincaré, 1897, pp. 63–70] and then in more detail in a lengthy article published in English translation in *The Monist*, 'On the Foundations of Geometry' [Poincaré, 1898]. This article, to which I now turn, contains a full exposition of Poincaré's first theory of dimension.

The central question of Poincaré's essay in *The Monist* is: how do we come to have knowledge of space and its properties and its features? Additionally, how is our notion of space related to geometry and the objects of geometry, as defined parts of mathematics? His answers are comprised in a theory about the genesis of our ideas of space and our knowledge of spatial geometry, which is an important expression of his conventionalist philosophy.[15] The theory has two interconnected parts: an empirical, genetic one and a rationalistic, mathematical one. First, he offers a psycho-physiological explanation of the origin of our mental concept of space and its corresponding geometry. In effect, he treats the first question from a 'subjectivist,' or probably more accurately 'intersubjectivist,' viewpoint.[16] It is a question of how each of us obtains knowledge of space; alternatively, how we obtain knowledge of 'our' space. The psycho-physiological explanation is concerned with sensations and their associations; it is a piece of associational psychology. It probably would not be difficult to criticize Poincaré's sensationalist theory of knowledge, but I shall refrain from doing so.[17] Second, he shifts his question to the domain of mathematics and develops a rationalistic explanation of spatial geometry and dimensionality in terms of the Lie theory of continuous groups.

---

[15] A little later E. Mach [Mach, 1901] [Mach, 1902] [Mach, 1903] [Mach, 1906] put forward an empirical and psycho-physiological theory of the origins of geometry. At the beginning of [Mach, 1901] he distinguishes his theory from Poincaré's.

[16] In his essay in *The Monist* [Poincaré, 1898] and his other essays dealing with the origins of our ideas of space and spatial geometry [Poincaré, 1895b] [Poincaré, 1903b] [Poincaré, 1906] [Poincaré, 1912] Poincaré very often treats the matter from a 'purely subjectivist' viewpoint. He seems to say that each individual subject independently creates or could create his own ideas of space and spatial geometry. However, he does use the nominative plural 'we' when describing how we create our spatial ideas. Indeed I think that he would wish eventually to take into account communications among subjects in the creation of spatial geometry, a geometry which is based on commonly agreed conventions. Hence, I think that in the long run he would wish to adopt an intersubjectivist viewpoint. Indeed in his important discussion of the objectivity of science [Poincaré, 1902d] [Poincaré, 1905a, chapitres X, XI] he argues for an intersubjectivist epistemology and ontology. Objective reality is what is common to many thinking beings and could be common to all. Moreover, in his last essays dealing with the genesis of spatial geometry [Poincaré, 1906] [Poincaré, 1912] he seems to strengthen his intersubjectivist position considerably. He even appears to move towards what one might describe as an 'objectivist' position by adopting an evolutionary epistemology and an argument from mathematical physics, though he would probably maintain that this position remains intersubjectivist. See the discussion later in this chapter.

[17] Cf. [Torretti, 1978, pp. 340–341] on this matter.

At the beginning of his inquiry into the origins of our ideas of space Poincaré argues that sensations per se, whether visual, tactile, or muscular, do not give us the notion of space directly, for by themselves they have no spatial character [Poincaré, 1898, pp. 1–5]. They do not produce the notion of our space by a direct causal relation, as a naive empiricist might imagine or claim. We have no direct knowledge of physical space. Indeed 'physical space' is not a part of Poincaré's philosophical vocabulary. In his view, however, there is a sensible space or representative space built up from visual, tactile, and motor sensations by a process of classification. This empirical space has none of the properties of true geometrical space; it is not infinite (or even unbounded), homogeneous, or isotropic [Poincaré, 1898, pp. 5–6]. Hence, his aim is not so much to investigate the empirical space, but rather to explain the origin and properties of the geometrical space, the purely conceptual space about which we can reason and which we can use in physical theories.[18]

In spite of what has just been stated, Poincaré does regard sensations as having a significant function in the creation and development of our notion of geometrical space. In a special way sensations occasion our minds to construct geometrical space.

Poincaré's view is that the mind forges a link between sensations and geometrical space through acts of classifying and analyzing certain changes in our sensations. The eventual result is that the mind produces the concept of geometrical space [Poincaré, 1898, pp. 6–9]. To begin, consider some examples. Suppose we see an object before us which then moves away from the centre of our vision. It may be that we can bring it back to the center of our retina by a movement of our eyes or body. In sensationalist terms we can reestablish our primitive sensations of the object. Poincaré gives the following more definite example, one of his favorites. Suppose there is a sphere before us with one hemisphere blue and the other red. If at first we see the blue hemisphere and then the sphere rotates so that the red one becomes visible to us, we can bring back the sensations of the blue one by making certain changes in our internal sensations, i.e., by moving around the sphere. In other words, we are able to correct an external change of sensations, a rotation of the sphere, by an internal change of sensations, by moving around the sphere in a compensating rotation. Often we can correct the external change that an object has undergone by means of an internal change for which we are responsible, but certainly not always. If instead of a blue and red sphere there had been a spherical vase filled with a blue liquid before us and this liquid was changed to red by a chemical reaction, then we would not have been able to bring our sensations back to their original state by means of a rotation. In general there are external changes of sensations, which are independent of our will and not accompanied by muscular sensations, and internal changes, which are voluntary and accompanied by muscular sensations.

[18] Sensible or representative space and geometrical space are the only two kinds of space Poincaré considers in his philosophy, Tactile space, visual space, and motor space are special forms of sensible space. In his philosophy there is no physical space per se. This exclusion is due in large part to his apparent identification of physical space with absolute space and his rejection of absolute space. He reasons that we can have no knowledge of physical space, i.e., absolute space. Cf. [Giedymin, 1977, pp. 280–281]. He advocates a relativistic philosophy of space. On these matters cf. [Poincaré, 1895b] [Poincaré, 1898] [Poincaré, 1903b] [Poincaré, 1906] [Poincaré, 1912].

Those external changes that we can correct by internal changes so as to reestablish our first impressions are called 'displacements'; the others that we cannot correct are called 'alterations.'

Furthermore, we proceed to classify those external changes that are displacements. This classification is crucial to Poincaré's theory. In the example of the rotating sphere he notes that if the hemispheres exhibited two other colors and then the sphere rotated as before, we could correct the external change of impressions, the displacement, by the same internal change as previously used. The internal change would be the same in that the muscular sensations would be the same, while the visual ones would not. Hence, in this case there is a variety of equivalent displacements. In geometrical terms, the class of all such equivalent displacements is supposed to correspond to a rotation of π radians or 180°. However, such geometrical terms are not yet available to Poincaré, since he is attempting to explain the origins of geometry and all its concepts. Similarly we may imagine that it is possible to distribute all the individual external displacements into classes. In effect, these are equivalence classes, to employ a mathematical term, which was not used in Poincaré's time but which would have been understandable to him. Two external displacements belong to the same equivalence class if they can be corrected by the same internal change, at least on the level of motor sensations. In a similar way we may consider internal displacements, i.e., internal changes that correspond to external displacements, and imagine that they are distributed into equivalence classes coextensive with those for the external displacements.[19]

Poincaré regards the individual displacements and especially the classes of displacements—which he also ambiguously calls 'displacements'—as fundamental to our creation of geometry. We learn to put the individual (external/internal) displacements into displacement classes by actively applying our minds to them. Essentially this double classification amounts to a process of abstracting pure motions, such as rotations and translations, from the sensations, while excluding incidental properties like color. Thus, in Poincaré's theory displacement classes on the level of sensations become identified with motions or transformations on the higher level of geometry. There are surely many difficulties attendant upon the theory, especially with respect to the precise recognition of individual displacements and the strict classification of them. The theory does not seem entirely plausible to me. Is it really likely that each of us comes to recognize more or less exact displacements through comparing our sensations, in particular, our muscular sensations? Is it likely that we develop our globally applicable theory of geometry out of such local phenomena? The situation is not clear. However, I shall not stop to criticize this empirical part of Poincaré's theory of geometry further.[20]

The key assumption of Poincaré's genetic theory, from which the remainder proceeds, is that each set of displacement classes (external and internal) forms a group in the mathematical sense [Poincaré, 1898, pp. 9–10]. In accordance with the mathematical view of his time he regards the closure property as the essential feature

---

[19] Cf. [Poincaré, 1903b], where he is somewhat clearer about these classifications.

[20] Cf. the penetrating criticism of [Torretti, 1978, pp. 343–345].

of a group, though he implicitly assumes the other properties (existence of identity and of inverses; associativity follows for any group of transformations).[21] In their *Theorie der Transformationsgruppen* (*Theory of Transformation Groups*) they define a group of transformations by requiring the closure property [Lie & Engel, 1888, page 3]. They mention the existence of identity and of inverses [Lie & Engel, 1888, page 2], but they do not include these properties in the definition and, in fact, stress that a group need not automatically include an identity transformation and inverse transformations [Lie & Engel, 1888, page 22]. Only much later in the book do they mention the associative property and prove that any group of transformations has the property [Lie & Engel, 1888, page 402]. The attitude of Lie and Engel is fairly typical of the time when Poincaré was writing. In other words, he primarily assumes that one (external/internal) displacement class A followed by another B together yield a displacement class C.[22] Actually he speaks loosely as if the individual displacements formed a group, because he normally identifies a displacement class with some individual member in it. Yet, strictly speaking, the individual displacements do not form a group. These depend upon the sensations of individual objects, and even equivalent displacements may start and end with distinct sensations, so that it is not always possible to combine two individual displacements by succession. It would appear that Poincaré tacitly assumes that for any two displacement classes there will always be appropriate individual displacements that can be combined according to the closure property.[23] Thus, one must realize that the group property for displacement classes is complicated and not immediately evident. Nevertheless, Poincaré asserts that if the displacement classes did not form a group, there would be no geometry. In this paper, as in the earlier one of 1887, he holds the view that group theory is the essence and true foundation of geometry. He had an important confirmation of this view in the work of Lie on the group-theoretic foundations of geometry, which had been recently published [Lie, 1887] [Lie, 1890a] [Lie, 1890b] [Lie & Engel, 1893, Abtheilung II, especially Theorem 9, page 139, Abtheilung V].

A philosophical question immediately becomes relevant: What justifies the assertion that the displacement classes form a mathematical group? Poincaré argues that the assertion is neither the result of *a priori* reasoning nor empirical [Poincaré, 1898, pp. 9–12, 41–43].[24] He dismisses the idea that it results from *a priori* reasoning, because it is not evident or absolutely certain. The very complexity of the compensation of external by internal changes and of their composition in classes leaves the assertion open to doubt. Hence it cannot be *a priori*. Yet neither is the assertion empirical. If it were, it would be open to revision and then geometry would be liable to destruction in the event of its being falsified. Indeed some likely candidates for

---

[21] One may compare Lie and Engel on the definition of group.

[22] For the group theory in his essay Poincaré uses an additive notation read from left to right, $A + B = C$. However, I prefer to use a 'functional' notation read from right to left, $BA = C$, because in effect one is dealing with a group of transformations that is not Abelian. See page 170 later in this chapter.

[23] Cf. [Poincaré, 1903b, pp. 407–410].

[24] On this matter cf. [Poincaré, 1895b, pp. 645–646] [Poincaré, 1903b, pp. 422–427] [Poincaré, 1912, pp. 498–502].

being displacements do not obey the group law. In other words, there already exist apparent refutations. However, we do not accept these as refutations. Rather, we adjust them to fit into the displacement classes to satisfy the group law. By disregarding the 'non-displacement' elements we force them into the group-theoretic mold, so to speak. We lay down a convention; indeed, a rather strong one. It is not an arbitrary convention, but one that is in line with rough experience. Those external changes that seem to satisfy the defining property for displacements also go into classes that obey the group law approximately. In the final analysis we enforce the law by convention, so that the construction of geometry based on group theory can take place. We elevate the group closure law for displacements to the status of a conventional principle, which we choose not to refute.

Having arrived at the fundamental convention that the displacement classes form a group, Poincaré can proceed to developing his full theory of the origins of geometry in accordance with his strongly-held view that group theory is at the foundation of all geometry. He feels able to bring to bear some group-theoretic results of Lie and others. At this point the mathematical part of his theory becomes more prominent than the empirical part. Within the mathematical part one finds his first theory of dimension.

9.3 Poincaré's First Theory of Dimension

What are the properties of the group of spatial displacements? [25] Poincaré spends several pages of his essay answering this question. He bases his answer on empirical as well as mathematical arguments. However, he emphasizes that he is studying the *formal* properties of the group [Poincaré, 1898, pp. 12–13]. Any isomorphic group will have the same properties. Thus, he is really studying the properties of a group in abstraction, although the group fits a concrete situation. In effect, he is studying an abstract group. About this time the abstract approach to the theory of groups was rapidly gaining currency among mathematicians. [26] With good reason Poincaré follows this approach, although in so doing he does not take up the axiomatic approach. His insistence that he is examining the group of displacements from an abstract standpoint is crucial for his first dimension theory, as will be seen presently.

In brief, he arrives at the following conclusions about the group of displacements. First, the very existence of the group insures that the resulting space is homogeneous, isotropic, and unbounded—properties that distinguish it from the representative space [Poincaré, 1898, page 12]. Second, the group is continuous [Poincaré, 1898, pp. 14–16]. To come to this conclusion he analyses the situation in experience, argues that displacement classes form a physical continuum, and then

[25] From now on when I use the word 'displacements' I mean displacement classes rather than individual displacements (unless otherwise stated).

[26] Among others, A. Cayley, W. von Dyck, G. Frobenius, H. Weber, and W. Burnside were prominent in their advocacy of the abstract approach to group theory. Cf. [Wussing, 1969, pp. 171–191] [Wussing, 1984, pp. 231–254].

contends that we substitute a perfect mathematical continuum according to his earlier theory of continua [Poincaré, 1893] (see above).[27] Third, the group is a 6-parameter group of transformations [Poincaré, 1898, pp. 18–21]. Every sufficiently small displacement can be generated by six infinitesimal rotations and translations. Finally, he suggests that we choose by convention the Euclidean group of rigid (orientation-preserving) motions (or direct isometries) as the best representation for the group of displacements. This conventional choice is motivated by the fact that the Euclidean group possesses an invariant (or normal) subgroup, the subgroup of all translations [Poincaré, 1898, page 21]. The alternative non-Euclidean groups do not have such subgroups, although they are conceivable representations of the group of displacements. Hence, the geometry of space becomes nothing but the study of the formal properties of a certain 6-parameter continuous group, the Euclidean group. According to Poincaré we impose this group on nature, though all along the way experience guides us in making our choice of the best group to apply to nature.

Poincaré regards the abstract group concept as an *a priori* element of our understanding [Poincaré, 1898, pp. 5–6, 41–43]. This view is highly significant for his entire philosophical theory of the origin and elaboration of geometry. According to him the mind has the innate ability to construct many continuous groups, and not just the Euclidean group, and to apply some of them to experience. It is this *a priori* part of our understanding that assures him that geometry can be exact and certain. His view contrasts with Kant's. For Poincaré the *a priori* element relevant to spatial geometry lies in our understanding, while for Kant it is a pure *a priori* form of our outer sensibility.

From the viewpoint of the history of mathematics it is not surprising that Poincaré should have given the group concept such a fundamental role in his philosophy. Around the turn of the twentieth century mathematicians increasingly saw group theory pervading almost every branch of their subject, bringing about a new unification. Poincaré himself promoted this trend of unification through many of his mathematical works.[28] One can properly regard his philosophy of geometry in *The Monist* essay as a philosophical expression of this trend. More specifically, his essay amounts to a philosophical rendition of Lie's work on the foundations of geometry based on continuous group theory [Lie, 1887] [Lie, 1890a] [Lie, 1890b] [Lie & Engel, 1893, Abtheilung II, especially Theorem 9, page 139, Abtheilung V]. Towards the end of his essay he refers directly to Lie's foundational work [Poincaré, 1898, page 37]. Furthermore, he implicitly relies upon Lie's solutions of the Riemann–Helmholtz–Lie space problem by contemplating only the three classical 6-parameter continuous groups of rigid motions, the Euclidean, the Gauss–Lobachevskii–Bolyaian, and the Riemannian,[29] as possible representations for the group of displacements. With this historical background one can see why Poincaré in his essay puts forward the view that the group concept, as an *a priori* concept, forms a natural and sure foundation for geometry.

---

[27] [Torretti, 1978, pp. 345–348] has pointed out an important difficulty in Poincaré's reasoning.

[28] Cf. [Poincaré, 1908a].

[29] For Poincaré the 'Riemannian group' can refer to either the elliptical group or the spherical group. Cf. footnote 7 earlier in this chapter.

In addition, one should realize that Poincaré's bringing in *continuous* group theory as his base geometrical theory implies that he is effectively bringing in topology, as one would understand the theory today. Lie's theory of continuous groups has developed and grown extensively into modern topological group theory. For Poincaré the topology is in the background, but is certainly there.[30]

I shall now consider Poincaré's first dimension theory within his broader theory of the genesis of spatial geometry. With respect to a continuous group of transformations Poincaré emphasizes the distinction between the order and the degree of the group [Poincaré, 1898, pp. 21–22]. The order of a continuous group is the number of essential parameters needed to define its operations. The Euclidean group of motions is a 6-parameter group, so its order is six. The degree of a continuous group is the number of coordinates of the number-manifold, i.e., the dimension number of the manifold, on which the group acts. Poincaré looks upon the order of a group as a formal property. It is a primary characteristic of the group regarded as an abstract object. The degree is a material element of secondary importance. It is not a characteristic of the group regarded as an abstract object. Two isomorphic groups of transformations will always have the same order, but they can have distinct degrees, because they can act on different manifolds. From the standpoint of contemporary topological group theory one would say that an abstract topological group of a given (finite) number of parameters or order can act on distinct manifolds having different dimensions or degrees yielding different groups of transformations, different representations.

For the group of displacements the degree is yet to be determined. In the first place this group acts on our sensations, subjecting them to changes. However, if we think of the sensations as forming a sensible space or sensible manifold, then this manifold would have as many 'coordinates' or dimensions as we have nerve fibers. Hence, it would have an exceedingly large dimension number, certainly much larger than one would expect for geometry. Consequently, Poincaré suggests that a much simpler manifold be found to replace this manifold of sensations, in order to derive a manageable and recognizable space on which the group of displacements can act. In other words, one derives a new and simpler group of displacement transformations acting on a simpler manifold that is in a sense isomorphic to the group of displacement transformations acting on the manifold of sensations.

How does one find a replacement for the manifold of sensations? In answer, Poincaré says that we initially observe that certain displacements conserve certain sensible elements. He asserts that we then identify certain subgroups of the principal group of displacements and recognize that the subgroups conserve certain spatial or geometrical elements. These subgroups form well-defined types comprising equivalent subgroups, and their corresponding conserved elements yield appropriate simpler manifolds as possible replacements. He examines three specific types of subgroups and the three corresponding possible replacement manifolds [Poincaré,

[30] See the contemporary treatment of continuous group theory as used in geometry in Fano's 1907 *Encyklopädie* (*Encyclopedia*) article [Fano, 1907]. In modern terms a Lie continuous group or Lie group is a group that is also a differentiable manifold. A modern topological group is more general. Brouwer helped enable the development of topological groups from Lie groups (see Chapter 11).

1898, pp. 23–24]. In this way subgroups become the key to his idea of replacement manifolds and his first theory of dimension. Part of Lie's work on continuous groups and their subgroups (as described, for example, in [Lie & Engel, 1888, pp. 270–289]) must have provided the fundamental mathematical background for his idea of replacement manifolds that are simpler. Moreover, he must have known about some important studies specifically concerned with the subgroups of the Euclidean group. Thirty years before Poincaré's essay was published, Camille Jordan had given a detailed analysis of the subgroups of the Euclidean group [Jordan, 1868]. More recently Eduard Study had examined this group and its subgroups by using some of the tools of Lie's group theory [Study, 1891].[31]

The following are some preliminaries to Poincaré's idea for defining a simpler manifold to be used as a replacement for the manifold of sensations. As noted above, he identifies the group of displacements, $\mathfrak{D}$, with the Euclidean group of rigid (orientation-preserving) motions, $\mathfrak{E}$. Thus the group $\mathfrak{D}$ should consist, like $\mathfrak{E}$, of rotations, translations, and screws or twists (i.e., rotations about axes and translations along the axes), but relative to sensations. In terms of modern topological group theory $\mathfrak{E}$ has a group manifold that is homeomorphic to $\mathbb{R}^3 \times \mathbb{RP}^3$, the topological product of Euclidean 3-space and real projective 3-space. In general, each Euclidean motion can be regarded as a translation of a given rectangular coordinate system of $\mathbb{R}^3$ followed by a rotation about a straight line (axis) passing through the translated origin of the system. The rotations about axes passing through a common point (origin) can be mapped onto $\mathbb{RP}^3$ according to the following mapping, which is a homeomorphism:

$$x = \tan(\theta/2)\cos\alpha, \quad y = \tan(\theta/2)\cos\beta, \quad z = \tan(\theta/2)\cos\gamma,$$

where θ is the angle of rotation ($-\pi < \theta \leq \pi$) and $\cos\alpha, \cos\beta, \cos\gamma$ are the direction cosines of the axis of rotation. Hence, the group manifold of $\mathfrak{E}$ is homeomorphic to $\mathbb{R}^3 \times \mathbb{RP}^3$. What Poincaré does is define some manifolds from selected subgroups of the group $\mathfrak{D} = \mathfrak{E}$ (viewed abstractly) on which this group can act, thereby generating some groups of transformations (representations) directly from the principal group.

Consider now the rotation of a solid body about a fixed place. It leaves a small patch of color unmoved. From this situation and others like it one may conclude that all the rotations around a center conserve the center as a fixed place. In other words, a rotative subgroup about a center of rotation serves to define a fixed point. Different rotative subgroups fix distinct points. Hence, on the empirical level there are fixed sensations or fixed places, while on the mathematical level there are corresponding rotative subgroups.

Proceeding further on the mathematical level, one notes that each of the rotative subgroups is a 3-parameter continuous subgroup. Moreover, they are all conjugates or transforms of one another. For a given subgroup $\mathfrak{G}$ and a given operation D of a group one says that the set $\mathfrak{G}' = D\mathfrak{G}D^{-1}$ is a conjugate or transform of $\mathfrak{G}$. The conjugate $\mathfrak{G}'$ will also be a subgroup. If one takes a fixed rotative subgroup $\mathfrak{R}_0$

[31] Cf. also H. Wiener's papers in volumes 42, 43, and 45 of *Berichte über die Verhandlungen der Sächsischen Akademie der Wissenschaften zu Leipzig: Mathematisch-Physische Klasse.*

and forms all conjugates, $D\mathfrak{R}_0D^{-1}$ (where D represents an arbitrary operation of the group), one derives all the rotative subgroups. In fact, if one restricts the D to just the translations, one derives the conjugate class of all rotative subgroups. Given expressions involving displacements such as 'DED^{-1}' Poincaré would interpret them from left to right. (His notation includes the addition sign '+'.) However, it is more appropriate to read such expressions functionally from right to left in accordance with a theorem of Jordan [Jordan, 1868, page 171], to which Poincaré implicitly refers [Poincaré, 1898, page 17]. Jordan's theorem states that the conjugate DED^{-1} of E (where D, E are displacements, i.e., rotations, translations, or screws) is the same type of displacement as E except that the axis or direction of DED^{-1} is the axis or direction of E displaced by D. For example, if E is a rotation about an axis, then DED^{-1} is a rotation of the same angle about the axis of E displaced by D. Further, if one lets $\mathfrak{R}_0$ represent all the rotations about axes passing through a given point 0 (which one may designate as the origin), then by Jordan's theorem $D\mathfrak{R}_0D^{-1}$ represents all rotations about axes passing through $D(0)$, the point 0 displaced by D, and, in particular, $T\mathfrak{R}_0T^{-1}$, where T is a translation, represents all rotations about axes passing through $T(0)$, the point 0 translated by T. In Lie's terms these subgroups are all equivalent (gleichberechtigt) and belong to the same type (Typus). Poincaré thinks of this type or conjugate class as forming a space or manifold. Mathematically, the type or conjugate space of rotative subgroups is homeomorphic to the ordinary Euclidean 3-space. By a simple modern argument one can straightforwardly show that the conjugate space of rotative subgroups, $\Gamma = \{T\mathfrak{R}_0T^{-1} \mid T \text{ is a translation}; \mathfrak{R}_0, \text{ the set of all rotations about axes passing through } 0\}$, is homeomorphic to $\mathbb{R}^3$. Define the mapping

$$f : \Gamma \to \mathbb{R}^3 \text{ by } f : T\mathfrak{R}_0T^{-1} \mapsto T(0) = p;$$

define $\Delta \subset \Gamma$ to be open if and only if $f(\Delta) = U \subset \mathbb{R}^3$ is open in the usual topology. f is a homeomorphism. It is one-one (injective) because distinct rotative subgroups $T\mathfrak{R}_0T^{-1}$ are defined by distinct translations T, and it is onto (surjective) because for each point p of $\mathbb{R}^3$ there is a unique translation T such that $T(0) = p$. Moreover, the topology defined for Γ is equivalent to the usual topology for $\mathbb{R}^3$. Empirically, the type defines a space of ordinary points or places, the usual 3-dimensional space. One can calculate the dimension number of the space as the difference between the number of parameters of the group of displacements $\mathfrak{D} = \mathfrak{E}$ and the number of parameters of any rotative subgroup: $6 - 3 = 3$. Poincaré does not explain how the main group of displacements acts on the conjugate space, but, since the space is homeomorphic to ordinary Euclidean 3-space, he apparently thinks that the group action is indirectly definable in a straightforward manner. Consequently, one derives a reduced 3-dimensional manifold homeomorphic to Euclidean 3-space, on which the group $\mathfrak{D} = \mathfrak{E}$ can act, and a representation of $\mathfrak{D} = \mathfrak{E}$, which is similar to the usual one. It is possible to complete Poincaré's sketch of the construction of the conjugate space of rotative subgroups with a small amount of modern topological group theory. One defines the action of $\mathfrak{D} = \mathfrak{E}$ on $\Gamma = \{T\mathfrak{R}_0T^{-1}\}$ to be conjugation:

$$\mathfrak{E} \times \Gamma \to \Gamma; \quad (D, T\mathfrak{R}_0 T^{-1}) \mapsto DT\mathfrak{R}_0 T^{-1} D^{-1},$$

($DT\mathfrak{R}_0T^{-1}D^{-1} = U\mathfrak{R}_0U^{-1}$, where U is a translation such that $(DT)(0) = U(0)$.) It is easy to show that one has a topological transformation group $(\mathfrak{E}, \Gamma)$ that is a faithful representation of $\mathfrak{E}$. $\mathfrak{E}$ is transitive on Γ, i.e., for each pair of members of Γ there is a displacement of $\mathfrak{E}$ mapping one member of the pair to the other. Hence, by definition Γ is a homogeneous space. The action of $\mathfrak{E}$ on Γ by conjugation is equivalent to the usual action of $\mathfrak{E}$ on $\mathbb{R}^3$, since the following diagrams commute:

$$\begin{array}{ccccccc}
\mathfrak{E} \times \Gamma & \xrightarrow{1 \times f} & \mathfrak{E} \times \mathbb{R}^3 & \quad & (D, T\mathfrak{R}_0 T^{-1}) & \longmapsto & (D, T(0)) \\
\downarrow & & \downarrow & & \updownarrow & & \updownarrow \\
\Gamma & \xrightarrow{f} & \mathbb{R}^3 & & DT\mathfrak{R}_0 T^{-1} D^{-1} & \longmapsto & (DT)(0) \\
& & & & = U\mathfrak{R}_0 U^{-1} & & = U(0).
\end{array}$$

Hence, one can say that the representation $(\mathfrak{E}, \Gamma)$ is similar to the usual representation of the 6-parameter Euclidean group, $(\mathfrak{E}, \mathbb{R}^3)$.[32]

Any two rotative subgroups have elements in common. These form a 1-parameter subgroup, a rotative sheaf comprising the rotations about a fixed axis (an Abelian subgroup). These 1-parameter subgroups effectively define fixed straight lines in our space. All the rotative sheaves are conjugates of one another. If one takes a fixed sheaf $\mathfrak{S}_0$ and forms all conjugates, $D\mathfrak{S}_0D^{-1}$ (where D represents an arbitrary operation of the group), one derives the conjugate class of all rotative sheaves. In Lie's terms they all belong to the same type. According to Poincaré they form a space or manifold. Mathematically, this conjugate space of rotative sheaves is homeomorphic to an unusual 5-dimensional space. The conjugate space of rotative sheaves is homeomorphic to the unusual 5-dimensional space $\mathbb{R}^3 \times \mathbb{RP}^2$, the topological product of Euclidean 3-space and the real projective plane. Empirically, one finds the space of fixed straight lines or lines with fixed points on themselves. One can calculate the dimension number of the conjugate space as the difference between the number of parameters of the principal group and the number of parameters of any rotative sheaf: $6 - 1 = 5$. Again, Poincaré does not explain how the principal group of displacements acts on the conjugate space, though, knowing that the space is homeomorphic to the 5-dimensional space of fixed lines, he probably thinks that the group action is indirectly definable. In this way one derives an unusual representation of $\mathfrak{D} = \mathfrak{E}$. It is possible to complete Poincaré's sketch of the construction of the conjugate space of rotative sheaves with a small amount of modern topological group theory. One would define the action to be conjugation. I omit the details.

Finally, Poincaré considers a third type of subgroup to derive another reduced manifold. The subgroup is the helicoidal subgroup consisting of combinations of rotations about a fixed axis and translations parallel to the axis, i.e., screws or twists

---

32 Cf. [Torretti, 1978, pp. 350–351]). The original 1978 version contained an error, which Torretti corrected in [Torretti, 1984, page xiii]. I wish to thank J.J. Gray for some help with this formal version of Poincaré's sketchy construction.

(an Abelian subgroup). The helicoidal subgroups clearly are 2-parameter subgroups. They effectively define straight lines that can slide along themselves. They all are conjugates of one another. If one takes a fixed helicoidal subgroup $\mathfrak{H}_0$ and forms all conjugates, $D\mathfrak{H}_0D^{-1}$ (where D represents an arbitrary operation of the group), one derives the conjugate class or type of all helicoidal subgroups. These form a space. Mathematically, this conjugate space of helicoidal subgroups is homeomorphic to a 4-dimensional Plückerian space of lines. The conjugate space of helicoidal subgroups is homeomorphic to the unusual 4-dimensional space $\mathbb{R}^2 \times \mathbb{RP}^2$, the topological product of the Euclidean plane and the real projective plane. Empirically, it is also a space of straight lines. One can again calculate the dimension number of the space as the difference between the number of parameters of the principal group of displacements and the number of parameters of any helicoidal subgroup: $6 - 2 = 4$. Again, he probably thinks that the action of the group of displacements on the conjugate space is indirectly definable through its action on the homeomorphic space of straight lines. One thus derives another unusual representation of $\mathfrak{D} = \mathfrak{E}$. It is possible to complete Poincaré's sketch of the construction of the conjugate space of helicoidal subgroups with a small amount of modern topological group theory. One would again define the action to be conjugation. I omit the details.

Poincaré has thus shown how to derive three special manifolds on which the group of displacements can act: the 3-dimensional space of points related to the type of rotative subgroups, the 4-dimensional space of lines related to the type of helicoidal subgroups, and the 5-dimensional space of lines with fixed points on themselves related to the type of rotative sheaves. In his derivation Poincaré clearly has made a direct application of Plücker's equivalence principle for constructing spaces of different dimension numbers by choosing different basic elements.

In the second volume of his very first book, *Analytisch-geometrische Entwicklungen* (*Analytical-Geometrical Developments*), Julius Plücker (1801–1868)[33] had hinted at the construction of a variety of spaces and geometries through coordinate systems deriving from basic figures (i.e., points, lines, etc.) of ordinary geometry [Plücker, 1831, Vorrede, pp. 283–284]. He had suggested that the geometries would be equivalent according to a general unifying principle. This principle would be analogous to the Poncelet–Gergonne duality principle of projective geometry. (He had himself contributed to the foundation of the latter principle.) Then in his *System der Geometrie des Raumes in neuer analytischer Behandlungsweise* (*System of the Geometry of Space in a New Analytical Presentation*) he had briefly indicated how, in particular, it is possible to consider the ordinary 3-dimensional space of points as a 4-dimensional space of straight lines [Plücker, 1846, pp. 322–323; cf. Chapter 1]. From this remark Plücker later developed his line geometry during the final years of his life. His book *Neue Geometrie des Raumes gegründet auf Betrachtung der geraden Linie als Raumelement* (*New Geometry of Space Founded on Consideration of the Straight Line as the Space Element*) [Plücker, 1868] [Plücker, 1869] stands as the culmination of this development of his thought.[34] In full generality Plücker's

[33] For biographical information on Plücker cf.
`https://mathshistory.st-andrews.ac.uk/Biographies/Plucker/`.

[34] Cf. Plücker's earlier paper [Plücker, 1865].

equivalence principle means that from a given space, e.g., the ordinary Euclidean 3-space, it is possible to construct a whole infinity of different spaces with different dimension numbers by using different basic elements (points, lines, etc.) and thereby build different but equivalent geometries.[35] It is this principle that Poincaré has put to use in constructing the manifolds of points, lines, and fixed lines on which the Euclidean group can act in order to produce various groups of transformations (representations) and alternative geometries.

The geometrical ideas of Plücker directly inspired the young Sophus Lie. In his important early paper 'Über Komplexe, insbesondere Linien- und Kugelkomplexe, mit Anwendung auf die Theorie partieller Differentialgleichungen' ('Concerning Complexes, especially Line and Sphere Complexes, with Application to the Theory of Partial Differential Equations') [Lie, 1872] Lie follows in the footsteps of Plücker. He considers the equivalence principle to be of fundamental importance. He extends it when applying it to derive his theory of the contact transformation.

The works of both Plücker and Lie strongly influenced the work of Poincaré. One can easily detect their influence on Poincaré's first philosophical investigations into dimension. In his essay of 1898 Poincaré whole-heartedly adopts the group-theoretic view of geometry, particularly as Lie championed it. In this context he applies the Plücker equivalence principle in order to develop a special 'relativistic' view of dimension in spatial geometry.[36] Overall, one can view him as following a particular important tradition of geometrical thinking of the late nineteenth century. His philosophical views are closely bound up with this tradition.[37]

In his essay Poincaré declares [Poincaré, 1898, page 24] that we choose three as the dimension number of our space, because it is the smallest number and because the associated subgroup/element recommends itself, being connected with our empirical ideas of place and point. In this choice another conventional principle of Poincaré's philosophy is apparent. There are two aspects of the convention. First, one can regard it as a simple convention of language (cf. [Poincaré, 1898, pp. 65–67]). In the formal theory of the Euclidean group one can specify a particular kind of subgroup/element for the construction of a space or manifold for the group action. The denoting of a particular subgroup/element as basic is a simple convention of language. As such it is quite arbitrary. For example, one could treat the helicoidal subgroup and corresponding line as basic and so derive a 4-dimensional space. However, by taking the rotative subgroup and point as the basic element one obtains the more usual 3-dimensional space. Yet beyond this arbitrary linguistic aspect of the chosen convention there is another aspect that is more characteristic of Poincaré's conventionalism. According to him our experience directs us to choose the point as the basic element of space. We fix upon the point as the best choice of element through the application of the group of displacements (displacement classes) to rough

[35] Cf. [Clebsch, 1872].

[36] In his essay Poincaré also mentions the work of Karl Georg Christian von Staudt on the 'geometry of position' [Staudt, 1847] [Staudt, 1856] [Staudt, 1857] [Staudt, 1860], but does not seriously apply it to his work [Poincaré, 1898, pp. 34–37].

[37] Rougier [Rougier, 1920] and Giedymin [Giedymin, 1977] have stressed the significance of this geometrical tradition for understanding Poincaré's geometrical conventionalism.

experience. To be sure, it is difficult to determine the notion of point or even the less precise but more immediate notion of place in our experience. Poincaré devotes several pages of his essay to analyzing this difficulty [Poincaré, 1898, pp. 24–32]. However, in the end we are led to adopt the strong convention that associates the empirical notion of place with the rotative subgroup. This convention is similar to the one asserting that the displacement classes form a group. There are many difficulties involved in our experience. Nevertheless, the convention is genuinely useful, since it provides a bridge from rough experience to exact geometry.

One should clearly recognize that, in creating a variety of spaces of differing dimensions by Plücker's principle, Poincaré always remains within the confines of the 6-parameter Euclidean group or by implication within the confines of the 6-parameter non-Euclidean groups. These 6-parameter groups are *a priori* elements of his theory of dimension. He is only willing to contemplate spaces and geometries associated with these groups. Accepting Lie's solution of the Riemann–Helmholtz–Lie space problem as a bound for his philosophical research, he only contemplates three groups and geometries and their corresponding spaces.

I should stress that Poincaré does accept the idea that our space is or ought to be regarded as 3-dimensional, and not 4- or 5-dimensional. Unusual spaces under nonstandard interpretations are certainly possible, but they are not as convenient as the usual one for our geometrical deliberations. Ultimately, the 3-dimensionality of our space rests upon a reasonable convention in line with common experience.

In ascertaining the dimension of space from the manifolds constructed out of the subgroups of the 6-parameter group of displacements Poincaré states that he differs from Helmholtz and Lie on the philosophy of the foundations of geometry [Poincaré, 1898, pp. 40–41]. While both Lie and Helmholtz assume that space is a 3-fold number-manifold (Zahlenmannigfaltigkeit) as an axiom for the foundations of geometry,[38] Poincaré does not make this assumption. Indeed he definitely wishes to avoid making it; otherwise he would be caught in the circle in which Couturat [Couturat, 1896, pp. 659–660] had earlier accused him of being ensnared (cf. especially [Poincaré, 1898, page 40]). In effect, he would be assuming the dimension number of space as three before proceeding to derive it. To avoid this *petitio principii* he starts with the 6-parameter group of displacements. He regards this as a formal (or abstract) continuous group with six essential parameters. First, it is possible to determine its formal properties. Then afterwards, it is possible to construct the 3-dimensional manifold of conjugate rotative subgroups/points on which the group can act. In Poincaré's terms the degree, 3, of the group and its matter are determined subsequently. In modern terms one has a representation of the formal group, the 6-parameter Euclidean group acting on the specially constructed 3-dimensional space. The space is derived and not assumed in the genesis of geometry. Although other representations are possible, experience guides us to this particular one as the most appropriate.

[38] Cf. [Lie & Engel, 1893, pp. 438–439, 452, 461, 506].

In his essay of 1898 Poincaré appears confident that his first theory of dimension is adequately or is indeed correctly founded upon continuous group theory and that it adequately explains the possible dimensions assignable to our space. From the mathematical standpoint his explanation is certainly interesting. It is a significant and even clever application of elementary continuous group representation theory. However, from the philosophical standpoint some may not regard his explanation as entirely satisfactory. It is quite complicated and very indirect. Explaining the conventionally preferred 3-dimensionality of our space by means of a special representation of the Euclidean group of rigid motions acting on the conjugate space of rotative subgroups appears as a very roundabout way of explaining a seemingly simple fact about our known space. One may wonder if the explanation is entirely plausible. Also, while Poincaré explains the 3-dimensionality of space, he does not explain the concept of dimension itself. The first theory does not appear to explain what dimension is per se.[39] Poincaré himself may have began to see the need to improve it shortly after its publication in the essay in *The Monist* of 1898. On this issue see the next Section 9.4.

Poincaré calls his theory of the dimension number of space based on the continuous group of displacements a 'dynamical theory.' However, when investigating the notion of place as an empirical substitute for the mathematical concept of point, he also puts forward very tentatively a 'statical theory' of the dimension of space. In accordance with his general psycho-physiological theory we can perceive each place in space only inexactly. Our sensations of a place form something like a 'wafer.' A place corresponds to a fuzzy area in sensible space. We can imagine sensible space as a series or collection of overlapping wafers with contiguous wafers associated with one another. This picture is a generalization of his idea of the 1-dimensional physical continuum mentioned above (cf. [Poincaré, 1893]). Concerning the dimension of the perceptual 'wafer space' he says the following [Poincaré, 1898, page 29]:

> If these wafers were to form by their superposition only a sort of one-dimensional ribbon, we should recognise it, because the associations . . . obey a law that may be stated as follows: if A is associated at once with B, C, and D, D is associated with B or with C. This law would not be true if our wafers covered by their superposition a plane or a space of more than two dimensions. When I say, therefore, that all possible places constitute an aggregate of one dimension or of more than one dimension, I mean to say simply that this law is true or that it is false. When I say that they constitute an aggregate of two or three dimensions, I simply affirm that certain analogous laws are true.

This explanation of dimension based on a covering of a space by overlapping sensation-wafers is not entirely clear. In particular, the associative law for a 1-dimensional wafer space does not appear to be stated correctly. Perhaps the translator of Poincaré's original French manuscript (T.J. McCormack) introduced an error at

[39] Some years ago Professor R. Torretti had suggested to me that Poincaré's first theory of dimension should not be considered as a genuine theory of dimension, because it does not include any explanation of the concept of dimension. Indeed it is only a theory of dimension in the sense that it offers an explanation of the dimension number of our space.

this point or perhaps the author himself made a mistake. Still, one could probably devise a sequence of correct laws for distinguishing wafer coverings of spaces of various dimensions along the lines suggested. Be that as it may, Poincaré eventually comes to the conclusion that it is very difficult to provide a fully adequate explanation of the concept of dimension by means of a statical theory [Poincaré, 1898, page 32]:

> ... the introduction of a group, more or less complicated, appears to be absolutely necessary. Every purely statical theory of the number of dimensions will give rise to many difficulties, and it will always be necessary to fall back upon a dynamical theory.

9.4 Poincaré's Second Theory of Dimension

In spite of this last negative and somewhat 'pessimistic' remark Poincaré continued to work on resolving the difficulties of constructing a 'statical theory' of dimension in the years just after the publication of his essay in *The Monist*. The result was a new paper, which appeared in 1903, 'L'espace et ses trois dimensions' ('Space and its three dimensions') [Poincaré, 1903b],[40] containing a novel explanation of the concept of dimension itself. Several years later he then published a modified and extended version of the new dimension theory in an article on 'Pourquoi l'espace a trois dimensions' ('Why space has three dimensions') [Poincaré, 1912].

These important papers of 1903 and 1912 contain a substantial modification to Poincaré's philosophical theory of the genesis of spatial geometry.[41] The centerpiece of the revised theory is his second theory of dimension. In the two papers he distinguishes very clearly between the mainly mathematical problem of defining the dimension concept and the epistemological, physical, or cosmological problem of explaining why we attribute three dimensions to our space. The striking feature of the new theory for defining dimension is that it is explicitly a topological theory, based on a notion of cut (coupure). The topological approach offers a direct way of explaining the concept of dimension. In contrast, the group-theoretic approach of his first theory of dimension using a group of transformations and its subgroups did not provide an explanation of the dimension concept per se. Also, the detour via the 6-parameter group of displacements was an indirect way of solving the epistemological problem. Poincaré admits this fact in his paper of 1903 [Poincaré, 1903b, pp. 409–410]. The earlier theory lacks a certain intuitive appeal. It seems unlikely that the creation of our familiar conception of a 3-dimensional space follows so indirectly from a 6-parameter group. For these reasons Poincaré realized some time between 1898 and 1903 that a better route to dimension and spatial geometry was needed, so he devised his explicitly topological theory.

[40] In [Poincaré, 1902d, chapitre II] there is a preliminary version of the definition of dimension given in [Poincaré, 1903b].

[41] Cf. [Poincaré, 1906], which also contains a revision of his theory of the genesis of spatial geometry but without the new dimension theory.

In the years around this time he was publishing his fundamental mathematical papers on analysis situs, that is, topology, essentially based on homology theory and the fundamental group: 'Analysis situs' [Poincaré, 1895a], eventually to include five 'compléments' or 'supplements' [Poincaré, 1899] [Poincaré, 1900] [Poincaré, 1902c] [Poincaré, 1902e] [Poincaré, 1904] (English translation [Poincaré, 2010]). It was natural for him to turn to a topological explanation of dimension. In the first of these mathematical papers, 'Analysis situs,' he introduced the concept of cut (coupure) [Poincaré, 1895a, page 42] [Poincaré, 1953, page 225] [Poincaré, 2010, page 28] and used it effectively in this series of papers. In his philosophical theory of dimension of a few years later he was able to make the concept fundamental.

Another event happened that altered Poincaré's view of the foundations of geometry: the publication of David Hilbert's *Grundlagen der Geometrie* (*Foundations of Geometry*) in 1899 (first edition) [Hilbert, 1899]. Before the appearance of this important work Poincaré was seemingly unaware of the developments in the pure logic of the foundations of geometry stemming from the investigations of Pasch, the Italian geometers, and Hilbert. In his extended review of Hilbert's *Grundlagen der Geometrie* (*Foundations of Geometry*) of 1902 [Poincaré, 1902a] [Poincaré, 1903a] he gives clear evidence of surprise at discovering the new approach to geometry's foundations that is described so forcefully and succinctly in Hilbert's little masterpiece.[42] He seems amazed at the wide variety of geometries that are conceivable by using the logical or axiomatic-deductive method. As we know Poincaré had grown up with the group-theoretic approach to geometry of Helmholtz, Klein, and above all Lie. Now the logical approach offered entirely new possibilities, which he clearly recognizes in his review. Of course, he did not entirely give up the group-theoretic viewpoint once he had read and reviewed Hilbert's work. Nevertheless, in the last years of his career there was a discernible 'backing away' from it. In his papers on dimension of 1903 and 1912 he no longer sees group theory as providing the complete foundation for the genesis of geometrical knowledge. Yet in partly retreating from the group-theoretic approach, he did not thereby rush to adopt Hilbert's thoroughly logical and axiomatic approach, for he considered it to be too formal. Rather he appears increasingly to consider analysis situs or topology as the bedrock of geometry.

I shall now examine the details of Poincaré's second, topological theory of dimension. In his first paper on space and topological dimension of 1903 Poincaré provides a definition of dimension based on cuts that is only applicable to physical continua [Poincaré, 1903b, pp. 286–290], whereas in his second paper of 1912 he gives two closely related definitions based on cuts, the one for mathematical continua and the other for physical continua. In these papers he applies the given definitions to various continua. In both papers his final goal remains the same: to explain why we say our space is 3-dimensional.[43] For the present discussion I shall just consider the mathematical part of Poincaré's revised philosophical theory of spatial geometry. I

---

[42] [Freudenthal, 1960, page 23] has rightly stressed this important historical point concerning the development of Poincaré's philosophical ideas on geometry.

[43] Cf. also [Poincaré, 1906].

shall follow the more mathematical exposition of the later paper of 1912 [Poincaré, 1912, pp. 486–490].

In the paper of 1912 Poincaré mentions the 'paradoxical' result of Cantor on the equivalence of continua of different dimensions through one-one correspondences. This is the very first time he has drawn attention to this result of such vital importance in the history of dimension theory. He immediately 'explains' the paradox by pointing to the well-known fact that Cantor's correspondences are not continuous. He then gives a general statement of the invariance of dimension theorem [Poincaré, 1912, page 486]:

> It is therefore possible to *deform* a plane in such a way as to obtain a straight line, provided that this deformation is not continuous. On the contrary, this would be impossible with a continuous deformation. Thus the question of the number of dimensions is closely linked to the notion of continuity, and it would have no meaning for anyone who wished to exclude this notion.[Ch9, 3]

Poincaré never tackled the dimensional invariance problem himself, but he knew of Brouwer's results resolving various topological invariance questions. They had a short correspondence in December 1911 and January 1912. L.E.J. Brouwer, a young man who was fairly new to the field of topology, opened the correspondence with a letter to Poincaré dated a little before 10 December 1911 [Dalen, 2011a, pp. 112–114] [Dalen, 2011b, pp. 473–474]. In the letter Brouwer asks Poincaré's permission to publish something relative to Poincaré's mathematical work. He included with the letter three articles of his own, one being an offprint of his 'Beweis der Invarianz des n-dimensionalen Gebiets' ('Proof of the invariance of the n-dimensional domain') [Brouwer, 1912f] and another possibly being an offprint of his 'Beweis der Invarianz der Dimensionenzahl' ('Proof of the invariance of the dimension number') [Brouwer, 1911a].[44] Poincaré's reply to Brouwer is postmarked 10 December 1911 [Dalen, 2011a, pp. 114–115] [Dalen, 2011b, page 475] (see also Alexandroff's article [Alexandroff, 1972]). He answers Brouwer's technical concerns and questions and makes a very complimentary remark at the end [Dalen, 2011a, page 115] [Dalen, 2011b, page 475]:

> ... I am happy to have an opportunity to be in contact with a man of your merit.[Ch9, 4]

There are apparently only two other surviving letters in the correspondence, which are dated around 4 January 1912 [Dalen, 2011a, pp. 120–121] [Dalen, 2011b, pp. 493–494]. They cover other technical matters.

It is not clear what level of importance Poincaré ascribed to a proof of dimensional invariance. In the paper of 1912 he emphasizes that the coordinate definition of an n-dimensional mathematical continuum is quite satisfactory, indeed 'irreproachable'

[44] Dirk van Dalen, editor of the correspondence, does not think that this latter paper was one of the three, given the topic of discussion [Dalen, 2011a, page 112, footnote (85)] [Dalen, 2011b, page 473, footnote (2)]. Nonetheless, the dimensional invariance paper is closely related to the discussion and the three papers that van Dalen suggests were included.

or 'faultless' ('irréprochable'), from the mathematical point of view [Poincaré, 1912, pp. 486–487]. However, from the philosophical viewpoint, i.e., the viewpoint of the foundations of mathematical knowledge, he does not find the definition to be entirely satisfactory. Its primary fault is that it lacks intuitive appeal. It obscures the intuitive origin of the notion of a continuum of n dimensions and all the rich ideas contained in it. As a result he is led to propose a new and philosophically more satisfactory definition of n-dimensional continuum. Thus, all along his sole motive for defining the concept has been to improve his solution to the fundamental epistemological problem. He wants a definition with direct intuitive appeal, one with immediate explanatory power.

Poincaré prefaces his cut definition for mathematical continua with some simple motivating examples. A (simple) closed curve, a typical 1-dimensional continuum, can be cut into two pieces by any pair of points on the curve. We cannot pass from one piece to the other except by passing through one of the cut points. However, on a closed surface, a typical 2-dimensional continuum, the exclusion of finitely many isolated points is not sufficient to divide the surface into parts. It is always possible to go around the excluded points in order to pass from one position on the surface to any other. Rather we need 1-dimensional closed curves to produce divisions of the surface into parts. Similarly in 3-dimensional space we need 2-dimensional closed surfaces to obtain divisions of the space into parts. These examples suggest the following definition of dimension applicable to continua [Poincaré, 1912, page488]:

> A continuum has n dimensions when it is possible to divide it into several parts by means of one or more cuts which are themselves continua of $n - 1$ dimensions. The continuum of n dimensions is thus defined by the continuum of $n - 1$ dimensions; this is a definition by recursion.[Ch9, 5]

Poincaré is convinced of the natural and intuitive character of his definition. He cites two pieces of evidence for the correctness of the definition. First, he notes that it is in harmony with the traditional definitions of the basic elements of geometry that have been handed down to us from the ancients. These definitions, which can be found in Euclid's *Elements* and some of the writings of Aristotle, say that surfaces are the boundaries of solids; lines, the boundaries of surfaces; and points, the boundaries of lines (see Chapter 2). Second, he recognizes the close analogy between his inductive definition proposed for distinguishing continua of different dimension numbers and Riemann's definition of connectivity for classifying (2-dimensional) surfaces and its higher-dimensional generalizations. Poincaré's cuts are similar to cross cuts (Querschnitte, Riemann's term) or more exactly to loop cuts (Rückkehrschnitte, Carl Neumann's term). Indeed the Riemannian sphere of ideas is fundamental to much of topology, as Poincaré well knew from his own work in the field. Thus, the new definition of dimension fits in well with the best mathematical traditions. This fact suggests to him that he has discovered the natural origin of the notion of dimension. He can feel confident in using it in his revised theory of the genesis of geometry.

After giving the cut definition of dimension for mathematical continua, Poincaré immediately provides a modified cut definition that is directly applicable to physical continua [Poincaré, 1912, pp. 488–490]. In the rest of the paper he then applies his twin definitions to various continua. He determines the dimension numbers of physical continua that are especially important in his revised theory of the genesis of space and geometry: the tactile, visual, and motor spaces [Poincaré, 1912, pp. 490–498]. Then in a most interesting section he adopts the viewpoint of mathematical physics, leaving aside his usual psycho-physiological (subjectivist) viewpoint starting from sensations, and considers how it is possible to write the laws of celestial physics for various mathematical spaces (or continua) of different dimension numbers [Poincaré, 1912, pp. 499–502]. The alternative laws for the different spaces can be made parallel or equivalent, but he argues that a 3-dimensional mathematical space of points is still the best choice of space for which to write the laws of physics. Significantly he bases his brief argument on the 6-parameter Euclidean group. The principal aim of his entire paper is to show why one should regard our space as 3-dimensional, even though spaces with other dimension numbers are conceivable and indeed relevant to the origins of our spatial ideas and our usual spatial geometry.

Although his topological definition of dimension for mathematical continua is the essential kernel of a possible mathematical theory, Poincaré spent no time developing its mathematical consequences, but only applied it along with its twin definition for physical continua in the solution of his epistemological problem. He did not even specify the general types of mathematical continua to which it is supposed to apply.[45] Yet the cut definition itself is a simple brilliant idea—a product of genius in harmony with geometrical tradition. Perhaps not surprisingly, it is open to objections; it needs improvement in order to make it serve as the basis for a mathematical theory.[46] Still, it is inspired, and if Poincaré did not see fit to pursue its mathematical consequences, then the other master of topology of the beginning of the twentieth century, L.E.J. Brouwer, did (as we shall see in Chapters 11 and 12).

9.5 Concluding Remarks on Poincaré's Philosophy of Geometry

To complete my examination of Poincaré's theories of dimension I would like to make some remarks on his general philosophy of geometry, his geometrical conventionalism, which encompasses these theories.[47]

---

[45] Perhaps in line with his own work in analysis situs Poincaré thought that the domain of application of his definition should be limited more or less to topological manifolds.

[46] Frigyes Riesz [Riesz, 1909a, page 313] criticized Poincaré's theory as formulated in [Poincaré, 1903b] (see Section 10.2) and then Brouwer [Brouwer, 1913b] gently but carefully criticized it as formulated in [Poincaré, 1912].

[47] These few remarks do not constitute a full exposition of Poincaré's geometrical conventionalism. I again refer the reader to [Rougier, 1920], [Giedymin, 1977], and [Torretti, 1978, pp. 320–358] (or [Torretti, 1984, pp. 320–358]), [Zahar, 1997], and [Soltani, 2020].

Poincaré's geometrical conventionalism is a response to a problem of knowledge, one of enduring importance. What is the epistemological status of the geometry or geometries most closely related or applicable to space? In the first place he accepts geometry as an exact mathematical science, as a part of assured mathematical knowledge. He regards the geometry or the geometries most suitable for describing our space as exact. Yet, in his view experience cannot provide us with absolutely certain knowledge, because experimental results are always open to revision. Geometry, particularly the branch that applies to our space, cannot be an empirical science but must be related to empirical science. Spatial geometry must be based in and relate to experience. The question becomes: How do we account for the relationship between spatial geometry and experience, in view of the fact that there is an enormous logical gap between exact mathematics and inexact empirical knowledge? He rejects the Kantian rationalist solution based on a pure *a priori* form of our outer sensibility. Recognition of the equal status of Euclidean and the alternative non-Euclidean geometries led him to this rejection. Instead, he puts forward his geometrical conventionalism as a middle course between geometrical empiricism and Kantian rationalism. The precise connection between experience and spatial geometry is a matter of conventions. We *impose* a geometry on 'our space' through conventions, although we are guided to the various conventions by experience. We choose those conventions that are most appropriate or attuned to experience and lift them to the status of agreed principles, thereby protecting them from future refutation. Conventions are not empirically testable, though it is conceivable that for adequate reasons we might at some time replace an agreed convention by an alternative. Conventions serve as bridges from inexact empirical findings and results to precise mathematics. Poincaré's conventionalism in geometry is in a sense an attempt to 'reconcile' the certainty of mathematics with the fallibility of experience.

When discussing geometry in his philosophical writings Poincaré is always thinking of applied or applicable geometry as opposed to pure geometry, although he neither makes the pure/applied distinction nor uses the terms explicitly. He only considers the geometries most applicable to 'our space.' This is the space that we partly create from the analysis of psycho-physiological associations. He never deals with physical space. Indeed he rejects physical space as a concept in his philosophy on various grounds. Geometrical space is the space about which we can reason. A spatial geometry is a language employed in physical theories for discussing natural phenomena. There are alternative geometrical languages. At first they have equal status and are intertranslatable or interchangeable in a sense. Each can provide a way of viewing natural phenomena.

Poincaré's view of the nature of space and geometry arises from his fundamental epistemological assumptions. From the start he adopts a sensationalist and associational psychology that shapes his theory of the genesis of geometry. We do not know space directly. Certainly we cannot perceive physical space or absolute space. Beginning from a subjectivist position but eventually adopting an intersubjectivist position, he describes how we learn about spatial relations and proceed to create geometrical space. We can create a set of alternative geometries or geometrical languages. We then choose one by convention as a better formal representation for our space and a better mathematical language for our physical theories.

In his philosophical writings Poincaré frequently tries to show that alternative conventions and alternative geometries are genuinely viable. Primarily he argues for the viability of two sets of alternatives. First, he argues that spatial geometry can be a form of non-Euclidean geometry or it can be Euclidean geometry. Second, he argues that it is possible to conceive of our space as other than 3-dimensional.

With respect to the first alternatives of Euclidean and non-Euclidean geometries his convention that the displacement classes form a group is relevant. Actually the convention is only a partial one. We can express it more fully by incorporating it in the alternative conventions that the displacements form a Euclidean group or that they form one of the non-Euclidean groups of transformations. For Poincaré a geometry is nothing but the study of some group. The possible groups of spatial geometry are all suggested by the motions of solid bodies in nature. We choose the Euclidean group as the best for spatial geometry. In the essay in *The Monist* he says that we choose it, because unlike the other alternatives it contains an invariant subgroup [Poincaré, 1898, page 21]. Elsewhere he says that we choose this group and geometry, because it is the simplest in mathematical terms and because it conforms closely to our experience of solid bodies [Poincaré, 1891, page 774]. However, the various Euclidean and non-Euclidean geometries are all possible ways of describing natural phenomena. Moreover, they are in large part intertranslatable. It is possible to devise dictionaries specifying equivalent terms for the different geometries [Poincaré, 1891]. The matter of choice of one geometry is similar to the choice of a coordinate system. It is largely a matter of convenience.

With respect to the question of the dimension of our space Poincaré explores the possibilities for spaces of low dimension numbers in several places (in particular, [Poincaré, 1895b] [Poincaré, 1898] [Poincaré, 1903b] [Poincaré, 1912]). For example, in the essay in *The Monist* he examines from the viewpoint of group theory the cases of 3-, 4-, and 5-dimensional spaces using different basic elements. Yet according to Plücker's equivalence principle the geometries are equivalent. We can regard them as different ways of speaking about the same mathematical phenomena. Criteria of simplicity and conformity to aspects of our experience guide us to choosing a 3-dimensional space for the geometry most relevant to physical theories.

We can distinguish between the two principal sets of alternative conventions according to the group-theoretic terminology of Poincaré's essay in *The Monist* [Poincaré, 1898, pp. 21–22]. In the first case when we conventionally pick the Euclidean group over the non-Euclidean groups for spatial geometry, we choose the structure of the group of displacements, the basic form of geometry. In the second case when we conventionally choose a 3-dimensional space for geometry, we thereby choose the material of our space, the material on which the group can act. The conventions are of different types. The first is more fundamental. First we need to pick the structure of the group and then one of its representations as a group of transformations.

Poincaré often stresses that his conventions are not chosen arbitrarily. Certainly they are not all simple conventions of language as when one arbitrarily picks a word to stand for a complex linguistic expression. However, he offers only a few general acceptance/rejection rules to help in choosing conventions. Even then, he does not analyze these philosophically. His main rules for choosing one convention over its rivals are approximate conformity to rough experience and simplicity. These are only guides and fairly subjective. Experience yields only inexact results. Hence, we must interpret them, in order to derive a precise convention. But interpretations can vary. They can involve subjective elements. Simplicity is really quite vague. Poincaré thinks mainly of mathematical simplicity: the simplicity of one expression, equation, calculation, or theory in comparison with another. There is also an aesthetic aspect of simplicity. Yet surely simplicity is dependent upon the case at hand. To use Poincaré's own example, there are cases in which one coordinate system is more convenient and simpler to use than another, but the coordinate system which is simplest in one case need not be simplest in all cases. Similarly there may be good grounds for regarding a non-Euclidean geometry as the simplest for stating a physical theory (e.g., the theory of relativity). Also there may be cases in which a 4-dimensional space provides a simpler means for stating a physical theory than the usual 3-dimensional space. In general it seems clear that approximate conformity to experience and simplicity do not yield strict, universally applicable rules for choosing conventions. They only provide suggestions. For this reason they are not entirely satisfactory.

In his last two essays dealing with problems of space and dimension, 'La relativité de l'espace' ('The relativity of space') [Poincaré, 1906] and 'Pourquoi l'espace a trois dimensions' ('Why space has three dimensions') [Poincaré, 1912], Poincaré appears to modify his position on the conventional acceptance of the 3-dimensionality of space. This apparent modification is quite significant. In the essay of 1906 he outlines a revision of his theory of the genesis of space and geometry, putting the theory into the framework of an evolutionary epistemology. Accordingly, our spatial and geometrical ideas originate partly through the evolution of the species. As such they can be inherited. This new view has important consequences for the question of the dimension of our space [Poincaré, 1906, pp. 12–16]. He suggests that it might be possible for us to regard the world as 2- or 4-dimensional, but from the standpoint of survival such nonstandard views could have disastrous effects. According to him we have adapted to perceiving the world as 3-dimensional through our evolution as organisms. It has become a way of our living and a means for our survival. In his last essay on space and dimension he puts this point in the following way [Poincaré, 1912, page 498]:

> We see from this brief account what are the experimental facts that have led us to attribute three dimensions to space. In view of these facts, it was more convenient for us to attribute to it three rather than four or two dimensions; but this word 'convenient' may not be forceful enough here. A being that had attributed two or four dimensions to space would have found himself at a disadvantage in his struggle for life in a world such as ours ....[Ch9, 6]

There seems little doubt that in his late essays of 1906 and 1912 Poincaré has modified his views on the question of the dimension of our space and strengthened his position on the conventional character of the 3-dimensionality of our space. Through the evolutionary development of the species we have come to regard space as 3-dimensional for purposes of survival. Through the long history of our evolution we have reacted and adapted to the world in which we live. The 3-dimensionality of space has become a part of our response to this world. In his essay of 1912 he augments this evolutionary argument with the one based on mathematical physics [Poincaré, 1912, pp. 499–502]. In the end he maintains that there are very strong practical and theoretical reasons for choosing the convention that space is 3-dimensional and rejecting any alternative. At first sight it would seem that he has moved towards an 'objectivist' position concerning the 3-dimensionality of space, but I think that he would maintain that his position remains an intersubjectivist one, honed by evolutionary development. For Poincaré what we call 'objective reality' is in the final analysis what is common to many thinking beings and could be common to all [Poincaré, 1902d] [Poincaré, 1905a, chapitre XI].

One final and very important remark about Poincaré's philosophy of geometry is in order. As noted above, part of his aim has been to describe the significant role of experience in the origin and creation of geometry. Yet for him geometry is not an empirical science. It is not supposed to be open to refutation by experiments and tests. Geometry is an exact science and even the geometry or geometries most applicable to our space are exact. To guarantee this exactness he requires the existence of an *a priori* element or intuition. In his first theory of dimension this *a priori* element is the abstract group concept, as he explicitly declares [Poincaré, 1898, page 41]. In his second theory he is even bolder. He proclaims that there is an *a priori* intuition of continua of any number of dimensions that is related to analysis situs (topology) [Poincaré, 1912, pp. 502–504]. He claims he can detect its immediate influence in Hilbert's 'axioms of order' of the *Grundlagen der Geometrie* [Hilbert, 1899, §3. Die Axiomgruppe II: Axiome der Anordnung].[48] In this topological intuition lies the true bedrock of geometry. His own extensive research in the analysis situs or topology of manifolds or varieties of three or more dimensions has taught him some of its difficulties and significant pitfalls. Based on his own research experience he declares [Poincaré, 1912, page 504]:

> But there is an Analysis situs of more than three dimensions. I do not say that it is an easy science; I have devoted too much effort to it not to have taken account of the difficulties that are encountered in it. But nevertheless this science is possible and it does not rest exclusively on analysis. It could not be pursued successfully without continual appeal to intuition. Therefore, there is surely an intuition of continua of more than three dimensions, and if it demands more sustained attention than ordinary geometric intuition it is doubtless a matter of habit and also the effect of the rapidly increasing complication of the properties of the continua when the number of dimensions increases. . . .

[48] Cf. [Poincaré, 1902a] [Poincaré, 1903a].

> I shall conclude that there is in all of us an intuitive notion of the continuum of any number of dimensions whatever, because we possess the faculty to construct a physical and mathematical continuum; and that this faculty preexists in us before all experience, because without it experience properly speaking would be impossible and would be reduced to brute sensations, unsuitable for any organization, and because this intuition is only the awareness that we have this faculty. . . . [Ch9, 7]

Poicaré certainly had encountered difficulties in his own topological research. These motivated significant corrections that are found in the supplements.[49] Moreover, Poincaré left an extensive array of difficult problems for later topologists to try to solve. Just like Poincaré, topologists of the twentieth and twenty-first centuries have had to continue to grapple with the difficulties of higher-dimensional topology or even three-dimensional topology. One wonders how well founded the intuition for this geometry or topology is?

Indeed, Poincaré is none too clear about the precise nature of the basic intuition underlying analysis situs and associated with it. He is cautious about describing its powers. He is not so rash as to say that it leads us immediately or even quickly to topological truths and actual theorems. However, he claims that it does exist and that a mathematician must be trained in its use in order to discover new results. It is a necessity for progress in the difficult field of topology. It is clear that Poincaré's intuition must go beyond the sort of mental visualization that geometers presumably call upon as a heuristic device in their work. In effect, his philosophy appears to demand a faculty of intellectual intuition with several Kantian transcendental properties to serve as a touchstone for topological truth. In the final analysis it must account for the certainty of mathematics. After all, Poincaré was a kind of protointuitionist before Brouwer.

I have come to the end of my examination of Poincaré's thoughts on space, geometry, and dimension. This brilliant French mathematician occupies an unusual place in the modern history of dimension theory. His motive for attacking problems of dimension developed out of his philosophy rather than out of his mathematics. His two theories of dimension form integral parts of his philosophy of geometrical conventionalism. Unlike many of his contemporaries who thought about dimension, he did not have a strong interest in the fundamental problem of dimensional invariance. He was far more concerned with epistemological issues in the foundations of mathematics. Nonetheless, whatever his motives or interests, he had some extremely important things to say about dimension. Through Brouwer he exerted a significant influence on the subsequent development of topological dimension theory.

---

49 Poincare's work in analysis situs or topology is groundbreaking and fundamental. There are many assessments of his topological work, including [Alexandroff, 1972] [Sarkaria, 1999] [Gray, 2013, pp. 427–466] [Poincaré, 2010, pp. vii–xvi].

Chapter 10
Riesz, Baire, Fréchet: New Approaches to Dimension

10.1 Introduction

In this chapter I shall examine the work of three other mathematicians who investigated the problems of dimension during the first decade of the twentieth century. These are the Hungarian Frigyes Riesz, and the Frenchmen René-Louis Baire and Maurice Fréchet.

10.2 Frigyes Riesz

The Hungarian mathematician Frigyes (Friedrich) Riesz (1880–1956)[1] is best known for his brilliant contributions to the development of abstract and functional analysis. However, at the very beginning of his career he had a lively interest in set-theoretic topology. His early work in this field had some effect on his later work in analysis. His interest in topology mainly stemmed from his studies at Göttingen, where he came under the influence of Hilbert.

Riesz made a brief attack on the topological problem of dimension in a research note published in the *Comptes rendus* (*Proceedings*) of the Academy of Sciences, Paris in 1905, 'Sur les ensembles discontinus' ('On discontinuous sets') [Riesz, 1905a]. In the first part of this note he tries to prove a variant of a theorem of Ludovic Zoretti [Zoretti, 1905b, page 12]. Riesz' theorem states that [Riesz, 1905a, page 652]:

> *Every [bounded] discontinuous set situated in a plane is part of a continuous curve without multiple point situated in this plane.*[Ch10, 1]

[1] There is a brief biography of Riesz included in [Riesz, 1960, pp. 19–20]. There are also online biographies: `https://mathshistory.st-andrews.ac.uk/Biographies/Riesz/` and `https://en.wikipedia.org/wiki/Frigyes_Riesz`.

D. M. Johnson, *Theories of Dimension from Ancient to Modern Times*, Sources and Studies in the History of Mathematics and Physical Sciences, https://doi.org/10.1007/978-3-032-02682-8_10

In his note Riesz implicitly assumes that the sets under discussion are bounded. He lays down the following definitions as a basis for his theorem. He calls a (bounded) set of points in the plane or Euclidean n-space (implicitly given the usual topology) 'connected' ('d'un seul tenant') if the set cannot be separated (divided) into two (nonempty) subsets having neither points nor limit points in common [Riesz, 1905a, page 650]. This is a weak form of connectedness. For example, the set of points belonging to the union of open intervals $(0, 1) \cup (1, 2)$ and the set of rational points in the closed interval $[0, 1]$ form connected sets in Riesz' sense. He then calls a (bounded) set that has no connected subset a 'discontinuous set' [Riesz, 1905a, page 650]. Such a discontinuum is strongly disconnected. Both it and its closure will be 0-dimensional according to the later dimension theory of Urysohn and Menger.[2] Consequently, Riesz' theorem states that a (bounded) discontinuum in the plane is contained in a simple Jordan arc (i.e., a homeomorphic image of an interval). The proof that Riesz gives in his note is not correct, for he misapplies a result of René-Louis Baire [Baire, 1899, page 94]. Shortly after the note appeared Zoretti [Zoretti, 1906] (1906) demolished the proof with a simple counterexample.[3] Notwithstanding the error in the proof, the asserted theorem is true and provable. Moreover, it can be extended to (bounded) discontinuous sets contained in n-dimensional Euclidean spaces ($n \geq 2$), as Riesz says in his note [Riesz, 1905a, page 652].[4] Thus he draws the following correct conclusion [Riesz, 1905a, page 652]:

> It follows from our theorem that *from the viewpoint of* analysis situs *all [bounded] perfect discontinuous sets are equivalent (homeomorphic).*[Ch10, 2]

In the second part of his short research note Riesz proceeds to apply his results in a partial analysis of the dimension concept for Euclidean spaces. He postulates certain requirements for a possible dimension function $d(E)$ that would assign dimension numbers to sets E [Riesz, 1905a, page 652]. (1) If $E \subset E'$, then $d(E) \leq d(E')$; (2) given the union of two sets, $E_1 \cup E_2$, one never has both $d(E_1 \cup E_2) > d(E_1)$ and $d(E_1 \cup E_2) > d(E_2)$; (3) the topological product (ensemble complexe, in Riesz' terminology) of two sets with dimensions m and n has dimension $m + n$; (4) the dimension of a set is an invariant of one-one continuous (topological) transformations of the containing space; (5) the dimension of the (closed) segment $[0, 1]$ is 1. These postulates for a dimension function are not consistent with later dimension theory. In the topological dimension theory initiated by Brouwer, Urysohn, and Menger

[2] Obviously a Riesz discontinuum is totally disconnected in the modern sense that no connected subset contains more than one point. Moreover, it is clear from the definition that the closure of a Riesz discontinuum is also a Riesz discontinuum. Inasmuch as a Riesz discontinuum is a bounded subset of a Euclidean space, its closure is compact. Hence, by a standard theorem of dimension theory (for separable metric spaces) stating that every compact totally disconnected space is 0-dimensional [Hurewicz & Wallman, 1948, pp. 20–22] (a space is 0-dimensional if each of its points has arbitrarily small neighborhoods with empty boundaries), the closure of a Riesz discontinuum is 0-dimensional. Thus any Riesz discontinuum is 0-dimensional, being a subset of its 0-dimensional closure.

[3] On Zoretti's work on curves in topology see Section 13.4.

[4] For proofs of Riesz' theorem and its generalization see [Schoenflies, 1908b, pp. 257–259], [Denjoy, 1910], [Moore & Kline, 1919], and [Antoine, 1921, pp. 302–304].

Riesz' postulates (2) and (3) do not hold.[5] Nonetheless, Riesz does provide a novel approach to dimension at an early date. He calls a set that is homeomorphic to an n-dimensional rectangle (rectangular parallelepiped) a 'simple set of n dimensions' ('ensemble simple à n dimensions'); a point is a simple set of dimension 0. In general, he calls a set 'n-dimensional' if it contains subsets that are dense in simple sets of n dimensions and does not contain any subset that is dense in a simple set of $n + 1$ dimensions. Though this general definition of dimension satisfies the given postulates, he is well aware that it is quite a difficult problem to ascertain the dimension numbers of arbitrary sets in Euclidean spaces. However, in the relatively simple case of (bounded) discontinuous sets his special theorem shows that these sets must have dimension 0.

Riesz continued his investigations in set-theoretic topology by writing a long paper, first published in Hungarian in 1906 and 1907 [Riesz, 1907] and then in German translation in 1909: 'Die Genesis des Raumbegriffs' ('The Origin of the Space Concept') [Riesz, 1909a]. This is a remarkable paper by the young Riesz, but unfortunately it went largely unnoticed. In the paper he takes up Poincaré's problem of explaining the origin and meaning of our fundamental concepts of space and spatial (i.e., applied) geometry (cf. the previous Chapter 9). He was primarily influenced by Poincaré's paper of 1903 on space, geometry, and dimension [Poincaré, 1903b] as republished in *La Valeur de la Science* (*The Value of Science*) [Poincaré, 1905a, chapitres III, IV]. He adopts Poincaré's distinction between physical and mathematical continua. His principal aim is to characterize the concept of space. In seeking to achieve this aim, he offers a constructive theory that builds the space concept from our basic spatial ideas, then from physical continua that describe certain spatial relations, and finally from a mathematical continuum that provides an exact model for geometrical space. For the most part he leaves aside the psycho-physiological approach adopted by Poincaré and only brings in certain assumptions about our spatial ideas that are necessary to facilitate his essentially mathematical analysis of the concept of space. He founds his theory of the genesis of the space concept upon set-theoretic topology, and, thus, he mainly analyzes the topological properties of our space. He concludes that our geometrical space is a special type of mathematical continuum, or in modern terms a special type of topological space. He barely mentions other possible geometrical properties, for example, its possible projective, affine, and metric properties. Hence, he does not provide a full foundation for a metrical geometry of space. In making his analysis of the space concept, Hilbert's important paper on the topological foundations of plane geometry [Hilbert, 1902b] (preliminary version of [Hilbert, 1903]) clearly influenced him. Riesz' paper is in several ways similar to Maurice Fréchet's celebrated thesis 'Sur quelques points du Calcul Fonctionnel' ('On some points in the functional calculus') [Fréchet, 1906]. Written independently, they both operate from an abstract perspective and introduce

[5] Riesz' postulate (2) does not hold in Urysohn–Menger dimension theory. The union of the 0-dimensional sets of the rational numbers and the irrational numbers of the real line form a 1-dimensional set, viz., the real line itself. Furthermore, Riesz' postulate (3) does not hold in Urysohn–Menger dimension theory as a remarkable example of Pontrjagin and Tolstowa [Pontryagin & Tolstowa, 1931] shows.

general spaces as an axiomatic foundation for abstract point-set theory and topology. Unfortunately, Riesz' work did not influence the development of mathematics as immediately as did Fréchet's work.[6]

In his paper Riesz touches upon the problems of defining the concept of dimension and explaining the 3-dimensionality of our space. Near the beginning of his paper he correctly criticizes Poincaré's topological theory of dimension for physical continua [Riesz, 1909a, pp. 313–314]:

> However, if one tries to place the Poincaré dimension concept at the foundation of a further investigation of the space concept, then one soon runs into great difficulties. First of all in introducing his concept Poincaré does not take account of those properties that already belong to the naive dimension idea, e.g., that a system of a certain dimension shall not contain a system of higher dimension. According to the Poincaré definition a double cone will be a physical continuum of one dimension. Still, it is quite easy to remedy this mistake. The greatest difficulty is that it is hard to capture the dimension concept for mathematical continua; prior to a fundamental analysis of this concept it must not be imagined that the dimension concept for physical continua and that for mathematical continua are connected with one another in any way. Thus for the present the dimension concept cannot serve as a certain foundation for further investigations.[Ch10, 3]

Riesz' double cone counterexample is very pertinent. A few years later Brouwer [Brouwer, 1913b] independently put it forward as a criticism of Poincaré's definition. Riesz' more general critical assessment of Poincaré's theory shows that he is well aware of the difficulties of constructing an adequate theory of dimension. Consequently, he abandons the idea of putting the dimension concept at the foundation of spatial geometry. Instead he examines the space concept in relation to other physical and mathematical continua in order to determine some assumptions for charactezising its dimension number three [Riesz, 1909a, pp. 340–349]. Using a theory of n-fold order types, which he had developed earlier [Riesz, 1905b], he defines an n-dimensional domain [Riesz, 1909a, page 345][7] and then proposes that space is a certain type of 3-dimensional domain [Riesz, 1909a, pp. 346–349]. His

[6] Riesz' work was published in German translation from the original Hungarian, but in a less accessible journal. Also, I would conjecture that Riesz' philosophical approach lost him many readers, even though he had some good ideas in an abstract setting.

With the hindsight of later developments in basic topology, Riesz' work [Riesz, 1909a] has some significance. For example, it contains one of the first occurrences of the now well-accepted definition of connectedness and includes the idea of topological relativization [Riesz, 1909a, pp.320–321]. He states the definition of connectedness as: 'The mathematical continuum is said to be *connected* if it cannot be separated into two open subsets that are complementary sets of one another.' ('Das mathematische Kontinuum heiße *zusammenhängend*, wenn es nicht in zwei offene Teilmengen zerlegt werden kann, die Komplementärmengen für einander sind.') However, when he next had occasion to discuss connectedness, he changed his definition to another, which later topologists have not accepted [Riesz, 1909b]. Cf. [Thron, 1997].

[7] A Riesz n-dimensional domain is effectively the modern concept of a connected open set in Euclidean n-space.

analysis of the dimension concept remains superficial inasmuch as he neither seeks to define the concept per se nor tries to explain or 'prove' the 3-dimensionality of space.

In his two contributions Riesz did not make much progress on the problems of dimension. However, he did demonstrate a very good grasp of the difficulties involved in a topological analysis of the dimension concept. He made no further contributions to dimension theory. Rather he went on to develop functional analysis with enormous success.

10.3 René-Louis Baire

Shortly after Riesz examined general problems connected with dimension, René-Louis Baire (1874–1932),[8] most famous for his investigations concerning discontinuous functions, made a direct assault on the problem of the topological invariance of dimension. First, in a research note presented by Paul Appell to the Academy of Sciences, Paris, at the session of 11 February 1907 [Baire, 1907a] and then, in a fuller exposition appearing in April of that year [Baire, 1907b], he put forward a method for proving the invariance. The proposed method is direct and sound, resting on a derivation from the domain invariance and the Jordan separation theorems for n dimensions. However, the outlined method is incomplete, because the Jordan separation theorem, upon which everything rests, is left unproven at the end. Later developments in topology show that the Jordan separation theorem for n dimensions is hard to prove.

The following is a summary of Baire's line of thought. Baire subsumes the invariance of dimension for Euclidean spaces (G_i in his notation) under the following theorem [Baire, 1907b, page 96]:

> Theorem I. — *There can be no application [i.e., homeomorphism] between two closed sets: E of* G_n, *F of* G_{n+p} $(p \geq 1)$, *if F contains all the points of a spherical domain [i.e., a solid closed sphere] of* $n + p$ *dimensions.*[Ch10, 4]

Employing a method similar to that used by Jürgens [Jürgens, 1879] and Schoenflies [Schoenflies, 1899], he finds it easy to deduce this theorem from the following one expressing domain invariance [Baire, 1907b, page 96]:

> Theorem II. — *In* G_n *let two sets P and* Π *be applicable one on the other [i.e., homeomorphic], P being a spherical domain; then every nonboundary [i.e., interior] point of P has as an image a nonboundary point of* Π.[Ch10, 5]

Briefly, Baire's deduction of dimensional invariance, Theorem I, from domain invariance, Theorem II, proceeds as follows. Suppose that a closed set E in G_n is homeomorphic to a closed set F in G_{n+p} containing a solid closed sphere Σ of $n+p$

[8] [Dugac, 1976a] contains an extensive biography of Baire and a lengthy survey of his works. Cf. the short biography at `https://mathshistory.st-andrews.ac.uk/Biographies/Baire/`.

dimensions. Let B be the center of Σ and consider an n-dimensional planar section Σ_1 of Σ through B. In the homeomorphism between F and E the planar section Σ_1 is mapped onto a closed set E_1. According to Theorem II the interior point B of Σ_1 (relative to the n-dimensional plane used to form Σ) must be mapped to an interior point A of E_1. If one considers a sequence of points B_h in Σ approaching B such that each B_h does not belong to Σ_1 (i.e., each B_h is not in the plane of Σ_1), then according to the homeomorphism the images A_h of the B_h approach A. Inasmuch as A is not a boundary point of E_1, for a sufficiently large h an A_h becomes a point of E_1 and, hence, the corresponding B_h becomes a point of Σ_1. But this situation contradicts the assumption that no B_h belongs to Σ_1. Hence, Theorem I follows from Theorem II. In this way Baire simply reduces the problem of proving dimensional invariance to the problem of proving domain invariance as expressed in Theorem II.

Proving Theorem II for the case $n = 1$ in which G_1 is the real line and P is a closed interval is an easy matter [Baire, 1907b, page 97]. In contrast, proving the theorem for the cases $n \geq 2$ is much more difficult, as Baire realizes. However, he sees that it is possible to base the proof of the difficult cases upon a generalization of the Jordan curve theorem, the n-dimensional Jordan separation theorem. Briefly, this general theorem states that a 'simple closed surface,' i.e., a homeomorphic image of a hypersphere, in Euclidean n-space divides the space into two parts. Baire's more exact version of the separation theorem is as follows [Baire, 1907b, page 98]:

Theorem III. — *Let Σ be a simple closed surface in G_n ($n \geq 2$):*

1° *Σ has an interior I and an exterior E. Every point of I (of E) is the center of a spherical domain [i.e., a solid closed sphere] all of whose points belong to I (to E).*
2° *The distance between any two points of I is less than the least upper bound of the distances between pairs of points of Σ.*
3° *A continuous path [i.e., a continuous interval image] going from a point of I to a point of E contains at least one point of Σ.*
4° *If a simple closed surface Σ varies in a continuous manner as a function of a parameter t which takes all values in an interval $[a, b]$, then a fixed point M of G_n that is not on Σ for any value of t is either always on the interior of Σ or always on the exterior.* [Ch10, 6]

The fourth part of this theorem essentially means that if there is a deformation of a simple closed surface Σ and there is a fixed point M that never lies on any of the surfaces Σ_t of the deformation, then M is either always on the interiors or always on the exteriors of the Σ_t.[9] It is this last part of theorem III that Baire finds most useful in deducing theorem II. The key idea of his deduction of theorem II is as follows.

[9] In the fourth part of theorem III Baire suggests the notion of a deformation without giving precise details. One may formulate Baire's suggestion in modern terms in the following way. One has a deformation of a simple closed surface Σ or rather a family of such surfaces Σ_t, $t \in [a, b]$. (The interval could be the standard unit interval $[0, 1]$.) Consider the continuous mappings:

$$f_a : \Sigma_a \to \Sigma_a \subset G_n, \text{ the identity mapping;}$$
$$f_b : \Sigma_a \to \Sigma_b \subset G_n, \text{ a homeomorphism of } \Sigma_a \text{ onto } \Sigma_b.$$

Let P be a solid closed sphere of unit radius in G_n. The assumption is that P is homeomorphic to a set Π. One can think of the solid sphere P as a set of concentric spheres S_ϱ (where ϱ denotes the radius, $0 \leq \varrho \leq 1$) with S_1 the boundary sphere and S_0 the center of P. Each of the spheres S_ϱ in P has a homeomorphic image Σ_ϱ in Π. As one contracts the sphere S_1 to the point S_0 according to the obvious linear contraction along the radii, the spheres S_ϱ sweep out the solid closed sphere P and at the same time there is a contraction of Σ_1 to Σ_0 with the Σ_ϱ sweeping out the entire region Π.[10] Baire shows that every point of the interior of Σ_1 belongs to Π and that no image of a point of P belongs to the exterior of Σ_1. In other words, Π is identical with the union of Σ_1 and its interior; moreover, the interiors correspond and the boundaries correspond.[11] It is in this way that Baire deduces theorem II from theorem III.[12]

In his works of 1907 Baire shows clearly how an n-dimensional Jordan separation theorem (his theorem III) entails n-dimensional domain invariance (his theorem II), and, in turn, how domain invariance entails dimensional invariance in general (his theorem I). However, he does not prove his version of the Jordan separation theorem, but only promises to provide a proof in a future work [Baire, 1907b, page 99].

Baire was never able to give a full proof of the Jordan separation theorem. Around 1907 he suffered a marked deterioration in his health. He remained very ill for the rest of his life.[13] The decline in his health most likely prevented him from developing a proof of the general separation theorem. He probably would never have been able to prove the general theorem, even if his health had remained good. He always thought that it would be possible to prove the general theorem along the lines of a proof of the Jordan curve theorem in the plane [Baire, 1907a]. Developing a proof would merely

---

For a deformation there exists a continuous function

$$F : \Sigma_a \times [a, b] \to G_n$$

such that

$$F(x, a) = f_a(x)$$

and

$$F(x, b) = f_b(x).$$

Thus, the identity mapping $f_a : \Sigma_a \to \Sigma_a$ is homotopic to the homeomorphism $f_b : \Sigma_a \to \Sigma_b$ or Σ_a is deformable into Σ_b over G_n. The fourth part of Baire's theorem III asserts that if there is a deformation of simple dosed surfaces in G_n and there is some fixed point $M \in G_n$ not lying on any of the surfaces Σ_t of the deformation, i.e., for every $x \in \Sigma_a$ and $t \in [a, b]$, $F(x, t) \neq M$, then M is on the interior of every Σ_t or M is on the exterior of every Σ_t.

[10] The unit sphere S_1 is contractible to its center S_0 along its radii. Thus, there is a contraction of S_1 to S_0 (or S_1 is deformable into S_0), i.e., the identity mapping $S_1 \to S_1$ is homotopic to the constant mapping $S_1 \to S_0$. At the same time the homeomorphism of P onto Π induces a contraction of Σ_1, the homeomorphic image of S_1 in Π, to the point Σ_0, the (homeomorphic) image of S_0 in Π, i.e., the identity mapping $\Sigma_1 \to \Sigma_1$ is homotopic to the constant mapping $\Sigma_1 \to \Sigma_0$.

[11] Baire does not give a full proof that the interiors correspond and the boundaries correspond, but it is easy to fill in that gap.

[12] For the case in the plane cf. [Schoenflies, 1899] and the similar proof of Hausdorff [Hausdorff, 1914, pp. 378–379].

[13] For details about Baire's illness see [Dugac, 1976a, pp. 299–314].

be a matter of providing the necessary technical details.[14] In this opinion surely he underestimated the difficulties to be overcome in proving the general separation theorem. What he did do in his notes of 1907 was to offer a promising line of attack on the problem of dimensional invariance, though in doing so he actually reduced the problem to a deeper and much more difficult problem, that of proving an n-dimensional Jordan separation theorem. Later even Brouwer found this latter problem quite a challenge (see Chapter 12).

Quite apart from his work on dimensional invariance Baire made another mathematical contribution, which much later had a bearing on the development of dimension theory. In one of his important studies of discontinuous functions, 'Sur la représentation des fonctions discontinues, Deuxiéme partie' ('On the representation of discontinuous functions. Second part') [Baire, 1909], published in 1909 but written between 1904 and 1906, he introduces a special space which he calls the 'space of 0 dimension' ('espace à 0 dimension') [Baire, 1909, chapitres II, IV]. This space consists of the set of all sequences of positive integers

$$(a_l, a_2, \ldots, a_n, \ldots).$$

By defining a notion of limit element, he effectively gives the space a topological structure [Baire, 1909, pp. 105–106]. A (countable) sequence of elements of the space

$$A_\nu = [(a_1)_\nu, (a_2)_\nu, \ldots, (a_n)_\nu, \ldots],$$

(varying over the index ν), has the element

$$A_0 = [(a_l)_0, (a_2)_0, \ldots, (a_n)_0, \ldots]$$

as its limit element, if, for whatever n, there is an integer h such that for $\nu > h$ it is the case that

$$(a_i)_\nu = (a_i)_0$$

for $i = 1, 2, \ldots, n, \ldots$.[15] A subset of the space of 0 dimension is called a 'set of 0 dimension' [Baire, 1909, page 134].

[14] Cf. Baire's letters to Fréchet in [Gispert, 1980, pp. 94–95, 96–97], and cf. Baire's correspondence with Brouwer in [Dugac, 1976a, pp. 368–370] and more completely in [Dalen, 2011a, pp. 104–109, 111–112] [Dalen, 2011b, pp. 454–460, 469–470] (including commentary). In contrast, Brouwer realized that Baire underestimated the difficulties to be overcome in proving the general separation theorem (see Chapter 12).

[15] It is much easier to define the Baire space of 0 dimension as a metric space and then consider the resulting topology. One may define a metric or distance function for pairs of sequences of the space, $(a_1, a_2, \ldots), (b_1, b_2, \ldots)$. One defines it to have the value 0 if the two sequences are identical and the value $1/m$ if the two sequences are different and they differ at the mth place for the first time. Baire wrote the part of his paper [Baire, 1909] on the space of 0 dimension in 1904, before Fréchet introduced (E) classes or metric spaces in his thesis [Fréchet, 1906].

Most importantly Baire demonstrates that this space of 0 dimension is homeomorphic[16] to the set of all irrational numbers on the real line (implicitly given the usual subspace topology) and, moreover, that the space of 0 dimension is homeomorphic to the set of all points having all coordinates irrational in an n-dimensional Euclidean space (for any n) (implicitly given the usual subspace topology) [Baire, 1909, pp. 136–138]. The demonstration naturally depends upon continued fraction expansions for the irrational numbers. He also shows that the Cantor discontinuum is homeomorphic to a subset of the space of 0 dimension [Baire, 1909, 98–99, 136]. Thus the Cantor discontinuum is a special set of 0 dimension.

Baire realizes that there is a profound relation between an n-dimensional Euclidean space and the space of 0 dimension. They have similar properties, all deriving from their similar notions of limit element or limit point [Baire, 1909, pp. 132–133]. In modern terms one would say that they have common topological properties.[17] Thus to signal the similarity between the spaces he uses the symbol 'G_0' for his space of 0 dimension and the symbol 'G_n' for an n-dimensional Euclidean space. However, he recognizes an extremely important difference between G_0 and any G_n ($n \geq 1$). Whereas the Euclidean n-spaces are connected,[18] the space of 0 dimension (like the Cantor discontinuum) is strongly disconnected. G_0 can be given a decomposition into a finite or an infinite number of disjoint perfect (i.e., closed and dense-in-themselves) subsets [Baire, 1909, pp. 133–134]. It is this property of disconnectedness that leads him to call G_0 the 'space of 0 dimension.'

When introducing his space G_0 of sequences of positive integers Baire was principally interested in throwing further light on the domain of discontinuous functions. However, much later his space turned out to have an importance in the dimension theory developed by Urysohn and Menger during the twenties. Not only is the Baire space 0-dimensional in the sense of Urysohn and Menger, but also it is universal in the sense that every 0-dimensional set in a separable metric space is homeomorphic to a subset of the Baire space of 0 dimension.[19] Hence, Baire hit upon something significant both for his studies of discontinuous functions and for the later development of topological dimension theory.

10.4 Maurice Fréchet

In 1909 Maurice Fréchet (1878–1973)[20] began to take an active interest in problems concerned with dimension. Already in 1906 he had published his influential thesis, 'Sur quelques points du calcul fonctionnel' ('On some points in the functional calcu-

---

16 Baire uses the term 'application' for the modern term 'homeomorphism' (deriving from Poincaré).

17 Baire was, of course, writing before the birth of abstract general topology.

18 Baire in [Baire, 1909, page 133] just gives a proof that the unit interval [0, 1] is connected.

19 Cf. [Fréchet, 1910, pp. 154–155], [Sierpiński, 1921], [Urysohn, 1925a, page 77], and [Engelking, 1978, pp. 20–27]. See Sections 15.7, 16.7, and 16.9.

20 [Mandelbrojt, 1973] is a short biographical note concerning Fréchet. See also `https://mathshistory.st-andrews.ac.uk/Biographies/Frechet/`.

lus') [Fréchet, 1906], which provided an important starting point for the development of the theory of abstract spaces and modern abstract analysis. By 1909 he was aware of the interest among mathematicians in problems of dimension. For example, he knew about Baire's efforts to solve the invariance problem.[21] On 3 May 1909 Paul Appell presented a research note by Fréchet to the Academy of Sciences, Paris, giving a general definition of dimension applicable to a very wide class of abstract spaces [Fréchet, 1909]. From this note he developed a longer exposition of his theory of 'dimension types,' 'Les dimensions d'un ensemble abstrait' ('The dimensions of an abstract space') [Fréchet, 1910], which appeared in *Mathematische Annalen* during 1910. Fréchet's abstract approach to dimension resembles Riesz' approach [Riesz, 1905a]. Perhaps one should expect this resemblance. Both mathematicians were instrumental in founding twentieth-century analysis based on abstract spaces. While Riesz' interest in dimension was short-lived, Fréchet continued to take an interest in dimension theory for many years. He devoted much of his research effort during the 1920s to it. In his well-known book of 1928, *Les Espaces abstraits* (*Abstract Spaces*) [Fréchet, 1928], he gives a final summary of his ideas about dimension. Then he left the topic for other mathematical research, though his students and a few others continued to develop his ideas.

Fréchet's publications on dimension show that his prime interest always was his abstract theory of dimension types. However, two letters of 1909 from Baire to Fréchet [Gispert, 1980, pp. 94–95] reveal that when he was first developing his theory he tried to solve the problem of dimensional invariance by completing Baire's line of argument [Baire, 1907b] with a proof of Baire's theorem III or an equivalent. In a letter dated 1 June 1909 Baire writes to Fréchet [Gispert, 1980, page 95]:

> I am very happy that you have in hand the demonstration of my theorem III or an equivalent, and my wish is that you publish your demonstration as soon as possible. From the scientific point of view the interest in it should be great. . . .[Ch10, 7]

In a subsequent letter dated 2 July 1909 Baire seems to suggest that in the meantime Fréchet had experienced difficulties with completing his proof [Gispert, 1980, page 95]:

> I keenly wish that your solution to theorem III is completed; I know by personal experience that deceptions are very frequent in this type of question, but that the gaps also get filled in after the event, very fortunately.[Ch10, 8]

It seems that in the long run Fréchet was not able to complete his demonstration and thus to solve the invariance problem. He never published a proof. In his paper on dimension of 1910 he indicated that by that time he was not prepared to continue Baire's line of thought in print [Fréchet, 1910, pp. 165–166]. Yet he was convinced that dimension is a topological invariant and, hence, that Euclidean spaces of different

---

[21] Cf. a letter of Baire to Fréchet dated 15 February 1907 in [Gispert, 1980, page 94].

dimension numbers are strictly distinguishable. In fact, the distinguishability of Euclidean spaces is highly relevant to his theory of dimension types.[22]

Fréchet did succeed in producing an interesting general theory of dimension. I shall briefly examine some of its details, particularly as given in his first extended exposition in *Mathematische Annalen* [Fréchet, 1910].[23]

In creating his theory of dimension types, Fréchet builds a theory analogous to Cantor's theory of cardinals or powers depending on one-one correspondences [Fréchet, 1910, page 145]. He first puts his theory into the framework of (L) classes. In his thesis of 1906 he had introduced such classes as the most general kind of abstract space for his analytic investigations [Fréchet, 1906, pp. 5–6]. At the same time he had introduced (E) classes [Fréchet, 1906, page 30]—classes which Hausdorff later named 'metric spaces' ('metrische Räume') [Hausdorff, 1914, page 211]. A Fréchet (L) class consists of a set of elements endowed with a definition of limit for sequences of its elements satisfying two conditions: first, that a sequence of elements all of which are the same A converges to A and, second, that the limit of every subsequence of a sequence converging to an element A is also A. The requirements for being an (L) class are quite weak, so that the concept comprehends a wide variety of spaces.[24] Nevertheless, it is possible to define many of Cantor's point-set-theoretic concepts for (L) classes. Furthermore, Fréchet is able to define the notion of a homeomorphism between subsets of a given (L) class [Fréchet, 1910, page 146]. Two subsets E_1, E_2, belonging to an (L) class, are homeomorphic or 'they are images of one another,' if there exists a one-one correspondence between them such that their corresponding sequences have corresponding limits.[25]

Three fundamental definitions of Fréchet's theory of dimension types are the following [Fréchet, 1910, page 146]:

---

[22] Fréchet's attempts to prove dimensional invariance do not seem to be among his papers now kept in the Archives of the Académie des Sciences, Paris.

[23] I shall only give a brief treatment of Fréchet's work on dimension and refer the interested reader to detailed treatments of L.C. Arboleda [Arboleda, 1980] [Arboleda, 1981b] and H. Gispert [Gispert, 1980], which focus on extant letters of the time. When I originally was researching the work of Fréchet I met Luis Arboleda in Paris, and we discussed the mass of Fréchet Papers kept in the Archives of the Académie des Sciences, Paris. Luis Arboleda was most helpful in my work as was Hélene Gispert. Luis Arboleda put the Fréchet Papers in excellent order [Arboleda, 1981a]. I am grateful to both for their help in my earlier research. Cf. also [Taylor, 1982] [Taylor, 1985] [Taylor, 1987], which together give a broader treatment of Fréchet's mathematics.

[24] Fréchet's two conditions for defining (L) classes are really too weak. They do not adequately characterize the notion of limit. There are simple examples of (L) classes that are pathological. However, a further condition can be added to Fréchet's two conditions in order to define L^* spaces. L^* spaces characterize the notion of limit better than (L) classes. Cf. [Kuratowski, 1966, pp. 188–204].

[25] [Fréchet, 1910, page 146] speaks of a 'correspondance biunivoque qui est bicontinue' and says that he has borrowed the term 'homéomorphe' from Hadamard (who certainly would have gotten it from Poincaré). Fréchet's definition of homeomorphism for (L) classes closely resembles [Baire, 1909, page 136] definition of 'application' for Euclidean spaces. [Arboleda, 1980] and [Gispert, 1980, page 74] regard the work of Baire and Fréchet as significant in the development of a precise concept of homeomorphism for the domain of set-theoretic topology.

Definition III. If two sets are such that one of them G_1 can be considered as the image of [i.e., is homeomorphic to] the other G_2 or a subset of the other, we say that the first G_1 has a type of dimension less than or equal to that of the second G_2 and we write $dG_1 \leq dG_2$ or $dG_2 \geq dG_1$.

Definition IV. If one has both $dG_1 \leq dG_2$ and $dG_2 \leq dG_1$, we say that G_1 and G_2 have the same type of dimension, and we write $dG_1 = dG_2$ or $dG_2 = dG_1$.

Definition V. If one has $dG_1 \leq dG_2$, and it is certain that G_2 cannot be considered in any way as homeomorphic to G_1 or a subset of G_1, we say that the type of dimension of G_1 is less than (and not equal to) that of G_2 and we write

$$dG_1 < dG_2 \text{ or } dG_2 > dG_1.$$ [Ch10, 9]

The analogy between these definitions and the Cantorian definitions for cardinals is clear.

In a remark following his initial definitions Fréchet says that he is not concerned with the question of knowing whether for two given sets G_1, G_2 in an (L) class it is necessary that one of the relations

$$dG_1 < dG_2, \quad dG_1 = dG_2, \quad dG_1 > dG_2$$

holds, i.e., that the law of trichotomy holds [Fréchet, 1910, page 147]. Yet in a note later in the paper he gives a simple example, which shows that the law can fail [Fréchet, 1910, page 158]. Consider a circle γ and a planar set F consisting of the point $(0, 0)$ and all the open segments

$$0 < x < 1, \quad y = \frac{1}{n}, \quad n = 1, 2, 3, \ldots.$$

The dimension types of these planar sets, $d\gamma$, dF, are clearly incomparable. Hence, the order relation $<$ for dimension types of an (L) class is in general not a total (or linear) order, though it is a partial order.[26] This lack of a total order for dimension types contrasts with the existence of a total order for Cantor's cardinals, which is provable on the assumption of the axiom of choice or an equivalent. Somewhat surprisingly Fréchet does not pursue the consequences of this lack in his paper of 1910.

Fréchet points out another peculiar fact about his order for dimension types [Fréchet, 1910, page 147]. It is possible to have two sets G_1, G_2 such that $dG_1 = dG_2$, i.e., such that each set is homeomorphic to a subset of the other, and yet G_1 is *not* homeomorphic to G_2. Consider the sets $G_1 = [0, 1]$ and $G_2 = (2, +\infty)$. It is easy to see that G_1 is homeomorphic to a proper subset of G_2 and that G_2 is homeomorphic to a proper subset of G_1, so that $dG_1 = dG_2$, but it is certain that G_1 and G_2 are themselves not homeomorphic. Hence, in general $dG_1 = dG_2$ does not imply that G_1 and G_2 are homeomorphic. In Fréchet's theory of dimension types there is no

[26] Cf. [Fréchet, 1910, pp. 147–148].

exact equivalent of the so-called Cantor–Bernstein equivalence theorem in Cantor's theory of cardinals.[27] In spite of the lack of this exact theorem in the theory of dimension types, much later Stefan Banach [Banach, 1924] clarified the situation by proving an important related theorem, a generalized Cantor–Bernstein-style theorem directly applicable to Fréchet's theory. He proved that if $dG_1 = dG_2$ then each set G_i has a decomposition into two disjoint subsets, G_{i1}, G_{i2}, such that G_{11} and G_{21} are homeomorphic and G_{12} and G_{22} are homeomorphic.

The two 'deficiencies' just explained may give one cause to wonder if Fréchet's theory of dimension types adequately expresses our intuitive ideas about dimension. Perhaps one should consider the theory further before judging its adequacy.

Although Fréchet proposes a general theory of dimension types for arbitrary (L) classes, he proves hardly any general results in his paper of 1910. Rather he concentrates on applying his theory to Euclidean spaces, i.e., examining the dimension types of Euclidean spaces and their subsets. Indeed his theory is mainly useful for making comparisons between Euclidean spaces and various other sets and spaces. In this respect it closely resembles Cantor's theory of cardinals, after which it was modeled. Fréchet gives some remarkable results [Fréchet, 1910, pp. 148–159, 160–163], particularly concerning the dimension types of subsets of the line and plane. For example, he tries to show that there is an uncountable set of dimension types all less than dR_1, the dimension type of the line [Fréchet, 1910, pp. 148–150]. The demonstration unfortunately is not correct, but the result is.[28] He also gives an example of a dimension type greater than all the dimension types of the Euclidean spaces, an infinite dimension type [Fréchet, 1910, pp. 159–160], though he does not study infinite dimension types in any detail in the paper of 1910 (cf. [Fréchet, 1910, pp.164–165]).

At the end of his paper Fréchet makes some remarks about Euclidean spaces and dimensional invariance [Fréchet, 1910, pp. 165–168]. He easily recognizes that the dimension types of Euclidean spaces ('R_n' denoting a Euclidean n-space) are ordered as follows:

$$dR_1 \leq dR_2 \leq \cdots \leq dR_n \leq \cdots .$$

Earlier partial results on dimensional invariance show that

$$dR_1 < dR_2$$

and also

$$dR_2 < dR_3,$$

[27] The so-called Cantor–Bernstein equivalence theorem states that if each of two sets is equivalent to a subset of the other (i.e., there is a one-one correspondence between each and a subset of the other), then the sets themselves are equivalent (i.e., there is a one-one correspondence between them), so that their cardinals are equal.

[28] Paul Mahlo [Mahlo, 1911] corrected Fréchet's mistake.

though at the time of writing his paper he did not actually know that these inequalities had effectively been proved earlier.[29] Further, he deduces that there are two possibilities for refining the order of Euclidean spaces. Either

$$dR_1 < dR_2 < \cdots < dR_n = dR_{n+1} = \cdots = dR_{n,+k} = \cdots$$

or

$$dR_1 < dR_2 < \cdots < dR_n < dR_{n+1} < \cdots < dR_{n,+k} < \cdots .$$

The second case is equivalent to dimensional invariance in general. He thinks that it is the most likely case, but he is well aware that dimensional invariance has not been proved in general so far. As already mentioned, Fréchet had not been successful in his own efforts during 1909 to prove the invariance along the lines suggested by Baire [Baire, 1907a] [Baire, 1907b]. Assuming the invariance as proved, i.e., the second case, he can give the simple formula

$$dR_{n+p} = dR_n + dR_p .$$

Moreover, it is natural to let

$$dR_n = n;$$

Euclidean n-space has dimension type or simply dimension n. He then proceeds to show how to create an infinity of sets K_n having distinct dimension types strictly between $n - 1$ and n:

$$n - 1 < dK_n < n.$$

In other words, there are subsets of points belonging to Euclidean spaces with fractional dimension types. This result is remarkable and unexpected from the standpoint of our usual understanding of dimension. To be the first to consider the possibility of fractional dimensions is a noteworthy achievement of Fréchet.[30]

After a gap of a decade Fréchet continued his work on the theory of dimension types in several publications of the 1920s. At this time he got some help and cooperation from other topologists, particularly Paul (Pavel) Alexandroff, Wacław Sierpiński, and Kazimierz Kuratowski. During the first semester of the academic year 1923–1924 he presented a course on abstract topology including the theory of dimension types at the University of Strasbourg. Then during 1924 he began writing a book. Before its publication a preliminary version was distributed among prominent mathematicians for comment and criticism. The endproduct, published in 1928, was a thorough summary of the development of his theory of dimension

[29] Cf. [Fréchet, 1924].

[30] [Fréchet, 1910, page 155] criticizes Baire's use of the term 'space of 0 dimension,' because in his theory of dimension types there is an infinity of fractional dimension types less than 1 and less than or equal to the dimension type of Baire's space of 0 dimension.

types as well as his ideas on general topology up to that time: *Les Espaces abstraits* (*Abstract Spaces*) [Fréchet, 1928]. In his later work on dimension types he put the theory into the framework of separable metric spaces and tried to bring it into line with the rapidly developing topological dimension theory initiated by Urysohn and Menger (see Chapters 14 and 15).[31]

Probably most would agree that in his first papers of 1909–1910 and his later work of the twenties Fréchet created a theory of some interest that offered a new way of looking at problems of dimension. However, his theory has not enjoyed great 'popularity' among mathematicians. The topological theory initiated by Brouwer, Lebesgue, Urysohn, and Menger has received much more attention and has been developed with much greater vigor than Fréchet's theory. Certainly a drawback of Fréchet's special theory is the lack of a total order among the dimension types of Euclidean spaces and their subsets. A much more serious disadvantage of the theory as presented by Fréchet is that it begins with a general definition of dimension, but it does not proceed to encompass a coherent set of general theorems. His dimension types are always linked with Euclidean spaces or other special spaces. Thus the theory is a special one for making comparisons among spaces rather than a general dimension theory like that of Urysohn and Menger. I think it is fair to say that Fréchet's theory has never been in the mainstream of the development of topological dimension theory.[32]

10.5 Concluding Remark

In this chapter I have examined a diversity of approaches to problems of dimension by three important mathematicians, Riesz, Baire, and Fréchet. Through them new light was shed on the basic concept of dimension. Yet, up to 1910 no one had been able to conquer the fundamental general problem of the invariance of dimension. However, a complete solution was to come very soon (see the next chapter).

---

[31] Fréchet [Fréchet, 1928, pp. 110–111] gives the noteworthy result that for every set G having finite dimension type and comparable with finite-dimensional Euclidean spaces the dimension of G in the sense of Urysohn and Menger can be considered as the integer part of the dimension type of G.

[32] Arboleda in [Arboleda, 1980] discusses a correspondence between Fréchet and Kuratowski of 1933–1934 in which Kuratowski at one stage remarked that he did not consider the theory of Fréchet to be a theory of dimension!

Chapter 11
Brouwer: Development of Topology, Invariance of Dimension

11.1 Introduction

Mathematicians recognize the Dutch mathematician L.E.J. Brouwer (1881–1966)[1] as the first to construct a rigorous proof of the invariance of dimension. In fact, Brouwer published two proofs of dimensional invariance, the first in 1911 and the second in 1913, both full of deep insights. Moreover, in the course of developing these proofs he made highly significant contributions to the whole domain of topology. He injected an entire array of new methods into the field and raised its standard of rigor to a high level. His monumental work in topology of 1909–1913 stands as an enduring record of achievement. At the beginning of the twentieth century Brouwer stands with Poincaré as one of the two founders of modern topology. While Poincaré introduced homology and the fundamental group, forming the basis for modern algebraic topology, Brouwer introduced mapping degree and a variety of concepts and methods, including homotopic methods, to show invariances and lay out fundamental structures.

---

[1] Brouwer's full name was Luitzan Egbertus Jan Brouwer, although in the mathematical world he has always been known as L.E.J. Brouwer. Brouwer's works have been collected in two volumes [Brouwer, 1975] [Brouwer, 1976]. In addition, Dirk van Dalen edited and published some of his correspondence in English translation, in [Dalen, 2011a]. Accompanying this, van Dalen edited and published a great many online documentary materials, including many letters in their original languages in [Dalen, 2011b]. See the early biography of Brouwer, in which two experts summarize his life and his important work in logic and mathematics [Kreisel & Newman, 1969]. Most significantly, there is Dirk van Dalen's very extensive recent biography of Brouwer in two editions. The first edition is in two volumes [Dalen, 1999] [Dalen, 2005]. The second edition is slightly reduced in length and in a single volume, but it remains very detailed and complete [Dalen, 2012].

Brouwer's *Nachlass* has now become the Brouwer Archive, hosted in the North-Holland Archive in Haarlem, The Netherlands. See [Dalen et al., 2019, page 390].

D. M. Johnson, *Theories of Dimension from Ancient to Modern Times*, Sources and Studies in the History of Mathematics and Physical Sciences, https://doi.org/10.1007/978-3-032-02682-8_11

Brouwer's achievements in topology form a coherent unity. To put Brouwer's proofs of dimensional invariance into proper historical perspective, I shall begin by examining some of the origins of his topological ideas.[2]

11.2 First Mathematical Research Interests

From the beginning of his mathematical career Brouwer had two primary interests: intuitionism, his special brand of mathematical foundational thinking and constructivist mathematics, and geometry. His thesis, *Over de Grondslagen der Wiskunde* (*On the Foundations of Mathematics*) [Brouwer, 1907] [Brouwer, 1975, pp. 11–101], cf. [Stigt, 1979]), presented in February 1907, offers an excellent picture of his twin interests at the time. It is possible to discern in it many of the themes of his entire contribution to mathematics throughout his career. Topology per se is not explored in the thesis, rather the closely related domains of the Riemann–Helmholtz–Lie space problem and the Lie theory of continuous groups are investigated. Lie's continuous groups are essentially an early form of topological groups. Topology underlies their structure, even though at the time of writing his thesis the topology was not fully understood or explored. Brouwer's study of Lie continuous groups effectively motivated his investigations into the underlying topology.

Continuing from the geometrical work of his thesis, Brouwer made a full-scale attack on Hilbert's fifth problem: to build the Lie theory of finite continuous groups independently of conditions of differentiability.[3] He presented his first results at the Fourth International Congress of Mathematicians held in Rome, 6–11 April 1908, where he came in contact with some of the world's most distinguished mathematicians, among them Jacques Hadamard.[4] He conceived his problem in Lie group theory as the task [Brouwer, 1908, page 296]:

---

[2] Prof. Dr. Hans Freudenthal contributed enormously to our understanding of Brouwer's topological work through his edition of the second volume of the *Collected Works* [Brouwer, 1976]. Apart from his edition, I benefited greatly from his helpful comments on the earlier version of my work.

In addition, van Dalen's biography provides an extensive historical treatment of Brouwer's work on geometry and topology [Dalen, 2012, Chapters 4, 5, 8.2, 11, pp. 119–192, 285–290, 395–434]. The earlier version is in [Dalen, 1999, Chapters 4, 5, 8.2, pp. 122–196, 290–294] and [Dalen, 2005, Chapter 12, pp. 441–480]. I have found this treatment very valuable in revising my earlier work on Brouwer.

See [Rowe, 2018, 1.11, pp. 46–70]. on Brouwer's topological work related to its publication in *Mathematische Annalen* and Brouwer's relationship with its managing editor, Otto Blumenthal.

[3] Cf. [Hilbert, 1901] [Hilbert, 1902a] [Hilbert, 1935] [Gray et al., 2000]. For a summary of Brouwer's interest in and involvement with Lie groups and Hilbert's fifth problem see [Dalen, 2012, pp. 122–127] as well as [Dalen, 1999, pp. 126–130].

[4] Brouwer wrote to his teacher and *promotor* (i.e., his chief supervisor for the doctorate) D.J. Korteweg from Italy just after the Rome Congress. The letter is an interesting document in the Brouwer Archive; see [Dalen, 2011b, pp.234–235]. In it he assesses mathematics and mathematicians at the Congress. He has the highest praise for the French mathematicians, particularly mentioning Poincaré, Darboux, Picard, Borel, and Hadamard. Though he mentions Hadamard by name, it is not clear from the letter that he actually made his acquaintance there, although it is

Fig. 11.1 Brouwer as Newly Appointed Professor at the University of Amsterdam, 1912 [Brouwer Archive `https://noord-hollandsarchief.nl/`, reproduced in [Dalen et al., 2019, page 395]].

"To determine all finite continuous groups of an n-dimensional manifold"[Ch11, 1]

under the conditions of Hilbert's fifth problem. In his Rome lecture and later papers [Brouwer, 1909c] [Brouwer, 1910c] he set out to determine these groups for 1- and 2-dimensional manifolds. When discussing p-dimensional (i.e., p parameter) groups acting on n-dimensional manifolds, Brouwer points out that [Brouwer, 1908, page 297, footnote ([1])] (cf. [Brouwer, 1909c, page 247, footnote ***)]):

> Whether p is an invariant for every group remains open as long as the 'nonapplicability' of two spaces whose dimension numbers are different is unproven.[Ch11, 2]

very probable that he did. He certainly saw him and heard him lecture. Hadamard became very important in his development of topological methods, as I shall explain shortly.

Thus, in early 1908 he was well aware of the implications of the problem of dimensional invariance for his own work on Lie groups.

11.3 First Involvement with Topology

In dealing with Lie groups on 2-dimensional manifolds, Brouwer relied on the topological results of Arthur Schoenflies (see Section 8.7). Around the time of the Rome Congress he was making a thorough study of Schoenflies' planar analysis situs as contained in the series of articles entitled 'Beiträige zur Theorie der Punktmengen' ('Contributions to the Theory of Point Sets') [Schoenflies, 1903] [Schoenflies, 1904a] [Schoenflies, 1906a] and the *Bericht: Die Entwickelung der Lehre von den Punktmannigfaltigkeiten. Zweiter Teil* (*Report: The Development of the Theory of Point Manifolds/Sets. Second Part*) [Schoenflies, 1908b]. At Rome he was proud to say that his group-theoretic investigations were probably the first instance of an application of the Schoenflies results. There is further evidence of his deepening interest in the field of topology during 1908 in a lecture on 'Vlakke krommen en vlakke gebieden' ('Planar curves and planar domains'), which he delivered to the Wiskundig Genootschap, the Dutch Mathematics Society, in October 1908. In the lecture he presented a full review of topology in the style of Schoenflies.[5] The review was not overly critical, although he did point to certain deficiencies in Schoenflies' attempt to prove the invariance of closed curves.

Not long after he gave the October lecture he discovered some very serious flaws in the topological theory of Schoenflies.[6] In a letter to Hilbert of 14 May 1909 Brouwer gives his own account of his discovery of the radical inadequacies of Schoenfliesian topology [Brouwer, 1976, page 371] [Dalen, 2011b, page 275]:

> During the previous winter when I had already completed my second communication on finite continuous groups for submission to the editorial board of *Mathematische Annalen*, I discovered all of a sudden that the Schoenflies investigations concerning the analysis situs of the plane, on which I had relied in the fullest way, could not be taken as correct in all parts, so that my group-theoretic results also became doubtful.
>
> In order to achieve clarity again, it was first necessary to examine thoroughly the corresponding Schoenflies theory and to ascertain exactly which of his results could be completely trusted. In this way the enclosed work arose, which

[5] The 30-page manuscript of this lecture in Brouwer's hand (with emendations by Korteweg) is part of the Brouwer Archive. The review of Schoenfliesian topology is very complete. In particular, the problem of the invariance of dimension is discussed. The lecture is noted by title in the *Jaarverslag* (*Annual Report*) of the Wiskundig Genootschap (Dutch Mathematics Society) for 1908. See van Dalen [Dalen, 2012, page 139]; also [Dalen, 1999, page 142].

[6] These flaws were already mentioned briefly in Section 8.7. For excellent detailed accounts of Brouwer's discoveries of the flaws and the interactions with Schoenflies see Freudenthal's notes [Brouwer, 1976, pp. 367–370] and van Dalen's account in [Dalen, 2012, pp. 125–127, 137–148] as well as in [Dalen, 1999, pp. 128–130, 140–150].

> interferes in several parts of the Schoenflies theory and as it were reorganizes some of them.[Ch11, 3]

The enclosure was the first submitted version of his critical paper 'Zur Analysis Situs' ('On Analysis Situs') [Brouwer, 1910a], in which he laid out a set of unexpected and quite devastating counterexamples, including the first indecomposable continua and a curve which divides the plane into more than two domains but which is the exact boundary of each of these domains,[7] counterexamples that all but demolished Schoenflies' planar topology. At the same time as he submitted the paper to Hilbert he sent a copy to Schoenflies. Correspondences among Brouwer, Schoenflies, and Hilbert resulted, and it is clear from what remains of these as well as from the paper itself that Brouwer had attained a new mastery of the field of topology that was quite beyond Schoenflies.[8]

Schoenflies deserves some credit for providing Brouwer with a point-set topological framework and more importantly a set of significant problems that needed solutions. A fair amount of Brouwer's important subsequent research in topology had its problem source in Schoenflies' point-set topological work.[9]

Brouwer got to know the mathematician David Hilbert through an exchange of mathematical papers (offprints or separata). Then during the summer of 1909 he met him in person when Hilbert spent part of his summer vacation at the resort Scheveningen in The Netherlands. Brouwer's letter to his friend Carel Adama van Scheltema reveals his considerable admiration for Hilbert [Dalen, 2011a, page 56] [Dalen, 2011b, page 326]:

> This summer the world's most famous mathematician [David Hilbert] was in Scheveningen; through my work I had already been in contact with him, now I have again and again walked with him, and talked as a young apostle with a prophet. He was 46 [actually 47], but youthful in soul and body, a strong swimmer who eagerly climbed over walls and barbed wire fences.[Ch11, 4]

---

[7] Much has been written about indecomposable and related continua. For example, see [Kuratowski, 1968, pp. 204–208] [Ingram, 2006] [Kennedy, 2020].

Contemporary with Brouwer Arnaud Denjoy [Denjoy, 1910] published a similar example of such a curve a few months after [Brouwer, 1910a].

[8] Cf. Schoenflies' response [Schoenflies, 1910].

[9] Concerning Brouwer's critical theory [Brouwer, 1910a] there are eight letters and one postcard from Schoenflies to Brouwer dating from 1909 in the Brouwer Archive as well as four letters from Brouwer to Hilbert from 1909 in the Hilbert *Nachlass* at Göttingen. The remaining parts of the correspondences relating to [Brouwer, 1910a] seem to have been lost. It is beyond the scope of the present work to give a full account of Brouwer's 'transformation' from Schoenfliesian to Brouwerian topology. See the van Dalen references for some details.

11.4 Early Research Program in Topology

By mid-1909 Brouwer was deeply engrossed in the problems of topology that he came to understand and in many cases to define clearly for the first time. His work on Lie groups plus some research on transformations of surfaces and on vector fields (discussed below) had led him to the very boundary of topology. His critical evaluation of Schoenflies' work had shown him some serious pitfalls to be avoided when entering this treacherous field. In his inaugural lecture as privaat-docent (private-tutor; unpaid lecturer) in the University of Amsterdam, *Het Wezen der Meetkunde* (*The Nature of Geometry* [Brouwer, 1909d] [Brouwer, 1975, pp. 112–120]), delivered on 12 October 1909, one finds a clear sketch of the problem situation in topology as he saw it at the time. A letter to Hilbert, written three days later [Brouwer, 1976, pp. 375–376], supplements the lecture. Together they outline a program for research in topology, which in the next few years he proceeded to carry out with remarkable mathematical insight.

In his lecture Brouwer views geometry in the manner of Klein's *Erlanger Programm*, i.e., on the basis of groups of transformations, and concentrates on the group of analysis situs—the group of one-one continuous transformations, as he defines it.[10] The basic problem for this group is to find [Brouwer, 1909d, page 19] [Brouwer, 1975, page 118]:

> . . . a *classification for analysis situs of the sets of points of a space;* . . . [Ch11, 5]

but he sees this problem in relation to problems from higher branches of mathematics, from the theories of algebraic and complex functions and from automorphic functions. A part of the fundamental problem of analysis situs is to characterize different spaces of the same dimension, but [Brouwer, 1909d, page 20] [Brouwer, 1975, page 118]:

> An immediately related problem is, in how far spaces of different dimension are different for our group. Most probably this is always the case, but it seems extremely hard to prove, and probably it will remain an unsolved problem for a long time to come.[Ch11, 6]

Again one finds Brouwer pointing to the significance and difficulty of the problem of dimensional invariance. Yet in spite of its apparent difficulty, he came to solve it in less than a year, once he had acquired his basic topological tool: the degree of a continuous mapping.

In his inaugural lecture Brouwer naturally pays special attention to the multitude of problems surrounding the attempt to classify subsets of a 2-dimensional space, i.e., the attempt to construct a sound analysis situs of the plane. He points out

---

[10] Like several other mathematicians of this time, e.g., Hurwitz [Hurwitz, 1898] and Schoenflies [Schoenflies, 1908b, pp. 149–150], Brouwer includes one-one continuous transformations that are not necessarily *bi*continuous among the topological transformations. In contrast, Baire [Baire, 1907a] [Baire, 1907b] [Baire, 1909, page 136] restricts his 'applications' to modern homeomorphisms.

that many of the difficulties surrounding this attempt are caused by the fact that neither differentiable nor monotone functions are available in this domain, so that it is not possible to proceed by calculation. Hence, the investigator is forced to rely upon the few invariants which have been defined so far for dimension two: limit point, connectedness, domain, and boundary.[11] Jordan curves and arcs are also invariants. Indeed, just at this time Brouwer had completed his beautiful direct proof of the Jordan curve theorem [Brouwer, 1910b] and submitted it to Hilbert for *Mathematische Annalen* with his letter of 15 October 1909 [Dalen, 2011b, pp. 312–313].[12] In the lecture he stresses the difficulty of proving such a seemingly obvious theorem [Brouwer, 1909d, page 21] [Brouwer, 1975, page 119]:

> The fact that this theorem, which seems quite evident, requires an elaborate proof and that the numerous fruitless attempts to simplify it fill an extensive literature, throws a clear light on the limitations to which reasoning in this field is subjected.[Ch11, 7]

Even an apparently slight generalization of the Jordan theorem, the statement that a one-one continuous image of a closed curve (i.e., a closed and bounded planar point set that divides the plane into two domains and that is the boundary of each of these domains) is again a closed curve, has not been proved satisfactorily in his view. This is, of course, an indirect criticism of Schoenflies, who had tried to prove this statement [Schoenflies, 1908b, pp. 158–160]. Brouwer had given a more direct critique of the Schoenflies proof in his lecture to the Wiskundig Genootschap (Dutch Mathematics Society) of October 1908. A few years later he constructed his own proof [Brouwer, 1912e] [Brouwer, 1912g].

In his inaugural lecture he says that when we proceed to 3-space, [Brouwer, 1909d, page 21] [Brouwer, 1975, page 119]:

> No more are we certain that a closed Jordan surface, i.e., a one-one continuous image of a spherical surface, divides the three-dimensional Cartesian space into two domains.[Ch11, 8]

Nevertheless, not many months later he also proved the Jordan theorem in n dimensions (see Chapter 12).

Brouwer's inaugural lecture as privaat-docent (private-tutor; unpaid lecturer) and the letter to Hilbert of the same time characterize very well his interests in topology just prior to his greatest discoveries and, moreover, they reflect his approach to mathematics. He had arrived at the gateway to topology via other 'higher' branches of mathematics, such as Lie group theory and the theory of vector fields, and perceived the need to solve some of its fundamental problems. He took many of his problems from Schoenflies, clarifying them to be sure, but treating them as important open problems. Above all, his goal was to solve outstanding problems and not to construct a merely 'formal' branch of mathematics. However, in the course of his successful

[11] Brouwer seems to have in mind [Schoenflies, 1908b, pp. 151–158, 162–164] and related works by Schoenflies and others.

[12] Hausdorff included Brouwer's proof in his well-known text [Hausdorff, 1914, pp. 354–357].

conquest of the basic problems he was able to create a new area of topology—practically from nothing!

11.5 Early Results in Topology

I shall now turn from Brouwer's general topological program to specific problems and results that led him to develop topological methods and tools suitable for attacking the problem of dimensional invariance and related topological problems.[13] Near the beginning of 1909 he wrote two papers on related topological subjects: a paper on one-one continuous mappings of spheres into themselves and one on continuous vector fields on spheres.

The first of these papers [Brouwer, 1909a] [Brouwer, 1909b], presented to the Koninklijke Nederlandse Akademie van Wetenschappen, the Royal Dutch Academy at Amsterdam, by his teacher and *promotor* D.J. Korteweg on 27 February 1909, marks the start of a series on 'Continuous one-one transformations of surfaces in themselves.' Brouwer's Lie group investigations motivated this research, which led eventually to the plane translation theorem. The initial paper of the series contains the earliest version of the celebrated fixed point theorem [Brouwer, 1909b, page 797]:

> THEOREM 1. *A continuous one-one transformation in itself with invariant indicatrix of a singly [i.e., simply] connected, twosided, closed surface possesses at least one invariant point.*

In other words, a one-one continuous orientation-preserving mapping of a surface homeomorphic to a 2-dimensional sphere to itself has at least one fixed point—an important theorem that Brouwer proves by Schoenfliesian methods. Yet such a mapping for a 2-dimensional sphere that inverts the orientation does not necessarily have a fixed point. In contrast, the situation is exactly the reverse when we consider the 1-dimensional versions of these theorems [Brouwer, 1909b, page 798]. For a closed curve, the homeomorph of a circle (i.e., 1-dimensional sphere), an orientation-inverting mapping always has a fixed point, whereas an orientation-preserving one will not necessarily have a fixed point. Hence, one finds that the 1- and 2-dimensional cases—very special odd and even dimensions—are just the opposite of one another. A little later this distinction between odd and even dimensions became important in Brouwer's search for an invariance proof for dimension.

A month after submitting the paper on mappings of spheres Brouwer produced a significant paper 'On continuous vector distributions on surfaces' [Brouwer, 1909e] [Brouwer, 1909f] for communication to the Royal Dutch Academy (27 March 1909), the first of three on vector fields. His work in many ways parallels the great contributions of Poincaré on vector fields related to the qualitative theory of differential equations, exposited in four papers [Poincaré, 1881] [Poincaré, 1882] [Poincaré,

[13] For this section cf. [Dalen, 2012, pp. 130–137] [Dalen, 1999, pp. 133–140].

1885] [Poincaré, 1886]. However, a letter to Hadamard of 24 December 1909 shows that he was not at first aware of Poincaré's 'Mémoires' ('Memoirs') [Dalen, 2011a, pp. 62–63] [Dalen, 2011b, pp. 344–345]. In addition, while his investigations were initially independent of the great French mathematician's, they were also more general, since his outlook was more fully topological.

Brouwer's investigations into vector fields on the sphere (or a homeomorphic image of the sphere) have their origin in a paper by Peano [Peano, 1890c] (cf. Arzelà's paper [Arzelà, 1896]) containing an existence theorem for differential equations, which can be stated as follows: The differential equation

$$\frac{\mathrm{d}y}{\mathrm{d}x} = f(x, y),$$

where f is a continuous function of x and y, possesses at least one integral curve (or solution curve) through each point (x_1, y_1). Regarding the differential equation as defining a continuous vector field in the plane or (by stereographic projection) on the sphere, one sees that Peano's theorem guarantees the existence of solution curves that are tangent curves to the vector distribution. In his first paper on vector fields Brouwer uses the Peano theorem and Schoenflies-style topological arguments to deduce the following result [Brouwer, 1909f, page 856]:

> THEOREM 2. *A vector distribution anywhere [sic!; i.e., everywhere] univalent and continuous on a singly [i.e., simply] connected, twosided, closed surface must be zero or infinite in at least one point.*

That is, a continuous vector field on the homeomorphic image of a sphere must have a singularity at one point at least, a theorem akin to one of Poincaré's [Poincaré, 1881, page 405].

In his paper on vector fields Brouwer notes [Brouwer, 1909f, page 857] that there is a close connection between the theorem on singularities and the fixed point theorem for mappings of spheres. Yet, while they seem to be mutually deducible one from the other, this is not really the case. For example, if one takes the singularity theorem as proved and considers a one-one continuous transformation of the sphere, one can construct a vector field which corresponds to the transformation by joining each point P to its image P' with the shortest arc of the great circle containing both points and then regarding the length and direction of the arc as the vector. However, in order that the vector field be well-defined and continuous it is necessary to assume that no points P and P' are diametrically opposed. In general, though, one cannot assume this condition for arbitrary mappings of the sphere. Hence, the fixed point theorem does not follow from the singularity theorem. One will fail in a similar way if one attempts to deduce the singularity theorem from the fixed point theorem.

Brouwer was so perplexed by the logical independence of these two theorems apparently so closely related, that towards the end of 1909 he asked the advice of the distinguished French mathematician Jacques Hadamard (1865–1963)[14] (see the interchange of letters between Hadamard and Brouwer [Dalen, 2011a, pp. 60–63] [Dalen, 2011b, pp. 342–345]).[15] Brouwer had heard Hadamard talk at the Rome Congress of Mathematicians in April 1908 and was impressed by his mathematical abilities.[16] That Brouwer should now consult the eminent Frenchman proved to have very important consequences. The resulting correspondence and contacts between the two men were decisive in the formation of Brouwer's most brilliant topological ideas. In apparently the only letter of Hadamard to Brouwer that is extant, written some time before Brouwer's reply of 24 December 1909 [Dalen, 2011a, pp. 60–61] [Dalen, 2011b, page 342], Hadamard suggests a way of deriving the fixed point theorem from the singularity theorem. In reply Brouwer rightly criticizes Hadamard's incorrect argument [Dalen, 2011a, pp. 62–63] [Dalen, 2011b, pp. 344–345]. However, what is really significant in this exchange of letters is that Hadamard refers to Poincaré's 'Mémoires' ('Memoirs') on the curves defined by differential equations. Upon reading these, Brouwer started to grapple with the concept of index for vector fields. In turn, this led him to his most important topological tool: the degree of a continuous mapping.

11.6 Background to Brouwer's Concept of Degree of a Mapping

Before attempting to trace the route that Brouwer took in formulating his concept of mapping degree, I shall sketch the historical background to the concept.[17] The idea of index or winding number and its generalizations, principally developed by Cauchy, Kronecker, and Poincaré, are direct ancestors of Brouwer's powerful, even revolutionary topological concept. In addition, in formulating his concept Brouwer was guided by the idea of a Riemann surface, as discussed below. I shall first examine the background idea of circulation index/winding number and its generalization.

---

[14] For biographies of Hadamard see [Mandelbrojt & Schwartz, 1965] [Cartwright, 1965], and the full-length book [Maz'ya & Shaposhnikova, 1999]. See also [Mandelbrojt, 1953]. Hadamard's contributions to mathematics are important and extensive. He was highly honored during his lifetime.

[15] Brouwer also separately involved Korteweg. His letters to Korteweg and Hadamard are dated 24.XII,1909.

[16] The fact that Brouwer should consult Hadamard at this time suggests to me that the two did become acquaintances at the Rome Congress, although it is not absolutely certain that they did. It seems that they did not exchange any letters between April 1908 and the end of 1909. There are no letters from this period according to van Dalen in [Dalen, 2011a] or [Dalen, 2011b].

[17] There are excellent modern mathematical treatments of the mapping degree concept, some that include historical notes. See, for example, [Alexandroff & Hopf, 1935, Kapital XII] [Siegberg, 1981] [Outerelo & Ruiz, 2009] [Dinca & Mawhin, 2009] [Dinca & Mawhin, 2021]. I found [Outerelo & Ruiz, 2009] particularly useful in revising what I had written in an earlier version of this material, because it outlines the history through a set of problems.

Augustin Louis Cauchy (1789–1857)[18] was the first mathematician to take up the idea of winding number or index in an explicit way. The motivation for Cauchy to define such an index was the problem of ascertaining the number of real and complex roots of an equation that are enclosed within a given contour. Even before Cauchy the concept of index or characteristic was implicit in Gauss' first, third, and fourth proofs of the fundamental theorem of algebra (dated 1799, 1816, 1849, respectively; cf. [Ostrowski, 1920] [Siegberg, 1981, pp. 129–132]). Cauchy's broader problem is clearly related to the fundamental theorem of algebra. Cauchy first explored his concept of index in an important *Mémoire detaché* (*separate memoir*), presented to the Turin Academy on 27 November 1831 when he was in self-imposed exile from France (due to the July 1830 Revolution and change of government) [Cauchy, 1831a]. Unfortunately, the *Mémoire* was not published in a journal at the time, but only handwritten and lithographed, and then distributed to a few mathematicians and societies. However, Cauchy did publish a short résumé (of some 13 pages) of the *Mémoire* in the *Bulletin de Férussac* [Cauchy, 1831b], and through this medium his ideas received a wider circulation among the members of the mathematical public.

To see how Cauchy introduced the notion of index one supposes, following the résumé [Cauchy, 1831b], that one has a real function $f(x)$ that is continuous except that it may have certain values of x for which it becomes infinite. Let $x = a$ be one of the values for which the function is infinite. As x increases, passing through a, the function $f(x)$ in becoming infinite may pass from negative to positive, or from positive to negative, or it may not change sign. One defines the index of the function $f(x)$ at a as $+1$ in the first case, -1 in the second, and 0 in the third. Then one defines the *integral index* of $f(x)$ between the limits $x = x_0$ and $x = X > x_0$ as the sum of the indices for the function over the interval,[19] employing with Cauchy the notation:

$$\mathop{\mathbf{J}}_{x_0}^{X}((f(x))).$$

The definition in the published paper of 1837 [Cauchy, 1837a, pp. 176–177] is similar.

Suppose now that one wants to find the number of roots of the equation

$$f(z) = f(x + y\sqrt{-1}) = 0$$

contained within a contour (i.e., simple closed curve) that does not pass through any roots. Considering f as a complex (analytic) function

$$f(x + y\sqrt{-1}) = \phi(x, y) + \sqrt{-1}\chi(x, y),$$

where ϕ, χ are real functions of x, y, and then letting

[18] See [Belhoste, 1991] for a recent biography of Cauchy. There is an older biography [Valson, 1868], which is not considered reliable.

[19] In his work Cauchy tacitly assumes that the sum is finite.

$$\psi(x, y) = \frac{\phi(x, y)}{\chi(x, y)}$$

and expressing ψ in terms of arc length s around the contour in a positive sense from 0 to c, one finds by a chain of reasoning that the number m of roots of the given equation is:

$$m = \frac{1}{2} \overset{c}{\underset{0}{\boldsymbol{J}}} ((\psi(s))).$$

The index relation actually determines the winding number, i.e., the number of times that the f-image of the contour is wrapped around the origin of the ϕ, χ-plane. However, in his writings Cauchy does not explicitly interpret his index relation in such a succinct geometrical way, although it is clear from the way he defines the integral index that the ratio $\frac{\phi}{\chi}$ will be infinite every time the f-image passes through the ϕ-axis, so that $\frac{1}{2}$ the index will yield the winding number.

In his original *Mémoire detaché* (*separate memoir*) Cauchy's derivation of the index formula for the number of roots is based on his calculus of residues [Cauchy, 1831a, pp. 3–23]. He expresses the number m of roots in residues as in [Cauchy, 1831a, page 22, formulas (89)] (see [Cauchy, 1831b, page 121, formula (4)]):

$$m = \frac{1}{2\pi\sqrt{-1}} \int_0^c \frac{f'(z)}{f(z)} \frac{\mathrm{d}z}{\mathrm{d}s} \mathrm{d}s,$$

since the poles of the function $\frac{f'(z)}{f(z)}$ are the roots of the function $f(z)$ [Cauchy, 1831a, pp. 12–14]. This integral is indeed equal to the variation of the natural logarithm of $f(z)$, and from this fact Cauchy derives the index relation [Cauchy, 1831a, pp. 14–23].

In a pair of papers, which appeared in 1836, Charles Sturm (1803–1855)[20] and Joseph Liouville (1809–1882)[21] put forward other proofs of Cauchy's index relation [Sturm & Liouville, 1836] [Sturm, 1836]. Sturm and Liouville seem to have been motivated by Cauchy's summary of his work in the résumé [Cauchy, 1831b] or perhaps the original *Mémoire* (*Memoir*) [Cauchy, 1831a] if they had available a copy of the handwritten lithograph version. In Sturm's paper one finds a more geometrical interpretation of the Cauchy index idea, one that is closer to the usual intuitive notion of winding number. For example, see the simple geometrical figures they include [Sturm & Liouville, 1836, pp. 279, 287].

In an additional *Mémoire detaché* (*separate memoir*), presented to the Turin Academy on 16 June 1833, then distributed in handwritten and lithographed form [Cauchy, 1833], Cauchy gives a generalization of his results for two functions of two variables [Cauchy, 1833, page 16]. A later note in the *Comptes rendus* (*Proceedings*) of the Academy of Sciences, Paris of 1837 provides a brief summary of the *Mémoire detaché* (*separate memoir*) of 1833 [Cauchy, 1837b], with the generalized result

[20] Biographical material on Sturm may be found in [Loria, 1938] and in [Sturm, 2009, pp. 1–11].

[21] For an extensive excellent biography of Liouville see [Lützen, 1990]

being quoted as the Eighth theorem (Huitième théorème) [Cauchy, 1837b, pp. 273–274] (in the 1833 version [Cauchy, 1833, page 16])[22] and with an elementary proof.

In a longer paper of 1837, Cauchy provides a fuller exposition of the notion of index in a way that is independent of residues and integrals [Cauchy, 1837a]. He considers both functions of one variable [Cauchy, 1837a, pp. 176–195] and functions of two variables [Cauchy, 1837a, pp. 195–229]. In particular, he examines pairs of continuous real functions $u = F(x, y)$, $v = f(x, y)$ with continuous first partial derivatives for their simultaneous solutions within a given contour [Cauchy, 1837a, page 197]. The index, i.e., double the winding number, then gives a direct measure of the number of such solutions [Cauchy, 1837a, 10e Théorème, page 217].[23]

Cauchy's 1837 note in the *Comptes rendus* (*Proceedings*) of the Academy of Sciences, Paris [Cauchy, 1837b] drew an immediate critical comment from Charles Sturm and Joseph Liouville [Sturm & Liouville, 1837]. In their note they provide three examples [Sturm & Liouville, 1837, pp. 722–723]—counterexamples to Cauchy's stated theorem, and then offer a corrected theorem to Cauchy's faulty one. The authors use $P = F(x, y)$ and $Q = f(x, y)$ as notation for the functions of two variables, and let

$$R = \frac{\partial P}{\partial x} \cdot \frac{\partial Q}{\partial y} - \frac{\partial P}{\partial y} \cdot \frac{\partial Q}{\partial x},$$

(expressed in this equation with a slight update using the partial derivative notation), in order to state their theorem as [Sturm & Liouville, 1837, pp. 723–724]:

> Let us consider a closed contour ABC on which P and Q never both vanish, and let us furthermore admit that, in the interior of this contour, the values of x, y, which make P and Q have value 0, give to R a positive or negative value, but one different from zero.
>
> Among the solutions (x, y) of the equations $P = 0$, $Q = 0$, contained in the interior of ABC, some may correspond to a positive value, the others to a negative value of R. We shall denote by μ_1, the number of solutions of the first kind, and by μ_2 the number of solutions of the second kind.
>
> That being said, we shall let
>
> $$\Delta = 2(\mu_1 - \mu_2),$$
>
> where Δ, representing the excess of the number of times that the fraction $\frac{P}{Q}$ goes from positive to negative over the number of times it goes from negative to positive, vanishes, as we move over the entire contour ABC, going from positive x to positive y [i.e., as we trace the contour in a positive direction].[Ch11, 9]

The number $\mu_1 - \mu_2$ is now called the topological degree of the mapping (F, f) at the origin [Outerelo & Ruiz, 2009, page 5].

Cauchy returned to the subject of index in a paper of 1855, 'Sur les compteurs logarithmiques' ('On the logarithmic counters') [Cauchy, 1855]. In this work he essentially introduces the 'Argument Principle' in relation to the winding number.

---

[22] Given in modernized notation in [Outerelo & Ruiz, 2009, page 3].

[23] A slightly later work on the subject of winding number is [Ullherr, 1846].

I shall not give the details of Cauchy's treatment. For a modern summary of it see [Outerelo & Ruiz, 2009, pp. 5–7].

I shall now examine a closely related problem of finding roots, a problem that is also part of the development of the degree concept. The problem is one of finding algorithmically the number of real roots of a polynomial in one variable with real coefficients in a given closed interval on the real line (that is, an interval defined by two real numbers). The focus of this problem and its generalization to multiple polynomials and multiple equations is the real domain, whereas the focus of the problem of the Cauchy index just discussed is the complex domain. The problems are closely related in a broader context and come together in the development of the degree concept.

In a celebrated paper of 1829 [Sturm, 1829], followed by a complete exposition in a separate publication of 1835 [Sturm, 1835], Charles Sturm solved the problem. In his work Sturm first considers a sequence of polynomials associated with a univariate polynomial P and its derivative that are computed by a variant of Euclid's algorithm for polynomials [Sturm, 1835, pp. 275–277]. Then Sturm's theorem expresses the number of distinct real roots of P located in an interval in terms of the number of changes of signs of the values of the Sturm's sequence at the bounds of the interval [Sturm, 1835, pp. 276–277]. When applied to the interval of all the real numbers, it gives the total number of real roots of P.[24]

Sturm's work generated a great deal of interest and spurred further work by several important mathematicians. In the 1850s Charles Hermite (1822–1901)[25] [Hermite, 1852] [Hermite, 1853] [Hermite, 1856] and James Joseph Sylvester (1814–1897)[26] [Sylvester, 1853a] [Sylvester, 1853b] [Sylvester, 1853c] were publishing extensions of Sturm's work. For example, Sylvester in a paper of some 142 pages provides an extensive analysis of certain generalizations relative to Sturm's theorem dealing with two functions [Sylvester, 1853a]. In the other two papers he gives a relatively simple modification of Sturm's theorem [Sylvester, 1853b] [Sylvester, 1853c].[27]

The well-known German mathematician Leopold Kronecker (1823–1891)[28] had a continuing interest in Sturm's theorem [Kronecker, 1869b] [Kronecker, 1873]. Both his research papers and his frequently given lecture cycles on the theory of algebraic equations at Berlin (*Vorlesungen über die Theorie der algebraischen Gleichungen*, that is, *Lectures on the theory of algebraic equations*) give evidence of his sustained interest in the theorem and its generalizations. In his doctoral thesis, Cédric Vergnerie

[24] Cauchy was interested in this kind of problem, going back to 1813/1815 [Cauchy, 1815]. For a full treatment of the interplay between the ideas and papers of Cauchy on this topic and those of Charles Sturm and Joseph Liouville see the detailed analysis of Sinaceur in the paper [Sinaceur, 1992] and the book [Sinaceur, 1991, Première Partie, pp. 33–144]. Cf. [Eisermann, 2008] or [Eisermann, 2012].

[25] For biographical information see [Darboux, 1906].

[26] For biographical information see [Parshall, 1998] [Parshall, 2006].

[27] For a nineteenth-century summary of the work on Sturm's theorem and the later related work of Kronecker see [Weber, 1895, pp. 270–298] [Weber, 1898, pp. 301–340]. On the historical reception to Sturm's theorem and related work see [Sinaceur, 2009].

[28] For biographical material see [Weber, 1893], [Edwards, 1988], and Edwards in [Begehr et al., 1998, pp. 64–69].

[Vergnerie, 2017, Deuxième partie, pp. 173–317] offers a detailed, extensive analysis of the *Vorlesungen* (*Lectures*) that Kronecker gave over the 1870s, 80s, and 90s to show how the theorem figured in his teaching.[29]

Kronecker's primary aim in researching Sturm's theorem was to generalize it as a way to deal with the case of real zeros or roots of n real equations in n unknowns. Kronecker viewed Sylvester's work as a step in that direction [Kronecker, 1869a, page 167]. However, he saw that Sylvester had encountered difficulties in looking for a suitable generalization, as he clearly states in a paper of 1869 [Kronecker, 1869a, page 167]:

> The difficulties of generalization, which Mr. Sylvester mentions, may well lie mostly in the restrictions on the algebraic structures he has considered. As soon as I gained the insight that all relevant considerations belong exclusively to that general area, which for the case where $n = 2$ or 3, is called the "Geometry of Location,"[30] I found the simplest means to overcome the difficulties encountered.[Ch11, 10]

In this passage he claims that he has overcome the difficulties that Sylvester encountered. In clearing up these difficulties he was able to proceed to fulfill his prime goal of generalization, and this came through developing his general theory of 'characteristic' ('Charakteristik').[31]

Kronecker put forward his theory of characteristic in detail in 1869 in a pair of insightful, but quite difficult and demanding papers, 'Über Systeme von Functionen mehrer Variablen' ('On systems of functions of several variables') [Kronecker, 1869a] [Kronecker, 1869c] in March and August of 1869 (also presented on 10 May 1869 to the Academy of Sciences, Paris [Kronecker, 1869b]). A few years later in a paper of 1873 he explored his theory of characteristic in relation to various Sturm series arising out of the work of Sturm, Sylvester, Hermite, and Jacobi [Kronecker, 1873]. Then in 1878 he wrote an additional paper on characteristic theory [Kronecker, 1878b]. He also incorporated characteristic theory as a part of some of his lecture cycles on the theory of algebraic equations.

In exploring his general theory of characteristic Kronecker set himself the task effectively to generalize Cauchy's index from two to n dimensions. He needed to explore sets of functions in several variables and thereby consider higher-dimensional differentiable manifolds.

In the first paper Kronecker provides some historical background to his work [Kronecker, 1869a, pp. 167, 186–187]. He mentions Sturm's theorems and Hermite's extensions as motivating his work [Kronecker, 1869a, pp. 186–187]. Significantly, he also states that he communicated the first part of his results to his friend (and colleague at Berlin University) Karl Weierstrass, who stimulated him to go further

---

[29] Vergnerie gives a list of the lecture series from 1872 to 1891 [Vergnerie, 2017, pp. 44–45]. Cf. also his paper [Vergnerie, 2019].

[30] No doubt this is a reference to von Staudt [Staudt, 1847].

[31] Vergnerie's further analysis in his thesis [Vergnerie, 2017, Troisième partie, pp. 319–508] is intended to show Kronecker's connection to the work of Sylvester and to his general theory of 'characteristic'.

to take up new points of view, including Cauchy's work. In this way he was able to expand his work to provide a fuller treatment.

The following paragraphs give a short summary of Kronecker's theory of characteristic as exposited in his papers, while also drawing on the early textbook summary of the theory written by Heinrich Weber [Weber, 1898, pp. 323–332], some modern mathematical treatments found in the books of [Outerelo & Ruiz, 2009, pp. 7–11], [Dinca & Mawhin, 2009, pp. 3–16], and [Dinca & Mawhin, 2021, pp. 1–15], and Vergnerie's thesis [Vergnerie, 2017, Troisième partie, pp. 319–508] along with his paper [Vergnerie, 2019].

In the first two papers of 1869 as well as several related ones [Kronecker, 1873] [Kronecker, 1878a] [Kronecker, 1878b] Kronecker described a wide diversity of applications to topics central to nineteenth-century mathematics. It is quite clear from Kronecker's papers that he had a deep understanding of his generalized Cauchy concept in relation to other mathematical developments of his time. His papers have the flavor of well-developed nineteenth-century algebraic and analytic investigations of equations, though he may well have not understood fully the topological 'nuggets' hidden within them.[32]

In his first paper Kronecker gives an extended functional and geometrical explanation of the generalized index, or 'characteristic' ('Charakteristik') [Kronecker, 1869a, pp. 159–161] [Kronecker, 1895, pp. 177–179].[33] He begins with $(n + 1)$ real-valued functions

$$F_0, F_1, F_2, \ldots, F_n$$

(a function-system) in n real variables

$$z_1, z_2, z_3, \ldots, z_n,$$

and makes the following assumptions about these functions:

1. the functions take on both positive and negative values infinitely often over the n variables;
2. the functions are everywhere continuous;
3. the functions are differentiable in each of the n variables;
4. the functions are assumed finite, such that none of the $(n + 1)$ functional determinants [Jacobians]

$$\left|\frac{\partial F_i}{\partial x_j}\right|_{i=0,\ldots,k-1,k+1,\ldots,n,\text{ i.e., excluding one } k \text{ in the sequence } 0,\ldots,n;\ j=1,\ldots,n}$$

vanish simultaneously with the given functions for infinitely many value systems of the z_i.

Kronecker uses the symbol Δ_k to denote the functional determinant (Jacobian) of the n functions

[32] I think that Kronecker's work in this case fits well with his algorithmic view of mathematics, as Edwards characterizes his view [Edwards, 2009].

[33] [Weber, 1898, pp. 323–332] gives an excellent textbook treatment of the Kronecker characteristic for the case of 3 functions.

$$F_0, F_1, \ldots, F_{k-1}, F_{k+1}, \ldots, F_n,$$

taking partial derivatives in all n variables, while leaving out the column of partial derivatives of function F_k.

If one sets any chosen $(n-1)$ of the functions F equal to 0, then the variation of the variables bound a simple infinity or manifold. The associated value systems of n quantities z form a continuous path, and it is possible to speak of the value systems as 'points' ('Punkte'), and the continuous path as a 'line' ('Linie'). To designate the line in this situation, Kronecker sets

$$F_i = 0,$$

for all indices with the exception of two (h and k). He then fixes a direction for the path or line, and uses $[hk]$ for one direction and $[kh]$ for the opposite direction.

It should now be assumed that all the $\frac{1}{2}n(n+1)$ lines, which are formed from the function-system F, are closed lines and that the number of points determined by the n equations $F = 0$ is finite. He then considers the directed line $[hk]$ and where it cuts the $(n-1)$-fold manifold $F_i = 0$. On this domain $F_h \cdot F_k$ is negative, or $F_h \cdot F_k$ is positive. In the first case the intersection point of $[hk]$ with $F_h = 0$ will be called an outgoing (Austritt) point; in the second case the cut point will be called incoming (Eintritt) point.

Kronecker asserts that the total number of incoming points (say, E, for Eintritt) and outgoing points (say, A, for Austritt) is an even number ($E + A$ is even) [Kronecker, 1869a, page 161] [Kronecker, 1895, page 179]. If one subtracts the number of outgoing points from the number of incoming points and takes the half, $\frac{1}{2}(E - A)$, then the result is an even number, or an odd number, or zero. The fundamental theorem is that this number is a constant (or invariant) even as the indices h and k vary over $0, \ldots, n$. In fact, the number for the function-system is characteristic, and so should be called the 'characteristic' ('Charakteristik') of the system.

From Kronecker's lectures it is possible to take two examples and their accompanying diagrams to get a sense of his analysis of characteristic. Vergnerie in his thesis includes a much more detailed analysis [Vergnerie, 2017, pp. 347–359, 375–388].

Take the case of two real-valued functions of one real variable, $\phi(z)$ and $\psi(z)$, as Kronecker considered in his lectures of 1880–81 and 1890–91. Figure 11.2 is essentially from Kronecker's unpublished 1880–81 series (page 157, as reproduced in [Vergnerie, 2017, page 352]). The intersections labeled are typical. The calculation of characteristic is [Vergnerie, 2017, page 351]:

$$\mathit{characteristic}(\phi, \psi) = \frac{1}{2}(2 - 4) = -1.$$

As another example take the case of three curves representing three functions, each of two variables: $f(z_1, z_2)$, $\phi(z_1, z_2)$, $\psi(z_1, z_2)$, as Kronecker considered in his lectures of 1879. Figure 11.3 is from Kronecker's unpublished 1879 series (page 526, as produced in [Vergnerie, 2017, page 385]). Again, the intersections labeled are typical. The calculation of the characteristic is more complicated, and there are

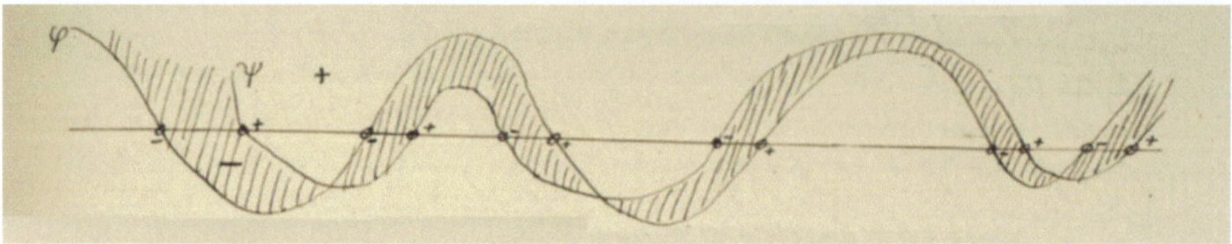

Fig. 11.2 Diagram for the Kronecker Characteristic: Two Real Functions of a Single Real Variable, $\phi(z)$, $\psi(z)$ [from Kronecker's 1880-81 Lecture Series, Collection Kurt Hensel, Bibliothèque de mathématique, IRMA Université de Strasbourg, reproduced in [Vergnerie, 2017, page 352]].

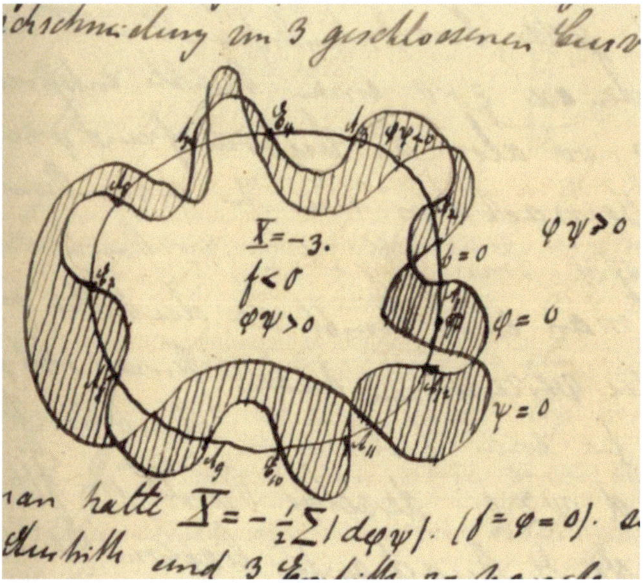

Fig. 11.3 Diagram for the Kronecker Characteristic: Three Real Functions of Two Real Variables, $f(z_1, z_2)$, $\phi(z_1, z_2)$, $\psi(z_1, z_2)$ [from Kronecker's 1879 Lecture Series, Collection Kurt Hensel, Bibliothèque de mathématique, IRMA Université de Strasbourg, reproduced in [Vergnerie, 2017, page 385]].

multiple interpretations, so I shall not go further into the matter; see [Vergnerie, 2017, pp. 380–385] for full details and analysis.

Kronecker's proof of the the invariance of the characteristic of a function-system takes a few paragraphs [Kronecker, 1869a, pp. 161–164] [Kronecker, 1895, pp. 179–182]. It is informal and, I think, does not provide sufficient details. (Perhaps, one could reasonably say that it is a little 'vague' or incomplete.) No doubt Kronecker had worked through many examples to arrive at his result, as examples from his later lecture series suggest. He gives a substitution procedure, substituting a newly introduced function ($\mathfrak{F}_k$) for one (F_k) in the original set forming the function-system, where the newly introduced function has special properties relative to a specific point.

The sign value (+1 or −1) at the specific point is called the 'character' ('Charakter') of the point. He gives a series of algebraic relations showing the connection between the point characters and the characteristic of the function-system. I shall not give the details (see Kronecker's original paper [Kronecker, 1869a, pp. 162–163] [Kronecker, 1895, pp. 180–181] and [Weber, 1898, pp. 323–332, especially page 328] (see also [Outerelo & Ruiz, 2009, pp. 8–9])).

Following from the work of Cauchy, Sturm, and Liouville before him, Kronecker is interested in the problem of determining the number of zeros of a set of equations and illuminating this problem with his concept of characteristic. Considering the function-system

$$F_0, F_1, F_2, \ldots, F_n,$$

he notes that the characteristic is equal to half the difference between the number of points for which

$$\Delta_0 \cdot F_0 < 0, \quad F_1 = F_2 = \cdots = F_n = 0$$

and the numbers of points for which

$$\Delta_0 \cdot F_0 > 0, \quad F_1 = F_2 = \cdots = F_n = 0$$

(where Δ_0 is the functional determinant (Jacobian) as defined earlier). If one denotes by the sequence

$$(\zeta_1, \zeta_2, \ldots, \zeta_n), \; (\zeta_1', \zeta_2', \ldots, \zeta_n'), \; \ldots$$

the totality of common value systems (points) for the equations

$$F_1 = 0, \; F_2 = 0, \ldots, F_n = 0,$$

then the characteristic of the system

$$(\Delta \cdot F_0', F_1, F_2, \ldots, F_n),$$

specifies half the excess of the number of points (ζ) lying inside of F_0' over the number of points lying outside of F_0'. Finally, if the domain $F_0' < 0$ completely encloses another domain $F_0'' < 0$, then the number of points (ζ) between the two boundaries $F_0' = 0$ and $F_0'' = 0$ is expressed as the difference of the two characteristics:

$$(\Delta_0 \cdot F_0', F_1, F_2, \ldots, F_n) \text{ and } (\Delta_0 \cdot F_0'', F_1, F_2, \ldots, F_n)$$

[Kronecker, 1869a, pp. 164–165] [Kronecker, 1895, pp. 182–183]. Hence, the result is a conceptual determination of the zeros of the function-system.

In Section V of his first paper of 1869 on characteristic Kronecker relates his concept of characteristic directly to the concept of winding number (Windungszahl). He also expresses characteristic as an integral, the Kronecker integral, and relates his integral to an integral of Gauss, highlighting it as a generalization [Kronecker, 1869a, pp. 167–171] [Kronecker, 1895, pp. 186–190].

For the winding number relation he considers the manifold formed of n real variables $x_1, x_2, \ldots x_n$ defined by the equations

$$x_1 = F_1(z_1, z_2, \ldots, z_n),\ x_2 = F_2(z_1, z_2, \ldots, z_n),\ \ldots,\ x_n = F_n(z_1, z_2, \ldots, z_n).$$

He notes that to each point z (with n coordinates) there corresponds a single point x (also with n coordinates), while, in general, to each x there may correspond several z. On the basis of these equations, from a well-defined $(n-1)$-fold manifold

$$F_0(z_1, z_2, \ldots, z_n) = 0,$$

there arises a well-defined $(n-1)$-fold manifold:

$$\Phi(x_1, x_2, \ldots, x_n) = 0.$$

This is exactly the set up for the concept of characteristic. If one thinks of the $(n-1)$-fold manifold $F_0 = 0$ as wrapped around an n-sphere according to the given functional mappings to yield the $(n-1)$-fold manifold $\Phi = 0$, then the meaning of characteristic as winding number emerges.[34]

In modern terms Kronecker's characteristic is the order of the origin with respect to the $(n-1)$-dimensional manifold $\Phi = 0$, i.e., it is the total number of signed intersections with the manifold of a ray emanating from the origin. However, Kronecker himself does not explicitly give this modern interpretation of characteristic.

Kronecker also offers an integral formula for the characteristic, a fairly complicated analogue of Cauchy's residue integral. To introduce the integral, he uses the following notations. Let dw be the positive (infinitesimal) element of the equation

$$F_0(z_1, z_2, \ldots, z_n) = 0,$$

the $(n-1)$-fold manifold, the element basis for the integral. Let $\bar{\omega}$ be the value of the integral:

$$\int \mathrm{d}w,$$

stretching over (the unit n-sphere)

$$z_1^2 + z_2^2 + \cdots + z_n^2 = 1.$$

Let S and $\mathfrak{S}$ be the positive values of

$$\sqrt{F_1^2 + F_2^2 + \cdots + F_n^2} \quad \text{and} \quad \sqrt{F_{01}^2 + F_{02}^2 + \cdots + F_{0n}^2},$$

respectively. Finally, let R be the functional determinant [Jacobian]:

[34] It should be noted that about this time Weierstrass was using the simple idea of winding number. Cf. [Weierstrass, 1927, pp. 71–72, 257–259].

$$R = \frac{1}{\bar{\omega}} \begin{vmatrix} 0 & F_{01} & F_{02} & \dots & F_{0n} \\ F_1 & F_{11} & F_{12} & \dots & F_{1n} \\ F_2 & F_{21} & F_{22} & \dots & F_{2n} \\ \dots & \dots & \dots & \dots & \dots \\ \dots & \dots & \dots & \dots & \dots \\ \dots & \dots & \dots & \dots & \dots \\ F_n & F_{n1} & F_{n2} & \dots & F_{nn} \end{vmatrix}.$$

On the basis of these notations he expresses the Kronecker integral as:

$$-\frac{1}{\bar{\omega}} \int \frac{R}{S^n} \mathrm{d}w.$$

The integral formula expresses the order in alternative terms. The order is measured by the total solid angle subtended by the manifold $\Phi = 0$ with respect to the origin or, speaking intuitively, it is the number of times $\Phi = 0$ 'covers' the unit sphere with center at the origin [Kronecker, 1869a, pp.169–171] [Kronecker, 1895, pp. 187–190]. In devising his formula, Kronecker took his cue from an integral that Gauss had used in potential theory [Gauss, 1813] [Gauss, 1840] [Gauss, 1867, pp. 7–9, 204–211].[35] Kronecker's 1878 paper [Kronecker, 1878b] expands on his earlier work. There he clearly states the connection between the characteristic expressed as an integral and winding number.

In his papers Kronecker then proceeds, applying the deep insights of an experienced master, to assemble a multitude of applications for his general characteristic. In particular, in his second paper of 1869 he links a special case of characteristic with Gauss' *curvatura integra* (Gaussian integrated curvature) [Kronecker, 1869c, pp. 688–698] [Kronecker, 1895, pp. 213–226]. Thus, he concludes this second paper with an expression of the conviction of having shown [Kronecker, 1869c, page 698] [Kronecker, 1895, page 226]:

> ... that the theory of the characteristic of function-systems is connected as closely with geometrical theories as with potential theory; and in these relationships of the concept of characteristic, which was originally developed from purely analytic principles, to other known theories, we may indeed recognize a test of the genuineness of the concept and a proof that the introduction of this concept into science is quite natural and necessary.[Ch11, 11]

Kronecker clearly assessed the positive value of his theory of characteristic by the diversity of its applications.[36]

---

[35] For a modern version of the Kronecker integral, following the lead of [Hadamard, 1910], see Alexandroff and Hopf in their treatise *Topologie I* (*Topology I*) [Alexandroff & Hopf, 1935, pp. 465–467].

[36] Émile Picard (1856–1941) was deeply interested in the problem of finding the roots common to two or more simultaneous equations and, hence, in Kronecker's characteristic theory and related integral, as is evidenced in several of his publications [Picard, 1889] [Picard, 1891a] [Picard, 1891b] [Picard, 1891c] [Picard, 1892] [Picard, 1893, Chapitre VII, pp. 193–207]. He took issue with Kronecker's characteristic, since it did not provide the exact number of such roots [Picard,

During the 1880s in a set of ground-breaking 'Mémoires' ('Memoirs') on the qualitative theory of differential equations related to work in celestial mechanics Henri Poincaré put the Cauchy index and Kronecker characteristic to good use in determining the distribution of singular points of differential equations and in their corresponding vector fields [Poincaré, 1881] [Poincaré, 1882] [Poincaré, 1885] [Poincaré, 1886].[37] Poincaré had been encouraged by Hermite's lectures at the École Polytechnique and the Sorbonne to study aspects of Sturm's theorem and Kronecker's characteristic. These 'Mémoires' ('Memoirs') were fundamental to Poincaré's own development of analysis situs or topology starting in 1895. As mentioned already, Poincaré's ideas on index/characteristic were an immediate influence on Brouwer as he was formulating his notion of mapping degree.

In his first 'Mémoire sur les courbes définies par une équation difféntielle' ('Memoir on the curves defined by a differential equation') Poincaré takes up the study of differential equations with the form [Poincaré, 1881]

$$\frac{\mathrm{d}x}{X} = \frac{\mathrm{d}y}{Y},$$

where X and Y are polynomials in x and y, and their defined solution curves or 'characteristics,' which he regards as projected on the sphere (taken as a model of the projective plane, the doubly covering sphere of the projective plane). For his qualitative theory Poincaré is concerned with the singularities of these equations where X and Y are simultaneously zero. For these special points with their related solution curve sets defined by the differential equations, he provides a case-by-case analysis, characterizing four types of singularities [Poincaré, 1881, Chapitre II, pp. 385–393], defining 'noeud' ('node'; German 'Knotenpunkt') [Poincaré, 1881, page 488], 'col' ('saddle point'; German 'Sattelpunkt') [Poincaré, 1881, page 488], 'foyer' ('focus'; German 'Strudelpunkt') [Poincaré, 1881, page 390], and 'centre' ('center'; German 'Wirbelpunkt') [Poincaré, 1881, page 391]. He then examines the possible distribution of these singularities [Poincaré, 1881, Chapitre III, pp. 394–409].

Midway through his work he announces the introduction of [Poincaré, 1881, page 400]:

> . . . a new consideration which will render us the greatest services.[Ch11, 12]

1891a, page 356], and proceeded to provide his own exact solution to the problem [Picard, 1891a] [Picard, 1891b] [Picard, 1891c] [Picard, 1892]. Kronecker responded to Picard's criticism in the last year of his life [Kronecker, 1891]. See the details of this minor controversy in [Vergnerie, 2020].

37 Christian Gilain has given a broad summary of the background and motivation for Poincaré's research into the qualitative theory of differential equations as well as the quantitative or analytical theory of differential equations [Gilain, 1991]. Jean Mawhin has provided an excellent historical analysis of Poincaré's early use of topology in this work on differential equations, while relating the matter to Kronecker's integral [Mawhin, 2000]. For a recent treatment of Poincaré index see [Aleksandrov, 2022].

the concept of index for vector fields related to the solution curve sets of the differential equations—Cauchy's index under a different guise. For this he considers the variation of the expression X/Y as a point moves around a 'cycle' or closed curve in one hemisphere in an assigned positive sense, i.e., the variation of the tangent direction around the cycle. Letting h be the number of times that the expression jumps from $-\infty$ to $+\infty$ and k be the number of times that it jumps from $+\infty$ to $-\infty$, he defines the index of the cycle (indice du cycle) as the integer

$$i = \frac{h - k}{2}.$$

This index is effectively the number of times that a variable vector tangent to the characteristics of the differential equation turns around as its initial point moves around the cycle. Actually, Poincaré's definition yields values for singularities that are opposite to the usual ones as used later. His index value for a cycle enclosing just one simple singular point is -1 for a node (noeud), focus (foyer), or center (centre), and is $+1$ for a saddle point (col) [Poincaré, 1881, page 401]. Indices for more complicated singular points can be other than 1. The four small figures given in Figure 11.4 (from [Liebmann, 1914, page 508]) illustrate the index concept for the four cases of (1) node (noeud; Knotenpunkt), (2) saddle point (col; Sattelpunkt), (3) center (centre; Wirbelpunkt), and (4) focus (foyer; Strudelpunkt). To determine the index for any of these singularities, draw a closed curve or cycle around the singularity and apply Poincaré's formula.

In the light of the concept of index an important fact emerges. This fact concerns the total number of singular points for the sphere, i.e., the total index [Poincaré, 1881, page 405]:

> *Corollary I.* — The total number of nodes and foci is equal to the total number of saddle points plus 2.[Ch11, 13]

This result was of special significance for Brouwer when he read it in late 1909. This is a first statement of the Poincaré–Hopf theorem [Hopf, 1927b] for the case of the sphere and of polynomial vector fields.

When investigating more general types of differential equations (first-order, higher-degree equations) in his third 'Mémoire' ('Memoir') [Poincaré, 1885], Poincaré has to place the solution curves on more general types of algebraic surfaces than the sphere. In this context he lays down a slightly different definition of index and offers a proof of a theorem giving a relation between the number of nodes, foci, and saddle points (N, F, C, respectively) on the surface—in reality, a highly significant theorem on total index [Poincaré, 1885, Chapitre XIII, page 208]:

$$N + F - C = 2 - 2p.$$

In this relation p is the topological genus of the (orientable) algebraic surface. In effect, one finds an important topological theorem arising in the field of differential equations.

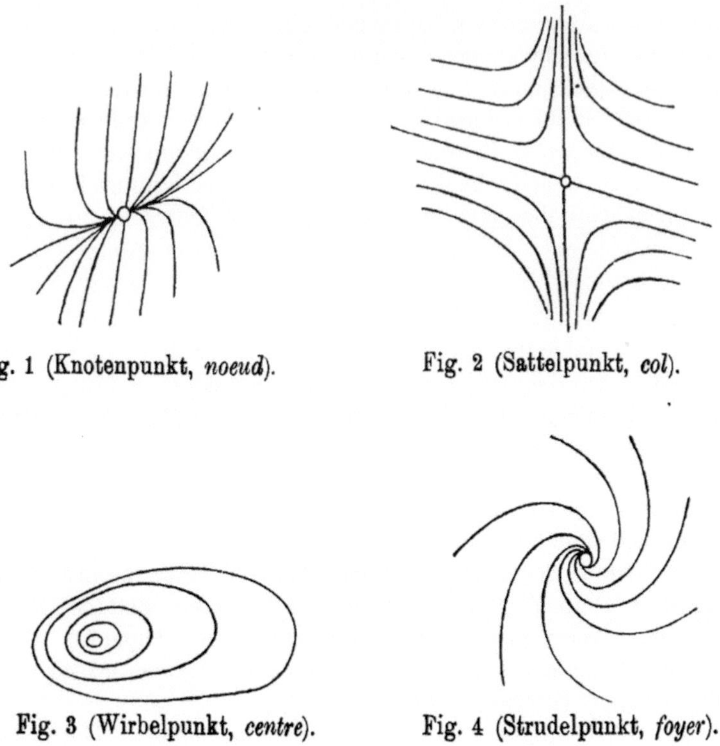

Fig. 11.4 Figures for Singularities in Vector Fields of Solution Curves and Poincaré Index [Liebmann, 1914, page 508].

In the fourth and final 'Mémoire' ('Memoir') of the series [Poincaré, 1886] dealing with second-order equations, Poincaré goes on to employ the more general Kronecker characteristic as a tool for analyzing the distribution of singular points [Poincaré, 1886, Chapitre XVIII]. What is clear from his profound investigations is the importance and usefulness of the Cauchy/Kronecker index. Poincaré's work in these 'Mémoires' ('Memoirs') is the basis for his work on analysis situs as exposited in his series of papers starting in the year 1895 [Poincaré, 1895a] [Poincaré, 1899] [Poincaré, 1900] [Poincaré, 1902c] [Poincaré, 1902e] [Poincaré, 1904] (English translation with historical commentary [Poincaré, 2010]). For an extensive treatment of the connections between Poincaré's 'Mémoires' ('Memoirs') and his analysis situs see [Mawhin, 2000].

A few years later Walther von Dyck sought to show the value of Kronecker's characteristic by taking it into the relatively new domain of analysis situs [Dyck, 1888] [Dyck, 1890].[38] Interestingly, in his two papers on the subject he does not mention Poincaré's topological work in the 'Mémoires' ('Memoirs').

[38] Cf. [Pont, 1974, pp. 131–152].

Jacques Hadamard clearly recognized the great significance of the Cauchy/Kronecker index. In 1910 he published a masterful exposition of the theory stemming from the concept with an emphasis on its powerful applications. This is his 'Note sur quelques applications de l'indice de Kronecker' ('Note on some applications of the index of Kronecker') [Hadamard, 1910], which appeared as an appendix, a kind of 'mathematical essay,' to Jules Tannery's textbook introduction to function theory. Tannery had been Hadamard's doctoral supervisor (1892), and Hadamard had a great respect for his teacher. His 'Note' appears in a teaching context, rather than a research context setting out new results, although certainly by 1910 he was a distinguished and much honored researcher.

Hadamard's expository 'Note' is remarkably similar to Brouwer's classic research paper defining the degree of a mapping, 'Über Abbildung von Mannigfaltigkeiten' ('Concerning the Mapping of Manifolds') [Brouwer, 1911b], which appeared a short time after the 'Note.' This resemblance is to be expected. The two men met in Paris around New Year 1910 and discussed the ideas that are basic to both works.[39] Brouwer's paper is a foundational piece of research, looking forward to new avenues of topological thinking, which he then proceeded to construct. In comparison, Hadamard's 'Note' or mathematical essay stands at the end of a great line of mathematical development, bringing together the elements of an important mathematical idea in a formal presentation. Brouwer's work is much more revolutionary.

In outline, Hadamard proceeds as follows in his 'Note.' He begins his exposition from a proof of the Jordan curve theorem (referencing L.D. Ames [Ames, 1905a] [Ames, 1905b]) based on the *order* of points inside and outside the simple closed curve. The *order* of a point P not on the curve C with respect to this curve is defined by means of the continuous variation of the argument (angle) of the vector PM when the point M traverses the curve C. Points outside the curve have order 0, and those inside have order ± 1 (the sign depending on the direction of traversing the curve) [Hadamard, 1910, I, pp. 437–440].

---

[39] It is not clear when the two works were written. Hadamard may have written a fair amount of his 'Note' in 1909, then interacted with Brouwer, and finalized his work for publication in 1910. He does refer to Brouwer and his work in the 'Note.' Given the hints and help from Hadamard, Brouwer then wrote his paper in 1910.

Hopf in a paper on vector fields from 1927 [Hopf, 1927b, pp. 225–226] briefly mentions the much earlier contacts between Brouwer and Hadamard. In fact, Brouwer asked Hopf to mention the earlier contacts between the two in a letter dated 21.12.1925 ([Dalen, 2011a, pp. 294–295] [Dalen, 2011b, pp. 1374–1375]; Hopf *Nachlass*, Zürich). A sentence from this letter reveals the following [Dalen, 2011a, page 294] [Dalen, 2011b, page 1374]:

> . . . The explanation is that the content of my work 'Ueber Abbildung von Mannigfaltigkeiten' ['Concerning the Mapping of Manifolds'] as well as of Hadamard's 'Note' found in the book by Tannery were discussed by Hadamard and me in Paris around Christmas 1909.[Ch11, 14]

In fact, the discussion must have taken place a little after New Year 1910. However, Brouwer and Hadamard were in correspondence around Christmas 1909 and New Year 1910 [Dalen, 2011a, pp. 60–61, 62–63, 67–69] [Dalen, 2011b, pp. 342, 344-345, 348–350].

For example, Hadamard did inform Brouwer of Poincaré's use of the Cauchy/Kronecker index.

In the next section of his 'Note' Hadamard covers the fundamentals of varieties or manifolds using the methods of simplicial decomposition for polyhedra and associated integrals [Hadamard, 1910, II, pp. 441–452]. Note that Brouwer may well have started using simplicial methods because of this basic work of Hadamard or his pointing to those methods. It is certainly possible that Hadamard set Brouwer on the track of simplicial topology.[40]

On this basis, Hadamard defines and analyzes in careful detail the *order* of a point with respect to a closed surface S in a space E_n of n dimensions [Hadamard, 1910, [III, pp. 452–465]. This order is the exact analogy with the order of a point with respect to the planar Jordan curve. For this analysis he relies on multiple integrals à la Kronecker.

He follows this with a treatment of the 'index of Kronecker' ('indice de Kronecker'), and in doing so he is passing from the definition of the order of a point with respect to a closed surface to the definition of the index of a system of functions on this surface [Hadamard, 1910, IV, pp. 465–469]. Given a surface S with points in the space of n dimensions, let $f_1, \ldots, f_n$ be a system of n continuous functions defined on the surface S and possibly elsewhere. The Kronecker index of the system of functions $f_1, \ldots, f_n$ is the order of the origin of the coordinates with the respect to the surface defined or generated by the points whose coordinates are $f_1, \ldots, f_n$. Following Poincaré, he recognizes the direct link between Kronecker index and the theory of vector fields.

He states and proves the fundamental property of index relative to the surface S and its enclosed volume V. Given the set of continuous functions $f_1, \ldots, f_n$, and supposing that the equations

$$f_1 = 0, \ \ldots, \ f_n = 0$$

never have a common solution in the interior of V, then the index of the system of functions on S is 0 (null). In the opposite case, the index is not necessarily 0 if the equations do have a common solution on the interior of V. The result is: if ω designates the index of the system of functions $f_1, \ldots, f_n$ on the surface S, then the inequality $\omega \lessgtr 0$ entails the existence of a solution common to the equations within volume V [Hadamard, 1910, page 466–467]. He concludes this section of his work with a statement and proof of the Poincaré–Bohl theorem [Poincaré, 1886, Chapitre XVIII, pp. 177–186] [Bohl, 1904] [Hadamard, 1910, pp. 467–468].

In the final section of the 'Note' Hadamard reviews significant applications, primarily topological ones, of the index concept [Hadamard, 1910, V, pp. 469–477]. This review demonstrates the power of the Kronecker index and shows the reader its wide applicability. I need only touch on a few of these applications.

He first considers a theorem of Schoenflies, related to the Jordan curve theorem in the plane and associated indices. Let two functions

$$X = f(x, y), \quad Y = g(x, y),$$

[40] For a contemporary treatment of the simplicial method cf. [Dehn & Heegaard, 1907].

continuous and defined on the circumference and the interior of a unit circle, be given, such that distinct points (x, y) are mapped to distinct points (X, Y) by f and g. Let C be the curve traced by point (X, Y) as (x, y) traces the unit circle. Then the given equations have a (unique) solution every time the specific point (X, Y) is contained in the interior of C [Hadamard, 1910, pp. 469–470].

Hadamard then assumes an n-dimensional version of the Jordan curve theorem, whereby a closed surface without double points divides the space of n dimensions into an exterior and an interior region. With this assumption, the theorem of Schoenflies is demonstrated in any number n of variables [Hadamard, 1910, pp. 470–471].

An immediate consequence of the Schoenflies theorem is the following important result. If two volumes V and V', each connected, in an n-dimensional space are in perfect and continuous correspondence (in the point-set-theoretic sense), then interior points of V correspond to interior points of V' and boundary points of V correspond to boundary points of V'. This is an important domain invariance result [Hadamard, 1910, pp. 471–472].[41] Hadamard's proof sketch depends on an n-dimensional Jordan separation theorem.

Further, he easily derives a vector distribution proof of the Brouwer fixed point theorem for n dimensions (one that Brouwer communicated to him in correspondence; see below and [Dalen, 2011a, pp. 62–63]). Consider the n-dimensional ball centered at the origin as the volume V. Then every perfect and continuous transformation of volume V in itself has at least one invariant or fixed point on the interior or on the frontier [Hadamard, 1910, pp.472–473].

In the last pages of his 'Note' Hadamard summarizes important topological results on vector fields and continuous transformations [Hadamard, 1910, pp. 473–477]. Again, these include the Brouwer fixed point theorem. Any perfect and continuous transformation of the surface of a sphere, if it preserves its orientation, leaves invariant at least one point. Hadamard credits Brouwer with the proof he gives.

Overall, in Hadamard's work one has a bold portrayal of the great sweep and power of the index concept. He masterfully draws out many consequences derived from a variety of mathematicians.

It is interesting to contrast the way Hadamard advances his exposition in the 'Note' with the way Brouwer proceeds in his mapping degree paper and his other classical works of 1910–1913. Although no one up to 1910 had proved an n-dimensional Jordan theorem—Brouwer published a proof of a version a year later [Brouwer, 1911d], Hadamard 'freely' accepts the 'theorem' in order to bring out the power of the index concept with respect to the Schoenflies and Brouwer theorems. In a similar vein he boldly proclaims in his work that the total index of a tangential vector field on a closed m-dimensional variety is an invariant of analysis situs, though without providing a proof [Hadamard, 1910, pp. 474–475].[42] Brouwer does not proceed in this way. His view of mathematics and his way of mathematical presentation require that conjectures be proved, in order for him to rely on them. Everything must be set

[41] The 'Schoenflies theorem' really goes back to Jürgens [Jürgens, 1878] [Jürgens, 1879]. On the deduction of domain invariance theorems from the Jordan curve and separation theorems cf. [Baire, 1907a] [Baire, 1907b] and Section 10.3.

[42] Cf. [Hopf, 1927b, page 225] [Hopf, 1966, page 183].

down in detail and with as much rigor as possible. As a result Brouwer's meticulous approach may look almost too careful. His line of thought is frequently quite tortuous. In contrast, Hadamard's way may seem somewhat loose, though it does permit him to give a wonderful description of the broad sweep of a mathematical development. He leaves room for filling in and revising or even correcting details. In the end one must give credit to Hadamard for his exposition of the index concept—shortly to become Brouwer's degree of a mapping—and then realize the greater depth and rigor of Brouwer's thought in this field of topology.

11.7 Defining the Concept of Degree of a Mapping

Returning to Brouwer's work, I shall retrace the crucial steps that he took towards defining the degree of a continuous mapping.[43] By the time of his letter to Hadamard of 24 December 1909 [Dalen, 2011a, pp. 62–63] [Dalen, 2011b, pp. 344–345] he was coming to grips with Poincaré's use of index for vector fields. In the letter he rephrases Poincaré's result quoted above to derive the fixed point theorem from the singularity theorem [Dalen, 2011a, page 62] [Dalen, 2011b, page 344]:

> *'If the singular points are finite in number, then each of them possesses a finite index, and the algebraic sum of all these indices is equal to* 2.*'*[Ch11, 15]

Taking a one-one continuous mapping of a sphere into itself, one can devise a better corresponding vector field. Let A be a particular point of the sphere that is not invariant, and let B be the image of A, and C, the image of B. Further, let M be an arbitrary variable point on the sphere with image M'. Take a little circle through M, M', and B, and define a vector at M as the tangent to that arc MM' that does not contain B. With this technique define a vector distribution for the whole sphere. There are two types of singular points possible for this distribution: the invariant points and the points A and B. If there were no invariant points of the transformation, then that supposition would contradict Poincaré's result. Since the index of point A is -1 and that of B is $+1$, their sum is 0, which is impossible. Thus, invariant points must exist.

Brouwer's argument in the letter to Hadamard is only a sketch. In order to carry it out fully he needed to prove the Poincaré result in full topological generality. In his second and third communications on vector fields he does this [Brouwer, 1910d] [Brouwer, 1910e] [Brouwer, 1910f] [Brouwer, 1910g].[44] In the second communication there is a version of the result developed from his own proof ideas [Brouwer, 1910e, page 734]:

[43] For a broader historical and mathematical treatment of the concept of degree cf. [Dieudonné, 1989, Part 2, Chapter I, pp. 167–181].

[44] These papers were communicated on 25 February and 28 May 1910, respectively.

Theorem 8. *A reduced distribution on a sphere possesses at least two radiating points.*[45]

However, it was Hadamard who indicated to Brouwer a simpler and more direct way of proving the result, Hadamard's 'belle démonstration directe' ('beautiful direct demonstration'), as Brouwer put it in a letter of 4 January 1910 [Brouwer, 1976, page 427] [Dalen, 2011a, page 69] [Dalen, 2011b, page 349]. Apparently Hadamard provided this better demonstration just before New Year 1910. Brouwer was then able to include it in his third communication on vector fields [Brouwer, 1910g, pp. 184–185].

Suppose one has a continuous tangential vector distribution on the sphere. Consider just one singular point surrounded by a simple closed curve. For a finite stereographic representation of the interior region of the simple closed curve (plus the curve itself) and the related part of the vector distribution on the interior region the total angle that a variable vector describes while its initial point traverses the projected curve in a positive sense is of the form $2k\pi$, k being the integer index of the singularity. For n singular points with indices $k_1, k_2, \ldots, k_n$ surrounded by a simple closed curve the total angle is $2\pi(k_1 + k_2 + \cdots + k_n)$ in the stereographic projection. Now take a great circle on the sphere that does not pass through any singular point. On the sphere, provided with an orientation, the total variation of the angle that a vector makes with respect to the tangent to the great circle as the vector goes once around the circle is $2h\pi$, h being an integer. With respect to the orientation and one hemisphere this angular variation will be reckoned as positive, but with respect to the other hemisphere it will be calculated as negative. Hence, for stereographic projections of the two hemispheres onto the plane the angular variation will be $2(1 + h)\pi$ for one hemisphere and $2(1 - h)\pi$ for the other, since a stereographic projection introduces an additional variation in the vector. Consequently, if in the first hemisphere there are m singularities with indices $k_1, \ldots, k_m$, so that $k_1 + k_2 + \cdots + k_m = 1 + h$, and in the second hemisphere there are $n - m$ singularities with indices $k_{m+1}, \ldots k_n$, so that $k_{m+1} + k_{m+2} + \cdots + k_n = 1 - h$, then by simple addition $k_1 + \cdots + k_n = 2$, i.e., the sum of the indices of the singular points for the entire sphere (the total index) is 2—an excellent generalization of Poincaré's result.

These straightforward arguments for the Poincaré total index theorem and the link between vector fields and the fixed point theorem reappear in a much more complex form in 'Über Abbildung von Mannigfaltigkeiten' ('Concerning the Mapping of Manifolds') [Brouwer, 1911b, pp. 107–115], where Brouwer generalizes the results to n-dimensional spheres. Of course, this great paper comes several months later in the development of his ideas on mapping degree. In fact, it marks the culmination of the first stage of his greatest research work in topology. One must first examine the initial part of this stage before considering the paper any further.

Hans Freudenthal [Freudenthal, 1975] (cf. [Brouwer, 1976, pp. 420–428]) has boldly asserted that the birthday of true Brouwerian topology was 1 January 1910. Given the documentary evidence from around that date, Freudenthal's designation

[45] A radiating point is a special kind of node around which all the vectors are directed straight outwards; cf. [Brouwer, 1910e, pp. 731–732].

of that date is very apt. At the time Brouwer was on holiday in Paris, staying with his brother, Aldert, at 6 Rue de l'Abbé de l'Epée in the Latin Quarter. He had gone there a day or two after Christmas 1909 and remained there until mid-January 1910. He used the time in Paris to explore his thoughts and new ideas on topology, jotting down discoveries in a student exercise book labelled 'Potentiaaltheorie en Vectoranalyse' ('Potential Theory and Vector Analysis'). Hence, on the morning of New Year's Day 1910 he was able to write a letter to Hilbert presenting a rough version of the mapping degree concept, which he had just arrived at through his feverish research activity [Brouwer, 1976, pp. 421–422] [Dalen, 2011a, pp. 65–67] [Dalen, 2011b, pp. 346–347]. Three days later on the 4th of January he wrote a similar letter to Hadamard [Brouwer, 1976, pp. 426–427] [Dalen, 2011a, pp. 67–69] [Dalen, 2011b, pp. 348–350].[46] A little later he then had a marvelous opportunity to discuss his novel topological ideas and conjectures directly with Hadamard. It seems clear that his stay in Paris was the starting point for his greatest period of research in topology. The student exercise book and the two letters testify to his restless activity at the time and the new direction of his research. Now I need to consider some of the details.

The first pages on topology of the student exercise book 'Potentiaaltheorie en Vectoranalyse' ('Potential Theory and Vector Analysis') (pages 7-15) show Brouwer working towards a general definition of index or degree while examining related theorems about vector distributions and mappings of n-spheres. He probably wrote these pages before he composed the letters to Hilbert and Hadamard, because the last page (15) contains a rough version of what he included in the letters: a definition of degree. Page 7 opens with a definition of index suitable for vector distributions on n-dimensional hyperspheres in terms of the (n-dimensional) solid angle that the vectors sweep out on the outer $(n-1)$-dimensional surface of a small volume.[47] There follows a proof (pages 7–10) of the theorem that the total index of a continuous (tangential) vector field on an n-dimensional hypersphere is 0 for odd n and 2 for even n—a powerful generalination of Poincaré's result [Poincaré, 1881, page 405]. A much refined version of this proof eventually appeared in the paper 'Über Abbildung von Mannigfaltigkeiten' ('Concerning the Mapping of Manifolds') [Brouwer, 1911b, pp. 107–112].

After the material on vector fields in the student exercise book Brouwer switches his attention to mappings, especially mappings of spheres and fixed point theorems (pages 10–15). The central issue is his famous general fixed point theorem (stated on page 12). A continuous mapping of a sphere with its interior to itself in n-space will always have an invariant point, but a continuous mapping of just an n-sphere to itself may fail to have invariant points in precisely two cases: (a) when n is even and the index (i.e., degree) of the mapping is -1 and (b) when n is odd and the index is

[46] It is not absolutely certain that Brouwer sent these two letters, but it is very likely that he did. The letter to Hilbert is not in the Hilbert *Nachlass* at Göttingen, although many of Brouwer's other letters to him are preserved there. Unfortunately, the papers of Hadamard were destroyed during the Second World War. In any case, the drafts of the letters are extremely important documents that shed some light on the development of Brouwer's topological ideas.

[47] On page 9 there is another definition of index in terms of vector field ideas.

+1. On the pages surrounding this theorem (pages 11–15) all the material concerns mappings of spheres and closed surfaces. No proof is given of the theorem. Rather the concepts relating to the theorem are examined. These pages mark the very important switch in Brouwer's ideas from index, a concept appropriate to vector distributions, to degree, a concept more suitable to mappings. On pages 11–12 he looks at the possibility of defining the 'index' of a mapping of a portion of a sphere to a sphere in terms of the algebraic number of times that a point of the sphere is covered by an image point of the mapping, but he realizes that there are certain difficulties with folds and boundaries in the mapping. On subsequent pages he considers algebraic approximations to mappings of spheres and closed surfaces. Finally, on page 15 he arrives at the definition of index, now rechristened 'degree' due to the polynomial functions used, and this he thought was good enough to put into the letters addressed to Hilbert and Hadamard.

A fundamental motivation for Brouwer's new definition of degree is from the theory of functions of a complex variable and their Riemann surfaces. In these middle pages of the student exercise book the main topic is mappings of spheres and algebraic approximations to them, and in the crucial letters to Hilbert and Hadamard Brouwer comments on mappings of 2-dimensional spheres in relation to rational complex functions. He perceived that two continuous mappings of one sphere to another can be continuously transformed into one another, i.e., in modern terms the mappings are homotopic, exactly when they have the same degree. In particular, all mappings of spheres that are of degree n can be continuously transformed into rational complex functions of the nth degree. Hence, the nth degree rational functions, which one can represent by Riemann surfaces with n sheets, are canonical models for the nth degree mappings of spheres.[48]

Thus, at the very beginning of 1910 Brouwer was in possession of a workable definition of degree, which he then refined in the next few months. His route to the concept had been through his investigations into mappings with fixed point theorems and into vector fields, and his attempt to connect the two subjects. Hadamard had wisely referred him to Poincaré's work on index and singularities in vector fields, and thence, he was impelled to transform the notion of index into his concept of degree. Excitedly, he wrote out his early version of the concept in the letters to Hilbert and Hadamard in the first days of 1910.[49] In the letter to Hadamard the definition is as

---

[48] Cf. [Brouwer, 1912c] [Brouwer, 1912d] [Brouwer, 1913a]. In modern terms one says that the nth degree mappings of spheres all belong to the same homotopy class. Cf. [Hopf, 1927a], an important generalization for n-spheres.

[49] There is an error in Brouwer's definition of degree as given in the letter to Hilbert of 1.1.1910. In this letter he specifies the use of ordinary homogeneous coordinates for the spheres. In the student exercise book at the top of page 15 he explains that each sphere is given a *stereographic projection* onto a plane with a point at infinity and then the extended plane is assigned a homogeneous coordinate system. However, by doing this the point at infinity derived from the stereographic projection is made to correspond to the line at infinity given by the homogeneous coordinates. In the middle of page 15 of the student exercise book he realizes this difficulty. There is also some crossing out that indicates a revision of his ideas. Hence, in the student exercise book at the bottom of page 15 and then in the letter addressed to Hadamard of 4.1.1910 he switches to

follows [Brouwer, 1976, page 426] [Dalen, 2011a, pp. 67–68] [Dalen, 2011b, page 348]:

> To determine this *degree*, let us introduce homogeneous coordinates (in the double sense), write x, y, z for the original sphere and ξ, η, ζ for the image, divide the sphere into a finite number of regions, and first consider the transformations defined by the relations:
>
> $$\xi : \eta : \zeta = f_1(x, y, z) : f_2(x, y, z) : f_3(x, y, z),$$
>
> where f_1, f_2, f_3 are polynomials, which, moreover, can be different for the different regions of the sphere. Let us call this transformation a polynomial transformation. Let us define an indicatrix [orientation] on the sphere: then each point P of the image *in general position* will appear a number of times r_P with positive indicatrix, and a number of times s_P with negative indicatrix. One then demonstrates that $r_P - s_P$ is a constant: it is the degree of the polynomial transformation.
>
> Let us return to a general single-valued and continuous transformation. This can be approximated by a sequence of polynomial transformations; one demonstrates that the last ones all have the same degree: thus it is the degree of the limit-transformation.
>
> The degree is always a finite whole number, positive or negative. The degree of a *one-one* transformation is $+1$, if the indicatrix remains the same, and -1, if it is reversed.[Ch11, 16]

Given in intuitive geometrical terms the degree is the algebraic number of times that the image of the first sphere covers points of the second. Brouwer immediately recognized that higher-dimensional generalizations of the degree concept are possible, and in his letters to Hilbert and Hadamard he set out the consequences of his new general idea for mappings of n-spheres and their fixed points.

Brouwer first published his fully elaborated definition of mapping degree explicitly in the paper 'Über Abbildung von Mannigfaltigkeiten' ('Concerning the Mapping of Manifolds') [Brouwer, 1911b, pp. 97–106], although he had already used the concept unnamed slightly earlier in his first paper on dimensional invariance (see below). In this fundamental paper on mapping degree he describes his new ideas with great precision and considerable succinctness. To introduce his concept of mapping degree properly he must show that his concept is well-defined in an appropriate topological context. He elaborates his context of topological manifolds by using and defining as necessary supporting objects and methods of simplicial complexes, simplicial (or barycentric) mappings, and simplicial (i.e., piecewise or simplexwise linear) approximations to continuous mappings. He needs to validate each step on his way to clarifying fully his definition of mapping degree. The mappings apply to n-dimensional manifolds, which Brouwer defines as connected complexes that are built from topological images of (closed) n-dimensional Euclidean simplexes

using homogeneous barycentric coordinates (homogeneous coordinates in the double sense) and to looking at the polynomial transformations region by region.

(called 'elements' by him), satisfying the essential requirement that the stars of their vertices have finitely many elements and form neighborhoods of their vertices.[50] He takes primarily a global view of degree in this paper. Considering first a mapping of a closed (i.e., having finitely many elements) orientable (zweiseitig) n-dimensional manifold μ to another such manifold μ' and simplicial approximations to the mapping, he defines the degree for 'ordinary points' ('gewöhnliche Punkte') P of μ' as the number p of image simplexes with a positive orientation less the number p' of image simplexes with a negative orientation covering P. This degree $p - p'$ is a well-defined integer (positive, negative, or zero), which in rough terms corresponds to the algebraic number of times that the image of μ covers the points of μ'. Further, he extends the degree concept to mappings from closed orientable manifolds to open or nonorientable manifolds, but it turns out that the degree is always 0 in these cases. The fundamental property of degree is that two continuous mappings of one manifold to another that can be continuously transformed into one another have the same degree; in other words, homotopic mappings possess identical degrees. Armed with this powerful weapon Brouwer began to conquer the most important topological problems of his time over the next few years.

11.8 First Solution of the Invariance of Dimension Problem

Having initially developed the degree concept for n-dimensional generalizations of the singularity theorem for vector fields and the fixed point theorem for mappings of spheres, Brouwer immediately started an assault on the refractory problem of dimensional invariance.[51] The last pages of the student exercise book 'Potentiaaltheorie en Vectoranalyse' ('Potential Theory and Vector Analysis') (pages 16–19) testify to this campaign. Also, some separate notes and jottings, mere odds and ends on

[50] Brouwerian manifolds are thus connected n-dimensional topological manifolds having locally finite simplicial subdivisions (triangulations). A Brouwerian manifold may be compact, in which case it has a finite simplicial subdivision, or it may be just locally compact, in which case it has a locally finite simplicial subdivision. Brouwer applies the term 'closed' ('geschlossen') to a finite simplicial manifold and the term 'open' ('often') to a locally finite simplicial manifold. These terms are still sometimes used. The general term 'manifold' ('Mannigfaltigkeit'; 'variété' in French) as used in geometry goes back to Gauss and Riemann, as noted in Chapter 5. However, Dyck [Dyck, 1888] [Dyck, 1890], Poincaré [Poincaré, 1895a] [Poincaré, 1899] [Poincaré, 1900] [Poincaré, 1904], Dehn and Heegaard [Dehn & Heegaard, 1907], Tietze [Tietze, 1908], and then Brouwer and Weyl [Weyl, 1913] were really the first to study manifolds from the viewpoint of modern topology. Cf. [Tietze & Vietoris, 1930, pp. 191–193]. Unfortunately, there has never been complete uniformity in terminology with respect to the concept of manifold, undoubtedly because of the great variety of possible manifolds and because of certain problems associated with their study (cf. the history of the later concept of homology manifold). Above all, cf. [Scholz, 1980] and [Scholz, 1999] on the history of the concept of manifold.

[51] For a broader historical and mathematical treatment of invariance of dimension and related topics cf. [Dieudonné, 1989, Part 2, Chapter II, pp. 182–196].

scraps of paper, show him making a variety of attacks on the invariance problem.[52] It seems most likely that these notes were written about the same time as the material in the student exercise book, that is, within the first six months of 1910. In most of these memoranda Brouwer uses the mapping degree concept as the main weapon for attacking the invariance problem. Frequently he considers mappings of spheres under the assumption that it is possible for Euclidean spaces of different dimensions to be homeomorphic, aiming to derive a contradiction. Occasionally he links the dimension problem to the n-dimensional Jordan separation theorem, thereby following the approach of Baire [Baire, 1907a] [Baire, 1907b] and some others (see Section 10.3 . In several notes he thinks of dimension and the invariance problem in terms of cuts ('coupuurs' in Dutch; 'coupures' in French) as Poincaré had done in his effort to define the concept of dimension [Poincaré, 1903b] [Poincaré, 1912] (see Chapter 9). However, he does not follow this path to dimension very far. Only later, in a paper published in 1913, did he succeed in proving the invariance of dimension by the cut approach, when he was able to couple it with Lebesgue's tiling principle [Lebesgue, 1911a] (see Chapter 12). By March 1910 Brouwer was able to report to Hilbert that he had a partial solution to the invariance problem, a solution directly related to the topological distinction between spheres of odd and even dimensions and their transformations, which he had discovered not long before [Brouwer, 1976, page 429] [Dalen, 2011a, pp. 69–70] [Dalen, 2011b, page 354]:

> . . . I am preparing an article for submission to the *Annalen* editorial board in which I solve the invariance problem of dimension, insofar as I show that at least spaces of even and odd dimension number cannot be mapped one-one continuously onto one another.[Ch11, 17]

The solution that Brouwer hints at in this letter is in reality *complete*, although strangely he did not think so at the time.[53] In any case, he saw the importance of his earlier work on transformations and vector distributions when using the mapping degree concept for the problem of dimensional invariance.

Brouwer must have wanted to make a name for himself by solving a longstanding problem of recognized significance for the mathematical community of his day. After he wrote to Hilbert in March 1910 he tried other ways of proving dimensional invariance. He sought a full proof and not just a partial solution to the problem. Finally, he discovered a solution that he could recognize as complete. It is one that is full of insight, employing his newly-developed mapping degree notion. He set it forth in a paper of merely five pages, 'Beweis der Invarianz der Dimensionenzahl' ('Proof of the Invariance of the Dimension Number') [Brouwer, 1911a], which he submitted to *Mathematische Annalen* in June 1910. This paper effectively swept away all previous attempts to prove the invariance of dimension. Unfortunately, in

[52] Brouwer commonly wrote mathematical notes on all sorts of bits of paper, used envelopes, old bills, etc. He willingly used anything that came to hand.

[53] The solution is complete, because one can compare an n-dimensional space with an $(n + h)$-dimensional space $(h > 0)$ by comparing the n-dimensional space with an $(n + 1)$-dimensional section of the higher-dimensional space, and it is known that these are topologically distinct according to Brouwer's results.

the Brouwer Archive there is not much documentary evidence to indicate how he arrived at this general proof. There are only the few notes in the student exercise book and the related jottings, and these do not reveal in any detail the exact path that he took in constructing the final proof.

Interestingly, in the Brouwer Archive there is a manuscript in Brouwer's hand that is directly related to the final proof, a manuscript entitled 'De invariantie van het aantal dimensies eener ruimte' ('The invariance of the number of dimensions of a space') (see Section 11.10 Appendix B; for a full transcription cf. [Freudenthal, 1977]). This is almost certainly a draft of a lecture that he gave to the Wiskundig Genootschap, the Dutch Mathematics Society, in October 1910, because the title is recorded in the *Jaarverslag* (*Annual Report*) of the Society for 1910. In this lecture he had his first opportunity to communicate in public his findings on invariance. The five-page draft of the lecture contains essentially the same proof as the published paper, but, as the lecture was delivered four months after the paper was submitted, it includes some improvements. In particular, the term 'simplicial approximation' ('simpliciale approximeering') and some terms and ideas related to vector fields appear in the draft of the lecture but not in the published paper. The lecture is also more informal. For example, it contains diagrams nicely illustrating Brouwer's thoughts in two dimensions. Further, there is a full review of the problem situation concerning dimensional invariance and attempted invariance proofs since Cantor's time (see Section 11.10 Appendix B). I shall examine some of the details of the published proof, while also referencing the early notes and the later lecture manuscript.

Brouwer's approach to the problem of demonstrating that an m-dimensional manifold (or Euclidean space[54]) and an $(m + h)$-dimensional manifold $(h > 0)$ cannot be put into one-one continuous correspondence, i.e., that the manifolds are not homeomorphic, is through the proof of an important lemma [Brouwer, 1911a, page 164]:

> *In a q-dimensional manifold if for a single-valued continuous mapping of a q-dimensional cube the maximum of the displacements [of the points] is less than half the side length, then there exists a concentric and homothetic cube which is contained entirely in the image set.*[Ch11, 18]

This is a weak form of domain invariance: the mapping satisfying the given conditions on displacements yields an image set that must contain a (smaller) q-dimensional cube. In their classic treatise *Topologie I* (*Topology I*) Alexandroff and Hopf call the lemma Brouwer's invariance principle [Alexandroff & Hopf, 1935, pp. 364–365]. It is a key to dimensional invariance. There is no trace of a proof of it among his notes from around this time, but, of course, it is proved in the manuscript of the lecture of

[54] Brouwer uses the term 'Mannigfaltigkeit' ('manifold') in [Brouwer, 1911a] without explicitly saying what he means by it. However, he surely means either a manifold simply in the sense of a Euclidean space (Zahlenraum) or a domain thereof (his references to other authors seem to suggest this interpretation) or a manifold in the sense of his [Brouwer, 1911b, pp. 97–98] (see above) (this is the concept he was using at the time and about to publish). In any case, the proof of invariance applies to both concepts of manifold.

October 1910. The proof in the published paper and that in the lecture of 1910 rest essentially on the mapping degree idea.

Letting K denote the original q-dimensional cube with side length $2a$ (cf. the 2-dimensional diagram from the lecture reproduced in Figure 11.5), and letting α (alpha) denote the continuous mapping of K under which the points of the cube are displaced by not more than a distance $b < a$, then to prove the lemma it must shown that the image set of K under the mapping α completely covers a concentric and homothetic cube K_1 with side length $2a_1 < 2(a - b)$ inside K. In the manuscript of the lecture Brouwer illustrates the situation concerning K and K_1 in two dimensions (which are thus simple squares) with a diagram, as reproduced in Figure 11.5.

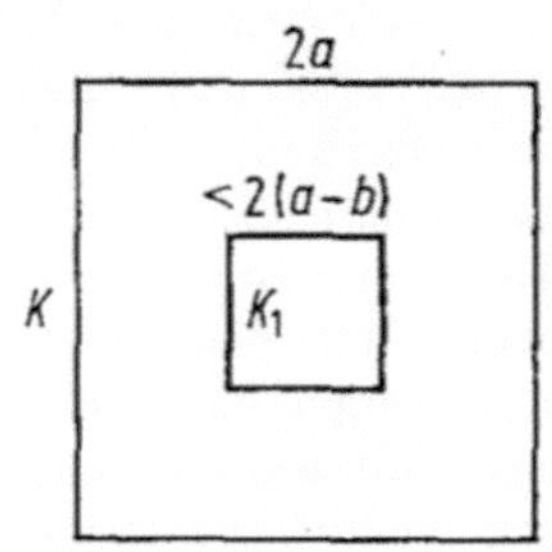

Fig. 11.5 Diagram from Brouwer's Lecture Manuscript: Brouwer's Key Lemma, the Invariance Principle in 2 Dimensions [Brouwer Archive `https://noord-hollandsarchief.nl/`].

It is interesting to note that in the lecture of 1910 Brouwer thinks in terms of the vector field associated with mapping α: the field of displacement vectors from points of K to their images under mapping α (or α-images). Thus b is the maximum length of the displacement vectors associated with the mapping α. The published proof does not contain this mode of thought.

Brouwer's method for proving the lemma is to consider simplicial mappings approximating the mapping α and the degrees of the various mappings. First, the cube K can be given a simplicial subdivision with a positive orientation by dividing it up into n^q little cubes (n, an arbitrary positive integer) and dividing each of these into $q!$ 'basic simplexes' ('Grundsimplexe') having positive orientations. In the lecture manuscript Brouwer provides a diagram to illustrate a simplicial subdivision for a 2-dimensional K (i.e., a triangulation of a square), reproduced in Figure 11.6.

Using these simplexes, he defines a simplicial mapping β related to mapping α, i.e., a simplexwise linear mapping based on the vertices of the basic simplexes and their α-images. In the lecture manuscript Brouwer provides a diagram as an aid to understanding the concept of a simplicial mapping, as reproduced in Figure 11.7. If β maps some of the simplexes to singular image simplexes that lie in $(q - 1)$-dimensional hyperplanes, it is possible to modify β slightly to eliminate

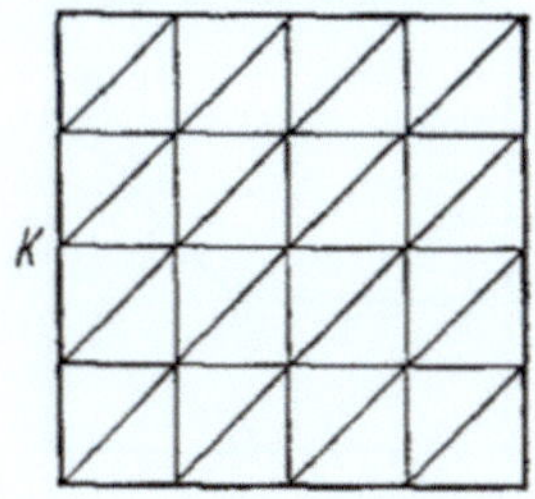

Fig. 11.6 Diagram from Brouwer's Lecture Manuscript: A Simplicial Subdivision (Triangulation of the Square) [Brouwer Archive `https://noord-hollandsarchief.nl/`].

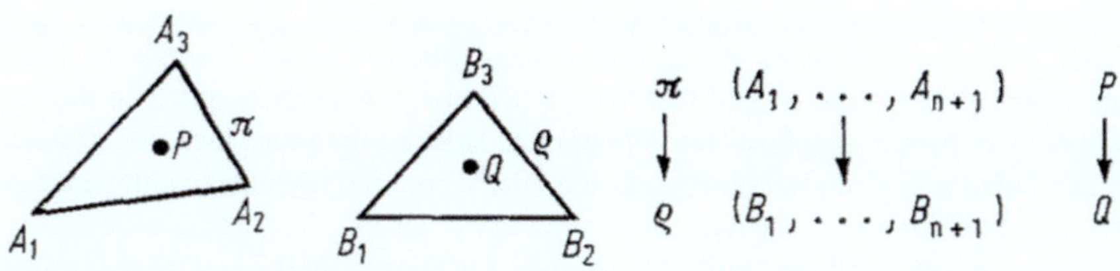

Fig. 11.7 Diagram from Brouwer's Lecture Manuscript: A Simplicial Mapping [Brouwer Archive `https://noord-hollandsarchief.nl/`].

these singular images, thereby deriving a modified simplicial mapping γ without them.[55]

Brouwer's main objective is to ascertain the degrees of the mappings γ and β, and, hence, of mapping α. He does not employ the term 'degree,' but rather he uses the concept unnamed, primarily in a local sense. Looking first at the mapping γ, he designates the set of interior points of the smaller cube K_1 that are not images of any of the edges of the basic simplexes by μ and notes that this set has the property that any two points of it can be connected by a polygonal path lying entirely in the set. It is clear that the γ-image of the boundary of the larger cube K cannot penetrate into the smaller cube K_1. Brouwer calls those points of μ that are not images of any points of the faces of the simplexes of K the 'ordinary points' ('gewöhnliche Punkte'). It is with respect to these points that he determines the degree of the mapping γ. For each ordinary point P he defines the integer c equal to the number p of positively-oriented image simplexes less the number p' of negatively-oriented image simplexes covering P: $c = p - p'$. To prove that c is the same number for every ordinary point he considers any two such points P_1 and P_2 and a variable point P which runs over a path in μ joining P_1 and P_2. In the lecture manuscript Brouwer provides a diagram to illustrate this situation in two dimensions, as reproduced in Figure 11.8. Brouwer uses 's' to denote the polygonal path joining P_1 and P_2.

[55] In the lecture manuscript Brouwer only operates with the mapping β and does not consider the modified mapping γ.

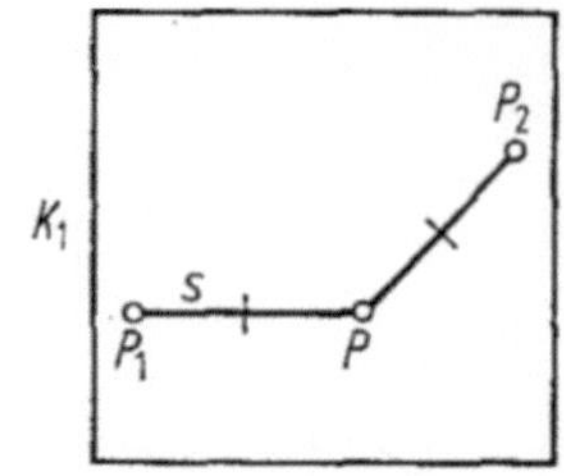

Fig. 11.8 Diagram from Brouwer's Lecture Manuscript: Joining Two Points by a Polygonal Path [Brouwer Archive `https://noord-hollandsarchief.nl/`].

The numbers p and p' for P can only change when P passes through the image of an inner $(q-1)$-dimensional face where two image simplexes meet. In the case where the two image simplexes are on opposite sides of the inner face, the numbers p, p', and also $p - p'$ do not change. In the case where the two image simplexes are on the same side of the inner face, both numbers p and p' either increase by one or decrease by one, but still the number $p-p'$ does not change. In the lecture manuscript Brouwer provides a pair of diagrams to illustrate the two cases for dimension two, reproduced in Figure 11.9.

Fig. 11.9 Diagram from Brouwer's Lecture Manuscript: Two Simplicial Cases for the Constancy of Number $c = p - p'$ [Brouwer Archive `https://noord-hollandsarchief.nl/`].

Hence, the number $c = p - p'$ is a well-defined constant for all ordinary points; it is the degree of mapping γ. By using approximating mappings γ, and a similar proof technique, it is possible to show that mapping β also has the well-defined degree c for all its associated ordinary points in K_1. By a simple homotopy argument it is easy to see that c has the value 1, because the degree of mapping β will not change suddenly if one transforms β continuously into the identity mapping by gradually moving the image points back to their original positions. Hence, under the mapping β all the points of K_1 are covered. Since one can approximate the original mapping α with any degree of accuracy by a β map based upon a sufficiently fine simplicial subdivision, mapping α has degree 1 and the cube K_1 must be covered by the image set of α. In this way Brouwer's invariance principle is completely proved.

To a topologist of today the original proof of the invariance principle will seem straightforward. In the case of Brouwer, he was introducing the concepts of simplicial mapping, simplicial approximation, and degree for the first time. Taken together, his ideas represent an enormous advance in topological thinking.

In the second (and last!) section of the paper Brouwer proceeds rapidly to a demonstration of the invariance of dimension. The approach of this part closely resembles several of his notes and jottings. When he was struggling to discover a full invariance proof the real stumbling block was without doubt the invariance principle. Once he had that, the rest was easy. In the second section of his paper he first assumes that the one-one continuous image of an m-dimensional domain (connected open set; Bereich) in an m-dimensional manifold is nowhere dense in a certain neighborhood of one of its points, so that one can have a one-one continuous correspondence between an m-dimensional cube K and a nowhere dense, connected, perfect set C in an m-dimensional manifold. This assumption leads to a contradiction with the fundamental lemma, because one can associate C with a nowhere dense part of K in such a way that one produces a continuous mapping of the cube K onto a nowhere dense subset of itself for which the maximum displacement of the points is less than half the side length. Hence, there follows [Brouwer, 1911a, page 165]:

> Theorem 1. *In an m-dimensional manifold the one-one continuous image of an m-dimensional domain contains a domain in an arbitrary neighborhood of any of its points.*[56] [Ch11, 19]

This is another weak form of domain invariance. It is the general case of the 2-dimensional theorem, which Jürgens [Jürgens, 1878] [Jürgens, 1879] had proved earlier. Jürgens had labored over his theorem with inadequate methods taken from real analysis. Brouwer's proof, constructed as it is with sophisticated topological tools, is a vast improvement. The evidence of progress could not be clearer.

The invariance of dimension is now a simple corollary. If one supposes that an m-dimensional manifold R_m contains the one-one continuous image B'_{m+h} of an $(m+h)$-dimensional domain B_{m+h} $(h > 0)$, then the image B'_m of an m-dimensional domain B_m, lying nowhere dense in B_{m+h}, would also be nowhere dense in R_m. This conclusion contradicts Theorem 1. Hence, there follows [Brouwer, 1911a, page 165]:

> Theorem 2. *An m-dimensional manifold cannot contain the one-one continuous image of a domain of higher dimension number.*[Ch11, 20]

Reversing the argument, one derives [Brouwer, 1911a, page 165]:

> Theorem 3. *In an m-dimensional manifold the one-one continuous image of a domain of lower dimension number is a nowhere dense point set.*[Ch11, 21]

Theorems 2 and 3 together express the invariance of dimension. Surely one must marvel at the elegance of Brouwer's proof. It is perhaps surprising to find such elegance at this stage in the history of topology.

---

[56] The proof of this theorem in the lecture manuscript is accompanied by two more diagrams illustrating the 2-dimensional case and also by some arrow diagrams denoting the mapping involved.

11.9 Concluding Remarks

To conclude the main part of this chapter I shall briefly review Brouwer's achievements thus far. Through his work on vector fields and mappings of spheres he came to define the degree of a continuous mapping, his principal weapon for attacking topological problems. In essence he was in possession of the concept by New Year's Day 1910, when he could write letters addressed to Hilbert and Hadamard about it. Then during the first half of 1910 he shaped his newly-found weapon in order to apply it to some important topological problems. In particular, he made a full-scale attack on the problem of dimensional invariance during this time and subdued it. Whence he wrote his paper 'Beweis der Invarianz der Dimensionenzahl' ('Proof of the Invariance of the Dimension Number'), which he submitted to *Mathematische Annalen* by June 1910. A month later he submitted his masterpiece, 'Über Abbildung von Mannigfaltigkeiten' ('Concerning the Mapping of Manifolds'), bringing together his most important general discoveries on degree and mappings. These two classical works are an exact record of his outstanding achievements in topology during the first half of 1910. They constitute an impressive demonstration of the fact that Brouwer must be classed with the greatest topologists, although the mathematical public did not recognize this fact for some time. Yet, these papers just mark the first stage of his great work in topology. Much more was to come. Part of this further work has a bearing on the dimension problem. In the next chapter I shall examine Brouwer's continuing progress on dimension and topology.

11.10 Appendix B: Transcription and translation of an excerpt from the manuscript, in Dutch, 'De invariantie van het aantal dimensies eener ruimte'

This section contains a transcription of an excerpt from the manuscript 'De invariantie van het aantal dimensies eener ruimte,' belonging to the Brouwer Archive, and a translation thereof. It was discovered through the assiduous efforts of the late Dr. W.P. van Stigt. It was unavailable to Prof. Dr. H. Freudenthal when he was preparing volume 2 of the Brouwer *Collected Works* [Brouwer, 1976], although he had attempted to gather together all relevant documents for his edition of the volume. A complete transcription is now available in [Freudenthal, 1977].

The manuscript provides an indication of how Brouwer first revealed his dimensional invariance proof in public. This manuscript in Brouwer's distinctive hand is almost certainly a draft of a lecture that Brouwer delivered to the Wiskundig Genootschap (Dutch Mathematics Society) in October 1910, because the title is given in the *Jaarverslag* (*Annual Report*) of the Wiskundig Genootschap for 1910. The lecture gave Brouwer his first opportunity to reveal his dimensional invariance proof in public. See the earlier Section 11.8 for further information about the lecture and manuscript.

The manuscript of the lecture consists of five sheets of paper, each approximately 21×33.6 cm. It includes many deletions and revisions, a typical feature of Brouwer's drafts. Nevertheless, the structure of the lecture is entirely clear and the words are all legible.

Inasmuch as the manuscript is very similar in content to his published paper 'Beweis der Invarianz der Dimensionenzahl' ('Proof of the Invariance of the Dimension Number') [Brouwer, 1911a], I decided not to include a full transcription and translation of the manuscript. However, Brouwer's review of the problem situation and earlier invariance proofs as contained in the manuscript is quite interesting. In the published paper the review is reduced to a short footnote [Brouwer, 1911a, page 161, footnote *)]. The following is a transcription kindly provided by Dr. W.P. van Stigt and a translation by me of the review in the manuscript.

Transcription:

[...]

We komen thans tot ons eigenlijk onderwerp, n.1. het probleem van de invariantie van her dimensie-aantal, dat zijn oorsprong heeft gekregen in de klassieke ontdekkingen van Peano en Cantor, nu ongeveer dertig jaar geleden. Peano heeft n.l. bewezen, dat, onafh[ankelijk], van p en q, een ruimte van p dimensies zich eenduidig en continu laat afbeelden op een ruimte van q dimensies, en Cantor, dat, eveneens onafh[ankelijk], van p en q, een ruimte van p dimensies zich één-éénduidig laat afbeelden op een ruimte van q dimensies.

Het oude dimensiebegrip werd daardoor rain of meer ondermijnd; vernietigd zou het echter eerst zijn, als bet gelukte, een ruimte van p dimensies 6n eeneenduidig n continu af te beelden op een ruimte van q dimensies. Men heeft dit van den beginne af zeer onwaardschijnlijk geacht, en reeds in 1878 werd door Thomae een bewijs geleverd van de onmogelijkheid van zulk een afbeelding. Kort daarop liet Lüroth echter zien, dat het bewijs van Thomae impliciet de stelling al bewezen onderstelt, zoodat men hiermee niet verder was gekomen. Sedert zijn nieuwe bewijzen van de onmogelijkheid gepubliceerd door Cantor, en door Netto. Maar Jürgens heeft laten zien, dat het bewijs van Cantor niet afdoende was, en Jürgens en nog duidelijker Schoenflies hebben hetzelfde gedaan voor het bewijs van Netto. In die bewijzen komen namelijk redeneeringen voor, die alleen geoorloofd zijn, als de afbeeldingen door analytische functies worden uitgedrukt. Zoo komt in het bewijs van Netto de volgende definitie voor: laat in R_n gegeven zijn een samenh[angende], verzameling. We zeggen, dat die m dimensies heeft, als op die verzameling in 't algemeen uit de incrementen van m der coördinaten die der overige volgen. En in 't vervolg van het bewijs wordt dan aangenomen, dat op grond van die definitie elke samenh[angende], verz[ameling], in R_n een zeker aantal dimensies, hoogstens gelijk aan n, bezit. Het geen natuurlijk alleen mag worden aangenomen, als die samenh[angende], verzameling door analytische functies wordt bepaald. En in het bewijs van Cantor wordt stilzwijgend aangenomen, dat de grens van een gebied in R_n $(n-1)$-dimensionaal is.

[...]

[Translation:

[...]

We now come to our proper subject, viz., the problem of the invariance of the dimension number, which had its origin in the classical discoveries of Peano and Cantor of about thirty years ago. For Peano has proved that, independently of p and q, a space of p dimensions can be mapped in a single-valued and continuous manner on a space of q dimensions, and Cantor has proved that, likewise independently of p and q, a space of p dimensions can be mapped in a one-one manner on a space of q dimensions.

The old dimension concept thereby became more or less undermined; however, it would be destroyed only if one succeeded in mapping a space of p dimensions in both a one-one and a continuous manner on a space of q dimensions. This has always seemed very unlikely, and already in 1878 a proof of the impossibility of such a mapping was supplied by Thomae. Shortly thereafter, however, Lüroth showed that the proof of Thomae implicitly assumed the theorem as proved, so that no progress had been made. Since then new proofs of the impossibility were published by Cantor, and by Netto. But Jürgens showed that the proof of Cantor was not conclusive, and Jürgens and even more clearly Schoenflies have done the same for the proof of Netto. For in the proofs there occur arguments which are only permissible when the mappings are expressed by analytic functions. Thus in the proof of Netto the following definition occurs: let a connected set be given in R_n. We say that this set has m dimensions, if in general (in the set) from the increments of m coordinates those of the remaining coordinates follow. And in the rest of the proof it is then assumed that on the basis of this definition every connected set in R_n has a certain number of dimensions, at most equal to n. Of course, this may only be assumed if the connected set is determined by analytic functions. And in the proof of Cantor it is tacitly assumed that the boundary of a domain in R_n is $(n-1)$-dimensional.

[...]

]

Chapter 12
New Ways of Proving Invariance of Dimension; Brouwer–Lebesgue Dispute

12.1 Introduction

Brouwer submitted his 'Beweis der Invarianz der Dimensionenzahl' ('Proof of the Invariance of the Dimension Number') [Brouwer, 1911a] to *Mathematische Annalen* in June 1910 and delivered his lecture 'De invariantie van her aantal dimensies eener ruimte' ('The Invariance of the Number of Dimensions of a Space') to the Wiskundig Genootschap, the Dutch Mathematics Society, in October of the same year.[1] Even before the paper was issued on 14 February 1911 two particular mathematicians residing outside of Holland were specially informed of its contents: Robert Fricke and Henri Lebesgue (1875-1941).[2] Brouwer himself told Fricke of his proof in a letter [Dalen, 2011b, page 370] because of its significance for the continuity method of proof in the theory of automorphic functions. As a result of Brouwer's friendly help, Fricke was able to refer to the invariance proof as a support for automorphic function theory in the second volume of the great treatise of Fricke and Klein, *Vorlesungen fiber die Theorie der automorphen Functionen* (*Lectures on the Theory of Automorphic Functions*) [Fricke & Klein, 1912, pp. vii–viii, 413–414]]. Otto Blumenthal (1876–1944)[3] told Lebesgue of Brouwer's new result. The resulting interaction between Lebesgue and Brouwer led to a dispute with significant mathematical consequences.[4] I shall now discuss the mathematics and the related history including the dispute in some detail.

[1] For a broader historical and mathematical treatment of topics related to this chapter cf. [Dieudonné, 1989, Part 2, Chapter II, pp. 182–196].

[2] For biographical material on Lebesgue cf. [Denjoy, 1946] [Lebesgue & Félix, 1974].

[3] For biographical material on Blumenthal cf. [Behnke, 1958] [Jongen & Krieg, 2000] [Butzer & Volkmann, 2006] [Rowe, 2018, Part I, pp. 1–82]. Blumenthal had a very tragic end to his life at the hands of the Nazis.

[4] On the Brouwer–Lebesgue dispute see also [Dalen, 2012, pp. 156–169] [Dalen, 1999, pp. 159–173] and [Rowe, 2018, Section 7.1, pp. 241–266]. The dispute is well covered in the collection of letters and notes of Brouwer and others that Dirk van Dalen carefully assembled in [Dalen, 2011a, pp. 73–100, 104–109, 111–112] [Dalen, 2011b, pp. 383–392, 395–397, 402–408, 412–421, 424–430, 433–434, 436–437, 454–460, 466–467, 469–470].

D. M. Johnson, *Theories of Dimension from Ancient to Modern Times*, Sources and Studies in the History of Mathematics and Physical Sciences, https://doi.org/10.1007/978-3-032-02682-8_12

12.2 Lebesgue's Initial Note on the Tiling Theorem and Invariance of Dimension, and Brouwer's Critical Reaction

During a holiday visit to Paris in the summer of 1910, Blumenthal, managing editor of *Mathematische Annalen*, met Lebesgue and apprised him of Brouwer's new result. In a letter to Hilbert of 27 October 1910 Blumenthal describes the meeting [Brouwer, 1976, page 438]:

> During the holidays we had a very lovely journey to Paris. Unfortunately, I did not see any mathematicians, as they were all still on holiday; i.e., except for Lebesgue, whose acquaintance I did make, for by chance he was in Paris. He is very interesting and told me that for some time already he has possessed not merely one but several proofs of the theorem of the invariance of dimension number, which Brouwer has now proved in the *Annalen*. He has sent me one of these proofs for the *Annalen* – one which seems very clever. I have not examined it in detail for the correctness of its execution but only for the correctness of its basic idea: one can leave the details to so perceptive a man. However, if you want to verify it more exactly, the work is at your disposal.[Ch12, 1]

The heart of the proof that Lebesgue sent is the geometrically intuitive tiling (or paving) principle (or theorem). The principle asserts that it is possible to cover a Euclidean n-space with sufficiently small closed sets (or little bricks) that cover the space, such that at most $n+1$ of the sets cover any one point in the space, and such that there exist points that are covered by $n+1$ of the sets. Blumenthal placed Lebesgue's three-page sketch of the proof, written as a letter to the editor and dated 14 October 1910 [Lebesgue, 1911a], just after Brouwer's proof in the *Annalen*. Thus, the two proofs appeared in print together in February 1911.

Brouwer was furious when he saw Lebesgue's proof. Although he at once recognized the tiling principle as a beautiful way to characterize dimension, he judged the proof to be utterly inadequate. In this way a dispute arose between these two giants in mathematics over the right way to attack the invariance problem. During 1911 the dispute was very bitter—as such mathematical disputes often can be. Without doubt Lebesgue's and Brouwer's personalities were not congenial. Nonetheless, beyond the level of mere personalities significant mathematics emerged. To put matters briefly, it is clear that Lebesgue had plenty of excellent intuitive ideas on dimension, which he revealed during 1911 in three short publications [Lebesgue, 1911a] [Lebesgue, 1911b][5] [Lebesgue, 1911c] and in private communications to Brouwer. Nevertheless, Brouwer could not accept what he regarded as Lebesgue's weak proofs. They certainly did not come up to the high standards of rigor that he was imposing on the new topology of the time. In his eyes the proofs were quite insufficient, and, given the counterexamples to Lebesgue's proof that Brouwer produced, I think that one should regard this judgment as correct. Nonetheless, Brouwer himself was spurred to some new attacks on the problem of dimension and on other topological questions

[5] This note is just a paragraph announcement of Lebesgue's tiling theorem in relation to Brouwer's *Annalen* article.

as well. In this way, Lebesgue's ideas produced immediate results through the efforts of his intellectual opponent!

In Volume 2 of Brouwer's *Collected Works* the editor Hans Freudenthal [Brouwer, 1976, pp. 438–452] discussed in some detail the Brouwer–Lebesgue dispute. After publication of that volume additional letters and documents were found, which provided more information about the back and forth of the dispute.[6] Additionally, in his biography of Brouwer Dirk van Dalen provided his detailed interpretation of the story [Dalen, 1999, Section 5.1, pp. 151–172] [Dalen, 2012, Section 5.1, pp. 149–169]. For completeness I shall explore the details of the dispute and the associated mathematics.

12.3 Mathematical Details of Lebesgue's Initial Note

In his short note 'Sur la non-applicabilité de deux domaines appartenant repectivement à des espaces à n et $n + p$ dimensions' ('On the non- applicability of two domains belonging to the spaces of n and $n + p$ dimensions, respectively') [Lebesgue, 1911a]—barely more than two printed pages—Lebesgue expresses the tiling principle in the following way [Lebesgue, 1911a, page 166]:

> *If each point of a domain D of n dimensions belongs to at least one of the closed sets $E_1, E_2, \ldots, E_p$, finite in number, and if these sets are sufficiently small, then there are points common to at least $n + 1$ of these sets.*[Ch12, 2]

For Lebesgue a domain of n dimensions in a Euclidean n-space is strictly a closed and bounded, hence, compact, and connected set that is the closure of its interior [Lebesgue, 1907, page 377] [Lebesgue, 1910, pp. 367, 377], or more generally (following Jordan) a set containing interior points (i.e., containing an open set) [Lebesgue, 1921, page 259]. To be a little more precise about the size of the E_i in relation to the domain D, one supposes that D contains an n-dimensional cube I of side length $2l$ ('intervalle' in Lebesgue's terminology), given by the inequalities: $0 \le x_i - x_i^0 \le 2l$ (the x_i are coordinates in n-space, and the x_i^0 are constants), and that each E_i can be contained within an n-dimensional cube of side length l, so that no E_i can stretch from one side of the cube I to the opposite side.

Lebesgue gives a straightforward but sketchy proof of the principle. He first assumes that each E_i is a union of finitely many little intervals or cubes (aligned with the coordinate axes and D). Initially, he considers just those E_i that contain points of the side of I defined at $x_1 = x_1^0$. Clearly the union e_1 of those E_i does not touch the opposite side $x_1 = x_1^0 + 2l$ and its boundary in the interior of I consists of a set of $(n - 1)$-dimensional hyperplane sections, some of the little cube faces. In

[6] I, with the help of Hélèn Gispert, edited most of those materials in the original version of my work [Johnson, 1981, Appendices B, C, pp. 187–220]. Dirk van Dalen then included all of the material and some additional material he found in his edition of Brouwer's correspondence [Dalen, 2011a] [Dalen, 2011b]. See the specific citations in this chapter.

particular, the boundary of e_1 within I contains a connected subset I_1, which extends from side $x_i = x_i^0$ to its opposite side $x_i = x_i^0 + 2l$, for each $i = 2, 3, \ldots, n$.

Lebesgue then considers the union e_2 of E_i that are not completely interior to e_1 and that have points in common with I_1 and also with the side of I at which $x_2 = x_2^0$. He claims that the intersection of e_2 and I_1 is bounded by a set of $(n-2)$- dimensional plane sections and, more importantly, that a certain part of this boundary, I_2, extends from $x_i = x_i^0$ to $x_i = x_i^0 + 2l$, for all $i = 3, 4, \ldots, n$.

Continuing in this way, he successively defines the sets $I, I_1, I_2, \ldots, I_n$. The basic claim is that every point of I belongs to at least one E_i, while every point of I_1 belongs to at least two E_i, and so forth. Finally, every point of I_n belongs to at least $n + 1$ E_i, that is, the tiling principle for the special case when the E_i are composed of little cubes. To derive the general case one merely needs to consider a sufficiently refined cubical subdivision of the space and cubical approximations to the E_i. Hence, according to Lebesgue the principle is entirely proved.

The invariance of dimension follows from the fact that the number $n + 1$ is characteristic of n-dimensional spaces. Indeed one can always choose E_i so that each point of an n-dimensional Euclidean space is covered by at most $n+1$ E_i. In the plane one can easily see the possibility of this choice in the pattern made by bricks (little closed square sets E_i) laid in the usual way, i.e., staggered row by row (see Figure 16.1). Alternatively, one can see the possibility in the pattern made by regular hexagonal tiles fit snugly together. In these two cases at most three bricks or tiles meet. Furthermore, it is easy to extend the brick-laying idea to higher dimensions, so that the 'Lebesgue order,' the number $n + 1$, obtains for cubical bricks in an n-dimensional space. This simple tiling or brick-laying idea must have been the leading one for Lebesgue.

In his paper he also makes an observation about Peano space-filling curves, which motivates the tiling principle [Lebesgue, 1911a, page 168].[7] Such curves covering n-dimensional domains always have multiple points of order at least $n+1$, i.e., points at which the curves cross at least $n+1$ times. However, it is possible to construct such curves, such that they do not have any multiple points of order greater than $n + 1$.[8]

One can see at a glance the beautiful intuitive appeal of Lebesgue's tiling principle. It is no wonder that Blumenthal admired its cleverness and wanted it to appear in the *Annalen*. The principle is a simple and direct route to dimensional invariance, at first sight surpassing Brouwer's ideas, which are elegant, but not as simple.

---

[7] Earlier Lebesgue had shown an interest in Peano curves in his [Lebesgue, 1904, pp. 44–45] and [Lebesgue, 1905b].

[8] Hilbert had noticed this possibility for the construction of space-filling curves in the plane [Hilbert, 1891]. For the later history of this topic see Section 13.8.

12.4 Brouwer's Contact with Lebesgue, with Blumenthal Acting as Intermediary

Notwithstanding the simplicity and immediate intuitive appeal of the principle, Brouwer quickly detected weaknesses and holes in Lebesgue's proof. It was clear to him that the proof was inadequate. Brouwer immediately wrote to Lebesgue to see if the latter could shore up what Brouwer regarded as a deficient proof, while at the same time he worked out his own proof of the tiling principle [Brouwer, 1976, pp. 449–451] (see below). In his various private notes he elaborates his assessment of Lebesgue's response [Dalen, 2011b, pp. 416–420]. In particular, he writes [Dalen, 2011b, page 416]:

> After Lebesgue's first letter it was certain that his paper was wrong and that he did not see the correct way (because he did not specialize the I_p).[Ch12, 3]

Thus, according to Brouwer Lebesgue was not able to do his duty. Once pressed to give adequate details to his proof, he could not do so.

Brouwer must have deeply resented the fact that in the *Annalen* Lebesgue's poor proof appeared next to his own deep one. To the casual reader it might even seem that Brouwer's proof using the mapping degree was unnecessarily complicated compared with Lebesgue's simple proof.[9] Earlier in March 1911 Brouwer decided to submit a 'Bemerkung zu dem Invarianzbeweise des Herrn Lebesgue' ('Remark on the invariance proof of Herr Lebesgue') [Dalen, 2011a, page 74] [Dalen, 2011b, page 383] pinpointing the error in Lebesgue's argument. In a reply that was intended for publication [Dalen, 2011a, page 76] [Dalen, 2011b, page 386] Lebesgue claimed that he had accepted certain 'evident facts' ('... j'ai annoncé que j'allais admettre des faits que je qualifiais de bien évidents ...' ; '... I stated that I was going to admit facts that I described as quite obvious ...') in his proof sketch, but that he would publish a complete proof through the Société Mathématique de France (Mathematical Society of France).

Blumenthal was reluctant to publish Brouwer's 'Bemerkung' ('Remark'). He did not want to involve the *Annalen* in a polemical dispute. As he told Hilbert in a letter of 14 March [Dalen, 2011a, page 75] [Dalen, 2011b, pp. 384–385], he did not like the 'unfriendly and unpleasant tone' ('unfreundlichen und unangenehmen Ton') of Brouwer's remarks. In fact, he was sympathetic to Lebesgue's position and wanted to give the French mathematician a chance to produce a complete demonstration of the tiling principle. Perhaps even Brouwer's own invariance proof was not without flaw. Consequently, Blumenthal wrote a long diplomatic letter, dated 25 March, to Brouwer requesting him to withdraw his note [Dalen, 2011a, pp. 76–78] [Dalen, 2011b, pp. 387–389].[10] He assured Brouwer that his priority rights were not being

[9] Such a casual reader was J. Pierpont who in his textbook on real functions [Pierpont, 1912, pp. 620–621] makes an even worse mess of Lebesgue's proof than Lebesgue himself.

[10] The note by Brouwer to which Blumenthal refers in his letters of 14 and 25 March 1911 [Dalen, 2011a, pp. 75–78] [Dalen, 2011b, pp. 384–385, 387–389] is most likely the 'Bemerkung' of March [Dalen, 2011a, ppage 74] [Dalen, 2011b, page 383]. I have assumed this is the case in the text

violated; Lebesgue was his successor and not his rival. However, Blumenthal went so far as to say that [Dalen, 2011a, page 77] [Dalen, 2011b, page 387]:

> *...your note is phrased in a very harsh form ...* Ch12, 4

The result was that Brouwer did withdraw his 'Bemerkung' ('Remark') (letters to Blumenthal [Dalen, 2011a, pp. 78–79] [Dalen, 2011b, page 390] and later to Hilbert [Brouwer, 1976, page 440] [Dalen, 2011a, page 96] [Dalen, 2011b, page 431]), although he insisted that Lebesgue ought to give his complete proof as soon as possible, preferably in the same volume of the *Annalen*. Nevertheless, Brouwer's private notes (likely dating from late June) [Dalen, 2011b, pp. 416–420] show that he was not genuinely convinced that Lebesgue was able to overcome the difficulties of the faulty proof and produce a cogent one. The notes reveal his plans, his course of action, to push Lebesgue really hard and to show his own superiority in these topological matters.

12.5 Brouwer's Criticism and Counterexample

Brouwer elaborated his criticism of Lebesgue's proof with both a confirming example in two dimensions and a counterexample in three dimensions, along with specific critical assertions and notes in letters to Blumenthal of 19 and 20 June and 2 and 8 July 1911 and related private notes ([Brouwer, 1976, pp. 446–452] [Dalen, 2011a, pp. 88–96] [Dalen, 2011b, pp. 412–421, 424–428]; cf. also [Brouwer, 1913b, page 151]). The difficulty lies with the I_i. Choosing I_1 (or, in fact, any I_i) merely with the provision that it be connected and extend from the sides at $x_i = x_i^0$ to the opposite sides at $x_i = x_i^0 + 2l$ $(i = 2, 3, \dots, n)$ will not assure the existence of an I_2 extending from $x_i = x_i^0$ to $x_i = x_i^0 + 2l$ $(i = 3, \dots, n)$; hence, the existence of the final I_n is not guaranteed.

For the confirming example let I be a simple square in two dimensions (see Figure 12.1, from the letter to Blumenthal of 19 June 1911). In this case in the boundary of e_1 there is only a single line (with little crossbars in Figure 12.1), a 1-dimensional space, which stretches from $x_2 = x_2^0$ to $x_2 = x_2^0 + 2l$. We choose this as I_1. (Other 1-dimensional spaces in the boundary of e_1, the cases with dashed lines, cannot extend from $x_2 = x_2^0$ to $x_2 = x_2^0 + 2l$.) Because I_1 is connected and stretches from $x_2 = x_2^0$ to $x_2 = x_2^0+2l$, two subsets of I_1, of which the first contains the subinterval that border $x_2 = x_2^0$, and the second one contains the subinterval that borders $x_2 = x_2^0+2l$, must have at least a point in common, so there exists a point I_2, giving the desired result.

For the counterexample let I be an ordinary cube in a 3-dimensional space, and divide it into 8^3 a little subcubes. For e_1 take the layer of 8^2 subcubes lying on the plane $x_1 = x_1^0$ plus the 4 subcubes on the next layer up, as indicated in the diagram in

above. However, there might have been other versions of this, which have been lost. Indeed the extant 'Bemerkung' is not especially severe, so perhaps Blumenthal's critical remarks do not apply to it. Hence, it is possible that Blumenthal could also be referring to other versions in his letters.

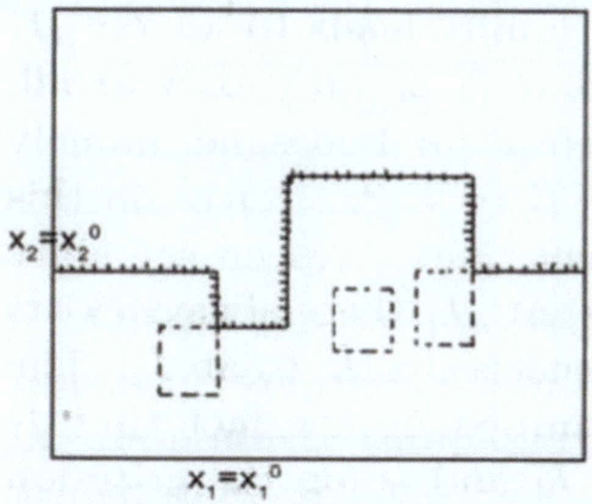

Fig. 12.1 Brouwer's Diagram for the Simple 2-Dimensional Case, a Square, with a Line Stretching across the Square [Brouwer Archive `https://noord-hollandsarchief.nl/`, reproduced [Brouwer, 1976, page 446]].

Figure 12.2 (from letter to Blumenthal of 19 June 1911), plus the whole third layer of 8^2 subcubes parallel to $x_1 = x_1^0$. Within I the boundary of this e_1 has two parts: one is the plane $x_1 = x_1^0 + \frac{3}{4}l$; the other is a rectangular cylinder or chimmey-shaped surface together with two planar collars. Each of these surfaces satisfies Lebesgue's requirement that it be connected and extend from $x_2 = x_2^0$ to $x_2 = x_2^0 + 2l$ and from $x_3 = x_3^0$ to $x_3 = x_3^0 + 2l$. Thus one is at liberty to choose the rectangular cylinder with collars as I_1. But under this choice if e_2 is composed of subcubes from the first layer next to $x_2 = x_2^0$ (the plane at the left side of the diagram) and if e_3 is composed of subcubes from the first layer next to $x_3 = x_3^0$ (the plane at the top of the diagram), then I_2 is at best a disconnected pair of polygonal lines, neither of which extends from $x_3 = x_3^0$ to $x_3 = x_3^0 + 2l$ (see the diagram in Figure 12.3, from letter to Blumenthal of 20 June 1911; Brouwer did not draw all the little faces of the subcubes of equal size and to scale), and I_3 does not exist at all! Hence, Lebesgue's proof collapses. It is really no proof at all.

Without question Brouwer's counterexample shows that the I_i as defined by Lebesgue do not yield the desired result. However, it seems clear that Lebesgue really intended that the I_i perform a different function. It would appear that I_1 should act like a separating boundary between the cube faces at $x_1 = x_1^0$ and $x_1 = x_1^0 + 2l$, while I_2 should act like a separating boundary in I_1 between the faces at $x_2 = x_2^0$ and $x_2 = x_2^0 + 2l$, and so forth. In fact, when Lebesgue later rewrote his proof, he claimed that this was his intention all along [Lebesgue, 1921, page 265]:

> In reality I have always thought of the separations into regions by the $I, I_1, \ldots$.[Ch12, 5]

Nevertheless, in the *Annalen* note he fails to define the I_i in a way that adequately carries out his intention.

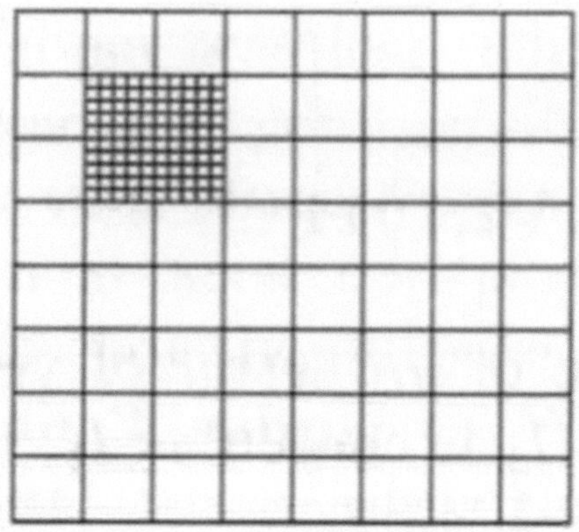

Fig. 12.2 Brouwer's Diagram of the First Layer of 8^2 Subcubes plus Shaded Area Indicating 4 Subcubes above First Layer [Brouwer Archive `https://noord-hollandsarchief.nl/`, reproduced [Brouwer, 1976, page 446]].

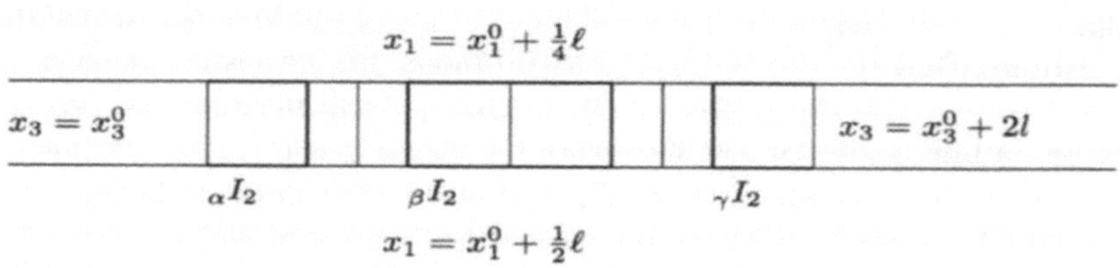

Fig. 12.3 A Section of the 3-dimensional Figure at the Plane $x_2 = x_2^0 + \frac{1}{4}l$ [Brouwer Archive `https://noord-hollandsarchief.nl/`, reproduced [Brouwer, 1976, page 448]].

12.6 Lebesgue's New Approach to Invariance of Dimension

Privately Lebesgue sent Brouwer a more detailed proof of his principle, but he did not want to publish this in *Mathematische Annalen*. At the same time he publicly embarked upon an entirely new approach to the problem of dimension in a research note presented by Émile Picard to the Academy of Sciences, Paris, at the session of 27 March 1911 [Lebesgue, 1911c].

In the first part of his research note in the *Comptes rendus* (*Proceedings*) [Lebesgue, 1911c, pp. 841–842] Lebesgue seeks to prove dimensional invariance by exhibiting certain separation properties in Euclidean spaces. The gist of his new proof sketch is as follows. Suppose that there is a homeomorphism (correspondance biunivoque et continue) between an $(n+1)$-dimensional coordinate (Euclidean) space E_{n+1} and an $(n+p+1)$-dimensional one E_{n+p+1} $(p > 0)$. Then a hypersphere S_n in the first space, dividing it into two parts, will have a homeomorphic image s_n in the second, which divides that space into regions. The next step is an important one in Lebesgue's argument, but he does not go very far towards explaining the ideas and concepts involved. Let σ_n be an n-dimensional polyhedron (variété polygonale), composed of hyperplanes, that 'approximates s_n' ('voisine de s_n et lui correspondant'). He says in a footnote [Lebesgue, 1911c, page 842] that σ_n can have self-intersections, so that in general it is not homeomorphic to s_n; otherwise, he reveals nothing about the way σ_n is defined. It seems clear that the rectilinear polyhedron σ_n does not divide E_{n+p+1} into regions; hence, its homeomorphic image Σ_n cannot divide E_{n+l} into regions. To complete his argument Lebesgue attempts to show the general proposition that every Σ_n close to (voisine de) S_n divides E_{n+1} into regions[11] by proving that an arbitrary radius OA of S_n extended into a straight ray D cuts every Σ_n close to S_n in an odd number of points; in other words, that the order of O with respect to Σ_n is odd. This completing step is crucial for the whole argument, but he only sketches an inductive proof for the general proposition, using the Jordan curve theorem as the basis for the induction $(n = 2)$. Overall, his argument for dimensional invariance founded on separation properties is very sketchy. It is reminiscent of Jordan's original proof of the closed curve theorem. It seems to rest on many topological 'facts' treated as though derivable from elementary geometry. Lebesgue's argument looks quite weak. However, Brouwer surprisingly recognized it as 'cogent' (!) (see below).

In the second part of his note in the *Comptes rendus* (*Proceedings*) [Lebesgue, 1911c, pp. 842–843] Lebesgue points to the close relationship between his sketch of a proof for dimensional invariance and the well-known Jordan curve theorem, and then he offers a remarkable generalization of the latter theorem. For the generalization he introduces the concept of 'variétés enlacées,' a notion of linked varieties or linked manifolds [Lebesgue, 1911c, page 842]:

> We say that two varieties T_n and T_p in the space E_{n+p+1} are *enlaced* [or *linked*] if, taking two polygonal varieties, t_n and t_p, homeomorphic to T_n and T_p and very close to them, in the course of a continuous deformation reducing t_n, say, to a point not on t_p, there is an odd number of crossings of t_p by t_n.[Ch12, 6]

Many years earlier in 1833 Gauss had introduced the idea of the linking of two closed or infinite curves using an integral formula [Gauss, 1867, page 605].[12] However, through the definition just given Lebesgue is trying to capture a more general notion,

[11] The proposition is supposed to apply to every Σ_n close to S_n and not just the particular Σ_n mapped from σ_n.

[12] Cf. Section 7.2 and particularly footnote 6 in that section; also cf. [Ricca & Nipoti, 2011].

that of the linking, modulo 2, of varieties. His definition is homotopic in character. One variety is contracted to a point. By means of the new term one can state the Jordan curve theorem as follows. For a simple closed curve in the plane there is a pair of points linked with the curve. Lebesgue's generalization is then [Lebesgue, 1911c, page 843]:

> *If there is an* [*n-dimensional*] *variety* T_n *in a space* E_{n+p+1}, *then there exists a* [*p-dimensional*] *variety* T_p *linked with* T_n.[Ch12, 7]

This existence theorem for linked varieties is an early ancestor of the great Alexander duality theorem, one of the most important research topics in the topology of the 1920s and 1930s. Though one must grant that Lebesgue discovered through his theorem a key to a brilliant treasure house of topological ideas, he seemingly did not fully understand the value of his find.[13]

Lebesgue's research note contains a brief sketch of a proof for the linking theorem, but the sketch is vague and imprecise at many points. Brouwer was soon to criticize this. Lebesgue then concludes [Lebesgue, 1911c, page 843]:

> In particular, one deduces from there that a T_n does not fill up E_{n+p+1} ($p \geqq 0$) and divides E_{n+p+1} into regions for $p = 0$ and only for this case. The invariance of the number of dimensions results from there in three ways.[Ch12, 8]

However, it is not completely clear from this statement how dimensional invariance does follow.

12.7 Brouwer's Critical Reaction, with Blumenthal Acting as Intermediary

Inevitably, Lebesgue's note in the *Comptes rendus* (*Proceedings*) of the Academy of Sciences, Paris added fuel to the dispute with Brouwer. As mentioned above, before its publication Brouwer had received from Lebesgue further elaborations of his proof of the tiling principle. He had found these quite unsatisfactory. After seeing the research note, probably near the end of April or the beginning of May, he decided upon crafting another critical 'Bemerkung' ('Remark') concerning Lebesgue's sketchy and insufficient proofs. He submitted this to Blumenthal on 9 May [Dalen, 2011a, pp. 80–82] [Dalen, 2011b, pp. 395–397].

The new 'Bemerkung' ('Remark') and the accompanying letter to Blumenthal reveal a great deal about Brouwer's attitude to Lebesgue and his view of the dispute as it had developed thus far. In the letter [Dalen, 2011a, pp. 80–81] [Dalen, 2011b, pp. 395–396] his condemnation of Lebesgue's further elaborations of the proof of the tiling principle is total: 'teeming with errors' ('wimmeln von Fehlschliüssen') and 'incurably wrong' ('unheilbar falsch'). In the 'Bemerkung' ('Remark') [Dalen,

[13] On the later developments associated with Lebesgue's ideas cf. Alexander's well-known paper [Alexander, 1922] and especially [Alexandroff & Hopf, 1935, Kapitel XI].

2011a, pp. 81–82] [Dalen, 2011b, pp. 396–397] he is equally critical of the purported deduction of dimensional invariance from the theorem on linked varieties as apparently given in the second half of Lebesgue's note in the *Comptes rendus* (*Proceedings*) of the Academy of Sciences, Paris. In making his criticism Brouwer interprets Lebesgue's vague statements in a particular way. He expresses the main ones as follows [Dalen, 2011a, page 81] [Dalen, 2011b, page 396]:

> 'For every regular closed manifold T_n lying in a [space] R_{n+p+1} there exists an arbitrarily small manifold T_p linked with T_n.' From this lemma it then follows that T_n is not everywhere dense in R_{n+p+1}.[Ch12, 9]

With regard to the $(n + p + 1)$-dimensional space R_{n+p+1} Brouwer sees two possibilities: either the space is a Cartesian (i.e., Euclidean) space or a 'regular' space of a more general type. In the first case he immediately points out that, inasmuch as T_n is a *closed* manifold and R_{n+p+1} is an open manifold, it is obvious that T_n is not everywhere dense in R_{n+p+1}. This is a trivial fact not dependent upon the linking theorem and certainly not significant for dimensional invariance. In the second case he detects a vicious circle, a hidden reliance on dimensional invariance! Hence, according to Brouwer's judgment there is no value in Lebesgue's linking theorem for the invariance problem at all.

Brouwer's strong attack on Lebesgue's deduction of dimensional invariance from the linking theorem at first glance seems devastating. Yet, almost certainly he was too hasty and too harsh when making it so quickly after the appearance of Lebesgue's research note. In fact, Brouwer's criticism rests on his particular interpretation of the French mathematician's (admittedly sketchy) presentation. Probably Lebesgue was thinking of Euclidean spaces, so that the first case is appropriate. Yet did Lebesgue make so trivial a blunder as Brouwer asserted, or did he mean something else? The question is difficult to decide in view of his imprecise text.[14] Brouwer at the time only had Lebesgue's sketchy note from which to work, so that it is perhaps not surprising that he found fault with the ideas expressed in it. However, he soon revised his opinion of Lebesgue's line of reasoning, though he remained critical of Lebesgue's ideas. In changing his opinion, he thereby achieved a deeper understanding of the topological situation involved.

It seems that Blumenthal did not find Brouwer's new 'Bemerkung' ('Remark') acceptable in the form in which it was submitted. When he and Brouwer met at Aachen in May or early June, they discussed it, and following that meeting Brouwer wrote a long letter in French—so that Lebesgue could not mistake his meaning—intended for publication in the *Annalen* [Dalen, 2011a, pp. 82–84] [Dalen, 2011b, pp. 402–404].

[14] Many years ago Prof. Dr. H. Fruedenthal suggested to me that Lebesgue might have intended his phrase [Lebesgue, 1911c, page 843] '... une T_n ne remplit pas E_{n+p+1} $(p \geqq 0)$ ...' to have the meaning 'a T_n is nowhere dense in E_{n+p+1} $(p \geqq 0)$' and not the interpretation that Brouwer placed upon it ('not everywhere dense'). If this alternative interpretation is the right one, then Lebesgue's assertion about T_n in the space E_{n+p+1} is not trivial. In view of this alternative, it should be possible to deduce dimensional invariance from the linking theorem, provided that the theorem is provable independently of the invariance.

12.8 Brouwer's New Insights and Further Reaction

In this letter dated 11 June Brouwer repeats his objections both against Lebesgue's 'third proof' of dimensional invariance based on the linking theorem and against his proof of the tiling principle. Moreover, he asserts quite surprisingly that his own proof of the invariance of dimension as given in *Mathernatische Annalen* [Brouwer, 1911a] is discernible in the first part of Lebesgue's note in the *Comptes rendus* (*Proceedings*) of the Academy of Sciences, Paris. When one examines how he relates his own demonstration to Lebesgue's sketch of a proof, there can be no doubt that this interpretation of the French mathematician's effort is overly generous. Supposing that there can be a homeomorphism between a q-dimensional space E_q and a $(q + h)$-dimensional space E_{q+h}, he considers, as in [Brouwer, 1911a], a q-dimensional cube K in E_q with center M mapped to a set k in E_{q+h} with m corresponding to M. By means of a small continuous deformation it is possible to transform k into a q-dimensional polyhedron p that will be nowhere dense in E_{q+h}. Hence, to p there corresponds a nowhere dense set P in E_q, which can be regarded as arising from a small deformation of the cube K. However, the existence of P is forbidden by Brouwer's fundamental invariance principle already proved [Brouwer, 1911a, page 164]. Every set π resulting from a sufficiently small deformation of K entirely covers a neighborhood around M. Furthermore, by modifying the argument and considering the boundaries of a π and p, it is clear that the boundary of the former separates M from infinity, while that of the latter cannot separate m from infinity. The foregoing is a liberal reformulation of the first part of Lebesgue's sketch for a proof. The boundary of K corresponds to Lebesgue's sphere S_n; the boundary of k, to his s_n; the boundary of p, to his σ_n; and the boundary of any π, to his arbitrary Σ_n. In the process of reconstructing Lebesgue's proof Brouwer has supplied the mainstay of his own demonstration, the invariance principle, and Lebesgue was nowhere close to a proof of that.[15]

At the end of Brouwer's letter in French of 11 June there is a review of the mathematical argument as he saw it [Dalen, 2011a, pp. 82–84] [Dalen, 2011b, pp. 402–404]. One cannot help feeling that Lebesgue must have found this review highly insulting if he ever read it. It is no wonder that Blumenthal, when he wrote to Brouwer on 14 June, said that some of Brouwer's expressions might need softening in the event of publication [Dalen, 2011a, pp. 85–86] [Dalen, 2011b, pp. 405–406]. In defense of Lebesgue Blumenthal suggested that the theorem on linked varieties might be intended as an independent theorem. The title of the note in the *Comptes rendus* (*Proceedings*) of the Academy of Sciences, Paris even points to this reading. In the letter he asks [Dalen, 2011a, page 85] [Dalen, 2011b, page 405]:

[15] On this part of the Brouwer–Lebesgue dispute cf. Brouwer's statement in his footnote *) in [Brouwer, 1911c, page 305] and also Freudenthal in [Brouwer, 1976, pp. 439–444]. The concepts of ϵ-transformation and ϵ-mapping can be used to recast Brouwer's version of Lebesgue's argument quite succinctly; cf. [Alexandroff & Hopf, 1935, pp. 364–365].

> Now I would like to know: Assuming the solution of the dimension problem, is the theorem on linked varieties correctly and rigorously proved? I do not venture to decide; it would almost be too beautiful if the statement were true.[Ch12, 10]

In what seems to be his answer to Blumenthal's letter of 14 June Brouwer [Brouwer, 1976, Y5, page 452] (probably written 15 June) reinforces his conclusion that there is a vicious circle lurking in Lebesgue's linking argument when it is used to support dimensional invariance. However, in answer to Blumenthal's question he says [Brouwer, 1976, page 452]:

> My position with regard to the second part of Lebesgue's note is this: He has proved a very beautiful theorem for three-dimensional space quite correctly, but for higher spaces he has only announced 'evident' extensions without having proved anything.[Ch12, 11]

Consequently, through Blumenthal's question and interest in the topological problems involved Brouwer started to think about linking as a concept independent of the problem of dimensional invariance. He continued to criticize Lebesgue's proof of the linking theorem (cf. letter of 8 July [Dalen, 2011a, pp. 94–96] [Dalen, 2011b, pp. 427–428], but gradually during the next few months he came to understand the depth of this circle of topological ideas.

By the time Brouwer wrote a letter to Blumenthal on 19 August [Dalen, 2011a, pp. 97–98] [Dalen, 2011b, pp. 433–434] (cf. [Dalen, 2011b, page 419]) he realized that providing an adequate definition of Lebesgue's notion of linked manifolds was a very deep problem, which he was not able to solve as yet. However, he felt that he had already succeeded in defining a narrower concept: the concept of linking for manifolds measured (gemessen) in a definite way, i.e., manifolds having simplexes with definite consistently-assigned barycentric coordinates.[16] From this restricted concept he was able to reconstruct Lebesgue's line of argument and, in particular, to deduce the result that a one-one continuous image of an n-dimensional sphere in $(n + 1)$-dimensional space determines *at least* two domains or, in other words, the first part of a restricted Jordan separation theorem for arbitrary dimension number.

About the time of writing the letter to Blumenthal he submitted his 'Beweis des Jordanschen Satzes für den n-dimensionalen Raum' ('Proof of the Jordan theorem for n-dimensional space') [Brouwer, 1911d] for the *Annalen*.[17] In this paper he credits Lebesgue with a proof of the first part of the Jordan theorem and, consequently, he does not include a proof of this part. By this action he undoubtedly wished to criticize Lebesgue indirectly, but nonetheless publicly. The Lebesgue 'proof' simply as given in the note in the *Comptes rendus* (*Proceedings*) of the Academy of Sciences, Paris hardly comes up to the standard that Brouwer expected in topology. At best the sketch of the proof requires essential 'Brouwerian' additions. In a letter of 26

[16] Cf. [Brouwer, 1911b, page 98]. The concept will not automatically be a topological invariant.

[17] Brouwer was interested in the n-dimensional Jordan theorem earlier, even before the contact with Lebesgue. His inaugural lecture of 1909 [Brouwer, 1909d] [Brouwer, 1975, page 119] and the student exercise book composed around the beginning of 1910 provide evidence of this interest.

August Blumenthal tried to persuade him to add a proof of the 'Lebesgue part' of the n-dimensional Jordan theorem to his paper [Dalen, 2011a, pp. 99–100] [Dalen, 2011b, pp. 436–437], but he refused. Moreover, he chose to include a further indirect 'criticism' in the paper in the *Annalen* by giving in a footnote his own rigorous demonstration of the first part of the Jordan theorem for the case of three dimensions, only following the approach of Lebesgue in his *Comptes rendus* (*Proceedings*) of the Academy of Sciences, Paris research note [Brouwer, 1911d, pp. 314–315]. A little later he then summarized his results on the Jordan separation theorem, including the 'credit' to Lebesgue, in a note in the *Comptes rendus* (*Proceedings*) of the Academy of Sciences, Paris [Brouwer, 1911f]. This kind of veiled criticism, which is nonetheless transparent to those who can genuinely understand the mathematics of the situation, is typical of one facet of Brouwer's personality. He seems to have relished engaging in subtle polemics.

Brouwer's final statement on Lebesgue's linking idea is in his important paper 'On looping coefficients' [Brouwer, 1912a] [Brouwer, 1912b], (1912c) (1912d) presented to the Koninklijke Nederlandse Akademie van Wetenschappen, the Royal Dutch Academy of Sciences at Amsterdam, on 24 February 1912. Instead of using Lebesgue's homotopy idea of contraction to a point for determining linking, Brouwer replaces it by a definition of linking founded on the concept of mapping degree. At the end of the paper there is yet another cryptic remark directed against Lebesgue [Brouwer, 1912b, pp. 121–122]. The Brouwer paper had some impact on topologists investigating ideas of linking and duality theorems during the 1920s and 1930s, but then the concept of linking became defined through homological terms and not by Lebesgue's or Brouwer's method.[18]

At the very end of the note on the tiling principle in *Mathematische Annalen* Lebesgue [Lebesgue, 1911a, page 168] had referred to Baire's method for attacking the dimension problem and related questions of topology [Baire, 1907a] [Baire, 1907b] (cf. Section 10.3). The Baire approach to dimension relies on a proof of domain invariance, which in turn rests on a Jordan separation theorem in n dimensions. In his note Lebesgue had suggested that completing Baire's line of reasoning would amount to no more than filling it out with a few technical details, although he knew that Baire had never done so.

Brouwer's first reaction to Lebesgue's reference to Baire's work, as given in his 'Bemerkungen' ('Remarks') of March and May [Dalen, 2011a, pp. 74, 81–82] [Dalen, 2011b, pp. 383, 396–397], was to assert that the theorems used by Baire to derive dimensional invariance, viz., the domain invariance theorem and the Jordan theorem, lie on a deeper level than the invariance of dimension itself. His first considered opinion was that they are much harder to prove. He put the three theorems, dimensional invariance, domain invariance, and Jordan separation, into a hierarchy ordered by increasing difficulty,[19] He repeated his critical assertion con-

[18] On these further developments cf. the works of J.W. Alexander, S. Lefschetz, E. van Kampen, and L. Pontryagin.

[19] Later Brouwer found a simple method for proving domain invariance directly from dimensional invariance [Brouwer, 1912f], so that dimensional and domain invariance are demonstrably on the 'same level of difficulty.'

cerning Lebesgue's estimate of the topological situation in a footnote to his own 'Beweis der Invarianz des n-dimensionalen Gebiets' ('Proof of the Invariance of the n-dimensional Domain') [Brouwer, 1911c, page 305], which he submitted to Hilbert on 14 July. Yet by the time he sent his 'Beweis des Jordanschen Satzes für den n-dimensionalen Raum' ('Proof of the Jordan Theorem for n-dimensional Space') [Brouwer, 1911d] to Blumenthal in late August he had come to the conclusion [Brouwer, 1911d, page 314] that it is possible to reconstruct a proof of domain invariance from the second part of Lebesgue's note in the *Comptes rendus* (*Proceedings*) of the Academy of Sciences, Paris and Baire's earlier work. In other words, he thought that Lebesgue's 'proof' of a part of the Jordan separation theorem could be joined with Baire's derivation of domain invariance from the separation theorem, and together they could be developed into a full proof of n-dimensional domain invariance.

Brouwer could be fairly sure that Lebesgue did not fully understand the topological situation concerning these theorems. Nevertheless, in the autumn of 1911 one finds him in correspondence with Baire, trying to discover the extent of Baire's knowledge and defending his priority rights over dimensional invariance and domain invariance [Dugac, 1976a, pp. 368–370] [Dalen, 2011a, pp. 104–115] [Dalen, 2011b, pp. 454–475].[20] He found that Baire had not been able to complete the line of reasoning of his papers of 1907 because of illness and because of not having friendly relations with Lebesgue for years. Brouwer had nothing to fear from Baire. Baire even would be likely to sympathize with him in the dispute with Lebesgue. Hence, in this correspondence one again sees Brouwer attempting to demonstrate his superiority over Lebesgue.

12.9 End of Brouwer's Direct Contact with Lebesgue

By the middle of July 1911 Brouwer's direct contacts with Lebesgue were at an end [Brouwer, 1976, page 440] [Dalen, 2011a, page 96] [Dalen, 2011b, pp. 429–431]. Afterwards he continued to make critical references directed at Lebesgue in print and in this way to carry on the battle for many years. On the level of personalities this was often Brouwer's style: uncompromising, anxious to show his pre-eminence in the mathematical fields he had made his own. Yet, on the mathematical level he must be regarded as the 'victor' (if that is not too strong a term). Without question he understood the situation in topology better than his adversary. Lebesgue and he were on quite different planes in this area of mathematics, greatly separated in distance. It was not until 1921 that Lebesgue came to publish a good detailed proof of the tiling principle [Lebesgue, 1921]; he never developed further the sketches in his note in the *Comptes rendus* (*Proceedings*) of the Academy of Sciences, Paris.

[20] Cf. Freudenthal in [Brouwer, 1976, pp. 441–443].

At this point I should like to reflect on another aspect of the Brouwer–Lebesgue dispute: the role of Blumenthal. Surely Blumenthal erred when he simply accepted Lebesgue's note on the tiling principle for the *Annalen*, without checking the proof—as he admitted himself. He merely took it for granted that someone like Lebesgue with a great reputation would have produced an adequate proof. He was mistaken in this opinion. Hence, he put the papers by Brouwer and Lebesgue sequentially in the *Annalen*. As a consequence Brouwer became extremely irritated when he saw the great disparity between his own proof of the invariance of dimension and Lebesgue's attempt. In Brouwer's view the stature of his new result—a breakthrough on an old problem—was diminished by the simultaneous publication of Lebesgue's sketchy note. As the dispute developed during 1911 one finds Blumenthal becoming increasingly sympathetic to Brouwer's views on the whole matter. Yet, the surviving documents show that he did not really understand deeply the topological issues involved. For example, his comprehension of the Jordan separation theorem was deficient. To his credit one should remember that he was at times helpful. After all it was he who persuaded Brouwer to consider Lebesgue's linking theorem independently of dimensional invariance, so leading Brouwer to an investigation of the problem of linking and related matters. In the final analysis I would say that, while Blumenthal was often diplomatic in his handling of the dispute on the personal side, he showed a lack of understanding of the deeper issues on the mathematical side.

12.10 Brouwer's Continuing Investigation of the Invariance of Dimension

At this point I shall return to Brouwer's investigations specifically directed towards the invariance of dimension. His continuing research eventually led to a second publication on the subject, which appeared in 1913. His notes and letters show that he devised a proof of Lebesgue's tiling principle a few days after he saw Lebesgue's note in *Mathematische Annalen*, that is, during the latter half of February 1911 (cf. [Brouwer, 1976, page 441] [Dalen, 2011b, pp. 416–420]). In fact, some of Brouwer's extant notes, jotted down on a few scraps of paper, are almost certainly a record of his progress towards this proof [Brouwer, 1976, pp. 449–451].

One part of these notes, probably composed first, reveals that he considered a proof based on a vector distribution over the n-dimensional cube, which is related to Lebesgue's e_i and their boundaries [Brouwer, 1976, page 449]. The existence of a singular point for the distribution on the interior of the cube would demonstrate that there is a point common to $n+1$ of the closed sets E_i. However, he did not bring this attempt at a proof to completion, apparently because the approach did not satisfy him.

In a longer series of notes he then moved towards the key idea of the proof that he ultimately published in 1913 [Brouwer, 1976, pp. 450–451]. First, he observed that if one has an n-dimensional cube I and the union e_1 of a collection of closed sets E_i covering one face of the cube, say at $x_1 = x_1^0$, but not touching the opposite face at

$x_1 = x_1^0 + 2l$, then the *projection* from $x_1 = x_1^0$ to $x_1 = x_1^0 + 2l$ of a certain connected part of the boundary of e_1, something like Lebesgue's I_1, will cover the opposite face with degree 1. In order to set up an inductive proof, a significant problem arises: how to extend this projection and covering idea to boundaries similar to Lebesgue's $I_2, I_3, \ldots$. Further, he realized that Lebesgue's principle is obvious in the case when one only needs to consider each first layer of little subcubes (in a cubical subdivision of I) next to each of the faces at $x_1 = x_1^0$, $x_2 = x_2^0, \ldots$. Then one has the problem of showing that the principle remains valid while adding little subcubes one at a time to one of the layers. Finally, he arrived at the fundamental idea of a good proof for the tiling principle based on suitable projections of Lebesgue-style boundaries I_i that all cover with degree 1. He knew that a different sequence of I_i than Lebesgue's had to be used in order to produce a sound inductive argument proving the existence of the final I_n. Thus, at the end of his notes one finds the following proof sketch [Brouwer, 1976, page 451]. For I_1 choose the *entire* set of boundaries of e_1, *except* those parts on the bottom face at $x_1 = x_1^0$ and on the vertical faces of the cube. Clearly, the projection of this I_1 upon $x_1 = x_1^0 + 2l$ covers the face on this plane with degree 1, and also under the same projection I_1's boundaries at the plane $x_2 = x_2^0$ cover the face at $x_1 = x_1^0 + 2l, x_2 = x_2^0$ with degree 1. If one now considers another projection that projects the face at $x_1 = x_1^0$ onto just the face at $x_1 = x_1^0 + 2l$, $x_2 = x_2^0$, then the boundaries of I_1 at $x_2 = x_2^0$ still cover that face with degree 1. Moreover, if one removes a little element from I_1, the boundaries of the modified I_1 still cover that face with degree 1. By removing little elements from I_1 one at a time, one eventually derives boundaries that are I_2 [21] and, hence, under the projection I_2 covers the face with degree 1. In essence, this is the crucial inductive step of Brouwer's paper of 1913, 'Über den natürlichen Dimensionsbegriff' ('On the natural dimension concept') [Brouwer, 1913b]. Thus, step by step, using appropriate projections, one eventually arrives at the conclusion that I_{n-1} covers a 1-dimensional edge with degree 1 and that I_n covers a vertex of the cube with degree 1. Consequently, I_n exists and yields a point common to $n+1$ E_i. At the end of his notes Brouwer put in the margin that everything should be divided into simplexes. Consequently, the paper of 1913 uses a simplicial subdivision as part of the argument.

Brouwer wrote out the full details of an induction step equivalent to that sketched in his private notes in a letter to Blumenthal of 2 July 1911 [Dalen, 2011a, pp. 92–94] [Dalen, 2011b, pp. 424–426]. The letter was intended for publication, but it never appeared. At the time of writing it he was still anxious to see whether Lebesgue could demonstrate his own principle (cf. [Dalen, 2011a, pp. 88–90] [Dalen, 2011b, pp. 412–420]).

It is interesting to discover that Brouwer conceived of at least one other proof of the tiling principle, if not more. He communicated this to Blumenthal in a letter of 19 August 1911 [Dalen, 2011a, pp. 97–98] [Dalen, 2011b, pp. 433–434]. In broad

[21] I_2 must be defined appropriately as the *entire* set of boundaries (within the cube I) of $e \cap I_1$, where e_2 is the union of E_i that are not completely interior to e_1 and that have points in common with I_1 and with $x_2 = x_2^0$. The rest of the I_i are defined in a similar way. Brouwer's note does not actually contain these definitions, but they seem to be intended from the context.

outline the proof looks especially simple, for it does not rely on the concept of degree of a mapping. However, he never chose to publish it. Possibly he could not see how to give it sufficient detail to convince himself of its correctness and rigor.

12.11 Brouwer's 1913 Paper Defining Dimension and Proving Its Invariance

Since by the beginning of 1912 a *mémoire étendu* (extended memoir) was still not forthcoming from Lebesgue, Blumenthal asked Brouwer to produce 'the absolutely shortest and best proof of the invariance of dimension' ('der absolut kürzeste und beste Beweis der Invarianz der Dimension') for the *Annalen* [Dalen, 2011a, pp. 122–123] [Dalen, 2011b, page 499]. He could even assure Brouwer that it would appear in the journal very quickly.[22] However, Brouwer did not comply with the request. It is not known why. Instead, when he eventually came to write a new paper on dimension incorporating a detailed proof of the tiling principle based on projections and the concept of degree of a mapping and also a definition of dimension in the manner of Poincaré's cut idea, he sent it to the *Journal für die reine und angewandte Mathematik* (*Journal for Pure and Applied Mathematics*).[23] The paper 'Über den natürlichen Dimensionsbegriff' ('On the natural dimension concept') [Brouwer, 1913b], a landmark in the history of topological dimension theory, is Brouwer's second and 'final' statement on the subject.

The paper of 1913 opens with a critical appraisal of Poincaré's definition of dimension by means of cuts (coupures) as given in 'Pourquoi l'espace a trois dimensions' ('Why space has three dimensions') [Poincaré, 1912] (see Section 9.4; on Poincaré and Brouwer on dimension also see [Katětov & Simon, 1997, pp. 119–123]). However, Brouwer's must have also known about the definition from Poincaré's earlier paper of 1903 [Poincaré, 1903b] or from information he received while in Paris, because he had already tried to use the cut approach for proving dimensional invariance during the first part of 1910.[24] Some might criticize Poincaré's definition of dimension for its lack of precision and rigor, since he was writing in a philosophical journal for a nonmathematical readership, but in his paper of 1913 Brouwer

[22] In his postcard [Dalen, 2011a, pp. 122–123] [Dalen, 2011b, page 499] Blumenthal does not say which proof he wanted Brouwer to submit to *Mathematische Annalen*. He may well have wanted the proof that Brouwer had offered him in his letter of 19 August [Dalen, 2011a, pp. 97–98] [Dalen, 2011b, pp. 433–434], rather than the proof based on the mapping degree concept that Brouwer had outlined in his letter of 2 July [Dalen, 2011a, pp. 92–94] [Dalen, 2011b, pp. 424–426] and which he later developed into his paper [Brouwer, 1913b].

[23] Freudenthal [Brouwer, 1976, page 444] states that there must have been subtle reasons for Brouwer's ceding his second paper on dimension to the *Journal für die reine und angewandte Mathematik* rather than his placing it in *Mathematische Annalen* as usual. Unfortunately, there seems to be no indication of these reasons in the Brouwer Archive.

[24] In view of the evidence of the student exercise book, Freudenthal [Brouwer, 1976, page 548] suggested that it is most likely that Brouwer learned about Poincaré's definition during his stay in Paris.

takes Poincaré's definition very seriously and submits it to a thorough mathematical appraisal. In a similar way Riesz had earlier assessed the definition [Riesz, 1909a, pp. 313–314] (see Section 10.2).

The main sentence of Poincaré's definition reads as follows [Poincaré, 1912, page 488]:

> A continuum has n dimensions when it is possible to divide it into several parts by means of one or more cuts that are themselves continua of $n-1$ dimensions ....[Ch12, 12]

Brouwer quotes this passage in German in his own paper, rather than in Poincaré's original French.

Brouwer critically examines three aspects of this definition at the beginning of his paper [Brouwer, 1913b, pp. 146–147].

First, he points out that, while the n-dimensional Jordan theorem might be used to justify the separations, this difficult theorem is nowhere mentioned by Poincaré. Hence, presumably it is not needed.

Second, he questions the meaning of the word 'continuum' ('continu,' 'Kontinuum') employed in the Poincaré definition. To what continua does the definition apply? Poincaré never provided a general abstract definition of the concept of continuum covering continua of all dimensions. When defining the concepts of a physical continuum and a mathematical continuum, he confined his attention to the 1-dimensional cases for the most part and based his definitions on arithmetical ideas [Poincaré, 1893] [Poincaré, 1902b, chapitre II]. When considering higher-dimensional continua, he always linked the concepts of continuum and dimension together. In his essay of 1912 [Poincaré, 1912, pp. 488–490] his question is: what is a continuum of n dimensions? However, Brouwer correctly understood that one really needs a definition of continuum prior to a definition of dimension of the type that Poincaré sought. In his paper of 1913 Brouwer rejects his own concept of manifold based on Euclidean simplexes [Brouwer, 1911b, pp. 97–98] as a meaning for 'continuum'. Using this concept brings one back to Euclidean spaces based on arithmetical ideas and the coordinate idea of dimension. In order to avoid a circle in the definitions, it would be necessary to characterize the manifold concept in a form that is independent of the coordinate representation. However, up to the time of Brouwer's writing no one had provided such a characterization. Consequently, he suggests the following abstract definition of continuum [Brouwer, 1913b, page 147]:

> '*A normal set* π (in *Fréchet*'s sense) *is called a continuum, if for every two of its elements,* m_1 *and* m_2*, there is a connected closed [and, hence, compact] set that is a subset of* π *and contains* m_1 *and* m_2*.*'[Ch12, 13]

In giving this definition of continuum he is following Schoenflies [Schoenflies, 1908b, page 117], who similarly defined the concept of a 'nonclosed continuum in the wider sense' for point sets in the plane or Euclidean n-space. In Brouwer's more general definition a Fréchet normal set or normal (E) class [Fréchet, 1906, pp. 18, 23–24, 30] is a complete separable metric space with no isolated points (using mainly Hausdorff's terminology in [Hausdorff, 1914]). Unfortunately, as he points out, this general abstract definition of continuum leads to immediate difficulties when applied to Poincaré's definition of dimension. A double cone, an obvious example of a continuum in the defined sense, turns out to be 1-dimensional, since it can be cut into two pieces by a single point.[25] Surely, one does not wish to regard a cone as 1-dimensional! Riesz had already raised this objection (see Section 10.2).

Third, Brouwer notes that the words 'one or more cuts' in Poincaré's definition cannot be correct, because several m-dimensional manifolds together can form an $(m + p)$-dimensional manifold.

In view of all these criticisms Brouwer leaves aside Poincaré's definition of dimension and puts forward one of his own. From a mathematical viewpoint his improved definition is more sophisticated and more complex. He is perhaps more conscious than Poincaré was of those set-theoretic foundations of topology that can support a definition of dimension. He intends that his definition apply to Fréchet normal sets, i.e., complete separable metric spaces with no isolated points. This represents an early use of Fréchet's ideas on abstract spaces. While abstract spaces are at present a commonplace in mathematics, this was not the case when Brouwer was writing his paper. He was one of the first to see the importance of the abstract approach in topology, although normally he did not advocate this style in mathematics per se.

12.12 Critical Problem in Brouwer's Definition; Brouwer's Early Correction; Urysohn's Later Criticism and Counterexample of 1923

Brouwer's definition is founded on a notion of separation among certain selected sets [Brouwer, 1913b, page 147]:

> Let π be any normal set, and let π_1, ϱ, and ϱ' be three subsets of π that are closed within π [i.e., closed relative to π] and have no points in common. Then ϱ and ϱ' are said to be *separated in π by π_1*, if every connected closed subset of π that has points in common with ϱ as well as with ϱ' also contains at least one point of π_1.[Ch12, 14]

There is a subtle difficulty in this definition of separation, which makes it 'wrong' as a sound logical support for Brouwer's intended definition of dimension and his principal dimension theorem (given below). It centers on the word 'abgeschlossene' ('closed' in English) in the 'wenn'-clause, or 'if'-clause in English, in the second sentence. As given, the definition is a 'weak' form of separation. If the word 'abgeschlossene' ('closed') is deleted, the definition becomes a 'strong' form of

[25] In defense of Poincaré one could point out that a double cone is not a manifold and that Poincaré was primarily interested in manifolds (varieties) in his topological work. However, limiting the Poincaré dimension definition to manifolds brings us back to Brouwer's criticism about the possibility of a logical circle in the definitions.

separation. That strong form is needed for his definition of dimension and his proof of the principal theorem on dimension. The weak and strong forms of separation relate to strong and weak forms of connectedness, respectively.

Documentary evidence shows that Brouwer knew about the difficulty with his definition of separation a short time after its publication in the paper of 1913. He sought to make corrections to deal with the problem, but at that time they were not well publicized. Then in a much more public way the difficulty came to the fore in 1923, when Paul (Pavel) Urysohn used a counterexample to pinpoint the difficulty in a precise mathematical way. First, I shall consider Brouwer's early recognition of the difficulty. Then, I shall examine the difficulty and the fixes for it when it became more public in 1923–1924.

In his edition of Brouwer's *Collected Works*, volume 2, Freudenthal carefully examined the matter and provided extensive and, I think, convincing evidence of Brouwer's early recognition and both his early and his later corrections of the difficulty [Brouwer, 1976, pp. 540–558]. The main evidence is [Brouwer, 1976, pp. 549–552]:

- In the *Collected Works* Freudenthal reprinted Brouwer's copy of the 1913 paper, revealing some of his own handwritten emendations [Brouwer, 1976, pp. 540–546].
- One emendation in light pencil (unfortunately not visible in the reprinted form in the *Collected Works*) crosses out the word 'abgeschlossene' ('closed') with an arrow drawn to a marginal note, 'zu streichen in Übereinstimmung mit S. 150 Fussnote *)' ('to strike [the word] in agreement with footnote *) on page 150'). The use of pencil strongly suggests that it dates from close to 1913.
- Another emendation on the word 'zusammenhängende' ('connected') consists of a short underlining in ink, which likely dates it from about 1923. Apparently related, there are words in ink in the margin, 'im Lennes'schen Sinne zusammenhängende' ('connected in Lennes' sense'). These ink markings are visible in the reprinted form in the *Collected Works*. It may be added that Brouwer used this notion of connectedness earlier, for example, in his paper of 1911 [Brouwer, 1911c, page 308].

  The modern, commonly used concept of connectedness originates from Lennes [Lennes, 1906] [Lennes, 1911, page 303], Riesz [Riesz, 1907] [Riesz, 1909a], and Hausdorff [Hausdorff, 1914, page 244]. Schoenflies used alternative definitions of connectedness and also of continuum in his work [Schoenflies, 1908b, pp. 108, 117], but these are not the modern ones and not now used to any extent.
- Much later in a letter to Hans Hahn of 4 August 1929, Brouwer asserted that he corrected his copy of the 1913 paper as early as March 1913 [Dalen, 2011b, page 1745]. Freudenthal found confirmation of Brouwer's later statement in the proof sheets Brouwer was correcting in 1913 to facilitate a second edition of Schoenflies' work on set theory. That edition was published in 1913, though it did not include a correction to Brouwer's 1913 paper that he asked Schoenflies to add.

I shall turn now to the later period of 1923–1924.[26]

At the 1923 annual meeting of the Deutsche Mathematiker-Vereinigung (German Association of Mathematicians), held in Marburg, 20–23 September 1923, the young Russian mathematician Paul (Pavel) S. Urysohn (1898–1924)[27] gave a talk highlighting difficulties in Brouwer's proof of his principal theorem on dimension and especially the underlying definition of separation. Key to his criticism was a significant counterintuitive example related to separation. Urysohn gave his talk on 21 September.[28] Urysohn was one of the later founders of dimension theory in the 1920s (see Chapter 14).

Brouwer did not attend Urysohn's talk, though he was at the meeting in Marburg lecturing about negative consequences of his intuitionistic mathematical program [Brouwer, 1923c]. He did hear about Urysohn's talk from other colleagues or most likely from Urysohn himself.

Just a few days after the meeting at Marburg Brouwer wrote a postcard to Urysohn in Moscow, dated 29 September 1923, clearly wanting to set the record straight with respect to Urysohn's criticism and perhaps trying to 'save his reputation' by mentioning his earlier correction in pencil (described above) to his 1913 paper, stating, 'This pencilled note, which clears up everything, must date back from many years ago. . . ' ('Deze potloodaanteekening, die alles opheldert, moet van vele jaren geleden afkomstig zijn. . . ') [Dalen, 2011a, page 262] [Dalen, 2011b, page 1194] [Dalen, 2012, pp. 404–405] [Dalen, 2005, page 451]. Brouwer promised to look into the matter further and expressed the hope that he might have already dealt with it by adding an 'Erratum' to a later publication. Brouwer often made such corrections—he was extremely careful in striving to get his mathematical publications exactly correct. Urysohn never received the postcard, so Brouwer summarized it in a later note to Urysohn of 29 November 1923 (Brouwer's summaries, given in Dutch). The matter shows that Brouwer was anxious to establish the correct historical record on his paper.[29]

Prompted by a request from Brouwer in Marburg, Urysohn wrote to Brouwer from Moscow on 24 October 1923, laying out a full explanation of his criticism and supporting counterintuitive example [Dalen, 2011a, pp. 256–260] [Dalen, 2011b, pp. 1182–1186] (cf. also [Brouwer, 1924b, pp. 636–637], [Brouwer, 1924a]).

---

[26] On this period see [Dalen, 2012, Chapter 11, pp. 395–434] [Dalen, 2005, Chapter 12, pp. 441–480], the related letters in [Dalen, 2011a] and [Dalen, 2011b] and also [Rowe & Felsch, 2019, Kapitel 5, pp. 165–194].

[27] For biographical material on Urysohn see [Alexandroff, 1925] and
`https://mathshistory.st-andrews.ac.uk/Biographies/Urysohn/`,
`https://en.wikipedia.org/wiki/Pavel_Urysohn`,
`https://dbpedia.org/page/Pavel_Urysohn`,
`https://www.encyclopedia.com/science/encyclopedias-almanacs-transcripts-and-maps/pavel-samuilovich-urysohn`.

[28] Cf. [Urysohn, 1923]. This short note does not mention this particular matter.

[29] Around this time he expressed some annoyance about the issue in a (draft) letter to Ludwig Bierberbach of October 1923 [Dalen, 2011b, page 1181].

I shall examine Urysohn's criticism and counterintuitive example more closely to understand it in relation to Brouwer's definition of separation [Dalen, 2011a, pp. 256–260] [Dalen, 2011b, pp. 1182–1186].[30] Much of Urysohn's criticism is directed at specific proof steps in the principal theorem on dimension in Brouwer's paper of 1913. The example is thus a counterexample directed at the proof. However, it is really only necessary to analyze Urysohn's criticism and counterintuitive example with respect to the definition of separation.

I shall provide an exposition of Urysohn's example in detail. The diagram in Figure 12.4 gives a picture of the geometrical situation. For the manifold choose the Euclidean plane π. Let a triangle (simplex) $E_1E_2E_3$ be defined by straight lines connecting the three points $E_1 = (0, -1)$, $E_2 = (3, 2)$, $E_3 = (-3, 2)$. Let π_1 consist of a set of six elements, the sinusoidal curve and five straight line segments:

$$\begin{array}{ll} (1)\ y = \sin^2 \frac{\pi}{x}, & 0 < x < 1; \\ (2)\ x = 0, & 0 \leqq y \leqq 1; \\ (3)\ y = 0, & -1 \leqq x \leqq 0; \\ (4)\ x = -1, & -2 \leqq y \leqq 0; \\ (5)\ y = -2, & -1 \leqq x \leqq 1; \\ (6)\ x = 1, & -2 \leqq y \leqq 0. \end{array}$$

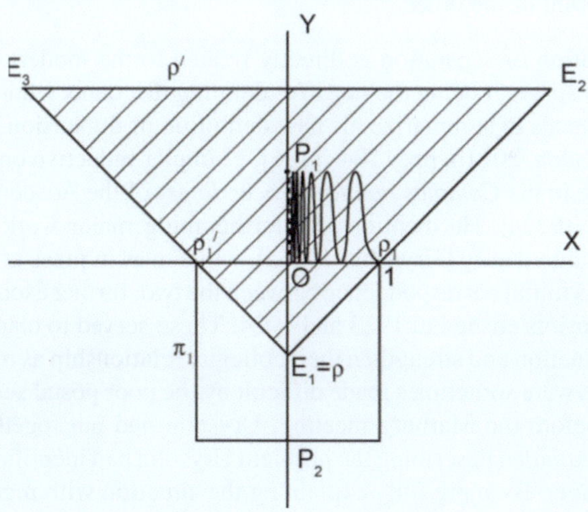

Fig. 12.4 Diagram of Urysohn's Counterintuitive Example [Letter of Urysohn to Brouwer, dated 24.X.1923, and reproduced in [Dalen, 2011a, page 208] and [Dalen, 2011b, page 1183]].

[30] Cf. the detailed discussion in [Dalen, 2012, Section 11.2, pp. 398–421] [Dalen, 2005, Section 12.2, pp. 444–467]).

The complete curve π_1 is a closed and bounded point set implicitly defining an inside domain and an outside domain. The curve is the common boundary of the two domains, though it is not a Jordan closed curve. Let π_2 be the set consisting of the two points $P_1 = (0, 1)$ and $P_2 = (0, -2)$. According to Brouwer's definition, curve π_1 separates the two sets point $\varrho = E_1$ and line segment $\varrho' = E_3E_2$ in the plane, π. That is, every connected closed subset of the plane, π, having points in common with $\varrho = E_1$ and $\varrho' = E_3E_2$ will contain at least one point of the curve π_1. Similarly, the set π_2 defined as consisting of the two points $P_1 = (0, 1)$ and $P_2 = (0, -2)$ separates the two one-point sets, $\varrho_1 = (1, 0)$ and $\varrho_1' = (-1, 0)$, on the curve π_1. That is, every connected closed subset of the curve π_1 having points in common with $\varrho_1 = (1, 0)$ and $\varrho_1' = (-1, 0)$ will contain at least one of the points $P_1 = (0, 1)$ and $P_2 = (0, -2)$. However, by dropping the condition of being closed, the two-point set π_2 will not separate the two one-point sets, ϱ_1 and ϱ_1', on the curve π_1, because the part of the curve π_1 consisting of the sinusoidal part (1) plus the modified vertical segment (2′) $x = 0, \quad 0 \leqq y < 1$ (*without point* P_1) plus the line segment (3) $y = 0, \ -1 \leqq x \leqq 0$ will be connected and not contain either point P_1 or P_2.

In his letter to Brouwer Urysohn gives a replacement for Brouwer's definition of separation [Dalen, 2011a, page 258] [Dalen, 2011b, page 1184]:

> 'ϱ and ϱ' are called separated in π by π_1, when $\pi - \pi_1$ can be split into two subsets, λ and λ', that contain ϱ and ϱ' and such that neither of them contains a boundary point of the other.' [Ch12, 15]

Urysohn's definition of separation is directly related to the modern definition of a connected set or space, it being the negative defining disconnectedness.

Urysohn proceeds to summarize his new definition of dimension [Dalen, 2011a, pp. 258–259] [Dalen, 2011b, pp. 1184–1186], a simpler inductive one, as published in his recent note in the *Comptes rendus* (*Proceedings*) of the Academy of Sciences, Paris [Urysohn, 1922a]. He mentions his forthcoming major work laying out his extensive dimension theory [Urysohn, 1925a], which was in press at the time.

Following this initial correspondence between the two, further exchanges of letters as well as of offprints ensued in 1923 and 1924. These served to clarify the broader mathematical situation and strengthen their collegial relationship as mathematicians. These exchanges were sometimes made difficult by the poor postal service in Russia.

Some time before the Marburg meeting, Urysohn had put together a paper for *Mathematische Annalen* describing the problem Urysohn had identified in Brouwer's original 1913 paper. By more fully explaining the situation with regard to the difficulty and its repair, Brouwer was able to persuade Urysohn to withdraw the paper.

Their correspondence shows the back and forth of the discussion of these topological issues. Urysohn with his close friend and fellow mathematician, Paul (Pavel) Alexandroff, did a considerable amount of traveling in Europe in 1924, particularly to Göttingen. Brouwer invited them to visit him in Holland, where they eventually stayed with him for several days in the summer. The photograph of the three with Brouwer between Alexandroff and Urysohn sitting by his hut in Blaricum provides a snapshot of their developing congenial friendship (see Figure 12.5, and [Brouwer, 1976, page 453] [Dalen, 2012, page 417] [Dalen, 2005, page 463]). Brouwer asked them to return to Holland the following year, but Urysohn tragically died in a swimming accident off the coast of Batz-sur-Mer, France on 17 August 1924—a great loss for mathematics. Fortunately, Alexandroff did an admirable job as literary executor in editing and publishing Urysohn's remaining papers.

Fig. 12.5 Brouwer with Alexandroff and Urysohn Sitting by his Hut in Blaricum [[Brouwer, 1976, page 453]; original courtesy of P. Alexandroff (deceased)].

As a result of Urysohn's calling attention to the difficulty with separation and the resulting faulty proof of the main dimension theorem, Brouwer issued a string of notes and papers seeking to correct the problem in a public way [Brouwer, 1923a] [Brouwer, 1923b] [Brouwer, 1924a] [Brouwer, 1924b] [Brouwer, 1924c] [Brouwer, 1924d]. The simple solution is to delete the word 'abgeschlossene' ('closed'), while at the same time to understand 'zusammenhängende' ('connected') as having its usual modern meaning. An equivalent way of amending the definition of separation is to change the final defining 'if'-clause, or 'wenn'-clause in the German, to the following [Brouwer, 1924d, page 253]:

> 'if π_1 determines a domain set [i.e., a union of connected open sets; thus an open set] in π containing ϱ, but containing no part of ϱ''.[Ch12, 16]

When correcting the error concerning separation in 1923–1924, Brouwer called the inclusion of the offending word, 'abgeschlossene' ('closed'), a 'slip of the pen' ('Schreibfehler') and argued that it is obvious from the text of the subsequent proof in the original paper [Brouwer, 1913b, page 150] that the word is a mistake [Brouwer, 1924c, page 312]. In the end, Urysohn agreed with him.

12.13 Brouwer's Definition of Dimension-Degree

I shall now examine Brouwer's fundamental definition of dimension-degree in detail. He first presents it in an inductive formulation [Brouwer, 1913b, pp. 147–148]:

> The expression '*π has the general dimension-degree n*,' where n denotes an arbitrary natural number, shall now mean that for every choice of ϱ and ϱ' there exists a separating set π_1 with the general dimension-degree $n - 1$, but that it is not the case that for every choice of ϱ and ϱ' there exists a separating set π_1 with a general dimension-degree less than $n-1$. Further, the expression '*π has the general dimension-degree zero, respectively, an infinite general dimension-degree*' shall mean that π contains no continuum as a part, respectively, that neither zero nor any natural number can be the general dimension-degree for π.[Ch12, 17]

This definition of dimension-degree is a global definition or a definition of dimension-degree in the large just as Poincaré's is.

Immediately after giving his definition of dimension-degree in an inductive form, Brouwer provides an equivalent recursive, noninductive formulation, a kind of game-theoretic one [Brouwer, 1913b, page 148]. It is this recursive formulation that he uses in proving Lebesgue's tiling principle; indeed his motive for providing it must have been to make the definition fit the proof. It can be paraphrased in the following way. One imagines two persons A and B subjecting the set π to a 'dimension-operation' ('Dimensionsoperation'), whereby they eventually reduce the set to a discontinuum (i.e., a set with no subcontinuum). First, A arbitrarily chooses two disjoint relatively closed subsets ϱ and ϱ' of π; after which B picks a relatively closed subset π_1 of π separating ϱ and ϱ'. Then A arbitrarily chooses two disjoint relatively closed subsets ϱ_1 and ϱ_1' of π_1; whereupon B finds a relatively closed subset π_2 in π_1 that separates ϱ_1 and ϱ_1'. A and B continue this process until a discontinuum π_h appears. On the basis of this dimension-operation one says that π *has general dimension-degree n*, if, independent of A's choices for the sets ϱ_ν, and ϱ_ν', B can provide for the appearance of a set π_h such that $h \leqq n$ (i.e., π is at most of dimension-degree n; Brouwer-dim-degree$[\pi] \leqq n$) and if, independent of B's choices for the sets π_ν, A can insure that no set π_h appears for which $h < n$ (i.e., π is at least of dimension-degree n; not Brouwer-dim-degree$[\pi] < n$). If, on the contrary, there is no natural number n with the property that, independent of A's choices for the ϱ's, B can guarantee the appearance of a set π_h such that $h \leqq n$, then one says that π *has an infinite general dimension-degree*.

To his global definition Brouwer adjoins a local definition of dimension-degree, the dimension-degree of a set at a point [Brouwer, 1913b, page 148]:

> If at a point P of π neighborhoods exist that have the general dimension-degree m, but no neighborhoods exist that have a lower general dimension-degree, then we shall say that π *has the dimension-degree m at P*.[Ch12, 18]

Simple geometrical figures show that it is quite possible for sets to have different dimension-degrees at different points.[31] A set of this type may have a dimension-degree in the large which is greater than or equal to any of its dimension-degrees at its points. However, if a set has all its dimension-degrees at its points equal to its global dimension-degree, then Brouwer expresses this fact by saying that the set possesses a homogeneous dimension-degree ('die Menge *einen homogenen Dimensionsgrad besitzt*').

12.14 Comparing the Brouwer, Urysohn, and Menger Definitions of Dimension

Brouwer's definitions can be compactly stated in the following form:

Definition of Brouwer Dimension-Degree (Brouwer-dim-degree).
A set π in a complete separable metric space shall be said to have Brouwer dimension-degree, Brouwer-dim-degree[π], or to have Brouwer dimension-degree at a point P, Brouwer-dim-degree$_P$[π], according to the following conditions:

(0) Brouwer-dim-degree[π] $= 0$, if π contains no subcontinuum, i.e., it contains no connected, closed, and compact subset of more than one point;

($\leqq n$) Brouwer-dim-degree[π] $\leqq n$, if every pair of disjoint relatively closed subsets ϱ and ϱ' in π can be separated by a closed subset π_1 in π for which Brouwer-dim-degree[π_1] $\leqq n-1$;

($= n$) Brouwer-dim-degree[π] $= n$, if Brouwer-dim-degree[π] $\leqq n$ but not Brouwer-dim-degree[π] $\leqq n-1$;

(∞) Brouwer-dim-degree[π] $= \infty$, if not Brouwer-dim-degree[π] $\leqq n$, for any natural number n, including 0;

($= m$) Brouwer-dim-degree$_P$[π] $= m$, if there exist neighborhoods of P, $U(P)$, contained in π, such that Brouwer-dim-degree[$U(P)$] $= m$, but none exist such that Brouwer-dim-degree[$U(P)$] $\leqq m-1$.

---

[31] Such figures are easy to construct. One can consider the union of a 0-dimensional point set, a 1-dimensional curve, a 2-dimensional surface, and a 3-dimensional solid as a simple example.

Taken together these definitions constitute Brouwer's proposal for understanding the 'natural concept of dimension.' They are the foundation for a theory of dimension having a wide application to point sets in metric spaces. In the following paragraphs I shall assess them further.

Above all Brouwer's sophisticated mathematical definitions proposed as a characterization of the concept of dimension are a response to and an improvement on Poincaré's informal 'philosophical' definition using cuts. In the first place, one should recognize that Brouwer and Poincaré were in fundamental agreement on the need for a good definition of dimension. The need arose from the evident inadequacy of the coordinate notion of dimension made plain by Cantor's paradoxical discovery. Like Poincaré, Brouwer wanted a basic definition, a definition at the foundations of geometry. He did not want it to be founded upon difficult theorems of topology such as the Jordan–Brouwer separation theorem or the dimensional invariance theorem; rather it should be logically prior to these. However, beyond this level of agreement Brouwer looked to improve upon Poincaré's work. In the simplest case he wished to overcome the difficulty posed by the double cone. His definition easily passes this test. The cone obviously has Brouwer dimension-degree two as one would expect on intuitive grounds.

Poincaré justified his informal cut definition by asserting that it is consonant with the traditional definitions of classical geometry, which have come down from Euclid and with topological ideas of connectivity originating from Riemann. Brouwer sought a more definite mathematical justification for his definition in a proof that an n-dimensional manifold, in particular, an n-dimensional Euclidean space, is n-dimensional in the newly-defined sense, i.e., that it has Brouwer dimension-degree n. From this theorem the topological invariance of dimension follows directly. Poincaré took dimensional invariance more or less for granted (see Section 9.4). In complete contrast Brouwer's primary goal in his paper of 1913 is to demonstrate the n-dimensionality of an n-manifold with its corollary of dimensional invariance. He recognized this result as the ultimate justification of his definition.

The theorem on the dimensionality of manifolds is, in fact, the only one Brouwer proves in his paper of 1913. He never went beyond it by deducing other consequences from his definition.[32] Carrying out this type of deductive program remained to the dimension-theorists of the 1920s: P.S. Urysohn, K. Menger, P.S. Alexandroff, W. Hurewicz, L.A. Tumarkin, and others. It seems he thought that by proving his theorem of justification he had reached his 'epistemological' goal. To investigate other consequences would have been merely an exercise in formal or formalistic mathematics [Brouwer, 1928b, 953–954]. Thus, he did not develop a *formal dimension theory*. That he did not do this is characteristic of his style in mathematics. He avoided the formal aspect of mathematics. He was far more concerned with solving the outstanding topological problems of his time. A definition of dimension-degree coupled with a proof of Lebesgue's tiling principle offered a very attractive route to dimensional invariance, and so he explored this route with great success in his paper

[32] However, in his paper he does state the monotonicity theorem (without giving the simple proof) [Brouwer, 1913b, page 148]. This theorem asserts that the Brouwer dimension-degree of a set is greater than or equal to the Brouwer dimension-degree at any given point in the set.

of 1913. Beyond this he did not go. As a result, one cannot say that Brouwer was a forerunner of the full development of dimension theory in the 1920s. It was not that he was unable to pursue this course, but that he chose not to do so.

Brouwer's definition of dimension-degree (dimension) is but the first in a sequence of such definitions lying at the foundations of modern topological dimension theory. The second and third members of this sequence are the definitions of Paul (Pavel) S. Urysohn [Urysohn, 1922a] [Urysohn, 1922b] [Urysohn, 1925a] [Urysohn, 1926] (cf. the emendations in [Alexandroff, 1926a]) and Karl Menger [Menger, 1923] [Menger, 1924d] [Menger, 1928b]. In order to give a fuller appraisal of Brouwer's definition, it is necessary to compare it with those of Urysohn and Menger. Though having different external forms, these latter are equivalent for separable metric spaces.[33] Hence, we compare the Brouwer definition with the combined Urysohn–Menger definition given in Menger's formulation [Menger, 1928b, pp. 77–83] [Hurewicz & Wallman, 1948, page 24]:

Definition of Urysohn–Menger Dimension (Urysohn–Menger-dim).
A set π in a separable metric space shall be said to have Urysohn–Menger dimension at a point P, Urysohn–Menger-dim$_P[\pi]$, or to have Urysohn–Menger dimension, Urysohn–Menger-dim$[\pi]$, according to the following conditions:

(-1) Urysohn–Menger-dim$[\pi] = -1$ if and only if π is the empty set $\emptyset$;
($\leqq n$) Urysohn–Menger-dim$_P[\pi] \leqq n$, if P has arbitrarily small neighborhoods with boundaries relative to π, $bdry\ U(P)$, such that Urysohn–Menger-dim$[bdry\ U(P)] \leqq n-1$;
($\leqq n$) Urysohn–Menger-dim$[\pi] \leqq n$, if for each point P of π Urysohn–Menger-dim$_P[\pi] \leqq n$;
($= n$) Urysohn–Menger-dim$_P[\pi] = n$, if Urysohn–Menger-dim$_P[\pi] \leqq n$ but not Urysohn–Menger-dim$_P[\pi] \leqq n-1$;
($= n$) Urysohn–Menger-dim$[\pi] = n$, if Urysohn–Menger-dim$[\pi] \leqq n$ but not Urysohn–Menger-dim$[\pi] \leqq n-1$;
(∞) Urysohn–Menger-dim$[\pi] = \infty$, if not Urysohn–Menger-dim$[\pi] \leqq n$, for any natural number n, including 0.

The definitions of Brouwer and Urysohn–Menger are highly sophisticated, employing all the relevant tools of twentieth-century set theory and topology. As one would expect they both assign the 'correct' dimension numbers to the usual simple geometrical objects. Hence, one can say that on the level of classifying ordinary figures of geometry by dimension they are both reasonable definitions.

On other levels they diverge. Brouwer's most important definition of dimension is his global definition; his local definition is of secondary importance. In contrast, the Urysohn–Menger definition begins from local considerations, but it differs very markedly from Brouwer's local definition.[34] It is much closer to Brouwer's global

[33] Dimension theorists of the 1920s quickly noticed this equivalence. Cf. [Brouwer, 1924b] [Urysohn, 1926, pp. 327–328] [Menger, 1926b, pp. 121–122] [Menger, 1928b, pp. 83–84].

[34] Brouwer's local dimension-degree concept is completely different in content from the Urysohn–Menger local dimension concept according to Menger [Menger, 1928b, page 86]. Indeed, dimension theorists have hardly explored the consequences of Brouwer's local definition at all.

definition in terms of its implications in the large. In the case of compact metric spaces the two definitions are equivalent. However, for separable metric spaces, the type of space which dimension theorists of the 1920s came to use as a matrix for developing the theory, the two are not equivalent. The nonequivalence arises in the bases of the definitions. Brouwer's definition begins with discontinua; those sets that have no subcontinua are given dimension 0. The Urysohn–Menger definition starts by declaring that the empty set and only this set has dimension -1. Then a set will be 0-dimensional if each of its points has arbitrarily small neighborhoods with empty boundaries.[35] Research into the notions of connectedness and disconnectedness, especially by the Polish school of topology in the early 1920s, has shown that these two ideas of 0-dimensionality differ radically (see Section 13.9). Although it is true that for any set π, Urysohn–Menger-dim$[\pi] = 0$ implies Brouwer-dim-degree$[\pi] = 0$, the converse of this statement is by no means true. In 1921 Sierpiński [Sierpiński, 1921] gave an example of a peculiar set that turns out to have Brouwer dimension 0, but which in the later theory of Urysohn and Menger is 'weakly 1-dimensional.' About this same time Knaster and Kuratowski [Knaster & Kuratowski, 1921] provided similar examples. In a more spectacular way Urysohn in his magnum opus on dimension theory, 'Mémoire sur les multiplicités Cantoriennes' ('Memoir on Cantorian multiplicities') [Urysohn, 1926, page 323], exhibited a discontinuum, a set with Brouwer dimension-degree 0, which is infinite-dimensional in the sense of Urysohn and Menger. Hence, on the level of classifying more unusual examples, extraordinary examples generated by deeper topological research, the Brouwer and Urysohn–Menger definitions differ greatly. Furthermore, the two theories deducible from these definitions are strongly divergent. The local character of the Urysohn–Menger definition gives its corresponding theory greater power and greater subtlety than the Brouwer theory. In Karl Menger's well-known book *Dimensionstheorie* [Menger, 1928b, pp. 317–318] there are several examples illustrating this fact. A number of important theorems in the Urysohn–Menger dimension theory turn out to be false in the Brouwer theory and, moreover, there do not appear to be interesting replacement theorems for these in the Brouwer theory. Consequently, with respect to theoretical power, the Urysohn–Menger theory has definite advantages over a possible Brouwer theory.

In the course of history mathematicians have chosen the Urysohn–Menger definition over the Brouwer definition as a basis for developing dimension theory in the context of separable metric spaces.[36] Their reasons for this choice seem to have been the following. First, with respect to classifying ordinary geometrical figures by dimension number, both definitions serve equally well. However, for extraordinary examples the Urysohn–Menger definition appears to be superior, because it offers a more refined scale of dimensionality. It copes better with the extreme examples, the

[35] On the Urysohn–Menger concept of 0-dimensionality cf. [Menger, 1928b, pp. 202–214] and [Hurewicz & Wallman, 1948, chapter II].

[36] Separable metric spaces, i.e., metric spaces with countable dense subsets or equivalently with countable bases, have been the main spaces used for developing dimension theory [Hurewicz & Wallman, 1948, pp. 153–156]. However, recently more general topological spaces have been considered for dimension theory, e.g., [Pears, 1975] or [Engelking, 1978].

pathological examples or monsters, so to speak, which have emerged from topological research. Indeed, some of these extraordinary examples helped to generate the Urysohn–Menger theory. Second, with respect to theoretical power, the Urysohn–Menger theory also seems to be superior to the Brouwer theory. It yields a greater wealth of results. Judged as a whole the theory is more fruitful and aesthetically more pleasing. Hence, topologists have come down on the side of the Urysohn–Menger definition and theory. This is not to say that the Brouwer definition is wrong and the Urysohn–Menger definition is right. On the contrary, there is a certain amount of arbitrariness in the choice of the Urysohn–Menger theory over the Brouwer one. Neither is a completely 'natural' explanation of the dimension concept.[37] Theory appraisal is not absolute and for all time. Nevertheless, as a result of a reasonable methodological decision and perhaps partly as a result of historical accident mathematicians have chosen to develop further the Urysohn–Menger dimension theory instead of the Brouwer theory.[38]

12.15 Mathematical Details of Brouwer's 1913 Paper on Dimension

I shall now consider further Brouwer's paper of 1913. Brouwer states the principal theorem that he recognizes as justifying his dimension definition and demonstrating its correctness in the following way [Brouwer, 1913b, page 148]:

> **Dimension Theorem.** *An n-dimensional manifold has the homogeneous dimension-degree n*[**)].
>
> [Brouwer's footnote] [**)] Because the dimension-degree is clearly an invariant of analysis situs, the invariance of the dimension number is contained in the Dimension Theorem.[Ch12, 19]

A Brouwerian n-dimensional simplicial manifold[39] is uniformly of dimension n; it is of dimension n at every point. In particular, a Euclidean n-space uniformly has dimension n. Since Brouwer's concept of dimension-degree is topologically invariant, the invariance of dimension number immediately follows from this theorem, as Brouwer asserts in his footnote [**)].

To prove the theorem Brouwer uses the noninductive formulation of dimension in the large and breaks the proof into two parts by demonstrating, first, that an n-dimensional manifold has dimension that is *at most n* and, second, that it has

[37] Thus, the term 'natürlichen' in the title of Brouwer's paper 'Über den natürlichen Dimensionsbegriff' [Brouwer, 1913b] may not seem entirely appropriate.

[38] It is possible, of course, to go beyond both theories in the dimension theory of general spaces. In the last few paragraphs I have just touched upon the question of the appraisal of theories, an important methodological and historical question.

[39] Recall that a Brouwerian simplicial manifold is (in modern terms) a connected n-dimensional topological manifold having a locally finite simplicial subdivision (triangulation). Cf. Brouwer's original formulation in [Brouwer, 1911b, pp. 97–98].

dimension that is *at least* n. In addition, to overcome Urysohn's counterexample to the proof, it is necessary to use the corrected definition of separation, as Urysohn asserts in his letter to Brouwer of 24 October 1923 [Dalen, 2011a, pp. 258–259] [Dalen, 2011b, page 1184].

The first part of the proof [Brouwer, 1913b, page 149] is by far the easier part. Brouwer needs to show that for an n-dimensional manifold π, Brouwer-dim-degree$[\pi] \leqq n$. Thus, it is necessary to show that no matter what sets ϱ_ν and ϱ'_ν A chooses, B can choose sets $\pi_1, \pi_2, \ldots$, such that a set π_h with no subcontinuum arises with $h \leqq n$. The method by which B can guarantee the existence of π_h is to construct the sequence of chosen sets using appropriate simplicial subdivisions. Suppose that A has determined the relatively closed sets ϱ and ϱ' within the manifold π. Then B constructs a simplicial subdivision of π in such a way that simplexes ${}_\pi s_\varrho$, which contain points of ϱ, and simplexes ${}_\pi s_{\pi'}$, which contain points of π', never are the same and never have a common face, i.e., the subcomplexes containing the sets ϱ and ϱ' are disjoint. This construction is always possible, since the sets ϱ and ϱ' are closed in π. The resulting subcomplexes, in fact, form closed neighborhoods of the sets. Brouwer asserts that the set of all those $(n-1)$-dimensional faces of simplexes ${}_\pi s_\varrho$ that do not contain any points of ϱ forms a system of orientable (closed) $(n-1)$-dimensional pseudomanifolds, which may have elements in common. Certainly this system will form a closed set π_1 that separates ϱ and ϱ'.[40] In a similar way, when A has determined relatively closed sets ϱ_1 and ϱ'_1 within π_1, possibly within the same subcontinuum of π_1, B can construct an appropriate simplicial subdivision of π_1 to derive subcomplexes containing ϱ_1 and ϱ'_1 and, hence, a system of orientable (closed) $(n-2)$-dimensional pseudomanifolds that forms a set π_2 separating ϱ_1 and ϱ'_1. Continued in this way the process will certainly lead to a set π_h that contains no subcontinuum with $h = n$ or even possibly with $h < n$.

For the second part of the proof [Brouwer, 1913b, pp. 149–152], to demonstrate that it is not the case that Brouwer-dim-degree$[\pi] \leqq n - 1$, it is necessary to show that A can force the existence of the final π_n, so that the set π_h with no subcontinuum is in reality π_n. Brouwer reduces this problem to one of proving Lebesgue's tiling principle for the simplex. Thus, A considers just one n-dimensional element (i.e., a topological image of an n-simplex) $E_1E_2\ldots E_{n+1}$ in the manifold π and takes the

---

[40] Brouwer defines a pseudomanifold as follows [Brouwer, 1911c, pp. 305–306] [Brouwer, 1921, page 286]:

> By an *n-dimensional pseudomanifold* we understand a point set formed from n-dimensional *elements* [i.e., topological images of simplexes] ...joined together in such a way that every $(n-1)$-dimensional face is common to two and only two elements, while every point not contained in an $(n-1)$-dimensional face belongs to only one element and every other point belongs to only a finite number of elements. Finally, every two points of the point set can be connected by a simple arc of a curve that belongs entirely to the point set and that meets no $(n-p)$-dimensional face $(p > 1)$.[Ch12, 20]

Brouwer's concept of pseudomanifold has become a standard concept in topology. For Brouwer's assertion about pseudomanifolds given in the text cf. his paper [Brouwer, 1911c, page 306]. It is clear that the system will form a closed set π_1 separating ϱ and ϱ', since the associated complex forms a closed neighborhood of ϱ.

point E_1 as ϱ and the opposite $(n-1)$-dimensional face $E_2 \dots E_{n+1}$ as ϱ'. Further, A takes the part of π_1 that belongs to the edge E_1E_2 as ϱ_1 and the part of π_1 that belongs to the face $E_1E_3 \dots E_{n+1}$ as ϱ_1'; then A takes the part π_2 that belongs to the face $E_1E_2E_3$ as ϱ_2 and the part of π_2 that belongs to the face $E_1E_2E_4, \dots E_{n+1}$ as ϱ_2'; etc. It is necessary to prove that all the point sets $\pi_1, \pi_2, \dots, \pi_n$ chosen by B are nonempty. By giving the element $E_1E_2 \dots E_{n+1}$ an appropriate simplicial subdivision, it is possible to approximate arbitrarily closely the sets $\pi_1, \pi_2, \dots, \pi_n$ by corresponding simplicial complexes $\sigma_1, \sigma_2, \dots, \sigma_n$. Hence, if σ_n is nonempty, so is π_n. In this way one arrives at Brouwer's version of Lebesgue's principle, stated as a lemma (I have displayed the translation below in a format with labelled items, and I have given Brouwer's single paragraph format in the endnote at the back as in the original printed paper) [Brouwer, 1913b, page 150]:[41]

Lemma. Let

- σ *be a simplicially subdivided n-dimensional element with vertices* $E_1, E_2, \dots, E_{n+1}$*;*
- γ *be a fragment, formed of basic simplexes of* σ*, that contains every basic simplex bounded at* E_1 *but none bounded on* $E_2E_3 \dots E_{n+1}$*;*
- σ_1 *be the part of the boundary of* γ *lying within* σ*;*
- γ_1 *be a fragment, formed of basic simplexes of* σ_1*, that contains every basic simplex bounded on* E_1E_2 *but none bounded on* $E_1E_3 \dots E_{n+1}$*;*
- σ_2 *be the part of the boundary of* γ_1 *lying within* σ_1*;*
- γ_2 *be a fragment, formed of basic simplexes of* σ_2*, that contains every basic simplex bounded on* $E_1E_2E_3$ *but none bounded on* $E_1E_2E_4 \dots E_{n+1}$*;*
- σ_3 *be the part of the boundary of* γ_2 *lying within* σ_2*;*

etc.

Then none of the point sets $\sigma_1, \sigma_2, \sigma_3, \dots, \sigma_n$ *can vanish.*[Ch12, 22]

Clearly Brouwer's γ_i and σ_i are modified versions of Lebesgue's e_i and I_i, respectively.

In the paragraph appearing immediately after the statement of the Lemma, Brouwer makes the following remark [Brouwer, 1913b, page 151]:

> This lemma, to which *Lebesgue* already referred in Math. Annalen 70 as the invariance of the number of dimensions, the proof of which is there but which has a significant gap*), is immediately obvious if we apply the concept of *mapping degree* introduced by me in Math. Annalen 71.[Ch12, 23]

---

[41] For the Lemma Brouwer requires his notion of *fragment*, defined as follows [Brouwer, 1911c, page 306]:

> By an *n-dimensional fragment* we understand a point set formed from a finite number of elements belonging to one or several n-dimensional pseudomanifolds.[Ch12, 21]

The remark is clearly a criticism of Lebesgue and his faulty proof along with a boost for Brouwer's degree of a mapping concept [Brouwer, 1911b]. In his footnote *) Brouwer elaborates the lemma in Lebesgue's terms and gives a version of his counterexample directed against Lebesgue's proof.

Brouwer's short proof of the lemma [Brouwer, 1913b, page 152] is in essence the same one as found in his private notes of February 1911 (see above and [Brouwer, 1976, pp. 450–451]), only revised for a simplex instead of a cube. The objective is to prove by induction that the projection σ_ν, from the face $E_1E_2 \ldots E_\nu$ covers the face $E_{\nu+1}E_{\nu+2} \ldots E_{n+l}$ with degree 1 for all ν. Inasmuch as the basis is trivial Brouwer only gives the induction step. From the induction hypothesis for ν it simply follows that the projection of that part of the boundary of σ_ν lying in the face $E_1 \ldots E_\nu E_{\nu+2} \ldots E_{n+1}$ projected from the face $E_l \ldots E_\nu$ or from $E_1 \ldots E_{\nu+1}$[42] covers the face $E_{\nu+2} \ldots E_{n+1}$ with degree 1. Now if σ_ν is reduced step by step to γ_ν by removing some of its basic simplexes one at a time, then the part of the boundary of σ_ν lying in the face $E_! \ldots E_\nu E_{\nu+2} \ldots E_{n+l}$ is eventually transformed into $\sigma_{\nu+1}$ and yet the degree of the projection upon $E_{\nu+2} \ldots E_{n+l}$ remains the same. So $\sigma_{\nu+1}$ also covers $E_{\nu+2} \ldots E_{n+I}$ with degree 1. Consequently, because each σ_ν covers an appropriate $(n-\nu)$-dimensional face of the element σ by projection with degree 1, none of the σ_ν can vanish, and Lebesgue's lemma is demonstrated.

By employing some extra symbolism one can put Brouwer's proof of the tiling principle into a little more modern form. This transformation of the proof will give a some indication of its relation to homology. The following is a recasting of the crucial induction step.

Consider the projection p_ν of σ_ν from the face $E_1 \ldots E_\nu$ to the face $E_{\nu+1} \ldots E_{n.+1}$,

$$p_\nu : \sigma_\nu \longrightarrow E_{\nu+1} \ldots E_{n+1}.$$

By the induction hypothesis the degree of p_ν[43] is 1:

$$\deg(p_\nu) = 1.$$

Furthermore, one can 'extend' p_ν to projection p_ν^* to map the combinatorial boundary of σ_ν, denoted as $\partial\sigma_\nu$,

$$p_\nu^* : \partial\sigma_\nu \longrightarrow \partial E_{\nu+1} \ldots E_{n+1},$$

and the degree of this mapping must also be 1:

$$\deg(p_\nu^*) = 1.$$

Hence, if one restricts p_ν^* to the part of $\partial\sigma_\nu$ in the face $E_1 \ldots E_\nu E_{\nu+2} \ldots E_{n+1}$, denoted as $\partial'\sigma_\nu$, then the restriction mapping

---

[42] Initially the projection from the face $E_1 \ldots E_{\nu+I}$ adds nothing to covering the face $E_{\nu+2} \ldots E_{n+l}$.

[43] Note that σ_ν is a relative cycle, modulo the boundary of the element σ, and also that the face $E_{\nu+I} \ldots E_{n+l}$ is a relative cycle, so that the mapping degree is meaningful for p_ν.

$$p_\nu^* \mid \partial'\sigma_\nu \to E_{\nu+2} \dots E_{n+1}$$

also has degree 1:

$$\deg(p_\nu^* \mid \partial'\sigma_\nu) = 1.$$[44]

Consider now the projection mapping $p_{\nu+1}$ from the face $E_1 \dots E_\nu$ or from the face $E_1 \dots E_{\nu+1}$ to the face $E_{\nu+2} \dots E_{n+l}$, changing in this instance to $p_{\nu+1}^*$ to apply to $\partial'\sigma_\nu$,

$$p_{\nu+1}^* : \partial'\sigma_\nu \to E_{\nu+2} \dots E_{n+1}.$$

Then $p_{\nu+1}^*$ coincides with $p_\nu^* \mid \partial'\sigma_\nu$, since only the projection from $E_1 \dots E_\nu$ contributes to covering $E_{\nu+2} \dots E_{n+I}$, so that

$$\deg(p_{\nu+1}^*) = \deg(p_\nu^* \mid \partial'\sigma_\nu) = 1.$$

If one considers the boundary of an $(n - \nu)$-simplex s in σ_ν, then the projection of the boundary ∂s by $p_{\nu+l}^*$ collapses ∂s on the face $E_{\nu+2} \dots E_{n+1}$ and the degree is 0:

$$\deg(p_{n+1}^* \mid \partial s) = 0.$$

If, along with Brouwer, one thinks of deleting simplexes s one by one from σ_ν, then the resulting boundaries within the element σ, call them

$$\partial'\sigma_{\nu 1}, \partial'\sigma_{\nu 2}, \dots,$$[45]

will all be projected by $p_{\nu+1}^*$ with degree 1:

$$\deg(p_{\nu+1}^* \mid \partial'\sigma_{\nu i}) = 1.$$

Eventually one arrives at a boundary $\partial'\sigma_{\nu m}$ that is identical with $\sigma_{\nu+1}$. Hence, for the projection mapping

$$p_{\nu+1} : \sigma_{\nu+1} \to E_{\nu+2} \dots E_{n+1}$$

one has

$$\deg(p_{\nu+1}) = 1,$$

which is the desired result. Consequently, one can conclude by induction that none of the $\sigma_\nu (\nu = 1, \dots, n)$ vanish.

Brouwer's use of the operation of deleting $(n-\nu)$-simplexes s and thus of subtracting their combinatorial boundaries appears as a form of geometrical combinatorial topology. What one has in Brouwer's topological work overall is the use of simplicial decomposition, simplicial approximation, and simplicial mapping.

Brouwer's paper on dimension of 1913 is the last substantial contribution that he made to dimension theory. It marks the end of his most productive period of research in topology. After the brilliant work of 1909–1913 he published only a few relatively

[44] Clearly $\partial'\sigma_\nu$ and $E_{\nu+2} \dots E_{n+1}$ are relative cycles; hence, the concept of degree is applicable.

[45] All of these are relative cycles.

minor papers or corrections to previous papers on the subject. There are probably several reasons for this diminution in his research activity in topology after 1913.[46] Most importantly, one should realize that by 1913 Brouwer had solved nearly all of the specific topological problems that he had set himself in his inaugural lecture as *privaat-docent* (private-tutor; unpaid lecturer) at the University of Amsterdam of 12 October 1909 [Brouwer, 1909d] [Brouwer, 1975, pp. 112–120]. His inaugural address on becoming Professor at Amsterdam, delivered on 14 October 1912 [Brouwer, 1912h] [Brouwer, 1913c] [Brouwer, 1913d] [Brouwer, 1975, pp. 123–138], already had revealed a shift in his interests towards the philosophy of mathematics. His intuitionist ideas, which had earlier found expression in his thesis [Brouwer, 1907] [Brouwer, 1975, pp. 11–101], were beginning to dominate his thinking. Then the First World War intervened, virtually halting scientific activity in Europe. After the war Brouwer devoted most of his attention to developing extensive foundations for his intuitionistic mathematics. Yet in spite of the fact that he did not continue to do much research in topology, he still kept abreast of new developments in the field. Moreover, during the 1920s quite a few younger topologists came to Amsterdam to be close to the (by then) acknowledged master (see Section 15.7).

12.16 Concluding Remarks on Brouwer's Work on Dimension and Topology

To bring this chapter to a close I shall briefly review Brouwer's work in topology and dimension theory. His output in topology was amazing, particularly in his highly creative years 1910-1913.

The geometrical part of Brouwer's thesis of 1907 brought him close to the field of topology. His research into the Riemann–Helmholtz–Lie space problem and Lie's theory of continuous groups (in reality topological groups) in the thesis, and then a little later his research into Hilbert's fifth problem—to build the Lie theory of finite continuous groups independently of conditions of differentiability—led him to investigate closely related parts of topology. Three mathematicians significantly propelled his work forward: Arthur Schoenflies, Jacques Hadamard, and Henri Lebesgue.

Schoenflies had a decisive effect on the direction of Brouwer's early mathematical research. His contributions to planar topology were the immediate spur for Brouwer to make a direct attack on topological problems. By carefully working through Schoenflies' papers and the second part of his *Bericht* (*Report*) and thereby detecting significant difficulties and errors, Brouwer quickly achieved deeper insights into the field that were beyond Schoenflies. The Schoenflies material gave Brouwer a full array research problems, which he pursued to successful solutions over the years from 1910. Schoenflies unintentionally started him on the path towards reshaping the domain of topology in accordance with a much higher standard of rigor. Brouwer's

[46] On Brouwer's change of interests from topology to intuitionism cf. also W.P. van Stigt's work [Stigt, 1979].

critique of Schoenflies, given in the paper 'Zur Analysis Situs' ('On Analysis Situs') [Brouwer, 1910a], submitted on 14 May 1909, marks the beginning of his transition from Schoenfliesian topology to more rigorous Brouwerian topology.

Hadamard's helpful influence on Brouwer came at a crucial stage in his research. Hadamard started him on the path towards constructing his most important topological tool, the degree of a continuous mapping. Although Hadamard's influence was apparently of only short duration, the contact in Paris just after New Year's Day 1910 came at a most opportune time. It is regrettable that there is not a more detailed record of the mathematical interchanges between the two, but it is clear from Brouwer's praise of the French mathematician that he benefited from their conversations and correspondence. Fortunately, there are the closely related records of their achievements: Hadamard's 'Note sur quelques Applications de l'Indice de Kronecker' ('Note on some applications of the index of Kronecker') [Hadamard, 1910] and Brouwer's 'Über Abbildung von Mannigfaltigkeiten' ('Concerning the Mapping of Manifolds') [Brouwer, 1911b]. After these were produced, Brouwer soon proceeded to go beyond Hadamard in the field that became his specialty.

Lebesgue's influence on Brouwer had a different character from Schoenflies' and Hadamard's. By the time Lebesgue came into Brouwer's view Brouwer was already well advanced in his research program. Lebesgue's topological ideas, though clever and intuitively appealing, were not expressed with precision and rigor like Brouwer's. It is no wonder that Brouwer found them unsatisfactory and required them to be entirely reformulated according to his higher standards of rigor. After the attempts to force Lebesgue to prove the tiling principle adequately, it became apparent that the French mathematician was no match for Brouwer in his chosen field. If one recognizes that Brouwer and Lebesgue were not congenial personalities, then one should not be surprised that their dispute ended badly. Ultimately it must be acknowledged that Brouwer profited from Lebesgue's ideas. They provided him with a necessary clue to proving dimensional invariance in a new way and with the beginnings of other significant topological results. Lebesgue's influence was mainly negative; Brouwer reacted *against* the French mathematician. Nevertheless, through the medium of Brouwer Lebesgue's ideas had a positive effect on the development of topology.[47]

In his remarkably short papers Brouwer produced landmarks of rigor and precision. One should note one very simple aspect of the papers that is quite characteristic of Brouwer's style. The writing is very spare; there is hardly a word too many. His leanness of expression is in total contrast to Poincaré's mode of writing. Poincaré wrote expansively, filling many pages with explanations and examples, all very useful to the reader, to be sure. Not so Brouwer! He was a master of compression, especially when he was writing in German, as he did in his major papers. The syntax of German, particularly with its scope for attributive participial phrases and similar constructions, is conducive to putting many ideas into a few sentences. Brouwer continually applied this feature of German in his compact definitions. For example, examine his definitions of manifold [Brouwer, 1911b, pp. 97–98] and pseudoman-

[47] I should also mention that Poincaré's ideas had an influence on Brouwer in his defining 'dimension' in the manner of Poincaré.

ifold [Brouwer, 1911c, pp. 305–306]. His compressed mode of expression, hardly providing the reader with any explanations or examples as guidance to get through difficult definitions, theorems, and proofs, at first made his works fairly inaccessible to many mathematicians.

In his classic textbook on set theory, *Grundzüge der Mengenlehre* (*Fundamentals of Set Theory*), Hausdorff complains that [Hausdorff, 1914, page 468]:

> ... The brevity of Brouwer's publications, forcing the reader to supply many of his own completions, together with the lack of another detailed and correct presentation, is very much to be regretted.[Ch12, 24]

Hausdorff himself was not one to waste words, but he also knew how to write books that could help readers to learn the mathematical content. He seems to have found it difficult to fathom the depths of some of Brouwer's papers. Nonetheless, he did incorporate Brouwer's 'astonishingly short proof' ('den erstaunlich einfachen Beweis') of the Jordan curve theorem [Brouwer, 1910b] into his textbook [Hausdorff, 1914, pp. 354–357].

In contrast to Hausdorff, Hermann Weyl (1885–1955) found Brouwer's new topological work of immense value and applied it in his very influential 1913 book of lectures, *Die Idee der Riemannschen Fläche* (*The Concept of the Riemann Surface*), the first to give an abstract topological treatment of the concept of Riemann surface [Weyl, 1913].[48] In the forward and introduction to the book Weyl writes admiringly [Weyl, 1913, page V]:

> ... In much greater measure than is apparent from the citations [in this book], I have been encouraged by BROUWER's fundamental topological investigations that have appeared in recent years, and whose intellectual sharpness and concentration must be admired; and I secretly hope that something of the spirit that animates this researcher's work has also come to life in my book.[Ch12, 25]

Clearly Weyl has an appreciation of Brouwer's deep topological work. Later Weyl also came to appreciate and apply Brouwer's intuitionistic thinking, as evidenced in his book *Das Kontinuum* (*The Continuum*) [Weyl, 1918][49]

I should mention another aspect of Brouwer's topological work, because it touches on his philosophical view of mathematics. Although his papers in topology do not conform to the full strictures of his later program of intuitionism, they do have definite constructivist facets. The descriptions of simplicial mappings and degree [Brouwer, 1911b], for example, seem to be constructivist in the basic sense that they delineate mental constructions. In general, the kinds of geometrical arguments that Brouwer put forward in his papers have a definite constructivist tone. Geometry

---

[48] For a more recent treatment of Riemann surfaces cf. [Ahlfors & Sario, 1960].

[49] Another significant early application of Brouwer's results on the invariance of dimension and domain to mathematical physics, specifically to statistical mechanics, should be mentioned. In 1913 Arthur Rosenthal directly applied Brouwer's results in his proof of the impossibility of ergodic systems [Rosenthal, 1913]. Thus quite soon after publication, Brouwer's dimension-theoretic results found an important application outside the domain of pure mathematics. For an interesting history of the foundations of statistical mechanics and ergodic theory cf. [Brush, 1967].

is naturally prone to imaginative exercises in construction. Of course, this type of geometrical constructivism is not the real essence of Brouwer's intuitionist program, which was directed more towards set theory and analysis. But around 1920 he made the claim that, although he had used older nonintuitionistic methods in his earlier publications, he always had striven to prove only those results that could eventually be incorporated into the body of intuitionist mathematics [Brouwer, 1919b]. Hence, he at least intended to comprehend his topological results in his intuitionist program.

Unquestionably, Brouwer made enormous strides forward in topology, though he did not cover the entire field. Freudenthal [Brouwer, 1976, page 436] found it surprising that he never fully employed any of the homological tools that Poincaré had constructed earlier and that have subsequently become so useful. It is not clear why he did not take advantage of the methods of Poincaré, although he must have known about them. There seems to be no explanation for this 'omission' discernible in the Brouwer Archive. However, it should be added that at the beginning of the century Poincaré's ideas did not seem to be adequately founded.

Brouwer left some open ends in his papers. Not everything is proved. For example, he did not give a demonstration of the simplicial approximation theorem, but merely assumed it. Apparently he thought it was fairly obvious and could be left for others to prove.[50] Such omissions, together with the lack of explanations, necessitated a full exposition of the Brouwerian ideas and methods. However, this did not really become available until the appearance in 1935 of the great book *Topologie I* by Paul Alexandroff and Heinz Hopf [Alexandroff & Hopf, 1935]![51] In retrospect I think that it is fair to say that modern topology is forever indebted to the genius of topology of the first years of the twentieth century.

[50] Cf. [Alexander, 1915] [Alexander, 1926].

[51] The book is dedicated to Brouwer, 'L.E.J. Brouwer gewidmet'.

Chapter 13
Problems of Curves and Dimension in the Early Twentieth Century

13.1 Introduction

Point-set theory and more general set theory were fairly well established as fields of mathematics by the beginning of the twentieth century.[1] The ideas and results of Cantor and Dedekind and those developing them further were having considerable impact in the mathematical community. In particular, various mathematicians were taking point-set-theoretic approaches in seeking to solve geometric problems. Schoenflies was one of the prime promoters of point-set theory and general set theory, including advocating point-set-theoretic approaches to geometrical and topological problems, in the last years of the nineteenth century and into the first decade of the twentieth. Then through a set of brilliant counterexamples published in 1910 Brouwer showed the work of Schoenflies and some others to be significantly flawed [Brouwer, 1910a]. From a broader perspective the problems of defining curves, surfaces, continua, and also discontinuous sets in the plane and in Euclidean spaces were significant motivators for the development of point-set topology and also dimension theory in the first decades of the twentieth century. In this chapter I shall examine these geometrical topics in conjunction with the developing point-set theory and point-set topology.

13.2 Problems of Defining Curves and Surfaces

I have discussed Jordan's notion of curve and the Jordan curve theorem in Section 8.4 and Peano's space-filling curve in Section 8.5. Also, I have discussed Schoenflies' point-set-theoretic topological work in Section 8.7 and Brouwer's critique of Schoenflies' work in Section 11.3. However, at this point I need to extend my exami-

[1] For this work I shall not discuss the related issues of the paradoxes of set theory, the various proposed solutions and the attendant logical and mathematical work, and the strict formalization and axiomatization of set theory.

D. M. Johnson, *Theories of Dimension from Ancient to Modern Times*, Sources and Studies in the History of Mathematics and Physical Sciences, https://doi.org/10.1007/978-3-032-02682-8_13

nation of the problem of defining or characterizing the concepts of line or curve and surface to understand its important role in the development of dimension theory.[2]

Jordan's functional definition of a continuous curve according to parametric equations is simple, intuitive, and also useful for the mathematician of the time (see Section 8.4), though Peano's space-filling curve clearly is a strong mark against it, a 'refutation' of the definition, so to speak (see Section 8.5). Jordan's definition is on the face of it too broad.

Hans von Mangoldt (1854–1925) wrote a valuable review, partly methodological and partly historical, of 'The Concept of 'Line/Curve' and 'Surface' ' ('Die Begriffe „Linie" und „Fläche" ') as an article in the *Encyklopädie der Mathematischen Wissenschaften mit Einschluss ihrer Anwendungen* (*Encyclopedia of the Mathematical Sciences Including their Applications*), published in 1906 [Mangoldt, 1906]. The article is an indicator of the problem situation and the progress made by the beginning of the twentieth century on defining some basic geometrical problems, in particular, in relation to classical analysis and to the more recent point-set theory. After opening with a very broad history of the notion of curve, Mangoldt provides a useful summary of specific approaches to characterizing and defining these basic geometrical objects and concepts circa 1906. The methodological development he lays out in his article moves from concepts and definitions embedded in nineteenth-century function theory, mainly complex function theory, to concepts and definitions belonging to point-set theory and geometrical topology—increasingly influential theories at the beginning of the twentieth century.

Mangoldt begins his analysis of the concept of curve with the notion of analytic line or curve deriving from Weierstrass' treatment of the theory of complex functions [Mangoldt, 1906, pp. 132–135]. He considers the representation of an analytic curve by means of a set of equations to be a fundamental mathematical tool for its representation [Mangoldt, 1906, pp. 135–139]. For example, in the domain of three variables, x, y, z, he considers an analytic curve l, as defined by means of two analytic functions $f(x, y, z)$, $g(x, y, z)$, given as the set of value systems x, y, z satisfying the equations:

$$f(x, y, z) = 0; \quad g(x, y, z) = 0.$$

Analysis of these various analytic curves begins by looking for tangents, tangent planes, arc lengths, curvatures, and so forth, all objects with properties and features that are needed in the continuing development of analysis. The scope is classic analytic and/or algebraic curve theory. He also includes real analytic curves.

Mangoldt proceeds to bring in other ways of looking at the concept of curve, including alternatives such as the curve as the image of a function, thereby adding real functions to the mix [Mangoldt, 1906, page 139]. Further, he considers the notion of curve as the path or trajectory of a point [Mangoldt, 1906, pp. 139–143]. In this frame he discusses the Jordan curve theorem. He also considers non-analytic

[2] Cf. McGrew's thesis [McGrew, 1976]. While taking a different approach, the thesis covers material overlapping to some extent with material in this chapter. The thesis has a useful, detailed coverage of space-filling curves. I only recently came upon this thesis. On space-filling curves see the excellent book by Sagan [Sagan, 2012].

curves, such as curves without tangents at any point (curves defined by functions that may be everywhere continuous and nowhere differentiable). He gives an example of a non-analytic Jordan curve.

Finally, he introduces the set-theoretic or point-set-theoretic approach, first by considering a curve as a 'length without breadth' (harking back to Euclid, see Section 2.4) or as the 'boundary of a surface' (also looking back to an ancient notion, see Chapter 2) [Mangoldt, 1906, pp. 143–147], and then in more up-to-date fashion by defining the curve concept as an object satisfying the set-theoretic condition: *a perfect connected point-set without interior points* (*eine perfekte zusammenhängende Punktmenge ohne innere Punkte*) [Mangoldt, 1906, page 143]. For this condition he refers to Schoenflies [Schoenflies, 1898, pp. 201–202], but the reference should go back to Cantor [Cantor, 1883b, §10, pp. 572–576] [Cantor, 1883c, §10] [Cantor, 1932, §10, pp. 190–194] (sse Section 8.2).

He raises specific concerns about the set-theoretic definition. For example, it appears to be too broad, because it includes objects that are not continuous one-one reversible images of a straight line or a circle. He offers two examples to illustrate this point [Mangoldt, 1906, page 144]:

A. The set of all the points as a curve consisting of the circumference of a circle in the plane together with all the points on a fixed radius $\mathfrak{R}$; essentially this is the circumference of the circle plus a single radius (spoke) to its center;
B. The set of points as a curve that includes a variant of the usual topologist's sine curve, that is, the curve defined by $y = \sin\frac{1}{x}$ $[x \neq 0]$, together with the points lying on the y-axis between -1 and $+1$, plus a polygonal line connecting two chosen points, A and B, on the sine curve on opposite sides of the y-axis and not intersecting the other points or parts of the curve.

Each of these set-theoretic curves has a well-defined inside region $\mathfrak{I}$ and a well-defined outside region $\mathfrak{A}$, separated in some sense by the given curve. However, the curve of example A is not a Jordan curve. The boundary of the inside region $\mathfrak{I}$ is the circumference plus the radius (spoke). The boundary of the outside region $\mathfrak{A}$ is just the circumference. The curve is not the common boundary of the two regions defined as the common limit points of both regions. The curve of the second example B presents a unusual situation: it not possible to possible to connect a point $p = (0, y_1)$ on the open segment of the y-axis part of the curve, such that $-1 < y_1 < +1$, to a point in the interior region $\mathfrak{I}$ by means of a polygonal line consisting of straight line segments and remaining entirely within the interior region. This situation represents the problem of accessibility (Erreichbarkeit), a concept that Schoenflies introduced (see Section 8.7). The interior points on the segment of the y-axis are not accessible from the inside region $\mathfrak{I}$. Mangoldt explicitly mentions the point-set topological work of Schoenflies related to closed curves and Jordan curves and provides further details [Mangoldt, 1906, pp. 144–147].

In his article Mangoldt appears to be somewhat 'uncomfortable' with these kinds of unusual counterintuitive 'curves' that arise so easily in point-set theory. His article also reveals a lack of agreement among mathematicians at the time about what are to be considered appropriate curves and what definitions of the concept of curve are acceptable. Consequently, the article provides a variety of diverse definitions for possible curve concepts. Mangoldt reveals himself as very much imbued with the analytic viewpoint deriving from the function theory of the time. In contrast, the parallel development of topological curve and dimension theory, which he partially summarizes, indicates an attempt by others to take an alternative set-theoretic path to defining these fundamental geometrical objects.

In the same year as Mangoldt's article appeared, 1906, William H. Young (1863–1942) and Grace Chisholm Young (1868–1944) published their important monograph on *The Theory of Sets of Points* [Young & Young, 1906] (revised with additional historical material [Young & Young, 2013]). This book is more thoroughgoing in its treatment of set theory and point-set theory deriving from Cantor and Dedekind. In it they devote a full chapter to 'Curves' [Young & Young, 1906, Chapter X, pp. 219–232]. They define a curved arc or curve as follows [Young & Young, 1906, page 219]:

> Def. *A plane set of points, dense nowhere in the plane, such that, given any small norm e, and describing round each point of the set a small region of span less than e, these small regions generate a single region R_e, whose span does not decrease indefinitely, is called a curved arc, or shortly a curve.*

The Youngs examine branches of curves, endpoints, and closed curves. They consider Jordan curves and give their own proof of the Jordan curve theorem. The book also emphasizes measure theory, which is reflected in this chapter. Their background was from that part of analysis. Overall, the chapter on curves is a useful treatment from a point-set-theoretic point of view. Mangoldt's article and the Youngs' book offer a contrast in approaches prevailing at the time.

To gain some perspective on the mathematical developments over time related to defining curves and surfaces, I shall consider another article of a later date from the *Encyklopädie der Mathematischen Wissenschaften mit Einschluss ihrer Anwendungen* (*Encyclopedia of the Mathematical Sciences Including their Applications*). This is the article on 'Point Sets' ('Die Punktmengen') by Ludovic Zoretti and augmented and further edited by Arthur Rosenthal [Zoretti & Rosenthal, 1923], published in 1923. It clearly shows the mathematical progress made over the period from 1906 to 1923, some 17 years, as enabled by the use of point-set-theoretic methods. It contains a lot more on the point-set topology of curves and surfaces. Three sections of the article are of most interest for this work: 12. Linienhalfte Kontinua (Linear/Line-like Continua), 13. Die Begrenzung eines ebensen Gebietes (The Boundary of a Planar Domain), and 13a. Die Begrenzung eines n-dimensionalen Gebietes (The Boundary of an n-dimensional Domain) [Zoretti & Rosenthal, 1923, pp. 907–915, 916–929, 929–931]. Zoretti and Rosenthal state explicitly that they are considering the concept of line/curve in its most general form from a geometrical-set-theoretic standpoint [Zoretti & Rosenthal, 1923, page 907]. Their consideration of the concept of surface is from the same standpoint.

In section 12, Zoretti and Rosenthal note the close connection between the concept of linear continuum and the concept of line or curve. In this case one is on traditional ground, but by this time mathematicians are using many new methods of point-set theory. In the plane one can define a 'line' in three different ways: 1. as a linear continuum ('line without breadth' and so containing no surface element), 2. as a trajectory curve (Bahnkurve) or parametric curve (a continuous movement of a point), and 3. as the boundary of a domain. The authors immediately point out that the three are by no means equivalent. Only 3 is contained in 1. A significant part of the development of the mathematics in this period is directed towards the goal of understanding the logical relationships among the three, often building logical bridges among them using further definitions and explanatory theories.

I should again mention the work of Georg Cantor on the concept of continuum. Note that in a long section of one of his key papers on point-set theory of 1883, Georg Cantor analyzes the notion of continuum, taking a broad historico-philosophical viewpoint [Cantor, 1883b, §10; pp. 572–576] [Cantor, 1883c, §10, pp. 28–32] [Cantor, 1932, §10, pp. 190–194].[3] He offers his own definition of a continuum (Kontinuum) as a perfect, connected point set. A perfect point set is a point set that is equal to its derived set of limit points. A connected set is a kind of 'well-chained' point set, such that for every pair of points of the set, t, t', and every ε there is always a finite ordered chain of points belonging to the set, $t_1, t_2, \ldots, t_\nu$, whereby the distances between successive points in the chain, $\overline{tt_1}, \overline{t_1t_2}, \overline{t_2t_3}, \ldots, \overline{t_\nu t'}$, are all less than ε. For the most part later mathematicians did not adopt Cantor's notion of connectedness as such, but references to Cantor's broader ideas concerning continua persisted for some time.

Zoretti and Rosenthal in their article view linear continua of the plane as continua without interior points. Painlevé [Painlevé, 1900, page 16] and Zoretti [Zoretti, 1905b, page 8] give such sets the name of Cantorian line (Cantorsche Linie, ligne cantorienne). This designation lasted for some time in the literature.

The authors devote most of their section 12 to describing details of the three given ways to define 'line' (or 'curve') and the efforts to show some, normally partial, equivalences. They cite many authors and papers, evidencing the activity in the field. They add Zoretti's notion of a continuum irreducible between two points a and b (continuum irréductible, irreduzibles Kontinuum) to the mix as well (see Section 13.4). In section 13, they are concerned with closed curves, closed Jordan curves, the Jordan curve theorem, and related notions like Schoeflies' accessibility and illustrative examples and striking counterexamples. Brouwer's ground-breaking work comes up often (see the next Section 13.3). In section 13a (mainly added by Rosenthal) they discuss in brief the boundaries of higher-dimensional domains, or surfaces. Again, they reference Brouwer's work.

Overall, Zoretti and Rosenthal's article provides a conspectus of developments on defining curve and surface concepts using geometical-set-theoretic methods. It also lays bare some open problems in the area. These authors, writing around 1923, do not point to dimension theory to indicate how that theory might help solve identified problems. The theory of dimension, say, for 1- and 2-dimensional objects, was not sufficiently developed. The relevant work of Urysohn and Menger was only just emerging.

[3] Cantor's criticism of Bolzano was mentioned in Section 4.6.

13.3 Schoenflies' Point-Set Topology and Brouwer's Critique Revisited

In Section 8.7 I gave a summary of Schoenflies' topological work. His three long papers, his 'Beiträge zur Theorie der Punktmengen I, II, III' ('Contributions to the Theory of Point Sets I, II, III') [Schoenflies, 1903] [Schoenflies, 1904a] [Schoenflies, 1906a], and his synthesis of these in his *Bericht II* (*Report II*) [Schoenflies, 1908b] on *Die Entwickelung der Lehre von den Punktamannigfaltigkeiten. Zweiter Teil* (*The Development of the Theory of Point Manifolds/Sets. Second Part*) lay out in detail his work on the topology of closed curves and simple closed curves in the plane. As I have noted in the earlier Section 8.7, Schoenflies had most success with the converse to the Jordan curve theorem. For this he introduced the concept of accessibility: If M is a connected domain (open set) and t is one of its boundary points, then we say that t is accessible from M if for every point m of M there is a simple path (einfacher Weg) in M leading from m to t [Schoenflies, 1906a, pp. 296–297]. Consider Mangoldt's example B given in the last section of a closed curve constructed in part with the topologist's sine curve, which he seems to have found problematic as a curve. The points of the open segment of the y-axis on this curve are not accessible from the interior region or inner domain $\mathfrak{J}$ that the curve determines.

In his overall assessment of Schoenflies' topological work Brouwer was satisfied with the results on simple closed curves, but he found significant problems with the results on closed curves and domains of a general sort [Brouwer, 1910a, page 422] described in Section 11.3.

Brouwer's counterexamples are the real drivers for his criticism of Schoenflies' results. In this domain of topology the delicate balance between provable results or theorems and counterintuitive examples or counterexamples is critical for acquiring new mathematical knowledge. I need to examine Brouwer's counterexamples a little further, because they arise elsewhere as refutations of seemingly provable results.

As a first example Brouwer provides an example of a simply connected domain without a closed outer boundary curve [Brouwer, 1910a, page 423, Fig. 1]. The diagram reproduced in Figure 13.1 tells all. The boundary curve for domain J wraps around the little circle infinitely often. This example counters or contradicts one of

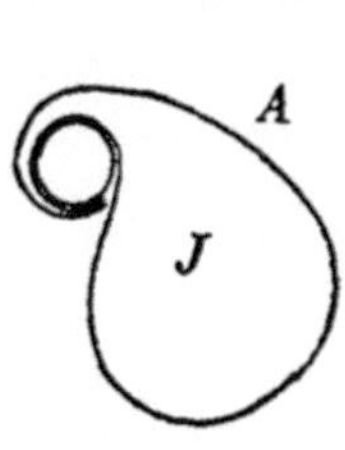

Fig. 1.

Fig. 13.1 Brouwer, Zur Analysis Situs [Brouwer, 1910a, page 423, Fig. 1].

Schoenflies' Theorems, among others [Schoenflies, 1908b, page 124]:

> XIII. For every bounded, simply connected domain $\mathfrak{G}$ there is a closed curve $\mathfrak{C}$ that is identical to its complete boundary or to a subset of the complete boundary.[Ch13, 1]

In Brouwer's example domain $\mathfrak{G} = J$, but the boundary of J is not the expected closed curve (see [Brouwer, 1910a, page 427]).

As a second example Brouwer provides a construction, step by step, of an elegant closed curve that can be divided (or decomposed) into two improper arcs [Brouwer, 1910a, pp. 423–425]. For this Brouwer gives the following definitions for proper and improper arcs [Brouwer, 1910a, page 423, footnote *)]:

> By an *improper arc* I understand an arc defined by two cuts that contains the completely closed curve, and in the opposite case I speak of a *proper arc*.[Ch13, 2]

Figure 13.2 shows the first steps of the construction. We begin with the outer boundary of the large rectangle, and we push black 'channel' or 'canal' areas step by step into the large rectangle, proceeding from the middle black channel on the right of the figure down the right side to form additional channels in succession, adding more boundary points to the outer curve. These black areas/channels plus the area outside the rectangle will form the outer domain. We also create red channel areas step by step on the inside of the large rectangle to create an inner domain. We proceed with this channel construction alternating between black and red, so that the large 'island' rectangle is taken up by more and more of the channels. The result at the limit is a boundary curve K common to an outer domain with black channels pushing into the rectangle together with the outer region and an inner red domain composed of the red channels. That is, the complement of the boundary curve K contains two domains, an inner red one and outer black one including the area outside the rectangle, and so K is a closed curve. Every point of the black domain within the rectangle is arbitrarily close to a point of the red domain, and every point of the red domain is arbitrarily close to a point of the black domain. Brouwer describes this construction very precisely. If we choose the two points Q_1 and Q_1' in the lower right corner of Figure 13.2 as cut points accessible from the inner domain, then we observe that they determine two improper arcs of the boundary curve. Brouwer uses this second example as a direct critique of Schoenflies' proof of another theorem (Theorem XIV), but I shall not go into the details [Brouwer, 1910a, page 427].

As a third example, related to the second, Brouwer offers an unusual example of a proper arc that can be divided (or decomposed) into two arcs that are identical with the entire arc [Brouwer, 1910a, page 425]. For this we have the beginnings of the construction in Figure 13.3. The figure results from taking two somewhat modified copies of Figure 13.2, slightly adjusted to let the red channel of the upper half join the red channel of the lower half, and so joining the two halves together as in Figure 13.3. Otherwise, the black and red channels are constructed step by step as in the previous example. The conclusion is that the set K formed by the boundaries of all red and black areas together with their limit points can be extended to form a closed

Fig. 13.2 Brouwer, Zur Analysis Situs [Brouwer, 1910a], Figure 2.

curve with the inclusion of the part of the lower line above which the red area begins. Thus, it can be regarded as an actual arc of that closed curve. However, as such it can be divided into two curves, both from the inner and from the outer area, which are equal, by closure, to the whole itself. This example serves as a broader criticism of Schoenflies' topological work, so Brouwer does not employ it in a specific way.

As a fourth example Brouwer offers an example of a closed curve that cannot be divided (or decomposed) into two proper curved arcs [Brouwer, 1910a, page 426]. The example in Figure 13.2 serves as the example. This is an indecomposable space or continuum, which Kuratowski defines as follows [Kuratowski, 1966, page 204]:

> Definition. A space $\mathfrak{X}$ is *indecomposable* if it is connected and is not the union of two closed connected sets different from $\mathfrak{X}$.

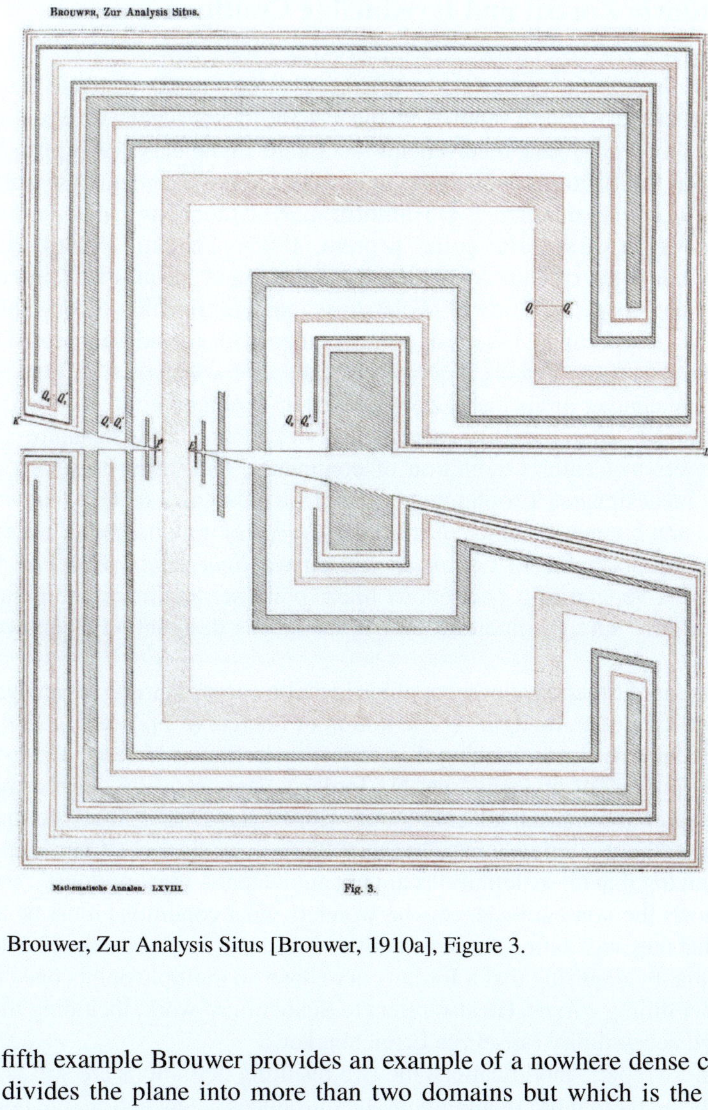

Fig. 13.3 Brouwer, Zur Analysis Situs [Brouwer, 1910a], Figure 3.

As a fifth example Brouwer provides an example of a nowhere dense connected set that divides the plane into more than two domains but which is the common boundary of all the domains [Brouwer, 1910a, page 427]. The Figure 13.3 provides such an example. This counterintuitive case is quite surprising!

Brouwer gives a very specific point-by-point critique of Schoenflies's topology over several pages [Brouwer, 1910a, pp. 427–434]. He summarizes his findings at the end, enumerating the counterexamples and stating some as yet unproven and so doubtful results [Brouwer, 1910a, page 434].

13.4 Ludovic Zoretti and Irreducible Continua

I shall consider next the work of Ludovic Zoretti[4] (1880–1948) related to the problem of defining curves. Zoretti studied at the Faculté des sciences de Paris (Faculty of Sciences of Paris) and received his doctorate in January 1905 with a thesis entitled, 'Sur les fonctions analytiques uniformes qui possèdent un ensemble parfait discontinu de points singuliers' ('On uniform analytic functions that possess a perfect discontinuous set of singular point' [Zoretti, 1905a] [Zoretti, 1905b]. His thesis committee consisted of Paul Appell, Paul Painlevé, and Jacques Hadamard. In the work he makes a special study of Weierstrass' analytic function theory and, as the title suggests, supporting methods of point-set theory for considering sets of singular points.

In the first chapter of his thesis Zoretti takes up problems concerned with perfect sets in the plane ('Les ensembles parfaits plans') [Zoretti, 1905b, Chapitre II, pp. 4–13]. He refers to Cantor's definition of continuum, and, using this as a point of departure, he designates a continuum in the plane either as a *surface* (*superficiel*) if it contains non boundary points, or as *linear* (*linéaire*) if all its points are boundary points. He adopts the term *Cantorian line/curve* (*ligne cantorienne*) or just *line* (*ligne*) for his version of a continuous linear point set (or linear continuum). His teacher Painlevé surely influenced him in using this designation [Painlevé, 1900, page 16].

For Zoretti's further development of his ideas on curves I turn to his analysis in his short paper 'La notion de ligne' ('The notion of line/curve') [Zoretti, 1909]. Again, Zoretti is concerned with defining this concept. He frames his analysis around the existing definitions of line or curve: (1) Jordan's parametrically defined trajectory, and (2) Cantor's definition of continuum scoped down to a linear continuum, a continuum without interior points, and thus 'ligne cantorienne' ('Cantorian line')

Recognizing that these definitions are not equivalent—the topologist's sine curve clearly shows the non-equivalence—he wonders what conditions must be added to each, so that they will define the same entities (i.e., have the same defined extension). He notes that by requiring that a Jordan curve have no multiple points one can avoid Peano space-filling curves. He also refers to Schoenflies' work, including the notion of all-sided accessibility (allseitige Erreichbarkeit).

With these typical modifications the two resulting definitions are still not equivalent. To bring them more in alignment, he introduces a new notion, the notion of a continuous set (or continuum) irreducible between two points [Zoretti, 1909, page 487]:

> Let E be a continuous set (a continuum), and a and b be two points of the set (continuum). I say that *this set (continuum) is* IRREDUCIBLE *between a and b if there does not exist any [proper] continuous subset ( ]proper] subcontinuum) of E that includes a and b.*[Ch13, 3]

[4] I could find only a small amount of biographical material on Zoretti. Cf. Wikipédia in French.

Zoretti immediately gives a few simple examples to illustrate his thinking: a line segment, an arc of a circle, a broken line without self intersections. In contrast, a surface continuum (continuous surface) is never irreducible between any two of its points.

Zoretti's objective in the paper is to demonstrate a selection of theorems about irreducible continua and to show connections with the other definitions of line/curve. He offers theorems connecting irreducible continua with Jordan curves, linear continua, and boundaries of domains.

Brouwer read Zoretti's paper with a careful eye and in 1910 published a critical note on the paper reinforced with counterexamples [Brouwer, 1910h]. Brouwer's note was dated Amsterdam, 5 June 1910. For example, in his note Brouwer points to one of Zoretti's claimed lemmas [Zoretti, 1909, page 489]:

> Let C be an irreducible continuum between a and b, and let c be any point of C; *then it is possible to divide C in a unique way into two continua having c as the only point in common* (and the two, by implication, are irreducible continua, one between a and c and the other between b and c)[Ch13, 4]

Brouwer's counterexample to this lemma is the topologist's sine curve, $y = \sin\frac{1}{x}$, $0 < x \leq 1$, completed by the limit points on the y-axis, $-1 \leq y \leq +1$, with $a = (0, 1)$, $b = (1, 0)$, and $c = (0, 0)$ [Brouwer, 1910h, page 565].

Brouwer bluntly asserts that the general structure of an irreducible continuum is not as simple as Zoretti supposes. Consequently, he singles out [Brouwer, 1910h, pp. 565–566] two results stated by Zoretti for direct criticism.

First, he considers Zoretti's assertion that [Zoretti, 1909, page 495]:

> . . . the external boundary of a domain can be divided into two continua having only two points in common a and b.[Ch13, 5]

This assertion is refuted by Brouwer's counterexample of a indecomposable continuum [Brouwer, 1910a, pp. 423–426], discussed in the previous section.

Second, he considers Zoretti's assertion that [Zoretti, 1909, page 496]:

> *The set of points that do not belong to an irreducible continuum form a continuum.*[Ch13, 6]

Brouwer counterexample to this assertion is his example of a closed curve that cannot be divided into two proper arcs [Brouwer, 1910a, page 426].

Zoretti's specific response to Brouwer, juxtaposed with Brouwer's note in the publication, though dated several months later as Grenoble, 14 October 1910, was short and rather weak [Zoretti, 1910a]. Yet, in the interval of time between the two notes Zoretti had continued to explore the topic of curves. He elaborated his new ideas on irreducible continua in a longer paper, 'Contribution à l'étude des lignes cantoriennes' ('Contribution to the study of Cantorian lines/curves'), published in 1913, but dated earlier at Grenoble, 11 July 1910 [Zoretti, 1913].[5] In the paper

[5] Cf. Zoretti's note in the *Comptes rendus* (*Proceedings*) of the Academy of Sciences, Paris from the session of 18 July 1910 [Zoretti, 1910c], which effectively is his response to a critical note of Janiszewski [Janiszewski, 1910b].

Zoretti sees the great value of applying point-set theory to geometrical problems of analyzing curves. He regards Cantorian lines or curves as providing a better perspective on this domain than Jordan curves. He gives two examples to illustrate the greater extension of Cantorian curves over Jordan curves [Zoretti, 1913, page 245]:

- The topologist's sine curve: $y = \sin \frac{1}{x}$, plus the segment of the y-axis between -1 and 1;
- The set constructed, beginning with two perfect nowhere dense sets on two segments $[0, 1]$ at right angles to one another, then drawing segments of length 1, thereby covering a unit square with a set of crosshatched lines having little open squares in the interstices.

Both of these examples are Cantorian lines, but certainly not Jordan curves.

Zoretti devotes most of the paper to analyzing irreducible continua, offering a variety of results on irreducible lines or curves [Zoretti, 1913, pp. 247–263], along with a few comments and results on tangents [Zoretti, 1913, pp. 263–265] and the extension of the definition of Cantorian lines to (3 or higher) space [Zoretti, 1913, pp. 265–268]. He includes an example with two irreducible sets having only two points in common and dividing the plane into three regions [Zoretti, 1913, page 260], à la Brouwer [Brouwer, 1910a] and Denjoy [Denjoy, 1910]. In fact, the paper includes quite a few counterintuitive examples to supplement and reinforce the analysis. This paper effectively is Zoretti's extended response to Brouwer's criticism.

13.5 Zygmunt Janiszewski

I shall now turn to the work of Zygmunt (or Sigismond) Janiszewski (1888–1920).[6] He is one of several Poles who made significant contributions to set theory, topology, logic, and the foundations of mathematics in the early years of the twentieth century. During that time Polish mathematics flourished due to the efforts of many mathematicians, including Wacław Sierpiński, Stefan Mazurkiewicz, Bronisław Knaster, Alfred Tarski,[7] and Kazimierz Kuratowski.[8]

For his higher mathematical education Janiszewski chose to study away from Poland in Zürich, Munich, and Göttingen with some of the great mathematicians of the time. He completed his higher mathematical education in Paris by studying for his doctorate at the Faculty of Sciences of the University of Paris with Lebesgue as his advisor. His thesis examining committee consisted of Poincaré, Borel, and Lebesgue.

[6] On Janiszewski's biography see Knaster's biographical note in [Janiszewski, 1962, pp. 5–13] and `https://mathshistory.st-andrews.ac.uk/Biographies/Janiszewski/`.

[7] His original family name was Tajtelbaum, but changed to Tarski in 1923.

[8] See, e.g., [Kuratowski, 1980], [Domoradzki & Stawiska, 2018], [Domoradzki & Stawiska, 2019].

At Paris Janiszewski became very much immersed in the problem situation concerning with curves and continua that the mathematics faculty and especially Zoretti had developed before him. Thus, for his doctoral research Janiszewski focused on problems of planar curves, Cantorian curves, and continua irreducible between two points. His papers frequently reference Zoretti, making clear their intellectual interactions and closely related mathematical interests.

Before completing his thesis Janiszewski produced three research notes for the *Comptes rendus* (*Proceedings*) of the Academy of Sciences, Paris [Janiszewski, 1910a] [Janiszewski, 1910b] [Janiszewski, 1911b]. His main thesis has the title *Sur les continus irréductibles entre deux points* (*On continua irreducible between two points*) [Janiszewski, 1911a] [Janiszewski, 1912a]. The thesis is remarkably clear in the way the mathematical material is presented.

In his thesis, following Hilbert, Schoenflies, and Zoretti, Janiszewski uses point-set-theoretic methods as his main tools to investigate curves and related geometrical/topological objects. He covers set-theoretic preliminaries at some length and in a well-delineated and clear fashion. Then he presents the results of his analysis of continua irreducible between two points, simple arcs, and point singularities. He is keenly aware of the mathematical limits Brouwer set for point-set theory through his counterexamples [Brouwer, 1910a]. Janiszewski carefully steers his analysis around these 'hazards' of topology.

He presents several counterintuitive examples of his own to clarify the bounds of his research results. One that is quite striking is an unusual irreducible continuum between two points A and B [Janiszewski, 1911a, pp. 35–36] [Janiszewski, 1912a, pp. 113–114], as depicted in Figure 13.4. Point A is in the center of the left half of the diagram and point B is in the lower right corner. The continuum is composed of (1) an uncountable set of concentric squares, such that the squares and their common diagonals intersect in a perfect nowhere dense set of points, and such that they are joined by spirals in a manner whereby each spiral admits one of them as an asymptotic square (left half of the figure); (2) a line asymptotically approaching a line of the form equivalent to the curve $y^2 = \sin \frac{1}{x}$ (right half of the figure).

This example offers a complex irreducible continuum, AB, which can have multiple points, such that the continuum/curve, if considered in the plane, divides the plane into many parts, which can contain an uncountable set of closed curves.

Knaster [Janiszewski, 1962, page 7] cites a widely recognized and quite useful result of Janiszewski, which is found in his chapter on irreducible continua in the thesis [Janiszewski, 1911a, page 45] [Janiszewski, 1912a, page 123]:

> Lemma. — *Let $\mathfrak{C}$ be a continuum, $\mathfrak{F}$ a closed subset of $\mathfrak{C}$, A a point of $\mathfrak{C} - \mathfrak{F}$ and $\mathfrak{A}$ the set of all points of $\mathfrak{C} - \mathfrak{F}$ that are well chained with A; then the derived set [set of limit/accumulation points] of $\mathfrak{A}$, which is a continuum, has points in common with $\mathfrak{F}$.*[Ch13, 7]

Janiszewski defines [Janiszewski, 1911a, page 51] [Janiszewski, 1912a, page 129] a simple arc AB as an irreducible continuum AB that can be divided into two continua, relative to every point M belonging to it, having only the point M in common. He asserts that for the plane this defined concept corresponds to what Schoenflies calls

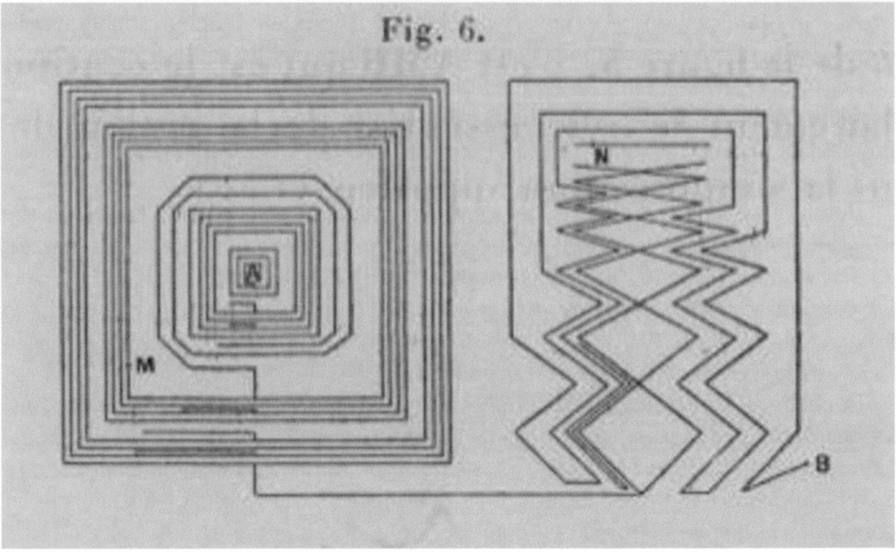

Fig. 13.4 Janiszewski's Unusual Example of a Continuum Irreducible Between Points *A* (Center Left Half) and *B* (Lower Right Corner), [Janiszewski, 1911a] [Janiszewski, 1912a], Figure 6.

a 'simple curved arc' ('einfacher Kurvenbogen') [Schoenflies, 1906a, page 305]. On this basis Janiszewski provides an array of theorems, including a necessary and sufficient condition that a continuum irreducible between two points be a simple arc [Janiszewski, 1911a, page 53] [Janiszewski, 1912a, page 131].

Following his thesis work, Janiszewski gave a short talk on 27 August 1912 at the Fifth International Congress of Mathematicians held in Cambridge, England, covering the concepts of 'line' and 'surface' [Janiszewski, 1912b]. This is a programmatic work laying out the problem situation of that time as he views it.

The presentation starts from his point-set-theoretic framework and approach to geometry. He is seeking a way to distinguish among these fundamental geometric objects, lines/curves and surfaces, as sets of points. In effect he apologizes for the negative character of his remarks. In simple cases one can easily distinguish between lines/curves and surfaces. Yet, in the complicated cases constructed in set-theoretic investigations, it is very much harder to make these distinctions. He queries, rhetorically: Are there continua that can be called neither lines/curves nor surfaces? For the case of two dimensions, the plane, he thinks that there is a definitive answer, asserting that every nowhere dense continuum defines a line (Zoretti's Cantorian line). However, for 3-dimensional space there is no universal, or universally accepted, definition. Generally, a point set being a nowhere dense continuum does not serve as a adequate definition.

In passing he mentions Fréchet's work on dimension type (see Section 10.4). However, he does not see Fréchet's theory as offering a good solution to the problem at hand.

Janiszewski offers two complex examples for which it is very difficult to decide if the constructed objects are curves or surfaces. He carefully lays out the constructions in his presentation [Janiszewski, 1912b, pp. 127–128]. In the end one is left with an open problem.

Janiszewski did further work in this domain (see [Janiszewski, 1962]), but it is not necessary to delve into it further. He did write a paper much later in 1920 with Kuratowski on indecomposable continua [Janiszewski & Kuratowski, 1920].

Janiszewski was one of the founders of the modern Polish school of mathematics. He was instrumental in the founding of the journal *Fundamenta Mathematicae*, primarily devoted to set theory, mathematical logic, and foundations of mathematics. He died quite young in January 1920 during the influenza epidemic. He was only 31.

13.6 Wacław Sierpiński's Triangle Curve—Every Point of the Curve is a Ramification Point

In this section and the next one I shall consider two counterintuitive examples that are important for the development of the topology of curves: the Sierpiński Triangle or Gasket and the Sierpiński Carpet. They are due to the influential Polish mathematician Wacław Sierpiński. Wacław Sierpiński (1882–1969)[9] was a highly productive mathematical researcher. His selected papers in topology and number theory span some three large volumes [Sierpiński, 1974] [Sierpiński, 1975] [Sierpiński, 1976]. Additionally, he produced several widely-read textbooks and monographs. He was a mathematical educator as well as a mathematical researcher. He was the doctoral supervisor for Stefan Mazurkiewicz and eventually for Kazimierz Kuratowski (after Janiszewski's death). After the First World War Sierpiński came to Warsaw University and, along with Zygmunt Janiszewski and Stefan Marurkiewicz, founded the Polish School of Mathematics. They started *Fundamenta Mathematicae* in 1920, soon to be a leading outlet for the publication of papers on set theory, topology, logic, and foundations of mathematics.

I shall consider first the Sierpiński Triangle curve, a curve that is both a Cantorian line/curve and a Jordan curve. Moreover, every point of this curve is what Sierpiński calls a *ramification point* or *branch point* (*point de ramification*). He defines a *ramification/branch point p* of a curve $\mathfrak{C}$ as a point of the curve for which there exist three [or more] continua, subsets of $\mathfrak{C}$, having in pairs the point p and only this point in common. Sierpiński described his unusual curve in a paper of 1915, which he extended in a Polish version of 1916 (also later translated into French) [Sierpiński, 1915] [Sierpiński, 1916a] [Sierpiński, 1975, pp. 99–106].

Sierpiński's extended exposition of the Triangle is wonderfully geometrical: It includes a sequence of eight figures (see Figures 13.5, 13.6, 13.7, 13.8) that explains the step-by-step construction. The construction reveals some of Sierpiński's

[9] For biographical material on Sierpiński see [Engelking, 1998] [Kuratowski, 1980, pp. 167–171].

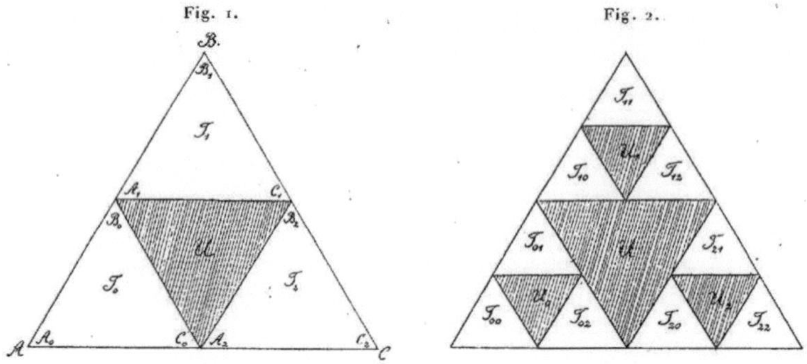

Fig. 13.5 Sierpiński's Triangle [Sierpiński, 1915], Figures 1 and 2.

geometrical intuition and thinking.[10] Begin with an equilateral triangle and join the midpoints of each side of the triangle by three straight lines, deriving four equilateral triangles. Delete the middle subtriangle, leaving the three outer subtriangles (Figure 13.5, left). Continue the construction within the three remaining subtriangles, deleting the middle subtriangles, thereby deriving nine subtriangles (Figure 13.5, right). Iterate this procedure. Sierpiński's Figures 1 and 2 in Figure 13.5 have iterations 1 and 2. At iteration i there will be a set of 3^i subtrianlges, all equilateral and congruent. Continuing the procedure as i goes to infinity, the intersection of all the sets obtained at each iteration, $S = \cap_{i\to\infty} S_i$, yields Sierpiński's Triangle.

The set S is closed and connected, and it contains at least one point, so it is a continuum. It also has an empty interior, so that it is a Cantorian line/curve. The geometric construction shows that S has the property of self-similarity of S: it contains copies of itself of many different sizes, including infinitely many copies when going to the limit.

In complementary fashion Sierpiński builds an actual curve C, which is piecewise linear. This construction starts with the structure given by the triangle (see Figure 13.6). In Figure 13.6, left, the bolded line $S'_0S'_1S'_2S'_3$, which connects four given points S'_0, S'_1, S'_2, S'_3, contains one edge from each of three subtriangles. Iterating this procedure, as shown in Figues 13.6, right, 13.7, 13.8, implies that at step i the line L_i will contain one edge belonging to each of the 3^i subtriangles of the level i iteration. The resulting curve C is a Jordan curve. Moreover, Sierpiński's Triangle and the Jordan curve C are provably identical [Sierpiński, 1915] [Sierpiński, 1916a] [Magrone, 2020, page 21].

Sierpiński proves that each point of the curve, except for the three vertices, A, B, C, is a ramification point of order 3. To make every point a ramification point, join together six Sierpiński's Triangles to obtain a hexagon or hexagonal figure or extended curve. In this way the three initial vertices will become finite ramification points.

[10] Cf. [Magrone, 2020] for a detailed exposition of Sierpiński's paper.

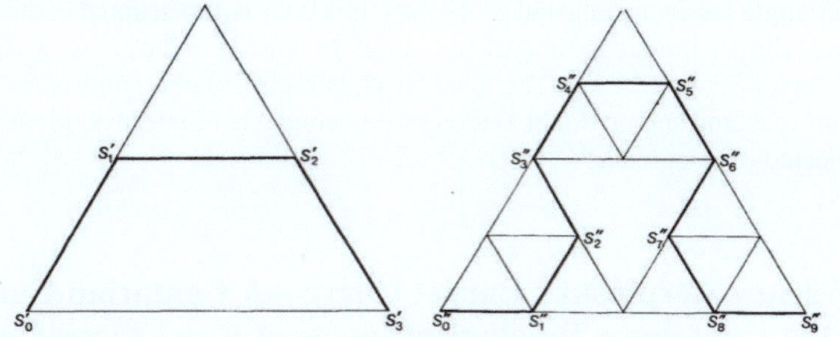

Fig. 13.6 Sierpiński's Triangle [Sierpiński, 1915], Figures 3 and 4.

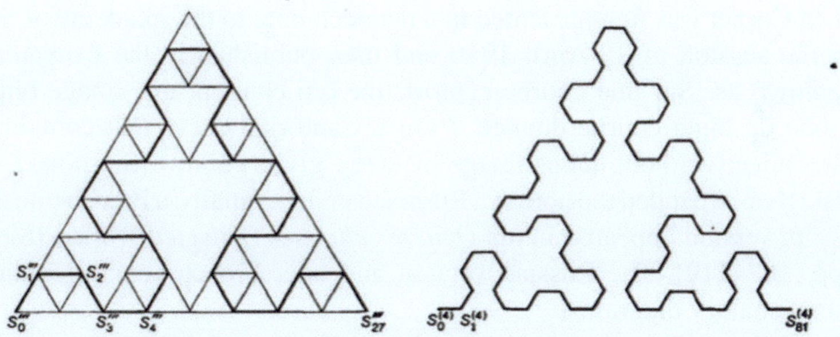

Fig. 13.7 Sierpiński's Triangle [Sierpiński, 1915], Figures 5 and 6.

Fig. 13.8 Sierpiński's Triangle [Sierpiński, 1915], Figures 7 and 8.

The Triangle has many unusual properties. It is both 1-dimensional in the sense of topological dimension theory and of fractional dimension in the sense of Hausdorff dimension (see Section 16.11). The Triangle construction yields some consequences that are far reaching and no doubt extend well beyond what Sierpiński himself might have expected or imagined.[11]

13.7 Wacław Sierpiński's Carpet Curve—A Cantorian Curve That Contains a Topological Image of Every Plane Curve

I shall now consider the other curve mentioned at the beginning of the previous section, also due to Sierpiński: the Sierpiński Carpet or Sierpiński Carpet Curve. The Sierpiński Carpet was first presented in a research note to the Academy of Sciences, Paris at the session of 25 April 1916 and then published in the *Comptes rendus* (*Proceedings*) as 'Sur une courbe cantorienne qui contient une image biunivoque et continue de toute courbe donnée' ('On a Cantorian curve that contains a one-one onto [bijective] continuous image of every given curve') [Sierpiński, 1916b]. Sierpiński then expanded the note in a Russian version. Finally, a French translation of the Russian version appeared in his *Oeuvres choisies* (*Selected Works*) [Sierpiński, 1975, pp. 107–119]. The Russian version and later French version contain quite useful, explanatory diagrams.

Sierpiński's special curve is a universal curve for bounded Cantorian curves in the plane. He constructs it in a manner related to Cantor's ternary set or discontinuum, i.e., Cantor's example of a perfect, nowhere dense set on a line segment. Figure 13.9 reproduces Sierpiński's two diagrams, Figure 1 and Figure 2, showing the first two steps of the construction.

First, one divides the large square into 9 subsquares and removes the interior points of the middle subsquare to yield Sierpiński's Figure 1. Then, one divides each of the the remaining 8 subsquares into 9 smaller subsquares and removes the interior points of the middle subsquare in each to derive Sierpiński's Figure 2. One continues with this process to construct the Sierpiński curve at the limit. The removal of the interior points from all the middle subsquares has the effect of removing all interior points from the resulting limit set. The result is a bounded Cantorian curve, a bounded planar set with no interior points.

The aim of Sierpiński's construction is to provide [Sierpiński, 1975, page 108]:

> *... an example of a plane curve C_0, simultaneously Cantorian and Jordanian, such that for any arbitrary plane Cantorian curve C, given in advance, there exists a one-one onto (bijective) continuous image C' all of whose points are points of the curve C_0.*[Ch13, 8]

[11] Cf. `https://mathworld.wolfram.com/SierpinskiSieve.html`;
`https://encyclopedia.pub/entry/32860`;
'Sierpiński gasket' article in the *Encyclopedia of Mathematics*.

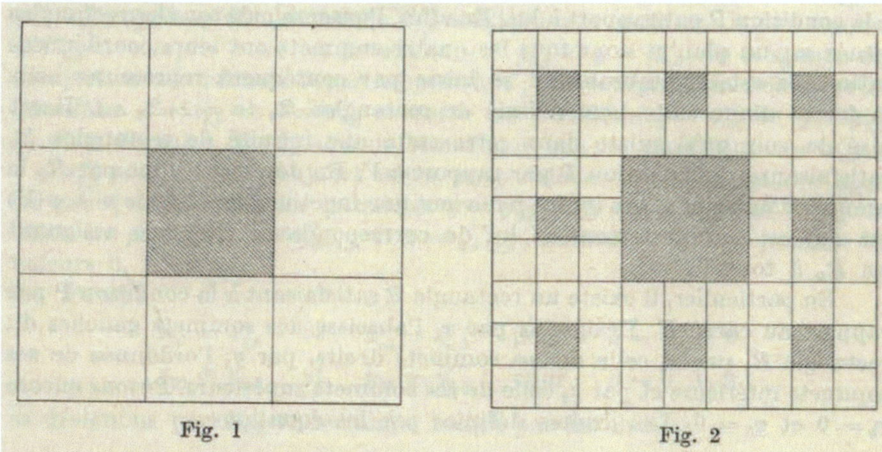

Fig. 13.9 Sierpiński's Carpet [Sierpiński, 1915], Figures 1 and 2.

In his note Sierpiński proceeds to prove this assertion and some additional facts about this remarkable universal plane curve. A few years later in 1926 Karl Menger developed a general 3-dimensional universal curve, the Menger Sponge, based on Sierpiński's Carpet and its construction (to be discussed in Section 15.7).

13.8 Stefan Mazurkiewicz, Hans Hahn: Characterizing Peano Curves

I shall consider next the topological work of Stefan Mazurkiewicz (1888–1945)[12] related to problems of curves and dimension theory. His research in this domain is broader and more extensive than that of Janiszewski, and it covers a longer period. Regrettably, Janiszewski died during the influenza epidemic in January 1920, as mentioned earlier, while Mazurkiewicz had a longer career in mathematics and so was able to make numerous contributions to topology as well as to other parts of mathematics. He died under the difficult circumstances at the end of the war in 1945.

Mazurkiewicz was from Warsaw, but, like his fellow Pole Janiszewski, he studied mathematics away from his native city of Warsaw in Kraków, then Munich and Göttingen. For his doctorate he studied in Lviv under the supervision of Wacław Sierpiński on the topic of (Peano) space-filling curves. Parts of his thesis work are to be found in the papers of 1913 [Mazurkiewicz, 1913a] [Mazurkiewicz, 1913b]

[12] For biographical information on Mazurkiewicz see Kuratowski's short biography in [Mazurkiewicz, 1969, pp. 9–18], [Pol, 1998], [Knaster, 1990], `https://mathshistory.st-andrews.ac.uk/Biographies/Mazurkiewicz/`.

[Mazurkiewicz, 1913c], while the paper of 1915 [Mazurkiewicz, 1915] is closely related.

Mazurkiewicz' early work on space-filling curves led him to his contributions towards the appropriately named Hahn–Mazurkiewicz Theorem (or Hahn–Mazurkiewicz–Sierpiński Theorem, as I shall discuss later). Deriving from his early research work he also had original ideas on dimension (to be discussed shortly).

Mazurkiewicz interacted with Janiszewski. Indeed, his first published paper exposits an alternative proof for a theorem of Janiszewski on irreducible continua [Mazurkiewicz, 1910, page 296] [Mazurkiewicz, 1969, page 27]:

> THEOREM. *Let C be a continuum, p and q two of its points; then there exists at least one irreducible continuum pq that is a subset of C.*[Ch13, 9]

Mazurkiewicz' second published paper, dating from 1913 and taken from his thesis, concerns 'punctiform' or hereditarily discontinuous sets, that is, sets that contain no continuum of more than a single point [Mazurkiewicz, 1913a]. Janiszewski had introduced the notion of punctiform sets in his Thesis [Janiszewski, 1911a, page 7] [Janiszewski, 1912a, page 85].[13] The objective of Mazurkiewicz' paper is to prove the following result concerning 'punctiform' sets in *n*-dimensional Euclidean space [Mazurkiewicz, 1913a] [Mazurkiewicz, 1969, page 30]:

> *Every continuum in an n-dimensional Euclidean space can be decomposed into two punctiform sets (i.e., not containing any subcontinuum).*[Ch13, 10]

Pol in his review of Mazurkiewicz' mathematical contributions [Pol, 1998, pp. 421–422] extols the originality of Mazurkiewicz' new method of proof. This result is just one indicator that various types of discontinuous sets were coming to the fore as definable objects of topological and dimension-theoretic research. I shall give more details about this development in Section 13.9 and in subsequent chapters

With two papers of 1913, 'On the arithmetization of continua' (in two parts; in Polish with French summary, 'Sur l'arithmétisation des contnuus'), and another of 1915, 'On the multiple points of the curves that fill a planar area' (in Polish with French summary, 'Sur les points multiples des courbes qui remplissent une aire plane'), Mazurkiewicz enters the domain of what was to become the Hahn–Mazurkiewicz Theorem and closely related dimension theory [Mazurkiewicz, 1913b] [Mazurkiewicz, 1913c] [Mazurkiewicz, 1915] [Mazurkiewicz, 1969, pp. 37–41, 42–45, 48–53]. The first two of this list are part of his thesis, and the third is closely related.

In the first paper of the three Mazurkiewicz states and proves the following [Mazurkiewicz, 1913b] [Mazurkiewicz, 1969, page 37]:

> THEOREM. *In order that an arbitrary continuum Γ of a Euclidean space of n dimensions be a Jordan curve (i.e., that the coordinates of the points of Γ be continuous, single-valued functions of a parameter t over a closed interval), it is necessary and sufficient*

[13] Cf. [Kuratowski, 1968, page 189].

I) *that* Γ *be bounded;*
II) *that for every pair of points* A, B *of* Γ *one can find a simple arc* (AB) *contained in* Γ *and connecting the two points;*
III) *that the diameter of* (AB) *approaches zero as the point* B *approaches* A.[Ch13, 11]

This theorem has a very clear intuitive geometrical flavor. Mazurkiewicz' proof of necessity of the three conditions is a direct analysis of the parametrization of Γ. His proof of sufficiency of the three conditions begins with an arbitrary Peano curve filling an n-dimensional unit cube containing Γ and then proceeds to show that the conditions yield a parameterized Jordan curve. Close examination of the proof demonstrates that by positing the given three conditions, Mazurkiewicz is able to expose the pieces of the space-filling curve construction. Thinking about the successive stages towards constructing the curve, one can intuitively 'see' the arcs connecting pairs of points as the process goes to the limit.

In the second paper of the three Mazurkiewicz replaces simple arcs with arbitrary continua for the statement and proof. In this version he demonstrates the following [Mazurkiewicz, 1913c] [Mazurkiewicz, 1969, page 42]:

> THEOREM. *A bounded continuum* Γ *of the Euclidean space of* m *dimensions is a Jordan curve if one can make every pair* a, b *of its points correspond to a continuum* $C(a, b)$ *containing* a *and* b*, being contained in* Γ *and such that the diameter of the continuum* $C(a, b)$ *approaches zero along with the distance between the points* a *and* b *approaching zero.*[Ch13, 12]

For this version one needs to substitute geometric intuition about continua $C(a, b)$ for geometric intuition about simples arcs. Taking the two results together from the two papers, he has shown in effect that locally connected continua are arcwise connected and coincide with continuous images of the interval—this is a very attractive characterization.

In his third paper of the three, published in 1915, Mazurkiewicz takes an excursion into dimension theory. His thesis supervisor Sierpiński had posed the problem that he analyzes [Mazurkiewicz, 1915] [Mazurkiewicz, 1969, pp. 48–53]. He refers to Lebesgue's deficient work on dimensional invariance and Lebesgue's stated theorem on the orders of the multiple points of space-filling curves as related to the tiling principle for dimension [Lebesgue, 1911a]. Mazurkiewicz asserts that while he knew the theorem to be true, he also knew from Brouwer's paper [Brouwer, 1911a] that Lebesgue's proof was incomplete. His statement of the Lebesgue theorem is as follows [Mazurkiewicz, 1915] [Mazurkiewicz, 1969, page 48]:

> ... theorem ...
>
> 1) a curve that covers a domain of n dimensions necessarily has multiple points at least of order $n + 1$, and
> 2) there exists a curve covering a domain of n dimensions that has no multiple points of order great than $n + 1$.[Ch13, 13]

In his paper Mazurkiewicz does not give general proofs for the two parts of Lebesgue's stated theorem, but limits himself to proofs for the plane. At this point the general case may have been too difficult for him to find an avenue to a proof.

Nevertheless, generalizing from the reasoning he uses in the limited proofs that he gives, he is able to offer a more general definition of dimension applicable to arbitrary closed sets in n-dimensional Euclidean space [Mazurkiewicz, 1915, page 118] [Mazurkiewicz, 1969, pp. 52–53]:

> Given a (closed) set A, one can represent A by considering the set of all continuous curves:
>
> $$J: \quad x_i = f_i(t),\ i = 1, 2, \ldots, n, \quad t \subset E,$$
>
> (where E is a closed linear discontinuous set). This set will be denoted by $Z(A)$.
>
> *Then there exists a unique integer d such that:*
>
> a) *all the curves of the set* $Z(A)$ *have multiple points of order at least* $d + 1$,
> b) *at least one of the curves of the set* $Z(A)$ *does not possess multiple points of order greater than* $d + 1$.
>
> The number d will be called *the dimension* of the set A.[Ch13, 14]

Mazurkiewicz immediately asserts four consequences of his definition of dimension:

1. dimension is an invariant of analysis situs;
2. punctiform sets have dimension $d = 0$, and, conversely, if $d = 0$, then the set is punctiform;
3. the dimension of plane continua is $d = 1$ or $d = 2$ follows if they do not contain or they do contain interior points;
4. if the Lebesgue theorem in the general case is true, then the n-dimensional cube has dimension $d = n$.

Mazurkiewicz is certainly aware that his development of his concept of dimension is incomplete. He says that it needs to be elaborated with many more provable properties to justify it. It certainly is not a fully developed theory. However, it has distinct similarities with the definitions of Brouwer, Urysohn, and Menger. Note that, in publishing the paper, he shows no awareness of Brouwer's definition and contribution of 1913 [Brouwer, 1913b].

Mazurkiewicz collected and elaborated his results on Jordan curves in a more extensive paper of 1920 intended as a systematic exposition, 'Sur les lignes de Jordan' ('On Jordan lines/curves'). It appeared in the very first volume of *Fundamenta Mathematicae* [Mazurkiewicz, 1920a] [Mazurkiewicz, 1969, pp. 76–113]. By then he knew of Hahn's equivalent work (shortly to be described below).

The following summary gives a few fundamentals of the paper. The assumed frame is the n-dimensional Euclidean space and sets and continua contained in that space. He does mention the possibility of generalizing the reasoning to more abstract spaces [Mazurkiewicz, 1920a, page 166]. The usual distance function between two points x and y, $\varrho(x, y)$, for Euclidean n-space is well defined. Given a set A in the n-space, the diameter (diamètre) δ of A, denoted as $\delta[A]$, is defined as the upper bound of the distances between pairs of points in the set A. The diameter may be finite or infinite (∞). The usual properties for distances between points and diameters of sets follow. A continuum is a closed, well-chained (bien enchaîné) or connected set containing more than one point. With regard to being well-chained he says that he is adopting Jordan's definition rather than Cantor's. Jordan's concept of being well-chained or connected for Euclidean spaces[14] is the direct ancestor of the modern commonly-accepted concept of connectedness for more abstract spaces, which Lennes, Riesz, and Hausdorff developed (see Section 12.12).

For further analysis Mazurkiewicz needs to define a more complex distance function between points relative to a continuum containing them within the Euclidean n-space. Given two points x and y belonging to a continuum A, he uses the notation

$$\mathfrak{C}(x + y, A)$$

to mean a (typical) subset of A that is closed and connected and contains x and y. In fact, any $\mathfrak{C}(x + y, A)$ will be a continuum provided it does not reduce to a single point x (so that $x = y$). He gives the following [Mazurkiewicz, 1920a, page 168]:

> ***Definition.*** Given A, a continuum containing the points x and y, I call *the distance between x and y relative to the continuum A* to be the lower bound of the numbers $\delta[\mathfrak{C}(x + y, A)]$.[15] I denote this number by the symbol $\varrho_A(x, y)$.[Ch13, 16]

He completes the definition with the following: if any $\mathfrak{C}(x + y, A)$ is not bounded, then:

$$\varrho_A(x, y) = \infty;$$

on the other hand, if there exists at least a bounded $\mathfrak{C}(x + y, A)$, then $\varrho_A(x, y)$ is the number defined by the conditions:

I. For every $\mathfrak{C}(x + y, A)$

$$\varrho_A(x, y) \leq \delta[\mathfrak{C}(x + y, A)],$$

II. There exists for every $\varepsilon > 0$ a $\mathfrak{C}(x + y, A)$ such that

$$\varrho_A(x, y) + \varepsilon > \delta[\mathfrak{C}(x + y, A)].$$

With this extended background of definitions Mazurkiewicz gives the [Mazurkiewicz, 1920a, page 170]:

---

[14] Jordan's definition [Jordan, 1893, page 25] is:

> We say that a perfect and bounded set is connected when it cannot be decomposed into several separate perfect sets.[Ch13, 15]

[15] Considering all the admitted $\mathfrak{C}(x + y, A)$.

Definition. Given a continuum A and a point a of A, I call *the oscillation of the continuum A at the point a* the number

$$\sigma_A(a) = \overline{\lim_{x,y \to a}} \varrho_A(x, y).^{\text{Ch13, 17}}$$

Given the assumed frame, the oscillation number satisfies the inequality:

$$\sigma_A(a) \geq 0.$$

A point a of continuum A is defined as being of *first type (premier genre)*, if

$$\sigma_A(a) = 0;$$

it is defined as being of the *second type* (*second genre*), if

$$\sigma_A(a) > 0.$$

He proves that the type of a point is an invariant of analysis situs.[16] Mazurkiewicz' principal result is [Mazurkiewicz, 1920a, page 191]:

Theorem V. *For a bounded continuum A to be a Jordan curve, it is necessary and sufficient that all the points of A are of first type.*[Ch13, 18]

The theorem has an attractive intuitive geometrical character. Having every point a of a continuum A be of first type with respect to the oscillation of A intuitively means that points arbitrarily close to a can be immediately connected to a by continua in A, in accordance with the defining limit process.

Over his career Mazurkiewicz made a great many significant contributions to topology and dimension theory. Pol's summary [Pol, 1998] shows the extensiveness and variety of his mathematical work.

I shall turn now to the work of Hans Hahn (1879–1934).[17] Hahn made important and extensive contributions to mathematics; his *Collected Works* span three volumes [Hahn, 1995] [Hahn, 1996] [Hahn, 1997]. He had a special interest in the theory of curves.[18] His independent contribution to the Hahn-Mazurkiewiicz Theorem complements the contribution of Mazurkiewicz. Beyond his mathematics, Hahn became a leading light and intellectual leader in the Vienna Circle, a group that sought to reform philosophy, make it more scientific, and provide foundations to philosophy of science.[19]

---

16 Cf. [Janiszewski, 1911a, page 25] and [Janiszewski, 1912a, page 103].

17 For biographical information on Hahn see [Mayrhofer, 1934] [Schmetterer & Sigmund, 1995], also `https://mathshistory.st-andrews.ac.uk/Biographies/Hahn/`, `https://en.wikipedia.org/wiki/Hans_Hahn_(mathematician)`. See Karl Popper's appreciation of Hahn as an educator and researcher in [Hahn, 1995, pp. 1–19].

18 Hahn's papers in this area of mathematics are reprinted in his *Collected Works* I [Hahn, 1995, pp. 283–438]. Cf. Hans Sagan's brief but excellent summary of Hahn's work in this area in [Hahn, 1995, pp. 275–282] and his excellent detailed book on space-filling curves [Sagan, 2012].

19 On the Vienna Circle see [Stadler, 2001] [Stadler, 2015] [Sigmund, 2017] [Edmonds, 2020]. For Hahn's contributions to the Circle see [Hahn, 1980].

In a paper of 1913, 'Über die Abbildung einer Strecke auf ein Quadrat' ('Concerning the mapping of a line segment onto a square'), Hahn offers a refined analysis of Peano space-filling curves—curves that are continuous images of a line segment [Hahn, 1913]. Several other mathematicians had already provided considerable analysis of such curves beyond Peano [Peano, 1890a], including Hilbert [Hilbert, 1891], E.H. Moore [Moore, 1900], Lebesgue [Lebesgue, 1904, pp. 44–45] [Lebesgue, 1905b],[20] and Sierpiński [Sierpiński, 1912a][21] (see also Section 8.5). In his paper Hahn investigates these curves with respect to their multiple points. He shows that for every continuous mapping of the line segment onto a square the collection of multiple points of order 2 (image points on the square with 2 pre-images from the line segment), or double points, is a set having the power of the continuum, while the set of points of order 3, or triple points, is a set that is everywhere dense in the square. Moreover, in constructing a space-filling curve covering a square it is not necessary to have any multiple points of order 4, although some examples of space-filling curves have multiple points of that order.

A natural mathematical question then arises: What characterizes these space-filling curves? Mazurkiewicz addressed this key question in his thesis and related papers. Hahn addressed the question too, and about the same time as Mazurkiewicz, without knowing at first of Mazurkiewicz' work.

Hahn first considered the characterization question in a talk he gave at the Versammlung deutscher Naturforscher und Ärzte (Conference of German Natural Scientists and Doctors) in September 1913 in Vienna. The title he gave his talk is appropriately, 'Über die allgemeinste ebene Punktmenge, die stetiges Bild einer Strecke ist' ('Concerning the most general planar point set that is the continuous image of a line segment') [Hahn, 1914a]. Hahn's lecture was then published in 1914. In that same year he presented to the Academy a more detailed analysis with some generalizations, which he published as 'Mengentheoretische Charakterisierung der stetigen Kurve' ('Set-theoretic characterization of the continuous curve') [Hahn, 1914b].

A background to Hahn's work is a characterization result that Schoenflies derived [Schoenflies, 1908b, Kapital VI], to which Hahn refers. Schoenflies had asked the same leading question that Hahn and Mazurkiewicz asked. So Schoenflies's work was a motivating background, but it was limited to planar continuous curves and not as insightful as the later work [Schoenflies, 1908b, X, page 237].

A key idea that Hahn added to the mix in solving this problem was his definition of a concept of 'zusammenhängend im kleinen' ('connectedness in the small' or 'local connectedness'). The definition itself adds geometrical insight towards understanding the problem and the whole problem situation.

Hahn starts from the notion of a continuous curve, the continuous image of a closed line segment [Hahn, 1914b, pp. 2433–2434]. Following Jordan, he defines a bounded, closed point set as connected if it cannot be the union of two closed sets

[20] Cf. [Sagan, 2012, Chapter 5, pp. 69–83] for a well-motivated exposition of Lebesgue's curve.

[21] Cf. [Sagan, 2012, Chapter 4, pp. 49–68] for a fine exposition of Sierpiński's curve and its extensions and generalizations.

without a common element. A bounded, closed and connected set defines a bounded continuum. On this basis he defines [Hahn, 1914b, pp. 2434–2425]:

> A set M is called *connected in the small [locally connected] at point P* if to every $\varepsilon > 0$ there corresponds an $\eta > 0$ having the following property: every point of M lying in the η-neighborhood of P [22] there is a continuum of M connecting the point with P and lying entirely in the ε-neighborhood of P.
>
> The set M is called *connected in the small [locally connected]*, if it is connected in the small [locally connected] at each of its points.[Ch13, 19]

Hahn's ε-η-style definition of being locally connected at a point P has an intuitive geometrical flavor suggestive of connecting all the points surrounding point P with a 'set of connecting strings' as continua emanating from 'hub' P. It provides the critical building block for showing that a space-filling curve can be the continuous image of a closed line segment.

On this basis Hahn shows that if a bounded continuum is locally connected, then every pair of its points can be connected by an arc contained entirely in the continuum [Hahn, 1914b, page 2436]. In more precise metrical terms he proves [Hahn, 1914b, pp. 2439–2440]:

> III. *If the bounded continuum M is locally connected, then to every* $\varepsilon > 0$ *there is an* $\eta > 0$ *having the following property: Every two points* P' *and* P'' *of* M*, whose distance [between them] is* $< \eta$*, are bound together by a continuous arc belonging to M that lies entirely in the* ε*-neighborhood of each of the two points* P' *and* P''*.*[Ch13, 20]

Hahn's principal result is the following theorem [Hahn, 1914b, page 2455]:

> XIV. *For a planar point set M to be a continuous image of a closed line segment it is necessary and sufficient that M be a bounded locally connected continuum.*[Ch13, 21]

Hahn goes on to generalize his results for Fréchet's normal (V) classes (or spaces) [Fréchet, 1906, pp. 18–30], with a defined metric (nearly metric spaces) [Hahn, 1914b, pp. 2456–2460, 2475–2489]. Additionally, he considers spatial curves [Hahn, 1914b, pp. 2460–2475].

The usual modern definitions of local connectedness are [Kuratowski, 1968, pp. 227, 230]:

> Definition. A topological space is said to be *locally connected at the point* p ... if every open neighborhood G of p contains a connected neighborhood of p ....
>
> Definition. A space which is locally connected at every point is said to be locally connected.

[22] i.e., the neighborhood consisting of points less than distance η from P.

These definitions accord with Hahn's definitions and with Mazurkiewicz's implicit ones.

In a paper published in 1920, 'Sur une condition pour qu'un continu soit une courbe jordanienne' ('On a condition for a continuum to be a Jordan curve'), Wacław Sierpiński offered an important result that is complementary to the results of Mazurkiewicz and Hahn [Sierpiński, 1920]. Sierpiński's necessary and sufficient condition for characterizing a Jordan curve or space-filling curve is: for every $\varepsilon > 0$ there exists a union of a finite number of continua of diameter $< \varepsilon$ representing the given continuum. Kuratowski sums up the results of Hahn, Mazurkiewicz, and Sierpiński as follows [Kuratowski, 1968, page 256]:

> THEOREM 2 (of HAHN–MAZURKIEWICZ–SIERPIŃSKI). *If $\mathfrak{X}$ is a non-empty continuum, the following conditions are equivalent.*
>
> (i) $\mathfrak{X}$ *is a continuous image of an interval,*
> (ii) $\mathfrak{X}$ *is the union of a finite number of continua of diameter* $< \varepsilon$ *for every* $\varepsilon > 0$,
> (iii) $\mathfrak{X}$ *is locally connected.*

13.9 Connectedness and Disconnectedness

Another important background area for the development of dimension theory is the research into various defining conditions and exotic examples of connectedness and disconnectedness, especially the analysis of diverse types of disconnected sets. This was a vigorous area of research in point-set topology in the first decades of the twentieth century, yielding a large collection of theorems, counterintuitive examples, and counterexamples that set logical and mathematical boundaries on the theory. Concerning this important research area I can only scratch the surface. Ryszard Engelking has provided a useful diagram showing a hierarchy of types of disconnected spaces (or sets), ordered by set inclusion, such that a space belonging to an inner set (depicted by a rectangle) is also a space belonging to any containing set (depicted by a rectangle) (see Figure 13.10), and has given a clear exposition of the topic in his book *Dimension Theory* [Engelking, 1978, pp. 31–37] [Engelking, 1995, pp. 24–29].

Polish mathematicians of the early twentieth century, including Wacław Sierpiński, Zygmunt Janiszewski, Stefan Mazurkiewicz, Bronisław Knaster, and Kazimierz Kuratowski played a considerable part in the investigation of these often rather unusual sets. I should add that a wide variety of European and American researchers in this domain also contributed to the mix of interesting results concerning connectedness and disconnectedness.[23]

[23] For example, the Austrian mathematician Leopold Vietoris (1891–2002) [Vietoris, 1921] also made an early study of the concept of connectedness à la Lennes, Riesz, and Hausdorff.

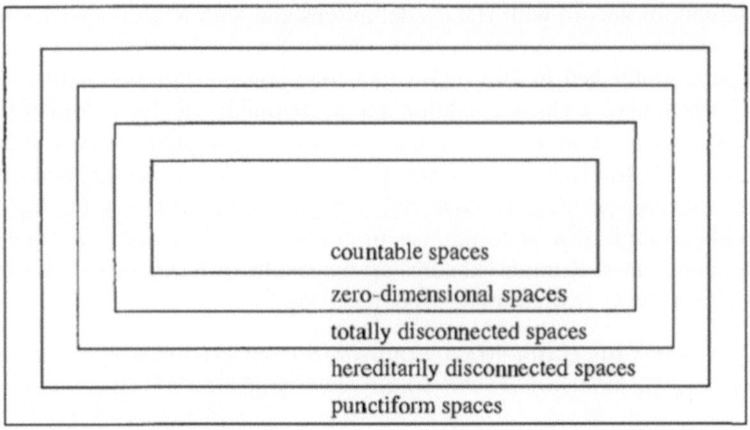

Fig. 13.10 Classification of Disconnected Sets, with Diagram of Sets of Spaces Ordered by Set Inclusion [Engelking, 1998, page 407].

To limit my survey concerning the topic of connectedness and disconnected sets (or spaces), I shall consider two papers by Sierpiński and one by Knaster and Kuratowski. These are representative of the research at the time.

First, I shall consider Sierpiński's paper, 'Sur les ensemble connexes et non connexes' ('Concerning connected and non-connected [disconnected] sets'), published in 1921 [Sierpiński, 1921]. This is a landmark paper, which many other topologists referenced in the years following its publication. In the paper Sierpiński uses the well-known standard definition of connected sets in an m-dimensional space, and, following Mazurkiewicz, calls two sets A and B in an m-dimensional space *separated* (*séparés*) if the following formulae hold:

$$AB = AB' = A'B = 0.$$ [24]

A set of points P will be *connected*, if there does not exist a decomposition of P into two separated sets.

Further, he says that two points a and b of a set P are *separated* in the set, if there exists a decomposition of P into two separated sets, such that one contains a and the other contains b.

A key concept that Sierpiński explores in the paper is the concept of a set being *dispersed*. He says that a set of points P is *dispersed* (*dispersé*), if it does not contain any connected set of more than one point. This notion is the same as Hausdorff's notion of *punkthaft* as defined in his classic book *Grundzüge der Mengenlehre*

[24] Juxtaposition of the letters denotes intersection, and '0' denotes the empty set.

(*Fundamentals of Set Theory*) [Hausdorff, 1914, page 322]. The modern term for this is *hereditarily disconnected* [Kuratowski, 1968, page 151].[25]

It is evident that a set P in which every pair of points is separated (totally disconnected) is dispersed or hereditarily disconnected. In other words, this statement means that every totally disconnected set is hereditarily disconnected (shown as an inclusion in Engelking's diagram). The following converse problem then naturally arises [Sierpiński, 1921, page 81]:

> *for two points of a dispersed set, are they necessarily separated in this set?* [Ch13, 22]

Sierpiński gives a negative response to this problem by describing an extended example in the plane, a point set bounded by a rectangle [Sierpiński, 1921, pp. 82–88].

At the same time Sierpiński considers the problem [Sierpiński, 1921, page 82]:

> Let P be a set in which every two points are separated in P, let a be a given point of P, and let ε be a given positive number, then can one always find a decomposition of P into two separated sets A and B such that A contains a and A is of diameter $< \varepsilon$? [Ch13, 23]

Sierpiński also uses his extended example to answer this question in the negative. He uses his example again to exhibit a punctiform G_δ set[26] in the plane that is not homeomorphic to any linear set. Thus, his geometrical example in the plane serves to give negative answers to all three of the problems considered. As noted, he elaborates the example over several pages [Sierpiński, 1921, pp. 82–88] with his conclusions stated as follows [Sierpiński, 1921, page 88]:

> We thus have *an example of a planar dispersed [hereditarily disconnected] set such that certain pairs of its points are not separated in this set.* [Ch13, 24]

> We thus have *an example of a planar punctiform set G_δ that is not homeomorphic with any linear set.* [Ch13, 25]

In this situation Sierpiński is using his preferred way of proceeding by offering a concrete geometric construction.

In slightly later terminology that Karl Menger introduced Sierpiński's first example is a (weakly) one-dimensional set that is hereditarily disconnected. (See papers of Menger [Menger, 1927d, page 707] and Hurewicz and Menger [Hurewicz & Menger, 1928, page 621]. See also Engelking's succinct summary [Engelking, 1998, page 406] and Kuratowski's characterization [Kuratowski, 1966, pp. 298–299].)

Further, Sierpiński proves [Sierpiński, 1921, page 89]:

---

[25] In relation to Engelking's diagram one has the following definitions. A set or space is *totally disconnected* if the set/space is not connected between any pair of points [Kuratowski, 1968, page 151]. A set/space is *zero-dimensional/0-dimensional* if each of its points lies in an arbitrarily small open neighborhood that is both closed and open, equivalent to the condition that the boundary of the neighborhood is empty [Kuratowski, 1966, page 277].

[26] A G_δ set is the intersection of countably many open sets.

> **Theorem:** *For a punctiform set P, situated in a space of m dimensions, to be homeomorphic to a linear set, it is necessary and sufficient that for every point p of P and for every positive number ε the set P be the sum [union] of two separated sets $P = A + B$* [27]*, such that A contains p and that the diameter of A be $< \varepsilon$.*[Ch13, 26]

Just a year or two later Paul Urysohn restated this theorem as a theorem concerning 0-dimensional sets [Urysohn, 1925a, page 76]:

> *For a punctiform set to be homomorphic to a linear set it is necessary and sufficient that it be of dimension 0.*[Ch13, 27]

Urysohn also interpreted Sierpiński's theorem equivalently in terms of the Baire null space [Urysohn, 1925a, page 77]. With these interpretations one may view Sierpiński as a immediate predecessor to dimension theory in Urysohn's formulation.

Sierpiński also proves [Sierpiński, 1921, page 94]:

> **Theorem:** *If P is a punctiform set, situated in a space of $m > 1$ dimensions, then the complement of P is a connected set.*[Ch13, 28]

In this paper Sierpiński provides a particularly good example of the systematic and relatively concrete geometrical exploration of a topic that he seems to prefer in his research.

Next, I shall consider another important paper by Sierpiński, 'Sur une propriété des ensembles frontières' ('Concerning a property of boundary sets'), published in 1922 [Sierpiński, 1922]. The motivation behind Sierpiński's paper was a conjecture of Mazurkiewicz that every planar set not containing any domain (i.e., not containing any set with only interior points and no boundary points) is homeomorphic to a nowhere dense planar set [Mazurkiewicz, 1920b, page 81]. Sierpiński not only proves Mazurkiewicz' conjecture, but he proves a significant generalization [Sierpiński, 1922, page 7]:

> **Theorem.** *Every boundary set in a space of m dimensions is homeomorphic to a nowhere dense set situated in the same space.*[Ch13, 29]

Sierpiński's quite concrete proof for this theorem yields a nowhere dense set whose closure is contained in the set of points with at most $n-1$ rational coordinates. In effect, he characterizes the n-dimensional subsets of the Euclidean n-space as the sets with a non-empty interior.[28]

Sierpiński mentions a connection between his result and Fréchet's theory of dimension types (cf. Section 10.4). By his construction he has effectively shown the existence of a closed nowhere dense set in the plane whose dimension type is the greatest of all those dimension types that are less than the dimension type of the plane. There is an analogous result for Euclidean m-dimensional space.

---

[27] '+' denotes union of sets.

[28] Cf. [Engelking, 1998, page 407].

Finally, I shall consider a paper by Knaster and Kuratowski, ‘Sur les ensembles connexes’ (‘Concerning connected sets’) from 1921 [Knaster & Kuratowski, 1921]. In this important paper Knaster and Kuratowski carry out a thorough analysis of connectedness and disconnectedness. They include several counterintuitive examples to focus on certain critical topological distinctions [Knaster & Kuratowski, 1921, pp. 241–254].

Of special importance from the paper is their startling example of a connected (and ‘biconnected’ [29]) set that becomes dispersed or hereditarily disconnected by the removal of a single point [Knaster & Kuratowski, 1921, pp. 213; 241–245; 254, Fig. 2]. See Figure 13.11. This is the Knaster–Kuratowski fan (or Cantor’s leaky tent or Cantor’s teepee). To construct the example in the plane, join point $(\frac{1}{2}, \frac{1}{2})$ by straight line segments to the Cantor middle-third set $\mathfrak{C}$ on the unit segment of the x-axis, [0, 1]. If a line segment contains an endpoint remaining in $\mathfrak{C}$ after the removal of one of the open ‘third’ segments, take the line segment points with rational ordinate values; in the complementary case relative to $\mathfrak{C}$, take the line segment points with irrational ordinate values. The collection of these points constitutes the example. The example is connected and biconnected, but becomes dispersed or hereditarily disconnected with the removal of point $(\frac{1}{2}, \frac{1}{2})$.

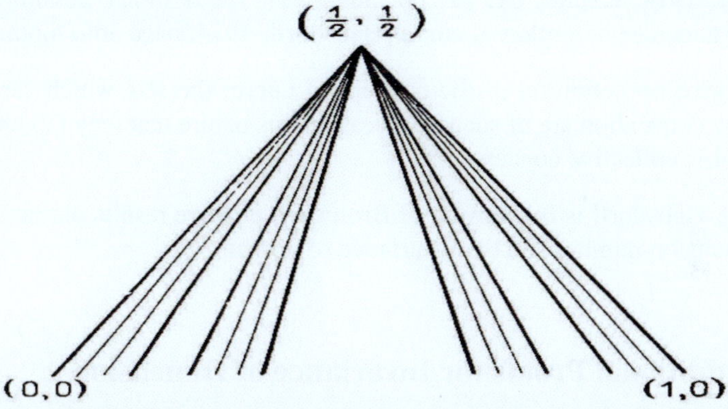

Fig. 13.11 Knaster and Kuratowski Example [[Knaster & Kuratowski, 1921], also reproduced in [Kuratowski, 1968, page 135], adjusted in size].

---

[29] Knaster and Kuratowski [Knaster & Kuratowski, 1921, page 214] call every connected set that is not a union of two disjoint connected (proper) subsets ‘biconnected’ (‘biconnexe’).

13.10 Growth of Set-Theoretic Topology

During the first two decades of the twentieth century both general set theory and point-set or set-theoretic topology grew rapidly as domains of mathematics. Mathematicians produced numerous research papers, monographs, and texts. Schoenflies's two-part *Bericht* (*Report*) [Schoenflies, 1900] [Schoenflies, 1908b] [Schoenflies & Hahn, 1913], Youngs' monograph [Young & Young, 1906], and Sierpiński's text of 1911 (in Polish) appeared, along with texts in function theory that featured chapters on point-set theory. Felix Hausdorff's (1868–1942)[30] treatise *Grundzüge der Mengenlehre* (*Fundamentals of Set Theory*) [Hausdorff, 1914] [Hausdorff, 2002] of 1914 is a considerable landmark. In this Hausdorff offers a beautiful synthesis of the diverse results of point-set theory through his theory of topological spaces. This theory joins together fundamental geometrical ideas and abstract analytical concepts of David Hilbert [Hilbert, 1903], Maurice Fréchet [Fréchet, 1906], and Hermann Weyl [Weyl, 2013]. Moreover, Hausdorff's text is eminently readable. The book was a wonderful inspiration for many students of the subject.

Felix Hausdorff was fully aware of the issues that space-filling curves and unusual Jordan curves and arcs brought to the fore in point-set topology. Thus, in one section of his treatise, he calls attention to the problem of defining the concept of curve [Hausdorff, 1914, Kapitel IX, §2, pp. 369–382]. He is rather pessimistic about finding an adequate definition of curve [Hausdorff, 1914, page 369, footnote]:

> We give no definition of the concept of curve; the sets which carry this name by convention are of such a heterogeneous nature that they fall under no reasonable collective concept.[Ch13, 30]

In contrast, Hausdorff is fully aware of Brouwer's positive results on the invariance of the dimension number and the invariance of domain.[31]

13.11 Additional Proofs for Invariance of Dimension

In the years just after the publication of Brouwer's two papers on dimension only a few mathematicians showed any interest in problems of dimension. Brouwer's second paper of 1913 was hardly noticed at all. Then suddenly in the year 1921 the topological theory of dimension again became a lively topic of research. In that year two young mathematicians, unknown to one another, Paul (Pavel) Urysohn

[30] For a full biography of Hausdorff cf. [Brieskorn & Purkert, 2018]. Also, cf. [Eichhorn, 1994] and `https://mathshistory.st-andrews.ac.uk/Biographies/Hausdorff/`.

[31] I may add that the English mathematician Eric H. Neville (1889–1961), moved by a desire to define the concept of planar curve produced a paper in 1918 (published 1920) on 'The Classification of Sets of Points,' which included a definition of dimension [Neville, 1920]. However, the paper is more philosophical than mathematical; by intention it includes no proofs. Indeed it fails to characterize dimension in any adequate way. Cf. the criticism of A. Rosenthal in the *Jahrbuch über die Fortschritte der Mathematik und ihrer Grenzgebiete*, 1916–18 46 (1923), 304–305.

and Karl Menger, began investigating problems of dimension and also problems of curves as based on notions of dimension, thereby initiating a vigorous growth of the theory of dimension (see Chapters 14 and 15). During the 1920s and 1930s many mathematicians contributed to developing the subject (see Chapter 16). At the same time several new proofs of dimensional invariance were proposed. Many regarded Brouwer's proofs as difficult, particularly as they depended upon his special concepts of combinatorial topology. Hence, it was natural that mathematicians should seek alternative proofs. In examining the growth of dimension theory after Brouwer's fundamental work, I shall consider the new proofs of dimensional invariance in this section, while leaving the broader developments to the later chaprers.

In 1921 Henri Lebesgue finally published a full and detailed proof of the tiling principle along with its immediate consequence, a proof of dimensional invariance [Lebesgue, 1921]. Thus, some ten years after the dispute with Brouwer (see Chapter 12) he at last put a satisfactory proof of the principle into print. The new combinatorial proof is not especially elegant. While it does avoid reference to special ideas of combinatorial topology, it does not improve much upon Brouwer's direct and insightful way of thinking about the principle.

In his paper of 1921 Lebesgue proceeds further to treat a variety of matters related to the tiling principle. For example, he seeks to deduce the domain invariance theorem from the principle. Yet his deduction is erroneous, as he later acknowledged [Lebesgue, 1924, page 96]:

> ... by a singularly thoughtless action, in effect, I have at one point in my reasoning admitted almost entirely the property to be demonstrated.[Ch13, 31]

Lebesgue never developed further the insightful ideas of his 1911 research note in the *Comptes rendus* (*Proceedings*) of the Academy of Sciences, Paris [Lebesgue, 1911c]. That is unfortunate. It seems that his efforts in topology only had limited success. His intuitive topological ideas were brilliant, but he may have needed other mathematicians to bring them to fruition and acceptable mathematical rigor.[32]

In the late 1920s, when the development of dimension theory was in full swing, Emanuel Sperner (1905–1980)[33] and Withold Hurewicz (1904–1956) produced solid proofs of the tiling principle and dimensional invariance. These proofs of Sperner [Sperner, 1928] and Hurewicz [Hurewicz, 1929] are extremely elegant and are probably about the simplest conceivable. They nearly reduce dimensional invariance to a topological triviality. They are also very similar, although they were constructed independently.

At the heart of Sperner's proof of the tiling principle lies a combinatorial proposition, which is interesting in its own right: the Sperner lemma. Briefly, Sperner's paper [Sperner, 1928] proceeds as follows. At the beginning he sets out to prove the tiling principle for the closed simplex: For a finite closed covering of an n-dimensional

---

[32] It should be mentioned that [Lebesgue, 1921, pp. 259–261] contains a proof of the so-called Lebesgue covering lemma. Cf. [Alexandroff & Hopf, 1935, pp. 101–102].

[33] For biographical information on Sperner cf.
`https://mathshistory.st-andrews.ac.uk/Biographies/Sperner/` and
`https://en.wikipedia.org/wiki/Emanuel_Sperner`.

simplex Σ by sufficiently small closed sets there is at least one point of the simplex that lies in at least $n+1$ of the sets. He calls such a point 'distinguished' ('ausgezeichnet'). When proceeding to show the existence of a distinguished point, he notes that it is possible to reduce the closed covering by taking unions of certain members, so that the reduced covering has precisely $n + 1$ members $\mathfrak{M}_1, \ldots, \mathfrak{M}_{n+1}$, and for each i the member of $\mathfrak{M}_i$ contains the vertex E_i of the simplex but contains no point of the opposite face s_i. Clearly, the distinguished points of the reduced covering must have been distinguished in the original covering, but the reduced covering may well have fewer such points. Yet, the reduced covering will still possess distinguished points, and this fact can be demonstrated as follows. Consider a simplicial subdivision of the simplex. A subsimplex of the subdivision will be called 'distinguished' if it has points in each of the $n + 1$ sets of the covering. If distinguished subsimplexes exist for every simplicial subdivision, then it is an easy matter to show that distinguished points exist for the closed covering by considering a sequence of simplicial subdivisions with the diameters of the subsimplexes of the successive simplicial subdivisions approaching zero. Thus, the main problem is to demonstrate the existence of at least one distinguished subsimplex σ_k for any given simplicial subdivision. This problem is solved by Sperner's lemma [Sperner, 1928, page 267]:

> Let a simplicial subdivision of Σ into the subsimplexes $\tau_1, \tau_2, , \ldots, \tau_p$ be given. A vertex of a τ_k then meets certain $\mathfrak{M}_i$, perhaps $\mathfrak{M}_{\nu_1}, \mathfrak{M}_{\nu_2}, \ldots, \mathfrak{M}_{\nu_q}$, where $\nu_1 < \nu_2 < \cdots < \nu_q$. We then associate the coordinate ν_1 with this vertex and deal similarly with all vertices of the subsimplexes. In this way a system of $n + 1$ coordinates is associated with each subsimplex τ_k, and a system of n coordinates is associated with each face. Then we assert:
>
> *The number of τ_k with which the system $(1, 2, 3, \ldots, n + 1)$, i.e., that with all different numbers, is associated is odd.*[Ch13, 32]

Sperner's proof of the lemma is inductive and has a simple combinatorial character. From the lemma the existence of a distinguished subsimplex for a simplicial subdivision and a closed covering follows immediately. Hence, the Lebesgue tiling principle is proved and from that the invariance of dimension. In addition, the last part of Sperner's wonderfully succinct paper contains a proof of the domain invariance theorem.

Hurewicz' paper [Hurewicz, 1929], like Sperner's, also provides a direct route to dimensional invariance. Yet, because it does not contain an explicit analogue of Sperner's lemma, it does not achieve quite the same degree of clarity in the proof. I shall say a little more about Hurewicz' paper in Section 16.4 when I discuss his broader contributions to the development of dimension theory.

Shortly after the appearance of the paper by Sperner the Polish topologists Knaster, Kuratowski, and Mazurkiewicz showed how to deduce Brouwer's fixed point theorem for the n-simplex from the Sperner lemma with a proof that is now regarded as classical [Knaster et al., 1929]. Hence, through the work of all of these authors a large and important part of Brouwer's topological work became provable by elementary means.

13.12 Concluding Remarks

Cantor and Jordan in the nineteenth century and mathematicians in the first two decades of the twentieth century explored various ways to define the concept of *curve*. Applying mathematical and logical tools, mathematicians developed and appraised the consequences of the various proffered definitions. Point-set theory and point-set topology provided the powerful methods needed to found and push this development forward.

At this point I shall mention three philosophical views of this historical matter. First, one might view the effort towards developing the concept of curve and its deductive consequences as an essentialist quest to find the unique definable concept. Second, one might consider the effort more liberally, as did Carnap and Menger, as looking for an *explication* or a set of *explications* of the concept and then appraising the deductive consequences for their mathematical value and fruitfulness. Third, one might view the effort à la Popper and Lakatos as developing theories of curves with provable consequences and using counterintuitive examples and counterexamples to lay out and delimit the theories more sharply. In this chapter I have described some of the interplay between proposed definitions of *curve*, their deductive consequences, and counterexamples, which in some cases offer refutations of purported results or at least narrow their scope. From the history just discussed, one could likely discern cases where the mathematicians involved took one or other of these views, either explicitly or implicitly. I shall leave further discussion of these philosophical views until Chapter 17.

In a way similar to the quest for a good definition of the concept of *curve*, mathematicians explored definitions of *connectedness* and *disconnectedness*. Counterexamples, in some instances subtle and complex, became the differentiators among the closely related concepts.

The quest to explain the phenomenon of space-filling curves yielded the Hahn–Mazurkeiwicz–Sierpiński Theorem based primarily on the notion of locally connectedness. In this case the quest for a sound mathematical *explanation* was a remarkable mathematical achievement.

I shall take up the main development of dimension theory of the 1920s and 1930s in the next chapters. I shall consider first the work of Urysohn (Chapter 14) and then the work of Menger (Chapter 15), followed by an examination of the broader development (Chapter 16), including the development of the special Hausdorff theory of dimension (Section 16.11) as well as other special theories.

Chapter 14
Urysohn's Theory of Curves and Dimension

14.1 Introduction

Paul Urysohn and Karl Menger were the prime developers of dimension theory in the early 1920s. As young mathematicians from different countries, they initially developed their differing, though remarkably similar, theories independently, each not knowing of the other's work. Each put together an extensive set of results. Then as various mathematicians sorted out the differences, the two theories became joined together, resulting in a reasonably unified theory derived from the efforts of several important mathematicians. Not surprisingly, the history exhibits ups and downs. In this chapter I shall discuss the work of Urysohn, and in the next, the work of Menger.

14.2 Mathematical Problems Motivating Urysohn's Research in Dimension

Paul (Pavel) Samuilovich Urysohn (1898–1924)[1] was born in Odessa (now a city of Ukraine) in 1898, but the family moved to Moscow in 1912. Urysohn entered Moscow University in 1915 with the intention of studying physics. However, he soon was drawn to mathematics, because he was charmed by the lectures and personalities of N.N. Luzin (or Lusin) and D.F. Egorov (or Egoroff). Luzin was a particularly dynamic mathematician at the university. It was he who persuaded Urysohn to stay on in order to study for a doctorate during 1919–1921. Yet, it was Egorov who set him the mathematical problems that spurred his developing a theory of dimension and related theory of curves. In his short life, cut short by a swimming accident, Urysohn

[1] For biographical information concerning Urysohn cf. [Alexandroff, 1925] [Neiman, 1972] [Alexandroff, 1950] [Arkhangelskii & Tikhomirov, 1998];
`https://mathshistory.st-andrews.ac.uk/Biographies/Urysohn/`;
`https://www.encyclopedia.com/science/dictionaries-thesauruses-pictures-and-press-releases/uryson-pavel-samuilovich`.

D. M. Johnson, *Theories of Dimension from Ancient to Modern Times*, Sources and Studies in the History of Mathematics and Physical Sciences, https://doi.org/10.1007/978-3-032-02682-8_14

Fig. 14.1 Pavel Samuilovich Urysohn (1898–1924) [[Urysohn, 1951a, frontispiece] and reproduced in [Arkhangelskii & Tikhomirov, 1998]].

still made many contributions to point-set topology, for example, his metrization theorem, along with his major and extensive contributions to dimension and curve theory.[2]

Initially, Urysohn's mathematical research was in analysis, but he switched to topology in 1921. In June of that year he was awarded his doctorate. He then was offered and accepted a post as lecturer at Moscow University. During this same

[2] Strictly with respect to publications Urysohn properly has priority over Menger. As a consequence I discuss his work first. Yet, it should be emphasized that both Urysohn and Menger started their researches on dimension theory in the same year, 1921. Menger did deposit a note [Menger, 1926e, pp. 1122–1123] [Menger, 1929b, pp. 411–412] with the Akademie der Wissenschaften (Academy of Sciences), Vienna, in the autumn of 1921, about a year before Urysohn published his first notes in the *Comptes rendus* (*Proceedings*) of the Academy of Sciences, Paris [Urysohn, 1922a] [Urysohn, 1922b] (September 1922). However, the note was only a deposition and not published at the time. Menger had to correct some errors in his research work. For my discussion of Urysohn's work I am relying on Alexandroff [Alexandroff, 1925] [Alexandroff, 1969] and especially his historical introduction in [Urysohn, 1951a, pp. 11–42], also [Alexandroff, 1979] [Alexandroff, 1980]; cf. [Neiman, 1972] and [Arkhangelskii & Tikhomirov, 1998].

month he had a conversation with his teacher Egorov, who proposed two research problems for him:

- First, to give an intrinsic topological definition of line or curve in general, which would be equivalent in the case of restriction to the plane to the known non-intrinsic definition of a Cantorian line;
- Second, to give an intrinsic topological definition encompassing a wide class of sets that could appropriately be regarded as surfaces.

Egorov put Urysohn directly into the problem situation surrounding these two problems, a problem area that I described in the previous Chapter 13 as well as in earlier chapters. In his lengthy 'Mémoires' ('Memoirs') Urysohn references virtually all the mathematicians whom I have already mentioned, including Cantor, Jordan, Brouwer, Zoretti, Janiszewski, Sierpiński, Hahn, Mazurkiewicz, and Hausdorff. His teacher Egorov must have posed the specific problems very clearly and forcefully. These problems immediately fired the imagination of the young Urysohn. Over the time of writing his big papers he must have analyzed the published mathematical literature very carefully to understand what was known and what needed more research. His meticulous study of the literature enabled him to develop his original ideas.

Urysohn's close friend, Paul (or Pavel) Alexandroff tells the continuation of the story from his memories of the immediate experience [Urysohn, 1951a, pp. 13–14]:[3]

> Pavel Samuilovich [Urysohn] immediately began thinking intensively about these problems. Very early on, he began to reflect on a general definition of dimension. He spent the entire summer of 1921 intensively seeking the 'real' ['actual'] definition, shifting from one version to another, constantly constructing examples to show why one version or another had to be rejected. These were two months of all-consuming reflection. At last, one morning in late August, he woke up with the complete, final, and now well-known inductive definition of dimension. This took place in the village of Burkovo near Bolshevo, on the bank of the river Klyazma. There a group of young (at the time) Moscow mathematicians—among them V.V. Stepanov, D.E. Men'shov, and N.K. Bari—were living in a dacha. That same morning, while we were swimming in the Klyazma, P.S. Urysohn told me about his definition of dimension, and in the course of that conversation, which lasted several hours, sketched an outline of the whole theory, including a number of theorems that were conjectures that no one knew how to approach at the time, but were proved one by one over the next several months. The whole program that he sketched at that time was realized over the winter of 1921/22. By the spring of 1922 the whole theory of dimension was complete. It turned out that there is a natural way of assigning to every point set R (every metric space) a certain non-negative number $\dim R$, which is the dimension of that set. The fruitfulness of the new definition results, of course, from the fact that it provided the basis for establishing many new properties of point sets, which in turn led to entirely new topological problems.[Ch14, 1]

[3] I am grateful to Professor Roger Cooke for providing me with the translation from the Russian.

14.3 Urysohn's Development of Dimension Theory

During the academic year 1921–1922 Urysohn gradually proved the theorems of his new dimension theory. He also gave a course of lectures on the topology of continua at Moscow University and used this as a medium to display his new results to his students, often just after he had proved them. Simultaneously, he announced his results in a series of notes to the Moscow Mathematical Society.[4] By the spring of 1922 his dimension theory was more or less complete. In September he had two notes outlining his theory, 'Les multiplicités cantoriennes' ('Cantorian multiplicities/manifolds/varieties') [Urysohn, 1922a] and 'Sur la ramification des lignes cantoriennes' ('On the ramification [branching] of Cantorian lines/curves') [Urysohn, 1922b] presented to the Académie des Sciences, Paris, by Lebesgue and published in the *Comptes rendus* (*Proceedings*) of the Academy of Sciences, Paris.

It was not until the academic year 1922–1923 that Urysohn wrote a full version of his dimension theory in his 'Mémoire sur les multiplicités Cantoriennes' (I) ('Memoir on Cantorian multiplicities/manifolds' (I)) [Urysohn, 1925a] [Urysohn, 1926]. The work bears the date 20 March 1923 [Urysohn, 1925a, page 64]. However, certain editorial changes were needed after this date. Urysohn and Brouwer had agreed on the issue of a 'slip of the pen' error in Brouwer's 1913 paper [Brouwer, 1913b].[5] In fact, Urysohn and Brouwer did not sort out the issue of the offending word 'abgeschlossene' ('closed') until later in 1923 and into 1924. Hence, referencing to Brouwer's work in the 'Mémoire' ('Memoir') needed to be modified accordingly. Meanwhile Urysohn tragically died in August 1924. Consequently, Paul Alexandroff, Urysohn's literary executor and close colleague and friend, had to undertake significant editing of the paper to prepare it for publication, including modifying some referencing to Brouwer's work. Alexandroff agreed to have Brouwer help with this editing. The two worked according to strict rules so as not to 'change,' 'modify,' or 'enhance' the history. Also, there were some publication delays with *Fundamenta Mathematicae*. The result was that the full work was not published until 1925 [Urysohn, 1925a] and 1926 [Urysohn, 1926].

My reporting of this part of the history may seem a bit labored. A little later Karl Menger raised a priority dispute centered on the dating of his early work and the dating of the work of Urysohn. The referencing to Brouwer in Urysohn's 'Mémoire' ('Memoir') and Brouwer's handling of it as part of the editing work with Alexandroff became an issue in the dispute [Menger, 1979c, pp. 245–247]. I think that van Dalen has thoroughly researched the question of the editing of Urysohn's 'Mémoire' ('Memoir') [Dalen, 2005, 12.4, pp. 470–480] [Dalen, 2012, 11.4, pp. 424–434], and I accept his findings. I shall return to the priority dispute that Menger opened in the next chapter.

[4] Alexandroff in [Urysohn, 1927a, page 3] says that these notes were delivered at the sessions of 16 October 1921, 20 November 1921, 19 February 1922, and 28 May 1922. He also gives their titles and brief descriptions of their contents.

[5] See Section 12.12 for a discussion of Brouwer's own corrections of the slip coming as early as 1913 and later in [Brouwer, 1924d] with his rewritten versions of the paper in [Brouwer, 1923a] [Brouwer, 1923b].

Note that the second part, 'Mémoire sur les multiplicités Cantoriennes (II). Les lignes Cantoriennes' ('Memoir on Cantorian multiplicities/manifolds (II). The Cantorian Lines/Curves') [Urysohn, 1927a], was published in 1927, and in a different journal. Paul Alexandroff had to undertake considerable editing of this work.

At this point I shall examine the theory as presented in the 'Mémoire' ('Memoir') (I).[6] At the beginning of the 'Mémoire' ('Memoir') Urysohn states his main problem [Urysohn, 1925a, page 32]:

> *To indicate the most general sets that still merit being called lines, surfaces etc.*[Ch14, 2]

In this context he also mentions Problem J_n: To give a purely geometrical definition of the n-dimensional Jordanian multiplicities or manifolds, where an n-dimensional Jordanian manifold is a set that is homeomorphic to an n-dimension cube. Further, he mentions Problem K_n: To give a purely geometrical definition of the n-dimensional Cantorian multiplicities or manifolds [Urysohn, 1925a, pp. 30–32]. By taking on this problem, Urysohn puts himself into the problem situation that Zoretti [Zoretti, 1909] [Zoretti, 1913] and Janiszewski [Janiszewski, 1911a] [Janiszewski, 1912a] [Janiszewski, 1912b] defined earlier (see the previous Chapter 13). Janiszewski's open problems must have been particularly attractive [Janiszewski, 1912b].

In his 'Mémoire' ('Memoir') he clearly intends that his research program encompass more than the pair of problems that Egorov had set him originally. Beyond seeking definitions of the concepts of curve and surface and the variants mentioned, he seeks a definition of dimension concept itself and its consequences. This general concept is the center of his attention in the first part of his 'Mémoire' ('Memoir'). Dimension theory becomes a frame for analyzing geometrical objects or well-defined point-sets of 0, 1, 2, 3, and n dimensions.

When approaching his problem Urysohn found Janiszewski's thesis *Sur les continus irréductibles entre deux points* (*On continua irreducible between two points*) [Janiszewski, 1911a] [Janiszewski, 1912a] especially important and useful (see Section 13.5). In the 'Mémoire' ('Memoir') he makes numerous references to it. Hausdorff's systematic treatise *Grundzüge der Mengenlehre* [Hausdorff, 1914] was another significant background work for Urysohn (cf. Section 13.10). In studying mathematics Urysohn was methodical in his careful detailed analysis of his immediate predecessors' work.

14.4 Urysohn's Methodological Principles and Framework for Developing Dimension Theory

Urysohn laid down three methodological principles or guidelines when attempting to solve his problem, which he states at the beginning of the 'Mémoire' ('Memoir') [Urysohn, 1925a, pp. 32–38]. First, he *desires that all his definitions be intrinsic*,

[6] Cf. [Urysohn, 1922a], [Urysohn, 1922b], and [Alexandroff, 1928a].

i.e., they should be based on the internal properties of the sets under consideration and not on their situations or positions in larger spaces.[7] He ascribes the fault of the known definition of a Cantorian curve, the fact that it only applies to the plane and not to 3- and higher-dimensional space, to its not being intrinsic. As noted earlier Egorov had emphasized the great importance of intrinsic definitions in his original discussion with Urysohn in June 1921 [Urysohn, 1925a, page 35]. Second, he attempts to *provide local definitions wherever possible* rather than global or 'integral' ones. Brouwer's main definition of dimension of 1913 was a global one. From the start Urysohn sought something different. Indeed, the local character of Urysohn's dimension definition distinguishes it from Brouwer's. It is this local character that makes it particularly fruitful. Third, Urysohn suggests that even though *closed sets are his principal concern* it is possible to *examine with profit their non-closed parts*. This suggestion certainly broadens the scope of his study.

In addition to laying down the three methodological principles for his research Urysohn places his theory within the *framework of the compact metric space*[8] and argues for the appropriateness of this frame [Urysohn, 1925a, pp. 38–42]. He had observed that compactness was the pivotal assumption in most of his arguments. Hence, it seemed natural to him to use the compact metric space as a base for his work rather than the more concrete Euclidean spaces. He boldly proclaims that the compact metric space is the 'natural domain of existence' ('domaine naturel d'existence') for intrinsic topology. He even goes further to assert that every mathematical theory possesses a natural domain of existence, while offering a few examples to substantiate his assertion. Although this methodological claim is open to question, it does mean that he puts his theory into the domain of abstract topology.[9] If one looks at Urysohn's subsequent work in general topology, one needs to be aware of the effect of his strong methodological views on this research.

14.5 Urysohn's Definition of Dimension

Urysohn's definition of dimension, like Brouwer's before him, starts from an antecedent definition of separation.[10] In the case of Urysohn this is a local form of separation called 'ε-separation' [Urysohn, 1922a, page 441] [Urysohn, 1925a, page

[7] To be sure, in other topological investigations the position or situation of an entity in a space matters a great deal. The situation of an entity relative to its complement in a space impacts the research direction.

[8] See Hewitt's charming article on the role of compactness in analysis [Hewitt, 1960].

[9] If Urysohn means that the natural domains of mathematical theories should be unchanging, then his methodological claim seems too restrictive with respect to any future developments and so methodologically mistaken. Surely, 'natural domains' change in the longer run of history; they are stretched to fit the current demands of mathematical research. In the case of dimension theory Urysohn's 'natural domain' of the compact metric space was soon expanded by other dimension theorists of the 1920s to the separable metric space. That change proved to be not a big expansion.

[10] See my analysis in Section 12.12 of Urysohn's critique of Brouwer's definition of separation as it was first published in Brouwer's 1913 paper [Brouwer, 1913b].

65]. Let C be any set in the given compact metric space, and let x be a point in C. We say that a decomposition of C into mutually disjoint sets, $C = A \cup B \cup D$, is an ε-separation of the point x or that the set B ε-separates the point x in C if

(1) $x \in A$;
(2) $(A \cap \bar{D}) \cup (\bar{A} \cap D) = \emptyset$, where $\bar{A}$, $\bar{D}$ denote topological closures;
(3) $A \cup B \subset S(x, \varepsilon)$, where $S(x, \varepsilon)$ denotes the open sphere about x with radius ε.

Condition (2) is the opposite of the Lennes–Riesz–Hausdorff condition for connectedness, so connectedness is implicitly involved in Urysohn's separation concept. Indeed Urysohn informs us [Urysohn, 1925a, page 37] that the study of connectedness was a starting point for his research, although in the end it dropped out of his exposition, because connectedness and disconnectedness do not have a great deal to do with his dimension concept per se.[11]

Urysohn's fundamental definitions of dimension given in inductive form are as follows [Urysohn, 1925a, pp. 65–66]:

> Def. 1_0. *If for any ε the point x of C can be ε-separated by the empty set, this point will be said to be of dimension* 0 *(with respect to C):*
>
> $$\dim_x\ C = 0.$$
>
> Def. 2_0. *If all the points of C are of dimension* 0 *with respect to C, C will itself have dimension* 0*:*
>
> $$\dim C = 0.$$
>
> Higher dimensions are defined by induction. We suppose, in effect, that the points and the sets of dimension $< n$ have been defined. In this case we say that
>
> Def. 1_n. *A point x of C that is not of dimension $< n$ (with respect to C), but for any ε can be ε-separated by a set B of dimension $< n$, this point is of dimension n:*
>
> $$\dim_x C = n.$$
>
> Def. 2_n. *A set C only containing points of dimension $\leq n$ will be said to be of dimension n*
>
> $$\dim C = n,$$
>
> *if it contains at least one point x such that* $\dim_x C = n$.
>
> Thus, one defines dimension step by step for every natural n. Moreover, it can happen that a point x (or a set C) does not receive any dimension in this way; we say then that *this point* (or this set) *is of infinite dimension:*
>
> $$\dim_x C = \infty, \text{ or } \dim C = \infty.^{\text{Ch14, 3}}$$

In addition, it is possible to say that the empty set has dimension -1 by definition.

[11] Cf. [Urysohn, 1925a, pp. 74–79], [Menger, 1928b, pp. 197–209], and [Hurewicz & Wallman, 1948, pp. 20–23].

When first developing these definitions Urysohn may not have known of Poincaré's philosophical approach to dimension or Brouwer's paper of 1913, at least until he had completed much of his investigations and was writing up the results.[12] Certainly Urysohn arrived at his inductive definitions completely independently of Menger's work, which at the time was only in progress.

Although Urysohn's definition of dimension is couched in metrical terms (ε-separations), it is not difficult to show that the concept defined is not metric dependent but is topologically invariant [Urysohn, 1925a, page 68]. His primary interest was in erecting a *topological* theory of dimension.

14.6 Urysohn's Development of His Theory of Dimension

In his 'Mémoire' ('Memoir') (I) [Urysohn, 1925a] [Urysohn, 1926] Urysohn develops a wealth of theorems with full proofs and many ingenious examples and counterexamples.[13] He also formulates interesting conjectures and open problems.

An immediate consequence of his definition is that the defined dimension concept is a topological invariant [Urysohn, 1925a, Théorème II, page 68].

Given all the earlier mathematical work (see Section 13.9), he is able to lay out the relationships among various disconnected sets and his defined 0-dimensional sets [Urysohn, 1925a, pp. 74–79].

Quite naturally, Urysohn investigates the dimensions of Euclidean spaces and the dimensions of points, sets (primarily closed sets), and continua contained in those spaces. He obtains significant results for the Euclidean plane and Euclidean 3-space [Urysohn, 1925a, Ch. II, pp. 79–137], including a definition of a Cantorian line/curve and a pair of theorems concerning the dimensions of various types of points and subsets in these spaces [Urysohn, 1925a, pp. 90, 120]. The results he proves [Urysohn, 1925a, pp. 90–92] justify or validate the following definition of Cantorian line/curve [Urysohn, 1925a, page 93]:

> Def. *A continuum of dimension* 1 *is said to be a Cantorian line/curve.*[Ch14, 4]

Urysohn adds in a footnote that the definition applies to a 1-dimensional continuum situated in any Euclidean space of any dimension.

Urysohn develops the following result for a point-set of Euclidean 3-dimensional space [Urysohn, 1925a, page 120]:

> **Theorem A_3**. *Let C be a set of E_3. Every interior point of C is of dimension* 3 *(with respect to C); every boundary point that is not a limit point of interior points is of dimension* 0, 1 *or* 2*; finally, a boundary point that is a limit point of interior points can be of any dimension* ≤ 3.[Ch14, 5]

[12] However, Poincaré's 1912 philosophical paper is referenced in the published 'Mémoire' ('Memoir') [Urysohn, 1925a, page 38]. Brouwer's paper is also referenced [Urysohn, 1926, page 287].

[13] In the course of his early research Urysohn [Urysohn, 1925a, pp. 121–122] independently produced an example very similar to a celebrated example of Antoine [Antoine, 1921], Antoine's necklace.

The reasoning for this theorem involves linking, starting from Gauss' linking integral and then developing into linking for polyhedra [Urysohn, 1925a, pp. 96–120]. In effect, the theorem leads to and justifies or validates the following definition [Urysohn, 1925a, page 122]:

> Def 3_2. *A continuum K of dimension* 2 *is said to be a Cantorian surface if it remains connected after the removal of the points of an arbitrary closed punctiform [hereditarily discontinuous] set.*[Ch14, 6]

In the general case one has the definition [Urysohn, 1925a, page 124]:

> Def 3_n. *A continuum of dimension n is said to be a Cantorian manifold/multiplicity (of dimension n) if it remains connected after the removal of any closed set of dimension* $n - 2$.[Ch14, 7]

This definition is justified or validated much later in the work by the following theorem and its proof [Urysohn, 1926, pp. 309; 309–310]:

> **Theorem $\mathbf{A_n}$**. *Let C be a set of* E_n. *Every interior point of C is of dimension n (with respect to C); every boundary point that is not a limit point of interior points is of dimension* $0, 1, \ldots, n-2$ *or* $n-1$; *finally, a boundary point that is a limit point of interior points can be of any dimension* $\leq n$.[Ch14, 8]

When Urysohn came to proving this theorem and other general theorems concerning Euclidean n-space and also to proving that Euclidean n-space has dimension n in the sense of his definition, he experienced difficulties, according to Alexandroff [Urysohn, 1951a, pp. 14–16]. In attempting to overcome them he was led to the Lebesgue tiling principle, possibly without at first knowing Lebesgue's or Brouwer's earlier work, though he does cite the papers of these two mathematicians in the 'Mémoire' ('Memoir') [Urysohn, 1926, page 287]. This principle became the key to solving many of his problems [Urysohn, 1926, Chapitre V. Les espaces Euclidiens, pp. 286–316]. In Urysohn's terms it can be stated in the following way [Urysohn, 1926, pp. 287–288]. First, a finite system of closed sets will be said to have order n if some n of them have a nonempty intersection but every $n + 1$ of them have an empty intersection. Further, a closed set will be said to have order n with respect to ε if there exists a finite closed covering of order n for the set, such that members of the covering all have diameters $< \varepsilon$. Finally, a closed set will be said to have the *true order n* if the set is of order n with respect to any positive ε and also there exists a positive ε_1, such that the set is not of order $< n$ with respect to ε_1. The primary result is then: the necessary and sufficient condition for a closed set to be of Urysohn-dimension n is that its true order is $n + 1$ [Urysohn, 1926, page 301]. In other words, there is an equivalence between Urysohn's dimension concept and 'covering dimension' based on the tiling principle.

On this basis and some additional steps of reasoning Urysohn is able to justify or validate his broad definition of Cantorian manifold (multiplicity), Def 3_n. His definitions of Cantorian line/curve, Def 3_1, Cantorian surface, Def 3_2, and Cantorian manifold, Def 3_n, constitute Urysohn's primary solutions to the problems that his teacher Egorov originally posed. He was convinced of their correctness after investigating numerous examples and special cases. Moreover, he went beyond these basic definitions and validating theorems to examine other dimension-theoretic properties of sets.

Thus, Urysohn proves a selection of fundamental theorems about dimension. For unions of sets the following theorem holds [Urysohn, 1926, page 317]:

> **Theorem.** *For any sets L and M, one always has*
>
> $$\dim(L + M) \leq \dim L + \dim M + 1.$$ [Ch14, 9]

In the inequality the symbol + means set union ∪, so that the union of any two sets is less than or equal to the sum of their dimension numbers plus 1.

In the case of *closed sets* the union of a finite number of closed sets or the union of a countable number of closed sets (i.e., F_σ sets) of dimension $\leq n$ is still of dimension $\leq n$ [Urysohn, 1926, pp. 337, 341]:

> **Theorem II.** *The set-sum [set union] of a finite number or a countable infinity of sets $\mathfrak{F}_\sigma$ of dimension $\leq n$ is also an $\mathfrak{F}_\sigma$ of dimension $\leq n$.* [Ch14, 10]

(Compare the statements and proofs of this theorem and its corollaries in Hurewicz and Wallman's *Dimension Theory* [Hurewicz & Wallman, 1948, pp. 30–32].)

Urysohn considers the decomposition of sets into sets having dimension 0 in some detail [Urysohn, 1926, Chapitre VI, pp. 316–351] (see background in Section 13.9). He presents two results of particular interest.

First, he proves the following theorem [Urysohn, 1926, page 324]:

> **Theorem I.** *If a set of dimension n is decomposed into a sum [union] of sets of dimension* 0*, then this sum [union] is constituted of at least n* + 1 *terms.* [Ch14, 11]

(On this theorem compare the statements and proofs of this theorem and its corollary in Hurewicz and Wallman's *Dimension Theory* [Hurewicz & Wallman, 1948, pp. 32–33].)

Second, he sees his goal as proving a converse to this theorem. However, in his 'Mémoire' ('Memoir') (I) he openly acknowledges some frustration concerning his desired goal and what he actually could achieve in this research direction [Urysohn, 1926, page 324]:

> I did not manage to prove the converse theorem in the general case; I will establish it in the particular case of closed sets and, more generally, of sets $\mathfrak{F}_\sigma$. [Ch14, 12]

Accordingly, he shows the following theorem [Urysohn, 1926, page 345]:

> **Theorem III.** *Every $\mathfrak{F}_\sigma$ of dimension n is the union of n*+1 *sets of dimension* 0.
>
> ...

> **Corollary.** ... *The necessary and sufficient condition that a closed set (or an $\mathfrak{F}_\sigma$) be of dimension n is that one can decompose it into n + 1 sets of dimension* 0 *and that one cannot decompose it into n sets of this type.*[Ch14, 13]

He says that he tried to prove these results without the restriction to closed or F_σ sets, but that he was unable to do so [Urysohn, 1926, Chapitre VI §.13. Conjecture and Problème μ, page 350]. However, a little later others did succeed in demonstrating the unrestricted theorem.[14] (Also, compare the results in Hurewicz and Wallman's *Dimension Theory* [Hurewicz & Wallman, 1948, pp. 32–33].)

Near the end of the 'Mémoire' ('Memoir') (I) Urysohn discusses various decompositions of sets according to dimension. As an example he mentions a decomposition of the Euclidean plane in relation to an analogous (though differing in dimension number) result of Sierpiński [Urysohn, 1926, page 347]:

> *The Euclidean plane E_2 being an $\mathfrak{F}_\sigma$ of dimension* 2 *can be decomposed into three sets of dimension* 0, *but it cannot be decomposed into two sets of this species.* This is stated with respect to the following theorem of M. Sierpiński [Sierpiński, 1923]:
>
> *E_2 can be decomposed into three sets homeomorphic to linear sets, but it cannot be decomposed into two sets of this type.*[Ch 14, 14]

The much later part II of the 'Mémoire sur les multiplicités Cantoriennes' ('Memoire on Cantorian multiplicities/manifolds') covers Urysohn's study of Cantorian lines. Because of his untimely death he did not complete his editing of it. Consequently, his mathematical colleague and close confidant Paul Alexandroff edited and published the posthumous work [Urysohn, 1927a]. The two parts of the 'Mémoire' ('Memoir') constitute an outstanding achievement for such a young mathematician.

Urysohn was very ambitious in building his extensive theory of dimension. His dimension concept provided a workable frame in which he could develop a great many topological results. In some sense, he was overly ambitious even to think of classifying 'all' general sets in n-space. He could not really take in all of topology. In point of fact his mainly set-theoretic methods would not be strong enough to solve all topological problems, as later developments in topology show. His friend Paul Alexandroff moved on to use homological and algebraic topological means to solve problems related to dimension [Alexandroff, 1932a].

Urysohn went beyond these investigations in dimension and curve theory to make remarkable contributions to general topology—well known to all topologists. In this research he often collaborated with his friend Paul Alexandroff. This work produced a considerable extension and deepening of the abstract theory of metric spaces (Fréchet) and topological spaces (Hausdorff).[15] An important example of his further topological investigations carried out in collaboration with Alexandroff is their 'Mémoire sur les espaces topologiques compacts' ('Memoir on the compact topological

---

[14] Cf. Tumarkin [Tumarkin, 1925a] [Tumarkin, 1926b] [Tumarkin, 1928b] and Hurewicz [Hurewicz, 1927a].

[15] Cf. [Arboleda, 1979] for historical information on the early topological work of Urysohn and Alexandroff.

spaces') [Alexandroff & Urysohn, 1929]. In this lengthy work the authors develop a whole range of results, both theorems and counterexamples, on compact, bicompact, and locally compact spaces. Above all, Urysohn, together with Alexandroff, made considerable contributions to point-set topology during the 1920s and thereby transformed the subject into a rich domain of modern mathematics.

14.7 Concluding Remarks

As noted earlier in Section 12.12 Urysohn died on 17 August 1924 while swimming in rough seas off the coast of France at Batz-sur-Mer. Considering the fact that he only had three years to devote to topology, one surely must marvel at his mathematical genius and productivity. He made his mark in his chosen field with brilliance and passion. Urysohn bequeathed much to modern topology.

Chapter 15
Menger's Theory of Curves and Dimension

15.1 Introduction

I shall turn now to Karl Menger's investigations into dimension theory and closely related curve theory. There are many similarities between the early progress of Menger and the nearly simultaneous early progress of Urysohn on this topic, even though they worked independently of one another. This coincidence is unsurprising. They both started work from essentially the same problem: the problem of defining the concept of curve[1] and by implication the problem of defining the concept of dimension. Problems in mathematics and science are the principal drivers of progress.

15.2 Menger's Earliest Thoughts about Curves and Dimension

Karl Menger (1902–1985)[2] began thinking about the twin problems of defining the concepts of curve and dimension in the very same year as Urysohn, 1921.[3] Menger, the son of the well-known Austrian economist, Carl Menger (1840–1921), had

[1] Unlike Menger, Urysohn also started from the problem of defining the concept of surface. See Section 14.2.

[2] Cf. [Kass, 1996] [Kass, 1997] [Sigmund, 2017] [Gillies, 1981] [Leonard, 1998] [Beham, 2012]. Cf. [Menger, 1994, Introduction, pp. vii–xx] and Menger's autobiographical writings [Menger, 1979a, pp. 237–255] [Menger, 1994]. Cf. `https://mathshistory.st-andrews.ac.uk/Biographies/Menger/` and `https://www.iit.edu/applied-math/about/remembering-karl-menger`, 'Remembering Karl Menger.' The Karl Menger papers, 1872-2000, are held at Duke University `https://archives.lib.duke.edu/catalog/mengerkarl`.

[3] For my first historical discussion of Menger's work [Johnson, 1981, pp. 233–241] I relied on [Menger, 1926e] [Menger, 1928b] [Menger, 1929b], some unpublished documents and letters, and several conversations I was fortunate to have with Menger many years ago. I remember his helpfulness and kindness in discussing the history in which he played a significant part. Since the time of my first discussion Menger wrote some detailed reminiscences concerning his early investigations into curve and dimension theory and his relations with L.E.J. Brouwer, which have been published [Menger, 1979a, pp. 237–255] [Menger, 1994, Chapter IV, pp. 38–53]. Also, his

D. M. Johnson, *Theories of Dimension from Ancient to Modern Times*, Sources and Studies in the History of Mathematics and Physical Sciences, https://doi.org/10.1007/978-3-032-02682-8_15

Fig. 15.1 Karl Menger (1902–1985) [Photograph from MacTutor History of Mathematics Archive, St. Andrews University, U.K.].

entered Vienna University as a student during the autumn of 1920 with the intention of studying physics.[4] About this time Hans Hahn (1879–1934) was called to Vienna. He arrived there near the beginning of 1921, and he soon announced a seminar on the problems associated with curves, 'Neueres über den Kurvenbegriff' ('New [ideas] concerning the concept of curve'). As I have discussed in Section 13.8, Hahn had a strong research background and specific interests in curve theory as evidenced by his important papers of 1913 and 1914 [Hahn, 1913] [Hahn, 1914a] [Hahn, 1914b] analyzing Jordan curves and Peano space-filling curves using the concept of local connectedness. At the urging of his close friend Otto Schreier (1901–1929) Menger decided to attend the first session of the seminar, which occurred in late April or early May.[5]

In his later reminiscences Menger writes *in extenso* about his earliest foray into curve and dimension theory [Menger, 1994, pp. 38–53].[6] I shall quote from and comment on what he has written. First, he relates how he became so strongly engaged with the topic of curves [Menger, 1994, pp. 38–39]:

Selecta Mathematica [Menger, 2002] [Menger, 2003] have appeared. Additionally, for my current discussion the secondary sources mentioned in the previous footnote have been useful.

[4] Recall that Urysohn likewise had planned to study physics when he entered the university.

[5] According to a letter of Schreier to Menger dated 7 April 1926 the first session took place on 27 April 1921, but according to a letter of Hahn to Brouwer dated 10 April 1926 it took place in early May.

[6] Cf. [Menger, 1979a, pp. 237–241].

> In March of 1921, just after Hahn's arrival in Vienna, I read a notice that he would conduct a two-hour advanced seminar every Wednesday during the spring semester, entitled *Neueres über den Kurvenbegriff* (*News [New ideas] about the concept of curve*). At that time I was at the end of my first semester at the university and primarily a student of physics. I was doubtful that I would be able to follow the advanced mathematical discussions; and I had never seen Hahn before. But after some deliberation, I decided to try.
>
> The first meeting of the seminar allayed my doubts. Hahn went right to the heart of the problem. Everyone, he began, has an intuitive idea of curves: everyone calls a circle and a lemniscate (of the shape of the numeral 8) curves; no one so designates a circular disc or a cube. When speaking of curves one thinks of thread-like objects, wire models, and the like. But anyone who would try to make that idea precise, Hahn said, would encounter great difficulties. In this seminar we would examine attempts by several eminent geometers to define the curve concept, only to find that some of their definitions were too wide, others too narrow and still others altogether unsatisfactory, so that at the end of the seminar we should see that the problem was not yet solved.
>
> I was completely enthralled; and when after that short introduction, Hahn set out to develop the principal tools used in those earlier attempts—the basic concepts of Cantor's point set theory, all totally new to me—I followed with utmost attention.

In his reminiscences Menger then summarizes some of the basics of point-set theory and current views about curves along with some of the paradoxical results (probably derived to some extent from Hahn's presentation at the first meeting). He vividly describes his reaction to Hahn's presentation at the first meeting of the seminar [Menger, 1994, pp. 41–42]:

> I left the seminar room in a daze. Like everyone else, I used the word 'curve' and had an intuitive idea of curves—mental pictures associated with the term. Should I not be able to spell out articulately *how* I used the word and to describe clearly what I saw?
>
> After a weekend of complete engrossment in the problem I felt I had arrived at what appeared to be a simple and complete solution. I told the idea to my friend, the highly talented mathematician Schreier, who listened with great interest. The next day he said that he had given much thought to my attempt but could not find any flaw in it. "However," he added, "if an idea as simple as yours could solve the problem, why would several great geometers have given unsatisfactory solutions and why would Hahn say that after discussing all previous attempts we should see that the problem was still unsolved?" I admitted that this thought had also crossed my mind. Schreier continued, "I then looked into the standard work on set theory, *Grundzüge der Mengenlehre* by Felix Hausdorff, the greatest authority on the subject in Germany; and he claims that the sets traditionally called curves are so heterogeneous that

> they do not fall under any reasonable collective concept."[7] Moreover, he showed me that Ludwig Biebarbach in his booklet on differential calculus said [[Bieberbach, 1917, pp. 4–5]]: "Anyone trying to define the concept of curve comprehensively and in a way that fully renders our intuitive idea certainly would need a description as long as a tape worm and of Gordian [rat-king-like, maybe, king-rat's nest] entanglement."[8] "Yet as you will agree," I said in keeping with a general penchant of mine for simplicity, "one should never reason that an idea is too simple to be correct. I shall tell my solution to Hahn."
>
> There was at that time hardly any contact between students and professors at the University of Vienna. So, an hour before the beginning of the second seminar meeting, I had to muster all my courage before knocking at the door of Hahn's office to tell him that I thought I could solve the problem he had formulated the week before.

After giving a brief description of his basic idea, Menger continues by relating his engagement with Hahn [Menger, 1994, page 43]:

> Hahn, who hardly looked up from the book he was reading when I entered, became more and more attentive as I went on. At the end, after some thought, he said that this indeed would be a workable definition, and asked me where I had learned so much about point sets and topology. I replied that I was a physics student at the end of my first semester and had not heard about topology; but that my definition used only concepts defined the week before in the first meeting of the seminar, which was all that I had ever heard about point sets. Hahn had not realized that two hours of his excellent presentation of basic concepts were sufficient to make them operative even in the mind of someone totally unfamiliar with the field. He nodded rather encouragingly and I left.

Menger continues with a summary of the fundamentals of his new proposed definitions. Then he describes his feverish work as follows [Menger, 1994, pp. 45–46]:

> The more I thought about these problems, the more they fascinated me and the more eager I became to complete their clarification. But after long studies in unheated libraries during the cold winter of 1921 and weakened by post-war malnutrition and overwork, I succumbed, in May 1921, to a serious respiratory infection, continuing to work feverishly in the literal as well as in the

---

[7] See [Hausdorff, 1914, page 369, footnote] and Section 13.10, which gives the quotation from Hausdorff's book.

[8] The quotation is from Bieberbach introductory text on differential and integral calculus. Bieberbach's original German is [Bieberbach, 1917, pp. 4–5]: 'Wollte der Leser versuchen, den Kurvenbegriff so zu fixieren, daß er den vollen Inhalt unserer Vorstellung in mathematisch faßbarer Weise wiedergäbe, so würde er sicher nicht ohne Beschreibung von bandwurmartiger Länge und rattenköniggleicher Verschlungenheit auskommen.' The text contain the usual simple examples of functions defining well-known curves, but it also contains the Peano space-filling curve and curves that are everywhere continuous and nowhere differentiable. Thus, Bieberbach wanted to acquaint the student with the diversity of possible functions and curves.

Fig. 15.2 Karl Menger as a Young Man; Date Unknown [[Leonard, 1998, page 1], courtesy Rosemary Menger Gilmore, Chicago].

> metaphorical sense. Too ill to attend the last meetings of the seminar I wrote Hahn a letter summarizing my results. But in the fall of 1921, my bout of the 'Viennese disease' became so serious that I felt it might be years before I could elaborate my ideas. After being bed-ridden for several months I had to retire for almost a year to a sanatorium in Styria—one of the two Austrian counterparts of the *Magic Mountain* which Thomas Mann set in pre-war Switzerland.

While it was an ordeal to be isolated on the mountain, for Menger it was also a blessing in disguise. He had the opportunity to read more and to teach himself more mathematics. Moreover, he had considerable time to develop his new ideas. He was also fortunate in this early time to be cured of the respiratory disease.

One of the attractions of Menger's early thinking about curves and dimension is the basic geometrical and physical intuitions guiding him in building his theories. He writes the following [Menger, 1994, page 48]:

> I looked for clues in the physical world, even though nothing but well-defined geometric entities such as cubes, surfaces of spheres, and ellipses are unequivocally categorized as solids, surfaces, and curves. Material objects, for example an apple, its skin, and its stem, are so classified merely with a grain of

allowance, for all material objects unless they are considered as dispersed sets of molecules are 3-dimensional. Yet only objects such as metal balls, wooden blocks and pieces of rock are regarded as *typically* 3-dimensional or as solids. Sheet metal, paper and films approximate what we mean when speaking of 2-dimensional objects or surfaces. Wire, threads, and chalk-marks roughly represent our idea of 1-dimensional objects or curves. A simple experiment that I devised differentiates between these three classes of physical objects: remove from an object a point and all points sufficiently near to it. The dimension of the object determines the *tool* required for the performance of the experiment. In the case of a wooden block, a *saw* is needed in order to cut the solid along surfaces. If the object is a cardboard surface, then a pair of *scissors* suffices, and the cuts are along curves. If the object is a wire curve, then all that is needed are *pliers* with which we have to pinch the curve in dispersed points. Finally, if the object is dispersed like a pile of sand, then *no tool at all* is required since nothing has to be dissected.

The physical images and simple experiments are suggestive of an inductive definition of dimension and a direct relation to the concept of curve. Menger was able to transform his 'intuitions' about curves and dimension through the adoption of point-set-theoretic methods to develop his theories of dimension and curves.

15.3 Menger's Earliest Papers on Curves and Dimension Theory

At this point I shall examine the written record of what Menger did in the first years of his research. By June 1921 Menger had produced a short paper of three and one-half pages entitled 'Der Begriff der Kurve' ('The concept of curve').[9] Although it contains only an informal attempt to define the concept of a curve embedded in Euclidean 3-space, it does reveal the germ of Menger's later ideas on curve theory and implicitly on dimension theory. According to him the most important property of a curve is its 1-dimensionality. This view expressed in the paper leads him to call every closed connected 1-dimensional point set (in 3-space) a 'curve in the wider sense' ('Kurve im weiteren Sinne'), where a set is 1-dimensional provided each of its points has arbitrarily small neighborhoods with boundaries, each one of which intersects the set in finitely many points. This is a definition of regular curve in Menger's later conceptual framework [Menger, 1926a]. However, a marginal insertion, probably added later, questions whether 'countably many' might not be a better defining property than 'finitely many' for the intersection.[10] In the rest of the paper there is a brief comparison between the newly-defined concept of curve and the vague popular one, plus there is an indication of types of points on curves,

[9] This paper has never been published to my knowledge. The original is in the Brouwer Archive. The content may be roughly equivalent to what he sent as a summary to Hahn toward the end of Hahn's seminar.

[10] In the original German the definition reads as follows:

e.g., endpoints, knot points, and so forth. No theorems are proved. Though the paper is only of historical interest, being Menger's first written effort on the subject of curves and dimension, it does show him from the beginning using the concepts of neighborhood and boundary in seeking to solve the basic geometrical problem.

Menger continued to pursue his investigations. However, illness forced him to be away from Vienna for long periods and inevitably slowed his progress, as Menger indicates in his reminiscences quoted above. Because he wished to declare the priority of his ideas, fearing that his illness would prevent him from elaborating them, he completed another short paper 'Über die Dimensionalität von Kontinuen' ('Concerning the dimensionality of continua'), an 'Entwurf' ('sketch') [Menger, 1926e, pp. 1122–1123] [Menger, 1929b, pp.411–412], by the autumn of 1921 for deposit with the Akademie der Wissenschaften (Academy of Sciences), Vienna.[11]

> Eine abgeschlossene, zusammenhängende Punktmenge $\mathfrak{C}$ heißt eine abgeschlossene *eindimensionale* Pnnktmenge, wenn zu jedem ihrer Punkte P ein von $\emptyset$ verschiedener Abstand A existi[e]rt, derart, daß zu jedem Abstand $a < A$ mindestens eine offene, dreidimensionale Punktmenge $\mathfrak{M}_a$ mit dem Durchmesser a gehört, die P enthält und die Eigenschaft hat, daß $\mathfrak{C}$ mit der Begrenzung ihrer abgeschlossenen Hülle endlich viele Punkte gemeinsam hat.
>
> Additionally, the word 'endlich' ['finitely'] is underlined, a '?' is added, and the following words appear: auch *abzählbar* wäre eine Beschränkung! (Kontinuumproblem!)

This definition may be translated as:

> A closed connected point-set $\mathfrak{C}$ is called a closed *one-dimensional* point-set, if to each of its points P there exists a distance A different from 0, such that to each distance $a < A$ there is at least one open three-dimensional point-set $\mathfrak{M}_a$ with diameter a, that contains P, and that has the property that $\mathfrak{C}$ has finitely many points in common with the boundary of its closed hull.
>
> *countably many* would also be a limitation [defining condition] (continuum problem!)

The main definition is essentially a definition of a regular curve in 3-space. The alteration yields the definition of a rational curve [Menger, 1926a].

[11] The original of this paper, to be found in the Brouwer Archive, actually bears the title as I have given it rather than the one printed in [Menger, 1926e, page 1122] and [Menger, 1929b, page 411]. At the top of the original the following inscription appears:

> Dieses Blatt bildete den Inhalt des zur Wahrung der Priorität der Akademie der Wissenschaften übergebenen versiegelten Schreibens Nr 778/1921.
> 9. April 1926
> F. Becke
> Generalsekretär der Akademie d. Wissenschaften in Wien.

This inscription may be translated as:

> This paper formed the content of the sealed letter No. 778/1921, which was given to the Academy of Sciences to protect its priority.
> 9. April 1926
> F. Becke
> General Secretary of the Academy of Sciences in Vienna.

In the paper there is a revised version of his concept of curve [Menger, 1926e, page 1123] [Menger, 1929b, page 412]:

> We define: A continuum $\mathfrak{K}$ is called a 'curve,' if in each neighborhood $\mathfrak{U}$ of every point P of $\mathfrak{K}$ a neighborhood $\mathfrak{U}_1$ is contained, such that the intersection $\mathfrak{D}$ of $\mathfrak{K}$ with the boundary of $\mathfrak{U}_1$ contains no connected part [i.e., it is hereditarily disconnected].[Ch15, 1]

This is the definition that Menger chose as the basis for his subsequent curve theory. After it, a few theorems about curves follow, stated without proof. There are also tentative definitions of the concepts of surface (i.e., 2-dimensional continuum) and n-dimensional continuum. While this deposited paper is only a sketch of his ideas, it clearly shows that Menger was conscious of the possibilities of a inductive approach to dimension, though at the time he still did not know about Poincaré's or Brouwer's earlier efforts.

By the turn of the year 1921/1922 Menger had written a longer essay 'Über die Dimensionalität von Punktmengen' ('Concerning the dimensionality of point-sets') [Menger, 1929b, pp. 412–418], which filled in some of the details of the previous sketch. He submitted it to Hahn in his capacity as editor of the *Monatshefte für Mathematik und Physik* (*Monthly Journal for Mathematics and Physics*) in hopes that it would be published. In this work for the first time there are proofs of theorems connected with the proposed definitions. The first section is about the curve concept, while the second section is devoted to the more general topic of dimension.

At the beginning of the essay Menger references Brouwer's paper on domain invariance [Brouwer, 1911c] (he also indirectly references Brouwer's paper on invariance of dimension [Brouwer, 1911a]), in which Brouwer thinks of dimension as a certain property of point-sets invariant under one-one continuous mappings. This view is in contrast with Menger's inductive approach. The editor, Hahn, adds that Menger did not know Brouwer's paper on defining dimension inductively [Brouwer, 1913b] at this time [Menger, 1929b, pp. 412–413].

In the first section of his essay Menger defines *curve* as follows [Menger, 1929b, page 413]:

> We call a *curve* a continuum K with the property that if p is an arbitrary point of K and U_1 an open neighborhood of p, then there exists an open neighborhood of p, $U_2 < U_1$ [$U_2 \subset U_1$], whose boundary has no connected part in common with K [i.e., the intersection of K and the boundary of U_2 has no connected part].[Ch15, 2]

He shows that the one-one continuous image of a curve is a curve. Also, he shows that a set having an open neighborhood in 2-dimensional space is not a curve.

In the second section of his essay Menger defines 0-dimensional sets as follows [Menger, 1929b, page 416]:

> *Definition*: We call 0-*dimensional* a nonempty set that contains no connected part [i.e., that is hereditarily disconnected].[Ch15, 3]

With this definition he attempts to prove the following assertion [Menger, 1929b, page 416]:

> For a set M to be 0-dimensional it is necessary and sufficient that for each of its points p and for every open neighborhood U_1 of p there exists an open neighborhood of p $U_2 < U_1$ $[U_2 \subset U_1]$ with whose boundary M has no point in common.[Ch15, 4]

Unfortunately, this assertion is not correct as it stands. Only the sufficiency part is correct. Hahn soon detected the straightforward mistake in the proof of necessity and, thus, could not accept the paper for publication in the form submitted. However, the error involved in the proposed theorem touches upon subtle differences between hereditarily disconnected sets (those defined as 0-dimensional in the above definition) and 0-dimensional sets (those described in the condition of the assertion and later defined by Menger) for the non-compact case. The Polish topologists were especially interested in these subtleties,[12] but Menger was not yet aware of them.

Menger concludes his short paper with definitions of 1-dimensional set and n-dimensional set, developing them inductively from his definition of 0-dimensional set. By this time he had become acquainted with Poincaré's second paper on topological dimension [Poincaré, 1912]. Although Hahn found the paper flawed, overall it confirmed Menger in the view that dimension should be approached inductively.

15.4 Menger's Further Thoughts on Dimension

Further consideration of the definition of n-dimensional set included in the paper just discussed [Menger, 1929b, page 418] led Menger to clarify it. In a letter to Hahn of 15 February 1922 (original letter reproduced in full in [Dalen, 2011b, pp. 1104–1106]) he offers the following modified definitions for n-dimensional sets [Dalen, 2011b, page 1104] [Menger, 1926e, page 1124] [Menger, 1929b, pp. 418–419]:[13]

> A nonempty set M of R_m is called *n-dimensional*, if
>
> 1.) for each of its points P and every open neighborhood $U_1(P)$ an open neighborhood $U_2(P) < U_1(P)$ $[U_2(P) \subset U_1(P)]$ exists with whose boundary M has an at most $(n-1)$-dimensional set in common; and if M

[12] See Section 13.9. Cf. [Sierpiński, 1921], [Knaster & Kuratowski, 1921] and also [Menger, 1928b, pp. 197–209], [Hurewicz & Wallman, 1948, pp. 20–23], and [Engelking, 1978, pp. 31–37] [Engelking, 1995, pp. 24–29].

[13] The original of this letter is in the Brouwer Archive, reproduced in [Dalen, 2011b, pp. 1104–1106]. I have used the original for my transcription. Prof. Menger informed me long ago that his clarification of the definitions for n-dimensional sets had nothing to do with Hahn's detection of the error in the proof of the 'theorem' about 0-dimensional sets in the paper just discussed. However, his clarification alters the basis of his inductive definition and is logically, though not historically, related to the error.

2.) contains at least one point Q for which there exists an open neighborhood U, such that with the boundary of every open neighborhood $U_1(Q) < U$ $[U_1(Q) \subset U]$ M has a set in common that contains an $(n-1)$- dimensional part. The empty set is -1-dimensional.

A set is called *exactly (perhaps "homogeneously") n-dimensional* when it contains only points Q with property 2. (Every null-dimensional set is exactly null-dimensional.) Every part [subset] of an n-dimensional set is at most n-dimensional. The union of all exactly n-dimensional parts [subsets] of M is called the *n-dimensional kernel* of an n-dimensional set M.[Ch15, 5]

With these definitions Menger had arrived at his version of the definition of dimension (in this case stated for Euclidean spaces), which became a foundation for the dimension theory of the 1920s and 1930s. He had come to it independently of Brouwer's and Urysohn's work. However, much more had to be done in terms of applying the definition and elaborating a full theory. In the letter to Hahn he only mentions as a conjecture the characterization of n-dimensional sets in Euclidean n-spaces as precisely those sets that contain interior points. What was needed now was a body of substantial theorems with complete proofs. [14]

In spite of illness and the necessity of convalescing away from Vienna, Menger persevered with his research during 1922. By November he had put together another paper 'Über die Dimensionalität von Punktmengen' ('Concerning the dimensionality of point-sets') [Menger, 1929b, pp. 419–425], his most substantial work on dimension so far. At the time he was staying in Aflenz, southwest of Vienna. He sent the new paper to his friend Schreier, asking him to submit it to Hahn for the *Monatshefte für Mathematik und Physik*. Schreier checked it and presented it to Hahn on 29 or 30 November 1922.[15]

The paper contains Menger's definitions for the dimension concept and four theorems with full proofs. In particular, he introduces a new, insightful way of defining dimension. This is a definition of local dimension or dimension at a point based on neighborhoods of the given point and their boundaries [Menger, 1929b, page 420].[16] This leads to a 'dimensional structure theorem' [Menger, 1929b, pp. 420–421]:

Theorem II. If the set M is at least n-dimensional at its point m, then every neighborhood $U(m)$ contains an at least n-dimensional set in M of points from M at which M is at least n-dimensional.[Ch15, 6]

[14] Most of the rest of the letter to Hahn concerns the relationship between measure and dimension.

[15] I derived the information about the submission of this paper from a letter of Schreier to Menger dated 7 April 1926. Schreier offered help and advice to Menger in the very early stages of his research.

[16] According to Menger a set M is n-dimensional at the point p, if for every neighborhood $U(p)$ there exists a neighborhood $U_1(p) \subset U(p)$ such that the intersection of M with the boundary of $U_1(p)$ is exactly $(n-1)$-dimensional and, moreover, a neighborhood $V(p)$ exists, such that the intersection of M with the boundary of every neighborhood $U_2(p) \subset V(p)$ is at least $(n-1)$-dimensional,

When examining the paper for possible publication, Hahn found the proof of this theorem defective. In the proof Menger incorrectly extends the domain of application of the Heine–Borel covering theorem; in effect, he asserts a new covering theorem that is not true in all cases [Menger, 1929b, page 422]. Hence, the proof is not valid. The main theorem needs modification to make it correct. Menger was forced to rethink the paper before it could be accepted for publication. The entire situation with respect to this theorem is related to 'weakly 1-dimensional and n-dimensional sets.' [17] However, in spite of the defects of the proof it does contain the essence of a method, which Menger found useful in his investigations: the method of modification of a neighborhood in the vicinity of its boundaries. The theorem itself is the prime ancestor of Menger's dimensional structure theorems.

15.5 Menger's First Publications on the Theory of Dimension and Curves

All this preliminary work led to Menger's first publications. After the long bout of illness he returned to Vienna in April 1923. He had already begun to write a 'definitive' paper for the *Monatshefte für Mathematik und Physik* intended to be a full exposition of his ideas, while correcting his earlier mistakes. It is amazing how quickly Menger developed mathematically over the two years 1921–1923 in spite of his illness and being away from the university.

About this time, April or May 1923, he became acquainted with Brouwer's key paper of 1913 defining the dimension concept [Brouwer, 1913b] as well as Lebesgue's second paper on the tiling principle [Lebesgue, 1921]. By the end of 1923 he had completed the paper 'Über die Dimensionalität von Punktmengen. Erster Teil' ('Concerning the dimensionality of point-sets. First Part') [Menger, 1923]. It was submitted to the printer on 12 December 1923 and issued on 12 April 1924. In fact, an earlier version of this paper was submitted to the *Monatshefte* in November 1922, though not published as such then; it is printed in [Menger, 1929b, pp. 419–425]. To the first published part Menger added a second part in 1924 [Menger, 1924d] and eventually a third part in 1929 [Menger, 1929d].

The first published part is mainly concerned with the definition of curve and the resulting theory. In this work Menger defines *curve* (*Kurve*) in the following way [Menger, 1923, page148]:

> A continuum K of a metric space is called a *curve*, if for every point k of K and to every neighborhood $U(k)$ there exists a neighborhood (Umgebung) $U'(k) < U(k)$ $[U'(k) \subset U(k)]$ such that the intersection of K with the boundary of $U'(k)$ is hereditarily disconnected; whereby we call a set *hereditarily disconnected* if all of whose components consist of only single points.[Ch15, 7]

[17] Cf. [Sierpiński, 1921] and [Menger, 1928b, pp. 45–50, 135–148].

Menger notes the equivalence of his notion of *zusammenhangslos* with Hausdorff's *punkthaft* [Hausdorff, 1914, page 322] and Sierpiński's *dispersé* [Sierpiński, 1921], in modern terms *hereditarily disconnected* (see Section 13.9). Then he proves several interesting theorems about curves as defined. For example, he shows [Menger, 1923, page 152]:

> *V. A curve of [Euclidean n-space]* R_n $(n > 1)$ *contains no interior point.*[Ch15, 8]

Also, he shows [Menger, 1923, page 156]:

> *VIa. The curves of [Euclidean 2-space]* R_2 *are identical with the continua with no interior points.*[Ch15, 9]

At the end of this paper he introduces his general definition of dimension for an arbitrary metric space and its subsets [Menger, 1923, pp. 157–160]. He briefly discusses 0-dimensional sets and 1-dimensional sets, with curves as special cases, and concludes by indicating his plans for the second part. At the very end there is a note in which he says that he had only just become acquainted with the work of Urysohn through a report of Urysohn's lecture at Marburg on 21 September 1923 [Urysohn, 1923] (see Section 12.12) and Urysohn's two notes in the *Comptes rendus* (*Proceedings*) of the Academy of Sciences, Paris [Urysohn, 1922a] [Urysohn, 1922b].

On 17 December 1923 Menger deposited another paper with the Akademie der Wissenschaften (Academy of Sciences), Vienna, presumably in order to provide a certified academic record of the independence of his own work from that of Urysohn. Thus far Urysohn had only published his two *Comptes rendus* (*Proceedings*) notes and given his Marburg talk. Menger's short paper, 'Zur Theorie der Punktmengen II' ('On the theory of point-sets II') [Menger, 1929b, pp. 425–432], was transformed in the following months into the second part of his larger work.

Continuing to work hard on his research, Menger attained his doctorate on 23 June 1924. The examiners were Hans Hahn and his colleague Phillip Furtwängler. The dissertation consists of the printed version of the first part of 'Über die Dimensionalität von Punktmengen' ('Concerning the dimensionality of point-sets') [Menger, 1923] and another preliminary version of the second part of the work.[18]

In his research Menger strove to solve the difficulties involved in the dimensional structure theorem that he had tried to prove in the paper of November 1922. He had made definite progress towards a solution in the deposited paper of December 1923. By the time of achieving his doctorate he had a good solution, which he incorporated in the second part of his dissertation (pages 16–26).

Moreover, in a paper on 'Einige Überdeckungssätze der Punktmengenlehre' ('Some covering theorems of point-set theory') [Menger, 1924a] he proceeded to examine in detail a variety of covering theorems related to the one that he had mis-

[18] There is a copy of the dissertation in the Universitäts-Bibliothek, Wien (University Library, Vienna), with the call-mark D213 Menger. It bears the inscription: 'Hahn-Furtwängler 23. VI. 1924 Z1074 & 5975.'

takenly formulated within the proof of the earlier dimensional structure theorem.[19] This paper was read to the Akademie der Wissenschaften (Academy of Sciences), Vienna, on 3 July 1924. In addition, he presented some of these results at the annual meeting of the Deutsche Mathematiker-Vereinigung (German Association of Mathematicians), held in Innsbruck, 21–27 September 1924, where he gave a talk on 23 September [Menger, 1924b].

By the autumn of 1924 Menger had completed a final version of the second part of his larger work on dimension for the *Monatshefte* [Menger, 1924d]. He submitted this on 6 October 1924. By 30 December 1924 the printer's proofs of the second part were completed, and the second part was then issued as an offprint or preprint. However, because of publication difficulties with the *Monatshefte* arising from the financial crisis in Austria at the time it was not issued in the journal until 20 September 1926 [Brouwer, 1928b, page 957] [Menger, 1930, pp. 180–181].

Some time in the early part of 1924 Menger and Brouwer began corresponding.[20] There is a letter from Menger to Brouwer, dated 12 March 1924, thanking Brouwer for an offprint of his corrected paper on natural dimension [Brouwer, 1923b]. In the letter Menger continues by informing Brouwer of his own early work [Dalen, 2011a, pp. 266–267] [Dalen, 2011b, page 1205]:

> When I tried in 1921 to define curves and the dimension concept, I was in the first year of my university study and did not know your article in the *Journal für die reine und angewandte Mathemarik* 141 [[Brouwer, 1913b]] at all, in which the definition is essentially anticipated. But also later, after I found the publication when studying the relevant literature, I hoped that I could offer through my results at least a small supplement.[Ch15, 10]

In the letter Menger goes on to summarize a few of his results. He writes that he hopes to send Brouwer the printed form of the second part of his work [Menger, 1924d], which, as noted above, was only available in printed form at the very end of 1924 and then issued in the journal in 1926.

Menger and Urysohn were also in correspondence in early 1924. There is a friendly postcard from Urysohn to Menger, dated 22 March 1924 [Dalen, 2011b, page 1209]. Urysohn writes to say that he only received a letter from Menger a few days before and that Menger's offprint was not included with it—apparently the offprint was lost. He says mail often gets lost in Moscow when sent to the university, and so he gives Menger his personal address. He writes [Dalen, 2011b, page 1209]:

> It seems we found the definition of dimension almost simultaneously: I found mine in July 1921, [and] presented it to the Moscow Mathematical Society in October 1921.[Ch15, 11]

Thus, Urysohn dates his discovery as a time in July 1921. (Recall that Alexandroff in a much later publication dated the discovery in August 1921 (Section 14.2).) Urysohn references his two notes that appeared in the *Comptes rendus* (*Proceedings*) of the

[19] See the analysis in [Menger, 2002, pp. 144–145].

[20] The Brouwer-Menger correspondence is extensive, with some 35 items, dating from 12 March 1924 to 29 November 1928, as printed in [Dalen, 2011b].

Academy of Sciences, Paris, in September 1922, and mentions that the first part of his principal work is currently in press with *Fundamenta Mathematicae*.

Brouwer must have encouraged Menger to summarize his results on dimension for the Dutch Academy. Thus, even before Menger submitted the second part of his main paper in October 1924, Brouwer was able to communicate Menger's summary, 'Ueber die Dimension von Punktmengen' ('On the dimension of point-sets'), on 27 September to the Koninklijke Nederlandse Akademie van Wetenschappen, the Dutch Academy at Amsterdam [Menger, 1924c]. In this short note Menger gives his definition of dimension and states that he found it in the autumn of 1921. He says that only later did he come upon Brouwer's 1913 paper and then Urysohn's *Comptes rendus* (*Proceedings*) note of 4 September 1922 [Menger, 1924c, page 639]. The summary highlights some of Menger's results on dimension. He says that his more extensive paper to appear in the *Monatshefte* is in press.

This second part of Menger's main work on dimension, 'Über die Dimension von Punktmengen. II. Teil' ('Concerning the dimension of point sets. Part II') [Menger, 1924d],[21] comprises a relatively complete exposition of his dimension theory, though it is not as expansive as Urysohn's 'Mémoire' ('Memoir') (I) [Urysohn, 1925a] [Urysohn, 1926]. I shall briefly examine its contents.

Menger begins by giving his definition of dimension for metric spaces or Hausdorff topological spaces [Menger, 1924d, pp. 138–139]:

> *A set A is called n-dimensional if n is the smallest nonnegative integer with the following property:*
>
> *For each point a of A there exists a sequence of neighborhoods descending to a with at most* $(n-1)$*-dimensional boundaries; i.e., for each point a of A and every neighborhood* $U(a)$ *there exists a neighborhood* $U_1(a) < U(a)$ $[U_1(a) \subset U(a)]$ *with an at most* $(n-1)$*-dimensional boundary. The empty set and only this is* (-1)*-dimensional and at most* (-1)*-dimensional.*
>
> Moreover, a set that is n-dimensional or less than n-dimensional is called *at most n-dimensional*, while we designate a set that is not at most n-dimensional as *at least* $(n+1)$*-dimensional*.[Ch15, 12]

After deriving the trivial proposition from this definition that a subset of an n-dimensional set is at most n-dimensional, Menger explores three principal topics in the rest of the paper: (1) the topological invariance of his dimension concept, (2) the structure of n-dimensional sets, and (3) the dimension of sets in Euclidean n-space. Topological invariance is straightforward to prove, so he devotes most attention to the last two topics.

In the second section of his paper Menger takes up the dimensional structure of sets, the topic upon which he had foundered in his earlier paper of November 1922. Having resolved the difficulty over covering theorems, he now develops and achieves two main results. These are stated for totally complete sets (total vollstämdige Mengen), a generalization of compact sets. A closed set is called 'totally complete' if every

[21] Note that there is a minor change in the title between part I [Menger, 1923] and part II [Menger, 1924d].

bounded subset of it is compact [Menger, 1924d, page 144].[22] First, he proves that for a totally complete set A, which is at least n-dimensional at its point a, there cannot exist both a neighborhood $U(a)$ of a and a closed at most $(n-1)$-dimensional set M, such that M contains all the points of $A \cap U(a)$ at which A is at least n-dimensional [Menger, 1924d, page 144]. This is Menger's revised dimensional structure theorem given in a somewhat negative form. In his 'Mémoire' ('Memoir') [Urysohn, 1926, page 270] Urysohn states and proves essentially the same theorem in a more positive form. If F is a closed set (in a compact metric space) and $\mathfrak{N}_k(F)$ represents the set of points of F at which it is at least k-dimensional, then every point of $\mathfrak{N}_k(F)$ has the same dimension with respect to F and to $\overline{\mathfrak{N}}_k(F)$ (the closure of $\mathfrak{N}_k(F)$). Second, Menger proves in his paper another important theorem: the sum theorem. In essence, this can also be found in Urysohn's work [Urysohn, 1926, pp. 337, 341]. The sum theorem in Menger's formulation asserts that in a totally complete space the union of countably many at most n-dimensional closed or F_σ sets is at most n-dimensional [Menger, 1924d, pp. 147, 150]. Of course, Menger has his own approach to proving these results, which differs markedly from Urysohn's.

15.6 Menger's Publications from 1925 Developing the Theory of Curves

On 13 February 1925 Menger submitted a substantial paper on the 'Fundamentals of a theory of curves' ('Grundzüge einer Theorie der Kurven') to *Mathematische Annalen*. Due to the delays typical of the time it only appeared as a publication in 1926 [Menger, 1926a]. Still, well before that 1926 publication date Menger offered a summary of the work for presentation to the Dutch Academy, and Brouwer communicated it on 31 January 1925 [Menger, 1925]. It is likely that Brouwer encouraged Menger to write the summary for the Dutch Academy.

The longer paper in the *Annalen* has many more details. In the next paragraphs I shall consider some of these.

Menger opens his paper by briefly mentioning the recent attempts to define the concept of curve: the Jordan concept, the concept of an irreducible continuum, and Cantor's concept (see Chapter 13 for details). Then Menger considers what one would think of as curves, surfaces, and solids in ordinary life. Intuitively a curve is like a fine wire, a surface is like thin sheet metal, and a solid is like a piece of wood. This intuitive and very concrete thinking motivates his whole conception of dimension. Recall Menger's description in his later reminiscences as quoted in Section 15.2. I shall have more to say about Menger's view of intuitive thinking leading to formal mathematical definitions in my later philosophical analysis of this entire area of mathematics.

[22] According to the common usage of the time Menger uses the term 'compact' in the original sense of Fréchet [Fréchet, 1906, page 6]: a set is compact if every infinite subset has a limit point (accumulation point); the limit point need not belong to the set. Fréchet compactness is sometimes called 'relative countable compactness' or 'relative sequential compactness.'

For his formal treatment Menger defines *continuum* (*Kontinuum*) as follows [Menger, 1926a, page 278]:

> *In the usual way, we call a continuum a closed set containing more than one point, which cannot be split into two separate closed parts, unless one of the two is empty.*[Ch15, 13]

Relying on this definition, he gives the following formal definition of *curve* [Menger, 1926a, page 278]:

> *A continuum K is called a curve if for every point of K there exist arbitrarily small neighborhoods whose boundaries contain no continua. A continuum K, which is embedded in a space, is called a curve if for every point of K there exist arbitrarily small neighborhoods whose boundaries have no continua with K in common.*[Ch15, 14]

Menger's definition has two ways of looking at the neighborhoods and boundaries in the definition. The first view looks at the neighborhoods and boundaries relative to the curve itself (intrinsically, so to speak); the second view looks at these with respect to the curve and the space in which the curve is embedded (extrinsically, so to speak), though the space is left unspecified for the moment.

In his later book on curve theory, *Kurventheorie* (*Curve Theory*), of 1932, Menger has a more elaborate discussion of the curve concept and supporting or related concepts [Menger, 1932a, pp. 79–82]. He includes a number of explanatory examples. Essentially, the definitions are the same.

In the paper he connects his definition with the dimension theory he was developing at the same time. He states [Menger, 1926a, pp. 278–279]:

> In Euclidean (and even more general) spaces the curves are identical with the one-dimensional continua. Thereby, we call a non-empty set M *zero-dimensional* (or also totally disconnected),[23] if to every point of M there exist arbitrarily small neighborhoods with empty boundaries (respectively, if M is embedded in a space, they are disjoint with M); — and we call a non-empty and non-zero-dimensional set M *one-dimensional*, if to every point of M there exist arbitrarily small neighborhoods whose boundaries are zero-dimensional (respectively, if M is embedded in a space, they have zero-dimensional intersections with M)[Ch15, 15]

Menger immediately states that his defined concept of curve is topological: the topological images of curves are again one-dimensional; if they are closed—a requirement for his continua—they are curves. Typically, a closed interval of the straight line and simple arcs are curves. In contrast, square surfaces and in general higher-dimensional sets containing open sets are not curves. In Euclidean spaces R_n, $n \geq 3$, the complement of a curve is connected. In the plane the curves are identical with the nowhere dense continua, therefore, with curves in Cantor's sense.

[23] Cf. Section 13.9 and [Engelking, 1978, 1.4, pp. 31–37].

He defines the *order* or *ramification order* (*Ordnung*) of a point on a curve. A point p has finite order n, if n is the least integer with the property that there exist arbitrarily small neighborhoods of p (intrinsically speaking) whose boundaries have cardinally n. A point p may have *growing/increasing order* (*wachsender Ordnung*) or be of order w, if, as the neighborhoods become arbitrarily smaller, there always exist some for which the number n increases. Further, a point p may have order $\aleph_0$, whereby the points have arbitrarily small neighborhoods with countable boundaries. In some instances it is possible to have a point p with the order of the continuum, $\mathfrak{c}$, or be of continuous order, if there exist arbitrarily small neighborhoods with boundaries having the power of the continuum.

In the spirit of point-set topology to illustrate the distinguishing characteristics of defined concepts by using examples, counterexamples, and counterintuitive examples, Menger gives seven examples. I shall only describe the first six [Menger, 1926a, page 280]:

1. Curve is an interval or line segment; its endpoints have order 1;
2. Curve is an interval or line segment; its interior points have order 2;
3. Lemniscate curve (figure 8) has order 4 at the crossing point;
4. Curve defined in the plane R_2 by the equations in polar coordinates:

$$\varphi = \frac{\pi}{n}, \quad 0 \leq r \leq \frac{1}{n} \quad (n = 1, 2, \ldots),$$

   is a set of short spikes all emanating from the origin and lying in the upper half-plane; starting with length 1 at π radians, they become shorter and closer together with increasing n as one moves clockwise in the upper half-plane; the defined curve is of growing order;
5. Curve defined in the plane R_2 by the equations in polar coordinates:

$$\varphi = \frac{\pi}{n}, \quad 0 \leq r \leq 1 \quad (n = 1, 2, \ldots)$$
$$\varphi = 0, \quad 0 \leq r \leq 1$$

   is similar to the one just described, but having all spikes of length 1 and all emanating from the origin; the curve is of order $\aleph_0$ at all the points of the segment $\varphi = 0, 0 \leq r \leq 1$;
6. Curve in the plane R_2, having Cartesian coordinates, defined by joining the points of a perfect nowhere dense set of points on the unit interval of the X-axis [e.g., the Cantor middle third set] to the point $(0, 1)$ on the Y-axis by straight lines; this curve has continuous order at all its points;

Example 6 may at first sight seem a little surprising. However, one may note that arbitrarily small neighborhoods of any point can be chosen with boundaries that include a short continuous segment.

In the rest of the paper Menger explores several topological topics concerning curves as 1-dimensional objects and examines various types of them. This is one of his major papers. He includes a section on local connectedness (Zusammenhang im kleinen), a concept that his teacher Hahn had introduced [Menger, 1926a, pp. 294–300] (cf. Section 13.8).

15.7 Menger and Others with L.E.J. Brouwer in Amsterdam/Blaricum

During the course of 1925 several mathematicians, mainly topologists, gathered around Brouwer in Amsterdam or rather in Blaricum, his home near Amsterdam, where he did his main work. Brouwer was no longer making breakthroughs in topology, but he was strongly interested in the ongoing development of the subject. His friendly and very warm relations with Urysohn and Alexandroff and then with Menger show his passion for the subject and his concern, both intellectual and emotional, for the well being of the researchers associated with him, even though at the time his primary interest was in his intuitionism.

Regrettably, Urysohn had died in August 1924; he had expected to spend additional time with Brouwer in Blaricum in 1925. Nonetheless, in 1925 other mathematicians came to be near and work in various ways with Brouwer. Menger came to The Netherlands at the end of March, Paul Alexandroff arrived in May, and Leopold Vietoris (1891–2002) came during the autumn. All three became first assistants to Brouwer. In this role they had rather light duties, giving them time for their own research. Other mathematicians joined too. At this time there was effectively a kind of 'Dutch Topological School' centered around Brouwer and his home in Blaricum.[24]

In contrast to his reputation as a difficult person, partly derived from the fairly public disputes in which he was involved, Brouwer had a very warm side to his personality. Thus, he quickly became close to Urysohn and Alexandroff and was greatly saddened by the death of the former. He sought to comfort Alexandroff and Urysohn's family after the untimely death.

In a similar way early in his dealings with Brouwer, Menger experienced the warm side of the great topologist. Menger left Amsterdam for the summer of 1925 and went to Heidelburg. Brouwer then had the difficult task of delivering the sad news to Menger of the death of his mother, and with the news offering his condolences. In a letter of 3 July 1925, Menger thanks Brouwer for his kind words on the death of his mother [Dalen, 2011a, pp. 291–292] [Dalen, 2011b, pp. 1340–1341]. In his response of 8 July 1925 Brouwer is effusive [Dalen, 2011a, pp. 292–293] [Dalen, 2011b, pp. 1343–1344]:

> I thank you for your letter of the third of this month, which gives me in any case the relief that you have withstood the stroke of fate, which has struck you so suddenly. For that reason I have worried very much, because during your stay here I have felt strongly to what extent the aura of your mother irradiated your life. So I surmised how great your loss was, and I expected the crisis that

[24] Dirk van Dalen gives a wonderful, detailed description of this 'School' in his biography of Brouwer [Dalen, 2005, pp. 514–537] [Dalen, 2012, pp. 468–490].

would be unleashed in you by the sudden emptiness and the sudden necessity to assume a different spiritual form of breathing. But after you have weathered the first crisis, I am certain that you will find the necessary concentration and religious dedication to work your way through, and that the certainty about the wish in that direction of the dear departed, and also the lasting, serene memory of her will help you with that.[Ch15, 16]

Menger was so stricken with his loss that he had a nervous breakdown and had to take a cure for several weeks (as reported by van Dalen [Dalen, 2005, page 521] [Dalen, 2012, page 474]). He recovered and returned to The Netherlands for the academic year 1925–26. He even returned for a second academic year 1926-27, although with less enthusiasm about being there with Brouwer and the other researchers.

During the two academic years 1925–26 and 1926–27, while mainly in The Netherlands, Menger carried out considerable research on curve and dimension theory and delved into other topics as well. Just before the start of the first academic year he wrote a wide-ranging 'Report on Dimension Theory' ('Bericht über die Dimensionstheorie') for the *Jahresbericht der Deutschen Mathematiker-Vereinigung* (*Annual Report of the German Association of Mathematicians*). The exposition encompassed a broad history and a diversity of topics [Menger, 1926b]. He first sent it to Brouwer, as revealed in a letter from Menger to Brouwer of 16 August [Dalen, 2011b, pp. 1350–1351]. He then submitted it to the journal on 30 August 1925, and it was published the following year.

During 1925–26 Menger produced two full papers on curve theory, 'Über reguläre Baumkurven' ('Concerning regular tree-like curves'), submitted to *Mathematishe Annalen* near the end of 1925 on 2 December 1925 [Menger, 1927a], and 'Zur allgemeinen Kurventheorie' ('On general curve theory'), submitted to *Fundamenta Mathematicae* on 10 July 1926 [Menger, 1927b]. These appeared in the journals in 1927. Publishing the latter paper in a different journal was perhaps intended as a way to give Menger's ideas a wider circulation.

In addition during 1926, Menger offered several research notes to the Dutch Academy, which Brouwer duly communicated. Three of these notes related general spaces to Cartesian (or Euclidean) spaces [Menger, 1926c], communicated 27 February 1926, [Menger, 1926d], communicated 27 March 1926, and [Menger, 1926f], communicated 26 June 1926; a fourth covered the principal problem of the dimensional structure of spaces [Menger, 1926g], communicated 27 November 1926.

The pair of research notes, 'Allgemeine Räume und Cartesische Räume' ('General spaces and Cartesian spaces'), first and second communications [Menger, 1926c] and [Menger, 1926f], are of special note. In the first Menger proves that [Menger, 1926c, page 476]:

> *Every compact one-dimensional metric space is homeomorphic to a subset of the three-dimensional number space.*[Ch15, 17]

In the second note he presents his 'universal curve,' which is now often called the 'Menger sponge' [Menger, 1926f, page 1125]:

There exist universal one-dimensional sets that include all the bounded continuous curves of R_3.[Ch15, 18]

Menger's construction of the universal curve is based on Cantor's ternary set or discontinuum, with the construction generalized to three dimensions. It is a simple and quite direct generalization of Sierpiński's work on the Carpet Curve, which he cites [Sierpiński, 1916b] (see Section 13.7). Menger's curve, like Sierpiński's, has no interior points, only boundary points, as required for a curve. Menger gives a full exposition of the universal curve in his later book *Kurventheorie* (*Curve Theory*) [Menger, 1932a, pp. 345–360], including a stage in its construction, as in Figure 15.3.

Fig. 15.3 Menger Universal Curve or Menger Sponge [Menger, 1932a, diagram between pages 346 and 347]

Menger's research note on 'Das Hauptproblem über die dimensionelle Struktur der Räume' ('The principal problem concerning the dimensional structure of spaces') (communicated 27 November 1926) [Menger, 1926g] analyzes the dimensionality of sets of type R^k, defined as sets of points that are at least k-dimensional in a compact, semicompact (halbkompakte, a union of countably many compact spaces), or separable metric space. Urysohn had considered the problem of dimensional structure within his preferred frame of compact metric spaces and regarded it as especially difficult and important [Urysohn, 1926, page 285]. He devoted some pages of his 'Mémoire' ('Memoir') (I) to treating these kinds of sets defining a form of dimensional structure [Urysohn, 1926, pp. 269–286].

In the note Menger proves three structural theorems and a general covering theorem. His first theorem considers structure in compact and semicompact metric spaces, while his second theorem deals more generally with structure in separable metric spaces. I shall consider just these two results.

His first structure theorem is the following [Menger, 1926g, page 138]:

> **Theorem I.** *In a compact or semicompact metric space R, for every k, the set R^k of all points at which the space is at least k-dimensional is an at least k-dimensional set at each of its points. In particular, therefore, in a compact or semicompact n-dimensional space, the set $\overset{n}{R}$ of all points at which the space is n-dimensional is homogeneously n-dimensional (i.e., n-dimensional at each of tis points). The $\overset{n}{R}$ in a compact n-dimensional space R is identical with the union of all homogeneous n-dimensional subsets of R.*[Ch15, 19]

His second structure theorem is the following [Menger, 1926g, page 141]:

> **Theorem II.** *If R is a separable metric space and R^k is the set of all points, at which R is at least k-dimensional, then every nonempty set open in R^k is at least $(k-1)$-dimensional. In particular, therefore, in a separable n-dimensional [metric] space R the set $\overset{n}{R}$ is either n-dimensional or $(n-1)$-dimensional. Both cases can occur.*[Ch15, 20]

These theorems characterize the dimensionality of the selected sets defined by dimension in a space and provide some notion of dimensional structure for the space and selected subsets. These results became the basis of a chapter of his book of 1928 *Dimensionstheorie* (*Dimension Theory*) [Menger, 1928b, Kapital IV, pp. 126–154].

During the academic year 1925–26 Menger taught a course on dimension theory at Amsterdam. In giving the course he could draw on the background from his 'Report' and his own and others' work. The other mathematicians in Amsterdam attended when they could [Dalen, 2005, page 527] [Dalen, 2012, page 480].

During that time issues also arose concerning the origins, independence, and Menger's claimed priorities concerning ideas on dimension. These issues motivated him to write a 'defensive' note 'On the origin of my work on dimension and curve theory' ('Zur Entstehung meiner Arbeiten über Dimensions- und Kurventheorie') [Menger, 1926e]. There must have been a definite rivalry between Alexandroff, representing his friend, the deceased Urysohn, and Menger. Brouwer, the senior mathematician and mentor, assumed the role of arbiter in this matter to make sure that the history was as accurate as possible. For example, Brouwer made clear distinctions with respect to Menger's earliest notes as to whether the original versions were available at the time of the appearance of this note or not [Brouwer, 1976, page 567] [Dalen, 2005, pp. 527–529] [Dalen, 2012, pp. 480–482]. The whole matter put a tremendous strain on Menger in gathering his documentary evidence and in verifying his particular claims of priority. A letter of 10 April from him to Brouwer shows the situation fairly clearly from his point of view [Dalen, 2011a, pp. 297–298] [Dalen, 2011b, pp. 1398–1399]:

> I write to you only today, because the priority matter is now completely settled: not only all evidence mentioned by me, but also other material has been found that I myself had already forgotten. I have put all the material in a safe, and I will hand over personally all the originals to you. Today I only mention shortly the *officially* certified documents:[Ch15, 21]

Menger then lists four items to validate his claims. He had sought specific corroborating evidence and original documents from his teacher Hahn and his friend Schreier. Menger was so stressed by this matter that it apparently caused him to have a nervous collapse. Finally, he completed his note, and Brouwer communicated it at the meeting of the Dutch Academy on 29 May 1926.

For the academic year 1926–27 Menger was less enthusiastic about being out in Blaricum with Brouwer, and apparently spent less time there. Yet, he was still productive. He wrote two research notes for presentation to the Dutch Academy during 1927, and Brouwer communicated them: [Menger, 1927d] and [Menger, 1927e], both communicated 24 September 1927. The last one is about metrics and not dimension or curve theory. As mentioned above, during 1927 two of Menger's longer papers on curve theory written in 1926 appeared in print [Menger, 1927a] [Menger, 1927b]. He clearly wanted to become known in the mathematical research community.

The research note 'Zusammenhangsstufen und Cantorsche Mannigfaltigkeiten' ('Levels/degrees of connectedness and Cantorian manifolds') [Menger, 1927d] is important. Its significant content is covered more fully in a paper that Menger and Witold Hurewicz wrote together, 'Dimension und Zusammenhangsstufe' ('Dimension and levels of connectedness') [Hurewicz & Menger, 1928]. As stated above, Brouwer communicated Menger's research note to the Dutch Academy on 24 September 1927. However, Menger and Hurewicz submitted the main part of their fuller joint paper even earlier on 30 July; its added appendix ('Anhang') authored by Hurewicz alone was then submitted by Hurewicz on 15 October. Menger was, in fact, on the verge of leaving The Netherlands for a new position in Vienna. Hence, the collaboration between Menger and Hurewicz was rather natural. Hurewicz was to stay on at Amsterdam until into the 1930s. He was effectively finishing the work Menger apparently started on this topic while in the The Netherlands.

The motivation for this work is straightforward from a simple geometrical point of view. A rectangular surface is more strongly connected than a line segment, though it is more weakly connected than a cube. A pair of rectangles joined at only one point is more weakly connected than a single rectangle. The aim of the research is to bring out these kinds of levels of connectedness with a whole range of possibilities under the recursively defined concept of *n-level connected space* (*n-stufig zusammenhängender Raum*), having a clear analogy with the recursively defined dimension concept. The base case of the empty set and only this set is *null-level disconnected* (*nullstufig zusammenhangslos*) In the paper the authors refer to Sierpiński's special example (using updated terminology) of a (weakly) one-dimensional set that is hereditarily disconnected [Sierpiński, 1921, pp. 82–88] (see Section 13.9). The other part of the paper is aimed at relating these levels to Urysohn's general Cantorian manifolds. The interested reader can examine the details in the research note and paper.

As already mentioned above, for the academic year 1926–27 Menger was less enthusiastic about being in Amsterdam and Blaricum. Some time toward the end of September 1926 Menger went to Düsseldorf to attend the annual meeting (Jahresversammlung) of the German Association of Mathematicians (Deutsche Mathematiker-Vereinigung), which occurred 19–26 September. He gave a talk there on 22 September [Menger, 1927c]. Afterwards he went to Paris and visited the painter Mondriaan (or Mondrian). Menger tells of his visits to Düsseldorf and Paris in a letter to his girl friend Mitzi[25] in Vienna. He also expresses his unhappiness with his return to Amsterdam. He was quite frustrated. The letter is undated, but likely from near the beginning of October 1926 (English translation by Robert Leonard) [Leonard, 1998, page 2] [Leonard, 2010, page 110] (original letter in the Karl Menger papers, 1872–2000, Box 3, at Duke University):

> From there [Paris] I returned to Amsterdam after all. I am reading about the calculus of variations. Personally, I am occupied with geometry of all kinds, further with epistemology. I hope that I shall have the strength to put together my views about the problem of truth etc. In the last weeks, I have had so many ideas that I do not have any time at all to write them down, and every evening run away to distract myself (of course, I live in Amsterdam) in order not to overwork. For the rest, I am not in Vienna, but here. I cannot get used to living here, and I will give all my effort to leave here forever in the month of June. . . . [Ch.15, 22]

Indeed, Menger did leave. Near the end of 1926 his teacher Hahn wrote to him inviting him to take up the chair of geometry at the University of Vienna, as the occupant Kurt Reidemeister (1893–1971) had left for Königsberg. In 1927 he returned to Vienna to start lecturing in the autumn [Menger, 1979c, page 247]. He continued his mathematical research in his new position and soon joined the Vienna Circle, of which Hahn was a primary leader.

15.8 Menger's Reminiscences of Brouwer

In 1979, many years later, Menger published his reminiscences of Brouwer as an essay, 'My Memories of L.E.J. Brouwer' [Menger, 1979c]. Menger paints a very mixed picture of the man and the mathematician, contrasting both 'positive qualities' and 'negative qualities' of the Dutchman. Even after so many years, the reader will be able to sense the emotions welling up in Menger in writing this piece. In his biography of Brouwer Dirk van Dalen provides a very careful analysis of the intellectual and 'emotional' relationships between Menger and Brouwer. The van Dalen analysis is based on very extensive evidence, which he collected and edited [Dalen, 2005, pp. 521–537] [Dalen, 2012, pp. 474–490]. I think the analysis has considerable merit—it is certainly not one-sided.

---

[25] I have no further details about Mitzi.

For purposes of this history I would like to draw attention to a few points from Menger's essay on Brouwer, as illuminated by the light of van Dalen's careful analysis. As I have already described above, Brouwer showed considerable compassion for Menger when the latter lost his mother.

On the mathematical side Menger was at first disappointed with meetings with Brouwer, which were in the company of others and concerned administrative matters [Menger, 1979c, page 242]. In his essay he also disparages Brouwer's attitude to other's concerning his topological work [Menger, 1979c, page 242]:

> Brouwer ... talked a great deal about other mathematicians' reactions to his topological and intuitionistic work. He gave the impression of being oversensitive and somewhat testy. In particular, I remember him complaining about Hausdorff, who once at a mathematical meeting appeared to him to suggest that the proofs of his topological theorems were obscure and would have to be completely redone.[26]

However, in his essay Menger also describes Brouwer's genuine concerns for him and his work [Menger, 1979c, page 244]:

> In the late spring and the fall of 1925, I reported to him my latest results almost every week and immensely enjoyed these occasions. He never formulated problems to me; but he listened with an eagerness that I found very simulating. Through the perpetually changing expressions of his face he had the rare ability to be, as it were, even silent with intensity. He also showed concern for my then still unstable health and expressed warm sympathy when I suffered a bereavement. He warned me against overworking. ... I revered him with a deep affection during those months.

Yet, there are certainly negative points in Menger's essay. Brouwer began insisting that the papers on dimension theory submitted to him contain mutual references. According to Menger, this insistence enabled Brouwer to exert a meticulous watch over Urysohn's rights. It also put Brouwer more firmly into the historical record for the development of dimension theory overall. Menger had no objection to this [Menger, 1979c, pp. 244–247]. However, Menger goes on to say [Menger, 1979c, page 247]:

> Moreover, at about this time two grave events occurred. For weeks I had found alterations and insertions in Brouwer's characteristic handwriting in my manuscripts and proofsheets. When on one of these occasions I tried, of course with utmost restraint, to expostulate with Brouwer, he terminated the conversation after a short while; and the next day, a friend of his looked me up to tell me that Brouwer was infuriated by my remonstrations and that he declared that he would hurt me with all the harm that he could do to me if I should not go along with his suggestions. I did not repeat my objections or change my tone, but my feelings toward Brouwer became ambivalent.

[26] Cf. [Hausdorff, 1914, page 468] and Section 12.16.

Menger goes on to 'accuse' Brouwer of adding and changing the references specifically to Brouwer's work in Urysohn's papers [Menger, 1979c, page 247]:

> About the same time, by coincidence, it came to my attention that even in Urysohn's posthumously published work, references to Brouwer were being added to and changed without the alterations being in any way designated as such.

With respect to this passage, see Menger's very detailed footnote [Menger, 1979c, Footnote 17, pp. 253–254]. Following Urysohn's untimely death in August 1924 Alexandroff had to complete the preparation of his topological works for publication. In fact, Urysohn's 'Mémoire' ('Memoir') (I) expositing his dimension theory [Urysohn, 1925a] [Urysohn, 1926] was already in press, and the editor of *Fundamenta Mathematicae*, Sierpiński, had to be consulted. Alexandroff did include Brouwer in the editing, given the agreement Urysohn and Brouwer had struck on the latter's 'slip of the pen' over his initial inclusion of the word 'abgeschlossene' in the original paper on defining dimension [Brouwer, 1913b] when the two had interacted and met before the former's death (see Setion 12.12). Moreover, new printed versions of the 'Mémoire' ('Memoir') (I) had to be made—at Brouwer's expense. On this significant matter see van Dalen's very detailed analysis in his biography of Brouwer [Dalen, 2005, pp. 470–480] [Dalen, 2012, pp. 424–434]. There is perhaps some justice in Menger's 'accusation' and concern for historical accuracy, since the references to Brouwer in the final printed version of the 'Mémoire' are not marked off in any special way. It would have been better for historical accuracy if an editorial note had been added to Urysohn's 'Mémoire' on the approach taken to the final editing, though this would not have been typical for a mathematical work. On the matter of Menger's specific accusation, van Dalen asks and then answers the hypothetical question [Dalen, 2005, page 478] [Dalen, 2012, page 432]:

> Did Brouwer actually 'falsify the text,' as Menger later suggested? It is highly unlikely; after all he did not act without Alexandrov's [Alexandroff's] approval, and Alexandrov [Alexandroff] had been present and taken part in the discussion between Brouwer and Urysohn in July 1924. Those who have known Alexandrov [Alexandroff], know that he would not be a party to shady practices of this sort, and he was certainly not one to be pressured into such an act.[27]

Menger naturally also took an interest in Brouwer's intuitionistic mathematics while in Amsterdam. He came to interpret Brouwer's 'spreads' in Brouwer's working out of intuitionistic set theory [Brouwer, 1918] [Brouwer, 1919a] [Brouwer, 1919b] as analytic sets from descriptive set theory [Menger, 1979c, page 246]. This work led Menger to write a set of four notes, then published in the *Jahresbericht der Deutschen Mathematiker-Vereinigung* (*Annual Report of the German Association of Mathematicians*) [Menger, 1928a] (note I is translated as [Menger, 1979b]). There is

[27] Note that this whole matter also relates indirectly to Menger's defensive 'Zur Entstehung . . . ' note [Menger, 1926e].

a lot of subtlety in Brouwer's intuitionistic set theory, which Menger's interpretation does not fully capture [Dalen, 2005, pp. 531–532] [Dalen, 2012, pp. 484–485], but it is, nonetheless, of some interest.

15.9 Concluding Remarks

At this point I shall leave Menger's early development of his curve and dimension theory. There will be more on Menger in Chapter 16.

Menger developed his curve theory and dimension theory independently of Urysohn and about the same time as Urysohn. He started from the motivation of Hahn's problem of defining the notion of curve. In a very similar way Urysohn had started from the motivation of Egorov's remarkably similar problem. Menger was a new student of mathematics, so it is understandable that he made some errors in his earliest papers, which he had hoped to publish. Hahn was particularly understanding and a careful critic. Menger learned mathematics quickly and gained a lot of mathematical maturity in a short time. He soon was on the road to developing his theories for the mathematical community.

A distinctive characteristic of Menger's dimension and curve theory is his intuitive way of thinking of the basics of dimension. He intuitive approach provides great value. I shall consider it later.

No doubt, Menger was a rather impetuous young fellow. He was anxious to make a name for himself. Perhaps his conflicts with Brouwer were inevitable. Brouwer wanted to assure his own place in the history of mathematics. I shall cover more of this conflict in Chapter 16.

Chapter 16
Development of Theories of Dimension and Curves

16.1 Introduction

In this chapter I shall cover the development of dimension and curve theory following the initial contributions of Urysohn and Menger. After returning to Vienna in 1927 Menger continued to work on dimension and curve theory, as well as on other parts of mathematics. He also took an active role in the Vienna Circle and founded a Mathematical Colloquium. Alexandroff edited and promoted the work of his friend Urysohn. He also developed a program relating dimension theory to complexes, polyhedral geometry, and geometrical intuition. This work led him to a theory of homological dimension. Other mathematicians began to contribute to topological dimension theory and to a lesser extent curve theory. The most prominent were Witold Hurewicz, a brilliant mathematician from Poland, Lev Tumarkin, a student of Alexandroff, Georg Nöbeling, a student of Menger and then his collaborator, and Eduard Čech, an important Czech mathematician. Beyond the framework of point-set topology, others developed alternative theories of dimension, in particular, Felix Hausdorff, who a little earlier initiated the development of Hausdorff dimension based on measure theory in a paper of 1919, and Georges Bouligand, who offered a theory of Minkowski dimension.

Problems and theories of dimension also came to the fore in algebraic geometry and commutative algebra. The most prominent and significant modern definition in this domain of mathematics appeared in 1937: Krull's definition of dimension [Krull, 1937a] (cf. [Krull, 1938a]). The history of this form of dimension is beyond the scope of this book, though I include a short Appendix C as Section 16.14 giving some pointers to this history.

In this chapter I shall concentrate on topological bases for dimension and curves and then to some extent on the non-topological Hausdorff dimension, while giving the others a lighter treatment. Hausdorff dimension is significant. It eventually led to fractal theory. I shall cover the basics and early development of Hausdorff dimension; the history of fractal theory is beyond the scope of this work.

D. M. Johnson, *Theories of Dimension from Ancient to Modern Times*, Sources and Studies in the History of Mathematics and Physical Sciences, https://doi.org/10.1007/978-3-032-02682-8_16

I shall bring the mathematical history to a close, marking it with the publication of Hurewicz and Wallman's beautiful and marvelously concise book *Dimension Theory* of 1941, revised in 1948 [Hurewicz & Wallman, 1948]. In the next and final chapter I shall examine selected aspects of the long history of dimension and curve theory and consider some important philosophical implications.

16.2 Menger's Report on Dimension and Curve Theory and Summaries of Urysohn's and Menger's Results

By the mid 1920s there was considerable interest in dimension and curve theory in the mathematical world. Given that interest, it is not surprising that Menger wrote his 'Bericht über die Dimensionstheorie' ('Report on Dimension Theory') [Menger, 1926b] for the *Jahresbericht der Deutschen Mathematiker-Vereinigung* (*Annual Report of the German Association of Mathematicians*) (already mentioned the Report in Section 15.7). As is customary for these special reports covering a developing field written for the German Association, Menger's 'Report' is wide ranging, containing historical and mathematical background and copious references to mathematicians doing research and contributing to the growing mathematical literature. As mentioned in Section 15.7 Brouwer had reviewed it before submission. Menger submitted it on 30 August 1925, but it was not published until 1926.[1]

A little later two summaries of dimension and curve theory were written for the journal *Mathematische Annalen* (possibly Brouwer asked for these to be written). Alexandroff wrote a summary of Urysohn's main work [Alexandroff, 1928a], and Hurewicz wrote a summary of Menger's principal work [Hurewicz, 1928a]. These summaries are essentially expositions of the main results with proofs in selected cases. Alexandroff's summary was submitted on 3 April 1926, and Hurewicz's summary, on 30 August 1926. Both appeared in the journal in 1928.

Following are short summaries of these three works. They reveal a new level of interest in these fields of mathematical research.

As indicated, Menger's 'Report' is broad, covering quite a few areas of the field. Menger provides a fairly comprehensive, yet succinct summary, circa 1925 developments. He first reaches back to the recent history motivating the development of curve and dimension theory, the 'ältere Dimensionsbegriffe' (the 'older dimension concepts'). He recalls Cantor's result on the equal power of intervals and cubes of various dimensions (the one-one correspondences among the sets of their points), the Peano space-filling curves, and the various attempts to define curves (Cantor, Jordan, Zoretti, Janiszewski, and so forth). He cites Brouwer's invariance result. He mentions combinatorial topology à la Poincaré. He even mentions Fréchet's homeomorphism types and Brouwer's intuitionistic work on mapping theory. Beyond topology, he mentions Hausdorff's measure constant (Maßkonstante), i.e., Hausdorff dimension

[1] Menger wanted to dedicate the offprints of the 'Bericht' to Brouwer. Issues arose between Menger and Brouwer concerning their meeting about coming to an agreement for the exact wording of the dedication. For the unpleasant story see [Dalen, 2005, pp. 529–530] [Dalen, 2012, pp. 482–483].

[Menger, 1926b, pp. 114–118]. I have covered nearly all these topics motivating the new development in the previous chapters, and I shall cover Hausdorff dimension a little later in Section 16.11.

The distinguishing characteristic of the new theory of dimension and curves, giving it a new starting point, is the concentration on intrinsic structural properties of sets, that is [Menger, 1926b, page 118]:

> *Structural properties of the sets in themselves, without regard to any transformations of the entities.*[Ch16, 1]

Menger describes his intuitive thought experiment about distinguishing objects of varying dimensions, whether lines, surfaces, or solids [Menger, 1926b, pp. 118–120], which he devised very early in his own research, as mentioned in Section 15.2. In effect, it is an epistemological observation about separation properties in various dimensions that is worth noting, and I shall examine it further in Chapter 17.

Menger states the fundamental definitions for dimension, emphasizing the underlying notions of neighborhood and boundary, relying on Hausdorff's concept of topological space [Menger, 1926b, pp. 120–121]. He may at first seem a little free and easy at this point in using Hausdorff's basic notion of topological space while not imposing further restrictions, say, to metrizable/metric spaces or compact metrizable/metric spaces. However, various restrictions come up later in the 'Report.' He gives ample references to the immediate founders of the dimension concept, Poincaré, Brouwer, Urysohn, and himself, as well as others [Menger, 1926b, pp. 121–123].

He proceeds to lay out the basic properties and results for the dimension concept, giving some mathematical justifications for the new definition, asserting its topological invariance, examining 0-dimensional sets and their defining properties, and characterizing the dimension of set unions. He treats curves and surfaces, and the orders or ramifications of their points too [Menger, 1926b, pp. 123–130].

He then devotes a lengthy section (III) to general investigations of the formal structure of topological spaces in relation to dimension [Menger, 1926b, pp. 130–141]. At the beginning of this section he makes a very general methodological point that the structural properties depend on the behavior of certain sequences of neighborhoods that converge to given points and on the boundaries of these neighborhoods. In other words, the structural properties are reflected in equivalence relationships among certain sets of neighborhoods and the entire system of neighborhoods for the topology of the space. Menger's methodological point comprehends a broad topological view of dimension theory. He then analyzes a variety of structural properties. For this analysis, he adopts a general abstract view of a given property (Eigenschaft) and considers sets of neighborhoods having a given property (E-Körper von Ungebungen) and also sets of points (E-Punkte). I shall not go further into the details.

He also devotes a section (IV) to a further examination of curve theory and the wide variety of curves [Menger, 1926b, pp. 141–147]. Again, I shall not go into the details.

Menger's final section (V) is of special interest. It covers two open problems [Menger, 1926b, pp. 147–150]. First, he considers the possibility of deriving the concept of dimension from requirements (Ableitung des Dimensionsbegriffes aus

Forderungen), which means developing axiomatics for the dimension concept. Menger soon took up the problem of axiomatizing dimension theory and so offered a solution (see Section 16.3). Second, he mentions relating the area of dimension to Cartesian [Euclidean] spaces (Dimension und Cartesische Räume). In essence, Menger touches on the embedding problem, which translates into the question: Is every finite-dimensional compact (or separable) space homeomorphic to a subset of a Cartesian [Euclidean] space? Menger had already found a partial result in his universal curve, whereby every one-dimensional compact metric space is homeomorphic to a subset of the Cartesian [or Euclidean] space R_3 (see Section 15.7). Several other mathematicians as well as Menger soon sought solutions to the general embedding problem (see Sections 16.7 and 16.9). The soon-to-be-proven positive results form a kind of capstone for dimension theory.

Alexandroff's exposition of Urysohn's version of dimension theory is a succinct presentation very much in the style of Urysohn [Alexandroff, 1928a]. Given the closeness of their friendship, it seems as if Alexandroff could even *be* Urysohn in writing. I shall review it in brief.

In accordance with Urysohn's formulation, Alexandroff frames the theory in the domain of compact metrizable (or just metric) spaces and their subsets. Yet, given Urysohn's and Tychonoff's companion results on metrization [Urysohn, 1924a] [Urysohn, 1924b] [Urysohn, 1925d] [Tychonoff, 1926], he notes that every metrizable topological space satisfying the Second Countability Axiom[2] is homeomorphic to a subset of the same compact metric space, and every regular topological space[3] satisfying the Second Countability Axiom is metrizable.

Alexandroff proceeds to give the Urysohn-style definitions of dimension and the basic results of the theory [Alexandroff, 1928a, Kapitel I, pp. 32–46]. He then gives an analysis of the Lebesgue Covering Theorem [Alexandroff, 1928a, Kapitel II, pp. 46–54] and finally a treatment of the decomposition of n-dimensional spaces into 0-dimensional spaces [Alexandroff, 1928a, Kapitel III, pp. 54–63]. The Alexandroff summary presents the Urysohn theory in a nutshell.

Hurewicz' exposition of Menger's version of dimension theory is similarly succinct [Hurewicz, 1928a]. In addition to using Menger's papers as a basis for his exposition Hurewicz was able to rely on Menger's lectures given at Amsterdam University during 1925–26. After giving Menger's formulation of the definitions of dimension, Hurewicz states and proves the following main results:

- **Sum Theorem and Corollaries**
  [Hurewicz, 1928a, pp. 66–73]
- **Dimensional Structure Theorem for Compact Spaces**
  [Hurewicz, 1928a, pp. 73–76]
- **Dimensional Structure Theorem for Separable Spaces**
  [Hurewicz, 1928a, pp. 77–80]

---

[2] i.e., a space having a countable base.

[3] i.e., a topological space that satisfies the regular separation property, such that every neighborhood U of any point x contains a closed neighborhood $\overline{V}$ of the point x, i.e., $x \in \overline{V} \subset U$.

- **Decomposition Theorem**
  [Hurewicz, 1928a, pp. 81–83]
- **Dimensional Relationships of Cartesian n-Spaces R_n**
  [Hurewicz, 1928a, pp. 83–88]

For example, the Sum Theorem asserts that [Hurewicz, 1928a, page 66]:

> **Sum Theorem.** *In a compact metric space the sum* [*union*] *of countably many closed n-dimensional sets is n-dimensional.*[Ch16, 2]

In proving the Sum Theorem Hurewicz uses a method that Menger had employed: the method of modification of neighborhoods in the vicinity of their boundaries [Hurewicz, 1928a, page 71]. To investigate the structure of n-dimensional sets in a space, the objective is to examine sets of points in the space that are exactly n-dimensional or k-dimensional ($0 \leq k \leq n$) and unions of these. The Hurewicz summary presents the Menger theory in a nutshell.

16.3 Menger's Further Development of Dimension and Curve Theory

As noted earlier, Menger returned to his native Vienna in the latter part of 1927 to take up the position of Professor of Geometry at the University of Vienna, a position that Hans Hahn had been able to offer him. He was happy to leave The Netherlands and Brouwer behind. He continued his mathematical research in dimension and curve theory and also pursued research topics in other parts of mathematics. At Vienna he lectured, taught, and mentored others. He became closely involved with the Vienna Circle, and so he developed his philosophical ideas further and more broadly.[4] As a case in point, some of his intuitive ideas on dimension were relevant to epistemological notions discussed in the Circle. He founded and led a Mathematical Colloquium, which included many other mathematicians and which produced many significant results over several years [Dierker et al., 1998].[5] At Vienna he was a colleague of the famous logician Kurt Gödel (1906–1978).

During this later period of his research Menger mainly investigated two problem areas: first, the embedding problem [Menger, 1929a], and, second, the axiomatization or axiomatics of dimension theory [Menger, 1929c] [Menger, 1929d] [Kuratowski & Menger, 1930]. The embedding problem turned out to be a much bigger research problem, which eventually involved several other mathematicians, including Menger's student and then collaborator, Georg Nöbeling. I shall consider this area later in Sections 16.7 and 16.9.

---

[4] See the collection of Menger's essays, many of a philosophical nature, [Menger, 1979a], his own *Reminiscences* [Menger, 1994], and the excellent books on the Vienna Circle [Stadler, 2015] [Sigmund, 2017] [Edmonds, 2020], as well as the review by Gillies [Gillies, 1981].

[5] This volume includes some very useful introductory essays, pp. 1–76; the essay by Menger himself is of special interest, pp. 63–76.

Menger's main paper on the axiomatics of dimension theory is the third in his series, 'Über die Dimension von Punktmengen' ('On the dimension of point-sets'), 'Zur Begründung einer axiomatischen Theorie der Dimension' ('On the foundation of an axiomatic theory of dimension') [Menger, 1929d]. I shall not go into its details. Exploring the axiomatics in this case fits very well with the quest for formal logical precision. The Vienna Circle and other mathematicians, logicians, and philosophers of the time encouraged the development of axiomatic theories as foundations. The axiomatic thinking of David Hilbert (1862–1943) and others impelled many to explore the foundations of mathematics and specific areas of mathematics like algebra and also areas of physics. Axiomatics is a way to capture a given field with formal precision.[6]

To my knowledge, Menger's work on this topic has not generated much interest nor has it exerted much impact on the field. In contrast, in other areas and domains of mathematics and science developing axiomatics has been extremely important and valuable. For example, Eilenberg and Steenrod's *Foundations of Algebraic Topology* has exerted considerable influence in that part of topology by axiomatizing homology and cohomology based on thinking in categories and functors [Eilenberg & Steenrod, 1952].

Menger capped his work on dimension theory and curve theory with the publication of his two books, *Dimensionstheorie* (*Dimension Theory*) [Menger, 1928b] and *Kurventheorie* (*Curve Theory*) [Menger, 1932a]. Menger's *Dimensionstheorie* (*Dimension Theory*) is a full exposition of the field at the time, 1928, with unsolved problems listed at the end. Some parts of it also turned out to be controversial, sparking a dispute with Brouwer.[7] I shall touch on that topic in Section 16.6. Menger's *Kurventheorie* (*Curve Theory*) is a fine exposition, with no associated controversy (to my knowledge).

16.4 Witold Hurewicz and Dimension Theory

I shall now consider the work of Witold Hurewicz (1904–1956)[8] on dimension theory. Though of Polish background, Hurewicz chose to study higher mathematics at the University of Vienna and completed his doctorate there in February 1926. Hans Hahn was his supervisor. Hurewicz had a wonderful ability for developing and

---

[6] However, modern logic shows the limitations of axiomatics. Think of the results of Kurt Gödel, Alfred Tarski, Leon Henkin, and other modern mathematical logicians.

[7] Many years ago I talked with Menger about his book and wondered if it should be reprinted, at least for historical reasons. He was very reluctant to discuss the book as such, but he did provide me with offprints of his papers relating to the dispute.

[8] For biographical material on Hurewicz see the chronology in [Hurewicz, 1995, pp. xix–xx] and see [Lefschetz, 1957] [Borsuk, 1997] Samuel Eilenberg's reminiscences in [Hurewicz, 1995, pp. xlv–xlvi], and
https://www.encyclopedia.com/science/dictionaries-thesauruses-pictures-and-press-releases/hurewicz-witold,
https://mathshistory.st-andrews.ac.uk/Biographies/Hurewicz/.

presenting the simplest and most beautiful arguments in his mathematical papers. His papers are written with precision and clarity. His book *Dimension Theory*, written with Henry Wallman, is a mathematical gem [Hurewicz & Wallman, 1948].

Hurewicz went to Amsterdam in 1926 or 1927 to join the group of topologists and mathematicians surrounding Brouwer. He stayed there until 1936, when he left for the United States. He carried out nearly all of his research work on dimension theory during his time in Amsterdam. Towards the end of his stay in Amsterdam he turned his attention to algebraic topology, a field to which he eventually made many outstanding research contributions.

Hurewicz' research into dimension theory is deep and extensive, resulting in a number of significant papers that furthered its development in multiple directions. He was one of the prime movers in pushing the field from the domain of compact metric/metrizable spaces to the broader domain of separable metric/metrizable spaces. I shall only be able to consider a portion of his work in the field.[9]

In his early paper, 'Normalbereiche und Dimensionstheorie' ('Normal families and dimension theory'), submitted 28 September 1925, but only published in 1927, Hurewicz treated the dimension concept using his defined concept of a *normal family* (*Normalbereich*) [Hurewicz, 1927a]. In that paper he moved to using separable metric spaces. He solidified this move in his compactification theorem [Hurewicz, 1927c]. He carried out some ground-breaking research in his examination of infinite-dimensional spaces [Hurewicz, 1928b] and in his exploration of mappings of spaces of finite dimension in a set of five papers. In the course of his research he also put forward a simplified proof of the Lebesgue–Brouwer tiling theorem [Hurewicz, 1929]. I shall cover these topics in a little more detail below. In addition, he did work on the embedding theorem, which I shall cover in Sections 16.7 and 16.9 (where I treat both his work and the work of other mathematicians).

Hurewicz' early paper on normal families and dimension theory is a fine example of his ability to lay out his ideas in a clear and insightful way to reveal a fundamental well-defined structure [Hurewicz, 1927a]. Normal families become an alternative way to characterize dimension. In the paper one also sees him shifting the basis of dimension theory from compact metric spaces to separable metric spaces. Lev Tumarkin made this change of basis independently, as I will discuss in Section 16.5.

In the introduction to his paper Hurewicz provides a succinct definition of normal families of point-sets, the basis of his approach [Hurewicz, 1927a, pp. 736–737]:

> ... we generally assume a family $\mathfrak{N}$ of point-sets and denote by $\mathfrak{N}^*$ the family of all point-sets, such that all points of which [i.e., all points in member sets M belonging to the family $\mathfrak{N}^*$] have relative neighborhoods whose boundaries are sets of the family $\mathfrak{N}$. We then examine the relationship between the two families $\mathfrak{N}$ and $\mathfrak{N}^*$. The family $\mathfrak{N}$ from which we start is subject to the following conditions.
>
> 1. If M is a set of $\mathfrak{N}$, so is every subset of M.

[9] See [Engelking & Pol, 1995] for a more complete summary of Hurewicz' contributions to dimension theory.

2. If M is the union of a countable number of sets closed in M, all of which belong to the family $\mathfrak{N}$, then M also belongs to the family $\mathfrak{N}$.

We briefly refer to a family of point-sets that satisfies these two conditions as a *normal family*, and a main result of the present work states that the family $\mathfrak{N}^*$ derived from a normal family $\mathfrak{N}$ according to the above requirement [i.e., satisfying the two conditions] is also a normal family.[Ch16, 3]

Hurewicz uses families $\mathfrak{N}$ and $\mathfrak{N}^*$ of sets so that he can discuss dimensional properties over a set. The language is more global than local.

In Part 1 of the paper Hurewicz then examines 0-dimensional sets or spaces in some detail [Hurewicz, 1927a, pp. 737–753]. He then uses this material to introduce his novel theory of normal families in Part 2 [Hurewicz, 1927a, pp. 753–764]. The new theory gives a clean way to define dimension in the finite case.

Hurewicz states the basic definition of 0-dimensional set in the following way [Hurewicz, 1927a, page 737]:

> *The non-empty set M of a metric space is said to be* 0*-dimensional, if to every point of M there is a sequence of subsets of M contracting to the point whose boundaries in M are empty.*[Ch16, 4]

A simple consequence of the definition is that every subset of a 0-dimensional set is 0-dimensional. This definition can also be localized to a point.

He then introduces the following alternative terms to be used in his theory [Hurewicz, 1927a, page 747]:

> ... we say, if the set M is 0-dimensional at the point p, that M is *discontinuous* at p, or also that p is a *point of discontinuity* of the set M. We call the other points of the space *continuity points* of M and also say that the set M is continuous in them.[Ch16, 5]

He opens part 2 of his paper, the main part, by defining totally discontinuous sets related to families of sets [Hurewicz, 1927a, page 753]:

> We turn to the task of investigating the sets M having the following characteristic feature: At every point of M there is a sequence of sets open in M, whose boundaries in M have a certain property, or what amounts to the same thing, whose boundaries in M belong to a certain family $\mathfrak{B}$ of sets. We call such sets *totally discontinuous with respect to the family* $\mathfrak{B}$. The totally discontinuous sets with respect to the family $\mathfrak{B}$, which only contains the empty set, are the 0-dimensional sets.[Ch16, 6]

In this statement Hurewicz is defining a generic class of what he calls 'totally discontinuous sets M.' His defined property is 'global' relative to sets M. Typically, for each of these sets M, each of its points has a sequence of sets open in M with boundaries of a certain type that is defined according to the family $\mathfrak{B}$. He then narrows his attention to the totally disconnected sets with respect to the family $\mathfrak{B}$ that just contains the empty set. These sets are the 0-dimensional sets.

With this background he offers his fundamental definition of a normal family (see above) [Hurewicz, 1927a, page 754]:

> *We call a family $\mathfrak{N}$ of sets a normal family, if the following conditions are satisfied:*
>
> 1. *If M is a set from $\mathfrak{N}$, then so is every subset of M,*
> 2. *If M is the union of countably many sets closed in M that belong to the family $\mathfrak{N}$, then M also belongs to the family $\mathfrak{N}$.* [Ch16, 7]

He needs the nesting condition dealing with subsets of M and the countable union condition, since he is concerned with dimension-theoretic properties.

From previous reasoning he shows that the family of all separable 0-dimensional sets is a normal family. Also, the family of all countable sets is a normal family.

Because of condition 1 of his basic definition every normal family contains the empty set. Thus, every 0-dimensional set with respect to an arbitrary normal family is totally discontinuous.

Hurewicz proves a sequence of seven theorems relating normal families to dimension [Hurewicz, 1927a, pp. 751, 758, 759, 760, 761, 762]. They lay out dimension theory à la normal families, providing an interesting development of the subject.

On this basis Hurewicz arrives at an important dimension-theoretic consequence [Hurewicz, 1927a, page 760]:

> In our terminology, *any set that is totally discontinuous* (discontinuous in p) *with respect to the family of at most $(n-1)$-dimensional sets, where the family of (-1)-dimensional sets only contains the empty set*, is *at most n-dimensional* (at the point p). Those at most n-dimensional sets that are not also at most $(n-1)$-dimensional are called n-dimensional.[Ch16, 8]

Huewicz' main theorem relating dimension to his theory of normal families is [Hurewicz, 1927a, page 760]:

> **Theorem IV.** *The separable at most n-dimensional sets form a normal family.*[Ch16, 9]

Hurewicz' paper on normal families shows him shifting the basis of dimension theory from compact metric spaces to separable metric spaces. A further support for this desirable shift comes in his compactification theorem, presented in his paper of 1927, 'Ueber das Verhältnis separabler Räume zu kompakten Räumen' ('On the relation of separable spaces to compact space') as [Hurewicz, 1927c, page 425]:

> **Theorem.** *Every separable metric space R is homeomorphic to a subset of a compact metric space that has the same dimension.*[Ch16, 10]

Hurewicz proves that the separable space R is dense in the compact metric space.

At this point I shall briefly consider Hurewicz' brilliant research note dealing with infinite-dimensional spaces [Hurewicz, 1928b]. While this research note is short, it is replete with significant results and conjectures or announcements of significant results (cf. [Hurewicz & Wallman, 1948, pp. 48–51]).

Hurewicz opens his note by recalling the known result that for every natural number n all separable spaces that are at most n-dimensional can be represented as a sum or union of $n + 1$ 0-dimensional subspaces. This fact leads him to defining a specific kind of infinite-dimensional space, a *countable-dimensional* space, as a space that can be decomposed into or represented by countably many 0-dimensional spaces, or, equivalently, countably many finite-dimensional spaces. For formal completeness reasons, the empty set is also regarded as a countable-dimensional space.

Hurewicz immediately addresses the question of whether every separable space is countable-dimensional. According to the result of Urysohn [Urysohn, 1924b, page 303] every separable space is homeomorphic to a subset of the Hilbert infinite-dimensional space H,[10] so in Hurewicz' terms the question becomes whether the Hilbert space is countable-dimensional. Hurewicz' answer is negative. *The Hilbert infinite-dimensional space cannot be composed of a union of countably-many finite-dimensional subspaces*. Hurewicz shows this result for the compact Hilbert cube, K, subspace of H [Hurewicz, 1928b, pp. 916–918]. The cube K is defined as the product of countably-many segments satisfying the coordinate equations:

$$0 \leq x_n \leq \frac{1}{n}.$$

Hurewicz also introduces the concept of a countable-dimensional space that possesses no finite dimension, which he calls $\aleph_0$-dimensional [Hurewicz, 1928b, page 916]. He gives a couple of simple examples of $\aleph_0$-dimensional spaces [Hurewicz, 1928b, page 918]. One of these is formed of all the points of H, of which only finitely many coordinates are different from 0. This space is not compact. However, an example of a compact $\aleph_0$-dimensional space is the following. Let W_m $(m = 1, 2, \ldots)$ is be the set of all points of H satisfying the condition:

$$0 \leq x_i \leq \frac{1}{i} \quad (i = 1, 2, \ldots, m)$$
$$x_i = 0 \quad (i = m + 1, m + 2, \ldots)$$

The union of the finite-dimensional sets W_m is $\aleph_0$-dimensional and indeed compact [Hurewicz, 1928b, page 918].

The rest of Hurewicz's research note is rich in ideas. However, I shall not go into it further.

I shall now briefly consider Hurewicz' simple proof of the Lebesgue–Brouwer tiling principle or theorem, published in his paper 'Über ein topolgisches Theorem' ('Concerning a topological theorem') [Hurewicz, 1929]. I have already covered Sperner's wonderfully simple proof of the theorem, which relies critically on the Sperner combinatorial lemma [Sperner, 1928] (see Section 13.11). In that section I also mentioned Hurewicz' proof. The two proofs were developed and published about the same time, and they are closely related. Apparently Hurewicz only found out about Sperner's proof while correcting the proofs for his own paper, as indicated by a note added in proof to his published paper [Hurewicz, 1929, page 211].

[10] See [Hausdorff, 1914, pp. 287–289].

Hurewicz states the Lebesgue–Brouwer tiling theorem in his paper as follows [Hurewicz, 1929, page 210]:

> (A) *If one divides a bounded closed domain of the Euclidean n-dimensional number space R_n into a finite number of closed sets whose diameters are sufficiently small, then there are points that have at least $n+1$ of these sets in common.*[Ch16, 11]

He first considers a cube or cubic interval I^n in n-dimensional number space (instead of a simplex as with Sperner), subdivided into smaller intervals or little bricks that have no interior points in common, $Z_1, Z_2, \ldots, Z_m$. That is, the interval is subdivided in such a way that the closed bricks, Z_i, only overlap at boundary points. The cube is viewed as a subset of the domain originally mentioned. For the 2-dimensional case one has the usual layering of bricks as in Figure 16.1.

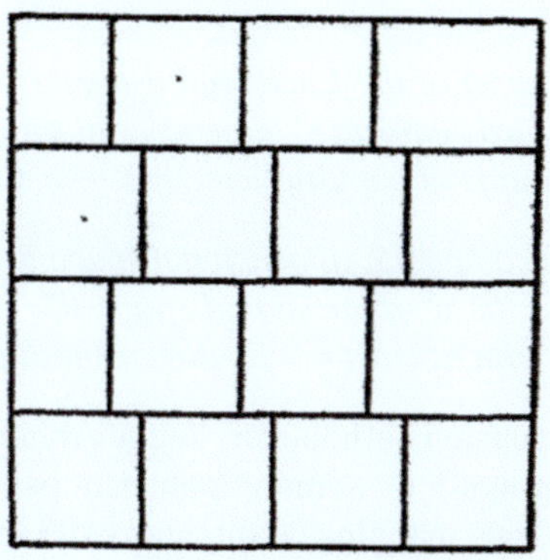

Fig. 16.1 Arrangement of Bricks for Tiling Principle [Hurewicz, 1929, page 213].

For a more general case, Hurewicz considers a new subdivision of the cube I as:

$$I = D_1 + D_2 + \cdots + D_k,$$

where each set D_i coincides either with one of the original subintervals Z_i or with a union of them, and each of the Z_i belongs to one of the complexes D_i. Then his aim is to prove [Hurewicz, 1929, page 214]:

> (A′) *Under the assumption that none of the interval-complexes D_i have points stretching between opposite sides of I, then there are points that $n+1$ of the sets D_i have in common.*[Ch16, 12]

This is just the Lebesgue–Brouwer tiling theorem for a special case. Hurewicz proves this in a combinatorial way. From it the general case follows easily.

Over the course of a few years from 1926 to 1930 Hurewicz wrote five significant papers on the behavior of mappings of finite-dimensional spaces. These are a pair of research notes [Hurewicz, 1926b] [Hurewicz, 1927b], a special appendix to his joint paper with Menger [Hurewicz & Menger, 1928, Anhang], a further paper on the Lebesgue–Brouwer tiling theorem [Hurewicz, 1930a], and final paper on dimension-raising mappings [Hurewicz, 1933b].

The second research note, from late 1926, 'Ueber stetige Bilder von Punktmengen. (Zweiter Mitteilung)' ('Concerning continuous mapping images of point-set. (Second communication)') is particularly important as it contains his well-known formulas on dimension-raising and dimension-lowering mappings [Hurewicz, 1927b]. He captures the formulas in three theorems [Hurewicz, 1927b, pp. 159, 161, 163]:

> **Theorem I.** *If the separable n-dimensional space R is mapped one-one and bicontinuously onto the n^*-dimensional space R^*, such that $n^* \geq n$ holds, then there are points in R^* that are images of at least $n^* - n + 1$ different points of R.*[Ch16, 13]

> **Theorem II.** *If a subset of the Euclidean n-dimensional space is mapped one-one and bicontinuously onto an n^*-dimensional space R^*, such that $n^* > n$ holds, then there are points of a multiplicity $\geq n^* - n + 2$ in R^*.*[Ch16, 14]

> **Theorem III.** *If the compact n-dimensional space R is one-one and bucontinuously mapped onto the n^*-dimensional space R^*, such that $n^* \leq n$, then there are points in R^* with at least $n - n^*$-dimensional pre-image sets.*[Ch16, 15]

The later paper 'Über dimensions-erhöhende stetige Abbildungen' ('On dimension-raising continuous mappings') is a more polished paper that deals only with dimension-raising continuous mappings [Hurewicz, 1933b].[11] I shall not go further into the research note with the three quoted theorems or the later paper.

The fourth paper in this set of five papers on mappings of finite-dimensional spaces, 'Ein Theorem der Dimensionatheorie' ('A theorem of dimension theory') was published in the American journal *Annals of Mathematics* [Hurewicz, 1930a]. The paper firms up results of Hurewicz' earlier papers on coverings and the intersection order that characterizes the Lebesgue–Brouwer tiling theorem [Hurewicz, 1926b] [Hurewicz & Menger, 1928, Anhang]. At the beginning of the paper he introduces a key conceptual construct, the 'branching' or 'refinement' ('Verzweigung') of a closed covering, whereby each of the closed sets F_i $(i = 1, 2, \ldots, m)$ in the closed covering is subdivided into smaller closed subsets $F_{i,k}$ $(i = 1, 2, \ldots, m; k = 1, 2, \ldots, n_i)$, the union of which is the original closed set F_i. In effect, the descending refinements of a given covering have the same intersection order. Applying this construct, Hurewicz is able to proceed to proving the following version of the Lebesgue–Brouwer tiling theorem [Hurewicz, 1930a, page 177]:

[11] Cf. the very short conference note [Hurewicz, 1932]).

(I) *For an n-dimensional subset of a compact space there is a sequence* $\sigma_1, \sigma_2, \sigma_3, \ldots$ *of infinitely fine coverings of order* $n+1$*, each of which is a branch/refinement of the immediately preceding one (and consequently also of all preceding ones).*[Ch16, 16]

I shall leave consideration of Hurewicz' work at this point, but I shall return later to his work on the embedding theorem in Section 16.9. Hurewicz made considerable, quite important contributions to dimension theory, and he went on to make contributions to algebraic topology, for example, through developing the higher homotopy groups.

16.5 Lev Tumarkin and Dimension Theory

I shall now turn to the work in dimension theory of the Russian mathematician Lev Abramovich Tumarkin (1904–1974).[12] In several ways his independent dimension-theoretic work parallels that of Hurewicz. They both sought to move the framework for developing dimension theory from compact metric/metrizable spaces to more general separable metric/metrizable spaces. Tumarkin studied at Moscow State University and graduated from that important institution in 1925. Tumarkin's interest in the problem situation surrounding general dimension arose in part due to his participation in the Alexandroff–Urysohn topological seminar at Moscow State University [Tumarkin, 1926b, page 57]. Tumarkin was a student of Alexandroff. The seminar with Alexandroff's mentoring[13] gave him the necessary background in topology and dimension theory for him to do research in these areas. He remained at Moscow State University as a researcher and teacher and administrator of mathematics for his entire career. His primary research domain was dimension theory.

His early work on dimension theory is covered in ten papers and research notes published between 1925 and 1928. From 1928 there is a large gap in the list of his research publications, with the list only resuming with works from 1951 and after [Alexandroff & Kolmogorov, 1964, pp. 162–163]. I have no explanation for this gap, but perhaps his teaching and administrative duties took precedence in those years.

Like Hurewicz, Tumarkin showed the way mathematically to move the framework for developing dimension theory from compact metric/metrizable spaces to more general separable metric/metrizable spaces. In a note honoring Tumarkin's 60th birthday Alexandroff and Kolmogorov characterize this move as follows, giving their colleague much praise, [Alexandroff & Kolmogorov, 1964, page 161]:

> Tumarkin began his scientific work, and obtained his first brilliant results in the field of topology, while still a student at the Moscow University. These results were as follows: he was able to prove by a simple and extremely clever

[12] For biographical material on Tumarkin see [Alexandroff & Kolmogorov, 1964], and [Alexandroff et al., 1975].

[13] Alexandroff had a reputation as a good teacher, and his texts reveal something of his good teaching ability and practices.

method that all the basic results in dimension theory, which were proved by P.S. Urysohn only for compact spaces, are also valid for arbitrary metrizable spaces with a countable base. This result was also obtained simultaneously and independently by the famous topologist W. Hurewicz (who was then also a very young mathematician, a pupil of Menger in Vienna). Now the Hurewicz–Tumarkin theorems are accepted as classical; in 1925, when they were proved for the first time, they seemed quite unexpected and at once brought fame to both young authors among mathematicians interested in dimension theory—an entirely new mathematical discipline at that time. Urysohn and Brouwer considered the possibility of the construction of a dimension theory as being inseparably bound up, if not necessarily with the compactness of space itself, then with some or other of its analogues or generalizations (completeness, absolute F_σ [countable union of closed sets]). Consequently, when Tumarkin's results were communicated to Brouwer, they made a profound impression on him.

Tumarkin initially published his results on extending the frame of dimension theory from compact to separable metric/metrizable spaces in a short research note, his first publication, communicated by Brouwer on 28 November 1925 to the Koninklijke Nederlandse Akademie van Wetenschappen, the Dutch Academy at Amsterdam, for its *Proceedings* [Tumarkin, 1925a]. Even before this date he had given a more extensive presentation of his results to the Moscow Mathematical Society on 4 October 1925. These results then appeared in 1926 as a longer publication, 'Beitrag zur allgemeinen Dimensionstheorie' ('Contribution to general dimension theory'), in the Russian journal *Matematiceskij sbornik*, bearing the place and date of Marphino, 23 July 1925,
[Tumarkin, 1926b]. The paper was supposed to appear in *Mathematische Annalen*, as Tumarkin remarks [Tumarkin, 1925a, page 996], and in its printed form it looks like an *Annalen* paper. However, for some reason it came out in the Russian journal.

Tumarkin forged a path to move the frame for dimension theory from compact to more general separable metric/metrizable spaces through his use of the supporting metrization results of Urysohn and Tychonoff and his application of his own generalization of the Heine–Borel–Lebesgue covering theorem. The result is effectually a rewriting or extension of Urysohn's dimension-theoretic results for separable metric/metrizable spaces, laid out in his long exposition [Tumarkin, 1926b]. The following are samples of his generalized results.

Tumarkin provides a generalized theorem concerning the decomposition of an n-dimensional (finite, positive n) set in a separable metric space into 0-dimensional sets [Tumarkin, 1926b, page 61]:

> Theorem III. *For a set (lying in a separable metric space) to have dimension* n $(n \geq 0)$ *it is necessary and sufficient that one can decompose the set into* $n + 1$ *and not fewer sets of dimension* 0.[Ch16, 17]

He also offers a generalized sum theorem [Tumarkin, 1926b, page 62]:

Theorem IV. *Let an arbitrary set M be given (in a separable metric space) and let* $L_1, L_2, \ldots, L_i, \ldots$ *be countably many at most* $(n-1)$*-dimensional subsets of M, that are closed in M. Then the union [of the subsets]* $\sum_{i=1}^{\infty} L_i$ *is at most* $(n-1)$*-dimensional.*[Ch16, 18]

Further, Tumarkin proves what he calls a 'Fundamental Theorem' ('Fundamentalsatz') [Tumarkin, 1926b, page 81]:

Theorem VIII_1. *Every set L lying in a separable complete*[14] *space R is contained in a* G_δ*-set [countable intersection of open sets]* Λ *of equal dimension to that of L (so that* $\Lambda \supset \overline{L}$*) such that:*

$$L \subset \Lambda \subset \overline{L} \subset R, \quad \Lambda \in \mathfrak{G}_\delta \text{ (rel. } R), \quad \dim \Lambda = \dim L.$$ [Ch16, 19]

Additionally, Tumarkin considers a universal space, which Urysohn had previously constructed, with the aim of applying his Fundamental Theorem to it. As background, note that Urysohn had offered a construction of a 'universal' separable metric space that contains an isometric image of every separable metric space in a short research note in the *Comptes rendus* (*Proceedings*) of the Academy of Sciences, Paris, 1925 [Urysohn, 1925b]. The note was published after Urysohn's death. A little later, Urysohn's colleague and friend Alexandroff edited a much more extensive treatment of this construction and related results, published in 1927 [Urysohn, 1927b]. Urysohn's construction is quite clever, but the details of the construction are not needed for the current discussion. I merely note that Tumarkin uses his Fundamental Theorem to yield the derived result [Tumarkin, 1926b, page 83]:

Every separable metric space R is congruent with a subset of a G_δ*-set* Λ *lying in a Urysohn universal space of equal dimension.*[Ch16, 20]

Overall, Tumarkin's exposition provides a significant extension of Urysohn's results to separable metric spaces [Tumarkin, 1926b] (along with the preliminary version [Tumarkin, 1925a]). As part of this effort Tumarkin gives extensive proofs.

During 1928 five of Tumarkin's papers were published, two jointly with other authors. In their assessment of Tumarkin's work Alexandroff and Kolmogorov consider one paper of special importance, 'Sur la structure dimensionnelle des ensembles fermés' ('On the dimensional structure of closed sets'), pointing to a significant result [Tumarkin, 1928a]. They write [Alexandroff & Kolmogorov, 1964, page 161]:

A very important achievement of Tumarkin (again obtained by him simultaneously with Hurewicz, and independently of him) is the theorem asserting that any compact space of dimension *n* contains a Cantor[ian] set [manifold] of the same dimension (i.e., a continuum of dimension *n*), the connectedness of which is not altered by removing from it an arbitrary set of dimension $\leq (n-2)$.

---

[14] A complete metric space is one in which every fundamental (Cauchy) sequence converges [Hausdorff, 1914, page 315].

Hurewicz' equivalent result with a full proof comes in his joint paper with Menger [Hurewicz & Menger, 1928, pp. 622–627] (cf. Section 15.7). Engelking gives a modern version of this result in the following formulation [Engelking, 1978, page 93]:

> **The Cantor[ian]-manifold theorem.** *Every compact metric space X such that [inductive dimension] ind X = n ≥ 1 contains an n-dimensional Cantor[ian]-manifold.*

Finally, I should mention an important longer paper of Tumarkin published in 1928, but submitted earlier on 10 June 1926, 'Über die Dimension nicht abgeschlossener Mengen' ('Concerning the dimension of non-closed sets') [Tumarkin, 1928b]. This paper again represents his effort to push beyond compact metric spaces, in this case not just to separable metric spaces, but also to *complete*[15] metric spaces. The paper of some 20 pages represents one of Tumarkin's last research efforts from 1928.

16.6 Menger's *Dimensionstheorie* (*Dimension Theory*) and the Brouwer–Menger Dispute

Menger published the first book on topological dimension theory in 1928, *Dimensionstheorie* (*Dimension Theory*) [Menger, 1928b]. His book on curve theory came out four year later in 1932, *Kurventheorie* (*Curve Theory*) [Menger, 1932a]. In comparison, Urysohn's lengthy articles fully expositing his dimension theory [Urysohn, 1925a] [Urysohn, 1926], together with his companion article on curve theory (theory of Cantorian lines) [Urysohn, 1927a], constitute a publication of book length.

In his book on dimension theory Menger includes several historical notes at the ends of various sections. As I discussed in Section 15.7, Menger was challenged earlier about the origins and priorities of his dimension theory work when he was in Amsterdam, resulting in the publication of a rather 'defensive' personal historical note, which included two of his earliest pieces. That note appeared in 1926 in the *Proceedings* of the Dutch Academy [Menger, 1926e]. Brouwer was involved with some of the editing of this note and communicated it to the Dutch Academy.

In his book Menger includes a long historical note at the end of the section that introduces the dimension concept [Menger, 1928b, Section II.2]. It is worth quoting *in extenso* and discussing its content, because the text soon became the proximate cause of an acrimonious dispute between Brouwer and Menger that lasted for years.

The dispute between Brouwer and Menger has been analyzed in great detail. Freudenthal in his notes to the related papers gave a very good analysis in his edition of Brouwer's *Collected Works*, volume 2 [Brouwer, 1976, pp. 564–567]. Then van Dalen in the first edition of his biography of Brouwer provided a very careful and quite detailed analysis [Dalen, 2005, pp. 643–671]. It seems to me that he sifted

---

[15] Cf. footnote 14.

through all the evidence he could find related to the matter. In the second edition he chose to give a simplified and much shorter summary of his analysis, though I do not think that he changed his mind about the facts and his overall analysis [Dalen, 2012, pp. 595–601]. In this work I shall limit myself to a short 'objective' (I hope) summary of the matter.

The following is the first part of Menger's historical note [Menger, 1928b, pp. 83–84]:

> **Historical Note.** In a note by me submitted in February 1922 to the Monatsheften für Math[ematik]. und Phys[ik]. [Monthly Journal for Mathematics and Physics] (printed in Proc[eedings]. Acad[emy]. Amsterdam *29*, page 1124 [Menger, 1926e]), it states literally: *"A non-empty set [M]*[16] *of the space R_m is called n-dimensional, if 1. to each of its points p and to every open neighborhood $U_1(p)$ there exists an open neighborhood $U_2(p) \subset U_1(p)$ with whose boundary M has an at most $(n-1)$-dimensional set in common; and if M 2. contains at least one point Q, for which an open neighborhood U exists such that M has a set in common with the boundary of every open neighborhood $U_1(Q) \subset U$ that contains an $(n-1)$-dimensional part. The empty set is (-1)-dimensional."* This definition can also be found in my works in the Monatsheften f[ür]. Math[ematik]. u[nd]. Phys[ik]. [Monthly Journal for Mathematics and Physics] (*33*, 1923, p.157ff., *34*, 1924, p. 138f. [Menger, 1923, pp. 157ff.] [Menger, 1924d, pp. 138f.]). The formally somewhat shorter form of definition listed above is also given there: *A space is called n-dimensional, if n is the smallest number, such that every point of the space is contained in arbitrarily small neighborhoods with at most $(n-1)$-dimensional boundaries.* (Further, for the intuitive-experimental derivation of this term see Math[ematische]. Annalen *95*, p.. 278 [Menger, 1926a, page 278] und Jahresber[icht]. d[er]. Deutschen Math[ematiker]. Vereinigung [Annual Report of the German Association of Mathematicians] *35*, p. 118 [Menger, 1926b, page 118].) Already in the submission from the fall of 1921 mentioned on page 77 [of this book] [printed at [Menger, 1926e, page 1123]] it says literally: *"A continuum K is called a curve if in every neighborhood U of every point p of K there is contained a neighborhood U_1, such that the intersection of K with the boundary of U_1 contains no connected part,"* wherein among other things the principle of recursion to higher dimensions is given.
>
> In September 1922, Urysohn (Comptes Rendus [Proceedings of The Academy of Sciences, Paris], *175*, p. 440 [Urysohn, 1922a, page 440]) published the following definition based on a compact metric space, in literal translation[17]: If C is a given subset of the space, p a given point of C, then one says of a subset B of C that it ε-separates the point p, *"if the set $C - B$ is the union of two sets A and D, such that 1. A and D are separated, i.e., that neither of the two sets contains a point or an accumulation point of the other,*

[16] omitted in this quotation, but in the original.

[17] from the French.

that 2. the set A contains the point p, and that 3. the sets A and C [B][18] *are contained in a sphere of radius* $< \varepsilon$. *If the point x of the set C for every* ε *can be* ε*-separated by an empty set, then C is called* 0*-dimensional at the point x. A set that is* 0*-dimensional at all its points is* 0*-dimensional. If the set C is not less than n-dimensional and the point x for every* ε *can be* ε*-separated by a set less than n-dimensional, then C is called n-dimensional at point x. If the set C in all of its points is* $\leq n$ *and in certain points has dimension n, that C is called n-dimensional."* As a result of not applying the definition of the boundary of the neighborhood, this definition is a little complicated but is equivalent to mine, since it is easy to show that any subset of the space that ε-separates the point p is nothing other than the boundary of a neighborhood of p, which is part of the sphere around p with radius ε, and that, conversely, the boundary of every neighborhood of p, which is part of this sphere, ε-separates the point p. So if a point p of a metric space is contained in arbitrarily small neighborhoods with at most $(n-1)$-dimensional boundaries, then for arbitrarily small ε it can be ε-separated by at most $(n-1)$-dimensional sets and vice versa. (See also Urysohn's great work on this definition in the Fundam[enta]. Math[ematicae]. *7* and *8*, 1925, 1926 [Urysohn, 1925a] [Urysohn, 1926]).[Ch16, 21]

This part of the text of Menger's historical note is very misleading. It suggests that Menger had priority of publication over Urysohn. This is certainly not the case. While both young men developed their new ideas on dimension and curve during 1921, Urysohn has priority of publication. As I have analyzed in Section 15.2 Menger did write down his ideas very early, including them in various notes and short papers intended for publication. However, Hahn and the editors of the *Monatshefte* (*Monthly Journal*) rejected these for publication, since they contained errors. They were deposited with the Vienna Academy. Two of these were included in the note on the origins of his work that Menger, when pressed, wrote in 1926 for the Dutch Academy [Menger, 1926e]. Additionally, there is a fuller publication of Menger's earliest notes and papers, which appeared later in 1929 [Menger, 1929b]. This came out because of the dispute with Brouwer (see below).

Menger's first *publication* on dimension theory was his 'Über die Dimensionalität von Punktmengen. (Erster Teil)' ('On the dimensionality of point-sets. First part') of 1923 [Menger, 1923]. Urysohn's first publications in this domain were his research notes appearing in the *Comptes rendus* (*Proceedings*) of the Academy of Sciences, Paris of 1922 [Urysohn, 1922a] [Urysohn, 1922b]. To be fairer to his Russian counterpart, Menger probably should have put Urysohn's work first in his historical note and given some explanation of his own early notes, drafts, and publications. Yet, there can be no doubt that Urysohn and Menger both started developing their ideas on dimension and curve theory in 1921.

Examining Menger's historical note further, he writes [Menger, 1928b, page 84]:

---

[18] There is an error in the quotation: the C should be B. Cf. [Urysohn, 1922a, page 441]. Urysohn's text is clearer.

> All of the following considerations and proofs are based exclusively on my concept of the dimension definition, which is based on the concept of neighborhood boundary.[Ch16, 22]

Menger had a wonderful intuitive approach to the concepts of dimension and curve, and he backed his approach with a concrete thought experiment. That is a great attraction for his work.[19]

In the next paragraphs of his historical note Menger mentions a series of forerunners or precursors (Vorläufer) of the modern dimension theory. He first mentions Euclid and Cantor as precursors [Menger, 1928b, pp. 84–85]. He might have also mentioned Peano at this point, but he does not.

Then Menger goes on to refer to Poincaré as another precursor and some of his last work on the concept of dimension [Menger, 1928b, page 85]:

> It was Poincaré who several times emphasized how unintuitive the arithmetic definition of dimension is and how little it touched the core of the problem, apart from its special nature. *"I am not saying that the arithmetization of mathematics is something bad, but I am saying that it does not exhaust the matter."* (Last Essays, German by Lichtenecker, Leipzig, 1913, page 64.) [Poincaré, 1963, pp. 25–44, especially page 29] And building on Euclid's ideas, he repeatedly emphasized the need for a recursive definition of the concept of dimension. *"I would like to base the determination of the number of dimensions on the concept of cut. . . . A continuum has n dimensions if it can be divided into several separate parts by making one or more cuts, which are themselves continua of* $n - 1$ *dimensions."* (loc. cit., pages 65f.).[Ch16, 23]

Naturally, Menger next considers Brouwer's work in his historical note with some three paragraphs [Menger, 1928b, pp/ 85–86]. In this case too, he regards Brouwer as a precursor. Menger's treatment of Brouwer as a precursor is, in my opinion, shocking. Menger had been a new researcher at Amsterdam during 1925–1927, and as I have discussed above Brouwer had mentored him while he was there. There are some early hints of a rift between the two. Still, as a very distinguished living mathematician, who had developed a major part of topology, I think that Menger owed Brouwer more than a little respect.

In the historical note Menger covers Brouwer's definitions of separation and dimension itself. He opens his treatment as follows [Menger, 1928b, page 85]:

> Brouwer subjected (Journal f[ür] d[ie] reine u[nd] angewandte Mathematik. 142, 1913, p. 146) Poincare's conceptual formulation to a critique and modification. He pointed out that, according to Poincaré's definition, a double cone, which can be divided into two separate parts by omitting a single point, can be designated as one-dimensional, contrary to ordinary intuition, and further that the concept of continuum and the words "one or several" in Poincaré's definition are not precise.[Ch16, 24]

[19] Menger's intuitive thinking and style, together with Hurewicz and Wallman's beautiful presentation of the overall theory in their book *Dimension Theory* [Hurewicz & Wallman, 1948], are what first attracted me to this area of mathematics.

Menger then quotes in full Brouwer's definitions of separation and dimension-degree, as given in the original 1913 publication [Brouwer, 1913b]. I have already discussed the issues in this case with regard to the inclusion of the word 'abgeschlossene' ('closed'), Brouwer's early recognition of the problem, and Brouwer's interaction with Urysohn in Sections 12.11, 12.12, and 12.13. Brouwer did not publish his corrections until 1923–1924[20], though the accumulated evidence shows that he knew of the problem much earlier, in 1913.

In his historical note Menger proceeds to write the following about Brouwer's work [Menger, 1928b, page 85]:

> This concept of the general dimension-degree already comes very close to the dimension concept in dimension theory. If one, namely, *first* replaces the Brouwer definition of separation with the following one that has since emerged: *"Two disjoint closed subsets A and B of the space R are called separated by the set C, if R – C can be decomposed into two disjoint relatively closed subsets of which one contains A and the other B as subsets"*, and if one *second* takes the beginning of the recursion not from the set without subcontinua but from the empty set as the (-1)-dimensional, then the concept of dimension-degree will be equivalent to the dimension concept. Namely, it is the case that, as we have seen (cf. page 117 and page 120 [of this book]), the theorem holds stating that in an n-dimensional space two disjoint closed subsets can be separated by an at least $(n-1)$-dimensional, but two disjoint closed subsets cannot be separated by a subset less than $(n-1)$-dimensional.[Ch16, 25]

Menger continues his discussion of notions of separation and Brouwer in the following way [Menger, 1928b, pp. 85–86]:

> *First*, as far as the intuitive idea of the separation of two sets is concerned, there can be various ways to make the concept precise in terms of form and content. We shall refer (p. 207 [of this book]) to Brouwer's original concept of separation (though also allowing for non closed separated sets) as *separating* [*Separieren*]. Brouwer also gave other definitions of separation (Proc. Acad. Amsterdam 27, 1924, p. 636), from which the following also comes, equivalent to the modern definition of separation, and very similar to Brouwer's original definition, but referring to the definition of the modern concept of connectedness, which has only recently emerged (cf. p. 197 [of this book]): *Two disjoint sets A and B of the space R are said to be separated by the set C, disjoint from A and B, if every set of R that is connected between A and B contains a point of C.* — *Second*, as far as the starting point of the recursion is concerned, one could actually develop a theory of dimension-degree (cf. p. 317 [of this book]), starting from the discontinuous sets, that would be consistent with one based on compact spaces and be identical to the dimension theory developed in this book, but would differ greatly from dimension theory based on arbitrary separable spaces (see p. 318 [of this book]). We will (p. 204 [of

[20] [Brouwer, 1923a] [Brouwer, 1923b] [Brouwer, 1924a] [Brouwer, 1924b] [Brouwer, 1924c] [Brouwer, 1924d].

this book]) get to know subsets of the plane that are one-dimensional in the sense of dimension theory, but do not contain any subcontinua, i.e., have a dimension-degree of zero. There are even (see p. 209 [of this book]) in the sense of dimension theory infinite-dimensional spaces with dimension-degree zero. — The concept of dimension-degree at a point is completely different in content from the concept of dimension at a point, which is fundamental to dimension theory.[Ch16, 26]

What Menger writes *technically* in his note about Brouwer's treatment of connectedness, separation, and dimension-degree appears to be correct. However, what he asserts *historically* about connectedness and separation is not fully accurate. As I have discussed in Section 12.12, the modern, commonly used concept of connectedness is due to Lennes [Lennes, 1906] [Lennes, 1911, page 303], Riesz [Riesz, 1907] [Riesz, 1909a], and Hausdorff [Hausdorff, 1914, page 244]. Brouwer knew and used this modern notion of connectedness as evidenced in a paper of 1911 [Brouwer, 1911c, page 308]. Brouwer also knew about the problem with the offending word 'abgeschlossene' ('closed') in his definition of separation very early, though it was only brought to the fore in publication later by Brouwer's interaction with Urysohn in 1923 and 1924. Perhaps without due consideration, Menger glosses over these fine points in his historical note. It may too much to demand of Menger in this work, but he may well have known something of this matter, having been in Amsterdam for two years around Brouwer and others working on the cluster of ideas related to dimension.

Menger concludes his historical note by mentioning Sierpiński, who implicitly introduced the concept of zero-dimensional sets [Sierpiński, 1921, page 89] (cf. Section 13.9), and Fréchet, who introduced his own special notion of 'dimension type' [Fréchet, 1910] (cf. Section 10.4). Menger even makes a brief mention of Hausdorff and Hausdorff dimension, based on measure theory [Hausdorff, 1919], which I shall cover below.

When Brouwer saw Menger's book and read his characterization of the history of dimension theory, he was outraged and reacted with considerable vigor. Hurewicz, who remained in Amsterdam with Brouwer after Menger left for Vienna, saw at close range Brouwer's reaction. According to van Dalen [Dalen, 2005, page 645], reporting on Hurewicz' witnessing of the mater: '... it appears that the situation in Amsterdam had definitely become unpleasant.... he [Brouwer] would fly into violent harangues on Menger and his views on dimension theory.' Hurewicz, of course, suffered some of the blowback. Thus, in a letter to his friend Menger, dated 10 October 1928, he writes [David M. Rubenstein Rare Book and Manuscript Library, Duke University, [Karl Menger papers, 1872-2000], Letter of Hurewicz to Menger, 10 October 1928] and [Dalen, 2005, page 646]:

What you write about Brouwer, about his character and his "morals", is nothing new for me. Nonetheless I find that your conduct with respect to him is incorrect and inappropriate [irregular, unlawful, unrechtmäßig]. As to the "matter" itself, in my opinion the historical survey in your book is not completely free from a subjective representation. Whether Brouwer is to be

> regarded as a mathematician or just as a precursor (as you portrayed him historically): the concept of dimension is a question on which in any case one can have differing opinions.
>
> ...
>
> Still it is precisely because of this and because you are aware of Brouwer's pathological sensitivity that you should have sent him the proofs and perhaps agreed to some changes that would not have caused you any great harm — yet you have to risk an argument, that will spoil your enjoyment of your beautiful book and, at best, will cost you an unreasonable expenditure of time and energy, which would really be worth spent on something better. A "life and death" argument over these issues! The fact that you did not think for a moment what an impossible situation I have to be in because of the argument between Brouwer and you, I take it very badly.[21] [Ch16, 27]

Brouwer immediate reaction in print to Menger's book came in a note appearing in the *Proceedings* of the Academy, Amsterdam, communicated on 29 September 1928: 'Zur Geschichtsschreibung der Dimensionstheorie' ('Historiography of Dimension Theory') [Brouwer, 1928b] (announced in [Brouwer, 1928a]). The note was short, but to the point.

Brouwer opens his note by declaring forthrightly what he thought that he had accomplished in dimension theory [Brouwer, 1928b, page 953]:

> I founded dimension theory in a paper appearing in 1913 "Concerning the natural dimension concept" (Journ[al]. f[ür]. Math[ematik]. 142. Pp. 146–152) for those species [sets] for which dimension has a natural meaning, namely, for the condensed [compact] metric species [spaces], by providing a *dimensional justification theorem*, i.e., defining a dimension concept that makes sense in the domain of condensed [compact] metric species [spaces], for which I was able to show that the same n-dimensional manifolds have dimension n.[Ch16, 28]

Given his current interest in furthering intuitionistic mathematics Brouwer's uses terms like 'Spezies' instead of 'Menge' for 'set' and 'kondensierte' instead of 'kompakt' for 'compact'. In these cases the meaning is virtually the same.

Brouwer summarizes the meanings of weak and strong connectedness and strong and weak separation, including references. He asserts [Brouwer, 1928a, page 953]:

> In my cited paper from 1913 the strong justification theorem is indeed formulated according to the words, but the proof following it clearly points to the weak justification theorem, so that clearly only that is meant, and the slip of the pen contained in the formulation of the theorem is immediately obvious to the thoughtful reader. That this slip of the pen remained for so many years with no published correction had, firstly, the general cause of being neglected with regard to easy-to-see slips of the pen and, secondly, a specific reason

[21] Hurewicz' letters of this time reveal that he wanted to leave Amsterdam, no doubt in part due to Brouwer's temperament and perhaps in part due to the dispute with Menger. However, he ended up staying on for many years, until 1936.

communicated in these Proceedings 27, p. 136 [p. 636] [Brouwer, 1924b, page 636, footnote 6].[Ch16, 29]

Brouwer next explains why he limited his work on dimension theory to the founding definitions and the justification theorem [Brouwer, 1928a, pp. 953–954]:

> In my cited paper of 1913 I limited myself to the foundation of dimension theory and put off the publication of further dimension-theoretic developments, on the one hand, because the proof of the justification theorem achieved the epistemological goal, and, on the other hand, because for further considerations (primarily those related to the sum theorem and the decomposition theorem) an intuitionistic development was not as close as the proof of the justification theorem.[Ch16, 30]

This explanation of his limited development of the theory may seem weak to a modern reader, but Brouwer took epistemological matters very seriously, and by then he was deeply involved with constructing his own intuitionistic mathematics and his defenses thereof. From his own special intuitionistic viewpoint he discerned orders of difficulty in proving the justification theorem versus other theorems like the sum and decomposition theorems. In his note he also points to his significant earlier result of 1911 proving the invariance of domain [Brouwer, 1911c] and to other fundamentals.

In the remaining sections of his short note Brouwer takes great pains to lay out carefully and in detail the priority of publication in dimension theory of Urysohn over Menger. He points to Urysohn's *Comptes rendus* (*Proceedings*) of the Academy of Sciences, Paris published notes of 1922 over Menger's deposition of 1921 with the Vienna Academy of Science and the note of 1922, which do not really count in terms of publication. Menger's first real publications come in 1923 and 1924.

Menger's historical note and Brouwer's responding historical note set off their full-blown dispute, which lasted for years. Hahn became involved as intermediary, because he knew Brouwer well and because he was Menger's doctoral supervisor and mentor in Vienna. Eventually, after much painful back and forth Menger published an 'Antwort auf eine Note von Brouwer' ('Answer to a Note of Brouwer') [Menger, 1930] in 1930. Much went on behind the scenes. Menger produced further publications. One was the careful publication of his early notes and draft papers [Menger, 1929b]. Another was self published in 1932 [Menger, 1932b]. Yet another appeared in 1933 [Menger, 1933]. I think that Menger was still 'seething' about Brouwer much later when writing his reminiscences of the great mathematician [Menger, 1979a, pp. 237–255] (see Section 15.8). For all the messy details see van Dalen's extensive treatment in the first edition of his biography of Brouwer [Dalen, 2005, pp. 643–671][22].

[22] I met and interviewed Menger in the 1970s and 80s near the end of his life. I may have been 'naive' then, but on reflection I think that he was still 'seething' about Brouwer and his own priorities. However, we never discussed the matter fully. He did provide me with offprints of his papers, some related to the dispute, and he hinted at other revealing items in his personal archive, which he never showed me.

16.7 Embedding Theorem for Dimension Theory: Menger's Assessment

The embedding theorem is a key result for the entire topological theory of dimension stemming from the research of the 1920s and 30s. In this Section I shall consider Menger's assessment of the problem situation and his proposed way to develop a proof in the general case. In the next Section 16.8 I shall examine Alexandroff's independent stream of research of the 1920s and 30s relating dimension to simplicial complexes (or complexes) and what he considered as geometrical intuition. Over the course of his pursuing his research goals and publishing his papers Alexandroff moved from a set-theoretic treatment of dimension theory to a combinatorial and algebraic topological approach to the theory. In Section 16.9 I shall cover the various proofs of the general embedding theorem that several mathematicians developed. Overall, these results relate dimension theory to more concrete geometrical spaces and intuitive geometrical ideas.

In Section 15.7 I discussed Menger's pair of research notes, 'Allgemeine Räume und Cartesische Räume' ('General spaces and Cartesian spaces'), first and second communications [Menger, 1926c] and [Menger, 1926f]. In the second of these Menger provides a construction of his universal curve, the so-called 'Menger sponge'—a 3-dimensional cube-like object with many holes or tunnels that can contain a homeomorphic image of every 1-dimensional curve (Figure 15.3). As I noted in the earlier section Menger's curve is an 'extension' or 'generalization' of Sierpiński's 2-dimensional Carpet Curve [Sierpiński, 1916b].

In the last part of the second note Menger sets out his initial thoughts about generalizations to higher dimensions related to his curve [Menger, 1926f, pp. 1127–1128]. First, he states two corollaries related to the curve construction [Menger, 1926f, page 1127]:

> *Every one-dimensional compact metric space is homeomorphic to a set in R_3 for which every point admits arbitrarily small intervals with whose boundary the set has a discontinuous intersection.* And: *Every compact one-dimensional space is homeomorphic to a subset of a surface of R_3.*[Ch16, 31]

Menger then goes on to state that investigating these ideas of a universal curve in higher dimensions is of interest. He proceeds to lay out the steps for constructing an n-dimensional Menger cube with holes or tunnels. For his higher-dimensional construction and his generalized results he refers to Sierpiński's fundamental work for two dimensions and also a result of Fréchet.

He concludes his note with the following [Menger, 1926f, page 1128]:

> In a subsequent communication it will be proved that every n-dimensional compact (and thus also every n-dimensional separable) metric space is homeomorphic to a subset of R_{2n+1}. Therefore, an example of a continuously

parametrizable continuum that contains a subset homeomorphic to every compact finite-dimensional space is obtained as the union [of sets] $\sum_{n=1}^{\infty} I_n$, where I_n $(n = 1, 2, \ldots)$ is an n-dimensional interval with side length $\frac{1}{n}$ containing a fixed point p.[Ch16, 32]

In his book *Dimensionstheorie* (*Dimension Theory*) Menger again considers the embedding problem for higher dimensions and regards the desired result as fundamental [Menger, 1928b, IX.3, pp. 295–296]:

> **Embedding Theorem.** *An n-dimensional compact space is homeomorphic to a subset of* R_{2n+1}.
>
> . . .
>
> **Theorem of Topological Characterization with the Coordinate Spaces.** *For a separable space to be characterizable with finitely-many coordinates, i.e., to be homeomorphic to a subset of a Cartesian space, it is necessary and sufficient that it be of finite dimension.*[Ch16, 33]

At this point in the book he gives a complete proof of the embedding of a 1-dimensional compact space into R_3, using some extensive dimension-theoretic reasoning with polyhedra [Menger, 1928b, pp.. 296–301]. However, he is only able to provide a proof sketch for the general result, while being hopeful that he can use the same type of reasoning for a full, rigorous proof for the general case [Menger, 1928b, pp. 301–303]. In the next few years other mathematicians put forward complete proofs.

16.8 Paul Alexandroff: Dimension, Complexes, and Geometrical Intuition

Before considering the proofs completing Menger's conjectured theorems, I shall examine another direction of research, having characteristics and results similar to embedding. The results can be and indeed were applied to proving formulations of the embedding theorem. This direction is Alexandroff's independent stream of research relating the general concept of dimension and the dimension of closed sets of arbitrary dimension to basic geometrical objects, simplicial complexes (complexes) and their corresponding point-sets defining polyhedra when considered as topological spaces, and 'geometrical intuition.' It relies to a great extent on simplicial approximation.

The Russian mathematician Paul Alexandroff (1896–1982)[23] began investigating the relation of the dimension concept to elementary geometrical objects, (simplicial) complexes, in the mid 1920s, apparently after the death of his colleague and close

[23] Sources of information on Alexandroff's life include his autobiographical notes [Alexandroff, 1979] [Alexandroff, 1980], `https://en.wikipedia.org/wiki/Pavel_Alexandrov`, and `https://mathshistory.st-andrews.ac.uk/Biographies/Aleksandrov/`. The Wikipedia article is useful, because it refers to Russian sources. Alexandroff was Urysohn's close colleague, friend, and literary executor. Cf. [Arhangel'skii & Dranishnikov, 1997]. Alexandroff was a very

friend Paul Urysohn. He wrote papers on the topic from late 1925 to 1932 and continued to include the results in other publications.

Three major influences played a part in the development of this stream of Alexandroff's topological research. These were, first, the fundamental topological ideas of Brouwer in Amsterdam, second, the new abstract algebraic approach of Emmy Noether in Göttingen as well as other mathematical ideas stemming from his association with Göttingen, and, third, the topological approaches and results of the group of American topologists at Princeton, New Jersey. Alexandroff was fortunate in being able to travel a great deal outside Russia in the 1920s and 1930s, enabling him to make good mathematical contacts that helped further his research.

Alexandroff's travel meant that he was often able to visit Brouwer in Amsterdam. Brouwer's topological work and especially the notion of geometrical complexes and simplicial approximations of mappings strongly impacted this stream of work.

A second influence on Alexandroff's development around this time was his close connection with Göttingen, its mathematicians, its style of mathematics, and its approach to mathematical theories. He frequently went to Göttingen and came under the mathematical spell of Emmy Noether (1882–1935),[24] who was changing the direction of the whole development of modern abstract algebra.[25] Noether emphasized the use of more abstract algebraic methods in topology. Alexandroff also met the brilliant German-Swiss topologist Heinz Hopf (1894–1971)[26] in Göttingen in 1925. They quickly developed a friendship, spent time together, and often travelled together. They soon began to collaborate in mathematics. Their collaboration eventually resulted in their seminal treatise *Topologie I* (*Topology I*) of 1935 [Alexandroff & Hopf, 1935].

It was highly significant for his mathematical development that Alexandroff spent an academic year, 1927–1928, in Princeton, getting to know and working with the American topologists, Oswald Veblen (1880–1960), James W. Alexander II (1888–1971), and Solomon Lefschetz (1884–1972). These mathematicians were more oriented to combinatorial and algebraic topology originating from Poincaré's work in analysis situs than to set-theoretic topology, and they had made important con-

---

good teacher, who inspired and trained many Russian mathematicians. The list of his textbooks on set theory, analysis, and topology is considerable. For aspects of his personal life and apparent homosexual relations with Urysohn and especially Kolmogorov see [Graham & Kantor, 2009, pp. 170–187].

[24] There is a vast amount of biographical material on Emmy Noether, since she had such an enormous impact on modern mathematics and abstract algebra. Thus cf., for example, [Brewer & Smith, 1981], [Dick, 1981], [Koreuber, 2015], [Rowe & Koreuber, 2020], [Rowe, 2021], `https://mathshistory.st-andrews.ac.uk/Biographies/Noether_Emmy/`, and `https://www.encyclopedia.com/science/dictionaries-thesauruses-pictures-and-press-releases/noether-amalie-emmy`.

[25] Emil Artin (1898–1962) was also part of this development.

[26] For biographical material on Hopf cf. his autobiographical notes [Hopf, 2001b], three biographical notes by Hilton, Alexandroff, and Eckmann in his *Collected Papers – Gesammelte Abhandlungen* [Hopf, 2001a, pp. 1219–1263], [Hirzebruch & Behnke, 1972], [Hilton, 1972], [Frei & Stammbach, 1999], `https://mathshistory.st-andrews.ac.uk/Biographies/Hopf/`, and `https://www.encyclopedia.com/science/dictionaries-thesauruses-pictures-and-press-releases/hopf-heinz`.

tributions to this domain. Alexandroff assimilated many ideas from the Princeton mathematicians into his framework of topological concepts. He got to know Veblen's Colloquium Lectures (1916) on *Analysis Situs*, published in 1922 [Veblen, 1922], and a little later Lefschetz' Colloquium book *Topology* of 1930 [Lefschetz, 1930]. He gained a deep understanding of Alexander's duality theorem. Going to Princeton in the United States gave him an added benefit and exposure to a broader set mathematical ideas. Moreover, his friend and collaborator Hopf was in Princeton during the same academic year.

Alexandroff first exposited his results in a series of research notes published in the *Comptes rendus* (*Proceedings*) of the Academy of Sciences, Paris in 1926, 1927, 1928, and 1930 [Alexandroff, 1926b] [Alexandroff, 1926c] [Alexandroff, 1927a] [Alexandroff, 1927b] [Alexandroff, 1927c] [Alexandroff, 1928b] [Alexandroff, 1928c] [Alexandroff, 1930a] [Alexandroff, 1930b]. At the same time he provided more details of his developing approaches, various methods, and diverse results in a sequence of six longer papers:

1. 'Simpliziale Approximationen in der allgemeinen Topologie' ('Simplicial approximations in general topology')), dated August 1925 and noted with the place Le Batz (Loire Inférieure), France, [Alexandroff, 1927d];
2. 'Über den allgemeinen Dimensionsbegriff und seine Beziehungen zur elementaren geometrischen Anschauung' ('Concerning the general concept of dimension and its relation to elementary geometrical intuition'), dated August 1926 and also noted with the place Le Batz (Loire Inférieure), France, [Alexandroff, 1928d];
3. 'Untersuchungen über Gestalt und Lage abgeschlossener Mengen beliebiger Dimension' ('Investigations concerning the form and position of closed sets of arbitrary dimension'), published in the American journal *Annals of Mathematics* and dated 15 January to 15 April 1928, with the place Princeton, N.[ew] J.[ersey], [Alexandroff, 1928e];
4. 'Zum allgemeinen Dimensionsproblem' ('On the general problem of dimension'), dated New York - Le Havre, Pentecost 1928 and Göttingen, 5 July 1928 [Alexandroff, 1928f];
5. 'Dimensionstheorie: Ein Beitrag zur Geometrie der abgeschlossenen Mengen.' ('Dimension Theory: A Contribution to the geometry of closed sets'), dated Göttingen, Mathematisches Institut, Winter 1930–31 [Alexandroff, 1932a];
6. 'Sur la notion de dimension des ensembles fermés' ('On the notion of dimension of closed sets'), extract of a letter to M. Fréchet [Alexandroff, 1932c].

I shall concentrate on the first two of these. Their approach is more set-theoretic and is founded on basic combinatorial topology. I shall touch on the others more lightly, while giving a few salient points from them. These papers, though important, are somewhat less relevant to the history of dimension theory.

Alexandroff's objective in his first paper 'Simpliziale Approximationen in der allgemeinen Topologie' ('Simplicial approximations in general topology') is to show [Alexandroff, 1927d, page 489]:

> ... that compact metrizable topological spaces can be approximated in a defined way by finitely-many connected simplicial forms (n-dimensional complexes); that, conversely, every space approximated in this way is compact and metrizable; that an n-dimensional space can be approximated by an (in the classical sense) n-dimensional complex and the n-dimensional complexes are at most n-dimensional spaces.[Ch16, 34]

In a standard way he defines an at most n-dimensional complex as a finite system of at most n-dimensional simplexes having appropriate neighboring and connecting relations among the simplexes and their sides and vertices. An n-dimensional complex must contain at least one n-dimensional simplex. For his general method of approximating topological spaces he defines a spectrum (Spektrum), a well-defined type of countable sequence of complexes. His principal theorem asserts that every space defined by an approximating spectrum is a compact metrizable topological space. Conversely, every compact metrizable space can be approximated by a spectrum; the dimension of the space is equal to the smallest number (n or ∞) that occurs as the dimension number of a spectrum defining this space [Alexandroff, 1927d, page 495].

At the beginning of his second paper 'Über den allgemeinen Dimensionsbegriff und seine Beziehungen zur elementaren geometrischen Anschauung' ('Concerning the general concept of dimension and its relation to elementary geometrical intuition') [Alexandroff, 1928d] Alexandroff makes a bold epistemological statement about dimension theory that asserts its validity as the correct and fundamentally applicable theory [Alexandroff, 1928d, page 617]:

> The development of dimension theory to date has fully justified the concept of dimension on which this theory is based as the correct concept of dimension: firstly, this concept itself is the simplest and most natural conceivable, and, secondly, it has been shown that those spaces or sets that should be a priori of dimension n (e.g., the Euclidean n-dimensional space R^n), the dimension n can also be obtained based on the general definition of dimension; and, thirdly, with the help of the general concept of dimensions, our knowledge of the topological structure of the geometrically interesting spaces, especially the Euclidean spaces and their closed subsets, is significantly deepened and increased.[Ch16, 35]

Alexandroff thinks that an important open question remains for dimension theory: whether the theory applies to much more general n-dimensional closed sets. As an example of a difficult case, he points to the indecomposable continuum that Brouwer originally constructed [Brouwer, 1910a]. It does turn out to be a curve according to dimension theory. However, can one recognize this example as a curve composed from intuitive elementary elements?

In response Alexandroff asserts that the question in its most general form can be answered in the affirmative; the goal of his paper is to show in the affirmative [Alexandroff, 1928d, page 618]:

> ... that *the general n-dimensional closed sets are geometrically completely characterized in that, for every $\varepsilon > 0$, each of these can be continuously deformed into an n-dimensional complex, such that during the entire deformation process no point is moved by more than an ε from its original location.*[Ch16, 36]

Continuing, he proceeds to formalize his positive result. He first defines a new term or verb 'to ε-transform' (ε-überführen'): a closed set F can be ε-transformed into closed set Φ, in case Φ is a one-one continuous image of F,

$$\Phi = \varphi(F),$$

such that for the mapping and the corresponding positions of F and Φ it is the case that no point x is more than an ε distant from it image point $\xi = \phi(x)$ [Alexandroff, 1928d, page 619].

Then Alexandroff gives a more precise statement of his positive assertion, splitting it into two theorems. He prefaces his two theorems with the following assumptions [Alexandroff, 1928d, page 620]

> *Let F be a given λ-dimensional closed set in [space] R (R may be as earlier either R^n, $n \geq \lambda$, or R^ω). Further, let ε be an arbitrary positive number.*[Ch16, 37]

Then he states the two theorems that express his positive result as [Alexandroff, 1928d, page 620]:

> I. *There exists a λ-dimensional complex K_ε^λ in R, into which F can be ε-transformed. This transformation can be achieved by means of a continuous deformation of the space R, whereby no point of the space during the deformation process will be more distant than ε from its original position.*
>
> *There exists, however, a positive number ε_0 of the type that for $\varepsilon \leq \varepsilon_0$, K_ε^λ in this assertion cannot be replaced by a complex of lower dimension.*
>
> II. *There exists in R a λ-dimensional complex $\tilde{K}_\varepsilon^\lambda$, a closed neighborhood $\overline{U}_\varepsilon$, and a continuous deformation Δ of the entire space R into itself, such that*
>
> 1. *all points of $\overline{U}_\varepsilon$ of F are distant by less than ε;*
> 2. *under this deformation $\overline{U}_\varepsilon$ maps into $\tilde{K}_\varepsilon^\lambda$;*
> 3. *during the deformation Δ no point of R is moved more than ε and (in the case where R is R^n) the points of R^n that are more than ε distant from F all remain fixed.*
>
> *There exists, however, a positive number $\tilde{\varepsilon}_0$ of the type that for $\varepsilon \leq \tilde{\varepsilon}_0$, $\tilde{K}_\varepsilon^\lambda$ in this assertion cannot be replaced by a complex of lower dimension.*[Ch16, 38]

At this point he thanks Heinz Hopf and credits Hopf's cycle of lectures given at Göttingen in the summer of 1926 [Alexandroff, 1928d, page 620] for helping him in this work.

He devotes the rest of his paper to proving these two theorems. A key foundation of his proof is a form of the Brouwer invariance principle [Alexandroff, 1928d, page 621]:

> **Brouwer's Invariance Principle.** *Let F be a closed λ-dimensional subset of the Euclidean [space] R^n or of the fundamental cube R^ω of Hilbert space. Then for sufficiently small ε F cannot be ε-transformed into any set Φ of lower dimension.*[Ch16, 39]

Earlier a form of the invariance principle had been a key to Brouwer's proof of the invariance of dimension (see Section 11.8).

In summary, Alexandroff's two theorems give a precise characterization of closed sets of any dimension, either finite or infinite, relative to small transformations into complexes. They succinctly achieve his goal through simplicial approximations and ε-transformations.

Alexandroff's third paper in the sequence, "Untersuchungen über Gestalt und Lage abgeschlossener Mengen beliebiger Dimension' ('Investigations concerniing the form and position of closed sets of arbitrary dimension'), marks a significant change in Alexandroff's thinking about his chosen problem domain [Alexandroff, 1928e]. It shows him moving away from purely set-theoretic methods to more combinatorial and algebraic topological methods. It was written January to April 1928 at Princeton and was received for publication on 16 April 2928. It was published in the American journal *Annals of Mathematics* and is by far the longest, some 87 pages. It appears as something of a work in progress, revealing his changing views on the matter (which continued to change afterwards). The American topologists had a strong impact on him. In the paper he recognizes the value of their methods and significant results in topology and dimension theory.

For this paper Alexandroff developed an extensive framework for combinatorial geometrical topology resulting in a unified presentation. The paper includes considerable topological machinery. In effect, Alexandroff presents his results afresh and extends them in a more abstract form of combinatorial topology, while continuing to rely at base on simplicial approximations and simplicial mappings. Brouwer's continuing influence is evident in the paper.

For purposes of this history of dimension I shall highlight just a few important items and results of this third paper. The material and results of Alexandroff's first two papers as covered above are sufficient for understanding the direction of his research for dimension theory. The interesting aspect of the third paper is his significant shift in conceptual basis.

A key part of this paper is Alexandroff's introduction of the notion of a 'projection spectrum' ('Projektionsspektrum') [Alexandroff, 1928e, pp. 107–108]. This concept of projection spectrum modifies his earlier notion of spectrum defined in his first paper [Alexandroff, 1927d, page 493]. He applies the concept in defining approximating spaces [Alexandroff, 1928e, page 108], especially in chapter 2 of the paper.[27] An important result of the chapter is that every compact metrizable space can be approximated by a projection spectrum of the same dimension as the space, and, conversely, every projection spectrum approximates a compact metrizable space, whose dimension does not exceed that of the spectrum [Alexandroff,

[27] The *Encyclopedia of Mathematics (Springer)* presents Alexandroff's concept in modern topological terms. The article also includes the following critical comment
`https://encyclopediaofmath.org/wiki/Projection_spectrum:`

1928e, Einleitung, pp. 108–109; Zweites Kapital, pp. 126–145; in particular, Satz III, page 132]. Another significant result of the second chapter is the homeomorphism theorem giving conditions under which two projection spectra approximate the same space [Alexandroff, 1928e, pp. 109, 134].

In the third chapter Alexandroff considers the problem of position (Lage) of closed sets in relation to their complements in their containing spaces [Alexandroff, 1928e, Drittes Kapital, pp. 145–155]. In this case he is thinking of J.W. Alexander's duality work [Alexander, 1915] [Alexander, 1922]. Early in the chapter he mentions the (two-part) paper of Lev Pontrjagin 'Zum Alexanderschen Dualitätssatz' ('On Alexander's duality theorem') [Pontrjagin, 1927a] [Pontrjagin, 1927b] in connection with the theorems he is proving. At the same place he might have also mentioned his own related research note of the same time 'Über die Dualität zwischen den Zusammenhangszahlen einer abgeschlossenen Menge und des zu ihr komplementären Raumes' ('Concerning the duality between linking numbers of a closed set and the space that complements it') [Alexandroff, 1927e]. Lev Semenovich Pontrjagin (1908–1988)[28] was a student of Alexandroff. One of his most significant contributions to topology is his generalization of Alexander duality. His research work in this domain culminated in his paper of 1931 'Über den algebraischen Inhalt topologischer Dualitätssätze' ('Concerning the algebraic content of topological duality theorems') [Pontrjagin, 1931].

A key summarizing result that Alexandroff gives in the paper is the following [Alexandroff, 1928e, page 148]:

> *The finite or infinite number of domains that defines an $n-1$-dimensional closed set in R^n is equal to the smallest number that appears as the linking number of an $n-1$-dimensional projection spectrum defining this set.*[Ch16, 40]

A little later in this chapter he specifically mentions Alexander's duality theorem [Alexandroff, 1928e, page 150].

Near the end of the third chapter he states and proves the following significant result [Alexandroff, 1928e, page 154]:

> GENERALIZED URYSOHN THEOREM: *The common boundary of two (or more) connected regions of R^n is an $n-1$-dimensional Cantor manifold, which, moreover, cannot be decomposed into a simply-connected $n-2$-dimensional closed subset.*[Ch16, 41]

> In the West, the concept of a projection spectrum is generally considered to have mainly historical importance, as the first version of the crucial concept of inverse system. For instance, the fundamental continuity axiom of abstract homology (cf. Steenrod–Eilenberg axioms [Eilenberg & Steenrod, 1952] is stated for complete limits of inverse systems of spaces. Picking out the upper limit succeeds . . . in representing a space by means of suitable nerves of coverings, but it combines badly with other topological constructions.

See also Alexandroff's later book [Alexandroff, 1960, page 41].

[28] For biographical material on the well-known mathematician cf.
`https://en.wikipedia.org/wiki/Lev_Pontryagin`,
`https://mathshistory.st-andrews.ac.uk/Biographies/Pontryagin/`, and
`https://www.rbth.com/history/336413-how-blind-boy-became-scientist`.

In this third paper Alexandroff presents a wide array of results following his new topological style and approach, as influenced by his recent close association with the Princeton topologists.

The fourth paper in the sequence, 'Zum allgemeinen Dimensionsproblem' ('On the general problem of dimension'), is a problem-setting and framework-setting paper [Alexandroff, 1928f]. He dated it Pentecost 1928 at New York and Le Havre [France], probably the time of his return trip to Europe from the U.S., and 5 July 1928 at Göttingen. It was delivered to the Society of Sciences at Göttingen (Gesellschaft der Wissenschaften zu Göttingen) on 6 July 1928. New ideas and new approaches were fresh in his mind.

The second paragraph of this paper sets the stage [Alexandroff, 1928f, page 25]:

> The general problem of the *geometrization* of dimension theory and thus of the entire set-theoretic topology involves even new questions that are not addressed at all by the approximation theorems alone, and that seem to be of essential importance for the entire theory.[Ch16, 42]

Alexandroff views the set-theoretic definition of the dimension concept as having a negative character [Alexandroff, 1928f, page 28]:

> *The set-theoretic dimension concept is a deepening and generalization of Jordan's negative definition of connectedness* ....[Ch16, 43]

Thus, he looks for a positive definition based on his new combinatorial topological approach and framework and provides the following based on cycles and homology [Alexandroff, 1928f, page 30]:

> **Definition of the geometric dimension of a space.** *Let F be any compact metric space. If r is the largest non-negative integer such that an r-dimensional non-identically vanishing cycle exists in F and is homologously zero there, then the number $r+1$ is called the geometric dimension of F. Even if for $r = 0$ is not a non-identically vanishing r-dimensional cycle in F homologous to zero, the geometric dimension of F is equal to zero; if non-identically vanishing r-dimensional boundary cycles occur in F for every r, the geometric dimension of F is by definition equal to infinity.* Finally, we shall assign the dimension -1 to the empty sets (as in set-theoretic dimensional theory).[Ch16, 44]

In the remainder of the paper he proves ten theorems based on this definition. I shall not go into the details of these.

Alexandroff first gave the content of the fifth paper in the sequence, 'Dimensionstheorie: Ein Beitrag zur Geometrie der abgeschlossenen Mengen.' ('Dimension Theory: A Contribution to the geometry of closed sets') [Alexandroff, 1932a], as a lecture in the summer of 1930 at Göttingen University. He then developed the paper from the lecture, dating it Göttingen, Mathematical Institute, Winter 1930–31. The paper was submitted in May 1931 to *Mathematische Annalen* and published in 1932. The lecture and resulting paper gave Alexandroff a chance to refine his new approach and present it in new places.

He opens the paper rather refreshingly with the following broad statement about the breadth and sweep of topology in the early 1930s [Alexandroff, 1932a, page 161]:

> The investigations of the last twenty years have melded several parts of the topology into a unified whole, the inner connections of which are now truly understood and carefully surveyed. The resulting piece of mathematical science contains not only the classic material of 'elementary' topology (as it was first compiled in Veblen's book *Analysis Situs*), but also the entire theory of decomposition and linking (from Jordan's theorem to the latest generalizations of Alexander's duality theorem), as well as large parts of the theory of continuous mappings, in particular, the degree of a mapping and the complete solution of the problem about the algebraic number of fixed points. Although the so-called classic problems of analysis situs—such as the main conjecture of combinatorial topology, the characterization of spherical manifolds (Poincaré's conjecture), the triangulation problem, the knot problem, etc.—are still unsolved, the one great success that has been achieved, because today you are standing in front of a clear and unified building, which extends far beyond the content of Veblen's book, can still be described as *elementary topology*.[Ch16, 45]

In this paragraph Alexandroff touches on nearly all parts of the topology of the time, seemingly looking forward to his planned wide-ranging treatise, to be written jointly with Heinz Hopf, *Topologie I* (*Topology I*).

A little later in the paper he states its aim [Alexandroff, 1932a, pp. 162–163]:

> *The aim of the present work is to show that dimension theory also belongs to 'elementary' topology and that it is also completely dominated by homology and intersection constructions.* This should hardly surprise anyone: in a series of papers that have appeared in recent years, it has gradually become clear that the shape properties of even general topological entities are only a limiting case of the corresponding properties of polyhedra, and consequently in principle can be attacked using the methods developed in polyhedral topology. In particular, this must, of course, also be the case for the concept of dimension; and it has actually been proven that the ordinary (Brouwer-Urysohn-Menger) dimension of a closed set F can be transferred over as the smallest r of an r-dimensional polyhedron.
>
> But even if a connection of the general concept of dimensions to the visual material of polyhedral topology was achieved in this way, next to nothing happened to geometrize the *theory* of dimension (not just the *definition* of dimension); this theory remained without any connection to the rest of topology—a theory which, according to its methods, belonged more to point-set theory, as it had developed within the framework of real functions, than to the actual analysis situs.
>
> *Only the actual application of combinatorial concepts of homology to the general set-theoretical structures allows us to determine that dimension theory is not a theory in itself, but merely the first paragraph of a general investi-*

> *gation, still in its infancy, of boundary and intersection (particularly linking) constructions with respect to closed sets and should be constructed as such.* Only this point of view gives us a correct perspective and leads to a natural and rich problem set. Its isolated position was only due to the lack of the correct geometrical basis in the previous purely set-theoretical development of dimension theory. However, this foundation, which is to be provided in the present work, means—as often happens—the end of the independent existence of a theory and its absorption into a more comprehensive chapter of mathematics.[Ch16, 46]

These paragraphs delineate Alexandroff's broad program to bring dimension theory into the fold of combinatorial/algebraic topology, into what he also calls 'elementary topology.' He expresses strong views about where dimension theory belongs as a domain in mathematics, namely, as part of homology and algebraic topology. His goals for the theory to be constructed in the paper are ambitious.

Given all the mathematical machinery required, I shall only give a short summary of this important paper. In brief the paper covers the following topics:

- Dimension and essential mappings (§1): Alexandroff uses Euclidean spaces R^n of arbitrary n dimensions as his basic domain of investigation. He assumes closed sets are bounded closed sets in these spaces, hence, compact sets. For Brouwer dimension he uses the covering definition based on the Lebesgue tiling principle. A mapping of a closed set F to an element of the space is 'essential' if the mapping is not homotopic to a constant mapping.
- Cycles and bounding (cycles) in closed sets (§2): Alexandroff sets out all the machinery for homology relative to Euclidean spaces and closed sets in them.
- Dimension as a homological invariant (§3): This section is key for Alexandroff. He sets out his principal theorem, given below.
- General dimension-theoretic theorems (§4): Alexandroff gives some further key results in this section as a way of developing dimension theory in a homological context.
- Dimension and closure (or closed sets) (§5): In this section he investigates closed Cantorian manifolds
- Concluding remarks (§6): In his remarks Alexandroff considers the axiomatization of dimension theory.

Alexandroff states his main result in §3 as [Alexandroff, 1932a, page 195]:

> **Principal Theorem.** *The dimension for every closed set F with respect to a variable modulus coincides with the Brouwer dimension:*
>
> $$\Delta(F) = dim\ F.$$[Ch16, 47]

He examines closed sets F in a Euclidean n-dimensional space R^n. The right-hand side of the equation, $\dim F$, means the Brouwer dimension of F according to the tiling principle. The left-hand side of the equation, $\Delta(F)$ refers to the Alexandroff [or Alexandroff-Čech] homology group with any (i.e., 'variable') coefficient domain (or coefficient group) relative to the closed set F. The *Encyclopedia of Mathematics (Springer)* under the entry 'Homological dimension of a space' provides a succinct explanation of the left-hand side of the equation (`https://encyclopediaofmath.org/wiki/Homological_dimension_of_a_space`): The left-hand side denotes the *homological dimension of a space* $X = R^n$, considering the space with respect to a coefficient group G, and is the largest integer n for which the Alexandroff [or Alexandroff-Čech] homology group $H_n(R^n, F; G)$ of some closed set $F \subset R^n$ is non-zero. The equation then states an equivalence of two views of dimension (see more extensive expositions in [Hurewicz & Wallman, 1948, pp. 108–152] and [Kuz'minov, 1968]).

The sixth paper in the sequence, 'Sur la notion de dimension des ensembles fermés' ('On the notion of dimension of closed sets'), was published in 1932 as an extract of a letter to M. Fréchet [Alexandroff, 1932c]. He wrote it in French and put it in a French journal no doubt to give wider circulation to his ideas about dimension. It is a summary of the ideas he developed over the last few years.

In course of his research over the period of the six papers Alexandroff moved from the set-theoretic theory of dimension (Brouwer, Urysohn, and Menger) to his own homological theory of dimension. Only a few mathematicians have continued developing his homological theory of dimension, while many others pursued other directions.

In concluding this section I direct the reader to Alexandroff's own revised and enhanced summary of his later view of dimension theory as given in his book written jointly with Heinz Hopf, *Topologie I* (*Topology I*) [Alexandroff & Hopf, 1935, pp. 363–376]. I may also mention his later introductory work on *Combinatorial Topology* [Alexandroff, 1956] [Alexandroff, 1957] [Alexandroff, 1960].

16.9 Embedding Theorem for Dimension Theory: Other Proofs

Georg Nöbeling (1907–2008)[29] offered one of the first full proofs of the general embedding theorem. Nöbeling was a student of Menger in Vienna and his doctoral work under him was an investigation of the general embedding theorem. A principal outcome from his investigation is the paper, published in 1931, 'Über eine n-dimensionale Universalmenge im R_{2n+1}' ('Concerning an n-dimensional universal set [space] in R_{2n+1}') [Nöbeling, 1931a]. His objective in the paper is to prove Menger's conjectured theorem for separable metric spaces stated as [Nöbeling, 1931a, page 71]:

> Theorem 1. *For arbitrary n, every n-dimensional space is homeomorphic with a subset of the* $(2n + 1)$*-dimensional Cartesian space* R_{2n+1}.[Ch16, 48]

[29] For biographical material see the article 'Georg Nöbeling zum 100. Geburtstag' and the Wikipedia entry for Georg Nöbeling.

He notes that Sierpiński had already proved this result for $n = 0$ [Sierpiński, 1921], and Menger had proved it for $n = 1$ and sketched a proof for $n \geq 2$ in his book [Menger, 1928b].

Nöbeling proves more. He shows that for every n-dimensional space there is a universal space, stating [Nöbeling, 1931a, page 71]:

> Theorem 2. *For every n, there exists an n-dimensional universal space.* Ch16, 49

Nöbeling's n-dimensional universal space is the set of all points of R_{2n+1} that have at most n rational coordinates. It has the 'universal' property that every n-dimensional space can be mapped to a subset of this restricted subspace of R_{2n+1}. Thus, he says [Nöbeling, 1931a, page 71]:

> We now want to prove that *the set of all those points of R_{2n+1} that have at most n rational coordinates is an n-dimensional universal space.* Ch16, 50

Over the next pages Nöbeling proves this result using geometrical arguments dealing with simplicial complexes and topological approximation methods [Nöbeling, 1931a, pp. 72–80]. In doing so, he cites the paper by Alexandroff of 1928, discussed above, 'Untersuchungen über Gestalt und Lage abgeschlossener Mengen beliebiger Dimension' ('Investigations concerning the form and position of closed sets of arbitrary dimension') [Alexandroff, 1928e, page 134][30] to support his argument.

Lev Pontrjagin and G. Tolstowa's joint paper on the embedding theorem, 'Beweis des Mengerschen Einbettungssatzes' ('Proof of the Menger embedding theorem'), arose out of their participation in Paul Alexandroff's Topology Seminar at the University of Moscow during the winter term of 1929–30 [Pontryagin & Tolstowa, 1931]. Alexandroff was their mentor in performing this research. At the beginning of their paper, which appeared in 1931, they state the embedding theorem for compact spaces [Pontryagin & Tolstowa, 1931, page 734]:

> Every n-dimensional compact metrizable space is homeomorphic to a subset of R^{2n+1}. Ch16, 51

Their proof very much depends on Alexandroff's methods and results in his 'Untersuchungen' ('Investigations') paper [Alexandroff, 1928e, Kapital II, pp. 126–145, especially page 134]. The proof is simple and combinatorial, apart from a necessary topological limit argument.

Near the end of their paper Pontrjagin and Tolstowa demonstrate a companion result [Pontryagin & Tolstowa, 1931, page 744]:

> Additional Result. *A continuous mapping f of an n-dimensional compact metric space F into a Euclidean space R of dimension $\geq 2n + 1$ can be transformed into a homeomorphism by an arbitrarily small continuous change.* Ch16, 52

[30] However, the page number Nöbeling gives, page 18, seems to be incorrect [Nöbeling, 1931a, page 77].

This companion result is very much in line with Alexandroff's thinking in this domain.

In a short paper printed just after Pontrjagin and Tolstowa's paper in *Mathematische Annalen* Nöbeling relates his universal space U_n^k, the set of all points with at most n rational coordinates in a k-dimensional Euclidean space R^k, to the Russians' result [Nöbeling, 1931b]. Given a compact space F of dimension $n < \frac{k}{2}$ and $f(F)$ a single-valued continuous image of F into R^k, Nöbeling asserts that [Nöbeling, 1931b, page 746]:

> *$f(F)$ can be transformed by an arbitrarily small change into a subset Φ of U_n^k that is homeomorphic to F.*[Ch16, 53]

Solomon Lefschetz (1884–1972)[31] is a very well-known American topologist—an intellectual giant in the early years of topology. He taught and performed research at Princeton University for many of his active years in mathematics. In 1930 he published his much-referenced book *Topology*, a landmark in combinatorial topology [Lefschetz, 1930].

Lefschetz examined the Menger embedding result in his paper of 1931 'On compact spaces' [Lefschetz, 1931]. As I noted above, Alexandroff had spent the academic year 1927–1928 in Princeton. One very likely result of his being there is that Alexandroff motivated Lefschetz to consider the embedding result and look for a proof.

In his paper Lefschetz introduces the aim of his research as follows [Lefschetz, 1931, page 521]:

> The chief purpose of the present paper is to give certain applications of the sequences of complexes approximating compact metric spaces introduced by Alexandroff.

In fact, Lefschetz refers to Alexandroff's 'Untersuchungen' ('Investigations') paper and the projection spectra that Alexandroff studied there [Alexandroff, 1928e, pp. 107–108].

Lefschetz makes an interesting methodological comment about his work in this case [Lefschetz, 1931, page 521]:

> The unavoidable link between dimensionality and combinatorial topology is the Lebesgue–Urysohn–Menger theorem on the order of covering sets. This theorem is all that we borrow from that theory. Aside from that, up to and including the universal space, we need only a few elementary properties of complexes. These are completely established in our Colloquium Lectures *Topology*, whose notations and terminology are used wherever practicable. Indeed had we obtained our basic lemma early enough the present paper would have been incorporated in Ch. VII of the book after §4.

[31] For biographical material on Lefschetz see [Griffiths et al., 1992], [Hodge, 1973], and `https://mathshistory.st-andrews.ac.uk/Biographies/Lefschetz/`.

Lefschetz proves Menger's theorem stated as [Lefschetz, 1931, pp. 527–528]:

> THE IMMERSION THEOREM. Every compact metric n-space can be mapped topologically on an S_{2n+1}[32] (Menger).

Lefschetz proceeds to construct a universal space, first giving the following explanation [Lefschetz, 1931, pp. 528–529]:

> The term "universal space" is now in general use in the following sense: Given a class $\{\mathfrak{R}\}$ of metric sets or spaces, if there exists a member $\mathfrak{R}_0$ of the class such that every $\mathfrak{R}$ can be topologically mapped on $\mathfrak{R}_0$, the latter is said to be a *universal* set or space for the whole class. The class to be considered here is that of all compact metric n-spaces and we propose to construct a *universal n-space* for that class.

He then provides a detailed construction of his universal n-space Λ for the designated class of all compact metric n-spaces [Lefschetz, 1931, pp. 529–533] and follows the construction with his proof that indeed Λ is a compact n-space that is universal. His space is similar to one Menger proposes in his book [Menger, 1928b, pp. 296–303] as a general construction, but Lefschetz uses simplexes instead of Menger's polytopes. Lefschetz' combinatioral-geometrical details are somewhat complicated, so I shall not go into them.

Lefschetz mentions Nöbeling's universal space, and he shows how to derive Nöbeling's universal space from his [Lefschetz, 1931, pp. 533–534]. That is a nice dovetailing of their results.

Finally, he considers universal spaces for separable metric spaces[33] and other metric spaces in the remainder of the paper [Lefschetz, 1931, pp. 534–538].

Given his extensive background in dimension theory, it is not surprising that Hurewicz became interested in investigating the embedding theorem. Characteristic of his intuitive mathematical thinking he developed deep insights into the problem area by defining and cleverly using a fundamental function space as a pivotal part of his work. He announced his results in two research notes, one in the *Proceedings* of the Academy, Amsterdam, communicated 28 March 1931 [Hurewicz, 1931a], and the other in the Announcements of the Academy of Sciences in Vienna some time later in 1931 [Hurewicz, 1931b]. Hurewicz' full exposition of his investigations then came in a longer research paper delivered to the Prussian Academy at Berlin on 9 February 1933 [Hurewicz, 1933a].

Hurewicz' paper 'Über Abbildungen von endlichdimensionalen Räumen auf Teilmengen Cartesischer Räume' ('Concerning mappings of finite-dimensional spaces onto subsets of Cartesian spaces') is, in my opinion, something of a tour de force dealing with the embedding problem, representing Hurewicz' profound thinking on

[32] Lefschetz' notation for an Euclidean/Cartesian space of $2n + 1$ dimensions.

[33] With regard to separable metric spaces Lefschetz [Lefschetz, 1931, page 534] would like to apply Hurewicz' useful result: *Every separable metric n-space can be mapped topologically on a compact metric n-space* [Hurewicz, 1927c], but he regards Hurewicz' proof in that paper as incomplete. However, around this time Hurewicz improved his proof in a further paper [Hurewicz, 1930c] and added more results.

the topic. He opens the paper stating a more general result [Hurewicz, 1933a, page 754]:

> *A compact metric space A of finite dimension n can be mapped to a subset of the Cartesian [space]* R_{n+m} *continuously and almost simply for every* $m > 0$, *such that those points that are assigned as images to more than one point of A, form an at most* $(n - m)$-dimensional *set.*[Ch16, 54]

By an 'almost simple' ('fastschlicht') mapping from a set A to a set B Hurewicz means one having the property that the pre-image set for each mapped point in B is a finite set in A. He understands a set of negative dimension to be the empty set. The Menger–Nöbeling result corresponds to the special case where $m = n + 1$ [Hurewicz, 1933a, page 755]:

> *Every n-dimensional compact space is homeophic with a subset of the [Cartesian or Euclidean space]* R_{2n+1}.[Ch16, 55]

A corollary of this result is [Hurewicz, 1933a, page 755]:

> The dimension of a compact space is the smallest number n with the property that the space can be mapped continuously and almost simply to a nowhere dense subset of R_{n+1}.[Ch16, 56]

He states his principal result as [Hurewicz, 1933a, page 755]:

> **Principal Theorem.** *An n-dimensional compact space A allows a continuous mapping onto a subset of* R_{n+m} $(m = 1, 2, \ldots)$, *such that for* $k = 2, 3, \ldots$ *those points of* R_{n+m} *that appear at least k times as image points form a maximum* $(n - (k - 1)m)$*-dimensional set.*[Ch16, 57]

Hurewicz proceeds to state a sharpened form of this theorem as follows [Hurewicz, 1933a, page 756]:

> *... Every continuous mapping f from A to a subset of* R_{n+m} *can be modified by an arbitrarily small change into a mapping* φ *that satisfies the conditions of the theorem. ...*[Ch16, 58]

To prove his Principal Theorem as well as the Menger–Nöbeling Embedding Theorem Hurewicz introduces a function space consisting of the set of continuous mappings or functions from one space to a second together with a metric for this set, thus forming a metric space. In advocating his method of using a function space he claims (rightly) [Hurewicz, 1933a, page 756]:

> This method avoids complicated convergence considerations and, especially for the Menger–Nöbeling case [where] $m = n + 1$, the proof can be reduced to just a few lines.[Ch16, 59]

The metric is defined in the following way. Consider two compact metric spaces A and B and the defined distances between pairs of points in each. Consider the set of all continuous mappings of the entire space A to a subset of B. For any two such mappings, f and f', define the distance between them as the maximum of the image point distances $\overline{f(x)f'(x)}$ taken over all the points of A. The space of functions with this metric forms a metric space, denoted as $\mathfrak{F}(A, B)$. That it is a metric space is easily shown. The space has the usual properties of a metric space, including in this case being a complete space in Hausdorff's sense that the Cauchy convergence criterion is satisfied.

With his method of defining the function space and the use of a covering theorem given in Menger's book [Menger, 1928b, page 155], Hurewicz proceeds to prove the Menger–Nöbeling Embedding Theorem in less than two pages [Hurewicz, 1933a, pp. 757–759]. For the proof of the theorem, A denotes the given n-dimensional compact metric space, and for B Hurewicz takes a a full sphere of R_{2n+1}, i.e., a sphere of R_{2n+1} with its interior. Considering the function space $\mathfrak{F}(A, B)$, he needs to show the following: if f is an arbitrary mapping from $\mathfrak{F}(A, B)$, then for every $\varepsilon > 0$ there is an ε-simple [ε-schlichte] mapping φ, such that $\overline{f\varphi} < \varepsilon$. An ε-simple mapping from $\mathfrak{F}(A, B)$ is one where for every two points of A having the same image point in B, the distance between these pre-image points is $< \varepsilon$. Using the covering theorem and some geometrical arguments (including the use of barycenters of image points with assigned weights), he provides a general construction for the approximating function φ and then for the embedding. He states at the end of the proof that he has proved a sharper result: to every continuous mapping f from A to a subset of R_{2n+1}, there exits a topological mapping differing from f by an arbitrary small amount [Hurewicz, 1933a, page 759].

In accompanying remarks he draws out some consequences, including a direct way to conceive Nöbeling's n-dimensional universal space [Hurewicz, 1933a, pp. 759–761].[34] In reviewing the paper in the *Zentralblatt* Nöbeling writes [Nöbeling, 1934, page 133]:

> The proof is very elegant and yields, in the case where $m = n + 1$, Menger's theorem on merely $1\frac{1}{2}$ pages.[Ch16, 60]

In the rest of his paper he is concerned with dimension-theoretic theorems about the function space $\mathfrak{F}(A, B)$, the proof of the general embedding theorem, and the mapping of n-dimensional spaces to subsets of R_{n-m} [Hurewicz, 1933a, pp. 761–768]. Some years later after he had come to the United States Hurewicz recast the ideas of this paper with its perceptive use of function spaces as a foundation in his book with Wallman *Dimension Theory* [Hurewicz & Wallman, 1948, pp. 55–66].

[34] Hurewicz added a correcting remark to his paper while in proof, dated July 1933, asserting that his proof is not merely an existence proof, but also that his proof contains a constructive method for developing the topological mapping from A to a subset of R_{2n+1}, so that the application of his method is comparable to Nöbeling's construction of a universal space [Hurewicz, 1933a, page 768].

At this point I would like to highlight three derived results that appear to validate the definitions of dimension due successively to Brouwer, Urysohn, and Menger, which together equivalently define a common concept and closely related theories.

First, Brouwer proved the invariance of dimension, and a little later he introduced his definition of dimension and coupled it with a second proof of the invariance. In defending the limited scope of his work on dimension *theory* (to Menger), he said he regarded his proofs as a validation of his defined concept from an epistemological point of view (see Section 16.6).

Second, the embedding theorem and associated universal spaces embody another validation of the recently defined concept of dimension. For general sets of dimension n one has a set of concrete models in the Euclidean/Cartesian space R_{2n+1}. These represent the abstract theories. In this way the embedding theorem is a kind of representation theorem.

Third, Alexandroff's results showing that compact n-dimensional metric spaces and closed compact subsets of Hilbert space can be transformed by small transformations or deformations into directly related complexes provide another validation, and a very interesting one at that. The results are very much in accord with his notion of geometric intuition, something he took very seriously.

Taken together, these three groups of results provide some assurance that the definitions of dimension that Brouwer, Urysohn, and Menger offered and the resulting equivalent dimension theories correctly represent a sound and 'intuitive' notion of dimension. I shall say more about this philosophical matter in the final Chapter 17.

16.10 Eduard Čech and Dimension Theory

Eduard Čech (1893–1960)[35] came later to the development of dimension theory, regarding the theory as a part of point-set or general topology. His main papers on the subject appeared in 1931, 1932, and 1933. He made his mark on the field by offering more general definitions and results in the larger context of topological spaces. Among topologists and analysts Čech is best known for the Čech–Stone (or Stone–Čech) Compatification, which he and Stone developed independently about the same time in 1937 [Čech, 1937] [Stone, 1937]. He also had a considerable impact on the development and organization of Czech mathematics in the mid-twentieth century. He was a Czech intellectual leader.

Čech's contributions to dimension theory come in a short research note in the *Comptes rendus* (*Proceedings*) of the Academy of Sciences, Paris 'Sur la théorie de la dimension' ('On the theory of dimension') [Čech, 1931], and two papers, 'Sur la dimension des espaces parfaitement normaux' ('On the dimension of perfectly

[35] For information on the life and work of Čech see [Katětov et al., 1993], [Balcar et al., 1993], and `https://mathshistory.st-andrews.ac.uk/Biographies/Cech/`. For assessments of his work in dimension theory and more generally topology see [Katětov, 1993] [Katětov & Simon, 1993] [Gray, 1994].

normal spaces') [Čech, 1932] and 'Příspěvek k teorii dimense' ('Contribution to Dimension Theory') [Čech, 1933].

The *Comptes rendus* (*Proceedings*) note merely announces his initial results, while the first paper provides the details. In the paper Čech defines dimension in the more general context of topological spaces. His definition is usually called the large inductive definition of dimension [Čech, 1968, page 73] [Katětov & Simon, 1993, page 132], akin to Brouwer's 1913 definition [Brouwer, 1913b] (corrected in [Brouwer, 1923b]).[36] He mainly investigates dimension for perfectly normal topological spaces. A normal space is a topological space R such that for every pair of disjoint subsets of R, closed in R, there are disjoint open subsets containing the two given subsets (a definition from Urysohn [Urysohn, 1925c, page 265]). A perfectly normal topological space is a topological space R that is normal and for which every open subset of R is of type F_σ, i.e., is a union of countably many closed subsets in R (also derived from Urysohn [Urysohn, 1925c, page 286]). His main results for perfectly normal spaces are (1) the sum theorem, asserting that a countable union of closed subsets, each of dimension $\leq n$ in such a space, is also of dimension $\leq n$, and (2) the subset theorem, asserting that any subset of a space of this type of dimension $\leq n$ is also of dimension $\leq n$ according to Čech's notion of large inductive dimension [Čech, 1968, pp. 75, 82] [Katětov & Simon, 1993, pp. 134, 141].

In the second paper, published in 1933, Čech defines the covering definition of dimension in the context of general topological spaces, examining its consequences and giving his critical assessment of the historical background [Čech, 1968, pp. 129–132] [Katětov & Simon, 1993, pp. 149–151].[37] The covering definition derives directly from Lebesgue's tiling theorem or tiling principle [Lebesgue, 1911a] [Lebesgue, 1921]. The definition certainly has an attractive intuitive appeal. In deriving consequences of the definition Čech mostly considers the covering dimension of normal or perfectly normal spaces, though sometimes of metric spaces. I shall not go into the details.

16.11 Constantin Carathéodory, Felix Hausdorff, and Hausdorff Dimension

I shall turn now from theories of dimension derived from or directly related to point-set or other branches of topology to other theories not related to topology *per se*. One of the most important of these is Hausdorff's dimension theory based in measure theory. Hausdorff laid out in detail his ground-breaking theory in a paper of 1919 [Hausdorff, 1919] (English translation as [Edgar, 2004, pp. 75–99]). The theory is distinctly measure-theoretic and not topological. The motivating background to this theory comes in an important earlier paper of 1914 by Carathéodory. I shall consider this first [Carathéodory, 1914] (English translation as [Edgar, 2004, pp. 47–72]).

---

[36] Cf. Engelking's treatment in [Engelking, 1978, Chapter 2, pp. 169–206].

[37] Cf. Engelking's treatment in [Engelking, 1978, Chapter 3, pp. 207–249] and Alexandroff and Hopf's treatment in [Alexandroff & Hopf, 1935, pp. 352–355, 363–376].

Constantin Carathéodory (1973–1950)[38] was of Greek origin, but he received his mathematical education at a variety of European educational institutions, and he spent most of his mathematical career and later life in Germany. He eventually became a professor of mathematics at Munich and was pretty much regarded as German.[39] His contributions to mathematics were extensive, covering real and complex function theory, conformal mapping, calculus of variations, and algebraic measure theory. He also contributed to parts of mathematical physics. His collected works extend to some five volumes [Carathéodory, 1957]. He also wrote an extensive set of mathematical texts.

Carathéodory's paper of 1914, 'Über das lineare Mass von Punktmengen—eine Verallgemeinerung des Längenbegriffs' ('On the linear measure of point-sets—a generalization of the concept of length') [Carathéodory, 1914] [Edgar, 2004, pp. 47–72], is a classic in measure theory, an abstract algebraic generalization of Lebesgue's ideas of measure.[40] Carathéodory opens his paper with the following paragraph [Carathéodory, 1914, page 404] [Edgar, 2004, page 47]:

> The idea of applying the path-breaking and extremely fruitful theories that Mr. Lebesgue has developed for the volume of point-sets to the concept of length is very tempting. It suffices to find an additive set function, whose value for every rectifiable curve is exactly equal to the conventional length of the curve, and that also retains the properties of ordinary Lebesgue measure for unions and intersections of sets.[Ch16, 61]

Carathéodory presents his exposition of a formal theory of measure in Chapter I of the paper. He begins the development with a proposed concept of outer measure characterized by three properties [Carathéodory, 1914, pp. 405–406] [Edgar, 2004, page 48]:

> We thus assume:
>
> I. To each point-set A of a q-dimensional space $\mathfrak{R}_q$ there is assigned a unique number μ^*A which can be zero, positive, or $+\infty$, and which is called *the outer measure of A*.
>
> II. For any subset B of A the following always holds
>
> $$\mu^* B \leq \mu^* A.$$
>
> III. If A is the set union of a finite or countably infinite sequence of point-sets $A_1, A_2, \ldots$, the following always holds
>
> $$\mu^* A \leq \mu^* A_1 + \mu^* A_2 + \cdots .$$

[38] Cf. the full-length biography of Carathéodory [Georgiadou, 2004] and the short synopsis of his life at `https://mathshistory.st-andrews.ac.uk/Biographies/Caratheodory/`.

[39] However, he was regarded with some suspicion under the National Socialist German Workers' Party (Nazi Party) regime.

[40] For example, cf. [Lebesgue, 1910].

Obviously, this last inequality is meaningful only if the sum on the right-hand side converges.[Ch16, 62]

He proceeds to give his fundamental definition of measurability as an equation or relation among the measurable entities [Carathéodory, 1914, page 406] [Edgar, 2004, page 49]:

Definition. A point-set A is said to be measurable if for any arbitrary point-set W of finite outer measure the following relation holds:

$$(1) \qquad \mu^* W = \mu^* AW + \mu^* (W - AW);$$ [41]

the measure μA of A is then defined by

$$\mu A = \mu^* A.$$ [Ch16, 63]

Carathéodory adds two clarifying remarks. First, one could drop the requirement that W should be finite, in which case if $\mu^* W = \infty$, then one of the terms in the equation must be infinite, so that (1) reduces to an identity. Second, the definition does not require the measure μA of a measurable point-set to be finite, so that one can draw conclusions about the measurability of point-sets without having to determine whether these sets have finite outer measures.

Carathéodory's treatment of the theory of measurability is really quite beautiful and is expressed in a succinct algebraic way! In the remainder of Chapter I of his paper Carathéodory provides an exposition of the measurability of various point-sets defined according to the usual algebra of sets and typical set-theoretic operations or constructions [Carathéodory, 1914, pp. 406–420] [Edgar, 2004, pp. 49–64]. He also provides a definition of inner measure of a point-set [Carathéodory, 1914, pp. 416–417] [Edgar, 2004, pp. 60–61].

In Chapter II Carathéodory defines and analyzes a concept of linear measure, which is a principal aim of his paper [Carathéodory, 1914, pp. 420–426] [Edgar, 2004, pp. 65–71]. He begins the chapter with his definition [Carathéodory, 1914, page 420] [Edgar, 2004, page 65]:

Let A be an arbitrary point-set in the q-dimensional space $(q > 1)$. We denote with $U_1, U_2, \ldots$ a sequence of finitely or countable infinitely many point-sets satisfying the following conditions:

a) The given point-set A is a subset of the union of the U_k,
b) The diameter d_k of each U_k is less than a given positive number ϱ.

We consider the *lower bound* of the sums

$$d_1 + d_2 + \cdots$$

---

[41] The juxtaposition 'AW' means intersection of sets A and W, so the formula in modern notation is: $\mu^* W = \mu^* AW + \mu^* (W - (A \cap W))$.

> of the diameters d_k for all sequences $U_1, U_2, \ldots$, satisfying the conditions a) and b), and we denote this lower bound, which, incidentally, can be $+\infty$, by $L_\varrho A$.
>
> If ϱ decreases, then the conditions that U_k must satisfy become more restrictive and the quantity $L_\varrho A$ cannot decrease. The following limit therefore *always* exists
>
> $$\lim_{\varrho=0} L_\varrho A = L^* A.$$
>
> *The quantity $L^* A$, which is uniquely defined for any point-set, will be called the outer linear measure of A.*[Ch16, 64]

In this definition he introduces a special kind of covering of typical point-sets A in q-dimensional space having sets $U_1, U_2, \ldots$ with diameters less than a small ϱ. This choice of covering is a critical part of his theory, and his intention is clear. Using more and more refined coverings of set A limited by small numbers $\varrho > 0$, then taking the lower bound of the sums of the diameters of the sets in these coverings for any given ϱ, and finally taking the limit, Carathéodory makes it clear that the limit exists and that intuitively this outer linear measure will actually measure A.

He then shows that his formal theory of Chapter I applies to his theory of linear measurability in a straightforward manner [Carathéodory, 1914, pp. 421–424] [Edgar, 2004, pp. 66–69]. He goes on to consider closed curves γ without double points, i.e., single-valued images of closed lines like circles. He shows that the linear measure of this type of curve γ is equal to the upper bound λ of the set of polygons inscribed in γ [Carathéodory, 1914, pp. 424–425] [Edgar, 2004, pp. 69–70]. Additionally, he shows that a point-set of finite linear outer measure has the usual Lebesgue measure zero [Carathéodory, 1914, pp. 425–426] [Edgar, 2004, pp. 70–71].

He ends his paper by stating that it is possible to develop a theory of p-dimensional measure in a q-dimensional space in an analogous way. He offers a path for building this higher-dimensional generalized theory by changing the choice of the basic diameters of sets for the covering. To make this change he proposes defining p-dimensional diameters for the smallest convex sets containing the original sets of the covering and using orthogonal projections onto p-dimensional linear manifolds [Carathéodory, 1914, page 426] [Edgar, 2004, page 71].[42] Carathéodory's theory of measure and linear measure is quite beautiful.

In the following years Carathéodory incorporated expositions of his theory in his lectures and books of lectures on real function theory [Carathéodory, 1918, pp. 237–258] [Carathéodory, 1927, pp. 237–258]. He dramatically revised his theory, giving it a more general algebraic form in his posthumously published book *Mass und Integral und ihre Algebraisierung* [Carathéodory, 1956] (in English translation *Algebraic Theory of Measure and Integration* [Carathéodory, 2010]).[43]

---

[42] However, introducing convex sets does complicate the development of the theory.

[43] See the *Encyclopedia of Mathematics (Springer)* for a short summary of Carathéodory's theory.

Felix Hausdorff (1868–1942)[44] developed his important work on measure and dimension directly from Carathéodory's 1914 paper. Hausdorff's paper of 1919, 'Dimension und äußeres Maß' ('Dimension and Outer Measure') [Hausdorff, 1919] [Hausdorff, 2001, pp. 19–43] [Edgar, 2004, pp. 75–99] is packed with interesting mathematical ideas, which have had a remarkable and lasting influence on the development of mathematics. In terms of historical ordering one must recognize that Hausdorff's work of 1919 comes a few years before the full development of topological dimension theory by Urysohn, Menger, and others in the 1920s and '30s. Only Brouwer's single paper with a topological basis comes before in 1913. Thus, in 1919 Hausdorff was striking out in a set-theoretic and measure-theoretic direction for developing a dimension theory, which comported with his background.[45]

Before examining Hausdorff's special theory presented in the paper of 1919 I need to provide some background for his work. Hausdorff was a master of set theory. His seminal book, *Grundzüge der Mengenlehre* (*Fundamentals of Set Theory*), is a full exposition of abstract set theory, transfinite cardinals and ordinals, point-set theory, metric spaces and topological spaces (which he introduced), mappings of functions, and content and measure theory [Hausdorff, 1914] [Hausdorff, 2002]. Mathematicians far and wide read the book—it was highly influential over many years, especially for its introduction of Hausdorff topological spaces.

The *Grundzüge der Mengenlehre* (*Fundamentals of Set Theory*) covers an extensive range of examples as well as important results and theorems. Included are analyses of counterintuitive examples like Cantor's ternary set or discontinuum and generalizations thereof (e.g., see [Hausdorff, 1914, Siebentes Kaptiel, §8., pp. 249–255]). The last, tenth chapter provides a synoptic treatment of content, measure, and integration theory [Hausdorff, 1914, Zehntes Kapitel, pp. 399–448]. Hausdorff was systematic in his study of set theory as well as in his exposition of the domain (see the general introduction in [Hausdorff, 2002, pp. 1–89]). Hausdorff must have been excited to see the abstracting work of Carathéodory's 1914 paper on measure. His extensive background gave him the tools needed to develop his measure and dimension work in 1919.

Hausdorff opens his 1919 paper rather modestly referencing Carathéodory's work [Hausdorff, 1919, page 157] [Edgar, 2004, page 75]:

---

[44] There is a full-length biography of Hausdorff [Brieskorn & Purkert, 2018]. See also [Eichhorn, 1994], [Brieskorn, 1996], [Brieskorn & Purkert, 2021], and
`https://mathshistory.st-andrews.ac.uk/Biographies/Hausdorff/`.
Hausdorff was a very distinguished mathematician who furthered the development of set theory. It is sad that at the end of career he was treated very badly by the Nazi regime and that he, along with his wife and his wife's sister, committed suicide in January 1942—a horrible end to a distinguished man's life.

[45] Cf. notes on and expositions of his theory in the following: [Brieskorn, 1996, pp. 149–252] [Brieskorn & Purkert, 2018, pp. 801–823] [Hausdorff, 2001, pp. 44–54] [Brieskorn & Purkert, 2021, pp. 372–390]. For a modern, abstract treatment of Hausdorff measure theory see [Rogers, 1998]. For full expositions of Hausdorff measure and dimension cf. [Edgar, 2008, Chapter 6, pp. 165–223] [Pesin & Climenhaga, 2009] [Falconer, 2014, Chapter 3, pp. 44–65].

> Mr. Carathéodory has developed an exceptionally simple and general measure theory that contains Lebesgue's theory as a special case, and that, in particular, defines the p-dimensional measure of a point-set in a q-dimensional space. In the following we shall add a small contribution to this work. After introductory considerations, which generalize the Carathéodory length measure in an obvious way and present an overview of the many analogous concepts of measure, we introduce an explanation of p-dimensional measure, which can immediately be extended to non-integer values of p, and suggests the possibility of the existence of sets of *fractional dimension*, and even of sets whose dimensions fill out the scale of positive integers to a more refined, e.g., logarithmic scale. Dimension thus becomes a sort of measure of gradation similar to the 'order' of convergence to zero, the 'strength' of convergence and related concepts.[Ch16, 65]

Hausdorff briefly compares his work to Fréchet's dimension types (see Section 10.4), claiming that Fréchet's assignment of dimension types to sets is not as natural as in his dimension theory. Then he states that he is limiting his work to considering simplest examples, namely, to certain linear nowhere dense, perfect sets (which typically exhibit abnormal behavior and appear in his treatment in a new light) and to planar and spatial sets arising from them by taking products [Hausdorff, 1919, page 157] [Edgar, 2004, pp. 75–76]. In this way he restricts his investigation to treating examples related to Cantor's ternary set or discontinuum and generalizations thereof.

Hausdorff begins the mathematical part his paper by setting out the five conditions of Carathéodory's measure theory for a set A to be measurable, the measure being denoted as $L(A)$, in a q-dimensional Euclidean space, concluding by giving the fundamental defining equation [Hausdorff, 1919, page 158] [Edgar, 2004, page 76]:

> A set A is said to be *measurable* (with respect to L) if for every set W the equation
> $$L(W) = L(AW) + L(W - AW)^{46}$$
> holds.[Ch16, 66]

This is Carathéodory's *outer measure* denoted as $L(A)$.

Hausdorff adds that we naturally confine our attention to those outer measures that are identical for congruent sets. If $L(B) = m^p L(A)$ whenever B is similar to A in proportion $m : 1$, then *we call L an outer measure of dimension p* [Hausdorff, 1919, page 158] [Edgar, 2004, page 76]. This last remark is key to his theory and prepares the reader for his treatment of various dimensions, including fractional ones.

He also gives ways to compare two outer measures. If for any set A two outer measures $L(A)$, $M(A)$ are simultaneously zero, positive, or infinite, then one calls them *measures of equal order*. Moreover, if $M(A)$ is always zero whenever $L(A)$ is finite (or if $L(A)$ is always infinite, whenever $M(A)$ is different from zero), then L is said to be *of lower order* and M *of higher order*. [Hausdorff, 1919, page 158] [Edgar, 2004, pp. 76–77].

---

[46] that is, $L(W) = L(A \cap W) + L(W - (A \cap W))$.

Following Carathéodory, Hausdorff considers any system $\mathfrak{U}$ of bounded point-sets U in a q-dimensional space, such that any given set A can be covered with a finite or countable number of the sets U *having arbitrarily small diameters* $d(U)$. Let a finite, nonnegative number $l(U)$ be assigned to every set U. Then, given a positive ϱ, it is possible to cover A by a union of U_n, where $d(U_n) < \varrho$. The lower bound of the sums $\sum l(U_n)$ arising from all possible sequences of sets may be denoted by $L_\varrho(A)$. It can be either zero, positive, or infinite. Clearly, as ϱ decreases, the constraints become stricter, L_ϱ does not decrease and has a finite or infinite limit,

$$\begin{aligned} L(A) &= \lim_{\varrho \to 0} L_\varrho(A) \\ &= \liminf_{\varrho \to 0} \sum l(U_n) \\ &\geq L_\varrho(A). \end{aligned}$$

This set function $L(A)$ satisfies four of Carathéodory's five conditions for being an outer measure, but needs further constraints on $l(U)$ to satisfy all five conditions. In the special case that the U are Borel sets (formed from open sets, or equivalently, from closed sets, through the operations of countable union, countable intersection, and relative complement), then all five conditions are satisfied and the set function $L(A)$ is an outer measure in Carathéodory's sense.

The set function $l(U) \geq 0$ may be thought of as defined for any bounded set, and it may have one of the following properties:

A) For the smallest closed set U_α containing U

$$l(U_\alpha) = l(U).$$

B) If one constructs a sphere of radius β around every point of U and if U_β denotes the set of interior points of these spheres, then one has for arbitrary $\varepsilon > 0$ and sufficiently small β

$$l(U_\beta) < l(U) + \varepsilon.$$

In case A) Hausdorff calls the set function $l(U)$ *closable* (*abschließbar*); in case B) he calls it *continuous* (*stetig*) [Hausdorff, 1919, page 160] [Edgar, 2004, page 78].

On this basis Hausdorff proves the following theorem [Hausdorff, 1919, page 161] [Edgar, 2004, page 79]:

> If $l(U) \geq 0$ is a closable or continuous set function defined for bounded U and if A is enclosed in finitely or countably many bounded sets U_n with diameters $< \varrho$, then
>
> $$L(A) = \liminf_{\varrho \to 0} \sum l(U_n)$$
>
> is an outer measure.[Ch16, 67]

The set function $l = l(U)$ is a basic element in Hausdorff's construction of his dimension theory. He offers functions in the following list as candidate functions l [Hausdorff, 1919, pp. 161–162] [Edgar, 2004, pp. 80–81]:

A) The diameter d, i.e., the upper bound of the distances between pairs of points u_0u_1 belonging to U.
B) The upper bound Δ of the surface area of a triangle $u_0u_1u_2$, whose vertices lie on U.
C) The upper bound Δ_p of the p-dimensional volume of a simplex $u_0u_1 \ldots u_p$, whose vertices lie on U. The first two functions are special cases of this one for $p = 1$ or $p = 2$.
D) The lower bound of the elementary geometry q-dimensional volume (in the sense of elementary geometry) of a sphere (or cube, parallelepiped, or simplex), which contains the set U.
E) The lower bound of the volume sum of *finitely* many spheres (or the like), enclosing U, i.e., the outer *volume* of U (in the sense of Jordan–Peano, not the outer Lebesgue measure).
F) The q-dimensional volume (in the Jordan or Lebesgue sense) of the smallest *convex* set U_c over U.
G) The upper bound of the p-dimensional volume of the orthogonal projection of U_c onto a p-dimensional plane. Carathéodory calls this the p-dimensional diameter of U. It is apparent that the functions of A) and F) earlier in the list are the special cases $p = 1$ and $p = q$, respectively.

These seven functions are all monotonic and continuous. Proofs of continuity for candidate functions of A) – E) reduce to elementary geometry, while the last two candidate functions of F) and G) require consideration of convex sets. Still, it is sufficient to use the more easily proven closability.

Note that every nonnegative, continuous, or monotonic function $\lambda(t)$ of $t \geq 0$, combined with a nonnegative, continuous, or monotonic set function $l(U)$, gives rise to a new composite set function $\lambda(l(U))$. There are other ways to combine other set functions, and there is a range of possibilities for defining outer measures in the sense of Carathéodory.

Hausdorff points out that for $p = 1, 2, \ldots$, one obtains the p-dimensional outer measure of Carathéodory by choosing the p-dimensional diameter for the set function $l(U)$ in the theorem above according to candidate function G). Since this definition requires a substantial amount of knowledge of the theory of convex sets, a simplification is certainly desirable. Thus, he begins to assess the advantages of alternative approaches according to the use of various geometrical figures.

For example, if one takes spheres instead of the set U, then one can obtain a p-dimensional outer measure L_p (in the sense defined near the beginning) defined as follows [Hausdorff, 1919, page 163] [Edgar, 2004, page 82]:

Definition 1. One covers A with finitely or countably many spheres K_n of diameters $d_n < \varrho$ and [then] one forms/constructs [the outer measure defined by the limit]

$$L_p(A) = \liminf_{\varrho \to 0} c_p \sum d_n^p,$$

where

$$c_p = \pi^{\frac{p}{2}} : 2^p \prod \left(\frac{p}{2}\right)$$

is the volume of a p-dimensional sphere of diameter 1.[Ch16, 68]

In fact, it is not necessary to consider the p-dimensional diameter at all, as L_p is certainly an outer measure by a previous argument (i.e., since spheres are Borel sets). Hausdorff then mentions possible consideration of other basic geometrical figures. For example, using cubes, parallelepipeds, or simplexes instead of spheres would result in a modification of the constant factor c_p. Yet, the corresponding outer measures would actually be of the same order in these cases.

Over the next few pages of his paper Hausdorff confines his attention to Definition 1 and shows (by proof sketch) that the cases for $p = 1$, $p = 2$, and $p = q$ give the usual expressions for length, area, and volume, at least for the simplest sets [Hausdorff, 1919, pp. 163–165] [Edgar, 2004, pp. 82–84]. These results tend to confirm that Definition 1 is correct, at least to an extent.

At this point in his paper he writes that it is tempting to extend the definition to *nonnegative positive values of p*—an extension that Carathéodory did not contemplate. *Consequently, he begins to consider the possibility of positive fractional numbers or positive real numbers as dimension numbers.* In this way *Hausdorff dimension* starts to take form. The question that must be asked first is whether this exercise is trivial in the sense that all sets may have outer measure 0 or ∞. Therefore, the problem arises to construct sets A for which $0 < L_p(A) < \infty$ and that are exactly of dimension p in this new extended sense.

One may consider this problem in a larger context. In the definition of L_p one may note that, apart from the numerical constant c_p, the pth power of the diameter d has played the role of the set function l. Can one possibly replace this with a function, such as $d^p \log \frac{1}{d}$, which tends to 0 as $d \to 0$ slower than d^p, but faster than any smaller power, and in this sense construct sets whose dimensions fall in between the scale of positive numbers? It is possible to make use of the freedom emphasized earlier in the paper and assign to every *function $\lambda(x)$ that is positive, continuous, increasing with x, and that converges to zero with* x an outer measure $L(A)$ defined in the following way [Hausdorff, 1919, page 166] [Edgar, 2004, page 84]:

> **Definition 2.** One covers A with finitely or countably many spheres K_n, whose diameters are $d_n < \delta$, and [then] one forms/constructs [the outer measure defined by the limit]
>
> $$L(A) = \lim_{\varrho \to 0} L_\varrho(A),$$
>
> where $L_\varrho(A)$ is the lower bound of the sums $\sum \lambda(d_n)$.[Ch16, 69]

Hausdorff indicates that the constraints imposed on $\lambda(x)$ are not absolutely necessary, but they are useful for the purpose of excluding zero or infinite $L(A)$.

If $0 < L(A) < \infty$, then one may say that A is of dimension $[\lambda(x)]$. In particular, denote the dimension $[x^p]$ by (p), thereby resolving the dimension $[x^p]$ to the power p. A set A of this dimension is considered of dimension p in the sense of the definition of dimension given near the beginning, but not vice-versa [Hausdorff, 1919, pp. 158, 166] [Edgar, 2004, pp. 76, 85].

It is useful to have comparisons among dimension functions. Thus, let $\lambda(x)$, $\mu(x)$ be two functions corresponding to the outer measures $L(A)$, $M(A)$ respectively, and let L, M be of the same order. Hausdorff considers the dimensions $[\lambda]$, $[\mu]$ to be equal. Every set of dimension $[\lambda]$ is also of dimension $[\mu]$ and vice-versa. If M is of higher order than L, then he calls $[\mu]$ the higher and $[\lambda]$ the lower dimension. A set of a given dimension has measure 0 for any higher dimension, and it has measure ∞ for any lower dimension. Naturally it is not necessary for two dimensions to be comparable in this sense.

Obviously, $L(A)$ only depends on the behavior of $\lambda(x)$ in an arbitrarily small neighborhood of the point $x = 0$, i.e., the dimension appears to be a characteristic of 'gradation' ('Abstufungsmerkmal'), like the 'order' of a function tending to 0 and similar concepts. If $\lambda(x) \leq \mu(x)$ for sufficiently small x, then $L(A) \leq M(A)$ holds for any set A. If $\mu = c\lambda$, $(c > 0)$, then $M = cL$. If μ/λ lies between two positive bounds, then the dimensions $[\lambda]$, $[\mu]$ are equal. If $\mu/\lambda \to 0$ for $x \to 0$, then $M \leq \varepsilon L$ for every positive ε, M vanishes for finite L, and $[\mu]$ is of higher dimension and $[\lambda]$ is of lower dimension. In particular, (q) is the higher dimension and (p) is the lower dimension whenever $q > p$. A function like $\mathrm{e}^{-1/x}$, which goes to zero faster than any x^p, results in an infinitely large dimension, which cannot be realized by any Euclidean point-set, whereas infinitely small dimensions can be realized, as will be discerned in the discussion below. The functions of the well-known logarithm scale

$$\lambda(x) = x^{p_0} \left(\frac{1}{l_1 \frac{1}{x}}\right)^{p_1} \left(\frac{1}{l_2 \frac{1}{x}}\right)^{p_2} \cdots \left(\frac{1}{l_k \frac{1}{x}}\right)^{p_k},$$

where the first nonvanishing p_i is assumed to be positive, are positive together with their derivatives for sufficiently small x, as is required by the initial assumptions. They define dimensions that may be denoted by

$$P = (p_0, p_1, \ldots, p_k).$$

These can be ordered lexicographically with Q higher than P if the first nonvanishing difference $q_i - p_i$ is positive. It is possible to divide these dimensions into two classes: The first class is:

$$\textit{Dimensions less than } (1): \quad \frac{\lambda(x)}{x} \to \infty.$$

In this case the first derivative $\lambda'(x) \to \infty$, and the second derivative λ'' is negative for small x. The function $\lambda(x)$ is convex upwards. The following holds for the exponents

either $p_0 = 0$ and the first nonvanishing $p_i > 0$,
or $0 < p_0 < 1$,
or $p_0 = 1$ and the next nonvanishing $p_k < 0$.

The second class is:

$$\textit{Dimensions greater than } (1): \quad \frac{\lambda(x)}{x} \to 0.$$

In this case the first derivative $\lambda'(x) \to 0$, and the second derivative λ'' is positive. The function $\lambda(x)$ is concave downwards.

either $p_0 = 1$ and the next nonvanishing $p_k > 0$,

or $p_0 > 1$.

The transition between the two regions occurs in dimension (1) when $\lambda(x) = x$.

This section (§8) of the paper marks the pivotal point where Hausdorff boldly ventures into fractional dimension theory and offers his initial analysis. In so doing he follows traditional ways of analyzing a new function.

To illustrate and partially validate his notion of fractional dimension Hausdorff constructs and analyzes in detail *linear sets* of dimension $[\lambda(x)]$ [Hausdorff, 1919, §§9–12, pp. 167–177] [Edgar, 2004, pp. 86–97]. To construct such linear sets the previous argument suggests that one should take $\lambda(x)$ to be *convex*, and to this end he gives an analysis of the convexity involved.[47] His primary goal at this point in his exposition is to show that there exist linear closed sets A of dimension $[\lambda(x)]$, that is, sets A with finite positive measure $L(A)$, where $L(A) = \lim_{\varrho \to 0} L_\varrho(A)$, with $L_\varrho(A)$ being the lower bound of $\sum \lambda(a_n)$ as one covers A with intervals of length $\alpha_n < \varrho$.

Hausdorff's construction follows a procedure for forming a generalized Cantor middle third set. For this he uses a sequence of positive numbers $\xi_0, \xi_1, \xi_2, \ldots$ with

$$\xi_0 > 2\xi_1, \ \xi_1 > 2\xi_2, \ \ldots$$

to define lengths of intervals to be considered in the process. Hausdorff's diagram from his paper, given in Figure 16.2, gives a broad idea of how the constriction proceeds [Hausdorff, 1919, page 169]. The idea is to construct a perfect, nowhere dense set A in closed interval $[0, \xi_0]$ using the ξ's for lengths of the intervals successively considered in the process. The β's in Figure 16.2 are used as labels for the open intervals to be deleted. More specifically, in the construction every rational dyadic number y between 0 and 1 is to be associated with or labels an open interval $\beta(y)$, and after deleting the open set (union) $B = \sum \beta(y)$ [i.e., $B = \bigcup \beta(y)$] from the original closed interval $[0, \xi_0]$, there remains a perfect, nowhere dense set A, which can be covered by 2^n intervals of length ξ_n for $n = 0, 1, 2, \ldots$ and whose measure $L(A)$ is at most equal to lim inf $2^n \lambda(\xi_n)$, since $\xi_n \to 0$. Hausdorff's detailed analysis over several pages leads to the result that *the constructed set A has dimension* $[\lambda(x)]$ [Hausdorff, 1919, page 172] [Edgar, 2004, page 91].

He follows the analysis with a selection of basic examples [Hausdorff, 1919, pp. 172–173] [Edgar, 2004, pp. 91–92]. To generate a set of dimension p with $0 < p < 1$ set $\lambda(x) = x^p$, which satisfies the requirements for all positive x. For example, to get dimension $p = 1/2$ choose length $\xi = 1/4$, so that one deletes the

---

[47] I shall not consider this further; see [Hausdorff, 1919, pp.167–168] [Edgar, 2004, pp. 86–87].

central half of each remaining interval. Moreover, for a given ξ where $0 < \xi < 1/2$ the dimension turns out to be

$$p = \log 2 : \log \frac{1}{\xi}.$$

For Cantor's particular middle third set ($\xi = \frac{1}{3}$) one finds the dimension to be:

$$p = \log 2 : \log 3 \approx 0.63093.$$

Hausdorff also gives a set of "infinitely small" dimension as well as a set that has dimension $[\lambda(x)]$ in every interval.

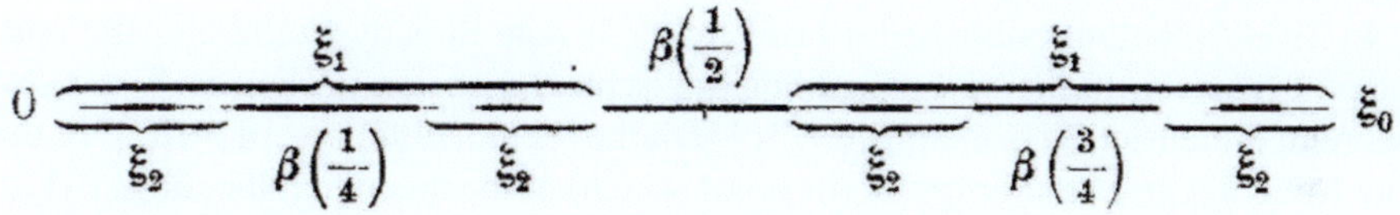

Fig. 16.2 Hausdorff's Diagram for Constructing a Linear Point-Set of Fractional Dimension [Hausdorff, 1919, page 169].

In two sections (§§11, 12) near the end of the paper Hausdorff analyzes further his constructed linear perfect set. Then in the final section (§13) he discusses *planar sets* of a prescribed dimension and devotes a sentence to *spatial sets* of given dimension.[48]

Hausdorff's theory of fractional dimension was a bold venture into a new measure-theoretic domain, and he grounded it in a classical analysis of functions. He published nothing further on this subject. However, material in his *Nachlaß* reveals that he had an interest in topological dimension theory for a long time. The commentator on the paper in his *Gesammelte Werke* (*Collected Works*), S.D. Chatterji, reports [Hausdorff, 2001, page 53]:

> There are several hundred pages of studies on the ongoing work on dimension theory of Hurewicz, Menger, P. Alexandroff, Urysohn, Kuratowski and others. NL Hausdorff: Kapsel 47: Fasz. 986 (written between 1930–1936) is a book length study of approx. 200 pages incorporating many of the results of the mathematicians mentioned.

[48] Many years before Carathéodory's and Hausdorff's papers Cantor had noticed the dependence of the content or measure of a point-set on the dimension of its containing space [Cantor, 1884a, pp. 389–390] [Cantor, 1932, page 258]. A unit square has content 0 in 3-dimensional Euclidean space, but it has content 1 in the 2-dimensional plane. Cantor thought that this observation of the relativity of content to containing space was indispensable for his research on the dimensions of various continua. However, he never published anything further on this subject. Cf. Chapters 6 and 8 and [Cantor, 1932, pp. 207–208]. Also cf. Cantor's 1884 article [Cantor, 1884b]. This observation was perhaps in the background when Hausdorff proposed a dimension theory founded on the concept of measure.

After the publication of Hausdorff's paper on fractional dimension several mathematicians investigated his innovative theory. Among the first and one of the most important was Abram Samoilovitch Besicovitch (1891–1970).[49] He did much to advance Hausdorff's theory through his extensive research and publications. His earlier papers of the 1920s and 1930s investigating the domain include the following: [Besicovitch, 1928] [Besicovitch, 1929] [Besicovitch, 1934] [Besicovitch, 1935a] [Besicovitch, 1935b] [Besicovitch, 1935c]. An analysis of Besicovitch's work in this domain is beyond the scope of this book.

There are provable connections between Hausdorff's dimension theory and more general measure theory and the topological theory of dimension of Brouwer, Urysohn, and Menger. In the 1930s several mathematicians sought to characterize these connections precisely. These included Georg Nöbeling, Lev Pontrjagin with Lev Schnirelmann, and Edward Szpilrajn.

In 1931 Nöbeling published a brief research note directly examining the connection, 'Hausdorffsche und mengentheoretische Dimension' ('Hausdorff and set-theoretic dimension') [Nöbeling, 1931c] [Dierker et al., 1998, pp. 180–181]. In the note Nöbeling gives a short proof (or proof sketch) of the theorem [Nöbeling, 1931c, page 24] [Dierker et al., 1998, page 180]:

> **Theorem.** *For every set M the Hausdorff dimension is at least equal to the set-theoretic dimension.*[Ch16, 70]

His proof is inductive on the set-theoretic dimension, dim M.

About the same time in 1931 Lev Pontrjagin and Lev Schnirelmann (1905–1938)[50] wrote a longer paper covering the topic, 'Sur une propriété métrique de la dimension' ('On a metric property of dimension') [Pontrjagin & Schnirelmann, 1932] (Translation into English [Edgar, 2004, pp. 133–141]). These authors were thinking generally about dimension theory in a topological context and aiming broadly to relate dimension theory to metrics and measure theory. They did not explicitly reference Hausdorff's work. They seem not to have been thinking in that direction. Their paper was published in 1932.

In the 1932 paper they state their aim as follows [Pontrjagin & Schnirelmann, 1932, page 156] [Edgar, 2004, page 133]:

> The aim of this note is to obtain the dimension of a compact metric space using a recovery process using 'standards of measure' ['metric standards,' 'étalons de mesure'], some of which make it possible to define dimension. This method is analogous to the measurement method used in elementary geometry. [Ch16, 71]

They use the Lebesgue tiling principle form of the definition of dimension with respect to closed covers for a compact metric space and its compact sets (while giving credit to Brouwer, Urysohn, and Menger) [Pontrjagin & Schnirelmann, 1932, page 156] [Edgar, 2004, page 134]:

---

49 For biographical material on Besicovitch's life and work cf. [Burkill, 1971] [Taylor, 1975].

50 For biographical information on Schnirelmann cf.
`https://mathshistory.st-andrews.ac.uk/Biographies/Shnirelman/`.

> A compact set F has dimension n, if n is the smallest integer such that there exists (with ε being a positive arbitrary number) a finite system of closed sets covering F, with diameters not exceeding ε and for which all the intersections of $n+2$ of them are empty.[Ch16, 72]

They use the symbol $N(\varepsilon)$ to define the minimum number of such closed sets necessary to cover F. The function $N(\varepsilon)$ takes positive values for all positive values of ε and grows indefinitely when ε tends to 0 (that is, unless set F contains only a finite number of points). This function obviously depends on the metric of the space F. This may be called the volume function of space F relative to the given metric, and it can be designated as $N_F(\varepsilon)$ to draw attention to F and its metric.

Further, let r be any positive real number. The authors say that r belongs to the first class if there exists a positive number c such that

$$N_F(\varepsilon) \geq \frac{c}{\varepsilon^r} \quad \text{(whenever } \varepsilon > 0);$$

in the opposite case they say that r belongs to the second class. The Dedekind cut determined by these two sets defines a non-negative number k ($\leq +\infty$) that they call the *metric order of the space* F. One can show without undue difficulty that

$$k = \liminf \left(-\frac{\log N_F(\varepsilon)}{\log \varepsilon} \right).$$

For the various metrics of F one obtains the various volume functions and the various metric orders. To support this statement the authors first give the example of a simple Jordan arc measured with the Euclidean metric and state that it has as a volume function (asymptotically) $\frac{c}{\varepsilon}$, where c is a constant. The metric order of such an arc is therefore equal to *one* (1). In contrast, they define a different metric of a simple arc in the following way: take a perfect everywhere discontinuous set in the k-dimensional Euclidean space R^k whose measure is positive. Draw a simple arc through such a set. This arc automatically has the metric of the Euclidean space R^k, its volume function is asymptotically $\frac{c}{\varepsilon^k}$, and its metric order is equal to k.

Thus, the metric order is not a topological invariant of F. However, one can obtain a topological invariant of the space F by considering the lower bound of the metric orders corresponding to various metrics that can be imposed on F without altering the topological properties of this space.

With this background the authors then state the main goal of the paper: to determine the topological invariant in question and *to show that it is equal to the Brouwerian dimension of the space* F. They state the principal result and corollary of the paper in the following form [Pontrjagin & Schnirelmann, 1932, page 157] [Edgar, 2004, page 135]:

FUNDAMENTAL THEOREM. The lower bound of the metric order

$$k = \liminf \left(-\log \frac{N_F(\varepsilon)}{\log \varepsilon} \right)$$

for all metrics of the compact space F is equal to its dimension.

COROLLARY. If the lower bound in question is finite, then it is necessarily an integer.[Ch16, 73]

The authors devote the rest of the paper to proving these results. Their method of proof is strikingly clever. It involves considering *all metrics* for a given space F to make it a metric space. Several mathematicians have found this approach of considerable value.

Building on the earlier results of Nöbeling, Pontrjagin, and Schnirelmann, Edward Szpilrajn (= Edward Marczewski) (1907–1976)[51] wrote a short paper 'La dimension et la mesure' ('Dimension and Measure') stating a connection between measure and dimension [Szpilrajn, 1937]. It appeared in 1937, a few years after the other papers.

In his paper of 1937 Szpilrajn first highlights differences between dimension and measure using a simple example. Then he employs ideas from papers of Nöbeling [Nöbeling, 1931c] and Pontrjagin and Schnirelmann [Pontrjagin & Schnirelmann, 1932] to offer his clarification of the mathematical situation, stated with a reference to the embedding theorem [Szpilrajn, 1937, pp. 81–82]:

> ... the relation between dimension and measure is not immediate: e.g., the set of points of the plane with two irrational coordinates is of dimension 0, but its surface measure[52] is positive. The divergence is also very natural, dimension being a topological notion and measure being a metric notion. However, the relation between them can easily be formulated in a very simple manner. In particular, taking advantage of the ideas of Nöbeling, Pontrjagin and Schnirelmann, one arrives at the theorem:
>
> *For a set (situated in a separable metric space) to be of dimension $\leq n$, it is necessary and sufficient that it be homeomorphic to a set of measure zero in or relative to an $(n + 1)$-dimensional space/set (situated in the cube of $2n + 1$ dimensions).*[Ch16, 74]

Proceeding to his formal treatment of the connection between dimension and measure, Szpilrajn provides a sequence of definitions and assumptions sufficient for rigorously stating and proving his results. He follows mainly Hausdorff along with the others who provided the background mathematics. I shall only touch on these lightly in order to give a statement of his principal result.

---

[51] Edward Szpilrajn changed his name to Edward Marczewski during the time of the Nazi takeover of Poland as a protection from the Nazi captors. From that time he has been known primarily by the latter name. For biographical information on Szpilrajn cf. [Kuratowski, 1980, pp. 189–192] and `https://mathshistory.st-andrews.ac.uk/Biographies/Marczewski/`. See also his *Collected Mathematical Papers* [Marczewski, 1996].

[52] I.e., its measure in or relative to the plane.

Let $\varphi(x)$ be a non-negative function of one non-negative, continuous, monotonically increasing variable, such that $\varphi(0) = 0$. Let Φ designate the class of all such functions of this type. Let E be a set belonging to a metric space. For each number $\varepsilon > 0$ let $L^{\varphi}_{\varepsilon}(E)$ be the lower bound of all the numbers $\sum_{n=1}^{\infty} \varphi[\delta(E_n)]$, where $E = E_1 + E_2 + \cdots$ is a decomposition of the E into a (countable) collection of sets E_n, all with diameter $[\delta(E_n)]$ less than ε. The limit of the number $L^{\varphi}_{\varepsilon}(E)$ for $\varepsilon \to 0$ will be designated as $L^{\varphi}(E)$.

In the important case where $\varphi(x) = x^p$ (for $p > 0$), the function $L^{\varphi}(E)$ will be designated by $L^{(p)}(E)$. Following Hausdorff, Szpilrajn takes the power p to be a dimension number.

In the Cartesian space of n dimensions, the class of the sets for which the function $L^{(n)}$ vanishes coincides with the class of the sets of Lebesgue measure zero (in or relative to the n-dimensional space). By analogy, one may call *sets of measure zero in or relative to p-dimensional space* the sets for which $L^{(p)}$ vanishes. Similarly, one may call the sets for which this function is finite *sets of finite measure in or relative to p-dimensional space*.

The function $L^{(1)}$ being equal to the linear sets of Lebesgue outer measure, one may call them sets of *linear measure*.

Additionally, one may designate by $L^{(0)}(E)$ (*0-dimensional measure of* E) the function equal to n for each finite set E composed of n points ($n = 0, 1, 2, \ldots$) and to $+\infty$ for each infinite set E.

Szpilrajn defines and uses comparative 'orders of functions' in a way close to the way Hausdorff uses them in his paper [Hausdorff, 1919, pp. 158, 166] (discussed above). Given two set functions $L(E)$ and $L^*(E)$, one says that the function L^* is *of a superior order* to the function L when the relation $L(E) < +\infty$ implies the relation $L^*(E) = 0$; one says that its order [the order of L^*] is *equal* to that of the function L when the relations $L(E)$ and $L(E) = +\infty$ are equivalent to the relations $L^*(E)$ and $L^*(E) = +\infty$, respectively.

Given these definitions and others as background and a sequence of intermediate proven results, Szpilrajn is able formally to assert his principal result at the end of his paper [Szpilrajn, 1937, page 89]:

> ***Theorem 5****. Let L^{φ} be a function of superior order to that of the function $L^{(n)}$, but of inferior or equal order to that of the function $L^{(n+1)}$. For a set E (situated in a separable metric space) to be of dimension $\leq n$, it is necessary and sufficient that it be homeomorphic to a set E^* (situated in [the* $(2n+1)$*-dimensional cube] I^{2n+1}) such that $L^{\varphi}(E^*) = 0$.*[Ch16, 75]

Szpilrajn associates his result with the embedding theorem (see Sections 16.7 and 16.9).[53]

In their publications of the 1930s Nöbeling, Pontrjagin and Schnirelmann, and Szpilrajn make clear the connections between dimension and measure. They reveal an interesting interplay between these two domains resting on the underlying geometry.

---

[53] For an exposition of Szpilrajn's result see [Hurewicz & Wallman, 1948, Chapter VII, pp. 102–107].

A beautiful outgrowth of the work on Hausdorff dimension is the development of fractal geometry and the theory of fractals. Benoit Mandelbrot (1924–2010)[54] was a primary early proponent of this theory, and he researched, wrote, and presented extensively on fractals. His book on the subject is well known (and rather exciting to read): *The Fractal Geometry of Nature* [Mandelbrot, 1983]. However, the work on fractal geometry and fractals is beyond the scope of this book.[55]

16.12 Hermann Minkowski, Georges Bouligand, and Dimension

There is another approach to analyzing sets that assigns them fractional dimensions, an approach that parallels the approach of Hausdorff. I shall consider it now in brief. Hermann Minkowski and Georges Bourligand are the principal mathematicians involved with this approach, and the work that I shall discuss comes from the beginning of the twentieth century to the 1930s, though their work has exerted an influence beyond this period.

Hermann Minkowski (1864–1909)[56] is most famous for his work on the geometry of numbers and the geometry of relativity theory, especially through his revolutionary essay 'Raum und Zeit' ('Space and Time') [Minkowski, 1909a] [Minkowski, 1909b]. He was a colleague and close friend of David Hilbert, who mourned his death at the young age of 44 in 1909.

Minkowski made a small, though significant contribution to the theory of fractional dimension through a single lecture. At the annual conference of the Deutsche Mathematiker-Vereinigung (German Association of Mathematicians) held in Aachen, Germany, 16–23 September 1900 he gave a lecture, 'Über die Begriffe Länge, Oberfläche und Volumen' ('Concerning the concepts of length, surface area, and volume'), which was then published in 1901 [Minkowski, 1901]. The theory contained there has an importance for fractional dimension theory.

In the lecture Minkowski first remarks that the concept of extended or multiple integrals for a manifold for the case $n = 3$ immediately yields the concept of volume of a solid body. This idea belongs to the elementary concepts of analysis, and he references a standard text, Camille Jordan's *Cours d'Analyse* (*Course of Analysis*) [Jordan, 1893, pp. 18–31]. He goes on to say [Minkowski, 1901, page 115]:

> Essentially more difficult than the introduction of the concept of volume is the founding of the *length of a curve* as the limit of the length of polygons suitably inscribed with respect to the curve or the founding of the *surface*

[54] For biographical material on Mandelbrot cf.
`https://mathshistory.st-andrews.ac.uk/Biographies/Mandelbrot/`.

[55] There are excellent texts on fractals, including [Edgar, 2008] [Pesin & Climenhaga, 2009] [Falconer, 2014].

[56] For biographical material on Minkowski cf. [Hilbert, 1910]
`https://mathshistory.st-andrews.ac.uk/Biographies/Minkowski/`
`https://www.encyclopedia.com/people/science-and-technology/mathematics-biographies/hermann-minkowski#2830902980`.

> *area of a curved surface* as the limit of the surface areas of polyhedra suitably constructed with respect to the surface.[Ch16, 76]

He then asserts that there is a better way forward [Minkowski, 1901, pp. 115–116]:

> However, these other concepts of length and surface area can also be developed solely by means of the concept of volume using a simple passing to a limit:
>
> Let C be a curve. One may think of each point of C as the center of a [closed] sphere with radius r, where r is a fixed positive quantity. The set of all these points of the space that lie in the interior or on the boundary of at least one of these spheres, defines the *domain with distance* $\leq r$ *from the curve* C. Let $V(r)$ be the volume of this region (in the case that it has a determined volume), then the limit of $\frac{V(r)}{\pi r^2}$ as r approaches zero (in the case that this limit exists) can be introduced as the *length of the curve* C. — Let F be a surface. One constructs in a corresponding way the *domain with distance* $\leq r$ *from* F. Let $V(r)$ be the volume of this domain, so the limit of $\frac{V(r)}{2r}$ as r approaches zero (assuming that the quantity $V(r)$ and this limit exist) can be introduced as the *surface area of the surface* F.
>
> It is obvious that initially the length of a straight line segment and the surface area of a plane triangle result in exactly the values usually assumed for this purpose; as a result, in *regular* cases, lengths and surface areas will have the same values in the sense just explained and in the usual sense.[Ch16, 77]

Note that Minkowski first takes the radius r as a single fixed positive quantity for all the spheres, and then he lets r vary and approach zero to derive the limit volumes and values $\frac{V(r)}{\pi r^2}$ for curve length and $\frac{V(r)}{2r}$ for surface area, assuming these limits exist. Intuitively this makes a great deal of sense. The factor $\frac{1}{\pi r^2}$ for curve length and the factor $\frac{1}{2r}$ for surface area each factor out the other dimensions to derive curve length and surface area, respectively.

In the rest of his lecture Minkowski considers some consequences for his new definitions and nascent theory particularly for convex surfaces and convex solids, but I shall not go into those. This is his only lecture on this topic.

Henri Lebesgue devoted a chapter of his well-known book of 1904, *Leçons sur l'intégration et la recherche des fonctions primitives* (*Lectures on integration and the search for primitive functions*) to 'Définition géométrique de l'intégrale' ('Geometric definition of the integral') [Lebesgue, 1904, Chapitre III, pp. 36–48]. In the book Lebesgue too is interested in the lengths of segments, areas of surfaces, and volumes of geometrical solids [Lebesgue, 1904, pp. 36–37], and he refers to Camille Jordan's work in two editions of his *Cours d'Analyse*, Georg Cantor's earlier work of 1884, and Hermann Minkowski's work of 1901. He specifically references Cantor, who much earlier stated the following [Cantor, 1884a, page 389]:

> I arrive at this general notion of *volume* or *magnitude* $\mathfrak{I}(P)$ of *any* set P contained in G_n by taking *each* point p, which belongs to P or to $P^{(1)}$ [the derived set of limit points of P], for the center of an n-dimensional full sphere

with radius ρ, which we will call $K(p, \rho)$. The smallest multiple [union] of all these full spheres $K(p, \rho)$ ..., denoted as:

$$\mathfrak{M}[K(p, \rho)],$$

(where ρ is a constant) constitutes for each value of ρ a set which is composed of continuous pieces in n dimensions and whose volume is determined according to the known rules by means of an n-tuple integral.

Let $f(\rho)$ be the value of this integral; $f(\rho)$ is a continuous function of ρ, which decreases with ρ; the limit of $f(\rho)$, when ρ converges to zero, serves as the definition of the volume $\mathfrak{I}(P)$; so that we have:

$$\mathfrak{I}(P) = \lim_{\rho=0} f(\rho).$$

I expressly point out that this value of the *volume* or the *magnitude* of any set P contained in a continuous plane space G_n with n dimensions is absolutely dependent on the same plane space G_n, of which P is considered as a component part, and particularly on the number n; so that, if we consider *the same set* P as a constituent part of another continuous plane space H_m, the value of the volume of P with respect to the space H_m is in general different from that, which relates to the same set P, considered as a constituent part of G_n.[Ch16, 78]

Cantor in this case essentially has the same idea of volume that Minkowski gives in his lecture, but he does not consider length of a segment or area of a surface. However, he emphasizes in an interesting way the relativity of the volume to the containing Euclidean space.[57]

Some years later, around 1924, the French mathematician Georges Bouligand (1889–1979)[58] was motivated through his research on the Dirichlet Principle and calculus of variations to examine problems of dimension. He reported his research and some ideas related to dimension in a paper 'Sur le problemè de Dirichlet' ('On the Dirichlet problem'), which was published in 1925 [Bouligand, 1925a]. Around this time he also submitted a short research note on dimension to the French Academy of Sciences, which was released at the session of 17 November 1924 and then published in the *Comptes rendus* (*Proceedings*) of the Academy of Sciences, Paris in 1925, 'Dimension, étendue, densité' ('Dimension, extension, density') [Bouligand, 1925b]. He proceeded with further research into his notions of dimension, publishing his results in a paper of 1927, 'Nombre dimensionnel et ensembles impropres dans le problème de Dirichlet' ('Dimension number and improper sets in the problem of

---

[57] I mentioned the relativity that Cantor recognized earlier in footnote 48 occurring near the end of paragraphs on Hausdorff in Section 16.11.

[58] For biographical material on Bouligand cf.
`http://serge.mehl.free.fr/chrono/Bouligand.html`
`https://en.wikipedia.org/wiki/Georges_Bouligand`.

Dirichlet') [Bouligand, 1927], and another in 1928, 'Ensembles impropres et nombre dimensionnel' ('Improper sets and dimension number') [Bouligand, 1928b].[59]

The second paper provides a fair summary of Bouligand's ideas on dimension and their application at this point, circa 1928. This paper and others of the same time appear to be exploratory in nature representing work in progress rather than a definitive theory of dimension. Still, it is of value to examine in brief the first part of the paper, which presents the basic mathematical content [Bouligand, 1928b, pp. 320–332] (translation into English in [Edgar, 2004, pp. 119–131]).

As background to his research he mentions Hausdorff's work ([Hausdorff, 1919], see Section 16.11) and Fréchet's work ([Fréchet, 1910], see Section 10.4). He notes that this earlier work is not fully topological. His own work will not be fully topological either. He calls it 'restricted topology of the first order' ('topologie restreinte du premier ordre'), saying that one considers in this case continuous and bijective transformations but restricted to functions with continuous first derivatives. For these transformations the rectifiability of a curve or of a surface is an invariant property. This restriction to 'smooth geometrical objects' is natural given his domain of application is Dirichlet's Principle and the calculus of variations.

The geometrical ideas of Cantor and Minkowski constitute a specific background to his work. He calls their construction of coverings for closed sets using open spheres to form open volumes the Cantor–Minkowski construction. Given a closed set E in Euclidean 3-space, one draws open spheres with radius ρ around each of the points of E, thereby forming a well-determined volume $f(\rho)$, which is an increasing function of ρ. This volume is also the set of points whose distance from E is less than ρ.

Given this construction, he considers the dimensional properties of the set of infinitesimal characteristics of the volume $f(\rho)$ as ρ tends to zero in the construction. For example, if the following are chosen, *a point*, *a line segment*, *a plane area*, then each of the $f(\rho)$ will have a well-defined infinitesimal order, whose value will be equal to 3, the given number of spatial dimensions, minus the number of dimensions (in the ordinary sense) of the set in question. From this basis he gives a tentative definition [Bouligand, 1928b, page 322] [Edgar, 2004, page 121]:

> ... *if $f(\rho)$ is of infinitesimal order α, the dimension number of the set being studied [E] will be* $3 - \alpha$.[Ch16, 79]

Yet, he immediately recognizes certain issues with respect to this definition. First, the function $f(\rho)$ of the construction may not always have a well-defined infinitesimal order as a number. Second, there will be an issue with regard to inhomogeneous sets E, such as the union of a disc joined with a line segment. The ambiguity of the situation leads him to move from the notion of dimension number to notions of *dimensional order* and *dimensional order of a local character*.

Over the next few pages of his paper he develops an analysis of his notion of dimensional order, though his analysis is none too clear. In the end he is much more definite when providing a characterization of the localized *dimensional order of a set*

[59] Cf. Bouligand's other related papers [Bouligand, 1928a] [Bouligand, 1928c] [Bouligand, 1928d] [Bouligand, 1928e].

E at one of its points Q using a volume function $f^Q(\rho)$ at the point. To be sure, not all sets E will be homogeneous in dimensional characteristics, so dimensional orders will vary over points of the set. In the homogeneous case he calls *E isodimensional* when the property of dimensional order holds for the same function $f(\rho)$ for all points Q of E [Bouligand, 1928b, pp.328–330] [Edgar, 2004, pp. 126–128].

An interesting part of his work is that he offers various geometrical constructions differing from the basic Cantor–Minkowski sphere construction, accompanying them with brief geometrical analyses. In particular, he examines the geometry of a cube/cell-counting or box-counting dimension [Bouligand, 1928b, pp. 327–328] [Edgar, 2004, page 126]. Let E be a closed and bounded set contained entirely in a larger (closed) cube, whose edge is taken as the unit of length. Subdivide the three sides of the containing cube into N equal parts to obtain N^3 equal cubical cells. Among the cells of the subdivision designate as *empty cells* (*mailles vides*) those that contain neither in their interior nor on their boundary points of E. Designate the others as *useful cells* (mailles utiles), so containing some points of E. Bouligand considers the total (open) volume of useful cells and compares geometrically this open volume with the open volume $f(\rho)$ generated in the Cantor–Minkowski construction.

Overall, Bouligand's work seems not to have been very influential, though it is relevant to aspects of fractional dimension theory. As a final piece on dimension theory Bouligand wrote a short expository monograph surveying *Les définitions modernes de la dimension* (*The modern definitions of dimension*), published in 1935 [Bouligand, 1935].[60]

16.13 Bringing the Story of Dimension and Curve Theory to a Close

In 1941 Hurewicz and Wallman published their beautiful book *Dimension Theory*, in which they presented with utmost clarity the dimension theory of separable metric spaces as a logically coherent and relatively complete structure of mathematics. A revised edition appeared in 1948. The book is a wonderful encapsulation of the theory, including not only the topological theory, but a summary of Hausdorff dimension and Alexandroff's homological theory.

Subsequently, topologists have refined the topological theory and explored the possibilities for extending the classical theory to more general spaces as in, for example, Pears in *Dimension Theory of General Spaces* [Pears, 1975].

Much more rapid growth has occurred in the area of Hausdorff dimension and the beautiful theory of fractals and fractal geometry. There have been many results proven and to some extent applied.

---

[60] For modern related work cf. [Tricot, 1981] (showing the equivalence of 12 different definitions of the same fractional dimension index: the logarithmic density) [Tricot, 1994] [Tricot, 1999] [Falconer & Troscheit, 2023] [Falconer et al., 2023].

For completeness to cover an entirely different aspect of dimension, I include Section 16.14 Appendix C: Note on Dimension in Algebraic Geometry and Commutative Algebra.

At this point I bring my story of dimension to a close. In the next chapter I shall make an assessment of dimension theory's origins, growth, and development, mainly from a philosophical point of view.

16.14 Appendix C: Note on Dimension in Algebraic Geometry and Commutative Algebra

So far in this book I have covered the early history of concepts and theories of dimension and then concentrated on the modern development of point-set and topological theories of dimension and of measure-theoretic (Hausdorff, fractal, Minkowski-Bouligand) theories of dimension. There are many other concepts and theories of dimension of recent date. In a very broad sense all are related, but they differ markedly in their specific details and in their domains of applicability. The 'quasi-taxonomy' in the nLab wiki lists many dimension concepts and theories. Also, there are two Wikipedia entries on dimension with ample references:

- `https://ncatlab.org/nlab/show/dimension`
- `https://en.wikipedia.org/wiki/Dimension`
- `https://en.wikipedia.org/wiki/Dimension_theory_(algebra)`

Concepts of dimension play an especially important central role in modern algebraic geometry and commutative algebra. Algebraic geometry at its base is geometry with dimension naturally a part. In this domain the dimension concept is applicable in algebraic terms to algebraic varieties, rings, and other algebraic and geometrical objects. The most prominent concept in this domain is *Krull dimension*.

Algebraic geometry and commutative algebra have experienced enormous growth in recent times. Hence, covering all the related dimension concepts and their developments as theories is infeasible and considerably beyond the scope of this book. Hence, in this appendix I shall just give a few pointers to the history of dimension in this domain.

The historical developments, problem shifts, and significant applications of concepts of dimension in modern algebraic geometry and commutative algebra, including Krull dimension, appear to be largely independent of the historical development of the topological dimension theories of Poincaré, Brouwer, Urysohn, and Menger. Yet, on the fundamental conceptual level there is a common geometrical and topological core underlying the two, at least by analogy or correspondence. The analogy is evidenced in the geometry of basic algebraic surfaces and varieties, the Krull dimension of topological spaces, and the Zariski topology applied to topological spaces for algebraic varieties and the spectra of rings (cf. [Kunz, 1985, page 39]).

For this domain I shall primarily reference the excellent book by David Eisenbud, *Commutative Algebra with a View Toward Algebraic Geometry* [Eisenbud, 1995]. The book has an entire Part II on Dimension Theory [Eisenbud, 1995, pp. 213–419], including an introductory chapter on the broad historical and conceptual development [Eisenbud, 1995, Chapter 8, pp. 215–226]. It contains many useful historical insights. In addition, Morris Kline's history of mathematics covers algebraic geometry [Kline, 1972, Chapter 39, pp. 924–946], and Jean Dieudonné's important historical works treat algebraic geometry extensively [Dieudonné, 1972] [Dieudonné, 1974] [Dieudonné, 1985]. Jeremy Gray's historical paper is also useful [Gray, 1997]. One can obtain further insights into the developments of algebraic and differential geometry from the nineteenth and early twentieth centuries by examining the extensive coverage given to these developments in the *Encyklopädie der mathematischen Wissenschaften mit Einschluss ihrer Anwendungen* (*Encyclopedia of the Mathematical Sciences, with the Inclusion of their Applications*), with 14 articles covering aspects of algebraic geometry, and 11 articles covering aspects of differential geometry, all written in the first three and a half decades of the twentieth century, [Akademie der Wissenschaften zu Göttingen, Leipzig, München und Wien, 1934] [Akademie der Wissenschaften zu Göttingen, Leipzig, München und Wien, 1927].

Additionally, the following is a list of some modern texts that present a variety of treatments of dimension in a useful way for this domain:

- Hartshorne, *Algebraic Geometry* [Hartshorne, 1977]; this standard text has ample information concerning dimension throughout;
- Shafarevich, *Basic Algebraic Geometry 1, 2* [Shafarevich, 2013a] [Shafarevich, 2013b]; this is a classic text in its third edition; it covers dimension throughout, especially [Shafarevich, 2013a, pp. 66–81] and contains a useful historical sketch [Shafarevich, 2013b, pp. 229–246];
- Cox, Little, and O'Shea, *Ideals, Varieties, and Algorithms: An Introduction to Computational Algebraic Geometry and Commutative Algebra* [Cox et al., 2013]; Chapter 9 covers the dimension of various types of variety in an elementary way [Cox et al., 2013, pp. 469–538];
- Kunz, *Introduction to Commutative Algebra and Algebraic Geometry* [Kunz, 1985]; Chapter II covers dimension concisely with a treatment of the topology [Kunz, 1985, pp. 39–62];
- Matsumura, *Commutative Ring Theory* [Matsumura, 1989]; Section 2.5 introduces dimension theory and Chapter 5 covers dimension theory [Matsumura, 1989, pp. 30, 83–104];
- Atiyah and Macdonald, *Introduction to Commutative Algebra* [Atiyah & Macdonald, 1969]; Chapter 11 covers dimension theory, including an interesting equivalence theorem [Atiyah & Macdonald, 1969, pp. 116–126, 121].

Eisenbud notes that for many centuries algebraic curves meant curves defined by algebraic equations with mainly real (and not complex) values. The work of Riemann in the mid nineteenth century changed that perspective [Eisenbud, 1995, page 217]:

> As far as I am aware, the early workers were not concerned with the collection of all complex points of a plane curve as a single geometric object, a surface. But that concern took the spotlight in the work of Riemann, as understood by [Alfred] Clebsch [1864] [Clebsch, 1864] and later authors, where algebraic curves, interpreted as "Riemann surfaces," arose as coverings of $C = R^2$ or better of the complex projective line, or "Riemann sphere" $C \cup \{\infty\}$. Now C could reasonably be called either the *complex plane* (from a topological point of view) or the *complex line* (from a complex-analytic point of view). Thus algebraic curves had reasonable claims to being called either one- or two-dimensional. Even the names, Riemann **surfaces** and algebraic **curves**, suggest a certain schizophrenia. We have grown used to the confusion, and speak happily in general of an n-dimensional complex manifold, which, as a topological space, has dimension $2n$.

In the later nineteenth century the study of algebraic functions involved the efforts of many mathematicians. An indication of these efforts is the lengthy report of Alexander von Brill and Max Noether on algebraic functions written for the Deutsche Mathematiker-Vereinigung (German Association of Mathematicians) and published in 1894 [Brill & Noether, 1894]. It covers an extensive history in 560 pages.

Eisenbud points out that mathematicians in the late nineteenth and early twentieth centuries used various definitions of dimension in this domain of mathematics [Eisenbud, 1995, pp. 217–219]. The definitions are scattered in the mathematical literature of the time: some for varieties, some for rings, and so forth; several being rather ad hoc. Their usage depended on mathematical context and the problems being investigated. In particular, he states [Eisenbud, 1995, pp. 217–218]:

> From these beginnings, and from the axiom, like that of the topologists, that affine d-space has dimension d, came the algebraic definition of dimension that was used early in this [twentieth] century: The dimension of an irreducible variety in affine r-space over a field k (initially $\mathbf{C}$) is the transcendence degree over k of the field of rational functions on X. This is by definition the quotient field of the domain $R = k[x_1, \ldots, x_r]/I$, where I is the ideal of all functions vanishing on X. It was only natural to define the dimension of R to be this same transcendence degree. This definition was still the accepted standard as late as 1935, as one may see from Krull's famous book *Idealtheorie*, published in that year.

In the last sentence Eisenbud points to the well-known modern algebraist Wolfgang Krull (1899–1971)[61] as one who used this definition in his influential monograph *Idealtheorie* (*Ideal Theory*) of 1935 [Krull, 1935, page 38].

[61] Krull was an important German algebraist. He studied at the University of Freiburg and eventually presented his doctoral thesis there in 1922. However, he also studied at Göttingen from 1920 to 1921 with Felix Klein and with Emmy Noether. Emmy Noether particularly influenced his work in algebra with the modern algebraic methods she promoted. He became an academic holding posts at Freiburg, Erlangen, and Bonn, and he published extensively, primarily in algebra. The following lists articles and sources of information about him:

Shortly after the publication of his monograph, Krull, in a paper of 1937 'Beiträge zur Arithmetik kommutativer Integritätsbereiche. III. Zum Dimensionsbegriff der Idealtheorie' ('Contributions to the arithmetic of commutative integral domains. III. On the dimension concept of ideal theory'), proposed a new definition of dimension for ideal theory (rings and ideals). The 'Krull dimension' is now well known and widely accepted [Krull, 1937a, pp. 745–746, 754].[62] Eisenbud states it in a modern setting as follows [Eisenbud, 1995, page 219]:

> **Definition.** *The* ***Krull dimension*** *(or simply the* ***dimension****) of a ring R, written* ***dim R****, is the supremum of the lengths of chains of distinct prime ideals in R.*

In the definition the length of the chain $P_r \supset P_{r-1} \supset \cdots \supset P_0$ involving $r+1$ distinct prime ideals is taken to be r.

This definition has become the standard one for rings and ideals and is used more generally in algebraic geometry and commutative algebra, though there are some alternatives. The background for the definition, as identified in Krull's references, comes in papers of Emmy Noether [Noether, 1923], Walter Rückert [Rückert, 1933], and F.S. Macaulay [Macaulay, 1934].

Mathematicians have analogously defined Krull dimension or (just) dimension more generally for topological spaces in the following way (to accord with algebraic geometry; cf. [Hartshorne, 1977, pp. 5–7], [Kunz, 1985, pp. 39–45]). A topological space X is called *irreducible* if it cannot be expressed as a union of two proper closed subsets. A subset Y of a topological space X is an *irreducible subset* if Y is an irreducible topological space with the subspace topology. Then Kunz states the following definition [Kunz, 1985, page 39]:

> Let X be a topological space . . . .
> **Definition . . . .** If $X \neq \emptyset$, the *Krull dimension* $\dim X$ (or combinatorial dimension) of X is the supremum of the lengths n of all chains
>
> $$X_0 \subset X_1 \subset \cdots \subset X_n \quad (X_{i+1} \neq X_i)$$
>
> of nonempty closed irreducible subsets X_i of X.

The empty topological space is assigned Krull dimension -1.

1. Retrospective on Krull in *Jahresbericht der Deutschen Mathematiker-Vereinigung* (*Annual Report of the German Association of Mathematicians*), volume 82(2) (1980), pp. 51–80; also in [Krull, 1999a, pp. 1–52]
2. Wolfgang Krull: *Gesammelte Abhandlungen/Collected Papers* [Krull, 1999a] [Krull, 1999b]
3. `https://mathshistory.st-andrews.ac.uk/Biographies/Krull/`
4. `https://en.wikipedia.org/wiki/Wolfgang_Krull`
5. [Bocklandt, 2021].

[62] This paper is one of an extended series of eight papers with an additional correction: [Krull, 1936a] [Krull, 1936b] [Krull, 1937a] [Krull, 1938a] [Krull, 1937b] [Krull, 1938b] [Krull, 1939a] [Krull, 1939b] [Krull, 1942]. Cf. also [Krull, 1938c] and the later paper [Krull, 1951].

The Krull dimension of an algebraic variety may be defined as its Krull dimension when the variety is regarded as a topological space. More explicitly, the Krull dimension for a variety may be given as follows [Artin, 2022, page 138]. First, a *chain* of closed subvarieties of a variety X is a strictly decreasing sequence

$$C_0 > C_1 > C_2 > \cdots > C_k$$

of closed subvarieties (of irreducible closed sets). The length of the chain is defined as k. The chain is *maximal* if it cannot be lengthened by inserting another closed subvariety and C_k is a point. It can be shown that all maximal chains have the same length. Then the Krull dimension of a variety is the length of maximal chains of subvarieties. (See further details in Artin's book [Artin, 2022, pp. 138–143].)

Eisenbud examines the definition of Krull dimension of rings further by providing motivating examples and mathematical arguments to relate the concept to other parts of mathematics [Eisenbud, 1995, pp. 219–226].[63] The following are some key points that Eisenbud presents in his book.

Krull dimension relates in a very clear way to the domain of vector spaces. One can think of the dimension of a vector space over a field k as the length of the longest chain of proper subspaces. In algebraic terms, an n-dimensional vector space corresponds to a polynomial ring $R = K[x_1, \ldots, x_r]$. Then an increasing sequence of subspaces, starting with 0, corresponds by analogy to the decreasing sequence of prime ideals

$$(x_1, \ldots, x_r) \supset (x_1, \ldots, x_{r-1}) \supset \cdots \supset (x_1) \supset 0.$$

Any field k has dimension 0, and the polynomial ring $k[x]$ has dimension 1.

Krull dimension is quite different from dimension according to Brouwer, Urysohn, and Menger for separable metrizable or metric spaces or more general spaces. Historically, they have had separate developments. However, Eisenbud also draws out a connection, a useful analogy between Krull's dimension and Menger's (and implicitly Urysohn's) topological dimension by giving the following argument [Eisenbud, 1995, page 219]:

> It is interesting to compare Krull's definition with Menger's . . . , although this is unhistorical in the sense that Krull does not mention the topological case or its pioneers. Krull's definition exploits the simple structure of algebraic varieties and replaces the closed sets that are the boundaries of small neighborhoods of p that occur in Menger's definition by maximal closed sets not containing any component *and* containing p. If we regard the set of primes of a ring as a topological space in the Zariski topology and apply Menger's definition, we get the same thing in many, but not all, cases; Krull's definition is in a certain sense more local.

To support the view that Krull's definition of dimension has merit in the domain of algebraic geometry and commutative algebra, Eisenbud provides a set of axioms characterizing the concept uniquely (as he claims) [Eisenbud, 1995, pp. 220–222].

---

[63] The rest of Part II of the book after the chapter being considered is a more formal treatment of dimension [Eisenbud, 1995, Chapters 9–16, pp. 227–419].

He also gives some additional alternative mathematical characterizations of dimension that are not Krull's but are useful in other contexts [Eisenbud, 1995, pp. 222–226]. He is not dogmatic about which ones are to be favored.

To conclude this appendix, I again draw attention to the nLab 'quasi-taxonomy' or list, `https://ncatlab.org/nlab/show/dimension`. It shows the richness of the domain of loosely related concepts of dimension. In some sense, they are based on our intuitive notion of dimension. The whole area of algebraic geometry and commutative algebra continues to be vigorously developed by many mathematicians. Dimension is an important part of that development.

Chapter 17
Historical, Mathematical, and Philosophical Perspectives

17.1 Introduction

My approach to the history of concepts and theories of dimension and objects that have assigned dimension numbers or exhibit forms of dimension, such as points, lines, surfaces, and solids, has been through the problems that have arisen concerning them and the solutions and theories that have been proposed to solve the problems. Over the years problems led to solutions and theories that drove the entire development. Problems and their solutions are the lifeblood of most human disciplines whether in the scientific or mathematical disciplines or in philosophy. Mathematicians and philosophers and to a significant extent physicists have been primary in treating problems of dimension and related geometrical objects.[1]

In treating problems I have sought to bring to the fore historical contexts and problem situations that surround and support these problems in various historical periods, and I have endeavored to consider them carefully. Historical contexts and problems situations have differed over the long period I have covered from ancient Greek times to various points in the modern era.

Additionally, the solutions, definitions, or theories often provide explanations of various phenomena and abstract situations concerning dimension. One prominent example of this is the explanation using local connectedness to characterize the phenomena associated with space-filling curves and Jordan curves.

With respect to the approach that I have taken, I have not been much concerned with the intuitive or mental ideas we may have of space and dimension or their origin as in our minds. That is a kind of psychological or epistemological problem, which I have not considered, at least not directly considered. Moreover, I have not considered an ontological view of the objects treated under the dimension concept. My approach has been methodological with respect to philosophy and mathematics rather than ontological, epistemological, or psychological.

[1] There is a related conceptual presentation of selected parts of the history that emphasizes the fractal side of the story. Cf. Stephen Lipscomb's works [Lipscomb, 2009] [Lipscomb, 2008].

D. M. Johnson, *Theories of Dimension from Ancient to Modern Times*, Sources and Studies in the History of Mathematics and Physical Sciences, https://doi.org/10.1007/978-3-032-02682-8_17

17.2 Problems of Dimension

In my introduction in Section 1.1 I identified the three main problems of dimension: (1) the problem of defining the concept of dimension itself and some closely related concepts, such as the concepts of curve and surface; (2) the problem of explaining the number of dimensions of physical space (perhaps including time); and (3) the problem of proving the invariance of the dimension numbers of mathematical spaces under some restricted type of mapping, e.g., homeomorphism. There are, of course, other related or subsidiary problems connected with this history, but I think that these are the main ones. How philosophers and mathematicians have treated these problems over many centuries has varied considerably. The philosophers and mathematicians who have worked on them have shaped them according to the way they understood them, the way they placed them in context, and the way they addressed and analyzed them.

I have mainly treated problems (1) and (3). I have only touched on problem (2) by mentioning that the ancient Greek mathematician and astronomer Ptolemy wrote something on it, now lost, and then by considering Bolzano's and Poicaré's treatments in Section 4.8 and in Chapter 9, respectively.

Bolzano's treatment of the three dimensionality of our space is interesting, but it is separate and distinct from his more general treatment of dimension with respect to the main geometrical objects, lines (curves), surfaces, and solids. I have concentrated on the latter, which is an instance of treating the definition problem (1).

Poincaré's treatment is perhaps more interesting, inasmuch as it relates to some of his other mathematics and his conventionalist philosophy. More importantly in terms of influence, it lies at the beginning of Brouwer's treatments of dimension; indeed it seems to have motivated Brouwer in part. Poincaré's treatment is more or less mathematical, but quite informal and philosophical. It needed a more rigorous formulation to be the foundation of a full mathematical theory of dimension. Brouwer, who admired Poincaré, gave it a stronger mathematical formulation. Essentially, Brouwer moved the problem situation from problem (2) to the definitional problem (1), and at the same time investigated and solved the invariance problem (3) in general for the first time.

Problem (2) on the nature of the dimensions of space (and time) is in an important sense related to modern physics, for example, as investigated in relativity theory or in string theory. However, I have determined that dimension as treated in modern physics as beyond the scope of this book.

A great portion of the work on dimension and the development of theories of dimension relates to the definition problem (1). It is only toward the end of the nineteenth century with the appearance of the counterintuitive results of Cantor, yielding a one-one mapping between the unit line segment and the square, and of Peano, giving a construction of space-filling curves that continuously map the line segment to fill the square, does the invariance problem (3) arise in full force. Brouwer solved that problem definitively in his work of 1910 and later and his publications of 1911 [Brouwer, 1911a] and 1913 [Brouwer, 1913b]. One has to add the 'interjections' of Lebesgue of 1911, including the intuitive tiling principle and other ideas, [Lebesgue, 1911a] [Lebesgue, 1911b] [Lebesgue, 1911c] and Brouwer's corrections of 1923 and 1924. See the full details in Chapters 11 and 12.

The whole matter of the invariance problem (3) was greatly simplified through the work of Sperner [Sperner, 1928] and Hurewicz [Hurewicz, 1929]. Sperner's lemma is an insightful simplification. For the details on Sperner's and Hurewicz' work see Sections 13.11 and 16.4.

17.3 Philosophical Reflections

Having covered the detailed history of dimension, I would like to consider again philosophical approaches to assessing the development. I shall consider the approaches of Carnap, Popper, and Lakatos in relation to the history, and reflect on the mathematical practices involved.

17.4 Rudolf Carnap: Words, Concepts, Explications

In Section 1.3 of the opening chapter of this book I summarized Rudolf Carnap's notion of *explication*. He introduced the idea in the 1940s and 1950s. The earlier work of Karl Menger strongly influenced his development of the idea of explication, but more on that matter below. Carnap succinctly states in a key text what he means by explication [Carnap, 1962, page 3]:

> By the procedure of *explication* we mean the transformation of an inexact, prescientific concept, the *explicandum*, into a new exact concept, the *explicatum*. Although the explicandum cannot be given in exact terms, it should be made as clear as possible by informal explanation and examples.
>
> The task of ***explication*** consists in transforming a given more or less inexact concept into an exact one or, rather, in replacing the first by the second.[2]

As noted in the earlier Section 1.3 Carnap puts forward criteria for appraising explications to determine if they are basically satisfactory, are fruitful, and have consequential value. These are the following [Carnap, 1962, page 7]:

1. The explicatum is to be *similar to the explicandum* . . . .
2. The characterization of the explicatum, that is, the rules of its use (for instance, in the form of a definition), is to be given in an *exact* form . . . .
3. The explicatum is to be a *fruitful* concept . . . .
4. The explicatum should be as *simple* as possible . . . .

[2] See Carnap's original statements on explication in [Carnap, 1956, pp. 7–8] [Carnap, 1962, Chapter 1] and some commentaries in [Gupta & Mackereth, 2023, pp. 7–9], [Leitgeb & Carus, 2023, Supplement D. Methodology], and [Cordes & Siegwart, 2024]. The last article gives an extensive and useful history and pre-history of explication.

The methodological ideas of Karl Menger as related to dimension theory directly influenced Carnap's development of the notion of explication. In the following few paragraphs I shall give the details about this connection. In terms of intellectual kinship both Menger and Carnap were members of the Vienna Circle during the 1920s and 1930s.

Menger's methodological ideas that form a basis for Carnap's notion can be traced back to his book *Dimensionstheorie* (*Dimension Theory*) [Menger, 1928b, II. Der Dimensionsbegriff, pp. 74–91] and to his article 'What is dimension?' [Menger, 1943] (the first is partly translated into English, and the second is reproduced, including a further explanation, in Menger's 1979 book [Menger, 1979a, pp. 207–216]).

In the 1928 book, having delineated the domain of set-theoretic geometry, i.e., point-set theory and point-set topology, in the first chapter, he introduces his general definition of dimension in the second as a basis for dimension theory. He opens the chapter with some general remarks about the dimension concept and related geometrical concepts as commonly used in ordinary life. The exposition is intuitive and revealing. He opens with his view of how informal terms are understood in ordinary life [Menger, 1928b, pp. 74–75]:

> The most important property of a spatial object is its dimension. In everyday life, curves, surfaces, and solids are considered to be the fundamentally different main classes of spatial objects, and the difference in dimensions is also the fundamental classification principle of all scientific geometry. . . . The dimension is something completely intuitive; it is the *core* of geometry.
>
> The terms 'curve', 'surface', 'solid' are taken from everyday life and are among the words that everyone uses, as one says, and that are associated with certain ideas. The fact that a certain individual person associates an idea with a certain word generally consists of the following: the person in question will, without thinking, describe certain entities with the word in question; of certain other things he will say, without thinking, that they can *not* be designated by the word in question. In the course of his life, especially in his childhood, the individual learns the use of words as names for certain things by being told: 'This is denoted by the word in question, [and] that is not.' Between these two classes there is a third class of things about which the individual is unsure whether he should describe them with the word in question or not. If we move on to considering the totality of people who use a word, we also see: certain things are referred to by all people with the word in question, while certain other things are not referred to by everyone with the word in question, and, moreover, there is another class of entities that we will call the *floating class*.
>
> For example, consider the word 'curve'. An arc, a circular line, a lemniscate [, each] is called a curve by every person who uses the word curve. A square surface, a spherical surface, a cubical solid [, each] is not called a curve by anyone. Because everyone who learns the words curve, surface, solid, learns that the word curve is used to describe the wire- and thread-like, line- and string-like forms, and the word surface is used to describe the plate-like

> forms, leaf- and skin-like forms, the word solid is used to name the block-like forms.[Ch17, 1]

He then goes on to consider difficult cases for which it may not be easy or not possible to ascribe a dimension. He seems to be referring to the exotic counterintuitive geometrical objects, those which mathematicians using point-set theory brought to the fore and in most cases specifically analyzed in some detail. This situation gives an opening to mathematicians to create a more formal inclusive dimension theory [Menger, 1928b, page 75]:

> However, a person who utters the word curve generally has nothing more than the ability to indicate things which are generally called curves and those which are not generally called curves, and the same is true of the words surface and solid. In everyday life, there is no classification of *all* spatial objects according to their dimensions: If you were to ask about the dimensions of certain complex objects that the modern geometer examines, you would get no answer or different answers from different people. Likewise, most people are unable to state precisely what they understand by a curve, a surface, a solid, i.e., to explain these words with the help of other words. The words curve, surface, solid fulfill a certain practical function for everyday life, of course, despite these circumstances, because the complicated objects of the floating [indeterminate] class with regard to the word dimension are rare in everyday life, and a completely precise delimitation of curves in general is not necessary for mutual understanding of people about certain special cases.
>
> Until the founding of dimension theory in the scientific domain, things were not much different there than in everyday life. In geometry, too, certain spatial objects were referred to as curves, others as surfaces, others as solids. But although the fundamental importance of this classification was clearly recognized — especially by outstanding mathematicians of all times — not all spatial objects were included in it. For the vast majority of spatial objects, a dimension could not be defined. There was no all-encompassing concept of dimension as a common property in mathematics.[Ch17, 2]

Menger seems almost boastful about the effectiveness of dimension theory in solving the ‘demarcation problem’ or ‘precision problem’ for classifying complicated geometrical objects.

Finally in this methodological part of the book, he considers requirements or criteria for defining more precise concepts that improve on informal ones and that serve in developing theories [Menger, 1928b, pp. 75–76]:

> If a word with which one associates an idea in everyday life is to be made precise in science by a definition, there is no reason to contradict the everyday use of the word, i.e., to exclude from the concept things that are generally referred to by the word in question, or to include in the concept things that are generally excluded from the designation with the word in question. A *formal requirement* for the strict definition of a word taken from everyday language is *that it represents the making precise and a completion of the use of the word,*

which, as we have seen, fluctuates and is incomplete in borderline cases, *with which it is not in contradiction*.

Such an addition and clarification can now take place in different ways, not only formally, but also in different ways (taking into account the floating classes of entities) in terms of content. For example, certain complicated objects, for which the individual has no knowledge when asked whether they are one-dimensional or different people give different answers, [and so] can both be counted among the one-dimensional objects and excluded from them, therefore, without the definition being in contradiction with common usage. Every commitment to a certain precision, therefore, contains a certain degree of *arbitrariness*, the justification of which can only be provided by the *fruitfulness* of the definition. The purpose of the word in everyday life is to help people understand each other; the purpose of a strict definition is to form the starting point of a deductive system. Definitions are dogmas, only the deductions from them are knowledge. It is therefore a *substantive requirement* of a definition in general *that it proves to be a source of knowledge by forming the starting point of a comprehensive, aesthetically perfect theory*.

The fulfillment of this *contentual*[3] requirement represents the only possible *validation of any definition*. If it is specifically a question of the definition of a term that is referred to with a name taken from everyday life, then the fulfillment of the formal requirement (which is secondary to the term itself), provides a *justification for the naming of the term*.

What has been said applies particularly to the concept of dimension. Already a consideration of special objects leaves a number of interesting and beautiful relationships to be considered with respect to a precise dimension concept. What is the dimension of the sum of two n-dimensional objects? What is the structure of n-dimensional objects, simultaneously with the distribution of points of different dimensional character? Do not the n-dimensional entities have certain properties under decomposition? How does the concept of dimension relate to the concept of connectedness among spatial objects? How does the dimension of an object behave under certain transformations? How do subsets of ordinary Cartesian [Euclidean] space behave under transformations?[Ch17, 3]

Clearly, for Menger the fruitfulness of a definition that is used as the starting point of a theory is a critically important criterion for the definition to be successful. So too with Carnap. Given that by 1928 dimension theory had developed to a significant degree, Menger is able to refer to key mathematical questions and to known answers, results, and theorems that are exposited in his book. Menger was part of this development along with several of his contemporaries as we have seen in Chapters 11, 12, 14, 15, and 16.

Appropriate to this history I would like to relate Carnap's (and effectively Menger's) notion of explication to selected examples from the history of dimension and curve theory to determine how well the examples fit with or conform to

---

[3] or *content-related*.

explication. First, I shall look again at the geometrical work of Bolzano. Then, I shall review Brouwer's work on dimension, Urysohn's work on dimension and curve, and Menger's work on dimension and curve. Finally, I would like to see how Hausdorff's measure-theoretic work on dimension may fit with the notion of explication.

In Chapter 4 I offered a detailed analysis of Bolzano's investigations into the geometrical concepts of line, surface, solid, and extension. His investigations took several turns in direction over his intellectual lifetime. With the benefit of hindsight his geometrical work looks very like an early emerging dimension and curve theory. Bolzano's quest for accurate and sound definitions of basic geometrical objects has the marks of being a search for explications of these concepts. He appears to be almost a pure 'explicationist,' to coin a word that seems appropriate. He surely imposed his own standards of 'adequacy,' 'rigor,' and 'success' on what he found. He was quite self critical in his analysis. Yet, we have only the record of his writings, many of which were unpublished in his lifetime. One might ask, how 'fruitful' were the results of his investigations? That question à la Carnap seems a little difficult to answer, since his output seems largely preliminary. We have no real legacy of Bolzano dimension theory or curve or surface theory. In other parts of mathematics, for example, as he presents in his *Rein analytischer Beweis* (*Purely Analytic Proof*) [Bolzano, 1817b], giving a more rigorous proof of the intermediate value theorem of analysis, he had more success with theory development. Nonetheless, his quest for defining the basic objects of geometry and implicitly dimension looks very much like explication, following the principles of his own philosophy. I shall return to Bolzano below with regard to Lakatos.

I consider now the main founders of dimension theory, Brouwer, Urysohn, and Menger. Brouwer was the first to prove the invariance of dimension in full generality [Brouwer, 1911a], and in his later paper of 1913 [Brouwer, 1913b] he proposed his important definition of dimension. His definition was based on a critical analysis of Poincaré's informal definition [Poincaré, 1903b] [Poincaré, 1912]. The main result of the paper and its proof partly arose from his dispute with Lebesgue (Chapter 12). Brouwer took the definition problem very seriously, but he also embedded it in the best point-set theory and topology available to him at the time, meticulously supporting it with carefully defined spaces and well-delineated definitions. As is known, he made an error in the initial publication of 1913, which I have argued he corrected soon after publication and later in more detail (Section 12.12). Thus, his definition was fully supported by theory and not a simple matter of definition. He did not spend much time drawing out consequences. He was satisfied with proving the invariance theorem afresh, with a proof based on Lebesgue's tiling principle. In summary, Brouwer was partly interested in the definition or explication problem, but more consumed by the invariance problem for dimension.

Urysohn and Menger both started with the definition problem for defining accurately and rigorously the concept of curve and by extension dimension. Their teachers and mentors, Egorov and Hahn, respectively, gave them the definition problem to research. Hence, they clearly started with the definition problem, an explication problem in Carnap's terms. However, each drew out many consequences of his respective definitions. Given all the ups and downs, they developed fruitful theories. Theory development became their goal. In many ways they conformed to what Menger described for definitions, but theory development was their bigger goal and achievement. So, the definition problem, the problem of explication, developed into larger and more interesting problems of theory development.

Hausdorff started from a rather different basis. He looked for a measure-theoretic way to define dimension for *exotic* discontinuous and also continuous or connected point sets. His examples to be analyzed are not the usual ones that an ordinary person might know or understand. They are sets like the Cantor discontinuum. Thus, he was not following the usual path of definition or explication that Menger described in the passage quoted above. Instead, he struck out on a different path to define and explicate dimension to apply to these exotic sets, which pop up in point-set theory. Hausdorff was the first in an important line of mathematicians investigating exotic sets using Hausdorff dimension and subsequent fractal dimension as a key to measure and understand them. The broad problem is a kind of explication and then theory development to explain complex cases.

17.5 Karl Popper: Problems and Theories, Conjectures and Refutations

The discussion of the previous section leads directly to considering Popper's views on this type of scientific development. He did not think that words and concepts were overly important.[4] In contrast, he did think that theories were vitally important. Theories develop out of problems and problem situations. They are conjectured answers to problems. They may be refuted, or, more likely, may need improvement and modification. In some cases they may be uninteresting in a larger historical context. However, in important cases they may provide explanations of one sort or another to problem situations or clusters of problems.

The examples of the previous section highlight some cases of definition and apparent explication in Carnap's sense. However, in the larger history of dimension theories and theories of geometrical objects, Popper's view appears to be a better fit for characterizing the developments overall or 'in the large,' so to speak. The full sequence of development in topological dimension theory, for example, from Brouwer and then Urysohn and Menger exhibits more of a Popperian pattern of development. The ups and downs and course corrections are significant parts of the scientific story line.

The development of Hausdorff dimension represents a different problem and theory development. It does not follow Menger's philosophical description of the development of dimension. It follows a different stream of overall development. The Hausdorff theory and subsequent fractal theory solve different problems. It represents a valid alternative development that needs to be treated in its own problem situation.

[4] He had no affinity with, really no liking for, the Oxford linguistic or analytic philosophers. He thought that there were real philosophical problems to be considered related to the sciences, including the social sciences, and these were not merely linguistic problems.

As an historian of mathematics I do not want to give too much of a rational reconstruction of the history, but the historical evidence points to a Popperian type of development of dimension and curve. The type can be seen in so many cases, especially in the current modern era, where there is so much available evidence.

Modern theories of dimension and curve have become parts of larger theories. The proof structure makes sense, and derives from the development. Urysohn and Menger dimension theory resides best in separable metric spaces. Urysohn, along with his colleague Alexandroff, specially developed related parts of topology. Much of this material is compactly summarized in Hurewicz and Wallman's book *Dimension Theory* [Hurewicz & Wallman, 1948], though without giving the history, except in a few notes.

One might also argue that some developments of dimension theories look like a Lakatosian research program [Lakatos, 1978a]. However, I shall not develop that thought further.

17.6 Imre Lakatos: Proofs and Refutations, and the Growth of Mathematical Knowledge

Certain parts of the development of curve and dimension theories appear to fit to an extent the Lakatos pattern of 'proofs and refutation' or the critical conjectural development of mathematics. I shall not give a detailed analysis according to this pattern but only point to two cases in outline.

First, some of Bolzano's work on precisely defining the basic objects of geometry, the 'line' (or 'curve'), the 'surface,' and the 'solid,' which he continued to consider over many years, appears to fit the Lakatos form of tentative conjectural development. The variety of definitions in his writings gives evidence of this way of proceeding: setting out a definition, finding a counterexample that shows that the definition is deficient, and then setting out a new alternative definition to deal with the deficiency or deficiencies, and so forth. I already pointed to this mode of thinking in Sections 4.5 and 4.9.

Second, the various strands of development of the notions of curve and of disconnected and/or zero-dimensional sets occurring around the beginning of the twentieth century appear to reveal a pattern fitting the Lakatos viewpoint. I covered these developments in Chapter 13. The early twentieth century was a time when point-set theory and point-set topology were coming into full flower, and mathematicians could offer very careful analyses of basic concepts. The theory provided the tools. See, for example, Figure 13.10 (assembled by Engelking), which offers a useful encapsulation of the analyses of all kinds of disconnected sets and the logical relationships among them. Overall, the back and forth of proposed definitions, the corresponding counterexamples and counterintuitive examples, together with related proven results and eventually theories provide clear evidence of this form of conjectural development.

17.7 Mathematical Practices

The modern development of dimension and curve theory beginning with Cantor and Dedekind in the 1870s is closely intertwined with the vigorous and extensive development of set theory, point-set theory, and point-set topology, and a little later with the development of combinatorial and then algebraic topology and measure theory. These broader mathematical theories with their particular methods and practices soon became part of the toolkit available to mathematicians. Thus, researchers who were especially interested in dimension and curve, such as Brouwer, Urysohn, and Menger, and several others, could adopt and adapt the methods and practices of these theories and also extend them in building detailed dimension and curve theories.

There are many historical cases that exemplify particular mathematical practices in action. I shall consider just three cases at this point, the first directly related to curve theory, the second related more generally to topology and various forms of topological thinking, and the third related to Hausdorff dimension.

In Section 13.8 I discussed the problem of characterizing Peano space-filling curves or Jordan curves. The key result was the Hahn–Mazurkiewicz–Sierpiński theorem. This provided an *explanation* of the phenomenon of space-filling curves. Mazurkiewicz and Hahn came to their results independently. *The crucial underlying explanatory concept was the concept of local connectedness.* Then slightly later Sierpiński provided a complementary result to round out the complete theorem and explanation. See the combined result at the end of Section 13.8.

Metrization has been an important research topic in point-set topology and is broadly related to curve and dimension theory: to determine the conditions or assumptions for a topological space to be metrizable, i.e., for it to be homeomorphic to a metric space. One of the first recognized metrization results was Urysohn's metrization theorem, published posthumously in 1925 [Urysohn, 1925d]. Andrei Tychonoff immediately generalized the Urysohn result in a paper of 1926 [Tychonoff, 1926]. The Tychonoff formulation is the one usually presented in modern texts. Munkres formulates the result in his text as follows [Munkres, 2000, page 215]:

> **Theorem . . . (Urysohn metrization theorem).** *Every regular space X with a countable base is metrizable.*

Munkres provides all the details, including the important supporting Urysohn lemma [Munkres, 2000, pp. 207–218]. What is significant to note is the typical mathematical practice of *searching for characterizing conditions*, in this case for metrization. The conditions form a type of explanation of the result. The later results of Nagata and Smirnov are in that same vein and involve paracompactness [Munkres, 2000, pp. 248–262].

There are many other examples of mathematicians searching for characterizing conditions in point-set topology, in dimension and curve theory, and in most other parts of mathematics. Hurewicz' analysis of his carefully defined normal families (Normalbereiche) to characterize aspects of dimension and 0-dimensional and n-dimensional sets is a good example (discussed in Section 16.4). Another example is Alexandroff's extensive search to characterize the relationship between the general concept of dimension and the dimension of closed sets of arbitrary dimension to basic geometrical objects, simplicial complexes and their corresponding point-sets defining polyhedra when considered as topological spaces, and 'geometrical intuition.' Section 16.8 covers Alexandroff's research direction and results.

In Section 16.11 I discussed Hausdorff dimension. Hausdorff was able to take the insightful and quite 'clever' measure theory of Carathéodory to create a dimension theory with sets of fractional dimension. In addition, he used some standard analysis of functions to capture dimension. In this case the mathematical practices were perhaps unusual, but quite fruitful. Hausdorff's pioneering work led to further developments, including fractal theory.

Earlier chapters of this book are replete with examples of the useful adoption and adaptation of mathematical practices applied to specific problems of dimension and curve. The interested reader might like to review the history in that light.

17.8 Embedding Theorems, Universal Spaces, and Representing Structures

In Section 15.7 I discussed Menger's universal curve, and in Sections 16.7 and 16.9 I considered in some detail the embedding theorem for dimension theory. Further, in Section 16.8 I analyzed Alexandroff's research program to forge a link between sets of Brouwer–Urysohn–Menger dimension and standard geometrical complexes, thereby showing a relation to geometrical intuition according to his view. This work led him eventually to offer a definition and theory of homological dimension, which he could relate to Brouwer–Urysohn–Menger dimension. The collection of all these results is significant. The collection typifies in diverse ways the mathematician's quest for an encompassing or representing model for a class of objects. The Menger cube or sponge provides a model for the collection of all possible bounded 1-dimensional continuous curves in 3-space. The embedding theorem offers a general model for n-dimensional spaces (n finite) in $(2n + 1)$-dimensional Euclidean space, the $(2n + 1)$-dimensional cube. The general cases of either bounded 1-dimensional curves or n-dimensional spaces (n finite) become represented concretely in well-known and well-understood spaces. These embedding theorems provide concrete representations and standard models. In a similar way Alexandroff's results show that general λ-dimensional (in this case λ can be either finite or infinite) closed sets or spaces can be transformed by small deformations into complexes of dimension λ but not less (cf. e.g., [Alexandroff, 1928d]). Thus, simplicial complexes become a standard model for λ-dimensional closed sets and spaces.

There are other similar mathematical results, for example, Nash's embedding theorem and Whitney's embedding theorem. The Nash embedding theorem states that every Riemannian manifold can be isometrically embedded into some Euclidean space. The strong Whitney embedding theorem asserts that any smooth real m-dimensional manifold (required to be Hausdorff and second-countable) can be smoothly embedded in the real Euclidean $2m$-space, $\mathbb{R}^{2m}$, if $m > 0$. I shall not go further into the details in these cases (cf. [Seshadri & Verma, 2016] for a useful exposition).

The broad pattern revealed in these examples is the mathematical quest for standard models to represent in some uniform manner a whole class of objects. They suggest a kind of completeness or closure in a particular domain.

17.9 Axiomatics and Foundations

In Sections 16.2 and 16.3 I mentioned Menger's interest in axiomatizing dimension theory, as a result of which he produced a set of axioms. I do not think that a lot has come from that direction of research in dimension theory. It appears that the early axiomatic efforts for dimension theory have largely lain dormant since Menger's contribution. However, in many domains of mathematics axiomatics and axiomatic thinking have proved important. For example, mathematicians have made and continue to make frequent and wide-ranging use of the abstracting axioms for general structures like groups, rings, fields, topological spaces, metric spaces, and the like. In the case of algebraic topology the axioms of Eilenberg and Steenrod for homology and cohomology have proven to be of considerable value [Eilenberg & Steenrod, 1952]. A great part of the value comes from the application of category theory in this effort of axiomatization

In modern mathematics there are many aspects and applications of axiomatic thinking and practice. Much more could be said about axiomatic theories, their characteristics (e.g., consistency, completeness or incompleteness, decidability, categoricity) and applications. Mathematicians have an extensive array of tools developed in mathematical logic by which to understand the depth of axiomatics in relation to mathematical theories. However, this topic is beyond the scope of this book.

17.10 Final Comments

I have come to the end of the history of dimension, curve, and related theories that I have chosen to cover in this book. With mathematics the story is never complete. New problems and new theories come to the fore, arising from previous and related work or from cross-fertilization, to excite the imagination of new researchers.

Taking a broad synoptic view of the history over centuries, one finds an extensive development of concepts and theories related to dimension. Initially, philosophers and mathematicians associated dimension with points, lines, surfaces, and solids, i.e., with standard 0-, 1-, 2-, and 3-dimensional geometrical objects and also with the real world. Then for important mathematical reasons consideration of higher-dimensional theories became necessary, with the full flowering of theories of n-dimensional and infinite-dimension spaces and geometries occurring in the nineteenth century. Counterintuitive examples and results forced a deeper examination and structuring of the continuum, resulting in the invariance theorem for dimension, new definitions, and new theories. In recent times the theories of Brouwer, Urysohn, and Menger have offered insightful ways to consider geometrical objects of $-1, 0, 1, 2, 3, \ldots$ and infinite dimensions. In addition, unusual examples and counterexamples have led to the creation of theories of more fine-grained fractional dimension, thus, Hausdorff dimension and Minkowski–Bouligand dimension. The synoptic view gives just an indication of the breadth and depth of the subject of dimension.

The history also reveals the names of a great many well-known mathematicians and philosophers who have been involved. The list includes, to name just a few, Aristotle and Euclid, representing the ancient Greeks, Bolzano, Cantor, Dedekind, Poincaré, Brouwer, Lebesgue, Urysohn, Menger, Hausdorff, and Hurewicz. To be sure, mathematicians belonging to a larger list helped solve the problems and create the full story.

In the case of dimension theory the older theories may have been fully exposited in texts, while Hausdorff dimension theory and fractal theory may continue to grow. There is considerable research activity going on in the mathematical world.

Bibliography

[Ahlfors & Sario, 1960] Ahlfors, L.V. & Sario, L. (1960). *Riemann Surfaces*. Princeton University Press. 282

[Akademie der Wissenschaften zu Göttingen, Leipzig, München und Wien, 1935] Akademie der Wissenschaften zu Göttingen, Leipzig, München und Wien (1898–1935). *Encyklopädie der mathematischen Wissenschaften mit Einschluss ihrer Anwendungen*. B.G. Teubner.

[Akademie der Wissenschaften zu Göttingen, Leipzig, München und Wien, 1927] Akademie der Wissenschaften zu Göttingen, Leipzig, München und Wien (1902–1927). *Encyklopädie der mathematischen Wissenschaften mit Einschluss ihrer Anwendungen*, volume III 3 D. Differentialgeometrie. B.G. Teubner. 422

[Akademie der Wissenschaften zu Göttingen, Leipzig, München und Wien, 1934] Akademie der Wissenschaften zu Göttingen, Leipzig, München und Wien (1903–1934). *Encyklopädie der mathematischen Wissenschaften mit Einschluss ihrer Anwendungen*, volume III 2 C. Algebraische Geometrie. B.G. Teubner. 422

[Aleksandrov, 2022] Aleksandrov, A.G. (2022). The Poincaré index and its Applications. *Universe*, 8(4), 223; 1–15. 224

[Alexander, 1915] Alexander, J.W. (1915). A proof of the invariance of certain constants of analysis situs. *Transactions of the American Mathematical Society*, 16(2), 148–154. 283, 389

[Alexander, 1922] Alexander, J.W. (1922). A proof and extension of the Jordan-Brouwer separation theorem. *Transactions of the American Mathematical Society*, 23(4), 333–349. 254, 389

[Alexander, 1926] Alexander, J.W. (1926). Combinatorial analysis situs. *Transactions of the American Mathematical Society*, 28(2), 301–329. 283

[Alexandroff, 1925] Alexandroff, P.S. (1925). Paul Urysohn (1898–1924). *Fundamenta Mathematicae*, 7, 138–140. 266, 321, 322

[Alexandroff, 1926a] Alexandroff, P.S. (1926a). Notes supplémentaires au "Mémoire sur les multiplicités Cantoriennes", rédigées d'après les papiers posthumes de Paul Urysohn. *Fundamenta Mathematicae*, 8(1), 352–359. 273

D. M. Johnson, *Theories of Dimension from Ancient to Modern Times*, Sources and Studies in the History of Mathematics and Physical Sciences, https://doi.org/10.1007/978-3-032-02682-8

[Alexandroff, 1926b] Alexandroff, P.S. (1926b). Sur la dimensions des ensembles fermés. *Comptes rendus hebdomadaires des séances de l'Académie des Sciences Paris*, 183, 640–642. Session of 18 October 1926. 385

[Alexandroff, 1926c] Alexandroff, P.S. (1926c). Sur les multiplicités cantoriennes et le théorème de Phragmèn-Brouwer généralisé. *Comptes rendus hebdomadaires des séances de l'Académie des Sciences Paris*, 183, 722–724. Session of 3 November 1926. 385

[Alexandroff, 1927a] Alexandroff, P.S. (1927a). Une définition des nombres de Betti pour un ensemble fermé quelconque. *Comptes rendus hebdomadaires des séances de l'Académie des Sciences Paris*, 184, 317–319. Session of 7 February 1927. 385

[Alexandroff, 1927b] Alexandroff, P.S. (1927b). Sur la décomposition de t'espace par des ensembles fermés. *Comptes rendus hebdomadaires des séances de l'Académie des Sciences Paris*, 184, 425–428. Session of 21 February 1927. 385

[Alexandroff, 1927c] Alexandroff, P.S. (1927c). Une généralisation nouvelle du théorème de Phragmén-Brouwer. *Comptes rendus hebdomadaires des séances de l'Académie des Sciences Paris*, 184, 575–577. Session of 7 March 1927. 385

[Alexandroff, 1927d] Alexandroff, P.S. (1927d). Simpliziale Approximationen in der allgemeinen Topologie. *Mathematische Annalen*, 96(1), 489–511. Dated August 1925; submitted 10 September 1925. 385, 386, 388

[Alexandroff, 1927e] Alexandroff, P.S. (1927e). Über die Dualität zwischen den Zusammenhangszahlen einer abgeschlossenen Menge und des zu ihr komplementären Raumes. *Nachrichten von der Gesellschaft der Wissenschaften zu Göttingen, Mathematisch-Physikalische Klasse*, 1927, 323–329. Read by R. Courant at Session of 25 November 1927. 389

[Alexandroff, 1928a] Alexandroff, P.S. (1928a). Darstellung der Grundzüge der Urysohnschen Dimensionstheorie. *Mathematische Annalen*, 98(1), 31–63. Submitted 3 April 1926. 325, 360, 362

[Alexandroff, 1928b] Alexandroff, P.S. (1928b). Sur l'homéomorphie des ensembles fermés. *Comptes rendus hebdomadaires des séances de l'Académie des Sciences Paris*, 186, 1340–1342. Session of 14 May 1928. 385

[Alexandroff, 1928c] Alexandroff, P.S. (1928c). Sur les frontières de domaines connexes dans l'espace à *n* dimensions. *Comptes rendus hebdomadaires des séances de l'Académie des Sciences Paris*, 186, 1696–1698. Session of 21 May 1928. 385

[Alexandroff, 1928d] Alexandroff, P.S. (1928d). Über den allgemeinen Dimensionsbegriff und seine Beziehungen zur elementaren geometrischen Anschauung. *Mathematische Annalen*, 98(1), 617–635. Dated August 1926; submitted 10 October 1926. 385, 386, 387, 437

[Alexandroff, 1928e] Alexandroff, P.S. (1928e). Untersuchungen über Gestalt und Lage abgeschlossener Mengen beliebiger Dimension. *Annals of Mathematics*. Second Series, 30(1), 101–187. Dated 15 January to 15 April 1928, Princeton, NJ. 385, 388, 389, 394, 395

[Alexandroff, 1928f] Alexandroff, P.S. (1928f). Zum allgemeinen Dimensionsproblem. *Nachrichten von der Gesellschaft der Wissenschaften zu Göttingen, Mathematisch-Physikalische Klasse*, 1928, 25–44. Dated New York - Le Havre, Pentecost 1928 and Göttingen, 5 July 1928.; Read at the session of 6 July 1928. 385, 390

[Alexandroff, 1930a] Alexandroff, P.S. (1930a). Sur la théorie de la dimension. *Comptes rendus hebdomadaires des séances de l'Académie des Sciences Paris*, 190, 1102–1104. Session of 12 May 1930. 385

[Alexandroff, 1930b] Alexandroff, P.S. (1930b). Analyse géométrique de la dimension des ensembles fermés. *Comptes rendus hebdomadaires des séances de l'Académie des Sciences Paris*, 191, 475–477. Session of 22 September 1930. 385

[Alexandroff, 1932a] Alexandroff, P.S. (1932a). Dimensionstheorie: Ein Beitrag zur Geometrie der abgeschlossenen Mengen. *Mathematische Annalen*, 106(1), 161–238. Göttingen, Mathematisches Institute, Winter 1930–31; Submitted 23 May 1931. 331, 385, 390, 391, 392

[Alexandroff, 1932b] Alexandroff, P.S. (1932b). *Einfachste Grundbegriffe der Topologie*. Springer-Verlag. 443

[Alexandroff, 1932c] Alexandroff, P.S. (1932c). Sur la notion de dimension des ensembles fermés. *Journal de Mathématiques Pures et Appliquées*, 11, 283–298. Extrait d'une lettre à M. Fréchet. 385, 393

[Alexandroff, 1950] Alexandroff, P.S. (1950). Pavel Samuilovich Uryson. *Uspekhi Matematicheskikh Nauk*, 5(1), 196–202. Article in Russian. Alexandroff = Aleksandrov. Uryson = Urysohn. 321

[Alexandroff, 1955] Alexandroff, P.S. (1955). The Present Status of the Theory of Dimension. *American Mathematical Society Translations,* Series 2, 1, 1–26.

[Alexandroff, 1956] Alexandroff, P.S. (1956). *Combinatorial Topology*, volume 1. Graylock Press. Alexandroff = Aleksandrov. 393

[Alexandroff, 1957] Alexandroff, P.S. (1957). *Combinatorial Topology*, volume 2. Graylock Press. Alexandroff = Aleksandrov. 393

[Alexandroff, 1960] Alexandroff, P.S. (1960). *Combinatorial Topology*, volume 3. Graylock Press. Alexandroff = Aleksandrov. 389, 393

[Alexandroff, 1961] Alexandroff, P.S. (1961). *Elementary Concepts of Topology*. Dover Publications, New York. English Translation of *Einfachste Grundbegriffe der Topologie* [Alexandroff, 1932b], translated by Alan E. Farley.

[Alexandroff, 1967] Alexandroff, P.S. (1967). *Einführung in die Mengenlehre und die Theorie der reellen Funktionen*. VEB Deutscher Verlag der Wissenschaften, fourth edition. 8

[Alexandroff, 1969] Alexandroff, P.S. (1969). Die Topologie in und um Holland in den Jahren 1920-1930. *Nieuw Archief voor Wiskunde*, 27, 109–127. 322

[Alexandroff, 1972] Alexandroff, P.S. (1972). Poincaré and topology. *Russian Mathematical Surveys*, 27(1), 157–168. Alexandroff = Aleksandrov. 178, 185

[Alexandroff, 1979] Alexandroff, P.S. (1979). Pages from an autobiography. *Russian Mathematical Surveys*, 34(6), 267–302. Alexandroff = Aleksandrov. 322, 383

[Alexandroff, 1980] Alexandroff, P.S. (1980). Pages from an autobiography. II. *Russian Mathematical Surveys*, 35(3), 315–358. Alexandroff = Aleksandrov. 322, 383

[Alexandroff et al., 1975] Alexandroff, P.S., Efimov, N.V., & Sklyarenko, E.G. (1975). In memory of Lev Abramovich Tumarkin. *Russian Mathematical Surveys*, 30(3), 135. 371

[Alexandroff & Hopf, 1935] Alexandroff, P.S. & Hopf, H. (1935). *Topologie I: Erster Band. Grundbegriffe der Mengentheoretischen Topologie Topologie der Komplexe· Topologische Invarianzsätze und Anschliessende Begriffsbildungen· Verschlingungen im n-Dimensionalen Euklidischen Raum Stetige Abbildungen von Polyedern.* Julius Springer. 9, 212, 223, 237, 254, 256, 283, 317, 384, 393, 400

[Alexandroff & Hopf, 2013] Alexandroff, P.S. & Hopf, H. (2013). *Topologie I: Erster Band. Grundbegriffe der Mengentheoretischen Topologie Topologie der Komplexe· Topologische Invarianzsätze und Anschliessende Begriffsbildungen· Verschlingungen im n-Dimensionalen Euklidischen Raum Stetige Abbildungen von Polyedern.* Springer-Verlag. 9

[Alexandroff & Kolmogorov, 1964] Alexandroff, P.S. & Kolmogorov, A.N. (1964). Lev Abramovich Tumarkin (on his 60th birthday). *Russian Mathematical Surveys*, 19(4), 161–163. 371, 373

[Alexandroff & Tumarkin, 1928] Alexandroff, P.S. & Tumarkin, L.A. (1928). Beweis des Satzes, dass jede abgeschlossene Menge positiver Dimension in einem lokal zusammenhängenden Kontinuum von derselben Dimension topologisch enthalten ist. *Fundamenta Mathematicae*, 11(1), 141–144. Lev = Leo.

[Alexandroff & Urysohn, 1928] Alexandroff, P.S. & Urysohn, P.S. (1928). Über nulldimensionale Punktmengen. *Mathematische Annalen*, 98(1), 89–106.

[Alexandroff & Urysohn, 1929] Alexandroff, P.S. & Urysohn, P.S. (1929). Mémoire sur les espaces topologiques compacts. *Verhandelingen der koninklijke Akademie van Wetenschappen te Amsterdam*, 14(1), 1–96. 332

[Ames, 1905a] Ames, L.D. (1905a). *An arithmetic treatment of some problems in analysis situs. A Dissertation.* Lord Baltimore Press–The Friedenwald Company. 227

[Ames, 1905b] Ames, L.D. (1905b). An arithmetic treatment of some problems in analysis situs. *American Journal of Mathematics*, 27, 343–380. 146, 227

[Antoine, 1921] Antoine, L. (1921). Sur l'homéomorphie de deux figures et de leurs voisinages. *Journal de mathématiques pures et appliquées*, 4(8), 221–325. 188, 328

[Arboleda, 1979] Arboleda, L.C. (1979). Les débuts de l'école topologique soviétique: Notes sur les lettres de Paul S. Alexandroff et Paul S. Urysohn à Maurice Fréchet. *Archive for History of Exact Sciences*, 20(1), 73–89. 331

[Arboleda, 1980] Arboleda, L.C. (1980). *Contribution à l'étude des premières recherches topologiques (d'après la correspondance et les publications de Maurice Fréchet (1904–1928)).* Thesis. Paris. 197, 201

[Arboleda, 1981a] Arboleda, L.C. (1981a). Rapport sur l'inventaire et l'analyse des papiers du Fonds-Fréchet dans les archives de l'Académie des Sciences de Paris. *Cahiers du séminaire d'histoire des mathématiques*, 2, 9–17. 197

[Arboleda, 1981b] Arboleda, L.C. (1981b). Les recherches de M. Fréchet, P. Alexandrov, W. Sierpiński et K. Kuratowski sur la théorie des types de dimensions et les débuts de la topologie générale. *Archive for History of Exact Sciences*, 24(4), 339–388. 197

[Argand, 1806] Argand, J.R. (1806). *Essai sur une manière de représenter les quantités imaginaires dans les constructions géométriques*. Privately Published. 74

[Argand, 1813] Argand, J.R. (1813). Essai sur une manière de représenter les quantités imaginaires dans les constructions géométriques. *Annales de Mathématiques pures et appliquées*, 4, 133–147. 74

[Argand, 1814] Argand, J.R. (1814). Réflexions sur la nouvelle théorie d'analyse. *Annales de Mathématiques pures et appliquées*, 5, 197–209. 74

[Argand, 1874] Argand, J.R. (1874). *Essai sur une manière de représenter les quantités imaginaires dans les constructions géométriques*. Gauthier-Villars. 74

[Argand, 1881] Argand, J.R. (1881). *Imaginary Quantities: Their Geometrical Interpretation*. Trans. A. S. Hardy. D. Van Nostrand. 74

[Arhangel'skii & Dranishnikov, 1997] Arhangel'skii, A. & Dranishnikov, A. (1997). P.[S]. Alexandroff and Topology: an introductory note. *Topology and its Applications*, 1(80), 1–6. 383

[Aristotle, 1939] Aristotle (1939). *On the Heavens*. Trans. W. K. C. Guthrie. Loeb Classical Library. 16

[Aristotle, 1984a] Aristotle (1984a). *The Complete Works of Aristotle: The revised Oxford translation*. J. Barnes and others, Eds., volume 1. Princeton University Press. 15, 20, 21

[Aristotle, 1984b] Aristotle (1984b). *The Complete Works of Aristotle: The revised Oxford translation*. J. Barnes and others, Eds., volume 2. Princeton University Press.

[Arkhangelskii & Tikhomirov, 1998] Arkhangelskii, A.V. & Tikhomirov, V.M. (1998). Pavel Samuilovich Urysohn (1898-1924). *Russian Mathematical Surveys*, 53(5), 875–892. xix, 321, 322

[Artin, 2022] Artin, M. (2022). *Algebraic Geometry: Notes on a Course*, volume 222 of *Graduate Studies in Mathematics*. American Mathematical Society. 425

[Arzelà, 1896] Arzelà, C. (1896). Sull'esistenza degli integrali nelle equazioni differenziali ordinarie. *Memorie della Reale Accademia delle Scienze dell'Istituto di Bologna*, 6(5), 131–140. 211

[Atiyah & Macdonald, 1969] Atiyah, M.F. & Macdonald, I.G. (1969). *Introduction to Commutative Algebra*. Addison-Wesley. 422

[Aull & Lowen, 1997] Aull, C.E. & Lowen, R., Eds. (1997). *Handbook of the History of General Topology*, volume 1. Kluwer Academic Publishers.

[Baire, 1899] Baire, R.L. (1899). Sur les fonctions de variables réelles. *Annali di Matematica Pura ed Applicata (1898-1922)*, 3(1), 1–123. 188

[Baire, 1906] Baire, R.L. (1906). Sur la représentation des fonctions discontinues. Première partie. *Acta mathematica*, 30(1), 1–48.

[Baire, 1907a] Baire, R.L. (1907a). Sur la non-applicabilité de deux continus à n et $n + p$ dimensions. *Comptes rendus hebdomadaires des séances de l'Académie des Sciences Paris*, 144, 318–321. Session of 11 February 1907. 191, 193, 200, 208, 229, 236, 258

[Baire, 1907b] Baire, R.L. (1907b). Sur la non-applicabilité de deux continus à n et $n + p$ dimensions. *Bulletin des Sciences Mathématiques*, 31(2), 94–99. 191, 192, 193, 196, 200, 208, 229, 236, 258

[Baire, 1909] Baire, R.L. (1909). Sur la représentation des fonctions discontinues. Deuxiéme partie. *Acta mathematica*, 32(1), 97–176. 194, 195, 197, 208

[Balcar et al., 1993] Balcar, B., Koutník, V., & Simon, P. (1993). Eduard Čech, 1893-1960. *Mathematica Slovaca*, 43(3), 381–392. 399

[Banach, 1924] Banach, S. (1924). Un théorème sur les transformations biunivoques. *Fundamenta Mathematicae*, 6(1), 236–239. 199

[Banks, 2005] Banks, E.C. (2005). Kant, Herbart and Riemann. *Kant-Studien*, 96(2), 208–234. 90

[Baumgarten, 1779] Baumgarten, A.G. (1779). *Metaphysica*. Impensis Coral. Herman. Hermderde. Halae Magdeburgicae., VII edition. 34

[Begehr et al., 1998] Begehr, H., Koch, H., Kramer, J., Schappacher, N., & Thiele, E.J. (1998). *Mathematics in Berlin*. Springer. 216

[Beham, 2012] Beham, B.J.K. (2012). *Die Genese des Mengerschen Dimensionsbegriffes im Spannungsverhältnis von Ökonomie, Mathematik und Philosophie*. PhD thesis, Universität Wien. 333

[Behboud, 2000] Behboud, A. (2000). *Bolzanos Beiträge zur Mathematik und ihrer Philosophie*. Institut für Philosophie, G. Grasshoff. 31

[Behnke, 1958] Behnke, H. (1958). Otto Blumenthal zum Gedächtnis. *Mathematische Annalen*, 136(5), 387–392. 245

[Belhoste, 1991] Belhoste, B. (1991). *Augustin-Louis Cauchy: A Biography*. Springer. 213

[Bellivier, 1956] Bellivier, A. (1956). *Henri Poincaré: ou, La vocation souveraine*. Gallimard. 158

[Beltrametti et al., 2009] Beltrametti, M.C., Carletti, E., Gallarati, D., & Bragadin, G.M. (2009). *Lectures on Curves, Surfaces and Projective Varieties: A Classical View of Algebraic Geometry*, volume 9 of *EMS Textbooks in Mathematics*. European Mathematical Society.

[Bendixson, 1883] Bendixson, I. (1883). Quelques theorèmes de la théorie des ensembles de points. Extrait d'une lettre adressée à M. Cantor à Halle. *Acta Mathematica*, 2, 415–429. 142

[Berg, 1965] Berg, J. (1965). *Bolzano's Logic*, volume 2 of *Stockholm Studies in Philosophy*. Almqvist & Wiksell. 31

[Bernstein, 1900] Bernstein, F. (1900). Ueber einen Schönflies'schen Satz aus der Theorie der Funktionen zweier reeller Veränderlichen. *Nachrichten von der Gesellschaft der Wissenschaften zu Göttingen, Mathematisch-Physikalische Klasse*, 1900, 98–102. 152

[Besicovitch, 1928] Besicovitch, A.S. (1928). On the fundamental geometrical properties of linearly measurable plane sets of points. *Mathematische Annalen*, 98(1), 422–464. 412

[Besicovitch, 1929] Besicovitch, A.S. (1929). On linear sets of points of fractional dimension I. *Mathematische Annalen*, 101(1), 161–193. 412

[Besicovitch, 1934] Besicovitch, A.S. (1934). Sets of fractional dimensions IV: On rational approximation to real numbers. *Journal of the London Mathematical Society*, 1(2), 126–131. 412

[Besicovitch, 1935a] Besicovitch, A.S. (1935a). On the sum of digits of real numbers represented in the dyadic system. (on sets of fractional dimensions II.). *Mathematische Annalen*, 110(1), 321–330. 412

[Besicovitch, 1935b] Besicovitch, A.S. (1935b). Sets of points of non-differentiability of absolutely continuous functions and of divergence of Fejér sums. (on linear sets of fractional dimensions III). *Mathematische Annalen*, 110, 331–335. 412

[Besicovitch, 1935c] Besicovitch, A.S. (1935c). On the density of certain sequences of integers. *Mathematische Annalen*, 110, 336–341. 412

[Bieberbach, 1917] Bieberbach, L. (1917). *Differential- und Integralrechnung, Band I: Differentialrechnung*, volume 1. B.G. Teubner. 336

[Biermann, 1990] Biermann, K. (1970–1990). Eugen Netto. In C.C. Gillispie (Ed.), *Dictionary of Scientific Biography*, volume 10 (pp. 24). Charles Scribner & Sons. 130

[Bocklandt, 2021] Bocklandt, R. (2021). A testimonial of troubled times. *Nieuw Archief voor Wiskunde*, 5(21), 181–184. 424

[Bohl, 1904] Bohl, P. (1904). Über die Bewegung eines mechnischen Systems in der Nähe einer Gleichgewichtslage. *Journal für die reine und angewandte Mathematik*, 127, 179–276. 228

[Boi, 1994] Boi, L. (1994). Die Beziehungen zwischen Raum, Kontinuum und Materie im Denken Riemanns, die Athervorstellung und die Einheit der Physik: Das Entstehen einer neuen Naturphilosophie. *Philosophia naturalis*, 31(2), 171–216. 90

[Bolzano, 0000a] Bolzano, B. (0000a). Anti-Euklid. *Unpublished manuscript in Handschriftensammlung der Österreichischen Nationalbibliothek, Wien.*, 3459:6.5. 55, 59, 64

[Bolzano, 0000b] Bolzano, B. (0000b). Geometrische Begriffe, die Jeder kennt und nicht kennt. Versuch einer Erhebung derselben ins deutliche Bewußtseyn. *Unpublished manuscript in Handschriftensammlung der Österreichischen Nationalbibliothek, Wien.*, 3459:3.b. 33, 55, 60, 63, 65, 67, 68

[Bolzano, 0000c] Bolzano, B. (0000c). Mathematische Gespräche. *Unpublished manuscript in Handschriftensammlung der Österreichischen Nationalbibliothek, Wien.*, 3459:1.d.

[Bolzano, 0000d] Bolzano, B. (0000d). Versuch einer Erklärung der Begriffe von Linie, Fläche und Körper. *Unpublished manuscript in Handschriftensammlung der Österreichischen Nationalbibliothek, Wien.*, 3459:3.c. 40, 48, 55, 56, 57, 61

[Bolzano, 1804] Bolzano, B. (1804). *Betrachtungen über einige Gegenstände der Elementargeometrie*. Karl Barth. 32, 37, 38, 39, 43

[Bolzano, 1810] Bolzano, B. (1810). *Beyträge zu einer begründeteren Darstellung der Mathematik: erste Lieferung*, volume 1. Im Verlage bey Caspar Widtmann. 40, 41, 42, 43

[Bolzano, 1816] Bolzano, B. (1816). *Der binomische Lehrsatz, und als Folgerung aus ihm der polynomische, und die Reihen, die zur Berechnung der Logarithmen und Exponentialgrößen dienen, genauer ans bisher erwiesen*. In der C. M. Enderschen Buchhandlung. 43

[Bolzano, 1817a] Bolzano, B. (1817a). *Die drey Probleme der Rectification, der Complanation und der Cubirung, ohne Betrachtung des unendlich Kleinen, ohne die Annahmen des Archimedes, und ohne irgend eine nicht streng erweisliche Voraussetzung gelöst; zugleich als Probe einer gänzlichen Umstaltung der Raumwissenschaft, allen Mathematikern zur Prüfung vorgelegt*. Paul Gotthelf Kummer. Prague. xvii, 33, 43, 49, 50, 51, 52, 53, 54

[Bolzano, 1817b] Bolzano, B. (1817b). *Rein analytischer Beweis des Lehrsatzes, dass zwischen je zwey Werthen, die ein entgegengesetztes Resultat gewähern, wenigstens eine reelle Wurzel der Gleichung liege*. Gottlieb Haase. Prague. 43, 49, 60, 123, 140, 433

[Bolzano, 1836] Bolzano, B. (1836). *Lebensbeschreibung des Dr. B. Bolzano: mit einigen seiner ungedruckten Aufsätze und dem Bildnisse des Verfassers*. Seidel. 31, 33, 35

[Bolzano, 1837] Bolzano, B. (1837). *Wissenschaftslehre*. J.E. v Seidel. Sulzbach. 31, 40, 43, 44, 45, 46, 47, 55, 56, 69

[Bolzano, 1843a] Bolzano, B. (1843a). *Über den Begriff des Schönen. Eine philosophische Abhandlung*. Prag. In Commission bei Borrosch et André. 49, 56, 60

[Bolzano, 1843b] Bolzano, B. (1843b). *Versuch einer objectiven Begründung der Lehre von den drei Dimensionen des Raumes*. Prag. In Commission bei Kronberger & Řiwanč. 60, 68, 69, 70

[Bolzano, 1845a] Bolzano, B. (1845a). Über den Begriff des Schönen. *Abhandlungen der königliche böhmische Gesellschaft der Wissenschaften*, V(3), 1–92. 49

[Bolzano, 1845b] Bolzano, B. (1845b). Versuch einer objectiven Begründung der Lehre von den drei Dimensionen des Raumes. *Abhandlungen der königliche böhmische Gesellschaft der Wissenschaften*, V(3), 201–215. 68, 69, 70

[Bolzano, 1851] Bolzano, B. (1851). *Dr. Bernard Bolzano's Paradoxien des Unendlichen.* Edited by F. Přihonsky. C. H. Reclam. Leipzig. 33, 55, 58, 60

[Bolzano, 1948a] Bolzano, B. (1948a). *Bernard Bolzano's Schriften. Band 5. Geometrische Arbeiten.* J. Vojtěch, Ed., volume 5. Královská česká společnost nauk v Praze.

[Bolzano, 1948b] Bolzano, B. (1948b). Über Haltung, Richtung, Kömmung und Schnörkelung bei Linien sowohl als Flächen samt einigen verwandten Begriffen. In J. Vojtěch (Ed.), *Bernard Bolzano's Schriften. Band 5. Geometrische Arbeiten*, volume 5 (pp. 139–183). Královská česká společnost nauk v Praze. xvii, 33, 55, 57, 58, 60, 61, 62, 63, 64

[Bolzano, 1948c] Bolzano, B. (1948c). Versuch einer objectiven Begründung der Lehre von den drei Dimensionen des Raumes. In J. Vojtěch (Ed.), *Bernard Bolzano's Schriften. Band 5. Geometrische Arbeiten*, volume 5 (pp. 51–65). Královská česká společnost nauk v Praze. 60, 68, 69, 70

[Bolzano, 1967] Bolzano, B. (1967). Anti-Euklid. K. Večerka, Ed. *Acta historiae rerum naturalium necnon technicarum*, 11, 203–216. 55, 59, 64

[Bolzano, 1969] Bolzano, B. (1969). *Bernard Bolzano-Gesamtausgabe*. E. Morscher; J. Berg, F. Kambartel, J. Loužil, B. van Rootselaar, and E. Winter, Eds. Frommann-Holzboog. Publication in progress, 1969–. 31, 40, 55

[Bolzano, 1972] Bolzano, B. (1972). *Theory of Science. Attempt at a Detailed and in the main Novel Exposition of Logic with Constant Attention to Earlier Authors.* R. George, ed. and trans. Blackwell. 31, 40, 44, 55

[Bolzano, 1975] Bolzano, B. (1975). Von der mathematischen Lehrart. In *Bernard Bolzano-Gesamtausgabe, Vol. II A 7, 1: Größenlehre* (pp. 46–97). Frommann-Holzboog. 40, 46, 47

[Bolzano, 1981] Bolzano, B. (1981). *Von der mathematischen Lehrart.* J. Berg, Ed. Frommann-Holzboog. 40, 46, 47

[Bolzano, 1989a] Bolzano, B. (1989a). Über den Begriff des Schönen. Eine philosophische Abhandlung. In *Bernard Bolzano-Gesamtausgabe, Vol. I, 18: Mathematisch-Physikalische und Philosophische Schriften 1842-1843* (pp. 87–217). Frommann-Holzboog. 49, 56, 60

[Bolzano, 1989b] Bolzano, B. (1989b). Versuch einer objectiven Begründung der Lehre von den drei Dimensionen des Raumes. In *Bernard Bolzano-Gesamtausgabe, Vol. I, 18: Mathematisch-Physikalische und Philosophische Schriften 1842-1843* (pp. 219–238). Frommann-Holzboog. 60, 68, 69, 70

[Bolzano, 2001a] Bolzano, B. (2001a). Geometrische Begriffe, die Jeder kennt und nicht kennt. Versuch einer Erhebung derselben ins deutliche Bewußtseyn. In *Bernard Bolzano-Gesamtausgabe, Vol. II A 11, 1: Vermischte mathematische Schriften 1832-1848 I* (pp. 151–169). Frommann-Holzboog. xvii, xvii, 33, 55, 60, 63, 65, 66, 67, 68

[Bolzano, 2001b] Bolzano, B. (2001b). Mathematische Gespräche. In *Bernard Bolzano-Gesamtausgabe, Vol. II A 11, 1: Vermischte mathematische Schriften 1832-1848 I* (pp. 107–112). Frommann-Holzboog.

[Bolzano, 2001c] Bolzano, B. (2001c). Versuch einer Erklärung der Begriffe von Linie, Fläche und Körper. In *Bernard Bolzano-Gesamtausgabe, Vol. II A 11, 1: Vermischte mathematische Schriften 1832-1848 I* (pp. 113–150). Frommann-Holzboog. 40, 48, 55, 56, 57, 61

[Bolzano, 2004a] Bolzano, B. (2004a). *The Mathematical Works of Bernard Bolzano.* S. Russ, editor and translator. Oxford University Press. 31, 32, 33, 34, 36, 38, 39, 40, 41, 42, 49, 50, 51, 53, 54, 58

[Bolzano, 2004b] Bolzano, B. (2004b). *On the Mathematical Method and Correspondence with Exner.* P. Rusnock and R. George, eds. and trans. Rodopi. 31, 40, 46, 47

[Bolzano, 2012] Bolzano, B. (2012). *Paradoxien des Unendlichen*, volume 630 of *Philosophische Bibliothek*. Felix Meiner Verlag. 33, 55

[Bolzano, 2014a] Bolzano, B. (2014a). *Bernard Bolzano: Theory of Science*. P. Rusnock and R. George, translators. Oxford University Press. 4 volumes. 31, 40, 43, 44, 45, 46, 47, 55, 56, 69

[Bolzano, 2014b] Bolzano, B. (2014b). *Paradoxes of the Infinite* (Routledge Revivals). Routledge. 33

[Bolzano, 20xxa] Bolzano, B. (20xxa). Anti-Euklid. In *Bernard Bolzano-Gesamtausgabe, Vol. II A 11, 2: Vermischte mathematische Schriften 1832-1848 II. In preparation*. Frommann-Holzboog. 55, 59, 64

[Bolzano, 20xxb] Bolzano, B. (20xxb). Paradoxien des Unendlichen. In *Bernard Bolzano-Gesamtausgabe, Vol. II A 11, 2: Vermischte mathematische Schriften 1832-1848 II. In preparation*. Frommann-Holzboog. 55, 58

[Bolzano, 20xxc] Bolzano, B. (20xxc). Uiber Haltung, Richtung, Kömmung und Schnörkelung bei Linien sowohl als Flächen samt einigen verwandten Begriffen. In *Bernard Bolzano-Gesamtausgabe, Vol. II A 11, 2: Vermischte mathematische Schriften 1832-1848 II. In preparation*. Frommann-Holzboog. 33, 55, 57, 58, 60, 61, 63

[Boniolo, 2003] Boniolo, G. (2003). Kant's Explication and Carnap's Explication: The redde rationem. *International Philosophical Quarterly*, 43(3), 289–298. 6

[Boole, 1854] Boole, G. (1854). *An Investigation of the Laws of Thought, on which are founded the Mathematical Theories of Logic and Probabilities*. Macmillan. Reprinted by Dover. 79

[Borsuk, 1997] Borsuk, K. (1997). Witold Hurewicz—Life and Work. In *Handbook of the History of General Topology*, volume 1 (pp. 79–84). Kluwer Academic Publishers. 364

[Bottazzini, 1994] Bottazzini, U. (1994). Geometry and "Metaphysics of Space" in Gauss and Riemann. In *Romanticism in Science* (pp. 15–29). Springer.

[Bottazzini & Tazzioli, 1995] Bottazzini, U. & Tazzioli, R. (1995). Naturphilosophie and its role in Riemann's mathematics. *Revue d'histoire des mathématiques*, 1(1), 3–38. 90

[Bouligand, 1925a] Bouligand, G. (1925a). Sur le problemè de Dirichlet. *Annales de la Société Polonaise de Mathématique*, 4, 59–112. 418

[Bouligand, 1925b] Bouligand, G. (1925b). Dimension, étendue, densité. *Comptes rendus hebdomadaires des séances de l'Académie des Sciences Paris*, 180, 245–248. Included at the session of 26 January 1925 from a session of 17 November 1924. 418

[Bouligand, 1927] Bouligand, G. (1927). Nombre dimensionnel et ensembles impropres dans le problème de Dirichlet. *L'Enseignement Mathématique*, (pp. 240–260). 419

[Bouligand, 1928a] Bouligand, G. (1928a). Sur l'aire d'un domaine plan. *Bulletin des Sciences Mathématiques*, 52, 55–63. 419

[Bouligand, 1928b] Bouligand, G. (1928b). Ensembles impropres et nombre dimensionnel. *Bulletin des Sciences Mathématiques*, 52, 320–344, 361–376. 419, 420

[Bouligand, 1928c] Bouligand, G. (1928c). Ordre de mesure et dimension des ensembles fermés. *Comptes rendus hebdomadaires des séances de l'Académie des Sciences Paris*, 187, 524–525. Session of 24 September 1928. 419

[Bouligand, 1928d] Bouligand, G. (1928d). Sur l'ordre de mesure d'un ensemble fermé. *Comptes rendus hebdomadaires des séances de l'Académie des Sciences Paris*, 187, 593–594. Session of 8 October 1928. 419

[Bouligand, 1928e] Bouligand, G. (1928e). Sur quelques points de topologie restreinte du premier ordre. *Bulletin de la Société Mathématique de France*, 56, 26–35. 419

[Bouligand, 1935] Bouligand, G. (1935). *Les définitions modernes de la dimension*, volume 274 of *Actualités scientifiques et industrielles*. Hermann & Cie. 420

[Brendan, 2024] Brendan, S. (2024). Karl Popper: Philosophy of Science. In *The Internet Encyclopedia of Philosophy*. The Internet Encyclopedia of Philosophy. ISSN 2161-0002, `https://iep.utm.edu/pop-sci/` (accessed 1 March 2024). 6

[Brewer & Smith, 1981] Brewer, J.W. & Smith, M.K., Eds. (1981). *Emmy Noether: a tribute to her life and work*. M. Dekker. 384

[Brieskorn, 1996] Brieskorn, E., Ed. (1996). *Felix Hausdorff zum Gedächtnis, Band I: Aspekte seines Werkes*, volume 1. Springer-Verlag. 404

[Brieskorn & Purkert, 2018] Brieskorn, E. & Purkert, W. (2018). *Felix Hausdorff, Gesammelte Werke, Band IB, Biographie*, volume IB. Springer. 316, 404

[Brieskorn & Purkert, 2021] Brieskorn, E. & Purkert, W. (2021). *Felix Hausdorff: Mathematiker, Philosoph und Literat*. Springer. 404

[Brill & Noether, 1894] Brill, A.v. & Noether, M. (1894). Die Entwicklung der Theorie der algebraischen Functionen in älterer und neuerer Zeit: Bericht erstattet der Deutschen Mathematiker-Vereinigung. *Jahresbericht der Deutschen Mathematiker-Vereinigung*, 3, I–XXIII, 109–566. 423

[Brill & Noether, 1911] Brill, A.v. & Noether, M. (1911). Jakob Lüroth. *Jahresbericht der Deutschen Mathematiker-Vereinigung*, 20, 279–299. 122

[Brouwer, 1907] Brouwer, L.E.J. (1907). *Over de grondslagen der wiskunde*. Maas & van Suchtelen. 204, 280

[Brouwer, 1908] Brouwer, L.E.J. (1908). Die Theorie der endlichen continuierlichen Gruppen, unabhängig von den Axiomen von Lie. In *Atti del IV. Congresso Internazionale dei Matematici. Roma*, volume II (pp. 296–303). 204, 205

[Brouwer, 1909a] Brouwer, L.E.J. (1909a). Over één-éénduidige, continue transformaties van oppervlakken in zichzelf. *Koninklijke Nederlandse Akademie van Wetenschappen te Amsterdam. Verslag, Series B Physical Sciences*, 17, 741–752. 210

[Brouwer, 1909b] Brouwer, L.E.J. (1909b). Continuous one-one transformations of surfaces in themselves. *Koninklijke Nederlandse Akademie van Wetenschappen te Amsterdam. Proceedings, Series B Physical Sciences*, 11, 788–798. 210

[Brouwer, 1909c] Brouwer, L.E.J. (1909c). Die Theorie der endlichen kontinuierlichen Gruppen, unabhängig von den Axiomen von Lie. (Erste Mitteilung.). *Mathematische Annalen*, 67(2), 246–267. 205

[Brouwer, 1909d] Brouwer, L.E.J. (1909d). *Het wezen der meetkunde*. P. Noordhoff. 208, 209, 257, 280

[Brouwer, 1909e] Brouwer, L.E.J. (1909e). Over continue vectordistributies op oppervlakken. *Koninklijke Nederlandse Akademie van Wetenschappen. Verslag, Series B Physical Sciences*, 17, 896–904. 210

[Brouwer, 1909f] Brouwer, L.E.J. (1909f). On continuous vector distributions on surfaces. *Koninklijke Nederlandse Akademie van Wetenschappen. Proceedings, Series B Physical Sciences*, 11, 850–858. 210, 211

[Brouwer, 1910a] Brouwer, L.E.J. (1910a). Zur Analysis Situs. *Mathematische Annalen*, 68(3), 422–434. xviii, xviii, xviii, 150, 154, 207, 281, 285, 290, 291, 292, 293, 295, 296, 297, 386

[Brouwer, 1910b] Brouwer, L.E.J. (1910b). Beweis des Jordanschen Kurvensatzes. *Mathematische Annalen*, 69(2), 169–175. 145, 209, 282

[Brouwer, 1910c] Brouwer, L.E.J. (1910c). Die Theorie der endlichen kontinuierlichen Gruppen, unabhängig von den Axiomen von Lie. (Zweite Mitteilung.). *Mathematische Annalen*, 69(2), 181–203. 205

[Brouwer, 1910d] Brouwer, L.E.J. (1910d). Over continue vectordistributies op oppervlakken II. *Koninklijke Nederlandse Akademie van Wetenschappen. Verslag, Series B Physical Sciences*, 18, 702–721. 230

[Brouwer, 1910e] Brouwer, L.E.J. (1910e). On continuous vector distributions on surfaces. (2nd communication.). *Koninklijke Nederlandse Akademie van Wetenschappen. Proceedings, Series B Physical Sciences*, 12, 716–734. 230, 231

[Brouwer, 1910f] Brouwer, L.E.J. (1910f). Over continue vectordistributies op oppervlakken III. *Koninklijke Nederlandse Akademie van Wetenschappen. Verslag, Series B Physical Sciences*, 19, 36–51. 230

[Brouwer, 1910g] Brouwer, L.E.J. (1910g). On continuous vector distributions on surfaces. (3rd communication.). *Koninklijke Nederlandse Akademie van Wetenschappen. Proceedings, Series B Physical Sciences*, 13, 171–186. 230, 231

[Brouwer, 1910h] Brouwer, L.E.J. (1910h). Sur les continus irréductibles de M. Zoretti. *Annales scientifiques de l'École Normale Supérieure*, 27, 565–566. 295

[Brouwer, 1911a] Brouwer, L.E.J. (1911a). Beweis der Invarianz der Dimensionenzahl. *Mathematische Annalen*, 70(2), 161–165. 178, 236, 237, 241, 243, 245, 256, 305, 340, 428, 433

[Brouwer, 1911b] Brouwer, L.E.J. (1911b). Über Abbildung von Mannigfaltigkeiten. *Mathematische Annalen*, 71(1), 97–115. Cf. 'Berichtigung'. *Mathematische Annalen* 71 (1911), 598; 'Berichtigung'. *Mathematische Annalen* 82 (1921), 286. 227, 231, 232, 234, 237, 257, 263, 275, 278, 281, 282

[Brouwer, 1911c] Brouwer, L.E.J. (1911c). Beweis der Invarianz des n-dimensionalen Gebiets. *Mathematische Annalen*, 71(3), 305–313. 256, 259, 265, 276, 277, 282, 340, 379, 381, 454

[Brouwer, 1911d] Brouwer, L.E.J. (1911d). Beweis des Jordanschen Satzes für den n-dimensionalen Raum. *Mathematische Annalen*, 71(3), 314–319. 229, 257, 258, 259

[Brouwer, 1911e] Brouwer, L.E.J. (1911e). Über Jordansche Mannigfaltigkeiten. *Mathematische Annalen*, 71(3), 320–327.

[Brouwer, 1911f] Brouwer, L.E.J. (1911f). Sur le théorème de M. Jordan dans l'espace à n dimensions. *Comptes rendus hebdomadaires des séances de l'Académie des Sciences Paris*, 153, 542–543. Session of 11 September 1911. 258

[Brouwer, 1912a] Brouwer, L.E.J. (1912a). Over omloopcoëfficiënten. *Koninklijke Nederlandse Akademie van Wetenschappen te Amsterdam. Verslag, Series B Physical Sciences*, 20, 1049–1057. 258

[Brouwer, 1912b] Brouwer, L.E.J. (1912b). On looping coefficients. *Koninklijke Nederlandse Akademie van Wetenschappen te Amsterdam. Proceedings, Series B Physical Sciences*, 15, 113–122. 258

[Brouwer, 1912c] Brouwer, L.E.J. (1912c). Over één-éénduidige, continue transformaties van oppervlakken in zichzelf V. *Koninklijke Nederlandse Akademie van Wetenschappen te Amsterdam. Verslag, Series B Physical Sciences*, 21, 300–309. 233

[Brouwer, 1912d] Brouwer, L.E.J. (1912d). Continuous one-one transformations of surfaces in themselves (5th communication.). *Koninklijke Nederlandse Akademie van Wetenschappen. Proceedings, Series B Physical Sciences*, 15, 352–361. 233

[Brouwer, 1912e] Brouwer, L.E.J. (1912e). Sur l'invariance de la courbe fermée. *Comptes rendus hebdomadaires des séances de l'Académie des Sciences Paris*, 154, 862–863. Session of 1 April 1912. 209

[Brouwer, 1912f] Brouwer, L.E.J. (1912f). Zur Invarianz des n-dimensionalen Gebiets. *Mathematische Annalen*, 72, 55–56. 178, 258

[Brouwer, 1912g] Brouwer, L.E.J. (1912g). Beweis der Invarianz der geschlossenen Kurve. *Mathematische Annalen*, 72(3), 422–425. 209

[Brouwer, 1912h] Brouwer, L.E.J. (1912h). Intuitionisme en formalisme. Inaugural lecture as professor in the University of Amsterdam, presented 14 October 1912. 280

[Brouwer, 1913a] Brouwer, L.E.J. (1913a). Sur la notion de classe de transformations d'une multiplicité. *Proceedings of the Fifth International Congress of Mathematicians. (Cambridge, 22–28 August 1912)*, II, 9–10. 233

[Brouwer, 1913b] Brouwer, L.E.J. (1913b). Über den natürlichen Dimensionsbegriff. *Journal für die reine und angewandte Mathematik*, 142, 146–152. 2, 180, 190, 250, 261, 262, 263, 264, 269, 270, 271, 272, 275, 276, 277, 278, 306, 324, 326, 340, 343, 345, 357, 378, 400, 428, 433

[Brouwer, 1913c] Brouwer, L.E.J. (1913c). Intuitionisme en formalisme. *Wiskundig tijdsehrift*, 9, 180–211. Inaugural lecture as professor in the University of Amsterdam, presented 14 October 1912. 280

[Brouwer, 1913d] Brouwer, L.E.J. (1913d). Intuitionism and Formalism. *Bulletin of the American Mathematical Society*, 20, 81–96. 280

[Brouwer, 1918] Brouwer, L.E.J. (1918). Begründung der Mengenlehre unabhängig vom logischen Satz vom ausgeschlossenen Dritten. Erster Teil: Allgemeine Mengenlehre. *Verhandelingen der Koninklijke Akademie van Wetenschappen te Amsterdam*, 12(5). 357

[Brouwer, 1919a] Brouwer, L.E.J. (1919a). Begründung der Mengenlehre unabhängig vom logischen Satz vom ausgeschlossenen Dritten. Zweiter Teil: Theorie der Punktmengen. *Verhandelingen der Koninklijke Akademie van Wetenschappen te Amsterdam*, 12(7). 357

[Brouwer, 1919b] Brouwer, L.E.J. (1919b). Intuitionistische Mengenlehre. *Jahresbericht der Deutschen Mathematiker-Vereinigung*, 28, 203–208. 283, 357

[Brouwer, 1921] Brouwer, L.E.J. (1921). Berichtigung. *Mathematische Annalen*, 82, 286. Correction to [Brouwer, 1911c]. 276

[Brouwer, 1923a] Brouwer, L.E.J. (1923a). Over het natuurlijke dimensiebegrip. *Koninklijke Nederlandse Akademie van Wetenschappen te Amsterdam. Verslag, Series B Physical Sciences*, 32, 881–886. 269, 324, 378

[Brouwer, 1923b] Brouwer, L.E.J. (1923b). Ueber den natürlichen Dimensionsbegrifff. *Koninklijke Nederlandse Akademie van Wetenschappen te Amsterdam. Proceedings, Series B Physical Sciences*, 26, 795–800. Communicated at the meeting of 24 November 1923. 269, 324, 345, 378, 400

[Brouwer, 1923c] Brouwer, L.E.J. (1923c). Die Rolle des Satzes vom ausgeschlossenen Dritten in der Mathematik. *Jahresbericht der Deutschen Mathematiker-Vereinigung*, 32, 67–68. Short note on Brouwer's presentation on 21 September 1923 at the Marburg Meeting of the D.M.-V., 20–25 September 1923. 266

[Brouwer, 1924a] Brouwer, L.E.J. (1924a). Opmerkingen over her natuurlijke dimensiebegrip. *Koninklijke Nederlandse Akademie van Wetenschappen te Amsterdam. Verslag, Series B Physical Sciences*, 33, 476–478. 266, 269, 378

[Brouwer, 1924b] Brouwer, L.E.J. (1924b). Bemerkungen zum natürlichen Dimensionsbegriff. *Koninklijke Nederlandse Akademie van Wetenschappen te Amsterdam. Proceedings, Series B Physical Sciences*, 27, 635–638. 266, 269, 273, 378, 381

[Brouwer, 1924c] Brouwer, L.E.J. (1924c). Zum natürlichen Dimensionsbegriff. *Mathematische Zeitschrift*, 21(1), 312–314. 269, 378

[Brouwer, 1924d] Brouwer, L.E.J. (1924d). Berichtigung zu der Abhandlung. *Journal für die reine und angewandte Mathematik*, 153, 253. 269, 324, 378

[Brouwer, 1928a] Brouwer, L.E.J. (1928a). Zur Geschichtsschreibung der Dimensionstheorie. *Koninklijke Nederlandse Akademie van Wetenschappen te Amsterdam. Verslagen*, 37, page 626. Summary and announcement of [Brouwer, 1928b]. 380, 381

[Brouwer, 1928b] Brouwer, L.E.J. (1928b). Zur Geschichtsschreibung der Dimensionstheorie. *Koninklijke Nederlandse Akademie van Wetenschappen te Amsterdam. Proceedings Series B Physical Sciences*, 31, 953–957. Communicated at the meeting of 29 September 1928. 272, 345, 380, 454

[Brouwer, 1975] Brouwer, L.E.J. (1975). *Collected Works 1. Philosophy and Foundations of Mathematics*. Edited by Arend Heyting. Elsevier. 203, 204, 208, 209, 257, 280

[Brouwer, 1976] Brouwer, L.E.J. (1976). *Collected Works 2. Geometry, Analysis, Topology, and Mechanics*. Edited by Hans Freudenthal. Elsevier. xviii, xviii, xviii, xviii, 156, 203, 204, 206, 208, 231, 232, 234, 236, 242, 246, 247, 249, 250, 251, 252, 256, 257, 259, 260, 261, 262, 265, 268, 269, 278, 283, 353, 374

[Browder, 1983] Browder, F.E., Ed. (1983). *The Mathematical Heritage of Henri Poincaré*, volume 39 - Part 2 of *Proceedings of Symposiaa in Pure Mathematics*. American Mathematical Society. 158

[Brush, 1967] Brush, S.G. (1967). Foundations of statistical mechanics 1845–1915. *Archive for History of Exact Sciences*, 4(3), 145–183. 282

[Burkert, 1972] Burkert, W. (1972). *Lore and Science in Ancient Pythagoreanism*. Harvard University Press. 12

[Burkill, 1971] Burkill, J.C. (1971). Abram Samoilovitch Besicovitch. 1891-1970. *Biographical Memoirs of Fellows of the Royal Society*, 17, 1–16. 412

[Butzer & Volkmann, 2006] Butzer, P. & Volkmann, L. (2006). Otto Blumenthal (1876–1944) in retrospect. *Journal of Approximation Theory*, 138(1), 1–36. 245

[Cajori, 1926] Cajori, F. (1926). Origins of fourth dimension concepts. *The American Mathematical Monthly*, 33(8), 397–406. 26

[Cantor, 1872] Cantor, G. (1872). Ueber die Ausdehnung eines Satzes aus der Theorie der trigonometrischen Reihen. *Mathematische Annalen*, 5(1), 123–132. 103, 140

[Cantor, 1874] Cantor, G. (1874). Ueber eine Eigenschaft des Inbegriffs aller reellen algebraischen Zahlen. *Journal für die reine und angewandte Mathematik*, 77, 258–262. 102

[Cantor, 1878] Cantor, G. (1878). Ein Beitrag zur Mannigfaltigkeitslehre. *Journal für die reine und angewandte Mathematik*, 1878(84), 242–258. 112, 116

[Cantor, 1879a] Cantor, G. (1879a). Ueber einen Satz aus der Theorie der stetigen Mannigfaltigkeiten. *Nachrichten von der Königlichen Gesellschaft der Wissenschaften und der Georg-Augusts-Universität zu Göttingen*, 1879, 127–135. 135

[Cantor, 1879b] Cantor, G. (1879b). Ueber unendliche, lineare Punktmannigfaltigkeiten 1. *Mathematische Annalen*, 15, 1–7. 140

[Cantor, 1880] Cantor, G. (1880). Ueber unendliche, lineare Punktmannigfaltigkeiten 2. *Mathematische Annalen*, 17, 355–358. 140

[Cantor, 1882] Cantor, G. (1882). Ueber unendliche, lineare Punktmannigfaltigkeiten 3. *Mathematische Annalen*, 20, 113–121. 102, 140, 152

[Cantor, 1883a] Cantor, G. (1883a). Ueber unendliche, lineare Punktmannigfaltigkeiten 4. *Mathematische Annalen*, 21, 51–58. 140, 141

[Cantor, 1883b] Cantor, G. (1883b). Ueber unendliche, lineare Punktmannigfaltigkeiten 5. *Mathematische Annalen*, 21, 545–591. 58, 140, 141, 142, 287, 289

[Cantor, 1883c] Cantor, G. (1883c). *Ueber unendliche, lineare Punktmannigfaltigkeiten, 5. Grundlagen einer allgemeinen Mannigfaltigkeitslehre. Ein mathematisch-philosophischer Versuch in der Lehre des Unendlichen*. B.G. Teubner. 140, 141, 287, 289

[Cantor, 1884a] Cantor, G. (1884a). De la puissance des ensembles parfaits de points. *Acta Mathematica*, 4(1), 381–392. 411, 417

[Cantor, 1884b] Cantor, G. (1884b). Ueber unendliche, lineare Punktmannigfaltigkeiten 6. *Mathematische Annalen*, 23, 453–488. 140, 142, 411

[Cantor, 1932] Cantor, G. (1932). *Gesammelte Abhandlungen.* A. Fraenkel and E. Zermelo, eds. Springer, Berlin. 58, 101, 102, 103, 112, 116, 135, 140, 141, 142, 152, 287, 289, 411

[Cantor & Dedekind, 1937] Cantor, G. & Dedekind, R. (1937). *Briefwechsel Cantor-Dedekind.* Jean Cavaillès and Emmy Noether, eds., volume 518 of *Actualités scientifiques et industrielles.* Hermann & Cie. xvii, 102, 103, 104, 105, 106, 107, 108, 109, 110, 111, 112, 113, 114, 115, 126, 132, 133, 134, 135

[Carathéodory, 1914] Carathéodory, C. (1914). Über das lineare Mass von Punktmengen—eine Verallgemeinerung des Längenbegriffs. *Nachrichten von der Gesellschaft der Wissenschaften zu Göttingen, Mathematisch-Physikalische Klasse*, 1914, 404–426. 400, 401, 402, 403

[Carathéodory, 1918] Carathéodory, C. (1918). *Vorlesungen über reelle Funktionen.* B.G. Teubner. 403

[Carathéodory, 1927] Carathéodory, C. (1927). *Vorlesungen über reelle Funktionen.* B.G. Teubner, second edition. 403

[Carathéodory, 1957] Carathéodory, C. (1954–1957). *Gesammelte mathematische Schriften. Herausgegeben im Auftrag und mit Unterstützung der Bayerischen Akademie der Wissenschaften.* C. H. Beck'sche. 5 volumes, 1954–1957. 401

[Carathéodory, 1956] Carathéodory, C. (1956). *Mass und Integral und ihre Algebraisierung*. Springer-Verlag. 403

[Carathéodory, 2010] Carathéodory, C. (2010). *Algebraic Theory of Measure and Integration.* American Mathematical Society. 403

[Carnap, 1956] Carnap, R. (1956). *Meaning and Necessity: A Study in Semantics and Modal Logic.* University of Chicago Press, second edition. 4, 429

[Carnap, 1962] Carnap, R. (1962). *Logical Foundations of Probability*. University of Chicago Press, second edition. 4, 5, 6, 429

[Cartwright, 1965] Cartwright, M.L. (1965). Jacques Hadamard, 1865–1963. *Biographical Memoirs of Fellows of the Royal Society*, 11, 75–99. 212

[Cauchy, 1815] Cauchy, A.L. (1815). Mémoire sur la détermination du nombre des racines réelles dans les équations algébriques. *Journal de l'École polytechnique*, X, 457–548. = [Cauchy, 1905, pp. 170–257]. 216

[Cauchy, 1821] Cauchy, A.L. (1821). *Cours d'analyse de l'École Royale Polytechnique; Par M. Augustin-Louis Cauchy ..., 1.re Partie. Analyse algébrique*. de l'Imprimerie royale. 123, 144

[Cauchy, 1831a] Cauchy, A.L. (1831a). *Mémoire sur les rapports qui existent entre le calcul des Résidus et le calcul des Limites, et sur les avantages qu'offrent ces deux nouveaux calculs dans la résolution des équations algébriques ou transcendantes.* Lithograph. Turin. Presented to the Turin Academy on 27 November 1831 and privately published. Handwritten, lithographed. = [Cauchy, 1974, pp. 183–261]. 213, 214

[Cauchy, 1831b] Cauchy, A.L. (1831b). Sur les rapports qui existent entre le calcul des résidus et le calcul des limites, et sur les avantages que présentent ces deux nouveaux calculs dans la résolution des équations algébriques ou transcendantes, formules extraites d'un mémoire présenté le 27 novembre 1831 à l'Académie des Sciences de Turin. *Bulletin des Sciences Mathématiques (Férussac)*, 16, 116–119, 119–128. = [Cauchy, 1958, pp. 169–172, 173–183]. 213, 214

[Cauchy, 1833] Cauchy, A.L. (1833). *Calcul des indices des fonctions*. Turin. Presented to the Turin Academy on 16 June 1833 and privately published. Handwritten, lithographed. 214, 215

[Cauchy, 1837a] Cauchy, A.L. (1837a). Calcul des indices des fonctions. *Journal de l'École Polytechnique*, 15(25), 176–229. = [Cauchy, 1905, pp. 416–466]. 213, 215

[Cauchy, 1837b] Cauchy, A.L. (1837b). Extrait d'une lettre de M. Cauchy sur un mémoire publié par luia Turin le 16 juin 1833 et relatif aux racines des équations simultanées. *Comptes rendus hebdomadaires des séances de l'Académie des Sciences Paris*, 4, 672–675. 214, 215

[Cauchy, 1847] Cauchy, A.L. (1847). Mémoire sur les lieux analytiques. *Comptes rendus hebdomadaires des séances de l'Académie des Sciences Paris*, 24, 885–887. 79

[Cauchy, 1855] Cauchy, A.L. (1855). Sur les compteurs logarithmiques. *Comptes rendus hebdomadaires des séances de l'Académie des Sciences Paris*, 40, 1009–1016. 215

[Cauchy, 1905] Cauchy, A.L. (1905). *Oeuvres complètes d'Augustin Cauchy, II Série*, volume 1. Gauthier-Villars. 456, 457

[Cauchy, 1958] Cauchy, A.L. (1958). *Oeuvres complètes d'Augustin Cauchy, II Série*, volume 2. Gauthier-Villars. 457

[Cauchy, 1974] Cauchy, A.L. (1974). *Oeuvres complètes d'Augustin Cauchy, II Série*, volume 15. Gauthier-Villars. 456

[Cavaillès, 1962] Cavaillès, J. (1962). *Philosophie mathématique*. Hermann. 102

[Cayley, 1844] Cayley, A. (1844). Chapters in the analytical geometry of (n) dimensions. *Cambridge Mathematical Journal*, 4, 119–127. 79

[Čech, 1931] Čech, E. (1931). Sur la théorie de la dimension. *Comptes rendus hebdomadaires des séances de l'Académie des Sciences Paris*, 193, 976–977. Session of 23 November 1931; also in [Čech, 1968, pp. 47–48]. 399

[Čech, 1932] Čech, E. (1932). Sur la dimension des espaces parfaitement normaux. *Bulletin international de l'Académie des Sciences de Bohême*, (pp. 38–55). Translated into French in [Čech, 1968, pp. 70–89] and into English as 'On the dimension of perfectly normal spaces' [Katětov & Simon, 1993, pp. 130–148]. 400

[Čech, 1933] Čech, E. (1933). Příspěvek k teorii dimense. *Časopis pro pěstování matematiky a fysiky*, 62(7), 277–291. Translated into French in [Čech, 1968, pp. 129–142] and into English as 'Contribution to Dimension Theory' [Katětov & Simon, 1993, pp. 149–160]. 400

[Čech, 1937] Čech, E. (1937). On bicompact spaces. *Annals of Mathematics*, 38(4), 823–844. 399

[Čech, 1968] Čech, E. (1968). *Topological Papers of Eduard Čech.* Edited by Miroslav Katětov, Josef Novák, and Alois Švec. Academia – Publishing House of the Czechoslovak Academy of Sciences. 400, 457

[Çevik, ND] Çevik, D. (N.D.). Meeting the metaphysics of geometry: The legacy of Herbart, Gauss and Riemann. *No journal given.* https://www.academia.edu/24975482/MEETING_THE_METAPHYSICS_OF_GEOMETRY_THE_LEGACY_OF_HERBART_GAUSS_AND_RIEMANN. 90

[Charpentier et al., 2010] Charpentier, É., Ghys, E., & Lesne, A., Eds. (2010). *The Scientific Legacy of Poincaré*, volume 36 of *History of Mathematics.* American Mathematical Society. 158

[Christianidis & Oaks, 2023] Christianidis, J. & Oaks, J. (2023). *The Arithmetica of Diophantus: A complete translation and commentary*. Routledge. 19

[Clebsch, 1864] Clebsch, A. (1864). Zur Theorie der algebraischen Flächen. *Journal für die reine und angewandte Mathematik*, 63, 14–26. 423

[Clebsch, 1872] Clebsch, A. (1872). Zum Gedächtniss an Julius Plücker. *Abhandlungen der Königlichen Gesellschaft der Wissenschaften zu Göttingen*, 16, 1–40. 173

[Clifford, 1882] Clifford, W.K. (1882). *Mathematical Papers*. Macmillan and Company. 89

[Cordes & Siegwart, 2024] Cordes, M. & Siegwart, G. (2024). Explication. In *The Internet Encyclopedia of Philosophy*. The Internet Encyclopedia of Philosophy. ISSN 2161-0002, https://iep.utm.edu/explicat/ (accessed 1 March 2024). 4, 429

[Couturat, 1896] Couturat, L. (1896). Études sur l'espace et le temps: De MM. Lechalas, Poincaré, Delbœuf, Bergson, L. Weber et Evellin. *Revue de Métaphysique et de Morale*, 4(5), 646–669. 162, 174

[Cox et al., 2013] Cox, D.A., Little, J., & O'Shea, D. (2013). *Ideals, Varieties, and Algorithms: An Introduction to Computational Algebraic Geometry and Commutative Algebra.* Springer, fourth edition. 422

[Crowe, 1967] Crowe, M.J. (1967). *A History of Vector Analysis: The Evolution of the Idea of a Vectorial System.* University of Notre Dame Press. 81, 84, 85

[Dalen, 1999] Dalen, D.v. (1999). *Mystic, Geometer, and Intuitionist. The Life of L. E. J. Brouwer. Volume 1: The Dawning Revolution.* Oxford University Press. 203, 204, 206, 210, 245, 247

[Dalen, 2005] Dalen, D.v. (2005). *Mystic, Geometer, and Intuitionist. The Life of L. E. J. Brouwer. Volume 2: Hope and Disillusion.* Oxford University Press. 203, 204, 266, 267, 268, 324, 350, 351, 353, 355, 357, 358, 360, 374, 379, 381

[Dalen, 2011a] Dalen, D.v., Ed. (2011a). *The Selected Correspondence of L. E. J. Brouwer*. Springer Science. xviii, 178, 194, 203, 207, 211, 212, 227, 229, 230, 231, 232, 234, 236, 245, 247, 249, 250, 254, 255, 256, 257, 258, 259, 261, 262, 266, 267, 268, 276, 345, 350, 353

[Dalen, 2011b] Dalen, D.v., Ed. (2011b). *Companion to The Selected Correspondence of L. E. J. Brouwer*. Springer Science. Online publication of 3185 pages of Brouwer correspondence and editorial material, dated August 16, 2011. xviii, 178, 194, 203, 204, 206, 207, 209, 211, 212, 227, 230, 231, 232, 234, 236, 245,

247, 249, 250, 254, 255, 256, 257, 258, 259, 260, 261, 262, 265, 266, 267, 268, 276, 341, 345, 350, 351, 353

[Dalen, 2012] Dalen, D.v. (2012). *L. E. J. Brouwer—Topologist, Intuitionist, Philosopher: How Mathematics is Rooted in Life*. Springer Science. 203, 204, 206, 210, 245, 247, 266, 267, 268, 324, 350, 351, 353, 355, 357, 358, 360, 375

[Dalen et al., 2019] Dalen, D.v., Jongbloed, G., Klop, J.W., van Mill, J., et al. (2019). L. E. J. Brouwer, fifty years later. *Indagationes Mathematicae*, 30, 387–402. xvii, 203, 205

[Darboux, 1906] Darboux, G. (1906). Charles Hermite. *La revue du mois*, 1, 37–58. 216

[Dauben, 1971] Dauben, J.W. (1971). The trigonometric background to Georg Cantor's theory of sets. *Archive for History of Exact Sciences*, 7(3), 181–216. 102

[Dauben, 1974] Dauben, J.W. (1974). Denumerability and dimension: the origins of Georg Cantor's theory of sets. *Rete*, 2, 105–134. 102

[Dauben, 1975] Dauben, J.W. (1975). The invariance of dimension: Problems in the early development of set theory and topology. *Historia Mathematica*, 2(3), 273–288. 122

[Dauben, 1979] Dauben, J.W. (1979). *Georg Cantor: His mathematics and philosophy of the infinite*. Harvard University Press. 101, 104, 118

[Dedekind, 1872] Dedekind, R. (1872). *Stetigkeit und irrationale Zahlen*. F. Vieweg & Sohn. 35, 103, 118

[Dedekind, 1931] Dedekind, R. (1931). *Gesammelte mathematische Werke*. Eds. R. Fricke, E. Noether, Ö. Ore, volume 2. Vieweg. 77, 134, 135

[Dehn & Heegaard, 1907] Dehn, M. & Heegaard, P. (1907). Analysis situs. In *Encyklopädie der Mathematischen Wissenschaften mit Einschluss ihrer Anwendungen*, volume III A B 3 (pp. 153–220). B.G. Teubner. 228, 235

[Denjoy, 1910] Denjoy, A. (1910). Continu et discontinu. *Comptes rendue hebdomadaire des séances de l'Académie des Sciences Paris*, 151, 138–140. Session of 11 July 1910. 154, 188, 207, 296

[Denjoy, 1946] Denjoy, A. (1946). Notice sur la vie et l'oeuvre de Henri Lebesgue 1875–1941. 245

[Dick, 1981] Dick, A., Ed. (1981). *Emmy Noether 1882–1935*. Springer. 384

[Diderot et al., 1765] Diderot, D. et al. (1751-1765). *Encyclopédie, ou dictionnaire raisonné des sciences, des arts et des mètiers*. André le Breton, Michel-Antoine David, Laurent Durand and Antoine-Claude Briasson. 29

[Dierker et al., 1998] Dierker, E., Sigmund, K., et al., Eds. (1998). *Karl Menger: Ergebnisse eines Mathematischen Kolloquiums*. Springer Vienna. 363, 412, 489

[Dieudonné, 1972] Dieudonné, J. (1972). The historical development of algebraic geometry. *The American Mathematical Monthly*, 79(8), 827–866. 422

[Dieudonné, 1974] Dieudonné, J. (1974). *Cours de géométrie algébrique*. Presses Universitaires de France. 422, 460

[Dieudonné, 1979] Dieudonné, J. (1979). The tragedy of Grassmann. *Linear and Multilinear Algebra*, 8(1), 1–14. 85

[Dieudonné, 1985] Dieudonné, J. (1985). *History of Algebraic Geometry: An Outline of the History and Development of Algebraic Geometry*. Belmont, CA: Wadsworth International Group. English translation of [Dieudonné, 1974]. 422

[Dieudonné, 1989] Dieudonné, J. (1989). *A History of Algebraic and Differential Topology, 1900-1960*. Birkhäuser. 230, 235, 245

[Dinca & Mawhin, 2009] Dinca, G. & Mawhin, J. (2009). *Brouwer Degree and Applications*. Self published. `https://www.ljll.fr/smets/ULM/Brouwer_Degree_and_applications.pdf`. 212, 218

[Dinca & Mawhin, 2021] Dinca, G. & Mawhin, J. (2021). *Brouwer Degree*. Birlhäuser. 212, 218

[Dirichlet, 1876] Dirichlet, P.G.L. (1876). *Vorlesungen über die im umgekehrten Verhältniss des Quadrats der Entfernung wirkenden Kräfte*. Edited by F. Grube. B.G. Teubner,. 135

[Dirichlet, 1887] Dirichlet, P.G.L. (1887). *Vorlesungen über die im umgekehrten Verhältniss des Quadrats der Entfernung wirkenden Kräfte*. Edited by F. Grube. B.G. Teubner, second edition. 135

[Domoradzki & Stawiska, 2018] Domoradzki, S. & Stawiska, M. (2018). Polish mathematicians and mathematics in World War I. Part I: Galicia (Austro-Hungarian Empire). *Studia Historiae Scientiarum*, 17. 296

[Domoradzki & Stawiska, 2019] Domoradzki, S. & Stawiska, M. (2019). Polish mathematicians and mathematics in World War I. Part II. Russian Empire. *Studia Historiae Scientiarum*, 18. 296

[Duda, 1979] Duda, R. (1979). The origins of the concept of dimension. *Colloquium Mathematicum*, 42(1), 95–110.

[Dugac, 1976a] Dugac, P. (1976a). Notes et documents sur la vie et l'œuvre de René Baire. *Archive for History of Exact Sciences*, 15(4), 297–383. 191, 193, 194, 259

[Dugac, 1976b] Dugac, P. (1976b). *Richard Dedekind et les fondements des mathématiques: avec de nombreux textes inédits*. Vrin. 101, 102

[Dvoretzky, 1948] Dvoretzky, A. (1948). A note on Hausdorff dimension functions. *Mathematical Proceedings of the Cambridge Philosophical Society*, 44(1), 13–16.

[Dyck, 1888] Dyck, W.v. (1888). Beiträge zur Analysis situs. I. *Mathematische Annalen*, 32(4), 457–512. 139, 226, 235

[Dyck, 1890] Dyck, W.v. (1890). Beiträge zur Analysis situs. II. *Mathematische Annalen*, 37(2), 273–316. 226, 235

[Edgar, 2004] Edgar, G.A., Ed. (2004). *Classics on Fractals*. Westview Press. 400, 401, 402, 403, 404, 405, 406, 407, 408, 409, 410, 412, 413, 419, 420, 485, 486

[Edgar, 2008] Edgar, G.A. (2008). *Measure, Topology, and Fractal Geometry*. Springer, second edition. 404, 416

[Edmonds, 2020] Edmonds, D. (2020). *The Murder of Professor Schlick: The Rise and Fall of the Vienna Circle*. Princeton University Press. 308, 363

[Edwards, 1988] Edwards, H. (1988). Kronecker's place in history. In *History and Philosophy of Modern Mathematics* (pp. 139–144). University of Minnesota Press. 216

[Edwards, 2009] Edwards, H.M. (2009). Kronecker's Algorithmic Mathemaitcs. *The Mathematical Intelligencer*, 2(31), 11–14. 218

[Eichhorn, 1994] Eichhorn, E. (1994). In Memoriam Felix Hausdorff (1868–1942). Ein biographischer Versuch. *Heldermann Verlag, Berlin*. 316, 404

[Eilenberg & Steenrod, 1952] Eilenberg, S. & Steenrod, N. (1952). *Foundations of Algebraic Topology*, volume 15 of *Princeton Mathematical Series*. Princeton University Press. 364, 389, 438

[Eisenbud, 1995] Eisenbud, D. (1995). *Commutative Algebra with a View Toward Algebraic Geometry*, volume 150 of *Graduate Texts in Mathematics*. Springer. 422, 423, 424, 425, 426

[Eisermann, 2008] Eisermann, M. (2008). The fundamental theorem of algebra made effective: an elementary real-algebraic proof via Sturm chains. *arXiv*, 2008.
https://arxiv.org/pdf/0808.0097. 216

[Eisermann, 2012] Eisermann, M. (2012). The fundamental theorem of algebra made effective: an elementary real-algebraic proof via Sturm chains. *The American Mathematical Monthly*, 119(9), 715–752. 216

[Encyklopädie, 1935] Encyklopädie (1898–1935). *Encyklopädie der mathematischen Wissenschaften mit Einschluss ihrer Anwendungen*. B.G. Teubner.

[Engel, 1911] Engel, F. (1911). Grassmanns Leben: Nebst einem Verzeichnisse der von Grassmann veröffentlichten Schriften und einer übersicht des handschriftlichen Nachlasses. In *Grassmanns Gesammelte mathematische und physikalische Werke*, volume 3.2. B.G. Teubner, Leipzig. 80, 84

[Engelking, 1978] Engelking, R. (1978). *Dimension Theory*. North-Holland Publishing Company Amsterdam. 195, 274, 311, 341, 348, 374, 400

[Engelking, 1995] Engelking, R. (1995). *Theory of Dimensions, Finite and Infinite*, volume 10 of *Sigma Series in Pure Mathematics*. Heldermann Verlag. 311, 341

[Engelking, 1998] Engelking, R. (1998). Wacław Sierpiński (1882–1969) His Life and Work in Topology. In *Handbook of the History of General Topology*, volume 2 (pp. 399–414). Kluwer Academic Publishers. xix, 299, 312, 313, 314

[Engelking & Pol, 1995] Engelking, R. & Pol, R. (1995). Hurewicz's contributions to dimension theory. In *Collected Works of Witold Hurewicz* (pp. xxi–xxviii). American Mathematical Society. 365

[Enriques, 1898] Enriques, F. (1898). Sulle ipotesi che permettono l'introduzione delle coordinate in una varieté a più dimensioni. *Rendiconti del Circolo Matematico di Palermo*, 12(1), 222–239.

[Erdmann, 1877] Erdmann, B. (1877). *Die Axiome der Geometrie: eine philosophische Untersuchung der Riemann-Helmholtz'schen Raumtheorie*. L. Voss. 98

[Ewald, 1996] Ewald, W.B. (1996). *From Kant to Hilbert: A Source Book in the Foundations of Mathematics. 2 volumes*. Oxford University Press. 102

[Falconer, 2014] Falconer, K.J. (2014). *Fractal Geometry: Mathematical Foundations and Applications*. John Wiley & Sons, third edition. 404, 416

[Falconer et al., 2023] Falconer, K.J., Fraser, J., & Käenmäki, A. (2023). Minkowski dimension for measures. *Proceedings of the American Mathematical Society*, 151(02), 779–794. 420

[Falconer & Troscheit, 2023] Falconer, K.J. & Troscheit, S. (2023). Box-counting dimension in one-dimensional random geometry of multiplicative cascades. *Communications in Mathematical Physics*, 399(1), 57–83. 420

[Fano, 1907] Fano, G. (1907). Kontinuierliche geometrische Gruppen. Die Gruppentheorie als geometrisches Einteilungsprinzip. In *Encyklopädie der Mathematischen Wissenschaften mit Einschluss ihrer Anwendungen*, volume III A B 4a (pp. 289–388). B.G. Teubner. 168

[Fearnley-Sander, 1979] Fearnley-Sander, D. (1979). Hermann Grassmann and the creation of linear algebra. *The American Mathematical Monthly*, 86(10), 809–817. 85

[Fearnley-Sander, 1982] Fearnley-Sander, D. (1982). Hermann Grassmann and the prehistory of universal algebra. *The American Mathematical Monthly*, 89(3), 161–166. 85

[Feigl, 1928] Feigl, G. (1928). Geschichtliche Entwicklung der Topologie. *Jahresbericht der Deutschen Mathematiker-Vereinigung*, 37, 273–286.

[Ferreirós, 1993] Ferreirós, J. (1993). On the relations between Georg Cantor and Richard Dedekind. *Historia Mathematica*, 20(4), 343–363. 102, 103, 118

[Ferreirós, 2006] Ferreirós, J. (2006). Riemann's Habilitationsvortrag at the Crossroads of Mathematics, Physics, and Philosophy. In *The Architecture of Modern Mathematics* (pp. 67–96). Oxford University Press. 86

[Ferreirós, 2007] Ferreirós, J. (2007). *Labyrinth of Thought: A History of Set Theory and Its Role in Modern Mathematics*. Birkhäuser, second revised edition. 75, 86, 101, 102, 103, 118

[Ferreirós, 2015] Ferreirós, J. (2015). *Mathematical Knowledge and the Interplay of Practices*. Princeton University Press. 8

[Ferreirós, 2024] Ferreirós, J. (2024). The Early Development of Set Theory. In E.N. Zalta & U. Nodelman (Eds.), *The Stanford Encyclopedia of Philosophy*. Metaphysics Research Lab, Stanford University, Winter 2024 edition. 102

[Ferreirós & Gray, 2006] Ferreirós, J. & Gray, J.J., Eds. (2006). *The Architecture of Modern Mathematics: Essays in History and Philosophy*. Oxford University Press.

[Fitzpatrick, 2008] Fitzpatrick, R. (2008). *Euclid's Elements of Geometry*. Published online.
`https://farside.ph.utexas.edu/books/Euclid/Elements.pdf`
(accessed 15 March 2024). 16

[Fleron, 1994] Fleron, J.F. (1994). A note on the history of the Cantor set and Cantor function. *Mathematics Magazine*, 67(2), 136–140. 141

[Folta, 1966] Folta, J. (1966). Bernard Bolzano and the foundations of geometry. *Acta historiae rerum naturalium necnon technicarum*, 2, 75–104. 31, 34, 36

[Folta, 1981] Folta, J. (1981). Life and scientific endeavour of Bernard Bolzano. In *Bolzano and the Foundations of Mathematical Analysis* (pp. 11–32). Society of Czechoslovak Mathematicians and Physicists. 31, 34, 36

[Fowler, 1987] Fowler, D.H. (1987). *The Mathematics of Plato's Academy*. Oxford University Press. 22

[Fraenkel, 1920] Fraenkel, A. (1920). Zahlbegriff und Algebra bei Gauss. Mit einem Anhang von A. Ostrowski in Göttingen: Zum ersten und vierten Gaussschen Beweise des Fundamentalgesetzes der Algebra. *Nachrichten von der Gesellschaft der Wissenschaften zu Göttingen, Mathematisch-Physikalische Klasse*, 1920, 1–58. 78

[Fraenkel, 1930] Fraenkel, A.A. (1930). Georg Cantor. *Jahresbericht der Deutschen Mathematiker-Vereignigung*, 39, 189–266. 101, 103, 117, 118

[Fréchet, 1906] Fréchet, M. (1906). Sur quelques points du calcul fonctionnel. *Rendiconti del Circolo Matematico di Palermo*, 22(1), 1–72. 146, 189, 194, 196, 197, 263, 310, 316, 347

[Fréchet, 1909] Fréchet, M. (1909). Une définition du nombre de dimensions d'un ensemble abstrait. *Comptes rendus hebdomadaires des séances de l'Académie des Sciences Paris*, 148, 1152–1154. Session of 3 May 1909. 196

[Fréchet, 1910] Fréchet, M. (1910). Les dimensions d'un ensemble abstrait. *Mathematische Annalen*, 68(2), 145–168. 195, 196, 197, 198, 199, 200, 379, 419

[Fréchet, 1924] Fréchet, M. (1924). Number of dimensions of an abstract set. In *Proceedings of the International Congress of Mathematicians, Toronto* (pp. 399–412). 200

[Fréchet, 1928] Fréchet, M. (1928). *Les Espaces abstraits*. Gauthier-Villars. 196, 201

[Frei & Stammbach, 1999] Frei, G. & Stammbach, U. (1999). Heinz Hopf. In *History of Topology* (pp. 991–1008). Elsevier Amsterdam. 384

[Freudenthal, 1955] Freudenthal, H. (1955). Neuere Fassungen des Riemann-Helmholtz-Lieschen Raumproblems. *Mathematische Zeitschrift*, 63, 374–405. 160

[Freudenthal, 1960] Freudenthal, H. (1960). Die Grundlagen der Geometrie um die Wende des 19. Jahrhunderts. *Semesterberichte Münster*, 7, 2–25. 158, 160, 177

[Freudenthal, 1964] Freudenthal, H. (1964). Lie groups in the foundations of geometry. *Advances in Mathematics*, 1(2), 145–190. 160

[Freudenthal, 1975] Freudenthal, H. (1975). The cradle of modern topology, according to Brouwer's inedita. *Historia Mathematica*, 2(4), 495–502. 231

[Freudenthal, 1977] Freudenthal, H. (1977). Een manuscript van Brouwer ['De invariantie van het aantal dimensies eener ruimte']. In T.J. Jong, R.H. de & P. Krijgsman (Eds.), *Tweehonderd jaar onvermoeide arbeid*, volume 2 (pp. 43–55). Mathematisch Centrum. 237, 242

[Fricke, 1904] Fricke, R. (1904). Beiträge zum Kontinuitätsbeweise der Existenz linear-polymorpher Funktionen auf Riemannschen Flächen. *Mathematische Annalen*, 59(4), 449–513. 156

[Fricke, 1905] Fricke, R. (1905). Neue Entwicklungen über den Existenzbeweis der polymorphen Funktionen. *Verhandlungen des dritten internationalen Mathematiker-Kongresses in Heidelberg vom 8. bis 13. August 1904*, (pp. 246–252). 156

[Fricke, 1913] Fricke, R. (1913). Automorphe Funktionen mit Einschluss der elliptischen Modutfunktionen. In *Encyklopädie der mathematischen Wissenschaften mit Einschluss ihrer Anwendungen*, volume II B 4 (pp. 349–470). B.G. Teubner. 156

[Fricke & Klein, 1897] Fricke, R. & Klein, F. (1897). *Vorlesungen über die Theorie der Automorphen Functionen. I*. B.G. Teubner. 156

[Fricke & Klein, 1912] Fricke, R. & Klein, F. (1912). *Vorlesungen über die Theorie der Automorphen Functionen. II*. B.G. Teubner. 156, 245

[Fritz, 1955] Fritz, K.v. (1955). Die APXAI in der griechischen Mathematik. *Archiv für Begriffsgeschichte*, 1, 13–103.

[Gauss, 1813] Gauss, C.F. (1813). Theoria attractionis corporum sphaeroidicorum ellipticorum homogeneorum. *Commentationes societatis regiae scientiarum Gottingensis recentiores*, 2. = [Gauss, 1867, pp. 1–22]. 223

[Gauss, 1828] Gauss, C.F. (1828). *Disquisitiones generales circa superficies curvas*. Typis Dieterichianis. 77

[Gauss, 1831] Gauss, C.F. (1831). Anzeige: Theoria residuorum biquadraticorum. commentatio secunda. *Göttingische gelehrte Anzeigen*, 1831 April 23. 75, 76, 89

[Gauss, 1832] Gauss, C.F. (1832). *Theoria residuorum biquadraticorum. Commentatio secunda*. Typis Dieterichchianis. 75, 89

[Gauss, 1840] Gauss, C.F. (1840). *Allgemeine Lehrsätze in Beziehing auf die im Verkehrten Verhältnisse des Quadrats der Entfernung wirkenden Anziehungs- und Abstossungs-Kräfte*. Leipzig. = [Gauss, 1867, pp. 195–242]. 223

[Gauss, 1849a] Gauss, C.F. (1849a). Beiträge zur Theorie der algebraischen Gleichungen. *Abhandlingen der Königlichen Gesellschaft der Wissenschaften zu Göttingen*, 4. 77, 89

[Gauss, 1849b] Gauss, C.F. (1849b). *Beiträge zur Theorie der algebraischen Gleichungen*. Dieterich. 77, 89

[Gauss, 1863] Gauss, C.F. (1863). *Werke*, volume 2. Königliche Gesellschaft der Wissenschaften, Göttingen. 75, 76, 78, 89

[Gauss, 1866] Gauss, C.F. (1866). *Werke*, volume 3. Königliche Gesellschaft der Wissenschaften, Göttingen. 77, 89

[Gauss, 1867] Gauss, C.F. (1867). *Werke*, volume 5. Königliche Gesellschaft der Wissenschaften, Göttingen. 125, 223, 253, 464

[Gauss, 1900a] Gauss, C.F. (1900a). Neue allgemeine Untersuchungen über die krummen Flächen. Dated 1825. In *Gauss Werke*, volume 8 (pp. 408–442). B.G. Teubner. 77

[Gauss, 1900b] Gauss, C.F. (1900b). *Werke*, volume 8. Königliche Gesellschaft der Wissenschaften, Göttingen. 78

[Gauss, 1917] Gauss, C.F. (1917). *Werke*, volume 10.1. Königliche Gesellschaft der Wissenschaften, Göttingen. 75, 77, 78

[Gauss, 1929] Gauss, C.F. (1929). *Werke*, volume 12. Königliche Gesellschaft der Wissenschaften, Göttingen. 76

[Gauss, 1933] Gauss, C.F. (1933). *Werke*, volume 10.2. Königliche Gesellschaft der Wissenschaften, Göttingen. Comprised of seven in-depth studies of selected areas of Gauss' mathematical work.

[Gauss, 2005] Gauss, K.F. (2005). *General Investigations of Curved Surfaces.* Edited Peter Pesic. Courier Corporation. 77

[Genocchi, 1884] Genocchi, A. (1884). *Calcolo differenziale e principii di calcolo integrale. Pubblicato con Aggiunte dal Dr. Giuseppe Peano*. Fratelli Bocca. 147

[Georgiadou, 2004] Georgiadou (2004). *Constantin Carathéodory*. Springer. 401

[Giannetto, 2023] Giannetto, E. (2023). Henry More and the vision of a fourth dimension. *AGON*, 36, 40–75. 26

[Giedymin, 1977] Giedymin, J. (1977). On the origin and significance of Poincaré's conventionalism. *Studies in History and Philosophy of Science Part A*, 8(4), 271–301. 158, 159, 160, 163, 173, 180

[Giedymin, 1982] Giedymin, J. (1982). *Science and Convention: Essays on Henri Poincaré's Philosophy of Science and The Conventionalist Tradition*. Pergamon Press. 158

[Giedymin, 1991] Giedymin, J. (1991). Geometrical and physical conventionalism of Henri Poincaré in epistemological formulation. *Studies in History and Philosophy of Science Part A*, 22(1), 1–22. 158

[Gilain, 1991] Gilain, C. (1991). La théorie qualitative de Poincaré et le problème de l'intégration des équations différentielles. *Cahiers d'histoire et de philosophie des sciences*, 34, 215–242. 224

[Gillies, 1981] Gillies, D. (1981). Karl Menger as a philosopher. *The British Journal for the Philosophy of Science*, 32(2), 183–196. 333, 363

[Gillies, 2023] Gillies, D. (2023). *Lakatos and the Historical Approach to Philosophy of Mathematics*. Cambridge University Press. 7

[Gispert, 1980] Gispert, H. (1980). Correspondance de Fréchet (1907-1926) et son apport à la théorie de la dimension (avec 3 lettres de Brouwer à Baire). *Cahiers du séminaire d'histoire des mathématiques*, 1, 69–120. 194, 196, 197

[Gispert-Chambaz, 1982] Gispert-Chambaz, H. (1982). *Camille Jordan et les fondements de l'analyse: Comparaison de la 1ère édition (1882-1887) et de la 2ème (1893) de son cours d'analyse de l'École Polytechnique*. Université de Paris-Sud, Départment de Mathématique. 143

[Goe, 1964] Goe, G. (1964). Kästner, forerunner of Gauss, Pasch, Hilbert. In H. Guerlac (Ed.), *Actes du dixième congrès international d'histoire des sciences : Proceedings of the tenth international congress of the history of science*, volume 1. Hermann. 35

[Goe, 1980] Goe, G. (1970–1980). Kästner, Abraham Gotthelf. In C.C. Gillispie (Ed.), *Dictionary of Scientific Biography*, volume 7 (pp. 206–207). Charles Scribner & Sons. 36

[Goursat, 1884] Goursat, É. (1884). Démonstration du théorème de Cauchy. *Acta Mathematica*, 4(1), 197–200. 144

[Goursat, 1900] Goursat, É. (1900). Sur la définition générale des fonctions analytiques, d'après Cauchy. *Transactions of the American Mathematical Society*, 1(1), 14–16. 144

[Graham & Kantor, 2009] Graham, L. & Kantor, J.M. (2009). *Naming infinity: A true story of religious mysticism and mathematical creativity*. Harvard University Press. 384

[Grassmann, 1844] Grassmann, H.E. (1844). *Die lineale Ausdehnungslehre ein neuer Zweig der Mathematik: dargestellt und durch Anwendungen auf die übrigen Zweige der Mathematik, wie auch auf die Statik, Mechanik, die Lehre vom Magnetismus und die Krystallonomie erläutert. Erster Theil*, volume 1. Otto Wigand. 81, 82, 83, 84

[Grassmann, 1845] Grassmann, H.E. (1845). Kurze Uebersicht über das Wesen der Ausdehnungslehre. *Archiv der Mathematik und Physik*, 6(1845), 337–350. 84

[Grassmann, 1862] Grassmann, H.E. (1862). *Die Ausdehnungslehre*. Enslin. 85

[Grassmann, 1896] Grassmann, H.E. (1894–1896). *Gesammelte mathematische und physikalische Werke*, volume 1. B.G. Teubner. 81, 84, 85

[Grassmann, 1904] Grassmann, H.E. (1902–1904). *Gesammelte mathematische und physikalische Werke*, volume 2. B.G. Teubner.

[Grassmann, 1911] Grassmann, H.E. (1911). *Gesammelte mathematische und physikalische Werke*, volume 3. B.G. Teubner. 80

[Grassmann, 1995] Grassmann, H.E. (1995). *A New Branch of Mathematics: The Ausdehnungslehre of 1844, and Other Works.* Trans. Lloyd C. Kannenberg. Open Court. 81, 82, 83, 84

[Grassmann, 2000] Grassmann, H.E. (2000). *Extension Theory.* Trans. Lloyd C. Kannenberg, volume 19. American Mathematical Society, Providence, RI, USA. 85

[Grassmann & Möbius, 1847] Grassmann, H.E. & Möbius, A.F. (1847). *Geometrische Analyse geknüpft an die von Leibniz erfundene geometrische Charakteristik*, volume 1. Weidmann'sche Buchhandlung. 84

[Grattan-Guiness, 1974] Grattan-Guiness, I. (1974). The Rediscovery of the Cantor-Dedekind Correspondence. *Jahresbericht der Deutschen Mathematiker-Vereinigung*, 76, 104–139. 102, 117, 118, 132

[Gray, 1994] Gray, J.J. (1994). Eduard Čech. *Mathematical Intelligencer*, 16(4), 48–49. 399

[Gray, 1997] Gray, J.J. (1997). Algebraic Geometry between Noether and Noether—a forgotten chapter in the history of Algebraic Geometry. *Revue d'histoire des mathématiques*, 3(1), 1–48. 422

[Gray, 2011] Gray, J.J. (2011). *Worlds Out of Nothing: A Course in the History of Geometry in the 19th Century*, volume 193 of *Springer Undergraduate Mathematics Series*. Springer. 98

[Gray, 2013] Gray, J.J. (2013). *Henri Poincaré: A scientific biography*. Princeton University Press. 158, 185

[Gray, 2017] Gray, J.J. (2017). Riemann om Geometry, Physics, and Philosophy—Some Remarks. In *From Riemann to Differential Geometry and Relativity* (pp. 97–110). Springer. 86

[Gray, 2019] Gray, J.J. (2019). *Simply Riemann*, volume 17. Simply Charly. 86

[Gray & Ferreirós, 2021] Gray, J.J. & Ferreirós, J. (2021). Epistemology of Geometry. In E.N. Zalta (Ed.), *The Stanford Encyclopedia of Philosophy*. Metaphysics Research Lab, Stanford University, Fall 2021 edition. `https://plato.stanford.edu/archives/fall2021/entries/epistemology-geometry/` (accessed 15 March 2024). 23, 98

[Gray et al., 2000] Gray, J.J., Rowe, D., et al. (2000). *The Hilbert Challenge*. Oxford University Press on Demand. 204

[Green, 1835] Green, G. (1835). On the determination of the exterior and interior attractions of ellipsoids of variable densities. *Transactions of the Cambridge Philosophical Society*, 5, 185–222. 74, 75

[Green, 1871] Green, G. (1871). *Mathematical Papers*. Macmillan. 75

[Griffiths et al., 1992] Griffiths, P., Spencer, D., & Whitehead, G. (1992). Solomon Lefschetz: 1884–1972. *National Academy of Sciences Biographical Memoirs*, 61, 269–313. 395

[Guggenheimer, 1977] Guggenheimer, H. (1977). The Jordan curve theorem and an unpublished manuscript by Max Dehn. *Archive for History of Exact Sciences*, 17(2), 193–200. 143, 146

[Guillén, 2024] Guillén, E.F. (2024). Bolzano's Theory of meßbare Zahlen: Insights and Uncertainties Regarding the Number Continuum. In *Handbook of the History and Philosophy of Mathematical Practice* (pp. 1–38). Springer.

[Guillén & Crippa, 2021] Guillén, E.F. & Crippa, D. (2021). The 1804 examination for the chair of Elementary Mathematics at the University of Prague. *Historia Mathematica*, 57, 24–54. 39

[Gupta & Mackereth, 2023] Gupta, A. & Mackereth, S. (2023). Definitions. In E.N. Zalta & U. Nodelman (Eds.), *The Stanford Encyclopedia of Philosophy*. Metaphysics Research Lab, Stanford University, Fall 2023 edition. `https://plato.stanford.edu/archives/fall2023/entries/definitions/` (accessed 15 March 2024). 4, 429

[Guthrie, 1962] Guthrie, W.K.C. (1962). *A History of Greek Philosophy: Volume 1, The Earlier Prosocratics and the Pythagoreans*. Cambridge University Press. 12, 13, 14

[Hadamard, 1910] Hadamard, J. (1910). Sur quelques applications de l'indice de Kronecker. In Jules Tannery, *Introduction à la Theorie des Fonctions d'une variable*, volume II (pp. 437–477). Librairie Scientifique à Hermann & fils, second edition. 223, 227, 228, 229, 281

[Hahn, 1913] Hahn, H. (1913). Über die Abbildung einer Strecke auf ein Quadrat. *Annali di Matematica Pura ed Applicata (1898-1922)*, 21(1), 33–55. 309, 334

[Hahn, 1914a] Hahn, H. (1914a). Über die allgemeinste ebene Punktmenge, die stetige Bild einer Strecke ist. *Jahresberichte der Deutschen Mathematiker-Vereinigung*, 23, 318–322. 309, 334

[Hahn, 1914b] Hahn, H. (1914b). Mengentheoretische Charakterisierung der stetigen Kurve. *Sitzungsberichte der Kaiserlichen Akademie der Wissenschaften in Wien. Mathematisch-Naturwissenschaftliche Klasse*, 123, 2433–2490. 309, 310, 334

[Hahn, 1980] Hahn, H. (1980). *Empiricism, Logic and Mathematics: Philosophical Papers*, volume 13 of *Vienna Circle Collection*. D. Reidel Publishing Company. 308

[Hahn, 1995] Hahn, H. (1995). *Gesammelte Abhandlungen / Collected Works*, volume I. Springer-Verlag. 308

[Hahn, 1996] Hahn, H. (1996). *Gesammelte Abhandlungen / Collected Works*, volume II. Springer-Verlag. 308

[Hahn, 1997] Hahn, H. (1997). *Gesammelte Abhandlungen / Collected Works*, volume III. Springer-Verlag. 308

[Hales, 2007a] Hales, T.C. (2007a). The Jordan curve theorem, formally and informally. *The American Mathematical Monthly*, 114(10), 882–894. 146

[Hales, 2007b] Hales, T.C. (2007b). Jordan's proof of the Jordan curve theorem. *Studies in logic, grammar and rhetoric*, 10(23), 45–60. 146

[Hamilton, 1853] Hamilton, W.R. (1853). *Lectures on Quaternions: Containing a Systematic Statement of a New Mathematical Method; of which the Principles Were Communicated in 1843 to the Royal Irish Academy; and which Has Since Formed the Subject of Successive Courses of Lectures, Delivered in 1848 and Sub Sequent Years, in the Halls of Trinity College, Dublin: With numerous Illustrative Diagrams, and with Some Geometrical and Physical Applications*. University Press by MH. 85

[Hartshorne, 1977] Hartshorne, R. (1977). *Algebraic Geometry*. Springer. 422, 424

[Hausdorff, 1914] Hausdorff, F. (1914). *Grundzüge der Mengenlehre*. von Veit. 97, 193, 197, 209, 263, 265, 282, 313, 316, 325, 336, 344, 356, 368, 373, 379, 404

[Hausdorff, 1919] Hausdorff, F. (1919). Dimension und äußeres Maß. *Mathematische Annalen*, 79, 157–179. xix, 379, 400, 404, 405, 406, 407, 408, 409, 410, 411, 415, 419

[Hausdorff, 2001] Hausdorff, F. (2001). *Gesammelte Werke, Band IV, Analysis, Algebra, und Zahlentheorie*, volume IV. Springer. 404, 411

[Hausdorff, 2021] Hausdorff, F. (2001–2021). *Gesammelte Werke*. Springer. 10 volumes, 2001–2021.

[Hausdorff, 2002] Hausdorff, F. (2002). *Gesammelte Werke, Band II, Grundzüge der Mengenlehre*, volume II. Springer. 316, 404

[Heath, 1910] Heath, T.L. (1910). *Diophantus of Alexandria: A study in the history of Greek algebra*. Cambridge University Press Archive. 19

[Heath, 1926] Heath, T.L. (1926). *The Thirteen Books of Euclid's Elements*. Cambridge University Press, second edition. 16

[Heath, 1949] Heath, T.L. (1949). *Mathematics in Aristotle*. Oxford: Clarendon Press. 15, 17, 18, 39

[Heath, 1956] Heath, T.L. (1956). *The Thirteen Books of Euclid's Elements*. Dover Publications, second edition. 16, 17, 18, 21

[Heath, 2016] Heath, T.L. (2016). *Mathematics in Aristotle*. (Reprint). Routledge. 17

[Heinzmann & Stump, 2024] Heinzmann, G. & Stump, D. (2024). Henri Poincaré. In E.N. Zalta & U. Nodelman (Eds.), *The Stanford Encyclopedia of Philosophy*. Metaphysics Research Lab, Stanford University, Summer 2024 edition. `https://plato.stanford.edu/archives/sum2024/entries/poincare/` (accessed 29 June 2024). 158

[Helmholtz, 1868a] Helmholtz, H.v. (1868a). Über die Thatsachen, die der Geometrie zum Grunde liegen. *Nachrichten von der Königlichen Gesellschaft der Wissenschaften zu Göttingen*, 9, 193–221. 97

[Helmholtz, 1868b] Helmholtz, H.v. (1868b). Über die thatsächlichen Grundlagen der Geometrie. *Verhandlungen des naturhistorisch-medicinischen Vereins zu Heidelberg*, 4(1866), 197–202. 97

[Helmholtz, 1883] Helmholtz, H.v. (1883). *Wissenschaftliche Abhandlungen*, volume 2. J A Barth. 97

[Henry, 2023] Henry, J. (2023). Henry More. In E.N. Zalta & U. Nodelman (Eds.), *The Stanford Encyclopedia of Philosophy*. Metaphysics Research Lab, Stanford University, Summer 2023 edition.
`https://plato.stanford.edu/archives/sum2023/entries/henry-more/` (accessed 8 April 2024). 26

[Herbart, 1822] Herbart, J.F. (1822). *Ueber die Möglichkeit und Nothwendigkeit, Mathematik auf Psychologie anzuwenden*. Bornträger. 91

[Herbart, 1824] Herbart, J.F. (1824). *Psychologie als Wissenschaft, neu gegründet auf Erfahrung, Metaphysik und Mathematik. Erster, synthetischer Theil.*, volume 1. A.W. Unzer. 91

[Herbart, 1825] Herbart, J.F. (1825). *Psychologie als Wissenschaft, neu gegründet auf Erfahrung, Metaphysik und Mathematik. Zweiter, analytischer Theil.*, volume 2. A.W. Unzer. 91, 92

[Herbart, 1828] Herbart, J.F. (1828). *Allgemeine Metaphysik, nebst den Anfängen der philosophischen Naturlehre. Erster historisch-kritischer Theil.*, volume 1. A.W. Unzer. 91

[Herbart, 1829] Herbart, J.F. (1829). *Allgemeine Metaphysik, nebst den Anfängen der philosophischen Naturlehre. Zweyter, systematischer Theil.*, volume 2. A.W. Unzer. 91, 92

[Herbart, 1912] Herbart, J.F. (1887–1912). *Sämtliche Werke: in chronologischer Reihenfolge*. Beyer. 19 volumes. 91, 92

[Hermite, 1852] Hermite, C. (1852). Sur l'extension du théorème de M. Sturm à un système d'équations simultanées. *Comptes rendus hebdomadaires des séances de l'Académie des Sciences Paris*, 35, 52–54. 216

[Hermite, 1853] Hermite, C. (1853). Remarques sur le théorème de M. Sturm. *Comptes rendus hebdomadaires des séances de l'Académie des Sciences Paris*, 36, 294–297. 216

[Hermite, 1856] Hermite, C. (1856). Sur le nombre des racines d'une équation algébrique comprises entre des limites données. Extrait d'une Lettre à M. Borchardt. *Journal für die reine und angewandte Mathematik*, 52, 39–51. 216

[Hestenes, 1996] Hestenes, D. (1996). Grassmann's vision. In *Hermann Günther Grassmann (1809–1877): Visionary Mathematician, Scientist and Neohumanist Scholar* (pp. 243–254). Springer. 85

[Hestenes, 2011] Hestenes, D. (2011). Grassmann's legacy. In *From Past to Future: Grassmann's Work in Context* (pp. 243–260). Springer. 85

[Hewitt, 1960] Hewitt, E. (1960). The rôle of compactness in analysis. *The American Mathematical Monthly*, 67(6), 499–516. 326

[Hilbert, 1891] Hilbert, D. (1891). Ueber die stetige Abbildung einer Line auf ein Flächenstück. *Mathematische Annalen*, 38(3), 459–460. 148, 149, 248, 309

[Hilbert, 1899] Hilbert, D. (1899). Grundlagen der Geometrie. In *Festschrift der Feier der Enthüllung des Gauss-Weber-Denkmals in Göttingen* (pp. 1–92). B.G. Teubner. 22, 23, 99, 177, 184

[Hilbert, 1901] Hilbert, D. (1901). Mathematische Probleme. *Archiv der Mathematik und Physik,* 3. Reihe, 1, 44–63, 213–237. 1, 204

[Hilbert, 1902a] Hilbert, D. (1902a). Mathematical Problems. *Bulletin of the American Mathematical Society*, 8(10), 437–479. 1, 204

[Hilbert, 1902b] Hilbert, D. (1902b). Ueber die Grundlagen der Geometrie. *Nachrichten von der Gesellschaft der Wissenschaften zu Göttingen, Mathematisch-Physikalische Klasse*, 1902, 233–241. 97, 189

[Hilbert, 1903] Hilbert, D. (1903). Ueber die Grundlagen der Geometrie. *Mathematische Annalen*, 56(3), 381–422. 97, 189, 316

[Hilbert, 1910] Hilbert, D. (1910). Hermann Minkowski. *Mathematische Annalen*, 68, 445–471. 416

[Hilbert, 1935] Hilbert, D. (1935). Mathematische Probleme. In *Gesammelte Abhandlungen,* Dritter Band: *Analysis· Grundlagen der Mathematik· Physik Verschiedenes nebst einer Lebensgeschichte* (pp. 290–329). Springer. 204

[Hilton, 1972] Hilton, P.J. (1972). Heinz Hopf. *Bulletin of the London Mathematical Society*, 4(2), 202–217. 384

[Hirzebruch & Behnke, 1972] Hirzebruch, F. & Behnke, H. (1972). In memoriam Heinz Hopf. *Mathematische Annalen*, 196, 1–7. 384

[Hodge, 1973] Hodge, W.V.D. (1973). Solomon Lefschetz. 1884–1972. *Biographical Memoirs of Fellows of the Royal Society*, 19, 433–453. 395

[Hopf, 1927a] Hopf, H. (1927a). Abbildungsklassen *n*-dimensionaler Mannigfaltigkeiten. *Mathematische Annalen*, 96(1), 209–224. 233

[Hopf, 1927b] Hopf, H. (1927b). Vektorfelder in *n*-dimensionalen Mannigfaltigkeiten. *Mathematische Annalen*, 96(1), 225–249. 225, 227, 229

[Hopf, 1953] Hopf, H. (1953). Vom Bolzanoschen Nullstellensatz zur algebraischen Homotopietheorie der Sphären. *Jahresbericht der Deutschen Mathematiker-Vereinigung*, 56, 59–76. 123

[Hopf, 1966] Hopf, H. (1966). Ein Abschnitt aus der Entwicklung der Topologie. *Jahresbericht der Deutschen Mathematiker-Vereinigung*, 68, 182–192. 229

[Hopf, 2001a] Hopf, H. (2001a). *Collected Papers – Gesammelte Abhandlungen.* Ed. B. Eckmann. Springer. 384

[Hopf, 2001b] Hopf, H. (2001b). Einige personliche Erinnerungen aus der Vorgeschichte der heutigen Topologie. In *Collected Papers – Gesammelte Abhandlungen* (pp. 1180–1191). Springer. 384

[Hurewicz, 1926a] Hurewicz, W. (1926a). Über eine Verallgemeinerung des Borelschen Theorems. *Mathematische Zeitschrift*, 24(1), 401–421.

[Hurewicz, 1926b] Hurewicz, W. (1926b). Über stetige Bilder von Punktmengen. *Proceedings of the Section of Sciences, Koninklijke Nederlandse Akademie van Wetenschappen te Amsterdam*, 29, 1014–1017. Communicated at the meeting of 29 May 1926. 370

[Hurewicz, 1927a] Hurewicz, W. (1927a). Normalbereiche und Dimensionstheorie. *Mathematische Annalen*, 96(1), 736–764. Submitted 28 September 1925. 331, 365, 366, 367

[Hurewicz, 1927b] Hurewicz, W. (1927b). Über stetige Bilder von Punktmengen (Zweite Mittelung). *Proceedings of the Section of Sciences, Koninklijke Nederlandse Akademie van Wetenschappen te Amsterdam*, 30(1), 159–165. Communicated at the meeting of 27 November 1926. 370

[Hurewicz, 1927c] Hurewicz, W. (1927c). Über das Verhältnis separabler Räume zu kompakten Räumen. *Proceedings of the Section of Sciences, Koninklijke Nederlandse Akademie van Wetenschappen te Amsterdam*, 30(3), 425–430. Communicated at the meeting of 26 March 1927. 365, 367, 396

[Hurewicz, 1928a] Hurewicz, W. (1928a). Grundriß der Mengerschen Dimensionstheorie. *Mathematische Annalen*, 98(1), 64–88. Submitted 30 August 1926. 360, 362, 363

[Hurewicz, 1928b] Hurewicz, W. (1928b). Über unendlich-dimensionale Punktmengen. *Proceedings of the Section of Sciences, Koninklijke Nederlandse Akademie van Wetenschappen te Amsterdam*, 31, 916–922. Communicated at the meeting of 26 May 1928. 365, 367, 368

[Hurewicz, 1929] Hurewicz, W. (1929). Über ein topologisches Theorem. *Mathematische Annalen*, 101(1), 210–218. Submitted 15 January 1928. xix, 317, 318, 365, 368, 369, 429

[Hurewicz, 1930a] Hurewicz, W. (1930a). Ein Theorem der Dimensionstheorie. *Annals of Mathematics*, (pp. 176–180). 370

[Hurewicz, 1930b] Hurewicz, W. (1930b). Über den sogenannten Produktsatz der Dimenstheorie. *Mathematische Annalen*, 102(1), 305–312. Submitted 18 October 1928.

[Hurewicz, 1930c] Hurewicz, W. (1930c). Über Einbettung separabler Räume in gleichdimensionale kompakte Räume. *Monatshefte für Mathematik und Physik*, 37, 199–208. 396

[Hurewicz, 1931a] Hurewicz, W. (1931a). Dimensionstheorie und Cartesische Räume. *Proceedings of the Section of Sciences, Koninklijke Nederlandse Akademie van Wetenschappen te Amsterdam*, 34, 399–400. Communicated at the meeting of 28 March 1931. 396

[Hurewicz, 1931b] Hurewicz, W. (1931b). Über Abbildungen von allgemeinen topologischen Räumen auf Teilmengen Cartesischer Zahlenräumen. *Anzeiger der Akademie der Wissenschaften in Wien* ***68***, *Mathematisch-naturwissenschaftlichen Klasse*, 12, 97–98. 396

[Hurewicz, 1932] Hurewicz, W. (1932). Über stetige Abbildungen topologischer Räume. In *Verhandlungen des Internationalen Mathematiker-Kongresses, Zürich 1932*, volume 2 (pp. 191).: Orell Füssli, Zürich. 370

[Hurewicz, 1933a] Hurewicz, W. (1933a). Über Abbildungen von endlichdimensionalen Räumen auf Teilmengen Cartesischer Räume. *Sitzungsberichte der Preussischen Akademie der Wissenschaften zu Berlin. Physikalisch-Mathematische Klasse*, 4, 754–768. Communicated at the meeting of 9 February 1933. 396, 397, 398

[Hurewicz, 1933b] Hurewicz, W. (1933b). Über dimensionserhöhende stetige Abbildungen. *Journal für die reine und angewandte Mathematik*, 169, 71–78. 370

[Hurewicz, 1995] Hurewicz, W. (1995). *Collected Works of Witold Hurewicz*. K. Kuperberg, Ed. American Mathematical Society. 364

[Hurewicz & Menger, 1928] Hurewicz, W. & Menger, K. (1928). Dimension und Zusammenhangsstufe; Anhang: Ein Additionssatz, Von Witold Hurewicz. *Mathematische Annalen*, 100(1), 618–627, 627–633. Submitted 30 July 1927; Anhang, submitted 15 October 1927; issue date December 1928. 313, 354, 370, 374

[Hurewicz & Wallman, 1948] Hurewicz, W. & Wallman, H. (1941; revised, 1948). *Dimension Theory*, volume 4 of *Princeton Mathematical Series*. Princeton University Press. 8, 57, 126, 188, 273, 274, 327, 330, 331, 341, 360, 365, 367, 377, 393, 398, 415, 435

[Hurwitz, 1898] Hurwitz, A. (1898). Über die Entwicklung der allgemeinen Theorie der analytischen Funktionen in neuerer Zeit. In *Verhandlungen des ersten internationalen Mathematiker-Kongresses in Zürich von 9. bis 11. August 1897* (pp. 91–112).: B.G. Teubner, Leipzig. 151, 208

[Iamblichus, 1988] Iamblichus (1988). *The Theology of Arithmetic: On the Mystical, Mathematical and Cosmological Symbolism of the First Ten Numbers*. R. Waterfield, Ed. and trans. Phanes Press. 12, 13

[Ingram, 2006] Ingram, W.T. (2006). A brief historical view of continuum theory. *Topology and its Applications*, 153(10), 1530–1539. 207

[Jacobi, 1834] Jacobi, C.G.J. (1834). De binis quibuslibet functionibus homogeneis secundi ordinis per substitutiones lineares in alias binas tranformandis, quae solis quadratis variabilium constant; una cum variis theorematis de tranformatione et determinatione integralium multiplicium. *Journal für die reine und angewandte Mathematik*, 12, 1–69. = [Jacobi, 1881, Volume 3, pp. 191–268]. 74

[Jacobi, 1881] Jacobi, C.G.J. (1881). *C.G.J. Jacobi's Gesammelte Werke*. Herausgegeben auf Veranlassung der Königlich Preussischen Akademie der Wissenschaften. Berlin,: G. Reimer. 74, 472

[Jacobson, 1883] Jacobson, J. (1883). Die Axiome der Geometrie und ihr „philosophischer Untersucher" Herr Benno Erdmann. *Altpreussische Monatsschrift*, 20, 301–341. 93

[James, 1999] James, I.M., Ed. (1999). *History of Topology*. Elsevier Amsterdam.

[Janiak, 2020] Janiak, A. (2020). Kant's Views on Space and Time. In E.N. Zalta (Ed.), *The Stanford Encyclopedia of Philosophy*. Metaphysics Research Lab, Stanford University, spring 2020 edition. `https://plato.stanford.edu/archives/spr2020/entries/kant-spacetime/` (accessed 15 March 2024).

[Janiszewski, 1910a] Janiszewski, Z. (1910a). Contribution à la géométrie des courbes planes générales. *Comptes rendus hebdomadaires des séances de l'Académie des Sciences Paris*, 150, 606–609. Session of 7 March 1910. Sigismond Janiszewski = Zygmunt Janiszewski. 297

[Janiszewski, 1910b] Janiszewski, Z. (1910b). Sur la géométrie de lignes cantoriennes. *Comptes rendus hebdomadaires des séances de l'Académie des Sciences Paris*, 151, 198–201. Session of 18 July 1910. Sigismond Janiszewski = Zygmunt Janiszewski. 295, 297

[Janiszewski, 1911a] Janiszewski, Z. (1911a). *Sur les continus irréductibles entre deux points*. PhD thesis, University of Paris. Sigismond Janiszewski = Zygmunt Janiszewski. xviii, 297, 298, 304, 308, 325

[Janiszewski, 1911b] Janiszewski, Z. (1911b). Sur les continus irréductibles entre deux points. *Comptes rendus hebdomadaires des séances de l'Académie des Sciences Paris*, 152, 752–755. Session of 20 March 1911. Sigismond Janiszewski = Zygmunt Janiszewski. 297

[Janiszewski, 1912a] Janiszewski, Z. (1912a). Sur les continus irréducible entre deux points. Thésis présentées à la Faculté des Sciences de Paris, 1911. *Journal de l'École Polytechnique*, 2(16), 79–170. Sigismond Janiszewski = Zygmunt Janiszewski. xviii, 297, 298, 304, 308, 325

[Janiszewski, 1912b] Janiszewski, Z. (1912b). Über die Begriffe „Linie" und „Fläche". In *Proceedings of the Fifth International Congress of Mathematicians, Cambridge*, volume 2 (pp. 126–128). 298, 299, 325

[Janiszewski, 1962] Janiszewski, Z. (1962). *Oeuvres choisies.* Edited by Karel Borsuk, Jan Jaworowski, Bronisław Knaster, and Kazimierz Kuratowski. Państwowe Wydawn. Naukowe. 296, 297, 299

[Janiszewski & Kuratowski, 1920] Janiszewski, Z. & Kuratowski, K. (1920). Sur les continus indécomposables. *Fundamenta Mathematicae*, 1(1), 210–222. 299

[Jarník, 1981] Jarník, V., Ed. (1981). *Bolzano and the Foundations of Mathematical Analysis*. Society of Czechoslovak Mathematicians and Physicists (Praha).

[Ji et al., 2017] Ji, L., Papadopoulos, A., & Yamada, S., Eds. (2017). *From Riemann to Differential Geometry and Relativity*. Springer.

[Johnson, 1977] Johnson, D.M. (1977). Prelude to dimension theory: The geometrical investigations of Bernard Bolzano. *Archive for History of Exact Sciences*, 17(3), 261–295. vii, 31, 34

[Johnson, 1979] Johnson, D.M. (1979). The problem of the invariance of dimension in the growth of modern topology, Part I. *Archive for History of Exact Sciences*, 20(2), 97–188. vii

[Johnson, 1981] Johnson, D.M. (1981). The problem of the invariance of dimension in the growth of modern topology, Part II. *Archive for History of Exact Sciences*, 25(2-3), 85–266. vii, 247, 333

[Johnson, 1987] Johnson, D.M. (1987). L. E. J. Brouwer's coming of age as a topologist. In E.R. Phillips (Ed.), *Studies in the History of Mathematics* (pp. 61–97). Mathematical Association of America. vii

[Jongen & Krieg, 2000] Jongen, H.T. & Krieg, A. (2000). Otto Blumenthal. *Mitteilungen der Deutschen Mathematiker-Vereinigung*, 8(4), 49–52. 245

[Jordan, 1868] Jordan, M.E.C. (1868). Mémoire sur les groupes de mouvements. *Annali di Matematica Pura ed Applicata (1867-1897)*, 2(1), 167–215, 322–345. 169, 170

[Jordan, 1881] Jordan, M.E.C. (1881). Sur la série de Fourier. *Comptes rendus hebdomadaires des séances de l'Académie des Sciences Paris*, 92, 228–230. 145

[Jordan, 1882] Jordan, M.E.C. (1882). *Cours d'analyse de l'École Polytechnique: Calcul différentiel*, volume 1. Gauthier-VIllars. 143

[Jordan, 1883] Jordan, M.E.C. (1883). *Cours d'analyse de l'École Polytechnique: Calcul intégral. Intégrales définies et indéfinies*, volume 2. Gauthier-VIllars. 143, 144

[Jordan, 1887] Jordan, M.E.C. (1887). *Cours d'analyse de l'École Polytechnique: Calcul intégral. Équations différentielles*, volume 3. Gauthier-Villars. 143, 144, 145, 146, 147, 150

[Jordan, 1892] Jordan, M.E.C. (1892). Remarques sur les intégrales définies. *Journal de mathématiques pures et appliquées*, 8, 69–100. 146, 147

[Jordan, 1893] Jordan, M.E.C. (1893). *Cours d'analyse de l'École Polytechnique: Calcul différentiel. Deuxième Édition, Entierement Refondue*, volume 1. Gauthier-Villars et fils, second edition. 144, 146, 307, 416

[Jordan, 1894a] Jordan, M.E.C. (1894a). *Cours d'analyse de l'École Polytechnique: Calcul intégral. Deuxième Édition, Entierement Refondue*, volume 2. Gauthier-Villars et fils, second edition. 144

[Jordan, 1894b] Jordan, M.E.C. (1894b). Question 60. *L'Intermédiaire des Mathématiciens*, 1, 23. 150

[Jordan, 1896] Jordan, M.E.C. (1896). *Cours d'analyse de l'École Polytechnique: Calcul intégral. Équations différentielles. Deuxième Édition, Entierement Refondue*, volume 3. Gauthier-Villars et fils, second edition. 144

[Julia, 1955] Julia, G. (1955). *Le livre du centenaire de la naissance de Henri Poincaré: 1854–1954*. Gauthier-Villars, Paris. 158

[Jürgens, 1878] Jürgens, E. (1878). Über eindeutige und stetige Abbildung von Mannigfaltigkeiten. In *Tageblatt der 5I. Versammlung Deutscher Naturforscher und Aerzte in Cassel* (pp. 137–140). L. Friederichsen & Co. 126, 229, 241

[Jürgens, 1879] Jürgens, E. (1879). *Allgemeine Sätze über Systeme von zwei eindeutigen und stetigen reellen Functionen von zwei reellen Veränderlichen*. B.G. Teubner. 127, 129, 131, 132, 191, 229, 241

[Jürgens, 1899] Jürgens, E. (1899). Der Begriff der n-fachen stetigen Mannigfaltigkeit. *Jahresbericht der Deutschen Mathematiker-Vereinigung*, 7, 50–55. 132, 136, 155

[Kant, 1746] Kant, I. (1746). *Gedanken von der wahren Schätzung der lebendigen Kräfte und Beurtheilung der Beweise derer sich Herr von Leibnitz und andere Mechaniker in dieser Streitsache bedienet haben, nebst einigen vorhergehenden Betrachtungen welche die Kraft der Körper überhaupt betreffen*. Martin, Eberhard Dorn, Königsberg. 28

[Kant, 1768] Kant, I. (1768). Vom dem ersten Grunde des Unterschiedes der Gegenden im Raume. *Königsberger Frag- und Anziegungsnachrichten*. 28

[Kant, 1781] Kant, I. (1781). *Critik [Kritik] der reinen Vernunft*. Hartknoch. 37, 42

[Kant, 1787] Kant, I. (1787). *Kritik der reinen Vernunft*. Hartknoch, second edition. 37, 42

[Kant, 1929] Kant, I. (1929). *Kant's Inaugural Dissertation and Early Writings on Space*. Trans. John Handyside. Chicago: Open Court.

[Kant, 1992a] Kant, I. (1992a). Concerning the ultimate ground of the differentiation of directions in space. In *Theoretical Philosophy, 1755–1770* (pp. 361–372). Cambridge University Press. = [Kant, 1992b, pp. 361–372.]. 28

[Kant, 1992b] Kant, I. (1992b). *Theoretical Philosophy, 1755–1770.* Translated and edited by David Walford and Ralf Meerbote. Cambridge University Press. 475

[Kant, 1998] Kant, I. (1998). *Critique of Pure Reason.* Translated and edited by Paul Guyer and Allen W. Wood. Cambridge University Press. 37

[Kant, 2012a] Kant, I. (2012a). *Natural Science.* Edited by Eric Watskins. Cambridge University Press. 475

[Kant, 2012b] Kant, I. (2012b). Thoughts on the true estimation of living forces and the assessment of the demonstrations that Leibniz and other scholars of mechanics have made use of in this controversial subject, together with some prefatory considerations pertaining to the force of bodies in general. In *Natural Science* (pp. 1–155). Cambridge University Press. = [Kant, 2012a, pp. 1–155]. 28

[Kass, 1996] Kass, S. (1996). Karl Menger. *Notices of the American Mathematical Society*, 43(5), 558–561. 333

[Kass, 1997] Kass, S. (1997). A tribute to Karl Menger. *Humanistic Mathematics Network Journal*, 1(16), 3. 333

[Kästner, 1746] Kästner, A.G. (1746). Gedanken zur Erläuterung des geometrischen Begriffs von dem Ursprunge einer Linie, aus der Bewegung eines Punktes. *Hamburgisches Magazin*, 4, 46–52. 36

[Kästner, 1782] Kästner, A.G. (1782). Occam hat Punkte, Linie und Fläsche sehr richtig erklärt. *Leipziger Magazin zur Naturkunde, Mathematik und Ockonomie*, (pp. 224–226). 36

[Kästner, 1792] Kästner, A.G. (1792). *Anfangsgründe der Arithmetik, Geometrie, ebenen und sphërischen Trigonometrie. und Perspektiv.* Vandenhoeck und Ruprecht, fifth edition. 34, 35, 36

[Kästner, 1800] Kästner, A.G. (1800). *Anfangsgründe der Arithmetik, Geometrie, ebenen und sphërischen Trigonometrie. und Perspektiv.* Vandenhoeck und Ruprecht, sixth edition. 34, 35, 36

[Katětov, 1993] Katětov, M. (1993). Dimension Theory. In *The Mathematical Legacy of Eduard Čech* (pp. 109–129). Springer. 399

[Katětov et al., 1993] Katětov, M., Novák, J., & Švec, A. (1993). Life and work of Eduard Čech. In *The Mathematical Legacy of Eduard Čech* (pp. 9–20). Springer. 399

[Katětov & Simon, 1993] Katětov, M. & Simon, P. (1993). *The Mathematical Legacy of Eduard Čech*. Springer. 399, 400, 457

[Katětov & Simon, 1997] Katětov, M. & Simon, P. (1997). Origins of Dimension Theory. In *Handbook of the History of General Topology*, volume 1 (pp. 113–134). Kluwer Academic Publishers. 31, 62, 262

[Kennedy, 2002] Kennedy, H.C. (2002). *Twelve Articles on Giuseppe Peano*. Peremptory Publications, San Francisco. 147

[Kennedy, 2020] Kennedy, J.A. (2020). A brief history of indecomposable continua. In *Continua* (pp. 103–126). CRC Press. 207

[Killing, 1892] Killing, W. (1892). Ueber die Grundlagen der Geometrie. *Journal für die reine und angewandte Mathematik*, 109, 121–186. 157

[Killing, 1893] Killing, W. (1893). *Einführung in die Grundlagen der Geometrie*, volume 1. F. Schöningh. 157

[Kim et al., 2024] Kim, A., Landerer, C., & Huemer, W. (2024). Johann Friedrich Herbart. In E.N. Zalta & U. Nodelman (Eds.), *The Stanford Encyclopedia of Philosophy*. Metaphysics Research Lab, Stanford University, Winter 2024 edition. `https://plato.stanford.edu/archives/win2024/entries/johann-herbart/` (accessed 15 March 2024). 90, 91

[Kirk & Raven, 1957] Kirk, G.S. & Raven, J.E. (1957). *The Presocratic Philosophers: A Critical History with a Selection of Texts*. Cambridge University Press, first edition. 12, 15

[Kirk et al., 1983] Kirk, G.S., Raven, J.E., & Schofield, M. (1983). *The Presocratic Philosophers: A Critical History with a Selection of Texts*. Cambridge University Press, second edition. 12

[Kirschner, 2021] Kirschner, S. (2021). Nicole Oresme. In E.N. Zalta (Ed.), *The Stanford Encyclopedia of Philosophy*. Metaphysics Research Lab, Stanford University, Fall 2021 edition.
`https://plato.stanford.edu/archives/fall2021/entries/nicole-oresme/` (accessed 15 March 2024). 25

[Klein, 1872] Klein, F. (1872). *Vergleichende Betrachtungen über neuere geometrische Forschungen: Programm zum Eintritt in die philosophische Facultät und den Senat der k. Friedrich-Alexanders-Universität zu Erlangen* [Das Erlanger Programm]. Deichert. 98

[Klein, 1882] Klein, F. (1882). *Ueber Riemanns Theorie der algebraischen Funktionen und ihrer Integrale*. B.G. Teubner. 139

[Klein, 1896] Klein, F. (1896). Über Arithmetisierung der Mathematik. Vortrag, gehalten in der öffentlichen Sitzung der K. Gesellschaft der Wissenschaften zu Göttingen. Geschäftliche Mitteilungen. *Zeitschrift für mathematischen und naturwissenschaftlichen Unterricht*, 27, 143–149. 99

[Klein, 1898] Klein, F. (1898). Gutachten, betreffend den dritten Band der Theorie der Transformationsgruppen von S. Lie anlässlich der ersten Vertheilung des Lobatschewsky-Preises. *Mathematische Annalen*, 50(4), 583–600. 96, 97

[Klein, 1921] Klein, F. (1921). *Gesammelte mathematische Abhandlungen*, volume 1. J. Springer. 96, 97

[Klein, 1922] Klein, F. (1922). *Gesammelte mathematische Abhandlungen*, volume 2. J. Springer. 99

[Klein, 1923] Klein, F. (1923). *Gesammelte mathematische Abhandlungen*, volume 3. J. Springer. 139, 156

[Klein, 1968] Klein, J. (1968). *Greek Mathematical Thought and the Origin of Algebra*. The M.I.T. Press. 19

[Kline, 1972] Kline, M. (1972). *Mathematical Thought from Ancient to Modern Times*. Oxford University Press. 422

[Knaster, 1990] Knaster, B. (1970–1990). Mazurkiewicz, Stefan. In *Complete Dictionary of Scientific Biography*. Oxford University Press. https://www.encyclopedia.com/science/dictionaries-thesauruses-pictures-and-press-releases/mazurkiewicz-stefan (accessed 15 April 2023 from Encyclopedia.com). 303

[Knaster & Kuratowski, 1921] Knaster, B. & Kuratowski, K. (1921). Sur les ensembles connexes. *Fundamenta Mathematicae*, 1(2), 206–255. xix, 274, 315, 341

[Knaster et al., 1929] Knaster, B., Kuratowski, K., & Mazurkiewicz, S. (1929). Ein Beweis des Fixpunktsatzes für n-dimensionale Simplexe. *Fundamenta Mathematicae*, 14(1), 132–137. 318

[Knorr, 1980] Knorr, W.R. (1980). On the early history of axiomatics: The interaction of mathematics and philosophy in Greek antiquity. In *Theory Change, Ancient Axiomatics, and Galileo's Methodology* (pp. 145–186). Springer. 22

[Koreuber, 2015] Koreuber, M. (2015). *Emmy Noether, die Noether-Schule und die moderne Algebra: zur Geschichte einer kulturellen Bewegung*. Springer-Verlag. 384

[Kraus, 2014] Kraus, L.B. (2014). *Der Begriff des Kontinuums bei Bernard Bolzano*. Academia Verlag. 31

[Krause, 1908] Krause, M. (1908). Enno Jürgens. *Jahresbericht der Deutschen Mathematiker-Vereinigung*, 17, 163–170. 126

[Kreisel & Newman, 1969] Kreisel, G. & Newman, M.H.A. (1969). Luitzen Egbertus Jan Brouwer. 1881-1966. *Biographical Memoirs of Fellows of the Royal Society*, 15, 39–68. 203

[Kronecker, 1869a] Kronecker, L. (1869a). Über Systeme von Functionen mehrer Variabeln. Erste Abhandlung. *Monatsberichte der Königlich Preussischen Akademie der Wissenschaften zu Berlin*, (pp. 159–193). Session of 4 March 1869; = [Kronecker, 1895, pp. 175–212]. 217, 218, 219, 220, 221, 223

[Kronecker, 1869b] Kronecker, L. (1869b). Sur le théorème de Sturm. *Comptes rendus hebdomadaires des séances de l'Académie des Sciences Paris*, 68, 1078–1082. Session of 10 May 1869. 216, 217

[Kronecker, 1869c] Kronecker, L. (1869c). Über Systeme von Functionen mehrer Variabeln. Zweite Abhandlung. *Monatsberichte der Königlich Preussischen Akademie der Wissenschaften zu Berlin*, (pp. 688–698). Session of 5 August 1869; = [Kronecker, 1895, pp.213–226]. 217, 223

[Kronecker, 1873] Kronecker, L. (1873). Über die verschiedenen Sturm'schen Reihen und ihre gegenseitigen Beziehungen. *Monatsberichte der Königlich Preussischen Akademie der Wissenschaften zu Berlin*, (pp. 117–154). Session of 17 February 1873; = [Kronecker, 1895, pp. 303–348]. 216, 217, 218

[Kronecker, 1878a] Kronecker, L. (1878a). Über Sturm'sche Functionen. *Monatsberichte der Königlich Preussischen Akademie der Wissenschaften zu Berlin*, (pp. 95–121). Session of 14 February 1878; = [Kronecker, 1897, pp. 37–70]. 218

[Kronecker, 1878b] Kronecker, L. (1878b). Über die Charakteristik von Functionen-Systemen. *Monatsberichte der Königlich Preussischen Akademie der Wissenschaften zu Berlin*, (pp. 145–152). Session of 21 February 1878; = [Kronecker, 1897, pp. 71–82]. 217, 218, 223

[Kronecker, 1891] Kronecker, L. (1891). Sur le nombre des racines communes à plusieurs équations simultanées. (Extrait d'une Lettre adressée à M. Picard.). *Comptes rendus hebdomadaires des séances de l'Académie des Sciences Paris*, 113, 1006–1012. Session of 28 December 1891. 224

[Kronecker, 1895] Kronecker, L. (1895). *Leopold Kronecker's Werke*, volume 1. B.G. Teubner. 218, 219, 220, 221, 223, 477

[Kronecker, 1897] Kronecker, L. (1897). *Leopold Kronecker's Werke*, volume 2. B.G. Teubner. 477, 478

[Krull, 1928] Krull, W. (1928). Primidealketten in allgemeinen Ringbereichen. In *Sitzungsberichte der Heidelberger Akademie der Wissenschaften. Mathematisch-naturwissenschaftliche Klasse. Jahrgang 1928. 7. Abhandlung*. Walter de Gruyter & Co. Submitted 22 May 1928.

[Krull, 1935] Krull, W. (1935). *Idealtheorie*. Number 4 in Ergebnisse der Mathematik und Ihrer Grenzgebiete. Julius Springer. 423

[Krull, 1936a] Krull, W. (1936a). Beiträge zur Arithmetik kommutativer Integritätsbereiche. *Mathematische Zeitschrift*, 41, 545–577. 424

[Krull, 1936b] Krull, W. (1936b). Beiträge zur Arithmetik kommutativer Integritätsbereiche. II. v-Ideale und vollständig ganz abgeschlossene Integritätsbereiche. *Mathematische Zeitschrift*, 41, 665–679. 424

[Krull, 1937a] Krull, W. (1937a). Beiträge zur Arithmetik kommutativer Integritätsbereiche. III. Zum Dimensionsbegriff der Idealtheorie. *Mathematische Zeitschrift*, 42, 745–766. 359, 424

[Krull, 1937b] Krull, W. (1937b). Beiträge zur Arithmetik kommutativer Integritätsbereiche. IV. Unendliche algebraische Erweiterungen endlicher diskreter Hauptordnungen. *Mathematische Zeitschrift*, 42, 767–773. 424

[Krull, 1938a] Krull, W. (1938a). Beiträge zur Arithmetik kommutativer Integritätsbereiche. III a. Eine Ergänzung von Beitrag III. *Mathematische Zeitschrift*, 43, 767. 359, 424

[Krull, 1938b] Krull, W. (1938b). Beiträge zur Arithmetik kommutativer Integritätsbereiche. V. Potenzreihenringe. *Mathematische Zeitschrift*, 43, 768–782. 424

[Krull, 1938c] Krull, W. (1938c). Dimensionstheorie in Stellenringen. *Journal für die reine und angewandte Mathematik*, 179, 204–226. 424

[Krull, 1939a] Krull, W. (1939a). Beiträge zur Arithmetik kommutativer Integritätsbereiche. VI. Der allgemeine Diskriminantensatz. Unverzweigte Ringerweiterungen. *Mathematische Zeitschrift*, 45, 1–19. 424

[Krull, 1939b] Krull, W. (1939b). Beiträge zur Arithmetik kommutativer Integritätsbereiche. VII. Inseparable Grundkörpererweiterung. Bemerkungen zur Körpertheorie. *Mathematische Zeitschrift*, 45, 319–334. 424

[Krull, 1942] Krull, W. (1942). Beiträge zur Arithmetik kommutativer Integritätsbereiche VIII. Multiplikative abgeschlossene Systeme von endlichen Idealen. *Mathematische Zeitschrift*, 48, 533–552. 424

[Krull, 1951] Krull, W. (1951). Jacobsonsche Ringe, Hilbertscher Nullstellensatz, Dimensionstheorie. *Mathematische Zeitschrift*, 54(4), 354–387. 424

[Krull, 1999a] Krull, W. (1999a). *Gesammelte Abhandlungen / Collected Papers*, volume 1. Walter de Gruyter. 424

[Krull, 1999b] Krull, W. (1999b). *Gesammelte Abhandlungen / Collected Papers*, volume 2. Walter de Gruyter. 424

[Krupski & Patkowska, 1996] Krupski, P. & Patkowska, H. (1996). Menger Curves in Peano Continua. *Colloquium Mathematicae*, 70(1), 79–86.

[Kunz, 1985] Kunz, E. (1985). *Introduction to Commutative Algebra and Algebraic Geometry*. Springer. 421, 422, 424

[Kuratowski, 1966] Kuratowski, K. (1966). *Topology. I.* Academic Press, New York; PWN–Polish Scientific Publishers, Warsaw. 197, 292, 313

[Kuratowski, 1968] Kuratowski, K. (1968). *Topology. II.* Academic Press, New York; PWN–Polish Scientific Publishers, Warsaw. xix, 207, 304, 310, 311, 313, 315

[Kuratowski, 1980] Kuratowski, K. (1980). *A half century of Polish mathematics: Remembrances and reflections*. Polish Scientific Publishers PWN (Wydawnictwo Naukowe PWN). 296, 299, 414

[Kuratowski & Menger, 1930] Kuratowski, K. & Menger, K. (1930). Remarques sur la théorie axiomatique de la dimension. *Monatshefte für Mathematik und Physik*, 37(1), 169–174. Received 20 November 1929; published 30 April 1930; issue date December 1930. 363

[Kuz'minov, 1968] Kuz'minov, V.I. (1968). Homological dimension theory. *Russian Mathematical Surveys*, 23(5), 1. 393

[Lagrange, 1797] Lagrange, J.L. (1797). *Théorie des fonctions analytiques*. l'Imprimerie de la République. 29

[Lakatos, 1964] Lakatos, I. (1963-1964). Proofs and refutations (i-iv). *The British Journal for the Philosophy of Science*, 14, 1–25, 120–139, 221–245, 296–342. 7, 52

[Lakatos, 1976] Lakatos, I. (1976). *Proofs and Refutations: The Logic of Mathematical Discovery*. Edited by J. Worrall and E. Zahar. Cambridge University Press. 7, 71

[Lakatos, 1978a] Lakatos, I. (1978a). *The Methodology of Scientific Research Programmes: Philosophical Papers, Volume 1*. Edited by J. Worrall and G. Currie. Cambridge University Press. 435

[Lakatos, 1978b] Lakatos, I. (1978b). *Mathematics, Science and Epistemology: Philosophical Papers, Volume 2*. Edited by J. Worrall and G. Currie. Cambridge University Press.

[Lakatos, 1978c] Lakatos, I. (1978c). Infinite regress and the foundations of mathematics. In J. Worrall & G. Currie (Eds.), *Philosophical Papers, Volume 2* (pp. 3–23). Cambridge University Press. 41

[Lakatos, 2015] Lakatos, I. (2015). *Proofs and Refutations: The Logic of Mathematical Discovery*. Edited by J. Worrall and E. Zahar. Cambridge University Press, new edition. 7, 52, 71

[Langsdorf, 1802] Langsdorf, K.C.v. (1802). *Anfangsgründe der reinen Elementar- und höhern Mathematik*. Palm. 36, 39

[Laugwitz & Laugwitz, 1995] Laugwitz, A. & Laugwitz, D. (1995). Impressions from Riemann's native country. *The Mathematical Intelligencer*, 17(3), 37–40. 86

[Laugwitz, 1996] Laugwitz, D. (1996). *Bernhard Riemann, 1826-1866: Wendepunkte in der Auffassung der Mathematik*. Springer. 86

[Laugwitz, 2008] Laugwitz, D. (2008). *Bernhard Riemann 1826–1866: Turning Points in the Conception of Mathematics*. Birkhäuser. 77, 86, 87

[Lebesgue, 1903] Lebesgue, H. (1903). Sur le problème des aires. *Bulletin de la Société Mathématique de France*, 31, 197–203. 150

[Lebesgue, 1904] Lebesgue, H. (1904). *Leçons sur l'intégration et la recherche des fonctions primitives*. Gauthier-Villars et Fils. 248, 309, 417

[Lebesgue, 1905a] Lebesgue, H. (1905a). Sur le problème des aires. *Bulletin de la Société Mathématique de France*, 33, 273–274. 150

[Lebesgue, 1905b] Lebesgue, H. (1905b). Sur les fonctions représentables analytiquement. *Journal de mathématiques pures et appliquées*, 1, 139–216. 248, 309

[Lebesgue, 1907] Lebesgue, H. (1907). Sur le probleme de Dirichlet. *Rendiconti del Circolo Matematico di Palermo*, 24(1), 371–402. 247

[Lebesgue, 1910] Lebesgue, H. (1910). Sur l'intégration des fonctions discontinues. *Annales scientifiques de l'École normale supérieure*, 27, 361–450. 247, 401

[Lebesgue, 1911a] Lebesgue, H. (1911a). Sur la non-applicabilité de deux domaines appartenant respectivement à des espaces à n et $n+p$ dimensions. *Mathematische Annalen*, 70(2), 166–168. 236, 246, 247, 248, 258, 305, 400, 429

[Lebesgue, 1911b] Lebesgue, H. (1911b). Sur la non-applicabilité de deux espaces d'un nombre différent de dimensions. *Comptes rendus des séances Société Mathématique de France*, 39, 20. 246, 429

[Lebesgue, 1911c] Lebesgue, H. (1911c). Sur l'invariance du nombre de dimensions d'un espace et sur le théorème de M. Jordan relatif aux variétés fermées. *Comptes rendus hebdomadaires des séances de l'Académie des Sciences Paris*, 152, 841–843. Session of 27 March 1911. 246, 252, 253, 254, 255, 317, 429

[Lebesgue, 1921] Lebesgue, H. (1921). Sur les correspondances entre les points de deux espaces. *Fundamenta Mathematicae*, 1(2), 256–285. 247, 251, 259, 317, 343, 400

[Lebesgue, 1924] Lebesgue, H. (1924). Sur le théorème de Schoenflies. *Fundamenta Mathematicae*, 6(1), 96–99. 317

[Lebesgue, 1926] Lebesgue, H. (1926). Notice sur la vie et les travaux de Camille Jordan (1838–1922): lue dans la séance du 4 juin 1923. *Mémoires de l'Académie des Sciences de l'Institut de France*, 58(2), xxxix–lxvi. 143, 144

[Lebesgue & Félix, 1974] Lebesgue, H. & Félix, L. (1974). *Message d'un mathématicien: Henri Lebesgue: pour le centenaire de sa naissance*. A. Blanchard. 245

[Lebon, 1912] Lebon, E. (1912). *Henri Poincaré: biographie, bibliographie analytique des écrits*. Gauthier-Villars. 158

[Lefschetz, 1930] Lefschetz, S. (1930). *Topology*, volume 12 of *Colloquium Publications*. American Mathematical Society. 385, 395

[Lefschetz, 1931] Lefschetz, S. (1931). On compact spaces. *Annals of Mathematics*, 32(3), 521–538. 395, 396

[Lefschetz, 1957] Lefschetz, S. (1957). Witold Hurewicz, in memoriam. *Bulletin of the American Mathematical Society*, 63(2), 77–82. 364

[Leitgeb & Carus, 2023] Leitgeb, H. & Carus, A. (2023). Rudolf Carnap. In E.N. Zalta & U. Nodelman (Eds.), *The Stanford Encyclopedia of Philosophy*. Metaphysics Research Lab, Stanford University, Summer 2023 edition. `https://plato.stanford.edu/archives/sum2023/entries/carnap/` (accessed 15 March 2024). 4, 429

[Lennes, 1906] Lennes, N.J. (1906). Curves in non-metrical analysis situs (abstract). *Bulletin of the American Mathematical Society*, 12, 284–285. 265, 379

[Lennes, 1911] Lennes, N.J. (1911). Curves in non-metrical analysis situs: With an application in the calculus of variations. *American Journal of Mathematics*, 33, 287–326. 265, 379

[Leonard, 1998] Leonard, R.J. (1998). Ethics and the excluded middle: Karl Menger and social science in interwar Vienna. *Isis*, 89(1), 1–26. xix, 333, 337, 355

[Leonard, 2010] Leonard, R.J. (2010). *Von Neumann, Morgenstern, and the creation of game theory: From chess to social science, 1900–1960*. Cambridge University Press. 355

[Lie, 1872] Lie, M.S. (1872). Über Komplexe, insbesondere Linien- und Kugelkomplexe, mit Anwendung auf die Theorie partieller Differentialgleichungen. *Mathematische Annalen*, 5, 145–256. 173

[Lie, 1880] Lie, M.S. (1880). Theorie der Transformationsgruppen I. *Mathematische Annalen*, 16(4), 441–528. 159

[Lie, 1887] Lie, M.S. (1887). Bemerkungen zu v. Helmholtzs Arbeit: Über die Tatsachen, die der Geometrie zu Grunde liegen. *Berichte über die Verhandlungen der Sächsischen Akademie der Wissenschaften zu Leipzig: Mathematisch-Physische Klasse 1886, Supplement*, 38, 337–342. 165, 167

[Lie, 1890a] Lie, M.S. (1890a). Ueber die Grundlagen der Geometrie (Erste Abhandlung). *Berichte über die Verhandlungen der Sächsischen Akademie der Wissenschaften zu Leipzig: Mathematisch-Physische Klasse*, 42, 284–321. 165, 167

[Lie, 1890b] Lie, M.S. (1890b). Ueber die Grundlagen der Geometrie (Zweite Abhandlung). *Berichte über die Verhandlungen der Sächsischen Akademie der Wissenschaften zu Leipzig: Mathematisch-Physische Klasse*, 42, 355–418. 165, 167

[Lie, 1924] Lie, M.S. (1924). *Gesammelte Abhandlungen.* Edited by F. Engel and P. Hegaard, volume V. B.G. Teubner. 159

[Lie & Engel, 1888] Lie, M.S. & Engel, F. (1888). *Theorie der Transformationsgruppen*, volume I. B.G. Teubner, Leipzig. 165, 169

[Lie & Engel, 1890] Lie, M.S. & Engel, F. (1890). *Theorie der Transformationsgruppen*, volume II. B.G. Teubner, Leipzig.

[Lie & Engel, 1893] Lie, M.S. & Engel, F. (1893). *Theorie der Transformationsgruppen*, volume III. B.G. Teubner, Leipzig. 165, 167, 174

[Liebmann, 1914] Liebmann, H. (1914). Geometrische Theorie der Differentialgleichungen. In *Encyklopädie der Mathematischen Wissenschaften mit Einschluss ihrer Anwendungen*, volume III D 8 (pp. 503–539). B.G. Teubner. xvii, 225, 226

[Liebmann, 1921] Liebmann, H. (1921). Johannes Thomae. *Jahresberichte der Deutschen Mathematiker-Vereinigung*, 30, 133–144. 125

[Lipschitz, 1986] Lipschitz, R. (1986). *Briefwechsel mit Cantor, Dedekind, Helmholtz, Kronecker, Weierstrass und anderen*. Springer. 102

[Lipscomb, 2008] Lipscomb, S.L. (2008). *Fractals and Universal Spaces in Dimension Theory*. Springer. 427

[Lipscomb, 2009] Lipscomb, S.L. (2009). The quest for universal spaces in dimension theory. *Notices of the American Mathematical Society*, 56(11), 1418–1424. 427

[Loria, 1887] Loria, G. (1887). La Definizioni di Spazio a *n* Dimensioni e l'ipotesi di Continuità del nostra Spazio secondo le Ricerche di Giorgio Cantor. *Giornale di matematiche*, 25, 97–108. 139

[Loria, 1938] Loria, G. (1938). Charles Sturm et son œuvre mathématique (1803-1855). *L'Enseignement Mathématique*, 37, 249–274. 214

[Lüroth, 1878] Lüroth, J. (1878). Über gegenseitig eindeutige und stetige Abbildung von Mannigfaltigkeiten verschiedener Dimensionen aufeinander. *Sitzungsberichte der physikalisch-medizinische Sozietät in Erlangen*, 10, 190–195. 122, 123, 124, 128

[Lüroth, 1899] Lüroth, J. (1899). Über gegenseitig eindeutige und stetige Abbildung von Mannigfaltigkeiten verschiedener Dimensionen aufeinander (Zweite Note). *Sitzungsberichte der physikalisch-medizinische Sozietät in Erlangen*, 31, 87–91. 122, 125, 155

[Lüroth, 1906] Lüroth, J. (1906). Über Abbildung von Mannigfaltigkeiten. *Mathematische Annalen*, 63(2), 222–238. 122, 125, 155

[Lützen, 1990] Lützen, J. (1990). *Joseph Liouville 1809–1882: Master of pure and applied mathematics*. Springer-Verlag. 214

[Lützen, 2001] Lützen, J., Ed. (2001). *Around Caspar Wessel and the Geometric Representation of Complex Numbers. Proceedings of the Wessel Symposium at The Royal Danish Academy of Sciences and Letters Copenhagen, August 11–15 1998. Invited papers*. The Royal Danish Academy of Sciences and Letters, C .A. Reitzels Forlag. 74

[Macaulay, 1916] Macaulay, F.S. (1916). *The Algebraic Theory of Modular Systems*, volume 19 of *Cambridge Tracts in Mathematics and Mathematical Physics*. Cambridge University Press.

[Macaulay, 1934] Macaulay, F.S. (1934). Modern algebra and polynomial ideals. *Mathematical Proceedings of the Cambridge Philosophical Society*, 30(1), 27–46. 424

[Mach, 1901] Mach, E. (1901). On physiological, as distinguished from geometrical, space. *The Monist*, 11(3), 321–338. 162

[Mach, 1902] Mach, E. (1902). On the psychology and natural development of geometry. *The Monist*, 12(4), 481–515. 162

[Mach, 1903] Mach, E. (1903). Space and geometry from the point of view of physical inquiry. *The Monist*, (pp. 1–32). 162

[Mach, 1906] Mach, E. (1906). *Space and geometry in the light of physiological, psychological and physical inquiry*. Open Court. 162

[Magrone, 2020] Magrone, P. (2020). Sierpiński's curve: a (beautiful) paradigm of recursion. *Slovenský Časopis pre Geometriu a Grafiku*, 17(33), 17–28. 300

[Mahlo, 1911] Mahlo, P. (1911). Über die Dimensionentypen des Herrn Fréchet im Gebiete der linearen Mengen. *Berichte über die Verhandlungen der Sächsischen Akademie der Wissenschaften zu Leipzig: Mathematisch-Physische Klasse*, 63, 319–347. 199

[Mancosu, 1999] Mancosu, P. (1999). *Philosophy of Mathematics and Mathematical Practice in the Seventeenth Century*. Oxford University Press, USA. 8

[Mancosu, 2008] Mancosu, P., Ed. (2008). *The Philosophy of Mathematical Practice*. Oxford University Press, Oxford. 8

[Mandelbrojt, 1953] Mandelbrojt, S. (1953). The mathematical work of Jacques Hadamard. *The American Mathematical Monthly*, 60(9), 599–604. 212

[Mandelbrojt, 1973] Mandelbrojt, S. (1973). Maurice Fréchet. *Comptes rendus hebdomadaires des séances de l'Académie des Sciences Paris*, 277, 73–76. 195

[Mandelbrojt & Schwartz, 1965] Mandelbrojt, S. & Schwartz, L. (1965). Jacques Hadamard (1865–1963). *Bulletin of the American Mathematical Society*, 71(1), 107–129. 212

[Mandelbrot, 1983] Mandelbrot, B.B. (1983). *The Fractal Geometry of Nature*. W.H. Freeman New York, revised and enlarged edition. 416

[Mangoldt, 1906] Mangoldt, H.C.F.v. (1906). Die Begriffe „Linie" und „Fläche". In *Encyklopädie der Mathematischen Wissenschaften mit Einschluss ihrer Anwendungen*, volume III A B 2 (pp. 130–152). B.G. Teubner. 286, 287

[Manin, 2006] Manin, Y. (2006). The notion of dimension in geometry and algebra. *Bulletin of the American Mathematical Society*, 43(2), 139–161.

[Marczewski, 1996] Marczewski, E. (1996). *Collected Mathematical Papers*. Polish Academy of Sciences, Institute of Mathematics. Edward Marczewski = Edward Szpilrajn. 414

[Matsumura, 1989] Matsumura, H. (1989). *Commutative Ring Theory*. Cambridge University Press. 422

[Mawhin, 2000] Mawhin, J. (2000). Poincaré's early use of analysis situs in nonlinear differential equation: variations around the theme of Kronecker's integral. *Philosophia Scientiae*, 4(1), 103–143. 224, 226

[Mayrhofer, 1934] Mayrhofer, K. (1934). Hans Hahn. *Monatshefte für Mathematik und Physik*, 41(2), 221–238. 308

[Mazur, 1997] Mazur, B. (1997). Conjecture. *Synthese*, (pp. 197–210). 8

[Mazurkiewicz, 1910] Mazurkiewicz, S. (1910). Sur la théorie des ensembles. *Comptes rendus hebdomadaires des séances de l'Académie des Sciences Paris*, 151, 296–298. Session of 25 July 1910; = [Mazurkiewicz, 1969, pp. 27–29]; Étienne Mazurkiewicz = Stefan Mazurkiewicz. 304

[Mazurkiewicz, 1913a] Mazurkiewicz, S. (1913a). Contribution à la théorie des ensembles. *Bulletin de l'Académie Polonaise des Sciences*, (pp. 46–55). Session of 7 January 1913; Corresponds to a chapter of Mazurkiewicz' thesis. = [Mazurkiewicz, 1969, pp. 30–36]. 303, 304

[Mazurkiewicz, 1913b] Mazurkiewicz, S. (1913b). O arytmetyzacji continuów. (Sur l'arithmétisation des continuus). *Comptes Rendus des Seances de la Société des Sciences et des Lettres de Varsovie*, 6, 305–311. In Polish with French summary. = [Mazurkiewicz, 1969, pp. 37–41]. 303, 304

[Mazurkiewicz, 1913c] Mazurkiewicz, S. (1913c). O arytmetyzacji continuów II. (Sur l'arithmétisation des continuus II). *Comptes Rendus des Seances de la Société des Sciences et des Lettres de Varsovie*, 6, 941–945. In Polish with French summary. = [Mazurkiewicz, 1969, pp. 42–45]. 304, 305

[Mazurkiewicz, 1914] Mazurkiewicz, S. (1914). O pewnej mnogości plaskiej, która ma z każdą prostą dwa i tylko dwa punkty wspólne. (Sur un ensemble plan qui a avec chaque droite deux et seulement deux points communs). *Comptes Rendus des Seances de la Société des Sciences et des Lettres de Varsovie*, 7, 382–384. In Polish with French summary. = [Mazurkiewicz, 1969, pp. 46–47].

[Mazurkiewicz, 1915] Mazurkiewicz, S. (1915). O punktach wielokrotnych krzywych wypeł niających obszar płaski. [Sur les points multiples des courbes qui remplissent une aire plane.]. *Prace Matematyczno-Fizyczne*, 26, 113–120. Original article in Polish; in French translation = [Mazurkiewicz, 1969, pp. 48–53]. 304, 305, 306

[Mazurkiewicz, 1920a] Mazurkiewicz, S. (1920a). Sur les lignes de Jordan. *Fundamenta Mathematicae*, 1(1), 166–209. = [Mazurkiewicz, 1969, pp. 76–113]. 306, 307, 308

[Mazurkiewicz, 1920b] Mazurkiewicz, S. (1920b). Sur un ensemble G_δ ponctiforme qui n'est homéomorphe à aucun ensemble linéaire. *Fundamenta Mathematicae*, 1(1), 61–81. 314

[Mazurkiewicz, 1969] Mazurkiewicz, S. (1969). *Travaux de topologie et ses applications*. PWN — Éditions scientifiques de Pologne. 303, 304, 305, 306, 484

[Maz'ya & Shaposhnikova, 1999] Maz'ya, V.G. & Shaposhnikova, T.O. (1999). *Jacques Hadamard: a universal mathematician*. American Mathematical Society. 212

[McGrew, 1976] McGrew, J.M. (1976). *The Origins of Connectedness im Kleinen*. PhD thesis, Michigan State University. 286

[Mendell, 2019] Mendell, H. (2019). Aristotle and Mathematics. In E.N. Zalta (Ed.), *The Stanford Encyclopedia of Philosophy*. Metaphysics Research Lab, Stanford University, Fall 2019 edition.

https://plato.stanford.edu/entries/aristotle-mathematics/ (accessed 15 March 2024). 22

[Menger, 1923] Menger, K. (1923). Über die Dimensionalität von Punktmengen. (Erster Teil). *Monatshefte für Mathematik und Physik*, 33(1), 148–160. Submitted to the printer, 12.12.1923; printer's proofs corrected by 28.1.1924; issued in the journal, 12.4.1924. 273, 343, 344, 346, 375, 376

[Menger, 1924a] Menger, K. (1924a). Einige Überdeckungssätze der Punktmengenlehre. *Sitzungsberichte der Akademie der Wissenschaften mathematisch-naturwissenschaftliche Klasse*, 133(2a), 421–444. Read at the session of 3 July 1924. 344, 485

[Menger, 1924b] Menger, K. (1924b). Einige Überdeckungssätze der Punktmengenlehre. *Jahresbericht der Deutschen Mathematiker-Vereinigung*, 33, 83. Presented at the annual meeting of the D.M.-V. on 23 September 1924 and appearing in [Menger, 1924a]. 345

[Menger, 1924c] Menger, K. (1924c). Über die Dimension von Punktmengen. *Koninklijke Nederlandse Akademie van Wetenschappen. Proceedings of the Section of Sciences*, 27, 639–643. Communicated at the meeting of 27 September 1924. 346

[Menger, 1924d] Menger, K. (1924d). Über die Dimensionalität von Punktmengen. II. Teil). *Monatshefte für Mathematik und Physik*, 34(1), 137–161. Submitted, 6.10.1924; submitted to the printer, 15.10.1924; printer's proofs corrected by 30.12.1924; offprints or preprints issued on or shortly after 30.12.1924; issued in the journal, 20.9.1926. 273, 343, 345, 346, 347, 375

[Menger, 1925] Menger, K. (1925). Grundzüge einer Theorie der Kurven. *Proceedings of the Section of Sciences, Koninklijke Nederlandse Akademie van Wetenschappen te Amsterdam*, 28, 67–71. Communicated at the meeting of 31 January 1925. 347

[Menger, 1926a] Menger, K. (1926a). Grundzüge einer Theorie der Kurven. *Mathematische Annalen*, 95(1), 277–306. Submitted 13 February 1925. 52, 338, 339, 347, 348, 349, 350, 375

[Menger, 1926b] Menger, K. (1926b). Bericht über die Dimensionstheorie. *Jahresbericht der Deutschen Mathematiker-Vereinigung*, 34, 113–150. Submitted 30 August 1925. 273, 351, 360, 361, 375

[Menger, 1926c] Menger, K. (1926c). Allgemeine Räume und Cartesische Räume. (Erste Mitteilung). *Proceedings of the Section of Sciences, Koninklijke Nederlandse Akademie van Wetenschappen te Amsterdam*, 29, 476–483. Communicated at the meeting of 27 February 1926; translated with commentary in [Edgar, 2004, pp. 103–111]. 351, 382

[Menger, 1926d] Menger, K. (1926d). Dimensionstheoretische Konsequenzen des Verhältnisses von allgemeinen Räume und Zahlenräumen. *Proceedings of the Section of Sciences, Koninklijke Nederlandse Akademie van Wetenschappen te Amsterdam*, 29, 648–649. Communicated at the meeting of 27 March 1926. 351

[Menger, 1926e] Menger, K. (1926e). Zur Entstehung meiner Arbeiten über Dimensions- und Kurventheorie. *Proceedings of the Section of Sciences, Koninklijke Nederlandse Akademie van Wetenschappen te Amsterdam*, 29, 1122–1124.

Communicated at the meeting of 29 May 1926. 322, 333, 339, 340, 341, 353, 357, 374, 375, 376

[Menger, 1926f] Menger, K. (1926f). Allgemeine Räume und Cartesische Räume. Zweite Mitteilung: „Ueber umfassendste *n*-dimensionale Mengen". *Proceedings of the Section of Sciences, Koninklijke Nederlandse Akademie van Wetenschappen te Amsterdam*, 29, 1125–1128. Communicated at the meeting of 26 June 1926; translated with commentary in [Edgar, 2004, pp. 103–104, 111–115]. 351, 382

[Menger, 1926g] Menger, K. (1926g). Das Hauptproblem über die dimensionelle Struktur der Räume. *Proceedings of the Section of Sciences, Koninklijke Nederlandse Akademie van Wetenschappen te Amsterdam*, 30, 138–144. Communicated at the meeting of 27 November 1926. 351, 352, 353

[Menger, 1927a] Menger, K. (1927a). Über reguläre Baumkurven. *Mathematische Annalen*, 96, 572–582. Submitted 2 December 1925. 351, 354

[Menger, 1927b] Menger, K. (1927b). Zur allgemeinen Kurventheorie. *Fundamenta Mathematicae*, 10(1), 96–115. Submitted 10 July 1926. 351, 354

[Menger, 1927c] Menger, K. (1927c). Die Haupttheoreme der Dimensionstheorie. *Jahresbericht der Deutschen Mathematiker-Vereinigung*, 36, 8–12. Short note on Menger's presentation on 22 September 1926 at the Düsseldorf Meeting of the D.M.-V., 20–24 September 1926. 355

[Menger, 1927d] Menger, K. (1927d). Zusammenhangsstufen und Cantorsche Mannigfaltigkeiten. *Proceedings of the Section of Sciences, Koninklijke Nederlandse Akademie van Wetenschappen te Amsterdam*, 30, 705–709. Communicated at the meeting of 24 September 1927. 313, 354

[Menger, 1927e] Menger, K. (1927e). Bemerkungen zur zweiten Untersuchung über allgemeine Metrik. *Proceedings of the Section of Sciences, Koninklijke Nederlandse Akademie van Wetenschappen te Amsterdam*, 30, 710–714. Communicated at the meeting of 24 September 1927. 354

[Menger, 1928a] Menger, K. (1928a). Bemerkungen zu Grundlagenfragen, I–IV. *Jahresbericht der Deutschen Mathematiker-Vereinigung*, 37, 213–226, 298–302, 303–308, 309–325. 357, 487

[Menger, 1928b] Menger, K. (1928b). *Dimensionstheorie*. B.G. Teubner. 6, 273, 274, 327, 333, 341, 343, 353, 364, 374, 375, 376, 377, 378, 383, 394, 396, 398, 430, 431

[Menger, 1929a] Menger, K. (1929a). Allgemeine Räume und Cartesische Räume III (Beweis des Fundamentalsatzes). *Proceedings of the Section of Sciences, Koninklijke Nederlandse Akademie van Wetenschappen te Amsterdam*, 32, 330–340. Communicated at the meeting of 28 January 1928. 363

[Menger, 1929b] Menger, K. (1929b). Zur Dimensions- und Kurventheorie: unveröffentlichte Aufsätze aus den Jahren 1921–23. *Monatshefte für Mathematik und Physik*, 36, 411–432. 322, 333, 339, 340, 341, 342, 343, 344, 376, 381

[Menger, 1929c] Menger, K. (1929c). Zur Frage nach der Herleitung des Dimensionsbegriffes aus Forderungen. *Anzeiger, Wiener Akademie der Wissenschaften, Mathematisch-Naturwissenschaftliche Klasse*, 66(10), 87–90. Presented by H. Hahn, 14 March 1929. 363

[Menger, 1929d] Menger, K. (1929d). Über die Dimension von Punktmengen III: Zur Begründung einer axiomatischen Theorie der Dimension. *Monatshefte für Mathematik und Physik*, 36, 193–218. Submitted 24 May 1929. 343, 363, 364

[Menger, 1930] Menger, K. (1930). Antwort auf eine Note von Brouwer. *Monatshefte für Mathematik und Physik*, 37, 175–182. 345, 381

[Menger, 1932a] Menger, K. (1932a). *Kurventheorie*. Edited with collaboration by Georg Nöbeling. B.G. Teubner. xix, 52, 348, 352, 364, 374

[Menger, 1932b] Menger, K. (1932b). Über die Hinweise auf Brouwer in Urysohns Mémoire. Im Selbstverlag, Self published. 381

[Menger, 1933] Menger, K. (1933). Eine Zuschrift von K. Menger an H. Hahn betreffend Mengers Antwort auf eine Note von Brouwer in Bd. 37 der Monatshefte. *Monatshefte für Mathematik und Physik*, 40, page 233. 381

[Menger, 1943] Menger, K. (1943). What is Dimension? *The American Mathematical Monthly*, 50(1), 2–7. 6, 430

[Menger, 1967] Menger, K. (1967). *Kurventheorie*. Edited with collaboration by Georg Nöbeling. American Mathematical Society, second edition. 52

[Menger, 1979a] Menger, K. (1979a). *Selected Papers in Logic and Foundations, Didactics, Economics*, volume 10 of *Vienna Circle Collection*. D. Reidel Publishing Company. 6, 333, 334, 363, 381, 430

[Menger, 1979b] Menger, K. (1979b). An Intuitionistic-Formalistic Dictionary of Set Theory. In *Selected Papers in Logic and Foundations, Didactics, Economics* (pp. 79–87). Springer. Translation with revisions of [Menger, 1928a, I, pp. 213–226]. 357

[Menger, 1979c] Menger, K. (1979c). My Memories of L. E. J. Brouwer. In *Selected Papers in Logic and Foundations, Didactics, Economics* (pp. 237–255). Springer. 324, 355, 356, 357

[Menger, 1994] Menger, K. (1994). *Reminiscences of the Vienna Circle and the Mathematical Colloquium*, volume 20 of *Vienna Circle Collection*. Kluwer Academic Publishers. 333, 334, 335, 336, 337, 363

[Menger, 2002] Menger, K. (2002). *Selecta Mathematica*. Edited by Bert Schweizer, Abe Sklar, Karl Sigmund, Peter Gruber, Edmund Hlawka, Ludwig Reich, Leopold Schmetterer, volume 1. Springer. 334, 345

[Menger, 2003] Menger, K. (2003). *Selecta Mathematica*. Edited by Bert Schweizer, Abe Sklar, Karl Sigmund, Peter Gruber, Edmund Hlawka, Ludwig Reich, Leopold Schmetterer, volume 2. Springer. 334

[Meschkowski, 1967] Meschkowski, H. (1967). *Probleme des Unendlichen: Werke und Leben Georg Cantors*. Springer. 101

[Meschkowski & Nilson, 1991] Meschkowski, H. & Nilson, W., Eds. (1991). *Georg Cantor: Briefe*. Springer. 102

[Milesi, 1892] Milesi, L. (1892). Sulla impossibile coesistenza della univocità e della continuità nella corrispondenza che si può stabilire fra due spazi continui ad un numero differente di dimensioni. *Rivista di Matematica*, 2, 103–106. 152

[Minkowski, 1901] Minkowski, H. (1901). Über die Begriffe Länge, Oberfläche und Volumen. *Jahresbericht der Deutschen Mathematiker-Vereinigung*, 9, 115–121. 416, 417

[Minkowski, 1909a] Minkowski, H. (1909a). Raum und Zeit. *Physikalische Zeitschrift*, 10, 104–111. 416

[Minkowski, 1909b] Minkowski, H. (1909b). Raum und Zeit. *Jahresbericht der Deutschen Mathematiker-Vereinigung*, 18, 75–88. 416

[Möbius, 1827] Möbius, A.F. (1827). *Der baryzentrische Kalkül*. Johann Ambrosius Barth, Leipzig. 81

[Moore, 1900] Moore, E.H. (1900). On certain crinkly curves. *Transactions of the American Mathematical Society*, 1(1), 72–90. xvii, 148, 149, 309

[Moore & Kline, 1919] Moore, R.L. & Kline, J.R. (1919). On the most general plane closed point-set through which it is possible to pass a simple continuous arc. *Annals of Mathematics*, (pp. 218–223). 188

[More, 1671] More, H. (1671). *Enchiridion Metaphysicum: sive, de Rebus Incorporeis Succincta & Inculenta Dissertatio*. F. Flesber, Guilielmum Morden, London. 26

[Morscher, 2018] Morscher, E. (2018). Bernard Bolzano. In E.N. Zalta (Ed.), *The Stanford Encyclopedia of Philosophy*. Metaphysics Research Lab, Stanford University, winter 2018 edition. `https://plato.stanford.edu/archives/win2018/entries/bolzano/` (accessed 15 March 2024). 31

[Mueller, 1969] Mueller, I. (1969). Euclid's Elements and the axiomatic method. *The British Journal for the Philosophy of Science*, 20(4), 289–309. 21, 23

[Mueller, 1981] Mueller, I. (1981). *Philosophy of Mathematics and Deductive Structure of Euclid's 'Elements'*. MIT Press. 21, 22, 23

[Munkres, 2000] Munkres, J.R. (2000). *Topology*. Prentise-Hall, second edition. 8, 123, 436

[Murzi, 2024a] Murzi, M. (2024a). Jules Henri Poincaré (1854–1912). In *The Internet Encyclopedia of Philosophy*. The Internet Encyclopedia of Philosophy. ISSN 2161-0002, `https://iep.utm.edu/poincare/` (accessed 29 June 2024). 158

[Murzi, 2024b] Murzi, M. (2024b). Rudolf Carnap (1891–1970). In *The Internet Encyclopedia of Philosophy*. The Internet Encyclopedia of Philosophy. ISSN 2161-0002, `https://iep.utm.edu/rudolf-carnap/` (accessed 1 March 2024). 4

[Musgrave & Pigden, 2023] Musgrave, A. & Pigden, C. (2023). Imre Lakatos. In E.N. Zalta & U. Nodelman (Eds.), *The Stanford Encyclopedia of Philosophy*. Metaphysics Research Lab, Stanford University, Spring 2023 edition. `https://plato.stanford.edu/archives/spr2023/entries/lakatos/`. 7

[Narasimhan, 2010] Narasimhan, R. (2010). Bernhard Riemann: Remarks on his life and work. *Milan Journal of Mathematics*, 78(1), 3–10. 86

[Neiman, 1972] Neiman, L. (1972). *The Joy of Discovery*. Detgiz, Moscow. Book in Russian. 321, 322

[Nerode & Greenberg, 2023] Nerode, A. & Greenberg, N. (2023). *Algebraic Curves and Riemann Surfaces for Undergraduates: The Theory of the Donut*. Springer.

[Netto, 1879] Netto, E.E. (1879). Beitrag zur Mannigfaltigkeitslehre. *Journal für die reine und angewandte Mathematik*, 86, 263–268. 130, 131

[Netto, 1882] Netto, E.E. (1882). *Substitutionentheorie und ihre Anwendung auf die Algebra*. B.G. Teubner. 130

[Netto, 1892] Netto, E.E. (1892). *The Theory of Substitutions and Its Application to Algebra*. Register Publishing Company. 130

[Netz, 2003] Netz, R. (2003). *The Shaping of Deduction in Greek Mathematics: A Study in Cognitive History*, volume 51 of *Ideas in Context*. Cambridge University Press. 22

[Neunhäuserer, 2021] Neunhäuserer, J. (2021). *Dimensionstheorie*. Springer.

[Neville, 1920] Neville, E.H. (1920). The classification of sets of points. *Acta mathematica*, 42(1), 63–93. 316

[Newton, 1704] Newton, I. (1704). *Opticks: or, a Treatise of the Reflexions, Refractions, Inflexions and Colours of Light. Also Two Treatises of the Species and Magnitude of Curvilinear Figures*. Royal Society of London. 27

[Nicomachus of Gerasa, 1926] Nicomachus of Gerasa (1926). *Introduction to Arithmetic*. M. L. D'Ooge, Trans. University of Michigan Studies. Macmillan. 12, 14

[Nöbeling, 1931a] Nöbeling, G.A. (1931a). Über eine n-dimensionale Universalmenge im $\mathbb{R}^{2n+1}$. *Mathematische Annalen*, 104, 71–80. Submitted 27 March 1930. 393, 394

[Nöbeling, 1931b] Nöbeling, G.A. (1931b). Bemerkungen zum Mengerschen Einbettungssatz. *Mathematische Annalen*, 105(5), 746–747. Submitted 8 June 1931. 395

[Nöbeling, 1931c] Nöbeling, G.A. (1931c). Hausdorffsche und mengentheoretische Dimension. *Ergebnisse eines Mathematischen Kolloquiums. Gesammelte Mitteilungen 1930/1931*, (pp. 24–25). = [Dierker et al., 1998, pp. 180–181]. 412, 414

[Nöbeling, 1934] Nöbeling, G.A. (1934). Review of Hurewicz, Witold, 'Über Abbildungen von endlichdimensionalen Räumen auf Teilmengen Cartesischer Räume'. *Zentralblatt für Mathematik und ihre Grenzgebiet*, 8, 133. 398

[Noether, 1923] Noether, E. (1923). Eliminationstheorie und allgemeine Idealtheorie. *Mathematische Annalen*, 90, 229–261. 424

[Nordmann & Boutroux, 2021] Nordmann, C. & Boutroux, P. (2021). *Henri Poincaré: son œuvre scientifique et sa philosophie*. Editions Le Mono. 158

[Oresme, 1968] Oresme, N. (1968). *Nicole Oresme and the Medieval Geometry of Qualities and Motions. Edited and translated by Marshall Clagett*. The University of Wisconsin Press. 25

[Osgood, 1900] Osgood, W.F. (1900). Ueber einen Satz des Herrn Schönflies aus der Theorie der Funktionen zweier reeller Veränderlichen. *Nachrichten von der Gesellschaft der Wissenschaften zu Göttingen, Mathematisch-Physikalische Klasse*, 1900, 94–97. 152

[Osgood, 1903] Osgood, W.F. (1903). A Jordan curve of positive area. *Transactions of the American Mathematical Society*, 4(1), 107–112. 150

[Ostrowski, 1920] Ostrowski, A. (1920). Über den ersten und vierten Gaussschen Beweis des Fundamentalsatzes der Algebra. In *Gauss Werke,* volume 10.2 (pp. 1–18). Julius Springer. 77, 213

[Otte, 1989] Otte, M. (1989). The ideas of Hermann Grassmann in the context of the mathematical and philosophical tradition since Leibniz. *Historia Mathematica*, 16(1), 1–35. 85

[Outerelo & Ruiz, 2009] Outerelo, E. & Ruiz, J.M. (2009). *Mapping Degree Theory*, volume 108. American Mathematical Society. 212, 215, 216, 218, 221

[Painlevé, 1900] Painlevé, P. (1900). *Notice sur les travaux scientifiques*. Gauthier Villars. 289, 294

[Parshall, 1998] Parshall, K.H. (1998). *James Joseph Sylvester: Life and work in letters*. Oxford University Press. 216

[Parshall, 2006] Parshall, K.H. (2006). *James Joseph Sylvester: Jewish Mathematician in a Victorian World*. Johns Hopkins University Press. 216

[Pascal, 1871] Pascal, B. (1871). De l'esprit géométrique et de l'art de persuader. In *Œuvres complètes*, volume 3 (pp. 163–182). Hachette. 41

[Pavlíková, 1985] Pavlíková, M. (1985). *Bolzanovo působení na pražské univerzitě*. Univerzita Karlova. 39

[Peano, 1887] Peano, G. (1887). *Applicazioni geometriche del calcolo infinitesimale*. Fratelli Bocca. 147

[Peano, 1890a] Peano, G. (1890a). Sur une courbe, qui remplit toute une aire plane. *Mathematische Annalen*, 36(1), 157–160. 147, 309

[Peano, 1890b] Peano, G. (1890b). Sulla definizione dell'area d'una superficie. *Atti della Accademia Naziohale dei Iincei, Rendiconti. Classe di Scienze Fisiche, Matematiche e Naturali*, 6, 54–57. 147

[Peano, 1890c] Peano, G. (1890c). Démonstration de l'intégrabilité des équations différentielles ordinaires. *Mathematische Annalen*, 37(2), 182–228. 211

[Peano, 1896] Peano, G. (1896). Réponse 60 (C. Jordan). *L'Intermédiaire des Mathématiciens*, 3, 39. 150

[Peano, 1973] Peano, G. (1973). *Selected Works of Giuseppe Peano, trans. and ed. Hubert C. Kennedy*. Toronto: University of Toronto Press. 147, 149

[Pears, 1975] Pears, A. (1975). *Dimension Theory of General Spaces*. Cambridge University Press. 274, 420

[Pesin & Climenhaga, 2009] Pesin, Y. & Climenhaga, V. (2009). *Lectures on Fractal Geometry and Dynamical Systems*, volume 52 of *Student Mathematical Library*. American Mathematical Society. 404, 416

[Petsche, 2009] Petsche, H.J. (2009). *Hermann Grassmann: Biography*. Springer Science. 80, 81, 82, 84, 85

[Petsche et al., 2010] Petsche, H.J., Kannenberg, L., Keßler, G., & Liskowacka, J. (2010). *Hermann Graßmann–Roots and Traces: Autographs and Unknown Documents*. Springer Science. 80

[Picard, 1889] Picard, É. (1889). Sur le nombre des racines communes à plusieurs équations simultanées. *Nouvelles annales de mathématiques: journal des candidats aux écoles polytechnique et normale 3e série*, 8, 5–13. 223

[Picard, 1891a] Picard, É. (1891a). Sur le nombre des racines communes à plusieurs équations simultanées. *Comptes rendus hebdomadaires des séances de l'Académie des Sciences Paris*, 113, 356–358. Session of 7 September 1891. 223, 224

[Picard, 1891b] Picard, É. (1891b). Sur la recherche du nombre des racines communes à plusieurs équations simultanées. *Comptes rendus hebdomadaires des séances de l'Académie des Sciences Paris*, 113, 669–672. Session of 16 November 1891. 223, 224

[Picard, 1891c] Picard, É. (1891c). Du nombre des racines communes à plusieurs équations simultanées. *Comptes rendus hebdomadaires des séances de l'Académie des Sciences Paris*, 113, 1012–1014. Session of 28 December 1891. 223, 224

[Picard, 1891d] Picard, É. (1891d). *Traité d'analyse: Intégrales simples et multiples. L'équation de Laplace et ses applications. Développements en séries. Applications géométriques du Calcul infinitésimal.*, volume 1. Gauthier-Villars et Cie.

[Picard, 1892] Picard, É. (1892). Sur le nombre des racines communes à plusieurs équations simultanées. *Journal de Mathématiques Pures et Appliquées*, 8, 5–24. 223, 224

[Picard, 1893] Picard, É. (1893). *Traité d'analyse. Fonctions harmoniques et fonctions analytiques. Introduction à la théorie des équations différentielles. Intégrales abéliennes et surfaces de Riemann.*, volume 2. Gauthier-Villars et Cie. 223

[Picard, 1896] Picard, É. (1896). *Traité d'analyse. Des singularités des intégrales des équations différentielles. Étude du cas où la variable reste réelle ; des courbes définies par des équations différentielles. Équations linéaires; analogies entre les équations algébriques et les équations linéaires.*, volume 3. Gauthier-Villars et Cie.

[Pierpont, 1912] Pierpont, J. (1912). *Lectures on the theory of functions of real variables*, volume 2. Ginn. 249

[Pincherle & Weierstrass, 1880] Pincherle, S. & Weierstrass, K.T.W. (1880). Saggio di una introduzione alla teoria delle funzioni analitiche secondo i principii del Prof. C. Weierstrass. *Giornali di matematiche*, 18, 178–254, 317–357. 98, 131

[Plücker, 1831] Plücker, J. (1831). *Analytisch-Geometrische Entwicklungen*, volume 2. G D Baedeker. 172

[Plücker, 1846] Plücker, J. (1846). *System der Geometrie des Raumes in neuer analytische Behandlungsweise, insbesondere die Theorie der Flächen zweiter Ordnung und Classe enthaltend.* W.H. Scheller. 79, 172

[Plücker, 1865] Plücker, J. (1865). On a New Geometry of Space. *Philosophical Transactions of the Royal Society of London*, 155, 725–791. 172

[Plücker, 1868] Plücker, J. (1868). *Neue Geometrie des Raumes gegründet auf die Betrachtung der geraden Linie als Raumelement. Erste Abteilung.* B.G.Teubner, Leipzig. 80, 172

[Plücker, 1869] Plücker, J. (1869). *Neue Geometrie des Raumes gegründet auf die Betrachtung der geraden Linie als Raumelement. Zweite Abteilung.* B.G. Teubner, Leipzig. 80, 172

[Poggi et al., 1994] Poggi, S., Bossi, M., & Straalen, B.v., Eds. (1994). *Romanticism in Science: Science in Europe, 1790-1840.* Springer.

[Poincaré, 1881] Poincaré, H. (1881). Mémoire sur les courbes définies par une équation différentielle (I). *Journal de mathématiques pures et appliquées*, 7, 375–422. 210, 211, 224, 225, 232

[Poincaré, 1882] Poincaré, H. (1882). Mémoire sur les courbes définies par une équation différentielle (II). *Journal de mathématiques pures et appliquées*, 8, 251–296. 210, 224

[Poincaré, 1885] Poincaré, H. (1885). Sur les courbes définies par les équations différentielles (III). *Journal de mathématiques pures et appliquées*, 1, 167–244. 211, 224, 225

[Poincaré, 1886] Poincaré, H. (1886). Sur les courbes définies par les équations différentielles (IV). *Journal de mathématiques pures et appliquées*, 2, 151–218. 211, 224, 226, 228

[Poincaré, 1887] Poincaré, H. (1887). Sur les hypothèses fondamentales de la géométrie. *Bulletin de la Société mathématique de France*, 15, 203–216. 159, 160

[Poincaré, 1891] Poincaré, H. (1891). Les Géométries non Euclidiennes. *Revue générale des Sciences pures et appliquées*, 2, 769–774. 159, 161, 182

[Poincaré, 1892] Poincaré, H. (1892). Correspondance sur les Géométries non Euclidiennes. Lettre. *Revue générale des Sciences pures et appliquées*, 3, 74–75. 161

[Poincaré, 1893] Poincaré, H. (1893). Le continu mathématique. *Revue de Métaphysique et de Morale*, 1(1), 26–34. 161, 167, 175, 263

[Poincaré, 1895a] Poincaré, H. (1895a). Analysis situs. *Journal de l'École Potytechnique*, 1(2), 1–121. 177, 226, 235

[Poincaré, 1895b] Poincaré, H. (1895b). L'espace et la géométrie. *Revue de Métaphysique et de Morale*, 3(6), 631–646. 161, 162, 163, 165, 182

[Poincaré, 1897] Poincaré, H. (1897). Réponse à quelques critiques. *Revue de Métaphysique et de Morale*, 5(1), 59–70. 162

[Poincaré, 1898] Poincaré, H. (1898). On the Foundations of Geometry. *The Monist*, 9(1), 1–43. 162, 163, 164, 165, 166, 167, 168, 169, 170, 173, 174, 175, 176, 182, 184

[Poincaré, 1899] Poincaré, H. (1899). Complément à l'analysis situs. *Rendiconti del Circolo Matematico di Palermo*, 13(2), 285–343. 177, 226, 235

[Poincaré, 1900] Poincaré, H. (1900). Second complément à l'analysis situs. *Proceedings of the London Mathematical Society*, 1(1), 277–308. 177, 226, 235

[Poincaré, 1902a] Poincaré, H. (1902a). Compte rendu: Hilbert, les fondements de la géométrie. *Bulletin des Sciences Mathématiques*, 26(2), 249–272. 159, 177, 184

[Poincaré, 1902b] Poincaré, H. (1902b). *La Science et l'Hypothèse*. E. Flammarion. 160, 161, 263

[Poincaré, 1902c] Poincaré, H. (1902c). Sur certaines surfaces algébriques; troisième complément à l'analysis situs. *Bulletin de la Société de France*, 30, 49–70. 177, 226

[Poincaré, 1902d] Poincaré, H. (1902d). Sur la valeur objective de la science. *Revue de Métaphysique et de Morale*, 10(3), 263–293. 162, 176, 184

[Poincaré, 1902e] Poincaré, H. (1902e). Sur les cycles des surfaces algébriques; quatrième complément à l'analysis situs. *Journal de mathématiques pures et appliquées*, 8, 169–214. 177, 226

[Poincaré, 1903a] Poincaré, H. (1903a). Correction to compte rendu: Hilbert, les fondements de la géométrie. *Bulletin des Sciences Mathématiques*, 27(2), 115. 159, 177, 184

[Poincaré, 1903b] Poincaré, H. (1903b). L'espace et ses trois dimensions. *Revue de Métaphysique et de Morale*, 11(3), 281–301, 407–429. 2, 161, 162, 163, 164, 165, 176, 177, 180, 182, 189, 236, 262, 433

[Poincaré, 1904] Poincaré, H. (1904). Cinquième complément à l'analysis situs. *Rendiconti del Circolo Matematico di Palermo*, 18(1), 45–110. 177, 226, 235

[Poincaré, 1905a] Poincaré, H. (1905a). *La Valeur de la Science*. E. Flammarion. 160, 162, 184, 189

[Poincaré, 1905b] Poincaré, H. (1905b). *Science and Hypothesis*. Science Press. 160

[Poincaré, 1906] Poincaré, H. (1906). La relativité de l'espace. *L'année psychologique*, 13(1), 1–17. 162, 163, 176, 177, 183

[Poincaré, 1908a] Poincaré, H. (1908a). L'avenir des mathématiques. *Revue générale des sciences pures et appliquées*, 19, 930–939. 167

[Poincaré, 1908b] Poincaré, H. (1908b). *Science et Méthode*. E. Flammarion. 160

[Poincaré, 1912] Poincaré, H. (1912). Pourquoi l'espace a trois dimensions. *Revue de Métaphysique et de Morale*, 20(4), 483–504. 2, 162, 163, 165, 176, 178, 179, 180, 182, 183, 184, 236, 262, 263, 341, 433

[Poincaré, 1913a] Poincaré, H. (1913a). *Dernières Pensées*. E. Flammarion. 160

[Poincaré, 1913b] Poincaré, H. (1913b). *The Value of Science*. Science Press. 161

[Poincaré, 1914] Poincaré, H. (1914). *Science and Method*. T. Nelson and Sons. 161

[Poincaré, 1953] Poincaré, H. (1953). *Oeuvres de Henri Poincaré*, volume VI. Paris. 177

[Poincaré, 1963] Poincaré, H. (1963). *Mathematics and Science: Last Essays*. Dover Publications. 161, 377

[Poincaré, 2010] Poincaré, H. (2010). *Papers on Topology: Analysis Situs and Its Five Supplements. Trans. John Stillwell*, volume 37 of *History of Mathematics*. American Mathematical Society. 177, 185, 226

[Pol, 1998] Pol, R. (1998). The Works of Stefan Mazurkiewicz in Topology. In *Handbook of the History of General Topology*, volume 2 (pp. 415–430). Kluwer Academic Publishers. 303, 304, 308

[Pólya, 1954] Pólya, G. (1954). *Mathematics and Plausible Reasoning: Induction and Analogy in Mathematics*, volume 1. Princeton University Press. 7

[Pólya, 1968] Pólya, G. (1968). *Mathematics and Plausible Reasoning: Patterns of Plausible Inference*, volume 2. Princeton University Press. 7

[Pólya, 2004] Pólya, G. (2004). *How to Solve It: A New Aspect of Mathematical Method*. Princeton University Press. 7

[Pont, 1974] Pont, J.C. (1974). *La topologie algébrique des origines à Poincaré*. Presses Universitaires de France. 226

[Pontrjagin, 1927a] Pontrjagin, L.S. (1927a). Zum Alexanderschen Dualitätssatz. *Nachrichten von der Gesellschaft der Wissenschaften zu Göttingen, Mathematisch-Physikalische Klasse*, 1927, 315–322. Read by R. Courant at session of 25 November 1927. 389

[Pontrjagin, 1927b] Pontrjagin, L.S. (1927b). Zum Alexanderschen Dualitätssatz. Zweite Mitteilung. *Nachrichten von der Gesellschaft der Wissenschaften zu Göttingen, Mathematisch-Physikalische Klasse*, 1927, 446–456. Read by R. Courant at session of 9 March 1928. 389

[Pontrjagin, 1930] Pontrjagin, L.S. (1930). Sur une hypothèse fondamentale de la théorie de la dimension. *Comptes rendus hebdomadaires des séances de l'Académie des Sciences Paris*, 190, 1105–1107. Session of 12 May 1930.

[Pontrjagin, 1931] Pontrjagin, L.S. (1931). Über den algebraischen Inhalt topologischer Dualitätssätze. *Mathematische Annalen*, 105, 165–205. 389

[Pontrjagin & Schnirelmann, 1932] Pontrjagin, L.S. & Schnirelmann, L. (1932). Sur une propriété métrique de la dimension. *Annals of Mathematics*, 33(1), 156–162. Dated Moscow, May 1931. 412, 413, 414

[Pontryagin & Tolstowa, 1931] Pontryagin, L.S. & Tolstowa, G. (1931). Beweis des Mengerschen Einbettungssatzes. *Mathematische Annalen*, 105(5), 734–745. Submitted 5 February 1931. 189, 394

[Popper, 1962] Popper, K. (1962). *Conjectures and Refutations: The Growth of Scientific Knowledge*. Routledge. 6

[Popper, 2002] Popper, K. (2002). *Conjectures and Refutations: The Growth of Scientific Knowledge*. Routledge, revised edition. 6

[Popper, 1972] Popper, K.R. (1972). *Objective Knowledge*. Oxford University Press. 7, 44

[Purkert & Ilgauds, 2013] Purkert, W. & Ilgauds, H.J. (2013). *Georg Cantor 1845–1918*, volume 1 of *Vita Mathematica*. Springer-Verlag. 101, 102

[Reich, 1973] Reich, K. (1973). Die Geschichte der Differentialgeometrie von Gauss bis Riemann (1828–1868). *Archive for History of Exact Sciences*, 11(4), 273–376. 77

[Ricca & Nipoti, 2011] Ricca, R.L. & Nipoti, B. (2011). Gauss' linking number revisited. *Journal of Knot Theory and Its Ramifications*, 20(10), 1325–1343. 253

[Riemann, 1851] Riemann, B. (1851). *Grundlagen für eine allgemeine Theorie der Functionen einer veränderlichen complexen Grösse*. Adalbert Rente. 92

[Riemann, 1857a] Riemann, B. (1857a). *Theorie der Abel'schen Functionen*. Georg Reimer. 93

[Riemann, 1857b] Riemann, B. (1857b). Theorie der Abel'schen Functionen. *Journal für die reine und angewandte Mathematik*, 54, 101–155.
Series consists of:
1. Allgemeine Voraussetzungen und Hülfsmittel für die Untersuchung von Functionen unbeschränkt veränderlicher Grössen;
2. Lehrsätze aus der analysis situs für die Theorie der Integrale von zweigliedrigen vollständigen Differentialien;
3. Bestimmung einer Function einer veränderlichen complexen Grösse durch

Grenz- und Unstetigkeitsbedingungen;
4. Theorie der Abel'schen Functionen. 93

[Riemann, 1868] Riemann, B. (1868). Über die Hypothesen, welche der Geometrie zu Grunde liegen.(Mitgetheilt durch R. Dedekind). *Abhandlungen der Königlichen Gesellschaft der Wissenschaften zu Göttingen*, 13, 133–152. 86, 88, 160

[Riemann, 1892] Riemann, B. (1892). *Bernhard Riemann's Gesammelte mathematische Werke und wissenschaftlicher Nachlass*. B.G. Teubner, second edition. 86, 87, 495

[Riemann, 1902] Riemann, B. (1902). *Nachträge*. M. Noether and W. Wirtinger, eds. B.G. Teubner. 495

[Riemann, 1953] Riemann, B. (1953). *Bernhard Riemann's Gesammelte mathematische Werke und wissenschaftlicher Nachlass und Nachträge*. H. Weber, M. Noether, W. Wirtinger, eds. Dover. Reprint of [Riemann, 1892] and [Riemann, 1902]. 87, 88, 89, 92, 93, 94, 95, 126, 160

[Riemann, 1990] Riemann, B. (1990). *Bernhard Riemann: Gesammelte Mathematische Werke, Wissenschaftlicher Nachlass und Nachträge. Collected Papers. Nach der Ausgabe von Heinrich Weber und Richard Dedekind neu herausgegeben von Raghavan Narasimhan*. Springer. 1, 86, 87, 88, 89, 92, 93, 94, 95, 126, 160

[Riemann, 2004] Riemann, G.F.B. (2004). *Collected Papers*. R. Baker, C. Christenson, H. Orde, eds. and trans. Kendrick Press, Heber City. 86, 87, 88, 89, 92, 93, 94, 95, 126, 160

[Riesz, 1905a] Riesz, F. (1905a). Sur les ensembles discontinus. *Comptes rendus hebdomadaires des séances de l'Académie des Sciences Paris*, 141, 650–653. Session of 23 October 1905. 187, 188, 196

[Riesz, 1905b] Riesz, F. (1905b). Über mehrfache Ordnungstypen. I. *Mathematische Annalen*, 61(3), 406–421. 190

[Riesz, 1907] Riesz, F. (1906, 1907). A térfogalom genesise. *Középiskolai Matematikai és Fizikai Lapok*, 15, 16, 97–122, 145–161. 189, 265, 379

[Riesz, 1909a] Riesz, F. (1909a). Die Genesis des Raumbegriffs. *Mathematische und naturwissenschaftliche Berichte aus Ungarn*, 24, 309–353. 180, 189, 190, 263, 265, 379

[Riesz, 1909b] Riesz, F. (1909b). Stetigkeitsbegriff und abstrakte Mengentehre. In *Atti del IV. Congresso Internazionale dei Matematici, Roma, 1908*, volume II (pp. 18–24). R. Accademia dei Lincei. 190

[Riesz, 1960] Riesz, F. (1960). *összegyűjtött munkái. Oeuvres Complètes. Gesammelte Arbeiten*. A. Csaszar, ed. Maison d'edition de l'Academie des Sciences de Hongrie. 2 volumes. 187

[Rogers, 1998] Rogers, C.A. (1998). *Hausdorff Measures*. Cambridge University Press. 404

[Rosenthal, 1913] Rosenthal, A. (1913). Beweis der Unmöglichkeit ergodischer Gassysteme. *Annalen der Physik*, 347(14), 796–806. 282

[Rougier, 1920] Rougier, L. (1920). *La philosophie géométrique de Henri Poincaré*. F. Alcan. 158, 173, 180

[Rowe, 2010] Rowe, D.E. (2010). Debating Grassmann's Mathematics: Schlegel versus Klein. *The Mathematical Intelligencer*, 32(1), 41–48. 80

[Rowe, 2018] Rowe, D.E. (2018). *Otto Blumenthal: ausgewählte Briefe und Schriften I: 1897-1918*. Springer. 204, 245

[Rowe, 2021] Rowe, D.E. (2021). *Emmy Noether–mathematician extraordinaire*. Springer. 384

[Rowe & Felsch, 2019] Rowe, D.E. & Felsch, V. (2019). *Otto Blumenthal: Ausgewählte Briefe und Schriften II: 1919–1944*. Springer-Verlag. 266

[Rowe & Koreuber, 2020] Rowe, D.E. & Koreuber, M. (2020). *Proving it her way: Emmy Noether, a life in mathematics*. Springer. 384

[Rückert, 1933] Rückert, W. (1933). Zum Eliminationsproblem der Potenzreihenideale. *Mathematische Annalen*, 107, 259–281. 424

[Rudolff & Stifel, 1554] Rudolff, C. & Stifel, M. (1554). *Die Coss Christoffs Rudolffs, Mit schönen Exempeln der Coss. Durch Michael Stifel Gebessert und sehr gemehrt*. Könoigsberg. At `https://www.e-rara.ch/zut/content/titleinfo/1238793` (accessed 8 April 2024). 26

[Rusnock, 1997] Rusnock, P. (1997). Remaking mathematics: Bolzano Reads Lagrange. *Acta Analytica*, 18, 51–72. 39

[Rusnock, 2000] Rusnock, P. (2000). *Bolzano's Philosophy and the Emergence of Modern Mathematics*, volume XXX of *Studien zur Österreichischen Philosophie*. Rodopi. 31

[Rusnock, 2020] Rusnock, P. (2020). Bolzano on Continuity. In S. Shapiro & G. Hellman (Eds.), *The History of Continua: Philosophical and Mathematical Perspectives* (pp. 187–216). Oxford University Press. Chapter 8. 31, 40, 48, 49, 56

[Rusnock & Šebestík, 2013] Rusnock, P. & Šebestík, J. (2013). The Beyträge at 200: Bolzano's Quiet Revolution in Philosophy of Mathematics. *Journal for the History of Analytical Philosophy*, 1(8), 1–14. 31, 40

[Rusnock & Šebestík, 2019] Rusnock, P. & Šebestík, J. (2019). *Bernard Bolzano: His Life and Work*. Oxford University Press, USA. 31, 33, 55, 70

[Russell, 1897] Russell, B. (1897). *An Essay on the Foundations of Geometry*. Cambridge University Press. 90, 96

[Sagan, 2012] Sagan, H. (2012). *Space-Filling Curves*. Springer. 131, 147, 148, 286, 308, 309

[Saks, 1937] Saks, S. (1937). *Theory of the Integral*. Monografie Matematyczne. Instytut Matematyczny Polskiej Akademii Nauk, Second Revised edition.

[Sarkaria, 1999] Sarkaria, K.S. (1999). The Topological Work of Henri Poincaré. In *History of Topology* (pp. 123–139). Elsevier Amsterdam. 185

[Schläfli, 1901] Schläfli, L. (1901). *Theorie der vielfachen Kontinuität*. Springer. 79

[Schläfli, 1950] Schläfli, L. (1950). *Gesammelte Mathematische Abhandlungen*, volume 1. Birkhäuser. 79

[Schläfli, 1953] Schläfli, L. (1953). *Gesammelte Mathematische Abhandlungen*, volume 2. Birkhäuser. 79

[Schlegel, 1878] Schlegel, V. (1878). *Hermann Grassmann: sein Leben und seine Werke*. F A Brockhaus. 80

[Schlesinger, 1912] Schlesinger, L. (1912). Über Gausss' Arbeiten zur Funktionentheorie. In *Gauss Werke,* volume 10.2 (pp. 1–222). Julius Springer. 77, 78

[Schmetterer & Sigmund, 1995] Schmetterer, L. & Sigmund, K. (1995). Hans Hahn — A Short Biography. In *Gesammelte Abhandlungen / Collected Works*, volume I (pp. 29–35). Springer-Verlag. 308

[Schoenflies, 1896] Schoenflies, A. (1896). Über einen Satz der Analysis Situs. *Nachrichten von der Gesellschaft der Wissenschaften zu Göttingen, Mathematisch-Physikalische Klasse*, 1896, 79–89. 146

[Schoenflies, 1898] Schoenflies, A. (1898). Mengenlehre. In *Encyklopädie der mathematischen Wissenschaften mit Einschluss ihrer Anwendungen*, volume I A 5 (pp. 184–207). B.G. Teubner. 139, 152, 287

[Schoenflies, 1899] Schoenflies, A. (1899). Über einen Satz der Analysis Situs. *Nachrichten von der Gesellschaft der Wissenschaften zu Göttingen, Mathematisch-Physikalische Klasse*, 1899, 282–290. 152, 191, 193

[Schoenflies, 1900] Schoenflies, A. (1900). *Die Entwickelung der Lehre von den Punktmannigfaltigkeiten. Bericht, Erstattet der Deutschen Mathematiker-Vereinigung*. B.G. Teubner. 148, 152, 316

[Schoenflies, 1902] Schoenflies, A. (1902). Über einen grundlegenden Satz der Analysis Situs. *Nachrichten von der Gesellschaft der Wissenschaften zu Göttingen, Mathematisch-Physikalische Klasse*, 1902, 185–192. 153, 154

[Schoenflies, 1903] Schoenflies, A. (1903). Beiträge zur Theorie der Punktmengen. I. *Mathematische Annalen*, 58(1), 195–234. 153, 154, 206, 290

[Schoenflies, 1904a] Schoenflies, A. (1904a). Beiträge zur Theorie der Punktmengen. II. *Mathematische Annalen*, 59(1), 129–160. 153, 154, 206, 290

[Schoenflies, 1904b] Schoenflies, A. (1904b). Ueber die geometrischen Invarianten der Analysis situs. *Nachrichten von der Gesellschaft der Wissenschaften zu Göttingen, Mathematisch-Physikalische Klasse*, 1904, 514–525. 153

[Schoenflies, 1906a] Schoenflies, A. (1906a). Beiträge zur Theorie der Punktmengen. III. *Mathematische Annalen*, 62(2), 286–328. 146, 153, 154, 206, 290, 298

[Schoenflies, 1906b] Schoenflies, A. (1906b). Die Beziehungen der Mengenlehre zur Geometrie und Funktionentheorie. *Jahresbericht der Deutschen Mathematiker-Vereinigung*, 15, 557–576. 150

[Schoenflies, 1907a] Schoenflies, A. (1907a). Ueber den allgemeinsten Begriff der ebenen stetigen Kurve. *Nachrichten von der Gesellschaft der Wissenschaften zu Göttingen, Mathematisch-Physikalische Klasse*, 1907, 28–49. 154

[Schoenflies, 1907b] Schoenflies, A. (1907b). Ueber den allgemeinsten Begriff der ebenen stetigen Kurve. Zweite Mitteilung. *Nachrichten von der Gesellschaft der Wissenschaften zu Göttingen, Mathematisch-Physikalische Klasse*, 1907, 299–320. 154

[Schoenflies, 1908a] Schoenflies, A. (1908a). Bemerkung zu meinem zweiten Beitrag zur Theorie der Punktmengen. *Mathematische Annalen*, 65(3), 431–432. 153

[Schoenflies, 1908b] Schoenflies, A. (1908b). *Die Entwickelung der Lehre von den Punktmannigfaltigkeiten. Bericht, Erstattet der Deutschen Mathematiker-Vereinigung. Zweiter Teil*, volume 2. B.G. Teubner. 146, 152, 154, 155, 188, 206, 208, 209, 263, 265, 290, 291, 309, 316

[Schoenflies, 1910] Schoenflies, A. (1910). Bemerkung zu dem vorstehenden Aufsatz des Herrn L. E. J. Brouwer. *Mathematische Annalen*, 68, 435–444. 207

[Schoenflies, 1925] Schoenflies, A. (1925). Über das eindeutige und stetige Abbild des Kreises (Jordankurve). *Jahresbericht der Deutschen Mathematiker-Vereinigung*, 33, 147–160. 146

[Schoenflies & Hahn, 1913] Schoenflies, A. & Hahn, H. (1913). *Entwickelung der Mengenlehre und ihrer Anwendungen. Erster Hälfte. Allgemeine Theorie der unendlichen Mengen und Theorie der Punktmengen*, volume I.1. B.G. Teubner, second edition. 152, 316

[Scholz, 1980] Scholz, E. (1980). *Geschichte des Mannigfaltigkeitsbegriffs von Riemann bis Poincaré*. Birkhäuser. 90, 235

[Scholz, 1982a] Scholz, E. (1982a). Herbart's influence on Bernhard Riemann. *Historia Mathematica*, 9(4), 413–440. 90, 91

[Scholz, 1982b] Scholz, E. (1982b). Riemanns frühe Notizen zum Mannigfaltigkeitsbegriff und zu den Grundlagen der Geometrie. *Archive for History of Exact Sciences*, (pp. 213–232). 92

[Scholz, 1992] Scholz, E. (1992). Riemann's vision of a new approach to geometry. In *1830–1930: A Century of Geometry* (pp. 22–34). Springer. 90

[Scholz, 1999] Scholz, E. (1999). The Concept of Manifold, 1850–1950. In *History of Topology* (pp. 25–64). Elsevier Amsterdam. 90, 235

[Schubring, 1996] Schubring, G., Ed. (1996). *Hermann Günther Grassmann (1809–1877): Visionary Mathematician, Scientist and Neohumanist Scholar*. Springer. 85

[Schubring, 2001] Schubring, G. (2001). Argand and the early work on graphical representation: New sources and interpretations. In *Proceedings of the Wessel Symposium at The Royal Danish Academy of Sciences and Letters, Copenhagen, August 11–15 1998. Invited papers* (pp. 125–146).: C .A. Reitzels Forlag. 74

[Schultz, 1789] Schultz, J. (1789). *Prüfung der Kantischen Critik der reinen Vernunft. Erster Theil*. GL Hartung. 37

[Schultz, 1791] Schultz, J. (1791). *Prüfung der Kantischen Critik der reinen Vernunft. Zweiter Theil*. GL Hartung.

[Šebestík, 1992] Šebestík, J. (1992). *Logique et Mathématique chez Bernard Bolzano*. Vrin. 31, 62

[Šebestík, 2017] Šebestík, J. (2017). Bolzano's Logic. In E.N. Zalta (Ed.), *The Stanford Encyclopedia of Philosophy*. Metaphysics Research Lab, Stanford University, Fall 2017 edition.
`https://plato.stanford.edu/archives/fall2017/entries/bolzano-logic/` (accessed 15 March 2024). 31

[Segre, 1904] Segre, C. (1904). On some tendencies in geometric investigations. *Bulletin of the American Mathematical Society*, 10(9), 442–468. 78

[Segre, 1912] Segre, C. (1912). Mehrdimensionale Räume. In *Encyklopädie der Mathematischen Wissenschaften mit Einschluss ihrer Anwendungen*, volume III C 7 (pp. 769–972). B.G. Teubner. 78

[Seifert & Threlfall, 1934] Seifert, H. & Threlfall, W. (1934). *Lehrbuch der Topologie*. B.G. Teubner.

[Seshadri & Verma, 2016] Seshadri, H. & Verma, K. (2016). The embedding theorems of Whitney and Nash. *Resonance*, 21, 815–826. 438

[Sextus Empiricus, 1936] Sextus Empiricus (1936). *Against the Physicists. Against the Ethicists.* R. G. Bury, trans. Loeb Classical Library: Harvard University Press. 15

[Shafarevich, 2013a] Shafarevich, I.R. (2013a). *Basic Algebraic Geometry 1: Varieties in Projective Space*. Springer, third edition. 422

[Shafarevich, 2013b] Shafarevich, I.R. (2013b). *Basic Algebraic Geometry 2: Schemes and Complex Manifolds*. Springer, third edition. 422

[Shapiro & Hellman, 2020] Shapiro, S. & Hellman, G., Eds. (2020). *The History of Continua: Philosophical and Mathematical Perspectives*. Oxford University Press.

[Siegberg, 1981] Siegberg, H.W. (1981). Some historical remarks concerning degree theory. *The American Mathematical Monthly*, 88(2), 125–139. 212, 213

[Sierpiński, 1912a] Sierpiński, W. (1912a). Sur une nouvelle courbe continue qui remplit toute une aire plane. *Bulletin international de l'Académie des sciences de Cracovie (Classe des sciences mathématiques et naturelles Série A)*, (pp. 462–478). 309

[Sierpiński, 1912b] Sierpiński, W. (1912b). *Zarys teorii mnogości [An outline of set theory]*. Biblioteka Matematyczno-Fizyczna, Skład Główny w Księgarni E. Wendego i S-ki, Warszawa, first edition.

[Sierpiński, 1915] Sierpiński, W. (1915). Sur une courbe dont tout point est un point de ramification. *Comptes rendus hebdomadaires des séances de l'Académie des Sciences Paris*, 160, 302–305. Session of 1 February 1915; expanded version in Polish 1916, translated into French in [Sierpiński, 1975, pp. 99–106]. xviii, xviii, xviii, xviii, xviii, 299, 300, 301, 303, 499

[Sierpiński, 1916a] Sierpiński, W. (1916a). O krzywej, której każdy punkt jest punktem rozgałęzienia (Sur une courbe dont tout point est un point de ramification). *Prace Matematyczno-Fizyczne*, 27, 77–86. Expanded version of [Sierpiński, 1915], translated into French in [Sierpiński, 1975, pp. 99–106]. 299, 300

[Sierpiński, 1916b] Sierpiński, W. (1916b). Sur une courbe cantorienne qui contient une image biunivoque et continue de toute courbe donnée. *Comptes rendus hebdomadaires des séances de l'Académie des Sciences Paris*, 162, 629–632. Session of 25 April 1916; expanded version in Russian 1916, translated into French in [Sierpiński, 1975, pp. 107–119]. 302, 352, 382

[Sierpiński, 1920] Sierpiński, W. (1920). Sur une condition pour qu'un continu soit une courbe jordanienne. *Fundamenta Mathematicae*, 1(1), 44–60. = [Sierpiński, 1975, pp. 308–321]. 311

[Sierpiński, 1921] Sierpiński, W. (1921). Sur les ensembles connexes et non connexes. *Fundamenta Mathematicae*, 2(1), 81–95. 195, 274, 312, 313, 314, 341, 343, 344, 354, 379, 394

[Sierpiński, 1922] Sierpiński, W. (1922). Sur une propriété des ensembles frontières. *Fundamenta Mathematicae*, 3(1), 7–13. 314

[Sierpiński, 1923] Sierpiński, W. (1923). Sur quelques propriétés topologiques du plan. *Fundamenta Mathematicae*, 4(1), 1–6. 331, 531

[Sierpiński, 1974] Sierpiński, W. (1974). *Oeuvres choisies*, volume I. Académie Polonaise des Sciences, Institut Mathématique, Warsaw. 299

[Sierpiński, 1975] Sierpiński, W. (1975). *Oeuvres choisies*, volume II. Académie Polonaise des Sciences, Institut Mathématique, Warsaw. 299, 302, 499

[Sierpiński, 1976] Sierpiński, W. (1976). *Oeuvres choisies*, volume III. Académie Polonaise des Sciences, Institut Mathématique, Warsaw. 299

[Sigmund, 2017] Sigmund, K. (2017). *Exact Thinking in Demented Times: The Vienna Circle and the Epic Quest for the Foundations of Science*. Hachette UK. 308, 333, 363

[Simon, 1981] Simon, P. (1981). Bernard Bolzano a teorie dimenze. *Pokroky matematiky, fyziky a astronomie*, 26(5), 248–258. 31, 62

[Sinaceur, 1991] Sinaceur, H.B. (1991). *Corps et modèles: essai sur l'histoire de l'algèbre réelle*. Vrin. 216

[Sinaceur, 1992] Sinaceur, H.B. (1992). Cauchy, Sturm et les racines des équations. *Revue d'histoire des sciences*, (pp. 51–67). 216

[Sinaceur, 2009] Sinaceur, H.B. (2009). L'œuvre algébrique de Charles François Sturm. In *Collected Works of Charles François Sturm* (pp. 13–24). Birkhäuser. 216

[Smith, 1875] Smith, H.J.S. (1875). On the integration of discontinuous functions. *Proceedings of the London Mathematical Society*, 6, 140–153. 141

[Soltani, 2020] Soltani, M. (2020). *Poincaré's philosophy of geometry and its reception in the twentieth-century philosophy of science*. PhD thesis, University of British Columbia. 159, 180

[Sperner, 1928] Sperner, E. (1928). Neuer Beweis für die Invarianz der Dimensionszahl und des Gebietes. *Abhandlungen aus dem Mathematischen Seminar der Universität Hamburg*, 6, 265–272. 317, 318, 368, 429

[Stäckel, 1901] Stäckel, P. (1901). Friedrich Ludwig Wachter. *Mathematische Annalen*, 54, 49–85. 75

[Stäckel, 1917] Stäckel, P. (1917). Gauss als Geometer. In *Gauss Werke*, volume 10.2 (pp. 1–123). Julius Springer. 77, 78

[Stadler, 2001] Stadler, F. (2001). *The Vienna Circle: Studies in the Origins, Development, and Influence of Logical Empiricism*. Springer. 308

[Stadler, 2015] Stadler, F. (2015). *The Vienna Circle: Studies in the Origins, Development, and Influence of Logical Empiricism*. Springer, revised edition. 308, 363

[Staudt, 1847] Staudt, K.G.C.v. (1847). *Geometrie der Lage*. Bauer und Raspe. 173, 217

[Staudt, 1856] Staudt, K.G.C.v. (1856). *Beiträge zur Geometrie der Lage*, volume 1. Fr. Korn'schen. 173

[Staudt, 1857] Staudt, K.G.C.v. (1857). *Beiträge zur Geometrie der Lage*, volume 2. Fr. Korn'schen. 173

[Staudt, 1860] Staudt, K.G.C.v. (1860). *Beiträge zur Geometrie der Lage*, volume 3. Fr. Korn'schen. 173

[Stepanoff & Tumarkin, 1928] Stepanoff, W. & Tumarkin, L.A. (1928). Über eine Erweiterung abgeschlossener Mengen zu Jordanschen Kontinuen derselben Dimension. *Fundamenta Mathematicae*, 12(1), 43–46.

[Stigt, 1979] Stigt, W.P.v. (1979). The rejected parts of Brouwer's dissertation on the foundations of mathematics. *Historia Mathematica*, 6(4), 385–404. 204, 280

[Stolz, 1881] Stolz, O. (1881). B. Bolzano's Bedeutung in der Geschichte der Infinitesimalrechnung. *Mathematische Annalen*, 18(2), 255–279. 50

[Stone, 1937] Stone, M.H. (1937). Applications of the theory of Boolean rings to general topology. *Transactions of the American Mathematical Society*, 41(3), 375–481. 399

[Struik, 1969] Struik, D.J. (1969). *A Source Book in Mathematics, 1200-1800*. Harvard University Press. 27

[Study, 1891] Study, E. (1891). Von den Bewegungen und Umlegungen: I. und II. Abhandlung. *Mathematische Annalen*, 39, 441–565. 169

[Sturm, 1829] Sturm, J.C.F. (1829). Analyse d'un Mémoire sur la résolution des équations numériques. *Bulletin Universel des Sciences et de l'Industrie*, 11, 419–425. = [Sturm, 2009, pp. 323–326]; complete exposition given in [Sturm, 1835]. 216

[Sturm, 1835] Sturm, J.C.F. (1835). Mémoire sur la résolution des équations numériques. *Mémoires présentés par divers savants à l'Académie Royale des Sciences*, 6, 273–318. = [Sturm, 2009, pp. 345–390]. 216, 501

[Sturm, 1836] Sturm, J.C.F. (1836). Autres démonstrations du même Théorème. *Journal de Mathématiques pures et appliquées*, 1, 290–308. 214

[Sturm, 2009] Sturm, J.C.F. (2009). *Collected Works of Charles François Sturm*. Edited by Jean-Claude Pont. Birkhäuser. 214, 501

[Sturm & Liouville, 1836] Sturm, J.C.F. & Liouville, J. (1836). Démonstration d'un théorème de M. Cauchy, relatif aux racines imaginaires des Équations. *Journal de Mathématiques pures et appliquées*, 1, 278–289. 214

[Sturm & Liouville, 1837] Sturm, J.C.F. & Liouville, J. (1837). Note sur un théorème de M. Cauchy relatif aux racines des équations simultanées. *Comptes rendus hebdomadaires des séances de l'Académie des Sciences Paris*, 4, 720–724. 215

[Sylvester, 1850] Sylvester, J.J. (1850). Additions to the articles in the September number of this journal, "on a new class of theorems," and on Pascal's theorem. *The London, Edinburgh, and Dublin Philosophical Magazine and Journal of Science*, 37(251), 363–370. 79

[Sylvester, 1851a] Sylvester, J.J. (1851a). On certain general properties of homogeneous functions. *Cambridge and Dublin Mathematical Journal*, 6, 1–17. 79

[Sylvester, 1851b] Sylvester, J.J. (1851b). An enumeration of the contacts of lines and surfaces of the second order. *The London, Edinburgh, and Dublin Philosophical Magazine and Journal of Science*, 1(2), 119–140. 79

[Sylvester, 1853a] Sylvester, J.J. (1853a). On a Theory of the Syzygetic relations of two rational integral functions, comprising an application to the Theory of Sturm's Functions, and that of the greatest Algebraical Common Measure. *Philosophical Transactions of the Royal Society of London*, 143, 407–548. 216

[Sylvester, 1853b] Sylvester, J.J. (1853b). On a remarkable modification of Sturm's theorem. *The London, Edinburgh, and Dublin Philosophical Magazine and Journal of Science*, 5(34), 446–456. 216

[Sylvester, 1853c] Sylvester, J.J. (1853c). Note on a remarkable modification of Sturm's theorem, and on a new rule for finding superior and inferior limits to the roots of an equation. *The London, Edinburgh, and Dublin Philosophical Magazine and Journal of Science*, 6(36), 14–20. 216

[Sylvester, 1912] Sylvester, J.J. (1904–1912). *The Collected Mathematical Papers of James Joseph Sylvester*. Cambridge University Press. 4 volumes. 79

[Szpilrajn, 1937] Szpilrajn, E. (1937). La dimension et la mesure. *Fundamenta Mathematicae*, 28(1), 81–89. Edward Szpilrajn = Edward Marczewski. 414, 415

[Taylor, 1982] Taylor, A.E. (1982). A study of Maurice Fréchet: I. His early work on point set theory and the theory of functionals. *Archive for History of Exact Sciences*, 27(3), 233–295. 197

[Taylor, 1985] Taylor, A.E. (1985). A study of Maurice Fréchet: II. Mainly about his work on general topology, 1909–1928. *Archive for History of Exact Sciences*, 34(4), 279–380. 197

[Taylor, 1987] Taylor, A.E. (1987). A study of Maurice Fréchet: III. Fréchet as analyst, 1909–1930. *Archive for History of Exact Sciences*, 37(1), 25–76. 197

[Taylor, 1975] Taylor, S. (1975). Abram Samoilovitch Besicovitch. *Bulletin of the London Mathematical Society*, 7(2), 191–210. 412

[Thomae, 1878] Thomae, J. (1878). Sätze aus der Functionentheorie. *Nachrichten von der Königliche Gesellschaft der Wissenschaften und der Georg-Augusts-Universität zu Göttingen*, 1878, 466–468. 125

[Thomas, 1939] Thomas, I., Ed. (1939). *Greek Mathematical Works I. Thales to Euclid.* Trans. I. Thomas. Loeb Classical Library. 13

[Thomas, 1941] Thomas, I., Ed. (1941). *Greek Mathematical Works II. Aristarchus to Pappus of Alexandria.* Trans. I. Thomas. Loeb Classical Library. 16, 19, 20

[Thomassen, 1992] Thomassen, C. (1992). The Jordan-Schönflies theorem and the classification of surfaces. *The American Mathematical Monthly*, 99(2), 116–130. 146

[Thornton, 2023] Thornton, S. (2023). Karl Popper. In E.N. Zalta & U. Nodelman (Eds.), *The Stanford Encyclopedia of Philosophy*. Metaphysics Research Lab, Stanford University, Winter 2023 edition.
`https://plato.stanford.edu/archives/win2023/entries/popper/` (accessed 15 March 2024). 6

[Thron, 1997] Thron, W. (1997). Frederic Riesz' contributions to the foundations of general ttopology. In *Handbook of the History of General Topology* (pp. 21–29). Kluwer Academic Publishers. 190

[Tietze, 1908] Tietze, H. (1908). Über die topologischen Invarianten mehrdimensionaler Mannigfaltigkeiten. *Monatshefte für Mathematik und Physik*, 19(1), 1–118. 235

[Tietze & Vietoris, 1930] Tietze, H. & Vietoris, L. (1930). Beziehungen zwischen den verschiedenen Zweigen der Topologie. In *Encyklopädie der Mathematischen Wissenschaften mit Einschluss ihrer Anwendungen*, volume III A B 13 (pp. 141–237). B.G. Teubner. 235

[Torretti, 1978] Torretti, R. (1978). *Philosophy of Geometry from Riemann to Poincaré*, volume 7 of *Episteme*. Kluwer. 157, 159, 160, 161, 162, 164, 167, 171, 180

[Torretti, 1984] Torretti, R. (1984). *Philosophy of Geometry from Riemann to Poincaré*, volume 7 of *Episteme*. D. Reidel Publishing Company. Revised, corrected edition, 1984. 159, 171, 180

[Torretti, 2019] Torretti, R. (2019). Nineteenth Century Geometry. In E.N. Zalta (Ed.), *The Stanford Encyclopedia of Philosophy*. Metaphysics Research Lab, Stanford University, fall 2019 edition.
`https://plato.stanford.edu/archives/fall2019/entries/geometry-19th/` (accessed 15 March 2024). 23, 73, 98, 159

[Tricot, 1981] Tricot, C. (1981). Douze définitions de la densité logarithmique. *Comptes rendus hebdomadaires des séances de l'Académie des Sciences Paris. Série I*, 293(11), 549. Session of 23 November 1981. 420

[Tricot, 1994] Tricot, C. (1994). *Curves and fractal dimension*. Springer. 420

[Tricot, 1999] Tricot, C. (1999). *Courbes et dimension fractale*. Springer, second edition. 420

[Tumarkin, 1925a] Tumarkin, L.A. (1925a). Zur allgemeinen Dimensionstheorie. *Proceedings of the Section of Sciences, Koninklijke Nederlandse Akademie van Wetenschappen te Amsterdam*, 28, 994–996. Lev = Leo. 331, 372, 373

[Tumarkin, 1925b] Tumarkin, L.A. (1925b). Ueber Zerlegungen kompakter metrischer Räume in zueinander fremde abgeschlossene Mengen. *Proceedings of the Section of Sciences, Koninklijke Nederlandse Akademie van Wetenschappen te Amsterdam*, 28, 1000–1001. Lev = Leo.

[Tumarkin, 1926a] Tumarkin, L.A. (1926a). Nouvelle démonstration d'un théorème de Paul Urysohn. *Fundamenta Mathematicae*, 1(8), 360–361. Lev = Leo.

[Tumarkin, 1926b] Tumarkin, L.A. (1926b). Beitrag zur allgemeinen Dimensionstheorie. *Matematiceskij sbornik*, 33(1), 57–86. Lev = Leo. 331, 371, 372, 373

[Tumarkin, 1928a] Tumarkin, L.A. (1928a). Sur la structure dimensionnelle des ensembles fermés. *Comptes rendus hebdomadaires des séances de l'Académie des Sciences Paris*, 186, 420–422. Session of 13 February 1928. Lev = Leo. 373

[Tumarkin, 1928b] Tumarkin, L.A. (1928b). Über die Dimension nicht abgeschlossener Mengen. *Mathematische Annalen*, 98, 637–656. Lev = Leo. 331, 374

[Tumarkin, 1928c] Tumarkin, L.A. (1928c). Ueber Dimension von Komponenten n-dimensionaler abgesehlossenen Mengen. *Matematiceskij sbornik*, 35(1), 133–138. Lev = Leo.

[Tychonoff, 1926] Tychonoff, A. (1926). Über einen Metrisationssatz von P. Urysohn. *Mathematische Annalen*, 95(1), 139–142. Tychonoff = Tichonoff. 362, 436

[Ullherr, 1846] Ullherr, J.C. (1846). Zwei Beweise für die Existenz der Wurzeln der höhem algebraischen Gleichungen. *Journal für die reine und angewandte Mathematik*, 31, 231–234. 215

[Urysohn, 1922a] Urysohn, P.S. (1922a). Les multiplicités Cantoriennes. *Comptes rendus hebdomadaires des séances de l'Académie des Sciences Paris*, 175, 440–442. Session of 4 September 1922. 268, 273, 322, 324, 325, 326, 344, 375, 376

[Urysohn, 1922b] Urysohn, P.S. (1922b). Sur la ramification des lignes Cantoriennes. *Comptes rendus hebdomadaires des séances de l'Académie des Sciences Paris*, 175, 481–483. Session of 18 September 1922. 273, 322, 324, 325, 344, 376

[Urysohn, 1923] Urysohn, P.S. (1923). Theorie der allgemeinen Cantorschen Kurven. *Jahresbericht der Deutschen Mathematiker-Vereinigung*, 32, 69. Short note on Urysohn's presentation on 21 September 1923 at the Marburg Meeting of the D.M.-V., 20–25 September 1923. 266, 344

[Urysohn, 1924a] Urysohn, P.S. (1924a). Über die Metrisation der kompakten topologischen Räume. *Mathematische Annalen*, 92, 275–293. 362

[Urysohn, 1924b] Urysohn, P.S. (1924b). Der Hilbertsche Raum als Urbild der metrischen Räume. *Mathematische Annalen*, 92, 302–304. 362, 368

[Urysohn, 1925a] Urysohn, P.S. (1925a). Mémoire sur les multiplicités Cantoriennes. *Fundamenta Mathematicae*, 7(1), 30–137. 195, 268, 273, 314, 324, 325, 326, 327, 328, 329, 346, 357, 374, 376

[Urysohn, 1925b] Urysohn, P.S. (1925b). Sur un espace métrique universel. *Comptes rendus hebdomadaires des séances de l'Académie des Sciences Paris*, 180, 803–806. Session of 2 February 1925 / 16 March 1925. 373

[Urysohn, 1925c] Urysohn, P.S. (1925c). Über die Mächtigkeit der zusammenhängenden Mengen. *Mathematische Annalen*, 94(1), 262–295. 400

[Urysohn, 1925d] Urysohn, P.S. (1925d). Zum Metrisationsproblem. *Mathematische Annalen*, 94(1), 309–315. 362, 436

[Urysohn, 1926] Urysohn, P.S. (1926). Mémoire sur les multiplicités Cantoriennes (suite). *Fundamenta Mathematicae*, 8(1), 225–351. 273, 274, 324, 328, 329, 330, 331, 346, 347, 352, 357, 374, 376

[Urysohn, 1927a] Urysohn, P.S. (1927a). Mémoire sur les multiplicités Cantoriennes II. Les lignes Cantoriennes. *Verhandelingen der Koninklijke Nederlandse Akademie van Wetenschappen te Amsterdam (Eerste Sectie)*, 13(4), 1–172. 324, 325, 331, 374

[Urysohn, 1927b] Urysohn, P.S. (1927b). Sur un espace métrique universel. *Bulletin des Sciences Mathématiques*, 51(2), 43–64, 74–96. Published posthumously; edited by P. S. Alexandroff. 373

[Urysohn, 1951a] Urysohn, P.S. (1951a). *Works on Topology and Other Areas in Mathematics*, volume 1. Gostehizdat, Moscow. Book in Russian. xix, 322, 323, 329

[Urysohn, 1951b] Urysohn, P.S. (1951b). *Works on Topology and Other Areas in Mathematics*, volume 2. Gostehizdat, Moscow. Book in Russian.

[Valson, 1868] Valson, C.A. (1868). *La vie et les travaux du baron Cauchy: membre de l'Académie des sciences*. Gauthier-villars. 213

[Veblen, 1905] Veblen, O. (1905). Theory on plane curves in non-metrical analysis situs. *Transactions of the American Mathematical Society*, 6(1), 83–98. 145, 146

[Veblen, 1922] Veblen, O. (1922). *Analysis situs*, volume 5 of *The Colloquium Lectures*. American Mathematical Society. 385

[Vergnerie, 2017] Vergnerie, C. (2017). *La théorie des caractéristiques dans les Vorlesungen über die Theorie der algebraischen Gleichungen de Kronecker: la fin du "cycle d'idées sturmiennes"?* PhD thesis, Sorbonne Paris Cité. xvii, xvii, 217, 218, 219, 220

[Vergnerie, 2019] Vergnerie, C. (2019). L'algébre sans les «fictions des racines»: Kronecker et la théorie des caractéristiques dans les Vorlesungen über die algebraischen Gleichungen. *Revue d'histoire des mathématiques*, 25(1), 1–107. 217, 218

[Vergnerie, 2020] Vergnerie, C. (2020). Une controverse entre Émile Picard et Leopold Kronecker. *Archive for History of Exact Sciences*, 74(2), 131–164. 224

[Verhulst, 2012] Verhulst, F. (2012). *Henri Poincaré: impatient genius*. Springer Science. 158

[Vietoris, 1921] Vietoris, L. (1921). Stetige Mengen. *Monatshefte für Mathematik und Physik*, 31, 173–204. 311

[Voss, 1911] Voss, A. (1911). Jakob Lüroth. *Sitzungsberichte der Mathematisch-Physikalischen Classe der Königliche Bayerischen Akademie der Wissenschaften zu München*, (pp. 21–33). 122

[Waerden, 1954] Waerden, B.L.v.d. (1954). *Science Awakening*. Noordhoff, Groningen. 12

[Waerden, 2012] Waerden, B.L.v.d. (2012). *Science Awakening I*. Springer Science.

[Wallis, 1685] Wallis, J. (1685). *A Treatise of Algebra, Both Historical and Practical*. London: Richard Davis. 26

[Walterhausen, 1856] Walterhausen, W.S.v. (1856). *Gauss zum Gedächtniss*. S. Hirzel. 77

[Weber, 1893] Weber, H. (1893). Leopold Kronecker. *Jahresbericht der Deutschen Mathematiker-Vereinigung*, Zweiter Band, 1891-92, 5–31. 216

[Weber, 1895] Weber, H. (1895). *Lehrbuch der Algebra*, volume 1. F. Vieweg & sohn, first edition. 216

[Weber, 1898] Weber, H. (1898). *Lehrbuch der Algebra*, volume 1. F. Vieweg & sohn, second edition. 216, 218, 221

[Weierstrass, 1870] Weierstrass, K. (1870). Über das sogenannte Dirichlet'sche Princip, Gelesen in der Königliche Akademie der Wissenschaften am 14. Juli 1870. *Monatsberichte der Königlich Preußischen Akademie der Wissenschaften zu Berlin*, (pp. 575). 135

[Weierstrass, 1880] Weierstrass, K. (1880). Zur Functionenlehre. *Monatsberichte der Königlich Preußischen Akademie der Wissenschaften zu Berlin*, (pp. 719–743). 131

[Weierstrass, 1894] Weierstrass, K. (1894). *Mathematische Werke. Herausgegeben unter Mitwirkung einer von der königlich preußischen Akademie der Wissenschaften eingesetzten Commission*, volume 1. Mayer & Müller. 131

[Weierstrass, 1895] Weierstrass, K. (1895). *Mathematische Werke. Herausgegeben unter Mitwirkung einer von der königlich preußischen Akademie der Wissenschaften eingesetzten Commission*, volume 2. Mayer & Müller. 131, 135, 141

[Weierstrass, 1927] Weierstrass, K. (1927). *Mathematische Werke. Herausgegeben unter Mitwirkung einer von der königlich preußischen Akademie der Wissenschaften eingesetzten Commission*, volume 7. Akademische Verlagsgesellschaft M. B. H. 98, 131, 140, 141, 222

[Weyl, 1913] Weyl, H. (1913). *Die Idee der Riemannschen Fläche*, volume 5 of *Mathematische Vorlesungen an der Universität Göttingen*. B.G. Teubner. 97, 235, 282

[Weyl, 1918] Weyl, H. (1918). *Das Kontinuum*. Veit Leipzig. 282

[Weyl, 1949] Weyl, H. (1949). *Philosophy of Mathematics and Natural Science*. Princeton University Press. 70

[Weyl, 2013] Weyl, H. (2013). *Die Idee der Riemannschen Fläche*, volume 5 of *Mathematische Vorlesungen an der Universität Göttingen*. Springer-Verlag. 316

[Whiteside, 1981] Whiteside, D.T. (1967-1981). *The Mathematical Papers of Isaac Newton*. Cambridge University Press. 27

[Wieleitner, 1925] Wieleitner, H. (1925). Zur Frühgeschichte der Räume von mehr als drei Dimensionen. *Isis*, 7(3), 486–489. 26

[Wilder, 1963] Wilder, R.L. (1963). *Topology of Manifolds*, volume 32 of *Colloquim Publications*. American Mathematical Society, revised edition.

[Wilkins, 2002] Wilkins, D.R., Ed. (2002). *Introduction to the Quadrature of Curves By Isaac Newton, Translated into English by John Harris*. Privately published. `https://www.maths.tcd.ie/pub/HistMath/People/Newton/Quadratura/HarrisIQ.pdf` (accessed 8 April 2024). 27

[Willard, 2004] Willard, S. (2004). *General Topology*. Dover Publications. 8

[Winter, 1969] Winter, E. (1969). *Bernard Bolzano. Ein Lebensbild. Bernard Bolzano-Gesamtausgabe. Einleitung*, volume E 1. Frommann-Holzboog. 31, 33, 34, 39, 55

[Wolff, 1743] Wolff, C. (1743). *Elementa Matheseos Universae*, volume 1. Henricum-Albertum Gosse, & Socios, novissima edition. 35

[Wussing, 1969] Wussing, H. (1969). *Die Genesis des abstrakten Gruppenbegriffes: ein Beitrag zur Entstehungsgeschichte der abstrakten Gruppentheorie*. Deutscher Verlag der Wissenschaften. 166

[Wussing, 1984] Wussing, H. (1984). *The Genesis of the Abstract Group Concept: A Contribution to the History of the Origin of Abstract Group Theory*. Translated by A. Shenitzer. MIT Press. 166

[Young & Young, 1906] Young, G.C. & Young, W.H. (1906). *The Theory of Sets of Points*. Cambridge University Press. 150, 288, 316

[Young & Young, 2013] Young, W.H. & Young, G.C. (2013). *The Theory of Sets of Points*. American Mathematical Society. 288

[Zahar, 1997] Zahar, E.G. (1997). Poincaré's philosophy of geometry, or does geometric conventionalism deserve its name? *Studies in History and Philosophy of Science Part B: Studies in History and Philosophy of Modern Physics*, 28(2), 183–218. 159, 180

[Zariski & Samuel, 1957] Zariski, O. & Samuel, P. (1957). *Commutative Algebra*, volume I. Springer.

[Zariski & Samuel, 1960] Zariski, O. & Samuel, P. (1960). *Commutative Algebra*, volume II. Springer.

[Zoretti, 1905a] Zoretti, L. (1905a). *Thésis presentées à la Faculté des sciences de Paris*. PhD thesis, Faculté des sciences de Paris. 1re Thése. Sur les fonctions analytiques uniformes qui possédent un ensemble parfait discontinu de points singulier. 2e Thése. Propositions donné par la faculté. 294

[Zoretti, 1905b] Zoretti, L. (1905b). Sur les fonctions analytiques uniformes qui possèdent un ensemble parfait discontinu de points singuliers. *Journal de mathématiques pures et appliquées*, 1(6), 1–51. 187, 289, 294

[Zoretti, 1906] Zoretti, L. (1906). Sur les ensembles discontinus. *Comptes rendus hebdomadaires des séances de l'Académie des Sciences Paris*, 142, 763–764. Session of 26 March 1906. 188

[Zoretti, 1909] Zoretti, L. (1909). La notion de ligne. In *Annales scientifiques de l'École Normale Supérieure*, volume 26 (pp. 485–497). 294, 295, 325

[Zoretti, 1910a] Zoretti, L. (1910a). Remarque au sujet de la remarque de M. Brouwer. *Annales scientifiques de l'École Normale Supérieure*, 27, 567. 295

[Zoretti, 1910b] Zoretti, L. (1910b). Sur les propriétés des lignes cantoriennes. *Comptes rendus hebdomadaires des séances de l'Académie des Sciences Paris*, 150, 1505–1506. Session of 6 June 1910.

[Zoretti, 1910c] Zoretti, L. (1910c). Sur la notion de ligne. *Comptes rendus hebdomadaires des séances de l'Académie des Sciences Paris*, 151, 201–203. Session of 18 July 1910. 295

[Zoretti, 1911] Zoretti, L. (1911). *Leçons sur le prolongement analytique professés au Collège de France*, volume 15. Gauthier-Villars.

[Zoretti, 1913] Zoretti, L. (1913). Contribution à l'étude des lignes cantoriennes. *Acta Methematica*, 36, 241–268. 295, 296, 325

[Zoretti & Rosenthal, 1923] Zoretti, L. & Rosenthal, A. (1923). Die Punktmengen. In *Encyklopädie der mathematischen Wissenschaften mit Einschluss ihrer Anwendungen*, volume II C 9a (pp. 852–1030). B.G. Teubner. 288

Appendix A
Original Language Texts of Quotations

Ch1, 1. Die hohe Bedeutung bestimmter Probleme für den Fortschritt der mathematischen Wissenschaft im Allgemeinen und die wichtige Rolle, die sie bei der Arbeit des einzelnen Forschers spielen, ist unleugbar. Solange ein Wissenszweig Ueberfluß an Problemen bietet, ist er lebenskräftig; Mangel an Problemen bedeutet Absterben oder Aufhören der selbstständigen Entwickelung. Wie überhaupt jedes menschliche Unternehmen Ziele verfolgt, so braucht die mathematische Forschung Probleme. Durch die Lösung, von Problemen stählt sich die Kraft des Forschers; er findet neue Methoden und Ausblicke, er gewinnt einen weiteren und freieren Horizont.

Ch2, 1. τὸ μὲν γὰρ ἓν στιγμή, τὰ δὲ δύο γραμμή, τὰ δὲ τρία τρίγωνον, τὰ δὲ τέσσαρα πυραμίς· ταῦτα δὲ πάντα ἐστὶ πρῶτα καὶ ἀρχαὶ τῶν καθ᾽ ἕκαστον ὁμογενῶν. καὶ ἀναλογιῶν δὲ πρώτη αὕτη ἐστὶν ἡ ἐν αὐτοῖς ὀφθεῖσα ἡ τὸ ἴσον μὲν ὑπερέχουσα, τέλος δὲ ἔχουσα ἐν τοῖς δέκα. ἔν τε ἐπιπέδοις καὶ στερεοῖς πρῶτά ἐστι ταῦτα· στιγμὴ γραμμὴ τρίγωνον πυραμίς· .... τὰ αὐτὰ δὲ καὶ ἐν τῇ γενέσει· πρώτη μὲν γὰρ ἀρχὴ εἰς μέγεθος στιγμή, δευτέρα γραμμή, τρίτη ἐπιφάνεια, τέταρτον στερεόν.

Ch2, 2a. δοκεῖ δέ τισι τὰ τοῦ σώματος "πέρατα, οἷον ἐπιφάνεια καὶ γραμμὴ καὶ στιγμὴ καὶ μονάς, εἶναι οὐσίαι, καὶ μᾶλλον ἢ τὸ σῶμα καὶ τὸ στερεόν.

Ch2, 2b. εἰσὶ δέ τινες οἳ ἐκ τοῦ πέρατα εἶναι καὶ ἔσχατα τὴν στιγμὴν μὲν γραμμῆς, ταύτην δ' ἐπιπέδου, τοῦτο δὲ τοῦ στερεοῦ, οἴονται εἶναι ἀνάγκην τοιαύτας φύσεις εἶναι.

Ch2, 3. ἔσται οὖν ἡ μὲν μονὰς σημείου τόπον ἐπέχουσα καὶ τρόπον ἀρχὴ μὲν διαστημάτων καὶ ἀριθμῶν, οὔπω δὲ διάστημα οὐδὲ ἀριθμός, ὡς τὸ σημεῖον ἀρχὴ μὲν γραμμῆς καὶ διαστήματος, οὔπω δὲ γραμμὴ οὐδὲ διάστημα· ἀμέλει οὔτε σημείῳ σημεῖον συντεθὲν πλεῖόν τι ποιεῖ, ἀδιάστατον γὰρ ἀδιαστάτῳ συντεθὲν διάστημα οὐχ ἕξει, ...

ἀδιάστατος ἄρα ἡ μονὰς καὶ ἀρχοειδής, πρῶτον δὲ διάστημα εὑρίσκεται καὶ φαίνεται ἐν δυάδι, εἶτ᾽ ἐν τριάδι, εἶτα ἐν τετράδι καὶ ἑξῆς ἐν τοῖς ἀκολούθοις· διάστημα γάρ ἐστι δυεῖν ὅρων τὸ μεταξὺ θεωρούμενον. πρῶτον δὲ διάστημα γραμμὴ λέγεται, γραμμὴ γάρ ἐστι τὸ ἐφ᾽ ἓν διαστατόν· δύο δὲ διαστήματα ἐπιφάνεια, ἐπιφάνεια γάρ ἐστι τὸ διχῇ διαστατόν· τρία δὲ διαστήματα στερεόν, στερεὸν γάρ ἐστι τὸ τριχῇ διαστατὸν

...

Ἔστιν οὖν σημεῖον ἀρχὴ διαστήματος, οὐ διάστημα δέ, τὸ δ᾽ αὐτὸ καὶ ἀρχὴ γραμμῆς, οὐ γραμμὴ δέ· καὶ γραμμὴ ἀρχὴ ἐπιφανείας, οὐκ ἐπιφάνεια δέ, καὶ ἀρχὴ τοῦ διχῇ διαστατοῦ, οὐ διχῇ δὲ διαστατόν. καὶ εἰκότως ἡ ἐπιφάνεια ἀρχὴ μὲν σώματος, οὐ σῶμα δέ, καὶ ἡ αὐτὴ ἀρχὴ μὲν τοῦ τριχῇ διαστατοῦ, οὐ τριχῇ δὲ διαστατόν. οὕτως δὴ καὶ ἐν τοῖς ἀριθμοῖς ἡ μὲν μονὰς ἀρχὴ παντὸς ἀριθμοῦ ἐφ᾽ ἓν διάστημα κατὰ μονάδα προβιβαζομένου, ὁ δὲ γραμμικὸς ἀριθμὸς ἀρχὴ ἐπιπέδου ἀριθμοῦ ἐφ᾽ ἕτερον διάστημα ἐπιπέδως πλατυνομένου, ὁ δὲ ἐπίπεδος ἀριθμὸς ἀρχὴ στερεοῦ ἀριθμοῦ ἐπὶ τρίτον διάστημα πρὸς τὰ ἐξ ἀρχῆς βάθος τι προσκτωμένου.

D. M. Johnson, *Theories of Dimension from Ancient to Modern Times*, Sources and Studies in the History of Mathematics and Physical Sciences, https://doi.org/10.1007/978-3-032-02682-8

Ch2, 4. ἐπεί φασι κινηθεῖσαν γραμμὴν ἐπίπεδον ποιεῖν, στιγμὴν δὲ γραμμήν, καὶ αἱ τῶν μονάδων κινήσεις γραμμαὶ ἔσονται· ἡ γὰρ στιγμὴ μονάς ἐστι θέσιν ἔχουσα.

Ch2, 5. τινὲς δ᾽ ἀπὸ ἑνὸς σημείου τὸ σῶμά φασι συνίστασθαι· τουτὶ γὰρ τὸ σημεῖον ῥυὲν γραμμὴν ἀποτελεῖν, τὴν δὲ γραμμὴν ῥυεῖσαν ἐπίπεδον ποιεῖν, τοῦτο δὲ εἰς βάθος κινηθὲν τὸ σῶμα γεννᾶν τριχῇ διαστατόν. διαφέρει δὲ ἡ τοιαύτη τῶν Πυθαγορικῶν στάσις τῆς τῶν προτέρων. ἐκεῖνοι μὲν γὰρ ἐκ δυεῖν ἀρχῶν, τῆς τε μονάδος καὶ τῆς ἀορίστου δυάδος, ἐποίουν τοὺς ἀριθμούς, εἶτ᾽ ἐκ τῶν ἀριθμῶν τὰ σημεῖα καὶ τὰς γραμμὰς τά τε ἐπίπεδα σχήματα καὶ τὰ στερεά· οὗτοι δὲ ἀπὸ ἑνὸς σημείου τὰ πάντα τεκταίνουσιν. ἐξ αὐτοῦ μὲν γὰρ γραμμὴ γίνεται, ἀπὸ γραμμῆς δὲ ἐπιφάνεια, ἀπὸ δὲ ταύτης σῶμα.

Ch2, 6. τῶν γὰρ φύσει συνεστώτων τὰ μέν ἐστι σώματα καὶ μεγέθη, τὰ δ᾽ ἔχει σῶμα καὶ μέγεθος, τὰ δ᾽ ἀρχαὶ τῶν ἐχόντων εἰσίν. Συνεχὲς μὲν οὖν ἐστι τὸ διαιρετὸν εἰς ἀεὶ διαιρετά, σῶμα δὲ τὸ πάντῃ διαιρετόν. Μεγέθους δὲ τὸ μὲν ἐφ᾽ ἓν γραμμή, τὸ δ᾽ ἐπὶ δύο ἐπίπεδον, τὸ δ᾽ ἐπὶ τρία σῶμα· καὶ παρὰ ταῦτα οὐκ ἔστιν ἄλλο μέγεθος διὰ τὸ τὰ τρία πάντα εἶναι καὶ τὸ τρὶς πάντῃ. Καθάπερ γάρ φασι καὶ οἱ Πυθαγόρειοι, τὸ πᾶν καὶ τὰ πάντα τοῖς τρισὶν ὥρισται· τελευτὴ γὰρ καὶ μέσον καὶ ἀρχὴ τὸν ἀριθμὸν ἔχει τὸν τοῦ παντός, ταῦτα δὲ τὸν τῆς τριάδος.

Ch2, 7.

Στοιχεῖα Εὐκλείδου α΄.

Ὅροι κγ΄.
α΄. Σημεῖόν ἐστιν, οὗ μέρος οὐθέν.
β΄. Γραμμὴ δὲ μῆκος ἀπλατές.
γ΄. Γραμμῆς δὲ πέρατα σημεῖα.

ε΄. Ἐπιφάνεια δέ ἐστιν, ὃ μῆκος καὶ πλάτος μόνον ἔχει.
ϛ΄. Ἐπιφανείας δὲ πέρατα γραμμαί.

Στοιχεῖα Εὐκλείδου ια΄

Ὅροι κγ΄.
α΄. Στερεόν ἐστι τὸ μῆκος καὶ πλάτος καὶ βάθος ἔχον.
β΄. Στερεοῦ δὲ πέρας ἐπιφάνεια.

Ch2, 8. εἴ γε μὴν γραμμὰς ἢ τὰ τούτων ἐχόμενα (λέγω δὲ ἐπιφανείας τὰς πρώτας) θήσει τις ἀρχάς, ταῦτά γ᾽ οὐκ εἰσὶν οὐσίαι χωρισταί, τομαὶ δὲ καὶ διαιρέσεις αἱ μὲν ἐπιφανειῶν αἱ δὲ σωμάτων (αἱ δὲ στιγμαὶ γραμμῶν), ἔτι δὲ πέρατα τῶν αὐτῶν τούτων· πάντα δὲ ταῦτα ἐν ἄλλοις ὑπάρχει καὶ χωριστὸν οὐδέν ἐστιν.

Ch2, 9. πάντες γὰρ διὰ τῶν ὑστέρων τὰ πρότερα δηλοῦσιν· τὸ μὲν γὰρ γραμμῆς, τὸ δ᾽ ἐπιπέδου, τὸ δὲ στερεοῦ φασι πέρας εἶναι.

Ch2, 10. Ἀλλὰ καὶ πρὸς τοῖσδε γινώσκοντί σοι πάντας τοὺς ἀριθμοὺς συγκειμένους ἐκ μονάδων πλήθους τινός, φανερὸν καθέστηκεν εἰς ἄπειρον ἔχειν τὴν ὕπαρξιν τυγχανόντων δὴ οὖν ἐν τούτοις

ὧν μὲν τετραγώνων, οἵ εἰσιν ἐξ ἀριθμοῦ τινος ἐφ᾽ ἑαυτὸν πολυπλασιασθέντος· οὗτος δὲ ὁ ἀριθμὸς καλεῖται πλευρὰ τοῦ τετραγώνου·

ὧν δὲ κύβων, οἵ εἰσιν ἐκ τετραγώνων ἐπὶ τὰς αὐτῶν πλευρὰς πολυπλασιασθέντων,

ὧν δὲ δυναμοδυνάμεων, οἵ εἰσιν ἐκ τετραγώνων ἐφ᾽ ἑαυτοὺς πολυπλασιασθέντων,

ὧν δὲ δυναμοκύβων, οἵ εἰσιν ἐκ τετραγώνων ἐπὶ τοὺς ἀπὸ τῆς αὐτῆς αὐτοῖς πλευρᾶς κύβους πολυπλασιασθέντων,

ὧν δὲ κυβοκύβων, οἵ εἰσιν ἐκ κύβων ἐφ᾽ ἑαυτοὺς πολυπλασιασθέντων, ἔκ τε τῆς τούτων ἤτοι συνθέσεως ἢ ὑπεροχῆς ἢ πολυπλασιασμοῦ ἢ λόγου τοῦ πρὸς ἀλλήλους ἢ καὶ ἑκάστων πρὸς τὰς ἰδίας πλευρὰς συμβαίνει πλέκεσθαι πλεῖστα προβλήματα ἀριθμητικά· . . . .

Ch2, 11. ...ἐπεὶ οὐκ ἔστι τι περιεχόμενον ὑπὸ πλειόνων ἢ τριῶν διαστάσεων. συγκεχωρήκασι δὲ ἑαυτοῖς οἱ βραχὺ πρὸ ἡμῶν ἑρμηνεύειν τὰ τοιαῦτα, μηδὲ ἓν μηδαμῶς διάληπτον σημαίνοντες, τὸ ὑπὸ τῶνδε περιεχόμενον λέγοντες ἐπὶ τὸ ἀπὸ τῆσδε τετράγωνον ἢ ἐπὶ τὸ ὑπὸ τῶνδε. παρῆν δὲ διὰ τῶν συνημμένων λόγων ταῦτα καὶ λέγειν καὶ δεικνύναι καθόλου καὶ ἐπὶ τῶν προειρημένων προτάσεων καὶ ἐπὶ τούτων τὸν τρόπον τοῦτον. ...

Ch2, 12. Ἀμέσου δ᾽ ἀρχῆς συλλογιστικῆς θέσιν μὲν λέγω ἣν μὴ ἔστι δεῖξαι, μηδ᾽ ἀνάγκη ἔχειν τὸν μαθησόμενόν τι· ἣν δ᾽ ἀνάγκη ἔχειν τὸν ὁτιοῦν μαθησόμενον, ἀξίωμα· ἔστι γὰρ ἔνια τοιαῦτα· τοῦτο γὰρ μάλιστ᾽ ἐπὶ τοῖς τοιούτοις εἰώθαμεν ὄνομα λέγειν. θέσεως δ᾽ ἡ μὲν ὁποτερονοῦν τῶν μορίων τῆς ἀντιφάσεως λαμβάνουσα, οἷον λέγω τὸ εἶναί τι ἢ τὸ μὴ εἶναί τι, ὑπόθεσις, ἡ δ᾽ ἄνευ τούτου ὁρισμός. ὁ γὰρ ὁρισμὸς θέσις μέν ἐστι· τίθεται γὰρ ὁ ἀριθμητικὸς μονάδα τὸ ἀδιαίρετον εἶναι κατὰ τὸ ποσόν· ὑπόθεσις δ᾽ οὐκ ἔστι· τὸ γὰρ τί ἐστι μονὰς καὶ τὸ εἶναι μονάδα οὐ ταὐτόν.

Ch3, 1. Quantitates Mathematicas non ut ex partibus quam minimis constantes, sed ut motu continuo descriptas hic considero. Lineæ describuntur ac describendo generantur non per appositionem partium sed per motum continuum punctorum, superficies per motum linearum, solida per motum superficierum, anguli per rotationem laterum, tempora per fluxum continuum, & sic in cæteris. Hæ Geneses in rerum natura locum vere habent & in motu corporum quotidie cernuntur. Et ad hunc modum Veteres ducendo rectas mobiles in longitudinem rectarum immobilium genesin docuerunt rectangulorum.

Considerando igitur quod quantitates æqualibus temporibus crescentes & crescendo genitæ, pro velocitate majori vel minori qua crescunt ac generantur, evadunt majores vel minores; methodum quærebam determinandi quantitates ex velocitalibus motuum vel incrementorum quibus generantur; & has motuum vel incrementorum velocitates nominando *Fluxiones* & quantitates genitas nominando *Fluentes*, incidi paulatim *Annis 1665 & 1666* in Methodum Fluxionum qua hic usus sum in Quadratura Curvarum.

Ch3, 2. Eine Wissenschaft von allen diesen möglichen Raumes-Arten, wäre ohnfehlbar die höchste Geometrie die ein endlicher Verstand unternehmen könnte. Die Unmöglichkeit, die wir bey uns bemerken, einen Raum von mehr als drey Abmessungen uns vorzustellen, scheinet mir daher zu rühren, weil unsere Seele ebenfalls nach dem Gesetze der umgekehrten doppelten Verhaltniss der Weiten die Eindrücke von draussen empfängt, und weil ihre Natur selber dazu gemacht ist, nicht allein so zu leiden, sondern auch auf diese Weise ausser sich zu würken.

Ch3, 3. Si l'on se représente qu'un *point* coule, il tracera une ligne; & une ligne qui couleroit engendreroit une surface, &c. Cette maniere de considérer la génération des dimensions ou des propriétés des corps, paroît être le premier fondement de la Géométrie moderne, c'est-à-dire, de la Géométrie analytique qui fait usage du calcul différenciel & intégral ....

Ch3, 4. DIMENSION ... c'est l'étendue d'un corps considéré en tant qu'il est mesurable, ou susceptible de mesure. ...

Ainsi, comme nous concevons que les corps sont étendus en *longueur*, *largeur*, & *profondeur* ou *épaisseur*, nous concevons aussi ces trois *dimensions* dans la matiere ; la longueur toute seule s'appelle *ligne* ; la longueur combinée avec la largeur prend le nom de *surface* : enfin la longueur, la largeur, & la profondeur ou l'épaisseur, combinées ensemble, produisent ce que l'on nomme un *solide*. ...

Ch3, 5. J'ai dit... qu'il n'étoit pas possible de concevoir plus de trois *dimensions*. Un homme d'esprit de ma connoissance croit qu'on pourroit cependant regarder la durée comme une quatrieme *dimension*, & que le produit du tems par la solidité seroit en quelque maniere un produit de quatre *dimensions*; cette idée peut être contestée, mais elle a, ce me semble, quelque mérite, quand ce ne seroit que celui de la nouveauté.

Ch4, 1. Mein besonderes *Wohlgefallen an der Mathematik* beruhte also eigentlich nur auf ihrem rein *speculativen Theile*, oder ich schätzte an ihr nur dasjenige, was zugleich *Philosophie* ist.

Ch4, 2. Eines der ersten philosophischen Bücher, die Bolzano etwa in seinem 16.Lebensjahre las, war Baumgartens Metaphysik, wo er mit aller Deutlichkeit einzusehen glaubte, daß viele Erklärungen verfehlt, manche Beweise, z.B. der des Satzes vom Grunde, das zu Beweisende schon voraussetzten, und mehrere ganz unrichtige Behauptungen (z.B. von der Zusammensetzung der Linie, Fläche und Körper aus einer endlichen Menge von Punkten) vorkommen.

Ch4, 3. Kästner bewies dort nämlich, was man sonst insgemein, weil es doch Jeder schon weiß, ganz übergeht; das heißt, er suchte dem Leser den Grund, auf welchem eines seiner Urtheile beruht, zum deutlichen Bewußtseyn zu bringen: und das war mir eben das Liebste.

Ch4, 4. Eine *stetige* Grösse (continuum) heißt, deren Theile alle so zusammenhängen, daß, wo einer aufhört, gleich der andere anfängt, und zwischen des einen Ende und des andern Anfange nichts ist, das nicht zu dieser Grösse gehörte.

Ch4, 5. *Die körperliche Ausdehnung, ein geometrischer Körper* (solidum; corpus) heißt eine solche Ausdehnung, die das, was sich innerhalb ihrer Gränzen befindet, überall, nach allen Seiten zu umgiebt. Die Ausdehnung der Körper an ihren Gränzen heißt *eine Fläche* (superficies) und die Ausdehnung der Fläche an ihren Gränzen, *eine Linie* (linea).

Ch4, 6. I. Dasjenige, was dem Puncte b in Beziehung auf a so zukömmt, daß es *unabhängig* ist von dem *bestimmten Puncte* a (qua praecise hoc est et non aliud); was folglich auch in der Beziehung auf einen *andern* Punct, z. B. α gleich vorhanden seyn kann: *genannt die Entfernung des Punctes* b *von* a.

II. Dasjenige, was dem Puncte b in Beziehung auf a so zukömmt, daß es *abhängig ist bloss von dem bestimmten Puncte* a; wovon nun getrennt werde, was schon in dem Begriffe der Entfernung liegt, d. h. was dem Puncte b auch in Rücksicht auf noch einen andern Punct zukommen kann: genannt *die Richtung, in welche* b *zu* a *liegt.*

Ch4, 7. Erkl.[ärung]. Ein Ding, welches alle jene, und nur jene Puncte enthält, die zwischen den zwey Puncten a und b liegen, heißt eine *gerade Linie zwischen* a *und* b.

Ch4, 8. Hier mangelt es zur Stunde noch an einer bestimmten Erklärung der wichtigen Begriffe: *Linie, Fläche, Körper*.

Ch4, 9. Car il y en a un [ordre], et c'est celui de la géométrie, qui est à la vérité inférieur en ce qu'il est moins convaincant, mais non pas en ce qu'il est moins certain. Il ne définit pas tout et ne prouve pas tout, et c'est en cela qu'il lui céde ; mais il ne suppose que des choses claires et constantes par la lumière naturelle, et c'est pourquoi il est parfaitement véritable, la nature le soutenant au défaut du discours. Cet ordre, le plus parfait entre les hommes, consiste non pas à tout définir ou à tout démontrer, ni aussi à ne rien définir ou à ne rien démontrer, mais à se tenir dans ce milieu de ne point définir les choses claires et entendues de tous les hommes, et de définir toutes les autres ; et de ne point prouver toutes les choses connues des hommes, et de prouver toutes les autres. Contre cet ordre pèchent également ceux qui entreprennent de tout définir et de tout prouver et ceux qui négligent de le faire dans les choses qui ne sont pas évidentes d'elles-mêmes.

Ch4, 10. In dem Reiche der Wahrheit, d.h. in dem Inbegriffe aller wahren Urtheile herrscht ein gewisser *objektiver*, von unserer zufälligen *subjektiven Anerkennung* derselben unabhängiger *Zusammenhang*, zu Folge dessen einige aus diesen Urtheilen die Gründe anderer, und diese die Folgen jener sind. Diesen objektiven Zusammenhang der Urtheile darzustellen, d.h. eine Menge von Urtheilen so auszuwählen und an einander zu reihen, daß jedes, das ein gefolgertes ist, auch als ein solches aufgeführt werde und umgekehrt, scheint mir der eigentliche *Zweck* zu sein, den wir bey einem wissenschaftlichen Vortrage verfolgen.

Ch4, 11. ... eine ächte Definition nur solche Merkmale des zu erklärenden Begriffs enthalten muß, die sein *Wesen* ausmachen, und ohne welchen er gar nicht gedacht werden kann ....

Ch4, 12. Die Logiker verstehen unter einer *Erklärung* (Definitio) in dieses Wortes eigentlichstem Sinne die *Angabe der nächsten* (zwei oder mehreren) *Bestandteile, aus welchen ein gegebener Begriff zusammengesetzt ist.*

Ch4, 13. Die erste Art der Betrachtungen über blosse Vorstellungen und Sätze ... sind die *Erklärungen*; worunter ich ... hier eben nichts Anderes verstehe, als Sätze, welche bestimmen, ob eine gewisse Vorstellung oder ein Satz einfach oder aus Theilen zusammengesetzt sey, und in dem letzteren Fälle, aus was für Theilen, und in welcher Verbindung derselben er bestehe.

Ch4, 14. Wenn wir ... schon bei dem Geschäfte des eigenen Nachdenkens in dem Streben nach Verdeutlichung unserer Gedanken zu weit gehen können: so ist es bei der Mittheilung unserer Gedanken an Andere um so gewisser, daß wir uns in Erklärungen und Zerlegungen verlieren können, welche in solchem Uebermaasse Tadel verbienen. ... Wir werden uns also in keiner Wissenschaft, selbst nicht in denjenigen, bei welchen es doch auf Uebung im Denken am Meisten abgesehen ist, zu einer vollständigen, bis auf die letzten einfachen Theille sich erstreckenden Erklärung aller Begriffe und Sätze, ja auch nur derer, die in den wesentlichen Lehren derselben vorkommen, anheischig machen.

Ch4, 15. Leicht zu erachten ist nun, daß unser Verfahren immer nur darin bestehen könne, vermittelst der schriftlichen Zeichen, welche den Lesern schon bekannt sind, gewisse Sätze über diejenigen Zeichen zu bilden, deren Bedeutung wir ihnen erst bekannt machen wollen. Dergleichen Sätze also, wie sie auch immer beschaffen seyn mögen, will ich *Verständigungen* nennen. Ich befasse somit unter dieser Benennung alle diejenigen, in einem Lehrbuche vorkommenden Sätze, die wir nur eben in der Absicht aufstellen, um unsern Lesern die Bedeutung, in der wir gewisse Zeichen gebrauchen, bekannt zu geben. Reichet ein solcher Satz zu seinem Zwecke bin, d.h. gibt er die Vorstellung, die wir mit unserm Zeichen verbunden wissen wollen, wirklich so vollkommen zu erkennen, daß sie mit keiner andern verwechselt werden kann: so mag man denselben auch eine *Bestimmung* dieser Bedeutung oder eine *Begriffsbestimmung* nennen. Andere geben ... solchen Bestimmungen auch den Namen *Erklärungen*, und zwar sowohl Erklärungen des *Zeichens*, als auch Erklärungen seiner *Bedeutung* oder seines *Bergriffes*, *Begriffserklärungen*.

Ch4, 16. Hier werden wir wohl thun, erst nachzusehen, ob unsere Vorstellung nicht zu der Classe der *zusammengesetzten* gehöre, und zwar von der Art, daß sie in Theile zerlegt werden kann, die sich im Einzelnen durch Zeichen, welche auch unseren Lesern schon bekannt sind, darstellen lassen. Ich dieser Fall vorhanden, und ist es uns gelungen, jene Bestandtheile unserer Vorstellung aufzufinden: so, wählen wir ein sehr sicheres Mittle, uns über dieselbe zu verständigen, wenn wir zuerst die Zeichen, die ihre einzelnen Bestandtheile ausdrücken, in eben der Folge nach einander hinstellen, in welcher die durch sie bezeichneten Vorstellungen in unserm Gemüthe auf einander folgen müssen, um die zusammengesetzte Vorstellung selbst zu erzeugen, und nun bemerken, die Vorstellung, welche die Leser durch Betrachtung jener Zeichen erhielten, sey eben diejenige, die wir mit unserm neuen Zeichen verbinden. Durch Anwendung dieses Mittels erfahren die Leser nicht nur den Begriff, den wir mit unserm Zeichen verbinden, sondern sie wissen nun auch, und zwar mit Klarheit, aus welchen (wenigstens nächsten) Bestandtheilen derselbe zusammengesetzt sey; d.h. ihre Vorstellung von demselben erhält nun *Deutlichkeit*. ... Da dieses durch die Verfahrungsarten ... nicht erreicht wird: so ist es billig, daß wir die gegenwärtige durch eine eigene Benennung auszeichnen. Sie mag denn die *Bestimmung eines Begriffes durch seine Thelie*, auch die *Zerlegung*, oder eine Erkläung im *engeren* oder strengwissenschaftlichen Sinne heißen. (Vgl. §. 555.) Beispiele solcher Erklärungen sind in den Lehrbüchern der Mathematik in grosser Menge zu finden.

Ch4, 17. ... das Word *Erklärung* für die Bezeichnung nur einer einzigen Art derselben aufzusparen, derjenigen nämlich, die uns mit der Bedeutung eines Zeichens dadurch bekannt macht, daß sie uns die Bestandteile, aus welchen der bezeichnete Begriff zusammengesetzt ist, angibt.

Ch4, 18. Es ist daher sehr gut, daß die Frage an deren richtiger Beantwortung uns so viel liegt, nicht sowohl diese ist, „aus welchen Theilen eine gewisse Vorstellung in der Art, wie sie *bisher* in unserm Bewussseyn vorkam, ja vielleicht auch sich *künftig* einfinden wird, bestehe?“ als vielmehr die

„welche Bestandtheile wir derselben beyzulegen haben, damit sie dem Zwecke, zu dem wir uns ihrer bedienen wollen, am Besten entspreche?"

Ch4, 19. ... in den meisten Fällen, wo wir Begriffe *in einem wissenschaftlichen Zusammenhange* aufstellen, wenig verschlägt, ob die Erklärung, die wir von ilnen aufführen, wirklich in allen ihren Theilen denjenigen Begriff darbiete, den wir mit dem zu seiner Bezeichnung gewählten Worte bisher verbunden hatten, wenn es nur jedenfalls ein zweckmässiger Begriff ist, und ein Begriff, der an dem Orte, wo wir ilm aufstellen, verdient, betrachtet und bei den Lehren, die wir über seinen Gegenstand vortragen sollen, zu Grunde gelegt zu werden ...

Ch4, 20. *Raumding* heißt überhaupt jedes *System* (jeder Inbegriff) *von Puncten* (sie mögen in endlicher oder unendlicher Menge vorhanden seyn).

Ch4, 21. Ein Raumding, zu dessen jedem Puncte es[,] anzufangen von einer gewissen Entfernung für alle kleineren abwärts, wenigstens *einen*, und höchstens nur eine *endliche* Menge von Puncten als Nachbarn gibt, heißt eine *Linie überhaupt* (Fig. 1-7.).

Ch4, 22. Ein Raumding, zu dessen jedem Puncte es, anzufangen von einer gewissen Entfernung für alle kleineren abwärts, wenigstens *eine*, und höchstens nur eine *endliche* Menge *getrennter Linien* voll Puncte gibt, heißt eine *Fläche überhaupt* (Fig. 13-19.).

Ch4, 23. Ein Raumding, zu dessen jedem Puncte es, anzufangen von einer gewissen Entfernung für alle kleineren abwärts, wenigstens *eine* durchaus zusammenhängende Fläche voll Puncte gibt, heißt *Körper überhaupt* (Fig. 20 und 21.).

Ch4, 24. Ein Raumding, dessen jeder Theil, der sich nach eben gegebener Erklärung als Linie ansehen lässt, mit dem noch übrigen Theile, der sich dann gleichfalls als Linie muß ansehen lassen, wenigstens *einen Punct* gemein hat, heißt *eine durchaus zusammenhängende Linie* (Fig. 2-7.).

Ch4, 25. Eine Linie ..., in der es Puncte gibt, die anzufangen von einer gewissen Entfernung für alle kleineren abwärts, nur *einen* Nachbarn haben, heißt eine begrenzte Linie (Fig. 2, 5, 6.).
Jene Puncte in ihr (z.B. e, e) heißen *Grenz-* (oder *End-*) *puncte*; die übrigen (z. B. m, m) *innere* Puncte.

Ch4, 26. Ein Raumding, dessen jeder Theil, der sich nach eben gegebener Erklärung als Fläche ansehen lässt, mit dem noch übrigen Theile, der sich dann gleichfalls als Fläche muß ansehen lassen, wenigstens *eine Linie* gemein hat, heißt eine einzige durchaus *zusammenhängende Fläche* (Fig. 14-19.).

Ch4, 27. Ein Raumding, dessen jeder Theil, der sich nach eben gegebener Erklärung als Körper ansehen lässt, mit dem noch übrigen Theile, der sich nun gleichfalls als Körper muß ansehen lassen, wenigstens *eine Fläche* gemein hat, heißt ein *einziger durchaus zusammenhängender Körper* (Fig. 21.).

Ch4, 28. Diese Erklärungen ... muß man nicht nothwendig inne haben, um die hier vorkommende Theorie verstehen und beurtheilen zu können. Gleichwohl schien es mir zweckmässig, ihnen hier einen Platz zu vergönnen; wäre es auch nur, um gelegenheitlich einen kleinen Vorbegriff von der Beschaffenheit jener gänzlichen *Umstaltung der Geometrie* zu geben, an der ich schon seit Jahren gearbeitet, bisher aber erst sehr wenige Proben dem Publico mit getheilt habe.

Ch4, 29. ... darf man sich durchaus nicht weigern, zuzugestehen, daß jeder ausgedehnte Raum, er sey nun Linie, Fläche oder Körper, auch aus einfachen Theilen, nämlich den Puncten, bestehe; denn er ist in der That nichts Anderes, als ein gewisser Inbegriff solcher Puncte.

Ch4, 30. Ein jeder Inbegriff von Punkten würde hiernächst ein *ausgedehntes Raumding* heißen, falls ein jeder Punkt ... mit andern verbunden ist, d. h. Nachbarn in diesem Raumdinge findet, die ihm so nahe, als man nur immer will, treten; oder noch anders, sobald sich für eine jede auch noch so kleine Entfernung Punkte im Raumdinge fänden, die diese Entfernung von ihm haben.

Ch4, 31. Wenn der Punct *i* in einem Raumdinge so liegt, daß keine auch noch so kleine Entfernung angeblich ist, von der behauptet werden könnte, *für diese und für alle kleineren Entfernungen*, die es nur überhaupt gibt, besitze *i* einen oder etliche Nachbarn: so sage ich, *i* stehe *isolirt* oder *vereinzelt* in diesem Raumdinge.

Ch4, 32. Die Bolzanosche Definition des Kontinuums (*Paradoxien §38*) ist gewiss nicht richtig; sie drückt einseitig bloss *eine* Eigenschaft des Kontinuums aus, die aber auch erfüllt ist bei Mengen, welche aus G_n dadurch hervorgehen, daß man sich von G_n irgendeine „isolierte" Punktmenge (man vgl. *Math. Ann.* Bd. 21, S. 51) entfernt denkt; desgleichen ist sie erfült bei Mengen, welche aus mehreren getrennten Kontinuis bestehen; offenbar liegt in solchen Fällen kein Kontinuum vor, obgleich nach Bolzano dies der Fall wäre.

Ch4, 33. *Lehrsatz*: Wenn eine in sich zurückkehrende Linie in einer Ebene liegt, und man verbindet einen Punct derselben Ebene der von ihr eingeschlossen wird mit einem anderen Puncte dieser Ebene, der aber von ihr nicht eingeschlossen wird, durch eine zusammenhängende Linie, so muß diese die zurückkehrende Linie schneiden.

Ch4, 34. Ein Ausgedehntes, dessen jeder Punct für jede hinlänglich kleine Entfernung der Nachbarn nur so viele hat, daß ihr Inbegriff, für eine jede dieser Entfernungen für sich allein betrachtet, noch kein Ausgedehntes darstellt, nenne ich ein *Raumding von einer einzigen oder einfachen Ausdehnung*, auch eine *Linie*. Ein Raumding, dessen jeder Punct für jede hinlänglich kleine Entfernung der Nachbarn so viele hat, daß ihr Inbegriff für eine jede dieser Entfernungen für sich allein betrachtet selbst noch ein Raumding von einfacher Ausdehnung darstellt, nenne ich ein *Raumding von zweifacher oder doppelter Ausdehnung*, auch eine *Fläche*. Ein Raumding endlich, dessen jeder Punct für jede hinlänglich kleine Entfernung der Nachbarn so viele hat, daß ihr Inbegriff für eine jede dieser Entfernungen für sich allein betrachtet schon ein Raumding von doppelter Ausdehnung darstellt, nenne ich ein *Raumding von dreifacher Ausdehnung* oder einen *Körper*.

Ch4, 35. Wohl dürfen wir also einem *jeden Ausgedehnten, das keine körperliche Ausdehnung hat*, in jedem seiner Puncte eine *Beschaffenheit* beilegen, welche wir uns als den *nächsten Grund* davon denken, warum diess Raumding in jedem seiner Puncte für eine jede hinlänglich kleine Entfernung aus dem ganzen Vorrathe von Puncten, die es für diese Entfernung überhaupt gibt, *nur eben jene und sonst keine anderen sich als Nachbarn aneignet*. Es werde mir verstattet, diese Beschaffenheit eines Raumdinges in jedem seiner Puncte *seine Haltung in diesem Puncte* zu nennen.

Ch4, 36. §.2. Eine *einfache in sich zurückkehrende Linie* ist eine solche, deren jeder Punct 2 Nachbarn hat, aber in keinem Punct Nachbarn besitzt, deren Entfernung grösser als eine gegebene ist. ***Et ut falsum vide fig.***

Ch4, 37. §. Eine Linie *heißt in sich zurückkehrend*, wenn 1, kein Punct derselben eine Entfernung von einem anderen hat grösser als eine gegebene **E**, wenn 2, jeder Punct für jede hinlänglich kleine Entfernung zwei Nachbarn hat; und wenn endlich 3, kein einzelner Punct, auch kein Inbegriff mehrerer, die nicht für sich selbst wieder eine Linie bilden, zu ihr hinzugefügt werden kann, ohne daß das neue Ganze, das so zum Vorschein kommt, aufhört eine Linie zu seyn und in jedem Punct für jede hinlänglich kleine Entfernung 2 Nachbarn zu haben.

Ch4, 38. §. Eine *Linie* heißt eine *einzige* in sich zurückkehrende Linie, wenn kein Theil derselben, der für sich allein schon eine in sich zurückkehrende Linie darstellt, weggelassen werden kann, ohne daß der Uiberrest aufhörte, eine in sich zurückkehrende Linie zu sein.
§. Eben so eine *einzige in sich zurückkehrende Fläche*.

Ch4, 39. §. Jede einfache in sich zurückkehrende Linie, welche in einer Fläche liegt, theilt diese in 2 Theile, welche sich dadurch unterscheiden, daß alle Puncte der Fläche, die ausserhalb dieser Linie liegen, entweder auf der Einen oder auf der ihr entgegengesetzten Seite der Linie liegen.

Ch4, 40. §. *Zwei Puncte in einer gegebenen Fläche heißen 1) auf derselben Seite oder 2) auf entgegengesetzten Seiten einer in dieser Fläche liegenden Linie*, je nachdem jede durch sie gelegte Ebene (höchstens mit Ausnahme eines oder mehrerer vereinzelt stehender Fälle) entweder 1) keine oder 2, 4, ... überhaupt *eine gerade Anzahl* von Durchschnittspuncten, oder 2) Ein oder 3 ... oder überhaupt eine *ungerade* Anzahl von Durchschnittspuncten mit jener Linie gemein hat, liegend in dem Stücke der Durchschnittslinie der Ebene mit der Fläche, das jene Puncte zu seinen Grenzpuncten hat.

Ch4, 41. *Es gibt Systeme von vier Puncten, in welcher keiner wie nicht an sich, so auch nicht durch sein Verhältniss zu den drei übrigen, so fern es durch einen reinen Begriff aufgefasst werden soll, bestimmt wird. Ist aber ein solches System von vier Puncten gegeben, so lässt sich ein jeder andere Punct und jeder Inbegriff von Puncten* (jegliches Raumding also) *durch blosse Begriffe, die dessen Verhältniss zu jenen vier Punctem ausdrücken, bestimmen.*

Ch4, 42. ... Gibt es endlich, wenn die vier Puncte a, b, c, d die angegebene Beschaffenheit haben, *keinen fünften*, der nicht durch einen bloßen Begriff seines Verhältnisses zu jenen schon bestimmt wäre: so folgt, daß jede andere aus a hervorgehende Richtung, welche auf zweien der nur eben gefundenen senkrecht aufsteht, mit der dritten entweder einerlei oder ihr entgegengesetzt ist; genau dasjenige, was in dem Lehrsatze von den drei Dimensionen des Raumes gemeint ist.

Ch5, 1. ... nous... observerons, en terminant, que les expressions a, a_b, a_{b_c}, qui désignent des lignes considérées par rapport à une, à deux, à trois dimensions, ne sont que les premiers termes d'une suite qui peut être prolongée indéfiniment.

Ch5, 2. Ich habe mich in diesem Herbst sehr viel mit der allgemeinen Betrachtung der krummen Flächen beschätigt, welches [sic] in ein unabsehbares Feld führt. ... Jene Untersuchungen greifen tief in vieles andere, ich möchte sogar sagen, in die Metaphysik der Raumlehre ein, und nur mit Mühe kann ich mich von solchen daraus entspringenden Folgen, wie z. B. die wahre Metaphysik der negativen und imaginären Grössen ist, losreissen. Der wahre Sinn des $\sqrt{-1}$ steht mir dabei mit grosser Lebendigkeit vor der Seele, aber es wird sehr schwer sein, ihn in Worte zu fassen, die immer nur ein vages, in der Luft schwebendes Bild geben können.

Ch5, 3. Ich werde die Beweisführung in einer der Geometrie der Lage entnommenen Einkleidung darstellen, weil jene dadurch die grösste Anschaulichkeit und Einfachheit gewinnt. Im Grunde gehört aber der eigentliche Inhalt der ganzen Argumentation einem höhern von Räumlichem unabhängigen Gebiete der allgemeinen abstracten Grössenlehre an, dessen Gegenstand die nach der Stetigkeit zusammenhängenden Grössencombinationen sind, einem Gebiete, welches zur Zeit noch wenig angebauet ist, und in welchem man sich auch nicht bewegen kann ohne eine von räumlichen Bildern entlehnte Sprache.

Ch5, 4. ... eine ausserordentliche Hoffnung setzte er aber auf die Ausbildung der Geometria situs, in der weite gänzlich unangebaute Felder sich befänden, die durch unsern gegenwärtigen Calcul noch so gut wie gar nicht beherrscht werden könnten.

Ch5, 5. Indem ich daher nun daran ging, die so gefundenen Resultate zusammenhängend und von Anfang an zu bearbeiten, so dass ich mich auch auf keinen in irgend einem Zweige der Mathematik bewiesenen Satz zu berufen gedachte, so ergab sich, dass die von mir aufgefundene Analyse nicht, wie mir Anfangs schien, bloss auf dem Gebiete der Geometrie sich bewegte; sondern ich gewahrte bald, dass ich hier auf das Gebiet einer neuen Wissenschaft gelangt sei, von der die Geometrie selbst nur eine specielle Anwendung sei. Schon lange war es mir nämlich einleuchtend geworden, dass die Geometrie keineswegs in dem Sinne wie die Arithmetik oder die Kombinationslehre als ein Zweig der Mathematik anzusehen sei, vielmehr die Geometrie schon auf ein in der Natur gegebenes (nämlich den Raum) sich beziehe, und dass es daher einen Zweig der Mathematik geben müsse, der in rein abstrakter Weise ähnliche Gesetze aus sich erzeuge, wie sie in der Geometrie an den Raum gebunden erscheinen. Durch die neue Analyse war die Möglichkeit, einen solchen rein abstrakten Zweig der Mathematik auszubilden, gegeben; ja diese Analyse, sobald sie, ohne irgend einen schon anderweitig erwiesenen Satz vorauszusetzen, entwickelt wurde, und sich rein in der Abstraktion bewegte, war diese Wissenschaft selbst. Der wesentliche Vortheil, welcher durch diese Auffassung erreicht wurde, war der Form nach der, dass nun alle Grundsätze, welche Raumesanschauungen ausdrückten, gänzlich wegfielen, und somit der Anfange in eben so unmittelbarer wurde, wie der der Arithmetik, dem Inhalte nach aber der, dass die Beschränkung auf drei Dimensionen wegfiel. Erst hierdurch traten die Gesetze in ihrer Unmittelbarkeit und Allgemeinheit ans Licht und stellten sich in ihrem wesentlichen Zusammenhänge dar, und manche Gesetzmässigkeit, die bei drei Dimensionen entweder noch gar nicht, oder nur verdeckt vorhanden war, entfaltete sich nun bei dieser Verallgemeinerung in ihrer ganzen Klarheit.

Ch5, 6. Den Uebergang des erzeugenden Elementes aus einem Zustande in einen andern nennen wir eine Aenderung desselben; und diese abstrakte Aenderung des erzeugenden Elementes entspricht also der Ortsänderung oder Bewegung des Punktes in der Geometrie.

...

„Unter einem Ausdehnungsgebilde erster Stufe verstehen wir die Gesammtheit der Elemente, in die ein erzeugendes Element bei stetiger Aenderung übergeht.“

...

Wir können eine solche Aenderung, durch die aus einem Element einer stetigen Form ein nächst angränzendes erzeugt wird, eine Grundänderung nennen, und werden dann sagen: „das *einfache Ausdehnungsgebilde* sei ein solches, das durch stetige Fortsetzung derselben Grundänderung hervorgeht.“

...

... nennen wir das einfache Ausdehnungsgebilde erster Stufe eine *Ausdehnungsgrösse* oder *Ausdehnung erster Stufe* oder eine *Strecke*.

Ch5, 7. Nehme ich nun, um zu den Verknüpfungen verschiedenartiger Strecken zu gelangen, zunächst zwei verschiedenartige Grundänderungen an, und lasse ein Element die erste Grundänderung (oder deren entgegengesetzte) beliebig fortsetzen und dann das so geänderte Element in der zweiten Aenderungsweise gleichfalls beliebig fortschreiten, so werde ich dadurch aus einem Element eine unendliche Menge neuer Elemente erzeugen kön-nen, und die Gesammtheit der so erzeugbaren Elemente nenne ich ein *System zweiter Stufe*. Nehme ich dann ferner eine dritte Grundänderung an, welche von jenem Anfängselemente aus nicht wieder zu einem Elemente dieses Systems zweiter Stufe führt, und welche ich deshalb als von jenen beiden ersten *unabhängig* bezeichne, und lasse ein beliebiges Element jenes Systems zweiter Stufe, diese dritte Aenderung (oder deren entgegensetzte) beliebig fortsetzen, so wird die Gesammtheit der so erzeugbaren Elemente ein System dritter Stufe bilden; und da dieser Erzeugungsweise dem Begrilfe nach keine Schranke gesetzt ist, so werde ich auf diese Weise zu Systemen beliebig hoher Stufen fortschreiten können.

...

... Es ist nämlich klar, dass das System zweiter Stufe der Ebene entspricht, und die Ebene dadurch erzeugt gedacht wird, dass alle Punkte einer geraden Linie nach einer neuen in ihr nicht enthaltenen Richtung (oder nach der entgegengesetzten) sich fortbewegen, wobei dann eben die Gesammtheit der so erzeugharen Punkte die unendliche Ebene bildet. Es erscheint somit die Ebene als eine Gesannntheit von Parallelen, welche alle eine gegebene Gerade durchschneiden; und es ist ersichtlich, dass, da diese Parallelen sich nicht schneiden, und auch die ursprüngliche Gerade nicht noch ein zweitesmal treffen, alle auf jene Weise erzeugten Punkte von einander verschieden sind und somit die Analogie eine vollständige ist. Ebenso gelangt man zu dem ganzen unendlichen Raume, als dem Systeme dritter Stufe, wenn man die Punkte der Ebene nach einer neuen, nicht in der Ebene liegenden Richtung (oder der entgegengesetzten) fortbewegt; und weiter kann die Geometrie nicht fortschreiten, während die abstrakte Wissenschaft keine Gränze kennt.

Ch5, 8. „Jedes System m-ter Stufe kann erzeugt gedacht werden durch je m unabhängige Aenderungsweisen desselben aus jedem beliebigen Element desselben, d. h. aus Einem solchen Elemente können alle übrigen durch jene Aenderungsweisen erzeugt werden.”

Ch5, 9. So erlebte denn *Graßmann* das, was den Verfasser einen neuen Werkes das Schmerzlichste sein muß: sein Buch fand nirgends Beachtung, es wurde in der Öffentlichkeit vollständig totgeschwiegen, niemand fand sich, der eine Besprechung geliefert, der es auch nur öffentlich getadelt hätte, und auch die Mathematiker, denen er das Werk zugeschickt hatte, äußerten sich zwar nicht unfreundlich, zum Teil sogar wohlwollend darüber, aber wirklich studiert hatte es keiner von ihnen.

Ch5, 10. *Erklärung*. Die Geammtheit der Grössen, welche aus einer Reihe von Grössen $a_1, a_2, \ldots, a_n$ numerisch ableitbar sind, nenne ich das aus jenen Grössen ableitbare *Gebiet* (das Gebiet der Grössen $a_1, \ldots, a_n$), und zwar nenne ich es ein Gebiet n-ter *Stufe*, wenn jene Grössen von erster Stufe (d. h. aus n ursprünglichen Einheiten numerisch ableibar) sind, und sich das Gebiet nicht aus weniger als n solchen Grössen ableiten lässt. Ein Gebiet, welches ausser der Null keine Grösse enhält, heisst ein Gebiet *nullter* Stufe.

Ch5, 11. Mit meinen Arbeiten steht es jetzt so ziemlich; ich habe Anfangs December meine Habilitationsschrift abgeliefert und musste dabei drei Themata zur Probevorlesung vorschlagen, von denen dann die Facultät eines wählt. Die beiden ersten hatte ich fertig und hoffte, dass man eins davon nehmen würde; Gauss aber hatte das dritte gewählt, und so bin ich nun wieder etwas in der Klemme, da ich dies noch ausarbeiten muss.

Ch5, 12. Da entschloss er sich plötzlich auf mein wiederholtes Bitten, „um die Sache vom Halse los zu werden", am Freitag nach Pfingsten Mittag das Colloquium auf den anderen Tag um halb elf anzusetzen und so war ich am Sonnabend um eins glücklich damit fertig.

Ch5, 13. Ich habe mir daher zunächst die Aufgabe gestellt, den Begriff einer mehrfach ausgedehnten Grösse aus allgemeinen Grössenbegriffen zu construiren.

Ch5, 14. Der Verfasser ist Herbartianer in Psychologie und Erkenntnisstheorie (Methodologie und Eidolologie), Herbart's Naturphilosophie und den darauf bezüglichen metaphysischen Disciplinen (Ontologie und Synechologie) kann er meistens nicht sich anschliessen.

Ch5, 15. Bei der Untersuchung der Functionen, welche aus der Integration vollständiger Differentialien entstehen, sind einige der analysis situs angehörige Sätze fast unentbehrlich. Mit diesem von Leibnitz, wenn auch vielleicht nicht ganz in derselben Bedeutung, gebrauchten Namen darf wohl ein Theil der Lehre von den stetigen Grössen bezeichnet werden, welcher die Grössen nicht als unabhängig von der Lage existirend und durch einander messbar betrachtet, sondern von den Massverhältnissen ganz absehend, nur ihre Orts- und Gebietsverhältnisse der Untersuchung unterwirft.

Ch5, 16. Fehlt dieses, so kann man zwei Grössen nur vergleichen, wenn die eine ein Theil der andern ist, und auch dann nur das Mehr oder Minder, nicht das Wieviel entscheiden: Die Untersuchungen, welche sich in diesem Falle über sie anstellen lassen, bilden einen allgemeinen von Massbestimmungen unabhängigen Theil der Grössenlehre, wo die Grössen nicht als unabhängig von der Lage existirend und nicht als durch eine Einheit ausdrüickbar, sondern als Gebiete in einer Mannigfaltigkeit betrachtet werden.

Ch5, 17. Wenn man, anstatt den Begriff als bestimmbar, seinen Gegenstand als veränderlich betrachtet, so kann diese Construction bezeichnet werden als eine Zusammensetzung einer Veränderlichkeit von $n + 1$ Dimensionen aus einer Veränderlichkeit von n Dimensionen und aus einer Veränderlichkeit von Einer Dimension.

Ch6, 1. Lässt sich eine Fläche (etwa ein Quadrat mit Einschluss der Begrenzung) eindeutig auf eine Linie (etwa eine gerade Strecke mit Einschluss der Endpunkte) eindeutig beziehen, so dass zu jedem Puncte der Fläche ein Punct der Linie und umgekehrt zu jedem Puncte der Linie ein Punct der Fläche gehört?

Ch6, 2. . . . da es sich von selbst verstünde, dass zwei unabhangige Veränderliche sich nicht auf eine zurückführen lassen.

Ch6, 3. ... dass Flächen, Körper, ja selbst stetige Gebilde von ρ Dimensionen sich eindeutig zuordnen lassen stetigen Linien, also Gebilden von nur *einer* Dimension, dass also Flächen, Körper, ja sogar Gebilde von ρ Dimensionen, dieselbe *Mächtigkeit* haben, wie Curven ....

Ch6, 4. Ich weiss nicht, ob mein Einwurf von wesentlicher Bedeutung für Ihre Idee ist, doch wollte ich ihn nicht zurtickhalten.

Ch6, 5. (A) «Eine nach ρ [correcting e as printed] Dimensionen ausgedehnte stetige Mannigfaltigkeit lässt sich eindeutig einer stetigen Mannigfaltigkeit von einer Dimension zuordnen, oder: (was nur eine andere Form desselben Satzes ist) die Puncte (Elemente) einer nach ρ Dimensionen ausgedehnten Mannigfaltigkeit lassen sich durch eine reelle Coordinate t so bestimmen, dass zu jedem reellen Werth von t im Intervalle $(0\ldots1)$ ein Punct der Mannigfaltigkeit, aber auch umgekehrt zu jedem Puncte der M. ein bestimmter Werth von t im Intervalle $(0\ldots1)$ gehört.»

Ch6, 6. (B) «Eine veränderliche Zahl e, welche alle *irrationalen* Zahlenwerthe des Intervalles $(0\ldots1)$ annehmen kann, lässt sich *eindeutig* einer Zahl x, welche *alle* Werthe dieses Intervalles ohne Ausnahme erhält, zuordnen.»

Ch6, 7. (C) «Eine Zahl f, welche alle Werthe des Intervalles $(0\ldots1)$ annehmen kann mit Ausnahme gewisser ε_ν, die an die Bedingungen gebunden sind: $\varepsilon_\nu < \varepsilon_{\nu+1}$ und $\lim \varepsilon_\nu = 1$ läisst sich einer stetigen Veränderlichen x eindeutig zuordnen, welche alle Werthe des Intervalles $(0\ldots1)$ ohne Ausnahme erhält.»

Ch6, 8. (D) «Eine Zahl y welche alle Werthe des Intervalles $(0\ldots1)$ mit alleiniger Ausnahme des Werthes 0 annehmen kann, lässt sich einer Zahl x eindeutig zuordnen, welche alle Werthe des Intervalles $(0\ldots1)$ ohne Ausnahme erhält.»

Ch6, 9. (F) «Einer Zahl z, welche alle Werthe eines Intervalles $(\alpha\ldots\beta)$ mit Ausnahme von α annehment kann, lässt sich eindeutig einer Zahhl u zuorden, die alle Werthe des Intervalls $(\alpha\ldots\beta)$ ohne Ausnahme erhält.»

Ch6, 10. (G) «Eine Zahl w, welche alle Werthe des Intervalles $(\alpha\ldots\beta)$ mit Ausnahme der beiden Endwerthe α, β erhält, lässt sich einer veränderlichen Zahl u eindeutig zuordnen, die alle Werthe des Intervalles $(\alpha\ldots\beta)$ annimmt.»

Ch6, 11. ... je le vois, mais je ne le crois pas.

Ch6, 12. Ihren Beweis habe ich noch einmal geprüft, und ich habe keine Lücke darin entdeckt; ich glaube gewiss, dass Ihr interessanter Satz richtig ist, und beglückwünsche Sie zu demselben.

Ch6, 13. Dabei fiel mir auf, dass alle in dieses Feld schlagenden Untersuchungen *ihrerseits* von einer unbewiesenen Voraussetzung ausgehen, die mir nicht als selbstverständlich, vielmehr einer Begründung bedürftig erschienen ist. Ich meine die Voraussetzung, dass eine ρ fach ausgedehnte stetige Mannigfaltigkeit zur Bestimmung ihrer Elemente ρ von einander unabhängiger reeller Coordinaten bedarf, dass diese Zahl der Coordinaten für eine und dieselbe Mannigfaltigkeit weder vergrössert noch verkleinert werden könne.

Ch6, 14. *Nun scheint es mir*, dass alle philosophischen oder mathematischen Deductionen, welche von jener irrthümlichen Voraussetzung Gebrauch machen, unzulässig sind. Vielmehr wird der Unterschied, welcher zwischen Gebilden von *verschiedener* Dimensionszahlen liegt, in ganz anderen Momenten gesucht werden müssen, als in der für characteristisch gehaltenen Zahl der unabhängigen Coordinaten.

Ch6, 15. ... dass die Dimensionenzahl einer stetigen Mannigfaltigkeit nach wie vor die erste und wichtigste Invariante derselben ist ....

Ch6, 16. « Gelingt es, eine gegenseitige eindeutige und vollständige Correspondenz zwischen den Puncten einer stetigen Mannigfaltigkeit A von a Dimensionen einerseits und den Puncten einer stetigen Mannigfaltigkeit B von b Dimensionen andererseits herzustellen, so ist diese *Correspondenz selbst*, wenn a und b *ungleich* sind, nothwendig eine *durchweg unstetige*. »

Ch6, 17. ... scheint mir bei Ihrem jetzigen Beweise die *anfängliche* Correspondenz zwischen dem Puncten des ρ-Strecks, deren Coordinaten sämmtlich irrational sind, und den Puncten des Einstrecks mit ebenfalls irrationaler Coordinate in gewissen Sinne (Kleinheit der Aenderungen) so stetig zu sein, wie möglich ...

Ch6, 18. ... zwingt Sie, eine grauenhafte, Schwindel erregende Unstetigkeit in der Correspondenz entreten zu lassen, durch welche Alles in Atome aufgelöst wird ....

Ch7, 1. ... *wenn auf einer Kugel zwei eindeutige und stetige Functionen gegeben sind, so gibt es stets Werthe, welche gleichzeitig von beiden Functionen in zwei Punkten der Kugel angenommen werden* ....

Ch7, 2. Eine zusammenhängende continuirliche Mannigfaltigkeit M_n von n Dimensionen kann durch eine oder mehrere Mannigfaltigkeiten von $n-2$ oder weniger Dimensionen ($M_\nu, M'_{\nu'}, M''_{\nu''}, \ldots$; $\nu, \nu', \cdots \leq n-2$) nicht in getrennte Stücke zerlegt werden.

Dabei muß allerdings vorausgesetzt werden, dass nicht die Anzahl der Mannigfaltigkeiten M_ν, $M'_{\nu'}, M''_{\nu''}, \ldots$ in jedem noch so kleinen Stücke einer continuirlichen Mannigfaltigkeit von $n-1$ Dimensionen abzäihlbar unendlich groß sei.

Ch7, 3. *Wenn zwei unabhängige reelle Veränderliche x_1 und x_2 als rechtwinklige Punktcoordinaten in der Ebene aufgefasst alle Stellen im Inneren und auf einem Kreise durchlaufen, wenn von ihnen zwei andere reelle Veränderliche y_1 und y_2 eindeutig und stetig abhängen und dabei dasselbe Werthepaar y_1, y_2 zu einer endlichen Anzahl von Werthepaaren x_1, x_2 gehört, so enthält, indem auch die Veränderlichen y_1 und y_2 in einer zweiten Ebene als rechtwinklige Punktcoordinaten angesehen werden, der von den Punkten y_1, y_2 gebildete Theil dieser Ebene ein zweifach ausgedehntes Stück der Ebene, etwa die ganze Fläche eines Kreises, in sich.*

Ch7, 4. *Wenn zwei eindeutige und stetige reelle Functionen von zwei reellen Veränderlichen dasselbe Werthepaar nicht wiederholt annehmen, so entspricht einer inneren Stelle des Gebietes der unabhängigen Veränderlichen eine innere Stelle im Gebiete der abhängigen Veränderlichen.*

Ch7, 5. *Kein Theil des dreifach ausgedehnten Raumes, welcher eine Kugel ganz enthält, kann auf irgend einen Theil der Ebene eindeutig und stetig abgebildet werden.*

Ch7, 6. „In einer Mannigfaltigkeit νten Grades wird jedes Gebilde νten Grades durch ein anderes von geringerem Grade begrenzt; in einer Mannigfaltigkeit $(\nu+1)$ten Grades fält jedes Gebilde νten Grades mit seiner Grenze zusammen".

Ch7, 7. ... der ... Inductionsschluss ist nur dann richtig, wenn auch die Unmöglichkeit einer eindeutigen und stetigen Abbildung einer stetigen Mannigfaltigkeit $n$ter Dimension auf einen unstetigen Theil einer zweiten stetigen Mannigfaltigkeit $n$ter Dimension, d.h. auf einen Theil ohne innere Punkte, nachgewiesen wird.

Ch7, 8. ... es scheint mir jedoch die Sache noch nicht ganz fertig gestellt zu sein.

Ch7, 9. ... so dankenswerth auch dieser mit Scharfsinn unternommene Beweisversuch mir erscheint, so kann ich dennoch gewisse Bedenken an demselben nicht verscheuchen und fürchte, daß es *nur* ein Versuch ist, der zur Klärung über die Sache aber gewiß beitragen würde.

Ch7, 10. « Eine stetige M_μ und eine stetige M_ν lassen sich, falls $\mu < \nu$, nicht *stetig* so einander zuordnen, dass zu jedem Elemente von M_μ *ein einziges* Element von M_ν und zu jedem Elemente von M_ν *ein oder mehrere* Elemente von M_μ gehören. »

Ch7, 11. . . . es wäre sehr verdienstlich, wenn diese ganze « Gebietslehre » ab ovo dargestellt würde, ohne die geometrische Anschauung zuzuziehen, und dabei müsste z.B. der Begriff einer von dem Puncte a nach dem Puncte b innerhalb des Gebietes G stetig führenden Linie recht bestimmt und deutlich definiert werden.

Ch7, 12. (III) »Hat man zwischen zwei stetigen Gebieten M_μ und M_ν eine solche Abhängigkeit, daß zu jedem Punkte z von M_μ *höchstens* ein Punkt Z von M_ν, zu jedem Punkte Z von M_ν *mindestens* ein Punkt z von M_μ gehört, und ist ferner diese Beziehung eine stetige, so daß unendlich kleinen Aenderungen von z, unendlich kleine Aenderungen von Z und auch umgekehrt unendlich kleinen Änderungen von Z unendlich kleine Aenderungen von z entsprechen, so ist $\mu \geq \nu$ «.

Ch8, 1. Nous avons apporté un soin particulier à l'établissement des théorèmes fondamentaux. Il n'en est aucun dont la démonstration ne soit subordonnée à certaines restrictions. Nous nous sommes efforcé d'apporter dans cette discussion, parfois dél icate, toute la précision et la rigueur compatibles avec un enseignement él émentaire.

Ch8, 2. *Soit **C** une ligne continue fermée et sans points multiples, ne contenant à son intérieur aucun point critique de la fonction* $f(z)$. *L'intégrale* $\int f(z)\,\mathrm{d}z$, *prise le long d'une ligne fermée et rectifiable quelconque **K** intérieure à **C**, est nulle.*

Ch8, 3. . . . toute courbe continue C divise le plan en deux régions, l'une extérieure, l'autre intérieure, cette dernière ne pouvant se réduire à zéro, car elle contient un cercle de rayon fini.

Ch8, 4. Pourrait-on signaler une courbe

$$x = f(t),\;\; y = \phi(t)$$

(f et ϕ étant des fonctions continues) dont l'aire fut indéterminée?

Ch8, 5. . . . was ist eine einfach geschlossene Linie, was ist eine Linie, insbesondere eine geschlossene Linie überhaupt, und sind alle oder nur gewisse geschlossene Linien in dem Ausspruch des Cauchy'schen Satzes zulässig?

Ch8, 6. Diese Einteilung der Punktmengen in Klassen bildet . . . die allgemeinste Grundlage der *Analysis situs*. Die Aufgabe der Analysis situs ist es, die Invarianten der einzelnen Klassen von Punktmengen aufzusuchen.

Ch8, 7. . . . daß *das umkehrbar eindeutige und stetige Abbild der Fläche eines Quadrats wieder ein einfach zusammenhängendes Fl ächenstück ist*, daß es nämlich aus der Gesamtheit aller Punkte besteht, die dem Innern einer *geschlossenen Curve* angehören.

Ch8, 8. Als eines der allgemeinsten Probleme aus der Theorie der Punktmengen kann man die Aufgabe bezeichnen, die grundlegenden Sätze der *Analysis Situs* mengentheoretisch zu formulieren und zu begründen und die Beziehungen darzulegen, die zwischen den *mengentheoretisch-geometrischen* und den *analytischen* Ausdrucksweisen derselben Begriffe und Sätze obwalten. Die paradoxen Resultate, wie sie z.B. in der eineindeutigen Abbildung der Continua und in der Peanoschen Kurve vorliegen, haben die naiven Vorstellungen der Analysis Situs gründlich zerstört. Um so mehr muß man verlangen, daß die Mengentheorie wiederum Ersatz schafft und die geometrischen Grundbegriffe in einer Weise definiert, die ihnen ihren natürlichen für die Analysis Situs charakteristischen Inhalt wieder zurückgibt. Ist auch die vielgeschmähte Anschauung keine Quelle des Beweises, so scheint es mir doch — wenigstens im Gebiet der Analysis Situs — ein Ziel der Forschung zu sein, den Inhalt der geometrischen Definitionen mit dem Anschauungs*inhalt* in Ubereinstimmung zu bringen.

Ch8, 9. Wir stehen mit dem, was den Inhalt der Analysis Situs ausmacht, grossenteils noch auf dem Boden der Anschauung, die zugleich als Beweisgrund dienen muss. Aber es dürfte nötig sein, auch auf diesem Gebiet nach grösserer Exactheit zu streben, und seine Sätze aus den geometrischen und mengentheoretischen Grundtatsachen abzuleiten. Dies allmählich zu tun, ist mein lebhafter Wunsch. Ich habe wieder Lust bekommen, hieriiber intensiver zu arbeiten, und hoffe, es wird mir mit der Zeit gelingen.

Ch9, 1. . . . la Géométrie n'est autre chose que l'étude d'un groupe . . . .

Ch9, 2. Sont-ce des faits expérimentaux, des jugements analytiques ou synthétiques *a priori*?

Ch9, 3. On peut donc *déformer* le plan de façon à obtenir une droite, pourvu que cette déformation ne soit pas continue. Cela serait impossible au contraire avec une déformation continue. Ainsi la question du hombre des dimensions est intimement liée à la notion de continuité et elle n'aurait aucun sens pour celui qui voudrait faire abstraction de cette notion.

Ch9, 4. . . . Je suis heureux d'avoir cette occasion d'entrer en rapport avec un homme de votre valeur.

Ch9, 5. Un continu a n dimensions quand on peut le décomposer en plusieurs parties en y pratiquant une ou plusieurs coupures qui soient elles-mêmes des continus à $n - 1$ dimensions. Le continu à n dimensions se trouve ainsi défini par le continu à $n - 1$ dimensions; c'est une définition par récurrence.

Ch9, 6. Nous voyons par ce rapide exposé quels sont les faits expérimentaux qui nous ont conduits à attribuer trois dimensions à l'espace. En présence de ces faits, il nous était plus commode de lui en attribuer trois que quatre ou que deux; mais ce mot de commode n'est peut-être pas ici assez fort; un être qui aurait attribué à l'espace deux ou quatre dimensions se serait trouvé dans un monde fait comme le nôtre, en état d'infériorité dans la lutte pour la vie . . . .

Ch9, 7. Mais il y a une Analysis situs à plus de trois dimensions; je ne dis pas que ce soit une science facile, j'y ai consacré trop d'efforts pour ne pas m'être rendu compte des difficultés qu'on y rencontre; mais enfin cette science est possible et elle ne repose pas exclusivement sur l'analyse; on ne saurait la cultiver avec fruit sans de continuels appels à l'intuition. II y a donc bien une intuition des continus à plus de trois dimensions et si elle exige une attention plus soutenue que l'intuition géométrique ordinaire, c'est sans doute une affaire d'habitude, et aussi l'effet de la complication rapidement croissante des propriétés des continus quand augmente le hombre des dimensions. . . .

Je conclurai que nous avons tous en nous l'intuition du continu d'un nombre quelconque de dimensions, parce que nous avons la faculté de construire un continu physique et mathématique; que cette faculté préexiste en nous à toute expérience parce que sans elle, l'expérience proprement dite serait impossible et se réduirait à des sensations brutes, impropres à toute organisation, que cette intuition n'est que la conscience que nous avons de cette faculté. . . .

Ch10, 1. *Tout ensemble discontinu, situé dans un plan, fait partie d'une courbe continue sans point multiple, située dans ce plan.*

Ch10, 2. II suit de notre théorème qu'*au point de vue de l'*analysis situs, *tous les ensembles parfaits discontinus sont équivalents (homomorphes).*

Ch10, 3. Wird man es aber versuchen, den Poincaréschen Dimensionsbegriff einer weiteren Untersuchung des Raumbegriffes zugrunde zu legen, so stösst man bald auf grosse Schwierigkeiten. Zunächst rechnet schon Poincaré bei der Einftihrung des Begriffes nicht mit den Merkmalen, die auch schon einer naiven Dimensionsvorstellung eigen sind, so z.B. dass ein System gewisser Dimension kein System höherer Dimension enthalten soil. Nach der Poincaréschen Definition wird ein Doppelkegel als physikalisches Kontinuum von einer Dimension sein. Doch 1ässt sich diesem Fehler noch leicht abhelfen. Die grösste Schwierigkeit liegt darin, dass der Dimensionsbegriff für mathematische Kontinua schwer zu fassen ist; vor einer gründlichen Analyse dieses Begriffes kann nicht daran gedacht werden, den Dimensionsbegriff für physikalische Kontinua und jenen für mathematische Kontinua irgend wie miteinander zu verbinden. Der Dimensionsbegriff kann somit den weiteren Untersuchungen einstweilen nicht als sichere Grundlage dienen.

Ch10, 4. Théorème I. — *Il ne peut y avoir application entre deux ensembles fermés:* E *de* G_n, F *de* G_{n+p} $(p \geq 1)$, *si* F *contient tous les points d'un domaine sphérique à* $n + p$ *dimensions.*

Ch10, 5. Théorème II. — *Soient, dans* G_n, *deux ensembles* P *et* Π *applicables l'un sur l'autre,* P *étant un domaine sphérique; tout point de* P *non frontière pour* P *a pour image un point non frontière pour* Π.

Ch10, 6. Théorème III. — Σ *étant une surface fermée simple de* G_n $(n \geq 2)$:

1° *Il y a pour* Σ *un intérieur* I, *un extérieur* E. *Tout point de* I *(de* E*) est centre d'un domaine sphérique dont tous les points appartiennent à* I *(à* E*).*
2° *La distance de deux points quelconques de* I *est moindre que la borne supérieure des distances de deux points de* Σ.
3° *Un chemin continu allant d'un point de* I *à un point de* E *contient au moins un point de* Σ.
4° *Si une surface fermée simple* Σ *varie d'une manière continue en fonction d'un paramètre* t *qui prend toutes les valeurs d'un intervalle* (a, b), *un point déterminé* M *de* G_n *qui ne se trouve sur* Σ *pour aucune valeur de* t *est, ou constamment à l'intérieur de* Σ, *ou constamment à l'extéieur.*

Ch10, 7. Je suis très heureux que vous ayez en main la démonstration de mon théorème III ou d'un équivalent et mon désir est que vous publiez le plus tôt possible votre démonstration. Au point de vue scientifique, l'intérêt en serait grand. . . .

Ch10, 8. Je souhaite vivement que votre solution du th. III se complète; je sais par expérience personelle que les déceptions sont assez fréquentes dans cet ordre de questions, mais les rides se comblent aussi après coup, fort heureusement.

Ch10, 9. Déf. III. Si deux ensembles sont tels que l'un d'eux G_1 puisse être considéré comme l'image de l'autre G_2 ou d'une partie de cet autre, nous dirons que le premier G_1 a un type de dimension inférieur ou égal à celui du second G_2 et nous écrirons $dG_1 \leq dG_2$, ou $dG_2 \geq dG_1$.

Déf. IV. Si l'on a à fois $dG_1 \leq dG_2$ et $dG_2 \leq dG_1$, nous dirons que G_1 et G_2 ont même type de dimension et nous écrirons $dG_1 = dG_2$ ou $dG_2 = dG_1$.

Déf. V. Si l'on a $dG_1 \leq dG_2$ et s'il est certain que G_2 ne peut être considéré d'aucune manière comme homéomorphe de G_1 ou d'une partie de G_1, nous dirons que le type de dimension de G_1 est inférieur (et non pas égal) à celui de G_2 et nous écrirons

$$dG_1 < dG_2 \text{ ou } dG_2 > dG_1.$$

Ch11, 1. „Alle endlichen continuierlichen Gruppen der n-dimensionalen Mannigfaltigkeit zu bestimmen"

Ch11, 2. Ob p für jede Gruppe eine Invariante ist, steht aus, solange die „Nichtapplizierbarkeit" zweier Räume, deren Dimensionenzahl verschieden, unbewiesen ist.

Ch11, 3. Als ich im vergangenen Winter meine zweite Mitteilung über endliche kontinuierliche Gruppen zur Zusendung an die Redaktion der Mathematischen Annalen schon fertig hatte, entdeckte ich auf einmal, dass die Schoenfliesschen Untersuchungen über Analysis Situs der Ebene, auf welche ich mich in ausgiebigster Weise gesfützt hatte, sich nicht in allen ihren Teilen aufrecht erhalten lassen, so dass auch meine gruppentheoretischen Ergebnisse in Frage kamen.

Um wieder Klarheit zu erlangen, war zunächst notwendig, die bezügliche Schoenfliessche Theorie gründlich durchzuarbeiten, und genau festzustellen, auf welche ihrer Resultate sich in vollem Vertrauen weiter bauen lässt. So entstand die hier eingeschlossene Arbeit, welche in mehrere Teile der Schoenfliesschen Theorie eingrifft [sic!], und einige gewissermassen neu gestaltet.

Ch11, 4. Dezen zomer was de eerste mathematicus van de wereld in Scheveningen, door mijn werk was ik al met hem in betrekking, nu heb ik telkens weer met hem gewandeld, en gepraat als een jong apostel met een profeet. Hij was 46, maar jong van ziel en lichaam, zwom krachtig en klom met liefde over muren en hekken van prikkeldraad.

Ch11, 5. … een *classificatie voor de analysis situs der in een ruimte gelegen puntverzamelingen;* …

Ch11, 6. Een hierbij terstond aansluitende vraag is dan, in hoeverre ruimten van verschillend dimensie-aantal voor onze groep verschillend zijn. Zeer waarschijnlijk is dit steeds het geval, doch bet bewijs schijnt uiterst moeilijk te leveren, en zal waarschijnlijk nog langen tijd een onopgelost probleem blijven.

Ch11, 7. Door het felt, dat deze zoo evident schijnende stelling een uitvoerig bewijs vereischt, en dat de vele niet geslaagde pogingen, het te vereenvoudigen, een uitgebreide litteratuur beslaan, wordt de beperktheid der redeneermiddelen op dit gebied duidelijk in het licht gesteld.

Ch11, 8. Evenmin heeft men nog zekerheid, dat de drie-dimensionale Cartesiaansche ruimte door een gesloten Jordansch oppervlak, d. w. z. een éénéénduidig continu beeld van den bol, in twee gebieden wordt verdeeld.

Ch11, 9. Considérons un contour fermé ABC sur lequel P et Q ne s'annullent jamais à la fois, et admettons de plus que, dans l'intérieur de ce contour, les valeurs de x, y qui annullent P et Q, donnent à R une valeur positive ou négative, mais différente de zéro.

Parmi les solutions (x, y) des équations $P = 0$, $Q = 0$, contenues dans l'intérieur de ABC, les unes pourront correspondre à valeur positive, les autres à une valeur négative de R. Nous désignerons par μ_1, le nombre des solutions de la première espèce, et par μ_2 le nombre des solutions de la seconde espèce.

Cela posé, on aura

$$\Delta = 2(\mu_1 - \mu_2),$$

Δ représentant l'excès du nombre de fois où la fraction $\frac{P}{Q}$ passe du positif au négatif sur le nombre de fois où elle passe du négatif au positif, en s'évanouissant, quand on parcourt d'un mouvement continu le contour entier ABC, en allant des x positives aux y positives.

Ch11, 10. Die Schwierigkeiten der Verallgemeinerung, deren Herr Sylvester Erwähnung thut, dürften wohl zumeist in der Beschränkung auf algebraische Gebilde liegen, die er festgehalten hat. Sobald ich die Einsicht gewann, dass alle bezüglichen Betrachtungen ausschliesslich jenem allgemeinen Gebiete angehören, welches für den Fall wo $n = 2$ oder 3 ist als „Geometrie der Lage" bezeichnet wird, ergaben sich mir die einfachsten Mittel zur Bewältigung der entgegenstehenden Schwierigkeiten.

Ch11, 11. … dass die Theorie der Charakteristik von Functionen-Systemen ebenso nahe mit geometrischen Theorieen zusammenhängt wie mit der Potentialtheorie; und man daft wohl in diesen Beziehungen des ursprünglich aus rein analytischen Principien entwickelten Begriffs der Charakteristik zu anderen bekannten Theorieen eine Probe für die Echtheit desselben erkennen, einen Beweis dafür, dass die Einführung dieses Begriffes in die Wissenschaft durchaus naturgemäss und nothwendig ist.

Ch11, 12. … une considération nouvelle qui nous rendra les plus grands services.

Ch11, 13. *Corollaire I.* — Le nombre total des noeuds et des foyers est égal au nombre total des cols plus 2.

Ch11, 14. … Die Erklärung ist diese, dass der Inhalt sowohl von meiner Arbeit „Ueber Abbildung von Mannigfaltigkeiten" wie von der im Buche von Tannery befindlichen Hadamardschen Note um Weihnaehten 1909 in Paris von Hadamard und mir durchgesproehen wurde.

Ch11, 15. *'Si les points singuliers sont en nombre fini, chacun d'eux possède un indice fini, et la somme algébrique de tous ces indices est égale à 2.'*

Ch11, 16. Pour déterminer ce *degré*, introduisons des coordonnées homogènes (à double sens), écrivons les x, y, z pour la sphère primitive et ξ, η, ζ pour l'image, divisons la sphère en un nombre fini de régions, et considérons d'abord les transformations définies par les relations:

$$\xi : \eta : \zeta = f_1(x, y, z) : f_2(x, y, z) : f_3(x, y, z),$$

où f_1, f_2, f_3 sont des polynôomes, qui d'ailleurs peuvent différer pour les differentes régions de la sphère. Appelons cette transformation une transformation polynomiale. Définissons une indicatrice sur la sphère: alors chaque point P de l'image *de position générale* apparaîtra un nombre r_P de fois avec indicatrice positive, et un nombre s_P de fois avec indicatrice négative. On démontre alors que $r_P - s_P$ est une constante: c'est le degré la transformation polynomiale.

Revenons à une transformation univoque et continue générale. Elle peut être approximée par une série de transformations polynomiales; on démontre que ces dernières ont toutes le même degré: c'est encore le degré de la transformation-limite.

Le degré est toujours un nombre entier fini, positif ou négatif. Le degré d'une transformation *bi*univoque est de +1, si l'indicatrice reste la même, et de −1, si elle est renversée.

Ch11, 17. ... präpariere ich zur Einsendung an die [An]nalenredaktion einen Artikel, in dem ich das Invarianzproblem der Dimension insoweit löse, dass ich zeige dass jedenfalls Räume gerader und ungerader Dimensionenzahl sich nicht stetig und ein-eindeutig auf einander abbilden lassen.

Ch11, 18. *Wenn in einer q-dimensionalen Mannigfaltigkeit bei einer eindeutigen und stetigen Abbildung eines q-dimensionalen Kubus das Maximum der Verrückungen kleiner ist als die halbe Kantenlänge, so existiert ein konzentrischer und homothetischer Kubus, der ganz in der Bildmenge enthalten ist.*

Ch11, 19. Satz 1. *In einer m-dimensionalen Mannigfaltigkeit enthält das eineindeutige und stetige Abbild eines m-dimensionalen Bereiches in beliebiger Nähe eines beliebigen seiner Punkte einen Bereich.*

Ch11, 20. Satz 2. *Eine m-dimensionale Mannigfaltigkeit kann nicht das eineindeutige und stetige Abbild eines Bereiches höherer Dimensionenzahl enthalten.*

Ch11, 21. Satz 3. *In einer m-dimensionalen Mannigfaltigkeit ist das eineindeutige und stetige Abbild eines Bereiches geringerer Dimensionenzahl eine nirgends dichte Punktmenge.*

Ch12, 1. Wir haben in den Ferien eine sehr schöne Reise nach Paris gemacht. Mathematiker freilich habe ich leider nicht gesehen, die waren noch alle in Ferien. D.h. ich habe doch die Bekanntschaft von Lebesgue gemacht, der zufällig in Paris war. Er ist sehr interessant und sagte mir, dass er schon seit 1anger Zeit nicht nur einen, sondern mehrere Beweise des Satzes von der Invarianz der Dimensionenzahl besitzt, den Brouwer jetzt in den Annalen bewiesen hat. Einen dieser Beweise, der sehr witzig aussieht, hat er mir für die Annalen eingeschickt. Ich habe ihn nicht genau auf Richtigkeit der Durchführung, sondern nur auf Richtigkeit der Idee angesehen: in Einzelheiten kann man sich doch auf einen so scharfsinnigen Mann verlassen. Wenn Sie aber noch genau nachprüfen wollen, steht die Arbeit Ihnen zur Verfügung.

Ch12, 2. *Si chaque point d'un domaine D à n dimensions appartient à l'un au moins des ensembles fermés* $E_1, E_2, \ldots, E_p$ *en nombre fini et si ces ensembles sont suffisamment petits, il y a des points communs au moins à* $n+1$ *de ces ensembles.*

Ch12, 3. Na Lebesgues' eerste brief stond vast, dat zijn artikel fout was, en dat hij den goeden weg (n.l. zonder specialiseering der I_p) niet zag.

Ch12, 4. ... *Ihre Note in sehr schroffer Form abgefasst* ist ...

Ch12, 5. En réalité, j'avais toujours pensé à des séparations en régions de $I, I_1, \ldots$.

Ch12, 6. Disons que deux variétés T_n et T_p de l'espace E_{n+p+1} sont *enlacées* si, t_n et t_p étant deux variétés polygonales correspondant à T_n et T_p et trés voisines de celles-ci, au cours de la déformation continue réduisant t_n, par exemple, à un point non situé sur t_p, il y a un nombre impair de traversées de t_p par t_n.

Ch12, 7. *Si dans un espace* E_{n+p+1} *on a une variété* T_n, *il existe une variété* T_p *enlacée avec* T_n.

Ch12, 8. En particulier, on déduit de là qu'une T_n ne remplit pas E_{n+p+1} ($p \geqq 0$) et divise E_{n+p+1} en régions pour $p = 0$ et seulement dans ce cas. L'invariance du nombre de dimensions résulte de là de trois manières.

Ch12, 9. „Zu jeder in einem R_{n+p+1} liegenden regulären, geschlossenen Mannigfaltigkeit T_n existiert eine beliebig kleine, mit T_n verschlungene Mannigfaltigkeit T_p." Aus diesem Hilfssatze wird dann gefolgert, dass T_n den R_{n+p+1} nicht überall dicht erfüllt.

Ch12, 10. Nun möchte ich aber wissen: Ist, die Lösung des Dimensionsproblems vorausgesetzt, der Satz über die Variétés enlacées richtig und streng bewiesen? Ich wage es nicht zu entscheiden: es wäre fast zu schön, wenn der Satz wahr wäre.

Ch12, 11. Meine Stellung dem zweiten Teile der Lebesgueschen Note gegenüber ist also diese, dass er für den dreidimensionalen Raum einen sehr schönen Satz ganz richtig bewiesen hat, dass er für höhere Räume aber nur „evidente" Erweiterungen ausgesprochen hat, ohne etwas zu beweisen.

Ch12, 12. Un continu a n dimensions quand on peut le décomposer en plusieurs parties en y pratiquant une ou plusieurs coupures qui soient elles-mêmes des continus à $n-1$ dimensions ....

Ch12, 13. „*Eine Normalmenge* (im *Fréchet*schen Sinne) π *heiße ein Kontinuum, wenn es für je zwei ihrer Elemente* m_1 *und* m_2 *eine zusammenhängende, abgeschlossene Menge gibt, welche Teilmenge von* π *ist und* m_1 *und* m_2 *enthält.*"

Ch12, 14. Es sei π irgendeine Normalmenge, π_1, ϱ, und ϱ' drei Teilmengen von π, welche innerhalb π abgeschlossen sind und keine gemeinsamen Punkte besitzen. Alsdann heißen ϱ und ϱ' *in* π *durch* π' *getrennt*, wenn jede zusammenhängende, abgeschlossene Teilmenge von π, welche sowohl mit ϱ wie mit ϱ' Punkte gemeinsam hat, auch von π_1 mindestens einen Punkt enthält.

Ch12, 15. 'ϱ und ϱ' heissen in π durch π_1 getrennt, wenn $\pi - \pi_1$ sich in zwei Teilmengen, λ und λ', spalten lässt, die ϱ beziehungsweise ϱ' enthalten und von denen keine einen Grenzpunkt der anderen enthält.'

Ch12, 16. „wenn π_1 in π eine ϱ enthaltende, aber ϱ' nicht enthaltende Gebietsmenge bestimmt".

Ch12, 17. Der Ausdruck: „π *besitzt den allgemeinen Dimensionsgrad* n", in welchem n eine beliebige natürliche Zahl bezeichnet, soll nun besagen, daß für jede Wahl von ϱ und ϱ' eine trennende Menge π_1 existiert, welche den allgemeinen Dimensionsgrad $n-1$ besitzt, daß aber nicht für jede Wahl von ϱ und ϱ' eine trennende Menge π_1 existiert, welche einen geringeren allgemeinen Dimensionsgrad als $n-1$ besitzt. Weiter soll der Ausdruck: „π *besitzt den allgemeinen Dimensionsgrad Null bzw. einen unendlichen allgemeinen Dimensionsgrad*" bedeuten, daß π kein Kontinuum als Teil enthält, bzw. daß zu π weder die Null noch irgendeine natürliche Zahl als ihr allgemeiner Dimensionsgrad gefunden werden kann.

Ch12, 18. Wenn zu einem Punkte P von π Umgebungen, welche den allgemeinen Dimensionsgrad m, aber keine Umgebungen, welche einen geringeren allgemeinen Dimensionsgrad besitzen, existieren, so werden wir sagen, daß π *in* P *den Dimensionsgrad* m *besitzt.*

Ch12, 19. **Dimensionssatz.** *Eine* n*-dimensionale Mannigfaltigkeit besitzt den homogenen Dimensionsgrad* n[**)].

[**)] Weil der Dimensionsgrad offenbar eine Invariante der Analysis Situs ist, so ist im Dimensionssatz die Invarianz der Dimensionenzahl enthalten.

Ch12, 20. Unter einer n*-dimensionalen Pseudomannigfaltigkeit* verstehen wir eine von derart aneinander schlissenden n-dirnensionalen *Elementen* ... gebildete Punktmenge, daß jede $(n-1)$-dimensionale Seite zwei und nur zwei Elementen gemeinsam ist, während jeder nicht in einer $(n-1)$-dimensionalen Seite enthaltene Punkt nur zu einem einzigen Elemente und jeder andere Punkt nur zu einer endliehen Zahl von Elementen gehört. Schließlich sollen je zwei Punkte der Punktmenge sich durch einen ganz der Punktmenge angehörigen und keine $(n-p)$-dimensionale Seite $(p>1)$ treffenden einfachen Kurvenbogen verbinden lassen.

Ch12, 21. Unter einem *n-dimensionalen Fragment* verstehen wir eine von einer endlichen Zahl von einer oder mehreren n-dimensionalen Pseudomannigfaltigkeiten angehörigen Elementen gebildete Punktmenge.

Ch12, 22. *Hilfssatz. Es sei σ ein simplizial zerlegtes n-dimensionales Element mit den Eckpunkten $E_1, E_2, \ldots, E_{n+1}$; γ ein aus Grundsimplexen von σ gebildetes Fragment, das alle an E_1, aber kein an $E_2E_3 \ldots E_{n+1}$ grenzendes Grundsimplex von σ enthält; σ_1 der innerhalb σ liegende Teil der Grenze von γ; γ_1 ein aus Grundsimplexen von σ_1 gebildetes Fragment, das alle an E_1E_2, aber kein an $E_1E_3 \ldots E_{n+1}$ grenzendes Grundsimplex von σ_1 enthält; σ_2 der innerhalb σ_1 liegende Teil der Grenze von γ_1; γ_2 ein aus Grundsimplexen von σ_2 gebildetes Fragment, das alle an $E_1E_2E_3$, aber kein an $E_1E_2E_4 \ldots E_{n+1}$ grenzendes Grundsimplex von σ_2 enthält; σ_3 der innerhalb σ_2 liegende Teil der Grenze von γ_2; usw. Alsdann kann von den Punktmengen $\sigma_1, \sigma_2, \sigma_3, \ldots, \sigma_n$ keine verschwinden.*

Ch12, 23. Dieser Hilfssatz, auf den schon *Lebesgue* in Math. Annalen 70 die Invarianz der Dimensionenzahl zurückgeführt hat, dessen Beweis daselbst aber eine wesentliche Lücke aufweist*), leuchtet unmittelbar ein, wenn wir den von mir in Math. Annalen 71 eingeführten Begriff des *Abbildungsgrades* heranziehen.

Ch12, 24. . . . Die den Leser zu vielen eigenen Ergänzungen zwingende Kürze der Brouwerschen Publikationen ist bei dem Mangel einer sonstigen einwandfreien und ausführlichen Darstellung sehr zu bedauern.

Ch12, 25. . . . In viel höherem Maße, als aus den Zitaten hervorgeht, bin ich dabei durch die in den letzten Jahren erschienenen grundlegenden topologischen Untersuchungen Brouwers, deren gedankliche Schärfe und Konzentration man bewundem muß, gefördert worden; und im stillen hoffe ich, daß etwas von dem Geist, der die Arbeiten dieses Forschers beseelt, auch in diesem meinem Buche lebendig geworden ist.

Ch13, 1. XIII. Für jedes geschränkte, einfach zusammenhängende Gebiet $\mathfrak{G}$ gibt es eine geschlossene Kurve $\mathfrak{C}$, die mit seiner vollen Grenze identisch oder aber eine Teilmenge der vollen Grenze ist.

Ch13, 2. Unter einem *uneigentlichen Kurvenbogen* möchte ich einen solchen durch zwei Schnitte definierten Kurvenbogen verstehen, welche[r] die ganz geschlossene Kurve enhält, und im entgegengesetzten Falle sprechen von einem *eigentlichen Kurvenbogen*.

Ch13, 3. Soient un ensemble continu E, a et b deux points de l'ensemble. Je dirai que *cet ensemble est* IRRÉDUCTIBLE *entre a et b s'il n'existe aucune portion continue de E qui contienne a et b.*

Ch13, 4. Soient C un continu irréductible entre a et b, c un point quelconque de C; *il est possible et d'une seule manière de décomposer C en deux continus ayant c pour seul point commun* (et qui, par suite, seront irréductibles l'un entre a et c, l'autre entre b et c).

Ch13, 5. . . . la frontière extérieure d'un domaine peut se décomposer en deux continus ayant seulement deux points communs a et b.

Ch13, 6. *L'ensemble des points qui n'appartiennent pas à un continu irréductible donné est un continuum.*

Ch13, 7. Lemme. — *Soit $\mathfrak{C}$ un continu, $\mathfrak{F}$ un sous-ensemble fermé de $\mathfrak{C}$, A un point de $\mathfrak{C} - \mathfrak{F}$ et $\mathfrak{A}$ l'ensemble de tous les points de $\mathfrak{C} - \mathfrak{F}$, bien enchainés avec A; l'ensemble dérivé de $\mathfrak{A}$, qui est continu, a des points communs avec $\mathfrak{F}$.*

Ch13, 8. . . . *un exemple de courbe plane C_0, cantorienne et jordanienne simultanément, et telle que C étant une courbe cantorienne plane arbitrairement donnée d'advance, il en existe une image biunivoque et continue C' dont tous les points sont des points de la courbe C_0.*

Ch13, 9. THÉORÈME. *Soit C un continu, p et q deux de ses points; il existe au moins un continu irréductible pq, sous-ensemble de C.*

Ch13, 10. *Tout ensemble continu tiré d'un espace euclidien à n dimensions peut être décomposé en deux ensembles punctiformes (c.-à-d. ne contenant aucun sous-ensemble continu).*

Ch13, 11. THÉORÈME. *Afin qu'un continu Γ arbitraire de l'espace euclidien à n dimensions soit une courbe de Jordan (c.-à-d. que les coordonnées des points de Γ soient des fonctions continues d'un paramètre t et univoques dans un intervalle fermé), il est nécessaire et il suffit*

I) *que Γ soit borné;*
II) *qu'on puisse pour tout couple de points A, B de Γ trouver un arc simple (AB) continu dans Γ et joignant les deux points;*
III) *que le diamètre de (AB) tende vers zéro lorsque le point B tend vers A.*

Ch13, 12. THÉORÈME. *Un continu borné Γ de l'espace euclidien à m dimensions est une courbe de Jordan si on peut faire correspondre à tout couple a, b de ses points un continu $C(a, b)$ contenant a et b, contenu dans Γ et tel que le diamètre du continu $C(a, b)$ tende vers zéro avec la distance entre les points a et b.*

Ch13, 13. ... théorème ...

1) une courbe qui remplit un domaine à n dimensions a nécessairement des points multiples d'ordre $n + 1$ au moine et
2) il y a des courbes remplissant un domaine à n dimensions qui n'ont pas de points multiples d'ordre supérieur à $n + 1$.

Ch13, 14. Etant donné un ensemble A (fermé), on prend en considération l'ensemble de toutes les courbes continues:

$$J: \quad x_i = f_i(t), \ i = 1, 2, \ldots, n, \quad t \subset E,$$

représentant A (où E est un ensemble fermé, linéaire, punctiforme). Cet ensemble sera designé par $Z(A)$.

Il existe alors un seul nombre entier d tel que:

a) *toutes les courbes de l'ensemble $Z(A)$ ont des points multiples d' ordre $d + 1$ au moins;*
b) *une au moins des courbes de l'ensemble $Z(A)$ ne possède pas de points multiples d'ordre supérieur à $d + 1$.*

Le nombre d sera appelé *la dimension* de l'ensemble A.

Ch13, 15. Nous disons qu'un ensemble parfait et borné est d'un seul tenant lorsqu'il ne peut être décomposé en plusieurs ensembles parfaits séparés.

Ch13, 16. ***Définition.*** A étant un continu et contenant les points x et y, j'appelle *distance entre x et y relative au continu A* la borne inférieure des nombres $\delta[\mathfrak{C}(x + y, A)]$. Je désignerai ce nombre par $\varrho_A(x, y)$.

Ch13, 17. ***Définition.*** A étant un continu et a un point de A, j'appelle *oscillation du continu A au point* a le nombre

$$\sigma_A(a) = \overline{\lim_{x,y \to a}} \varrho_A(x, y).$$

Ch13, 18. ***Théorème V.*** *Pour qu'un continu borné A soit une ligne de Jordan, il faut et suffit que tous les points de A soient de premier genre.*

Ch13, 19. Eine Menge M heißt *zusammenhängend im kleinen im Punkte P*, wenn zu jedem $\varepsilon > 0$ ein $\eta > 0$ von folgender Eigenschaft gehört: jeder in der Umgebung η von P liegende Punkt von M ist mit P verbunden durch ein ganz in der Umgebung ε von P liegendes Kontinuum von M.

Die Menge M heißt *zusammenhängend im kleinen*, wenn sie in jedem ihrer Punkte zusammenhängend im kleinen ist.

Ch13, 20. III. *Ist das geschränkte Kontinuum M zusammenhängend im kleinen, so gehört zu jedem $\varepsilon > 0$ ein $\eta > 0$ von folgender Eigenschaft: Je zwei Punkte P' und P'' von M, deren Abstand $< \eta$ ist, sind verbunden durch einen zu M gehörigen stetigen Kurvenbogen, der ganz in der Umgebung ε jedes der beiden Punkte P' und P'' liegt.*

Ch13, 21. XIV. *Damit eine ebene Punktmenge M stetiges Bild einer abgeschlossenen Strecke sei, ist notwendig und hinreichend, daß M ein geschränktes, im kleinen zusammenhängendes Kontinuum sei.*

Ch13, 22. *deux points d'un ensemble dispersé, sont-ils nécessairement séparés dans cet ensemble?*

Ch13, 23. P étant un ensemble dont tous deux points sont séparés dans P, a étant un point donné de P et ε un nombre positif donné, peut-on toujours décomposer P en deux ensembles séparés A et B de sorte que A contienne a et que le diamètre de A soit $< \varepsilon$?

Ch13, 24. Nous avons donc *un exemple d'un ensemble plan dispersé et tel que certains deux de ses points ne sont pas séparés dans cet ensemble.*

Ch13, 25. Nous avons ainsi *un exemple d'un ensemble plan G_δ punctiforme qui n'est pas homéomorphe avec aucun ensemble linéaire.*

Ch13, 26. **Théorème:** *Pour qu'un ensemble punctiforme P, situé dans l'espace à m dimensions, soit homéomorphe avec un ensemble linéaire, it faut et il suffit que pout tout point p de P et pour tout nombre positif ε l'ensemble P soit une somme de deux ensembles séparés $P = A + B$, tel que A contienne p et que le diamètre de A soit $< \varepsilon$.*

Ch13, 27. *Pour qu'un ensemble punctiforme soit homéomorphe à un ensemble linéaire il faut et il suffit qu'il soit de dimension* 0.

Ch13, 28. **Théorème:** *Si P est un ensemble punctiforme, situé dans l'espace à $m > 1$ dimensions, le complémentaire de P est un ensemble connexe.*

Ch13, 29. **Théorème.** *Tout ensemble frontière dans l'espace à m dimensions est homéomorphe d'un ensemble non dense, situé dans le même espace.*

Ch13, 30. Wir geben keine Definition des Begriffs Kurve; die Mengen, die herkömmlicherweise diesen Namen führen, sind von so heterogener Beschaffenheit, daß sie unter keinen vernünftigen Sammelbegriff fallen.

Ch13, 31. . . . par une singulière étourderie, en effet, j'ai admis presque entièrement la propriété à démontrer à un moment de mon raisonnement.

Ch13, 32. Gegeben sei eine simpliziale Zerlegung von Σ in die Teilsimplizes $\tau_1, \tau_2, , \ldots, \tau_p$. Ein Eckpunkt eines τ_k kommt dann in gewissen $\mathfrak{M}_i$ vor, etwa in $\mathfrak{M}_{\nu_1}, \mathfrak{M}_{\nu_2}, \ldots, \mathfrak{M}_{\nu_q}$, wo $\nu_1 < \nu_2 < \cdots < \nu_q$ sein soll. Dann ordnen wir diesem Eckpunkt die Koordinate ν_1 zu und verfahren so mit allen Eckpunkten der Teilsimplizes. Jedem Teilsimplex τ_k ist auf diese Weise ein System von $n+1$ Koordinaten zugeordnet und jeder Seite eines Teilsimplex ein System von n Koordinaten. Dann behaupten wir:

Die Anzahl der τ_k, denen das System $(1, 2, 3, \ldots, n + 1)$, d.h. das mit lauter verschiedenen Ziffern, zugeordnet ist, ist ungerade.

Ch14, 1. Павел Самуилович сразу же стал упорно думать над этими задачами. Очень скоро предметом его размышлений стало общее определение размерности; все лето 1921 г. прошло в напряженных попытках найти «настоящее» определение, причем Павел Самуилович переходил от одного варианта к другому, постоянно строя примеры, показывавшие, почему тот или иной вариант надо отбросить. Это были два месяца всепоглощающих размышлений. Наконец, в одно утро в конце августа Павел Самуилович проснулся с готовым, окончательным и всем теперь хорошо известным индуктивным опрелелением размерности. Это было в деревне Бурково вблизи Болшева, на берегу реки Клязьмы; там группа тогда молодых московских математиков—в том числе В. В. Степанов, Д. Е. Меньшов, Н. К. Бари—жила на даче. В то же утро, во время купанья в Клязьме П. С. Урысон рассказал мне свое определение размерности и тут же, во время этого разговора, затянувшегося на несколько часов, набросал план всего построения теории размерности с целым рядом теорем, бывших тогда гипотезами, за которое не известно было, как и взяться, и которые затем доказывались одна за другой в течение последующих месяцев. Вся набросанная тогда программа полностью осуществилась в течение зимы 1921/22 г.; к весне 1922 г. вся теория размерности была готова. Оказалось, что можно естественным образом поставить в соответсвие каждому точечому множеству R (каждому метрическому пространству) некоторое неотрицательное число dim R — размерность этого множества; все дело, конечно, в плодотворности нового определения, в том, что оно явилось источником установления целого ряда новых свойств точечных множеств, что в свою очередь повело к совершенно новым топологическим задачам.

Ch14, 2. *Indiquer les ensembles les plus généraux qui méritent encore d'être appelés lignes, surfaces etc.*

Ch14, 3. Déf.1_0. *Si le point x de C peut être, quel que soit ε, ε-séparé par l'ensemble vide, ce point sera dit de dimension* 0 *(par rapport à C):*

$$\dim_x C = 0.$$

Déf. 2_0. *Si tous les points de C sont de dimension* 0 *par rapport à C, C aura lui-même la dimension* 0*:*

$$\dim C = 0.$$

Les dimensions supérieures seront définies par induction. Supposons, en effet, que les points et les ensembles de dimension $< n$ soient déjà définis. En ce cas nous dirons que

Déf. 1_n. *Un point x de C qui n'est pas de dimension $< n$ (par rapport à C) mais peut être, quel que soit ε, ε-séparé par un ensemble B de dimension $< n$, ce point est de dimension n:*

$$\dim_x C = n.$$

Déf. 2_n. *Un ensemble C ne contenant que des points de dimension $\leq n$ sera dit de dimension n*

$$\dim C = n$$

s'il contient au moins un point x tel que $\dim_x C = n$.

On définit ainsi, de proche en proche, la dimension pour tout n naturel. Il peut d'ailleurs arriver qu'un point x (ou un ensemble C) ne reçoive de cette manière aucune dimension; nous dirons alors *ce point* (ou cet ensemble) *est de dimension infinie:*

$$\dim_x C = \infty, \text{ ou } \dim C = \infty.$$

Ch14, 4. Déf. *Un continu de dimension* 1 *sera dit une ligne Cantorienne.*

Ch14, 5. **Théorème $\mathbf{A_3}$.** *Soit C un ensemble de E_3. Tous point intérieur de C est de dimension* 3 *(par rapport à C); tout point frontière non limite de points intérieurs, de dimension* 0, 1 *ou* 2*; enfin un point frontière limite de points intérieurs peut être d'une dimension quelconque* ≤ 3.

Ch14, 6. Déf 3_2. *Un continu K de dimension* 2 *sera dit une surface Cantorienne s'il reste connexe après la suppression des points d'un ensemble punctiforme fermé arbitraire.*

Ch14, 7. Déf 3_n. *Un continu de dimension n sera dit une multiplicité Cantorienne (de dimension n) s'il reste connexe après la suppression d'un ensemble fermé de dimension $n-2$ quelconque.*

Ch14, 8. **Théorème $\mathbf{A_n}$.** *Soit C un ensemble de E_n. Tous point intérieur de C est de dimension n (par rapport à C); un point-frontière non limite de points intérieurs, de dimension* $0, 1, \ldots, n-2$ *ou* $n-1$*; enfin, un point-frontière limite de points intérieurs peut être d'une dimension quelconque* $\leq n$.

Ch14, 9. **Théorème.** *Quels que soient les ensembles L et M, on a toujours*

$$\dim(L + M) \leq \dim L + \dim M + 1$$

Ch14, 10. **Théorème II.** *L'ensemble-somme d'un nombre fini ou d'une infiniité dénombrable d'ensembles $\mathfrak{F}_\sigma$ de dimension $\leq n$, est, lui aussi, un $\mathfrak{F}_\sigma$ de dimension $\leq n$.*

Ch14, 11. **Théorème I.** *Si un ensembles (quelconque) de dimension n est décomposé en une somme d'ensembles de dimension* 0*, cette somme est constituée d'au moins $n+1$ termes.*

Ch14, 12. Je ne suis arrivé à démontrer ce théorème inverse dans le cas général; je l'etablirai dans le cas particulier des ensembles fermés et, plus généralement, des ensembles $\mathfrak{F}_\sigma$.

Ch14, 13. **Théorème III.** *Tout $\mathfrak{F}_\sigma$ de dimension n est la somme de $n+1$ ensembles de dimension* 0.

...

Corollaire. ... *La condition nécessaire et suffisante pour qu'un ensemble fermé (ou un $\mathfrak{F}_\sigma$) soit de dimension n, consiste en ce qu'on puisse la décomposer en $n+1$ ensembles de dimension* 0 *et qu'un ne puisse pas le décomposer en n ensembles de cette sorte.*

Ch 14, 14. *Le plan Euclidien E_2 étant un $\mathfrak{F}_\sigma$ de dimension* 2*, il peut être décomposé en trois ensembles de dimension* 0*, mais il ne peut être décomposé en deux ensembles de cette espèce.* Cet énoncé est à rapproacher du théorème suivant de M. Sierpiński [Sierpiński, 1923]:

E_2 peut être décomposé en trois ensembles homéomorphes à des ensembles linéarires, mais il ne peut être décomposé en deux ensembles de cette sorte.

Ch15, 1. Wir definieren: Ein Kontinuum $\mathfrak{K}$ heißt Kurve, wenn in jeder Umgebung $\mathfrak{U}$ jedes der Punkte P von $\mathfrak{K}$ eine Umgebung $\mathfrak{U}_1$ enthalten ist, so daß der Durchschnitt $\mathfrak{D}$ von $\mathfrak{K}$ mit der Begrenzung von $\mathfrak{U}_1$ keinen zusammenhängenden Teil enthält.

Ch15, 2. Wir nennen *Kurve* ein Kontinuum K von der Eigenschaft, daß, wenn p ein beliebiger Punkt von K und U_1 eine offene Umgebung von p ist, eine offene Umgebung von p $U_2 < U_1$ existiert, mit deren Begrenzung K keine zusammenhängenden Teil gemein hat.

Ch15, 3. *Definition*: Wir nennen *nulldimensional* eine nichtleere Menge, welche keinen zusammenhängenden Teil enthält.

Ch15, 4. Damit eine Menge M nulldimensional sei, ist notwendig und hinreichend, daß für jeden ihrer Punkte p zu jeder offenen Umgebung U_1 von p eine offene Umgebung von p $U_2 < U_1$ existiert, mit deren Begrenzung M keinen Punkt gemein hat.

Ch15, 5. Eine nicht leere Menge M des R_m heißt *n-dimensional*, wenn

1.) zu jedem ihrer Punkte P und jeder offenen Umgebung $U_1(P)$ eine offene Umgebung $U_2(P) < U_1(P)$ existiert, mit deren Begrenzung M eine höchstens $(n-1)$-dimensionale Menge gemein hat; und wenn M
2.) mindestens einen Punkt Q enthält, für den eine offene Umgebung U existiert derart, daß M mit der Begrenzung jeder offenen Umgebung $U_1(Q) < U$ eine Menge gemein hat, die einen $(n-1)$-dimensionalen Teil enthält. -1-dimensional ist die leere Menge.

Genau (ev.[entuell] „homogen") n-dimensional heiße eine Menge, die nur Punkte Q mit der Eigenschaft 2. enthält. (Jede null-dimensional Menge ist genau nulldimensional.) Jeder Teil einer n-dimensionalen Menge ist höchsten n-dimensional. *n-dimensionaler Kern* einer n-dimensionalen Menge M heiße die Vereinigung aller genau n-dimensionalen Teile von M.

Ch15, 6. *Satz II.* Ist die Menge M in ihrem Punkte m mindestens n-dimensional, so enthält jede Umgebung $U(m)$ eine in M mindestens n-dimensionale Menge von Punkten aus M, in denen M mindestens n-dimensional ist.

Ch15, 7. Ein Kontinuum K eines metrischen Raumes heißt *Kurve*, wenn für jeden Punkt k von K und zu jeder Umgebung $U(k)$ eine Umgebung $U'(k) < U(k)$ existiert, so daß der Durchschnitt von K mit der Begrenzung von $U'(k)$ zusammenhangslos ist. Dabei nennen wir *zusammenhangslos* eine Menge, deren sämtliche Komponenten nur je einen Punkt enthälten.

Ch15, 8. *V. Eine Kurve des R_n $(n > 1)$ enthält keine inneren Punkt.*

Ch15, 9. *VIa. Die Kurven des R_2 sind identisch mit den Kontinua ohne inneren Punkt.*

Ch15, 10. Als ich 1921 die Kurven und den Dimensionsbegriff zu definieren versuchte, war ich im ersten Jahr meines Universitätsstudiums und hatte keinerlei Kenntnis von Ihrer Abhandlung, verehrter Herr Professor, im *Journal für die reine und angewandte Mathematik* 142, in welcher die Definition im Wesentlichen vorweggenommen ist. Aber auch nachdem ich später beim Studium der einschlägigen Literatur den Aufsatz gefunden hatte, hoffte ich durch meine Resultate wenigstens eine kleine Ergänzung desselben zu bieten.

Ch15, 11. Wie es scheint, haben wir fast gleichzeitig die Dimensionsdefinition gefunden: die meinige habe ich Juli 1921 gefunden, October 1921 der Moskauer Mathematischen Gesellschaft vorgetragen.

Ch15, 12. *Eine Menge A heißt n-dimensional, wenn n die kleinste ganze, nicht negative Zahl ist von folgender Eigenschaft:*

Zu jedem Punkt a von A existiert eine auf a sich zusammenziehende Folge von Umgebungen mit höchstens $(n-1)$-dimensionalen Begrenzungen; d.h. zu jedem Punkt a von A und zu jeder Umgebung $U(a)$ existiert eine Umgebung $U_1(a) < U(a)$ mit höchstens $(n-1)$-dimensionaler Begrenzung. (-1)-dimensional und höchstens (-1)-dimensional ist die leere Menge und nur diese.

Dabei heißt *höchstens n-dimensional* eine Menge, die n-dimensional oder weniger als n-dimensional ist, während wir eine Menge, die nicht höchstens n-dimensional ist, als *mindestens $(n+1)$-dimensional* bezeichnen wollen.

Ch15, 13. *Dabei bezeichnen wir in üblicher Weise als Kontinuum eine mehr als einen Punkt enthaltende abgeschlossene Menge, die sich nicht spalten läßt in zwei fremde abgeschlossene Teile, es sei denn, daß einer von beiden leer ist.*

Ch15, 14. *Ein Kontinuum K heißt Kurve, wenn zu jedem Punkt von K beliebig kleine Umgebungen existieren, deren Begrenzungen keine Kontinua enthalten. Ein Kontinuum K, welches in einen Raum eingebettet ist, heißt Kurve, wenn zu jedem Punkt von K beliebig kleine Umgebungen existieren, deren Begrenzungen keine Kontinua mit K gemein haben.*

Ch15, 15. In Euklidischen (und noch allgemeineren) Räumen sind die Kurven identisch mit den eindimensionalen Kontinua. Dabei nennen wir eine nicht leere Menge M *nulldimensional* (oder auch total zusammenhanglos), wenn zu jeden Punkt von M beliebig kleine Umgebungen mit leere (bzw.

wenn M in einen Raum eingebettet ist, mit zu M fremden) Begrenzungen existieren; — und wir nennen eine nicht leere und nicht nulldimensionale Menge M *eindimensional*, wenn zu jeden Punkt von M beliebig kleine Umgebungen existieren, deren Begrenzungen nulldimensional sind (bzw. wenn M in einen Raum eingebettet ist, mit M nulldimensionale Durchschnitte haben).

Ch15, 16. Ich danke Ihnen für Ihren Brief vom 3.dieses Monats, der mir jedenfalls die Beruhigung bringt, daß Sie den Schicksalsschlag, der Sie so jäh getroffen hat, überstanden haben. Ich war deswegen in grosser Sorge, eben weil ich während Ihres Aufenthalt hier sehr stark empfunden hatte, in wie hohem Masse die Sphäre Ihrer Mutter Ihr Leben durchstrahlte. Ich ahnte also den Umfang Ihres Verlustes, und erwartete die Krise, die die plötzliche Leere, und die plötzliche Notwendigkeit eine andere geistige Atemform anzunehmen, bei Ihren auslösen würde. Nachdem aber diese erste Krise überstanden ist, bin ich sicher, daß Sie die nötige Konzentration und religiöse Hingabe finden werden, um sich weiter durchzuarbeiten, und daß die Sicherheit des dahin gehenden Wunsches der lieben Verstorbenen, sowie die Ihren von ihr bleibende abgeklärte Erinnerung Ihnen dabei helfen wird.

Ch15, 17. *Jeder kompakte eindimensionale metrische Raum ist homöomorph mit einer Teilmenge des dreidimensionalen Zahlenraumes.*

Ch15, 18. *Es existieren umfassendste eindimensionale Mengen*, darunter sogar beschränkte, stetig durchlaufbare Kurven des R_3.

Ch15, 19. **Theorem I.** *In einem kompakten oder halbkompakten metrischen Raum R ist für jedes k die Menge R^k aller Punkte, in denen der Raum mindestens k-dimensional ist, eine in jedem ihrer Punkte mindestens k-dimensionale Menge. Insbesondre ist also in einem kompakten oder halbkompakten n-dimensionalen Raum, die Menge $\overset{n}{R}$ aller Punkte, in denen der Raum n-dimensional ist, eine homogen n-dimensionale (d.h., in jedem ihrer Punkte n-dimensionale) Menge. Die Menge $\overset{n}{R}$ ist in einem kompakten n-dimensionalen Raum R identisch mit der Summe aller homogen n-dimensionalen Teilmengen von R.*

Ch15, 20. **Theorem II.** *Ist R ein separabler metrischer Raum und R^K die Menge aller Punkte, in denen R mindestens k-dimensional ist, dann ist jede nicht-leere in R^k offene Menge mindestens $(k-1)$-dimensional. Insbesondere ist also in einem separabeln n-dimensionalen Raum R die Menge $\overset{n}{R}$ entweder n-dimensional oder $(n-1)$-dimensional. Beide Fälle können sich ereignen.*

Ch15, 21. erst heute schreibe ich Ihnen, da nunmehr die Prioritätsangelegenheit restlos erledigt ist: Es haben sich nicht nur alle von mir erwähnten Beweisstücke, sondern auch noch andere gefunden, die ich selbst schon vergessen hatte. Ich habe das gesamte Material in einem Safe deponiert und werde Ihnen alle Originale persönlich übergeben. Erwähnen möchte ich heute nur kurz die *amtlich* beglaubigten Stücke:

Ch.15, 22. Von dort bin ich also doch wieder nach Amsterdam. Ich lese über Variationsrechnung. Persönlich bin ich mit Geometrie aller Arten beschäft-igt, ferner mit Erkenntnistheorie. Ich hoffe ich werde mich dazu aufschwingen, meine Absichten über das Wahrheitsproblem etc[.] zusammenzustellen. Seit einige Wochen ist mir so vielerlei eingefallen, so daß ich überhaupt nicht dazu Komme etwa nicht zuschreiben und jeden Abend, davonlaufe, um mich zu zerstreuten (ich wohne natürlich in Amsterdam) und nicht zu überarbeiten. Im übrigens fluche ich, daß ich nicht in Wien bin, sondern hier. Ich kann mich hier nicht einleben und will mir alle Mühe geben, daß ich mit Juni für immer weggehe. . . .

Ch16, 1. *Struktureigenschaften der Mengen an sich, ohne Rücksicht auf irgendwelche Transformationen der Gebilde.*

Ch16, 2. **Summensatz.** *In einem kompakten metrischen Raum ist die Summe von abzählbar vielen abgeschlossenen n-dimensionalen Mengen n-dimensional.*

Ch16, 3. ... legen wir allgemein einen Bereich $\mathfrak{N}$ von Punktmengen zugrunde und bezeichnen mit $\mathfrak{N}^*$ den Bereich aller Punktmengen, auf deren sämtliche Punkte sich Relativumgebungen zusammenziehen, deren Begrenzungen Mengen des Bereiches $\mathfrak{N}$ sind. Wir untersuchen dann die Beziehung zwischen den beiden Bereichen $\mathfrak{N}$ und $\mathfrak{N}^*$. Der Bereich $\mathfrak{N}$, von dem wir angehen, wird dabei folgenden Bedingungen unterworfen.

1. Ist M eine Menge aus $\mathfrak{N}$, so auch jede Teilmenge von M.
2. Ist M Summe von abzählbar vielen in M abgeschlossenen Mengen, die alle zum Bereich $\mathfrak{N}$ gehören, dann gehört auch M zum Bereich $\mathfrak{N}$.

Einen Bereich von Punktmengen, welcher diesen beiden Bedingungen genügt, bezeichnen wir kurz als *Normalbereich*, und ein Hauptresultat der vorliegenden Arbeit besagt, daß der nach der obigen Vorschrift aus einem Normalbereich $\mathfrak{N}$ abgeleitete Bereich $\mathfrak{N}^*$ ebenfalls ein Normalbereich ist.

Ch16, 4. *Die nicht leere Menge M eines metrischen Raumes heißt nulldimensional, wenn auf jeden Punkt von M eine Folge von Teilmengen von M sich zusammenzieht, deren Begrenzungen in M leer sind.*

Ch16, 5. ... sagen wir, wenn die Menge M im Punkte p nulldimensional ist, daß M in p *unstetig* oder auch, daß p *Unstetigkeitspunkt* der Menge M sei. Die anderen Punkte des Raumes nennen wir *Stetigkeitspunkte* von M und sagen auch, die Menge M sei *in ihnen stetig*.

Ch16, 6. Wer wenden uns der Aufgabe zu, die Mengen M von folgender Beschaffenheit zu untersuchen: Auf jeden Punkt von M zieht sich eine Folge von in M offenen Mengen zusammen, deren Begrenzungen in M eine gewisse Eigenschaft haben, oder, was auf dasselbe hinausläuft, deren Begrenzungen in M einem gewissen Bereich $\mathfrak{B}$ von Mengen angehören. Wir nennen solche Mengen *total unstetig in bezug auf den Bereich* $\mathfrak{B}$. Die total unstetigen Mengen hinsichtlich des Bereiches $\mathfrak{B}$, welcher bloß die leere Menge enthält, sind die nulldimensionalen Mengen.

Ch16, 7. *Einen Bereich $\mathfrak{N}$ von Mengen nennen wir einen Normalbereich, wenn folgende Bedingungen erfüllt sind:*

1. *Ist M eine Menge aus $\mathfrak{N}$, so auch jede Teilmenge von M.*
2. *Ist M Summe von abzählbar vielen in M abgeschlossenen Mengen, die zum Bereich $\mathfrak{N}$ gehören, dann gehört auch M zum Bereich $\mathfrak{N}$.*

Ch16, 8. *Höchstens n-dimensional* (im Punkte p) heißt in unserer Terminologie *jede Menge, die total unstetig* (unstetig in p) *in bezug auf den Bereich der höchstens $(n-1)$-dimensionalen Mengen, wobei der Bereich der (-1)-dimensionalen Mengen bloß die leere Menge enthält.* Jene höchstens n-dimensionalen Mengen, die nicht auch höchstens $(n-1)$-dimensional sind, heißen n-dimensional.

Ch16, 9. **Theorem IV.** *Die separablen höchstens n-dimensionalen Mengen bilden einen Normalbereich.*

Ch16, 10. **Theorem.** *Jeder separable metrische Raum R ist homöomorph mit einer Teilmenge eines kompakten metrischen Raumes, welche dieselbe Dimension besitzt wie R.*

Ch16, 11. (A) *Zerlegt man ein beschränktes abgeschlossenes Gebiet des Euklidischen n-dimensionalen Zahlraumes R_n in endlich viele abgeschlossene Mengen, deren Durchmesser hinreichend klein sind, so gibt es Punkte, die mindestens $n+1$ dieser Mengen gemeinsam sind.*

Ch16, 12. (A′) *Unter der Voraussetzung, daß keiner der Intervallkomplexe D_i Punkte auf zwei gegenüberliegenden Randseiten von I besitzt, gibt es Punkte, die $n+1$ der Mengen D_i gemeinsam sind.*

Ch16, 13. **Theorem I.** *Ist der separable n-dimensionale Raum R eindeutig und beiderseits stetig auf den n^*-dimensionalen Raum R^* abgebildet, wobei $n^* \geq n$ gilt, dann gibt es in R^* Punkte, die Bilder von mindestens $n^* - n + 1$ verschiedenen Punkten von R sind.*

Ch16, 14. **Theorem II.** *Wird eine Teilmenge des Euklidischen n-dimensionalen Raumes eindeutig und beiderseits stetig auf einen n^*-dimensionalen Raum R^* abgebildet, wobei $n^* > n$ gilt, dann treten in R^* Punkte von einer Vielfachheit $\geq n^* - n + 2$ auf.*

Ch16, 15. **Theorem III.** *Ist der kompakte* n*-dimensionale Raum* R *eindeutig und stetig auf den* n^**-dimensionalen Raum* R^* *abgebildet, wobei* $n^* \leq n$*, dann gibt es in* R^* *Punkte, mit mindestens* $n - n^*$*-dimensionalen Urbildmengen.*

Ch16, 16. (I) *Zu einer* n*-dimensionalen Teilmenge eines kompakten Raumes gibt es eine Folge* $\sigma_1, \sigma_2, \sigma_3, \ldots$ *von unendlich fein werdenden Überdeckungen von der Ordnung* $n+1$*, deren jede eine Verzweigung der unmittelbar vorangehenden (und folglich auch aller vorangehenden) ist.*

Ch16, 17. Satz III. *Damit eine (in einem metrischen separablen Raume gelegene) Menge die Dimension* n $(n \geq 0)$ *besitzt, ist notwendig und hinreichend, dass man diese Menge in* $n + 1$ *und nicht weniger Mengen von Dimension Null zerlegen kann.*

Ch16, 18. Satz IV. *Es sei* M *eine beliebige (in einem metrischen separablen Raume gelegene) Menge und* $L_1, L_2, \ldots, L_i, \ldots$ *abzählbarviele, in* M *abgeschlossene, höchstens* $(n-1)$*-dimensionale Teilmengen von* M*. Dann ist auch die Vereinigungsmenge* $\sum_{i=1}^{\infty} L_i$ *höchstens* $(n-1)$*-dimensional.*

Ch16, 19. Satz VIII_1. *Jede in einem separablen vollständigen Raume* R *gelegene Menge* L *ist enthalten in einer* G_δ*-Menge* Λ *der gleichen Dimension wie* L *(wobei* $\Lambda \supset \overline{L}$*):*

$$L \subset \Lambda \subset \overline{L} \subset R, \quad \Lambda \in \mathfrak{G}_\delta \text{ (rel. } R), \quad \dim \Lambda = \dim L.$$

Ch16, 20. *Jeder separable metrische Raum* R *ist kongruent mit einer Teilmenge einer im Urysohnschen Universalraume* U *liegender* G_δ*-Menge* Λ *der gleichen Dimension wie* R*.*

Ch16, 21. **Historisches.** In einer von mir im Februar 1922 bei den Monatsheften für Math. und Phys. eingereichten Note (abgedruckt Proc. Acad. Amsterdam *29*, S. 1124), heißt es wörllich: „*Eine nicht leere Menge [M] des Raumes* R_m *heißt* n*-dimensional, wenn 1. zu jedem ihrer Punkte* p *und jeder offenen Umgebung* $U_1(p)$ *eine offene Umgebung* $U_2(p) \subset U_1(p)$ *existiert, mit deren Begrenzung* M *eine höchstens* $(n-l)$*-dimensionale Menge gemein hat; und wenn* M *2. mindestens einen Punkt* Q *enthält, für den eine offene Umgebung* U *existiert, derart, daß* M *mit der Begrenzung jeder offenen Umgebung* $U_1(Q) \subset U$ *eine Menge gemein hat, die einen* $(n-1)$*-dimensionalen Teil enthält.* (-1)*-dimensional ist die leere Menge.*" Diese Definition des Dimensionsbegriffes findet sich auch in meinen Arbeiten in den Monatsheften f. Math. u. Phys. (*33*, 1923, S.157ff., *34*, 1924, S. 138f.). Daselbst ist auch die oben angeführte formal noch etwas kürzere Definitionsform angegeben: *Ein Raum heißt* n*-dimensional, wenn* n *die kleinste Zahl ist, so daß jeder Punkt des Raumes in beliebig kleinen Umgebungen mit höchstens* $(n-1)$*-dimensionalen Begrenzungen enthalten ist.* (Vgl. ferner für die anschaulich-experimentelle Ableitung dieses Begriffes Math. Annalen *95*, S. 278 und Jahresber. d. Deutschen Math. Vereinigung *35*, S. 118). Bereits in der S. 77 erwähtten Hinterlegung vom Herbst 1921 heißt es wörtlich: „*Ein Kontinuum* K *heißt Kurve, wenn in jeder Umgebung* U *jedes der Punkte* p *von* K *eine Umgebung* U_1 *enthalten ist, so daß der Durchschnitt von* K *mit der Begrenzung von* U_1 *keinen zusammenhängenden Teil enthiält,*" worauf a. a. O. das Prinzip des rekursiven Aufstieges zu höheren Dimensionen angegeben wird.

Im September 1922 veröffentlichte Urysohn (Comptes Rendus *175*, S.440) in wörtlicher Übersetzung folgende Definition, der ein kompakter metrischer Raum zugrunde gelegt ist: Ist C eine gegebene Teilmenge des Raumes, p ein vorgelegter Punkt von C, so sagt man von einer Teilmenge B von C, daß sie den Punkt p ε-aussondert. „*wenn die Menge* $C - B$ *Summe zweier Mengen* A *und* D *ist, so daß 1.* A *und* D *getrennt sind, d. h. daß keine der beiden Mengen einen Punkt oder Häufungspunkt der anderen enthält, daß 2. die Menge* A *den Punkt* p *enthält und daß 3. die Mengen* A *und* C *[B] in einer Kugel vom Radius* $< \varepsilon$ *enthalten sind. Wenn der Punkt* x *der Menge* C *für jedes* ε *durch eine leere Menge* ε*-ausgesondert werden kann, dann heißt* C *im Punkt* x *nulldimensional. Eine Menge, die in allen ihren Punkten nulldimensional ist, heißt nulldimensional. Wenn die Menge* C *im Punkt* x *nicht weniger als* n*-dimensional ist und der Punkt* x *für jedes* ε *durch eine weniger als* n*-dimensionale Menge* ε*-ausgesondert werden kann, dann heißt* C *im Punkt* x n*-dimensional. Wenn die Menge* C *in allen ihren Punkten eine Dimension* $\leq n$ *und in gewissen Punkten die Dimension* n *besitzt, so heißt* C n*-dimensional.*" Diese infolge der Nichtverwendung des Begriffes der

Umgebungsbegrenzung ein wenig verwickelte Definition ist mit der meinen äquivalent, denn man bestätigt leicht, daß jede Teilmenge des Raumes, welche den Punkt p ε-aussondert, nichts anderes ist, als die Begrenzung einer Umgebung von p, die Teil der Sphäre um p mit dem Radius ε ist, und daß umgekehrt die Begrenzung einer jeden Umgebung von p, die Teil dieser Sphäre ist, den Punkt p ε-aussondert. Wenn also ein Punkt p eines metrischen Raumes in beliebig kleinen Umgebungen mit höchstens $(n-1)$-dimensionalen Begrenzungen enthalten ist, so kann er für beliebig kleine ε durch höchstens $(n-1)$-dimensionale Mengen ε-ausgesondert werden und umgekehrt. (Vgl. über diese Definition ferner Urysohn's große Arbeit in den Fundam. Math. *7* und *8*, 1926, 1926).

Ch16, 22. Allen folgenden Überlegungen und Beweisen liegt ausschließlich meine auf den Begriff der Umgebungsbegrenznng gestützte Form der Dimensionsdefinition zugrunde.

Ch16, 23. Es war Poincaré, der mehrmals hervorhob, wie unanschaulich diese arithmetische Dimensionsdefinition sei und wie wenig sie, abgesehen von ihrer speziellen Natur, den Kern des Problems berühre. *„Ich will nicht sagen, daß die Arithmetisierung der Mathematik etwas Schlechtes sei, ich sage, daß sie die Sache nicht erschöpft."* (Letzte Gedanken, Deutsch von Lichtenecker, Leipzig 1913, S. 64.) Und er betonte wiederholt, an Euklids Gedanken anknüpfend, die Notwendigkeit einer rekursiven Definition des Dimensionsbegriffes. *"Ich möchte die Bestimmung der Anzahl der Dimensionen auf den Begriff des Schnittes aufbauen. ... Ein Kontinuum besitzt n Dimensionen, wenn man es in mehrere getrennte Teile zerlegen kann dadurch, daß man einen oder mehrere Schnitte führt, die selbst Kontinua von* $n-1$ *Dimensionen sind."* (a. a. O. S. 65 f.).

Ch16, 24. Brouwer hat (Journal f. d. reine u. angew. Matbem. *142*, 1913, S.146) diese Poincarésche Begriffsbildung einer Kritik und Abänderung unterzogen. Er weist darauf bin, daß der Poincaréschen Definition zufolge ein Doppelkegel, welcher ja durch Weglassung eines einzigen Punktes in zwei getrennte Teile zerlegt werden kann, im Widerspruch zu aller Anschauung als eindimensional bezeichnet werden miüßte und daß ferner der Begriff des Kontinuums und die Worte „ein oder mehrere" in der Poincaréschen Definition nicht präzisiert sind.

Ch16, 25. Dieser Begriff des allgemeinen Dimensionsgrades kommt dem der Dimensionstheorie zugrundeliegenden Dimensionsbegriff bereits sehr nahe. Wenn man nämlich *erstens* die Brouwersche Trennungsdefinition durch die folgende inzwischen entstandene ersetzt: *„Zwei fremde abgeschlossene, Teilmengen A und B des Raumes R heißen durch die Menge C getrennt, wenn R – C in zwei fremde relativabgeschlossene Teilmengen zerfält, von denen die eine A, die andere B als Teilmenge enthält"*, und wenn man *zweitens* nicht von den Mengen ohne Teilkontinua als den nulldimensionalen, sondern wie in der Dimensionstheorie von der leeren Menge als der (-1)-dimensionalen die Rekursion ihren Ausgang nehmen laßt, dann wird der Begriff des Dimensionsgrades mit dem Dimensionsbegriff äquivalent. Es gilt nämlich, wie wir (vgl. S. 117 und S. 120) sehen werden, der Satz, daß in einem n-dimensionalen Raum je zwei fremde abgeschlossene Teilmengen durch eine höchstens $(n-1)$-dimensionale, aber nicht je zwei fremde abgeschlossene Teilmengen durch eine weniger als $(n-1)$-dimensionale Teilmenge trennbar sind.

Ch16, 26. Was *erstens* die anschauliche Vorstellung des Trennens zweier Mengen betrifft, so ist dieselbe übrigens formell und inhaltlich verschiedener Präzisierungen fähig. Wir werden (S. 207) den ursprünglichen Brouwerschen Begriff des Trennens (allerdings unter Zulassung auch von nicht abgeschlossenen trennenden Mengen) als *Separieren* bezeichnen. Andere Trennungsdefinitionen gab Brouwer (Proc. Acad. Amsterdam 27, 1924, S. 636), von dem auch die folgende mit der modernen äquivalente Trennungsdefinition stammt, die Brouwers ursprünglicher Definition sehr ähnlich ist, sich allerdings auf den inzwischen erst entstandenen modernen Zusammenhangsbegriff (vgl. S. 197) beruft: *Zwei fremde Mengen A und B des Raumes R heißen durch die zu A und B fremde Menge C getrennt, wenn jede zwischen A und B zusammenhängende Menge von R einen Punkt von C enthält.* — Was *zweitens* den Ausgangspunkt der Rekursion betrifft, so könnte man (vgl. S. 317) tatsächlich auch, von den diskontinuierlichen Mengen ausgehend, eine mit der Anschauung übereinstimmende Theorie des Dimensionsgrades entwickeln, die hinsichtlich kompakter Räume mit der in diesem Buch entwickelten Dimensionstheorie identisch wäre, sich aber hinsichtlich beliebiger separabler Räume von der Dimensionstheorie stark unterscheiden würde (vgl. S. 318). Wir werden (S. 204) Teilmengen der Ebene kennen lernen, die im Sinne der Dimensionstheorie eindimensional sind, aber keine Teilkontinua enthalten, also den Dimensionsgrad Null haben. Es gibt sogar (vgl. S. 209) im Sinne der Dimensionstheorie unendlichdimensionale Riäume mit dem Dimensionsgrad Null. — Der Begriff des Dimensionsgrades in einem Punkt ist von dem für die Dimensionstheorie grundlegenden Begriff der Dimension in einem Punkt inhaltlich völlig verschieden.

Ch16, 27. Was Du über Brouwer schreibt, über seinen Charakter und seine „Moral", is für mich nichts neues. Trotzdem finde ich Dein Vorgehen ihm gegenüber unrichtig und unrechtmäßig. Was den „Tatbestand" selbst anbetrifft, so ist meiner Ansicht dis historische Übernicht in Deinen Buche von einer subjektiven Darstellung nicht vollkommen frei: Ob Brouwer als Mathematiker oder nur als Vorläufer (so wie Du ihn dargestellt historisches): Dimensionsbegriffes zu gelten hat, ist eine Frage über die man jedenfalls verschieben Meinung sein kann.

...

Gerade deswegen aber und weil Du Brouwers krankhafte Empfindlichkeit kernst, hälbst Du ihn die Korrektur schicken und eventuell man-chen Änderungen, die Dir doch keinen großen Schaden bringen könten, zustimmen sollen, — habt es auf einen Streit ankommen zu lassen, den Dir die Freude an Deinem wunderschönen Buch trüben wird und Dich bestenfalls einer ungebusen Aufwand von Zeit und Energie kosten wird, die wirklich einen besseren Sache wert wären. Ein Streit auf „Leben und Tod" weg dieser Sächenlichkeiten! Dass Du Dir keinen Augenblick gedacht hat, in welch unmögliche Lage bin eurem Streit zwischen Brouwers und Dir ich können musste, dass nehme ich Dir sehr übel.

Ch16, 28. Die Dimensionstheorie habe ich in einem 1913 erschienenen Aufsatz „Ueber den natuerlichen Dimensionsbegriff" (Journ. f. Math. 142. S. 146–152) fuer diejenigen Spezies, fuer welche die Dimension einen natuerlichen Sinn besitzt, naemlich fuer die kondensierten metrischen Spezies, begruendet, indem ich einen *dimensionellen Rechtfertigungssatz* brachte, d.h. einen im Bereich der kondensierten metrisch en Spezies sinnvollen Dimensionsbegriff definierte, fuer den ich zeigen konnte, dass die n-dimensionalen Mannigfaltigkeiten die Dimension n besitzen.

Ch16, 29. In meinem zitierten Aufsatz vom Jahre 1913 wird zwar schon der starke Rechtfertigungssatz dem Wortlaute nach formuliert, der darauf folgende Beweis bezieht sich aber aufs deutlichste auf den schwachen Rechtfertigungssatz, so dass ersichtlich nur dieser gemeint ist, und der in der Formulierung des Satzes enthaltene Schreibfehler jedem mitdenkenden Leser sofort offenbar wird. Dass dieser Schreibfehler lange Jahre hindurch keine gedruckte Verbesserung erfuhr, hatte erstens die allgemeine Ursache der Fahrlaessigkeit muehelos einzusehenden Schreibfehlern gegenueber, zweitens einen in diesen Proceedings 27. S. 136 [S. 636] mitgeteilten speziellen Grund.

Ch16, 30. In meinem zitierten Aufsatz von 1913 habe ich mich auf die Begruendung der Dimensionstheorie beschraenkt und auf die Veroeffentlichung weiterer dimensionstheoretischer Entwicklungen verzichtet, einerseits weil mit dem Beweis des Rechtfertigungssatzes das gestellte erkenntnistheoretische Ziel erreicht war, andererseits weil fuer die anschliessenden Betrachtungen (in erster Linie diejenigen welche sich um den Summensatz und den Zerlegungssatz gruppieren) nicht wie fuer den Beweis der Rechtfertigungssatzes eine intuitionistische Verwirklichung nahe lag.

Ch16, 31. *Jeder eindimensionale kompakte metrische Raum ist homöomorph mit einer Menge des R_3, zu deren sämtlichen Punkten beliebig kleine Intervalle existieren, mit deren Begrenzungen die Menge diskontinuierliche Durchschnitte hat.* Und: *Jeder kompakte eindimensionale Raum ist homö-omorph mit einem Teil einer Fläche des R_3.*

Ch16, 32. In einer folgenden Mitteilung wird bewiesen werden, dass jeder n-dimensionale kompakte (und vermutlich sogar jeder n-dimensionale separable) metrische Raum homöomorph ist mit einer Teilmenge des R_{2n+1}. Ein stetig durchlaufbares Kontinuum, welches ein topologisches Bild jedes kompakten endlichdimensionalen Raumes als Teil enthält, ist dann z. B. die Menge $\sum_{n=1}^{\infty} I_n$, wobei I_n $(n = 1, 2, \ldots)$ ein einen festen Punkt p enthaltendes n-dimensionales Intervall mit der Seitenlänge $\frac{1}{n}$ ist.

Ch16, 33. **Einbettungssatz.** *Ein n-dimensionaler kompakter Raum ist mit einer Teilmenge des R_{2n+1} homöomorph.*

...

Theorem von der topologischen Charakterisiernng der mit Koordinaten beschreibbaren Räume. *Damit ein separabler Raum mit endlich-vielen Koordinaten beschreibbar, d. h. mit einer Teilmenge eines Cartesischen Raumes homöomorph sei, ist notwendig und hinreichend, daß er eine endliche Dimension besitze.*

Ch16, 34. ... daß kompakte metrisierbare topologische Räume sich auf eine bestimmte Weise durch aus endlich-vielen Simplexen zusammengesetzte Gebilde (n-dimensionale Komplexe) approximieren lassen; daß, umgekehrt, jeder auf diese Weise approximierte Raum kompakt und metrisierbar ist; daß dabei ein n-dimensionaler Raum sich durch (im klassischen Sinne) n-dimensional Komplexe approximieren läßt und n-dimensionale Komplexe höchstens n-dimensionale Räume approximieren.

Ch16, 35. Die bisherige Entwicklung der Dimensionstheorie hat im vollen Maße den dieser Theorie zugrunde liegenden Dimensionsbegriff als *den richtigen* Dimensionsbegriff gerechtfertigt: erstens ist dieser Begriff selbst der denkbar einfachste und natürlichste, zweitens hat es sich gezeigt, daß diejenigen Raüme bzw. Mengen, die a priori von der Dimension n sein sollten (z. B. der Euklidische n-dimensionale Raum R^n), auch auf Grund der allgemeinen Dimensionsdefinition die Dimension n erhalten, drittens ließ sich mit Hilfe des allgemeinen Dimensionsbegriffs unsere Erkenntnis des topologischen Aufbau der geometrisch interessanten Raüme, vor allem der Euklidischen Raüme und ihrer abgeschlossenen Teilmengen, wesentlich vertiefen und vermehren.

Ch16, 36. ... daß *die allgemeinen n-dimensionalen abgeschlossenen Mengen geomet-risch dadurch völlig charakterisiert sind, daß sie sich für jedes $\varepsilon > 0$ in einen n-dimensionalen Komplex stetig deformieren lassen, so daß während des ganzen Deformationsprozesses kein Punkt um mehr als ε von seiner versprünglichen Lage entfernt wird.*

Ch16, 37. *Es sei F eine in R gelegene abgeschlossene λ-dimensionale Menge (R ist dabei wie früher R^n, $n \geq \lambda$, oder R^ω). Es sei ferner ε eine beliebige positive Zahl.*

Ch16, 38. I. *Es existiert ein in R gelegener λ-dimensionaler Komplex K_ε^λ, in den F sich ε-überführen läßt. Diese Überführung läßt sich mittels einer stetige Deformation des Raumes R erreichen, wobei während dieses Deformationsprozesses kein Raumpunkt sich um mehr als ε von seiner ursprünglichen Lage entfernt.*

Es existiert aber eine positive Zahl ε_0 von der Art, daß, für $\varepsilon \leq \varepsilon_0$, K_ε^λ in dieser Behauptung nicht durch einen Komplex niedrigerer Dimension ersetzt werden kann.

II. *Es existieren ein in R gelegener λ-dimensionaler Komplex $\tilde{K}_\varepsilon^\lambda$, eine abgeschlossene Umgebung $\overline{U}_\varepsilon$ von F und eine stetige Deformation Δ des ganzen Raumes R in sich, so daß*

1. *alle Punkte von $\overline{U}_\varepsilon$ von F um weniger als ε entfernt sind;*
2. *bei dieser Deformation $\overline{U}_\varepsilon$ in $\tilde{K}_\varepsilon^\lambda$ übergeht;*
3. *während der Deformation Δ kein Punkt von R um mehr als ε verschoben wird und (im Falle, wo R der R^n ist) die von F um mehr als ε entfernten Punkte von R^n überhaupt fest bleiben.*

Es existiert aber eine positive Zahl $\tilde{\varepsilon}_0$ von der Art, daß für $\varepsilon \leq \tilde{\varepsilon}_0$, $\tilde{K}_\varepsilon^\lambda$ in dieser Behauptung nicht durch einen Komplex niedrigerer Dimension ersetzt werden kann.

Ch16, 39. **Brouwersches Invarianzprinzip.** *Es sei F eine abgeschlossene λ-dimensionale Teilmenge des Euklidischen R^n oder des Fundamentalquaders R^ω des Hilbertschen Raumes. Dann läßt sich F für ein hinreichend kleines ε in keine Menge Φ niedrigerer Dimension ε-überführen.*

Ch16, 40. *Die endliche oder unendliche Anzahl von Gebieten, die eine $n-1$-dimensionale abgeschlossene Menge im R^n bestimmt, ist gleich der kleinsten Zahl, die als Zusammenhangszahl eines diese Menge definierenden $n-1$-dimensionalen Projektionsspektrums auftritt.*

Ch16, 41. Verallgemeinerter Urysohnscher Satz: *Die gemeinsame Grenze zweier (oder mehrerer) zusammenhängender Gebiete des R^n' ist eine $n-1$-dimensionale Cantorsche Manniqfaltigkeit, die überdies durch keine einfach-zusammenhängende $n-2$-dimensionale abgeschlossene Teilmenge zerlegt werden kann.*

Ch16, 42. Das allgemeine Problem der *Geometrisierung* der Dimensionstheorie und somit der ganzen mengentheoretischen Topologie involviert aber neue Fragestellungen, die durch die Approximationssätze allein noch garnicht angegriffen werden, und die für die ganze Theorie von einer wesentlichen Bedeutung zu sein scheinen.

Ch16, 43. *Die mengentheoretische Dimensionsbegriff ist eine Vertiefung und Verallgemeinerung der negativen Jordanschen Zusammenhangsdefinition* ....

Ch16, 44. **Definition der geometrischen Dimension eines Raumes.** *Es sei F ein beliebiger kompakter metrischer Raum. Wenn r die größte nicht negative ganze Zahl ist von der Beschaffenheit, daß ein r-dimensionaler nicht identisch verschwindender Zyklus in F existiert und daselbst homolog Null ist, so heißt die Zahl $r+1$ die geometrische Dimension von F. Wenn selbst für $r=0$ kein nicht identisch verschwindende r-dimensionale Zyklus in F homolog Null ist, ist die geometrische Dimension von F gleich Null; wenn für jedes r nicht identisch verschwindende r-dimensionale berandende Zyklen in F auftreten, ist die geometrische Dimension von F definitionsgemäß gleich Unendliche.* Endlich werden wir (wie in der mengentheoretischen Dimensionstheorie) der leeren Mengen die Dimension -1 zuschreiben.

Ch16, 45. Die Untersuchungen der letzten zwanzig Jahre haben mehrere Teile der Topologie zu einem einheitlichen Ganzen zusammenschmelzen lassen, welches man in seinen inneren Zusammenhängen heute wirklich versteht und übersieht. Das so entstandene Stück der mathematischen Wissen-schaft enthält nicht nur den klassischen Stoff der „elementaren" Topologie (so wie er zum erstenmal in Veb-lens Buch „Analysis Situs" zusammengestellt worden ist), sondern auch die ganze Zerlegungs- und Verschlingungstheorie (vom Jordanschen Satz bis zu den neuesten Verallgemeinerungen des Alexanderschen Dualitätssatzes), sowie große Teile der Theorie der stetigen Abbildungen, insbesondere den Abbildungsgrad und die vollständige Lösung des Problems über die algebraische Anzahl von Fixpunkten. Obwohl die sogenannten klassischen Probleme der Analysis Situs — so die Hauptvermutung der kombinatorischen Topologie, die Charakterisierung der sphärischen Mannigfaltigkeiten (die Poincaréschen Vermutung), das Triangulierbarkeitsproblem, das Knotenproblem usw. — noch immer ungelöst sind, ist der erreichte Erfolg groß, denn man steht heute vor einem klaren und einheitlichen Gebäude, welches weit über den Inhalt des Veblenschen Buches hinausragt und dennoch als *elementare Topologie* bezeichnet werden kann.

Ch16, 46. *In der vorliegenden Arbeit soll nun gezeigt werden, daß auch die Dimensionstheorie zu dieser „elementaren" Topologie gehört, daß auch sie durchaus von den Homologie- und Schnittkonstruktionen beherrscht wird.* Das dürfte kaum jemanden überraschen: in einer Reihe von Arbeiten, die in den letzten Jahren erschienen sind, wurde allmählich klar gemacht, daß die gestaltlichen Eigenschaften selbst der allgemeinen topologischen Gebilde nur ein Grenzfall der entsprechenden Eigenschaften der Polyeder sind, und infolgedessen prinzipiell mit den Methoden, die sich in der Polyedertopologie ausgebildet haben, angegriffen werden können. Insbesondere muß das natürlich auch für den Dimensionsbegriff der Fall sein, und es wurde tatsächlich bewiesen, daß die gewöhnliche (Brouwer-Urysohn-Mengersche) Dimension einer abgeschlossenen Menge F die kleinste Zahl r ist ein r-dimensionales Polyeder übergeführt werden kann.

Wenn aber auf diese Weise ein Anschluß des allgemeinen Dimensionsbegriffes an das anschauliche Material der Polyedertopologie auch erbracht wurde, so geschah zur Geometrisierung der Dimensions*theorie* (nicht allein der Dimensions*definition*) so gut wie gar nichts; diese Theorie blieb ohne jeden Zusammenhang mit der übrigen Topologie — eine Theorie, die ihren Methoden nach mehr zu der Punktmengenlehre, wie sie sich im Rahmen der reelle Funktionen ausgebildet hatte, als zur eigentliche Analysis Situs gehörte.

Erst die tatsächliche Anwendung des kombinatorischen Homologiebegrifes auf die allgemeinen mengentheoretischen Gebilde erlaubt uns festzu-stellen, daß die Dimensionstheorie gar keine Theorie für sich, sondern lediglich der erste Paragraph einer noch in ihren Anfängen stehenden allgemeinen Untersuchung der Berandungs- und Schnitt- (insbesondere Verschlingungs-) Konstruktionen in abgeschlossenen Mengen ist und als solche auch aufgebaut werden soll. Erst dieser Standpunkt gibt uns eine richtige Perspektive und führt zu einer natürlichen und reichen Problematik. Nur durch das Fehlen der richtigen geometrischen Grundlage in der bisherigen rein mengentheortischen Entwicklung der Dimensionstheorie war ihre isolierte Stellung bedingt; diese Grundlage, die in der vorliegenden Arbeit gegeben werden soll, bedeutet aber — wie öfters — das Ende der selbständigen Existenz einer Theorie, ihr Aufgehen in einem umfassenderen Kapitel der Mathematik.

Ch16, 47. **Hauptsatz.** *Die Dimension nach variablem Modul stimmt für jede abgeschlossene Menge F mit der Brouwerschen Dimension überein:*

$$\Delta(F) = \dim F.$$

Ch16, 48. Satz 1. *Für beliebiges n ist jeder n-dimensionale Raum mit einer Teilmenge des $(2n+1)$-dimensionalen Cartesischen R_{2n+1} homöomorph.*

Ch16, 49. Satz 2. *Für jedes n gibt es einen n-dimensionalen Universalraum.*

Ch16, 50. Wir wollen nun beweisen, daß *die Menge aller derjenigen Punkte des R_{2n+1}, die höchstens n rationale Koordinaten haben, ein n-dimensionaler Universalraum ist.*

Ch16, 51. Jeder n-dimensionale kompakte metrisierbare Raum ist einer Teilmenge des R^{2n+1} homöomorph.

Ch16, 52. Zustaz. *Eine stetige Abbildung f eines n-dimensionalen kompakten metrischen Raumes F in einen Euklidischen Raum R von einer Dimension $\geq 2n+1$ kann durch eine beliebig kleine stetige Abänderung in eine Homöomorphie verwandelt werden.*

Ch16, 53. *$f(F)$ läßt sich durch eine beliebig kleine Abänderung in eine zu F homöomorphie Teilmenge Φ von U_n^k überführen.*

Ch16, 54. *Ein kompakter metrischer Raum A von einer endlichen Dimension n läßt sich für jedes $m > 0$ auf eine Teilmenge des Cartesischen R_{n+m} stetig und fastschlicht abbilden, derart, daß jene Punkte, die mehr als einem Punkt von A als Bilder zugeordnet sind, eine höchstens $(n-m)$*-dimensionale *Menge bilden.*

Ch16, 55. *Jeder n-dimensionale kompakte Raum ist homöomorph mit winer Teilmenge des R_{2n+1}.*

Ch16, 56. Die Dimension eines kompakten Raumes ist die kleinste Zahl n mit der Eigenschaft, daß sich der Raum stetig und fastschlicht auf eine nirgends dichte Teilmenge des R_{n+1} abbilden läßt.

Ch16, 57. **Haupttheorem.** *Ein n-dimensionaler kompakter Raum A gestattet eine derartige stetige Abbildung auf eine Teilmenge des R_{n+m} $(m = 1, 2, \ldots)$, daß für $k = 2, 3, \ldots$ jene Punkte des R_{n+m}, die mindestens k mal als Bildpunke auftreten, eine höchstens $(n-(k-1)m)$-dimensionale Menge bilden.*

Ch16, 58. ...*jede stetige Abbildung f von A auf eine Teilmenge des R_{n+m} sich durch eine beliebig kleine Abänderung in eine die Bedingungen des Theorems befriedigende Abbildung φ überführen läßt.*...

Ch16, 59. Durch diese Method werden komplizierte Konvergenzbetrachtungen vermeiden, und insbesondere für den Menger-Nöbelingschen Fall $m = n+1$ wird sich der Beweis auf wenige Zeilen reduzieren lassen.

Ch16, 60. Der Beweis ist höchst elegant und liefert für den Fall $m = n+1$ den obigen Mengerschen Satz auf knapp $1\frac{1}{2}$ Seiten.

Ch16, 61. Der Gedanke, die bahnbrechenden und äußerst fruchtbaren Theorien, die Herr Lebesgue für den Inhalt von Punktmengen entwickelt hat, auf den Begriff der Länge zu übertragen, liegt sehr nahe: es genügt eine additive Mengenfunktion zu finden, deren Wert für jede rektifizierbare Kurve gleich der gewöhnlichen Länge dieser Kurve ist, und die im übrigen bei der Bildung von Vereinigungs- und Durchschnittsmengen die Eigenschaften des gewöhnlichen Lebesgueschen Maßes besitzt.

Ch16, 62. Wir setzen also voraus:

I. Jeder beliebigen Punktmenge A eines q-dimensionalen Raumes $\mathfrak{R}_q$ ist eine Zahl $\mu^* A$ eindeutig zugeordnet, die Null, positiv, oder gleich $+\infty$ sein kann, und die *das äußere Maß von A* heißt.

II. Für eine Teilmenge B von A ist stets

$$\mu^* B \leq \mu^* A.$$

III. Ist A die Vereinigungsmenge einer Folge von endlich oder abzählbar unendlich vielen Punktmengen $A_1, A_2, \ldots$, so ist stets

$$\mu^* A \leq \mu^* A_1 + \mu^* A_2 + \cdots.$$

Diese letzte Ungleichheit hat offenbar nur dann eine Bedeutung, wenn die Summe rechterhand konvergiert.

Ch16, 63. *Definition. Eine Punktmenge A soll meßbar heißen, wenn für jede willkürliche Punktmenge W von endlichem äußeren Maße die Relation gilt*

$$\mu^* W = \mu^* AW + \mu^*(W - AW); \quad (1)$$

das Maß μA von A wird dann durch die Gleichung definiert

$$\mu A = \mu^* A.$$

Ch16, 64. Es sei A eine beliebige Punktmenge im q-dimensionalen Raume ($q > 1$); mit $U_1, U_2, \ldots$ bezeichnen wir eine Folge von endlich oder abzählbar unendlich vielen Punktmengen, die folgenden beiden Bedingungen genügt:

a) die gegebene Punktmenge A ist Teilmenge der Vereinigungsmenge der U_k,
b) der Durchmesser d_k eines jeden U_k ist kleiner als eine vorgeschriebene positive Zahl ϱ.

Wir betrachten die *untere Grenze* der Summen

$$d_1 + d_2 + \cdots$$

der Durchmesser d_k, für alle Folge $U_1, U_2, \ldots$, die den Bedingungen a) und b) genügen, und bezeichnen diese untere Grenze, die übrigens gleich $+\infty$ sein kann, mit $L_\varrho A$.

Nimmt ϱ ab, so werden die Bedingungen, denen die U_k genügen müssen, enger, und die Zahl $L_\varrho A$ kann sich nicht verkleinern. Es existiert also *stets* der Limes

$$\lim_{\varrho=0} L_\varrho A = L^* A.$$

Die Zahl $L^* A$, die für jede beliebige Punktmenge eindeutig definiert ist, soll das äußere lineare Maß von A genannt werden.

Ch16, 65. Herr Carathéodory hat eine hervorragend einfache und allgemeine, die Lebesguesche als Spezialfall enthaltende Maßtheorie entwickelt und damit insbesondere das p-dimensionale Maß einer Punkt-menge im q-dimensionalen Raume definiert. Hierzu geben wir im folgenden einen kleinen Beitrag. Nach einleitenden Betrachtungen, die das Carathéodorysche Längenmaß in naheliegender Weise verallgemeinern und einen Überblick über die reiche Fülle analoger Maßbegriffe gestatten, stellen wir eine Erklärung des p-dimensionalen Maßes auf, die sich unmittelbar auf nicht ganzzahlige Werte von p ausgedehnen und Mengen *gebrochener Dimension* als möglich erscheinen läßt, ja sogar solche, deren Dimensionen die Skala der positiven Zahlen zu einer verfeinerten, etwa logarithmiischen Skala ausfüllen. Die Dimension wird so zu einem Graduierungsmerkmal wie die „Ordnung" des Nullwerdens, die „Stärke" der Konvergenz und verwandte Begriffe.

Ch16, 66. Meßbar (bezüglich L) heißt eine Menge A, wenn für jede Menge W die Gleichung gilt

$$L(W) = L(AW) + L(W - AW).$$

Ch16, 67. Ist $l(U) \geq 0$ eine für beschränktes U definierte, abschließbare oder stetige Mengenfunktion und schließt man A in endlich oder abzählbar viele beschränkte Mengen U_n mit Durchmessern $< \varrho$ ein, so ist

$$L(A) = \liminf_{\varrho \to 0} \sum l(U_n)$$

ein äußeres Maß.

Ch16, 68. **Definition 1.** Man schließe A in endlich oder abzählbar viele Kugeln K_n von Durchmessern $d_n < \varrho$ ein und bilde

$$L_p(A) = \liminf_{\varrho \to 0} c_p \sum d_n^p,$$

wobei

$$c_p = \pi^{\frac{p}{2}} : 2^p \prod \left(\frac{p}{2}\right)$$

das Volumen einer p-dimensionalen Kugel von Durchmesser 1 ist.

Ch16, 69. **Definition 2.** Man schließe A in endlich oder abzählbar viele Kugeln K_n ein, deren Durchmesser $d_n < \varrho$ sind, und bilde

$$L(A) = \lim_{\varrho \to 0} L_\varrho(A),$$

wo $L_\varrho(A)$ die untere Grenze der Summen $\sum \lambda(d_n)$ ist.

Ch16, 70. **Satz.** *Für jede Menge M ist die Hausdorffsche Dimension mindestens gleich der mengentheoretischen Dimension.*

Ch16, 71. Le but de cette note est d'obtenir la dimension d'un espace métrique compact à l'aide d'un procédé de recouvrement par des «étalons de mesure» dont le nombre permet de définir la dimension. Cette méthode est analogue à la méthode de mesure employée en géométrie élémenitaire.

Ch16, 72. Un ensemble compact F a la dimension n, si n est le plus petit entier tel qu'il existe (ε étant un nombre arbitraire positif) un système fini d'ensembles fermés recouvrant F, de diamètres ne surpassant pas ε et dont toutes les intersections de $n + 2$ sont vides.

Ch16, 73. Théorème Fondamental. La borne inférieure de l'ordre métrique

$$k = \liminf \left(-\log \frac{N_F(\varepsilon)}{\log \varepsilon}\right)$$

pour toutes les métriques de l'espace compact F est égale à sa dimension.

Corollaire. Si la borne inférieure en question est finie, elle est necéssairement un nombre entier.

Ch16, 74. ... la relation entre la dimension et la mesure n'est pas immédiate: p. ex. l'ensemble des points du plan à deux coordinées irrationnelles est de dimension 0, tandis que sa mesure superficielle est positive. Cette divergence est d'ailleurs bien naturelle, la dimension étant une notion topologique et la mesure une notion métrique. Cependant la relation entre elles se laisse formuler d'une façon assez simple. Notamment, en profitant des idées de MM. Nöbeling, Pontrjagin et Schnirelmann, on parvient au théorème:

Pour qu'un ensemble (situé dans un espace métrique séprable) soit de dimension $\leq n$. it faut et it suffit qu'il soit homéomorphe à un ensemble de mesure $(n+1)$-dimensionnelle nulle (situé dans le cube à $2n+1$ dimensions).

Ch16, 75. ***Théorème 5**. Soit L^{φ} une fonction d'un ordre supérieur à celui de la fonction $L^{(n)}$, mais inférieur ou égal à celui de la fonction $L^{(n+1)}$. Pour qu'un ensemble E (situé dans un espace métrique séparable) soit de dimension $\leq n$, il faut et il suffit qu'il soit homéomorphe à un ensemble E^* (situé dans I^{2n+1}) tel que $L^{\varphi}(E^*) = 0$.*

Ch16, 76. Wesentlich schwieriger als die Einführung des Volumenbegriffs ist die Begründung der *Länge einer Curve* als Grenze der Länge von Polygonen, die der Curve geeignet eingeschrieben sind, und der *Oberfläche einer krummen Fläche* als Grenze der Oberfläche von Polyedern, die in Bezug auf die Fläche geeignet construiert sind.

Ch16, 77. Man kann jedoch diese anderen Begriffe Länge und Oberfläche auch allein aus dem Begriffe des Volumens mittels eines einfachen Grenzüberganges entwickeln:

Es sei C eine Curve. Um jeden Punkt von C als Mittelpunkt denke man sich eine Kugel mit dem Radius r abgegrenzt, unter r eine feste positive Größe verstanden. Die Menge aller derjenigen Punkte des Raumes, welche in das Innere oder die Begrenzung von wenigstens einer dieser Kugel zu liegen kommen, definiert uns den *Bereich der Entfernung $\leq r$ von der Curve C*. Es sei $V(r)$ das Volumen dieses Bereichs (falls ihm ein bestimmtes Volumen zukommt), so kann der Grenzwert von $\frac{V(r)}{\pi r^2}$ für ein nach Null abnehmendes r (falls dieser Grenzwert existiert), als die *Länge der Curve C* eingeführt werden. — Es sei F eine Fläche. Man construiere in entsprechender Weise den *Bereich der Entfernung $\leq r$ von F*. Es sei $V(r)$ das Volumen dieses Bereichs, so kann der Grenzwert von $\frac{V(r)}{2r}$ für ein nach Null abnehmendes r (vorausgesetzt, daß die Größe $V(r)$ sowie dieser Grenzwert existiert), als die *Oberfläche der Fläche F* eingeführt werden.

Es ist einleuchtend, daß hierbei zunächst die Länge einer geradlinigen Strecke und die Oberfläche eines ebenen Dreiecks genau mit den gewöhnlich dafür angenommenen Werten sich ergeben; infolgedessen werden überhaupt in *regulären* Fällen Längen und Oberflächen in dem eben erklärten und andererseits in dem üblichen Sinne die gleichen Werte vorstellen.

Ch16, 78. J'arrive à cette notion générale de *volume* ou de *grandeur* $\mathfrak{J}(P)$ d'un ensemble *quelconque* P contenu dans G_n en prenant *chaque* point p, qui appartient à P ou à $P^{(1)}$, pour centre d'une sphère pleine à n dimensions au rayon ρ, que nous appellerons $K(p,\rho)$. Le plus petit multiple de tous ces sphères pleines $K(p,\rho)$ ... savoir:

$$\mathfrak{M}[K(p,\rho)],$$

(où ρ est une constante) constitue pour chaque valeur de ρ un ensemble qui se compose de pièces continues à n dimensions et dont le volume se détermine d'après les règles connues au moyen d'une intégrale n-tiple.

Soit $f(\rho)$ la valeur de cette intégrale; $f(\rho)$ est une fonction continue de ρ, qui diminue avec ρ; la limite de $f(\rho)$, lorsque ρ converge vers zéro, me sert de définition du volume $\mathfrak{J}(P)$; en sorte, que nous avons:

$$\mathfrak{J}(P) = \lim_{\rho=0} f(\rho).$$

Je fais remarquer expressément que cette valeur du *volume* ou de la *grandeur* d'un ensemble quelconque P contenu dans un espace continu plan G_n à n dimensions est absolument dépendante de l'espace plan G_n même, duquel P est considéré comme une partie composante, et particuliérement du nombre n; de sorte que, si l'on considère *le même ensemble P* comme une partie constituante d'un autre espace continu plan H_m la valeur du volume de P par rapport à l'espace H_m est en général différente de celle, qui se rapporte au même ensemble P, considéré comme partie constitutive de G_n.

Ch16, 79. ... *si $f(\rho)$ est d'ordre infinitésimal α, le nombre dimensionnel de l'ensemble étudié sera* $3-\alpha$.

Ch17, 1. Die wichtigste Eigenschaft eines Raumgebildes ist seine Dimension. Kurven, Flächen, Körper gelten im täglichen Leben als die grundsätzlich verschiedenen Hauptklassen der Raumgebilde, und ebenso ist die Verschiedenheit der Dimension das fundamentale Klassifikationsprinzip jeder wissenschaftlichen Geometrie. ... Die Dimension ist etwas völlig Anschauliches, sie ist der *Kern* der Geometrie.

Die Bezeichnungen „Kurve“, „Fläche“, „Körper“ sind dem täglichen Leben entnommen und gehören zu jenen Worten, mit denen jeder, der sie ausspricht, eine gewisse Vorstellung verbindet. Die Tatsache, daß ein bestimmter einzelner Mensch mit einem gewissen Wort eine Vorstellung verbindet, besteht im allgemeinen in Folgendem: Gewisse Wesenheiten wird der betreffende Mensch, ohne sich zu besinnen, mit dem betreffenden Wort bezeichnen; von gewissen anderen Wesenheiten wird er, ohne sich zu bedenken, sagen, daß sie mit dem betreffenden Wort *nicht* zu bezeichnen sind. Der einzelne lernt ja im Laufe seines Lebens, zumal in seiner Kindheit, den Gebrauch der Worte als Bezeichnungen für gewisse Wesenheiten dadurch, daß man ihm sagt: „Dies wird mit dem betreffenden Worte bezeichnet, dies nicht.“ Zwischen diesen beiden Klassen schwebt eine dritte Klasse von Wesenheiten, bei denen der einzelne unsicher ist, ob er sie mit dem betreffenden Wort bezeichnen soll oder nicht. Wenn wir zur Betrachtung der Gesamtheit der ein Wort gebrauchenden Menschen übergehen, so sehen wir ebenfalls: Gewisse Wesenheiten werden von allen Menschen mit dem betreffenden Wort bezeichnet, gewisse Wesenheiten werden von allen mit dem betreffenden Wort nicht bezeichnet, und überdies gibt es noch eine Klasse von Wesenheiten, die wir als die *schwebende Klasse* bezeichnen wollen.

Betrachten wir beispielsweise das Wort „Kurve“. Ein Bogen, eine Kreislinie, eine Lemniskate wird von jedem Menschen, der das Wort Kurve gebraucht, als Kurve bezeichnet. Eine Quadratfläche, eine Kugeloberfläche, ein Wiirfelkörper wird von niemandem Kurve genannt. Denn jeder, der die Worte Kurve, Fläche, Körper lernt, erfährt, daß das Wort Kurve zur Zusammenfassung der draht- und fadenförmigen, der strich- und sehienenartigen Gebilde, das Wort Fläche zur Bezeichnung der plätten-, blätter- und hautartigen Gebilde, das Wort Körper zur Benennung der blockartigen Gebilde verwendet wird.

Ch17, 2. Mehr allerdings als die Fähigkeit, Dinge anzugeben, die allgemein als Kurven bezeichnet werden, und solche, die allgemein nicht als Kurven bezeichnet werden, besitzt ein Mensch, der das Wort Kurve ausspricht, gemeinhin nicht, und ebenso steht es mit den Worten Fläche und Körper. Eine Klassifikation *sämtlicher* Raumgebilde nach ihrer Dimension gibt es im täglichen Leben nicht: Auf die Frage nach der Dimension gewisser verwickelter Gebilde, die der moderne Geometer untersucht, erhielte man bei verschiedenen Menschen keinen Bescheid oder voneinander abweichende Antworten. Ebensowenig sind die meisten Menschen imstande, präzis anzugeben, was sie unter einer Kurve, einer Fläche, einem Körper verstehen, d. h. diese Worte mit Hilfe anderer Worte zu erklären. Eine gewisse praktische Funktion für das tägliche Leben erfüllen die Worte Kurve, Fläche, Körper natürlich trotz dieser Umstände, denn die verwickelten Gebilde der hinsichtlich des Wordes Dimension schwebenden Klasse sind im täglichen Leben selten, und eine völlig präzise Umgrenzung der Kurven überhaupt ist für die gegenseitige Verständigung der Menschen über gewisse spezielle Fälle nicht erforderlich.

Nicht viel anders als im tägliehen Leben stand es bis zur Begründung der Dimensionstheorie in der wissenschaftliehen Raumlehre. Auch in der Geometrie wurden gewisse Raumgebilde als Kurven, andere als Flächen, andere als Körper bezeichnet. Aber obwohl die fundamentale Bedeutung dieser Klassifikation — gerade von hervorragenden Mathematikern aller Zeiten — klar erkannt wurde, so gingen doch nicht im entferntesten alle Raumgebilde in dieselbe ein. Für die ungeheure Mehrzahl der Raumgebilde konnte eine Dimension nicht definiert werden. Es existierte kein allumfassender Dimensionsbegriff als Gemeingut der Mathematik.

Ch17, 3. Wenn ein Wort, mit dem man im täglichen Leben eine Vorstellung verbindet, in der Wissenschaft durch eine Definition präzisiert werden soIl, so besteht kein Anlaß, sich mit dem tägliehen Gebrauch des Wortes in Widerspruch zu setzen, d. h. Dinge, welche allgemein mit dem betreffenden Wort bezeichnet werden, aus dem Begriff auszuschließen, oder Dinge, welche allgemein von der Bezeichnung mit dem betreffenden Wort ausgeschlossen werden, in den Begriff aufzunehmen. Eine *formale Forderung* an die strenge Definition eines der täglichen Sprache entnommenen Wortes ist also, *daß sie die Priäzisierung und Ergänzung des*, wie wir sahen, in Grenzfällen schwankenden und unvollständigen *Wortgebrauches darstellt, welche mit demselben nicht in Widerspruch tritt.*

Eine derarlige Ergänzung und Präzisierung kann nun aber in verschiedener Weise, und zwar nicht nur formal, sondern (mit Rücksicht auf die schwebenden Klassen von Wesenheiten) auch inhaltlich in verschiedener Weise erfolgen. Beispielsweise können gewisse verwickelte Gebilde, für welche der einzelne auf die Frage, ob sie eindimensional seien, keinen Bescheid weiß oder verschiedene Menschen verschiedene Antworten geben, sowohl zu den eindimensionalen Gebilden gerechnet als auch von ihnen ausgeschlossen werden, ohne daß die Definition deshalb in Widerspruch zum allgemeinen Sprachgebrauch tritt. Jede Festlegung auf eine bestimmte Präzisierung enthält also ein gewisses Maß von *Willkür*, deren Rechtfertigung ausschließlich durch die *Fruchtbarkeit* der Definition geliefert werden kann. Der Zweck des Wortes im täglichen Leben ist die Verständigung der Menschen untereinander; der Zweck einer strengen Definition ist es, den Ausgangspunkt eines deduktiven Systems zu bilden. Definitionen sind Dogmen, nur die Deduktionen aus ihnen sind Erkenntnisse. Es ist demnach eine *inhaltliche Forderung* an eine Definition überhaupt, *daß sie sich als Erkenntnisquelle erweise dadurch, daß sie den Ausgangspunkt einer umfassenden, ästhetisch vollkommenen Theorie bildet.*

Das Erfülltsein dieser *inhaltlichen* Forderung stellt die einzige mögliche *Recktfertigung einer jeden Definition* dar. Handelt es sich speziell um die Definition eines Begriffes, der mit einem dem täglichen Leben entnommenen Namen bezeichnet wird, so liefert das Erffültsein der (für den Begriff an sich sekundären) *formalen* Forderung eine *Rechtfertigung für die Benennung des Begriffes.*

Ganz besonders gilt das Gesagte für den Dimensionsbegriff. Bereits eine Betrachtung spezieller Gebilde läßt eine FülIe von interessanten und schönen Beziehungen für einen präzisen Dimensionsbegriff vermuten. Welches ist die Dimension der Summe zweier n-dimensionaler Gebilde? Wie steht es mit der Struktur der n-dimensionalen Gebilde, zumal mit der Verteilung der Punkte verschiedenen dimension eilen Verhaltens? Besitzen die n-dimensionalen Gebilde nicht gewisse Zerlegungseigenschaften? In welcher Beziehung steht der Dimensionsbegriff zum Begriff des Zusammenhanges von Raumgebilden? Wie verhält sich die Dimension eines Gebildes bei gewissen Transformationen desselben? Wie steht es speziell mit den Teilgebilden des gewöhnlichen Cartesischen Raumes?

Name Index

D. M. Johnson, *Theories of Dimension from Ancient to Modern Times*, Sources and Studies in the History of Mathematics and Physical Sciences, https://doi.org/10.1007/978-3-032-02682-8

Subject Index

D. M. Johnson, *Theories of Dimension from Ancient to Modern Times*, Sources and Studies in the History of Mathematics and Physical Sciences, https://doi.org/10.1007/978-3-032-02682-8

MIX
Papier aus verantwortungsvollen Quellen
Paper from responsible sources
FSC® C105338

FSC
www.fsc.org

If you have any concerns about our products,
you can contact us on
ProductSafety@springernature.com

In case Publisher is established outside the EU,
the EU authorized representative is:
Springer Nature Customer Service Center GmbH
Europaplatz 3, 69115 Heidelberg, Germany

Printed by Libri Plureos GmbH
in Hamburg, Germany